WITHDRAWN

Industrial Fire Hazards Handbook

Industrial Fire Hazards Handbook

Second Edition

A Guide to Fire Protection in Industry

 National Fire Protection Association
Quincy, Massachusetts

Jim L. Linville, *Editor*
John S. Petraglia, *Associate Editor*
Ann E. Coughlin, *Copyeditor*
Louise Grant, *Composition*

Dedicated to Gordon P. McKinnon, an example to aspire to.

The Editors

First Printing, September 1984
Copyright © 1984
National Fire Protection Association, Inc.
All rights reserved
NFPA No. SPP-57A
ISBN 0-87765-283-X
Library of Congress No. 84-61186
Printed in the United States of America

CONTENTS

PART ONE: INTRODUCTION
  Chapter   1 Industrial Fire Risk Management 1
  Chapter   2 Life Safety in Industrial Occupancies 11
  Chapter   3 Plant Emergency Organization and Training 21

PART TWO: FIRE HAZARDS IN MAJOR INDUSTRIES
  Chapter   4 Electric Generating Plants 39
  Chapter   5 Nuclear Power Plants 69
  Chapter   6 Bulk Grain Handling 101
  Chapter   7 Vegetable and Animal Oil Processing 125
  Chapter   8 Paints and Coatings Manufacturing 143
  Chapter   9 Rubber Products 169
  Chapter 10 Mining 193
  Chapter 11 Aluminum and Nonferrous Metals Processing 233
  Chapter 12 Wood Products 249
  Chapter 13 Furniture Manufacturing 275
  Chapter 14 Pulp and Paper Manufacturing 289
  Chapter 15 Paper Products Manufacturing 315
  Chapter 16 Printing and Publishing 331
  Chapter 17 Textile Manufacturing 349
  Chapter 18 Clay Products Plants 375
  Chapter 19 Plastics Products 393
  Chapter 20 Motor Vehicle Assembly 415
  Chapter 21 Shipyards 441
  Chapter 22 Food Processing 465
  Chapter 23 Semiconductor Manufacturing 481

PART THREE: SPECIAL PROCESS FIRE HAZARDS
  Chapter 24 Welding and Cutting 495
  Chapter 25 Spray Finishing and Powder Coating.................. 521
  Chapter 26 Dipping and Coating Processes....................... 553
  Chapter 27 Heat Processing Equipment............................ 571
  Chapter 28 Oil Quenching.. 597
  Chapter 29 Molten Salt Baths...................................... 609
  Chapter 30 Machine Tool Processes................................ 627
  Chapter 31 Chemical Processes..................................... 651
  Chapter 32 Solvent Extraction 673
  Chapter 33 Lumber Kilns and Agricultural Dehydrators and
                     Dryers.. 693
  Chapter 34 Grinding and Milling Operations 707

Chapter 35 Storage and Filling of Aerosol Products.............. 729
Chapter 36 Radioactive Materials................................. 743

PART FOUR: GENERAL OCCUPANCY FIRE HAZARDS
Chapter 37 Flammable and Combustible Liquids Handling and
          Storage 767
Chapter 38 Handling and Storage of Industrial Gases............. 795
Chapter 39 Liquefied Petroleum Gases at Industrial Plants....... 827
Chapter 40 Computer Centers..................................... 843
Chapter 41 Laboratories 855
Chapter 42 Boiler-Furnaces 871
Chapter 43 Fluid Power Systems.................................. 895
Chapter 44 Refrigeration Systems 907
Chapter 45 Air Moving Equipment................................. 919
Chapter 46 Materials Handling Systems 933
Chapter 47 Electrical Installations in Industrial Locations..... 947
Chapter 48 Industrial Storage Practices......................... 971
Chapter 49 Industrial Waste Control............................. 1021
Chapter 50 Industrial Housekeeping Practices.................... 1049
          Index... 1065

Preface

For more than 80 years the National Fire Protection Association has been working to protect people and property from fire, and part of that effort has been to serve the needs of industry through the publication of its codes, standards, and recommended practices, many of which apply to industrial processes. The INDUSTRIAL FIRE HAZARDS HANDBOOK is *not* intended to summarize or consolidate these documents or to discuss them in any detail. Rather, it approaches industrial fire protection from a different perspective; its purpose is to provide a broad yet thorough introduction to major industries and industrial processes, with emphasis on the fire hazards that accompany them. This HANDBOOK, therefore, is intended to complement the NFPA codes and standards, and help the user intelligently apply them. It is meant to be a basic reference book which will serve a broad audience, including the fire service, insurance companies, regulatory authorities, and fire science students as well as those directly involved in industrial fire protection.

Until now, there has been no one source for the general information provided in this HANDBOOK, though the need for it has been evident. Briefly, the book is designed to outline major industries and industrial processes, identify the fire hazards associated with them, and explain the methods used to eliminate or control the hazards. Illustrations have been included to expand the material presented in the text.

The professionals who contributed to this HANDBOOK brought hundreds of years of cumulative experience to the task; their chapters include the latest advancements in industrial fire protection and prevention. In addition, the NFPA Industrial Fire Protection Section's Executive Committee and Pulp & Paper—Wood Products Committee offered valuable help.

As a body, the National Fire Protection Association is not responsible for the contents of this HANDBOOK, as the Association's full membership has had no opportunity to review the contents prior to publication. The editors would appreciate receiving comments and suggestions that will improve the contents, as future editions of the HANDBOOK will be revised and expanded to keep the information current.

Preface to the Second Edition

The first edition of the INDUSTRIAL FIRE HAZARDS HANDBOOK sought to describe many of the manufacturing processes used in industry, identify the fire hazards associated with those processes, and detail the methods used to control and eliminate them. This new edition represents the NFPA's continuing effort to provide a basic reference book for those charged with protecting life and property in industry. While the organizational structure of the first volume has been retained, significant new material has been added throughout the text. The new chapters include discussions of innovative technology and manufacturing processes that have become mainstream in today's economy, as well as common processes found to be of growing importance. The general updating of the text reflects a thorough revision of the material to include state-of-the-art industrial processes and hazard protection and prevention methods. Several chapters from the first edition which were found to be of lesser interest to the fire protection community have been deleted. The new and revised material has been reviewed by industrial experts and NFPA staff for technical accuracy, current usage, and readability; parenthetical metric equivalents have been added alongside all conventional measurements for convenience; numerous cross-references have been added to all chapters to direct readers to further discussions and related chapters; and the bibliographies have been extensively revised and updated.

**Industrial
Fire
Hazards
Handbook**

Part One
INTRODUCTION

Industrial Fire Risk Management

Peter K. Schontag, P.E., S.F.P.E.

As modern industry continues to expand and change, fire risk management becomes more complex and difficult. New processes and products bring new fire hazards. Power generating units, computers, and production machines have become even larger, and more expensive, and their loss has a greater impact on production. Greater values are concentrated in single buildings. Storage is piled higher and higher.

Fire detection and prevention equipment is hard pressed to keep pace with the new hazards. As a result, the risk of very large losses is increased — losses which can threaten the existence of the entire plant or business organization. Maintaining these risks within reasonable bounds is a major challenge to management.

Fire risk management doesn't just happen. It is the result of corporate policy and related programs. Good organization, with responsibilities clearly assigned and specific duties spelled out, will result in implementation of effective programs. The two primary ways to manage fire risk are to prevent fires and to limit or control their size.

This chapter outlines the organization, procedures, and supervision needed to achieve these two goals. Although intended primarily

---

Peter K. Schontag is a fire protection engineer with Rolf Jensen & Associates, Inc.

for industrial properties, the general principles outlined here also apply to other facilities including mercantile establishments, institutions, and large residential and office buildings.

The introductory sections are concerned with evaluating loss possibilities, general administration and planning. The sections which follow cover guard or watch service, fire prevention and fire equipment inspections, and private fire brigade organization. Training is covered in Chapter 3.

Evaluating Loss Possibilities

The fire prevention and control manager will devote much attention to evaluating the likelihood of fire or explosion in any part of the property, and the probable extent of the resulting damage.

Two specific figures are useful for an appraisal of the problem:

1. The first figure is the maximum foreseeable or maximum possible (MFL or MPL) loss with normal fire protection out of service.
2. The second figure is the probable maximum loss (MPL) with normal protection in service.

Work sheets are a useful tool in evaluating loss possibilities. Pertinent data would include a listing of building areas, construction, occupancy, fire protection, approximate value, and relative importance.

Once the loss possibilities have been evaluated, the corporate risk manager or a higher level of management can decide what corrective action should be taken to:

a. Reduce the risk (probability of fire occurring).
b. Control the risk (extent of ensuing loss).
c. Transfer the risk (provide insurance to cover losses).
d. Assume part or all of the risk (through self-insurance or large deductibles).
e. A combination of two or more of the above.

Frequently, action in more than one of these areas is needed; although the fire prevention and control manager will usually be responsible for only a and b.

The Fire Prevention and Control Program

An effective program receives its driving force and continuing motivation from top management, but strong interest extending down through the various levels of management and supervision to the individual employees is needed for the program to succeed.

The objectives of a fire prevention and control program can be stated very simply as follows:

1. To plan and construct low hazard buildings, processes, and equipment.
2. To provide fire control and suppression equipment where needed.
3. To maintain the equipment in readiness.
4. To educate and train employees in loss prevention and proper action in emergencies.

The details of the organization and program needed to carry out these objectives will vary with each property. Both safety to life and to property are the vital considerations. Management has to develop the program best suited to their individual needs. The program must then be communicated in a forceful way to those at all levels. This is no easy task. It takes the talent of a skilled communicator to generate enthusiasm.

The Fire Prevention and Control Manager

In small properties the president, manager, or other chief executive officer can manage the details of the loss prevention program. Usually, however, an administrator reporting directly to the top management is appointed to carry out the program. The administrator may be given various titles, such as loss prevention manager, director of property conservation, chief of plant protection, or fire chief.

The concept of a single administrator to evaluate and manage the control of all risks to the facility has been adopted by some plants, particularly very large ones. At other plants, fire protection, safety, and plant security are handled in separate departments.

The fire prevention manager (or other administrator of the program) should have an understanding of fire protection and the ability to work well with other managers.

The major responsibilities include:

1. Evaluating and analyzing loss possibilities and recommending corrective action.
2. Providing and supervising systems and procedures for loss prevention and control.
3. Working with other departments in matters affecting loss control.
   (a) *with production departments:* concerning new processes, new materials, changes affecting production, housekeeping, operational hazards, and cutting and welding supervision.
   (b) *with engineering and maintenance:* concerning new construction, fire protection installations, impairments to protection, and maintenance and inspection of fire protection equipment.
   (c) *with personnel:* concerning fire safety education, and personnel changes affecting emergency organization.
   (d) *with insurance:* concerning insurance company inspections, reports, losses, and fire protection problems.

Chapter 1

(e) *with the public fire department:* concerning plant layout and hazards, and the location and type of fire protective equipment (particularly the location of sprinkler valves and the areas they control):

4. Administering the plant emergency organization and its training.
5. Assuming administrative direction of loss control operations in emergencies.
6. Interpreting laws, codes, and standards dealing with industrial loss control.
7. Assisting management in contacts with public regulatory agencies and in public relations.

PLANNING NEW FACILITIES

Decisions made during the planning stage largely determine the degree of fire risk the facility will present after it is built.

The important considerations are in the following areas:

1. Safety to life.
2. Protection of property.
3. Continuity of operations.

Top management (and public regulatory agencies) determine the general level of acceptable risks in these areas. Careful planning and good liaison among managers, architects, engineers, and public officials are needed to design a facility that will achieve the desired level of firesafety.

Early review of plans by regulatory agencies and insurance interests is highly desirable to avoid expensive changes after construction has started.

Planning considerations which affect fire risk include, but are by no means limited to, the following:

1. Access to the site.

    a. Adequacy of the water supply.
    b. Fire exposure from surrounding property.
    c. Exposure to natural hazards such as flood, earthquake and heavy snow.
    d. Availability and adequacy of the public fire department.
    e. The social environment of the neighborhood.

2. Planning the facility.

    a. Limiting size of areas at risk by use of firewalls and fire separations.
    b. Use of fire resistant materials.
    c. Provision of adequate emergency exits.

d. Provision of approved automatic sprinklers, hydrants, and other fixed protection systems.
e. Provision of safe and reliable building services.
f. Safe location of hazardous processes.

Applicable NFPA fire protection standards and NFPA *Fire Protection Handbook* chapters furnish invaluable guidance in planning a firesafe facility.

FIRESAFETY DURING BUILDING CONSTRUCTION

Buildings undergoing construction or alterations are unusually vulnerable to fire. Wooden forms, packing materials, scrap lumber, and other combustibles accumulate from place to place. Cutting and welding sparks, heating salamanders, temporary electric wiring, roofer's tar kettles, and discarded smoking materials furnish ready ignition sources. Valuable, difficult-to-replace equipment is stored in unprotected wooden construction shanties.

With sprinkler protection not yet in service, or impaired due to construction alterations, any fire is likely to cause serious damage and costly delays to the project.

Management should make one person responsible for the fire prevention and control of the construction project.

Important areas of responsibility include:

1. Expediting the installation of sprinklers, hydrants, hose, and an adequate water supply.
2. Prompt and safe disposal of rubbish.
3. Supervising cutting and welding operations.
4. Safeguarding temporary heaters.
5. Safeguarding valuable equipment in storage.
6. Providing extinguishers and fire hose.
7. Providing guard and alarm service.
8. Safeguarding the handling of flammable liquids.
9. Guarding against wind damage.
10. For buildings undergoing demolition, keeping sprinklers in service as long as possible (at least until after valuable contents are removed).

GUARD SERVICE

Guards are an important part of the plant protection organization. They protect the property against trespassers and incendiarists, control the orderly movement of people during working hours, and are responsible for the overall safety of the property during nonoperating periods.

The duties of guards may be supplemented, or in some cases replaced in part, by various approved protective systems.

The property or security manager supervises guard service or designates a responsible person such as the fire prevention and control man-

SAMPLE ONLY — NO ONE BLANK CAN BE DESIGNED TO FIT ALL CONDITIONS. USE THIS BLANK AS A BASIC GUIDE IN DEVELOPING YOUR OWN FORM. ITEMS THAT DO NOT APPLY CAN BE OMITTED. OTHER ITEMS CAN BE EXPANDED AS DESIRED, AND NEW ITEMS CAN BE ADDED. FOR ASSISTANCE, CONSULT THE NEXT FACTORY MUTUAL ENGINEER WHO VISITS YOUR PROPERTY.

FIRE PREVENTION INSPECTION

Factory Mutual System

INSTRUCTIONS TO INSPECTOR: FILL OUT FORM WHILE MAKING INSPECTION. SEND COMPLETED FORM TO YOUR SUPERVISOR FOR NECESSARY ACTION. REPORT SHOULD BE HELD FOR REVIEW BY THE NEXT FACTORY MUTUAL ENGINEER.

PLANT	LOCATION	DATE

VALVE INSPECTIONS

PHYSICALLY TRY LOCKED VALVES AT LEAST MONTHLY AND UNLOCKED VALVES WEEKLY. IN ADDITION VISUALLY INSPECT ALL LOCKED VALVES WEEKLY. RECORD BOTH WEEKLY AND MONTHLY INSPECTIONS.

ALL INSIDE AND OUTSIDE VALVES CONTROLLING SPRINKLERS OR FIRE PROTECTION WATER SUPPLIES ARE LISTED BELOW. CHECK CONDITION OF VALVE AS FOUND. PHYSICALLY "TRY" GATE VALVES INCLUDING NONINDICATING AND INDICATOR POST GATE VALVES. DO NOT REPORT A VALVE OPEN UNLESS YOU PERSONALLY HAVE TRIED IT. FM APPROVED PIVA'S (POST-INDICATOR-VALVE ASSEMBLIES), IBV'S (INDICATING BUTTERFLY VALVES) AND STANDARD OUTSIDE SCREW & YOKE VALVES DO NOT HAVE TO BE TRIED BUT SHOULD BE VISUALLY CHECKED AT CLOSE RANGE.

NO.	VALVE LOCATION	AREA CONTROLLED	OPEN	SHUT	LOCKED	SEALED
1						
2						
3						
4						
5						
6						
7						
8						
9						
10						
11						
12						
13						
14						
15						
16						
17						
18						
19						
20						

THE FACTORY MUTUAL RED TAG ALERT SYSTEM IS USED TO GUARD AGAINST DELAYED REOPENING OF VALVES. FACTORY MUTUAL RED TAGS SHOULD BE USED EVERY TIME A SPRINKLER CONTROL VALVE IS CLOSED. WHEN THE VALVE IS REOPENED THE 2 INCH DRAIN SHOULD BE FLOWED WIDE OPEN TO BE SURE THERE IS NO OBSTRUCTION IN THE PIPING. THE VALVE SHOULD THEN BE RELOCKED.

WERE ANY VALVES OPERATED SINCE THE LAST INSPECTION	☐ Yes	☐ No
WERE FACTORY MUTUAL RED TAGS USED	☐ Yes	☐ No
WAS THE VALVE REOPENED FULLY AND A FULL FLOW 2 IN. DRAIN TEST MADE BEFORE THE VALVE WAS RELOCKED OR RESEALED	☐ Yes	☐ No

COMMENTS:

264 (8-79)ENGINEERING PRINTED IN USA

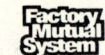

FIGURE 1.1. Factory Mutual inspection blank.

ager to administer it. Guards should be given detailed, specific instructions as to their duties. The addresses and telephone numbers of the proper management officials to notify for emergencies of various kinds should be posted at guard headquarters or other convenient locations.

INSPECT THESE ITEMS AT LEAST WEEKLY

SPRINKLERS	Auto-Matic Sprinklers	ANY HEADS DISCONNECTED OR NEEDED Yes ☐ No ☐				OBSTRUCTED BY HIGH PILING Yes ☐ No ☐			
		HEAT ADEQUATE TO PREVENT FREEZING (NOTE BROKEN WINDOWS, ETC.) Yes ☐ No ☐				Water Pressure	LB. AT YARD LEVEL		
	COMMENTS								

| DRY PIPE VALVES | VALVE ROOM PROPERLY HEATED | No. 1 Yes ☐ No ☐ | No. 2 Yes ☐ No ☐ | No. 3 Yes ☐ No ☐ | No. 4 Yes ☐ No ☐ | No. 5 Yes ☐ No ☐ | No. 6 Yes ☐ No ☐ | No. 7 Yes ☐ No ☐ | No. 8 Yes ☐ No ☐ |
| | AIR PRESSURE | No. 1 Lbs. | No. 2 Lbs. | No. 3 Lbs. | No. 4 Lbs. | No. 5 Lbs. | No. 6 Lbs. | No. 7 Lbs. | No. 8 Lbs. |

WATER SUPPLIES	FIRE PUMP	TURNED OVER Yes ☐ No ☐	GOOD CONDITION Yes ☐ No ☐
		AUTO. CONTROL TESTED Yes ☐ No ☐	FUEL TANK FULL Yes ☐ No ☐
		PUMP ROOM PROPERLY HEATED AND VENTILATED Yes ☐ No ☐	PRIMING TANK FULL Yes ☐ No ☐
	TANK OR RESERVOIR	FULL Yes ☐ No ☐	HEATING SYSTEM IN USE Yes ☐ No ☐
		TEMPERATURE AT COLD WATER RETURN (SHOULD BE 42°F MINIMUM)	CIRCULATION GOOD Yes ☐ No ☐

MFL WALL FIRE DOORS	CONDITION	OBSTRUCTED Yes ☐ No ☐	BLOCKED OPEN Yes ☐ No ☐

OTHER ITEMS

INSPECT THESE ITEMS AT LEAST MONTHLY

MANUAL PROT	EXTINGUISHERS	CHARGED Yes ☐ No ☐	ANY MISSING	ACCESSIBLE Yes ☐ No ☐	ATTENTION NEEDED (Give Location)
	INSIDE HOSE	IN GOOD CONDITION Yes ☐ No ☐		ACCESSIBLE Yes ☐ No ☐	
	YARD HYDRANTS & HOSE	CONDITION NO. 1 / NO. 2	NO. 3 / NO. 4	NO. 5 / NO. 6	
		HYDRANTS DRAINED Yes ☐ No ☐		REMARKS:	

OCCUPANCY	GENERAL ORDER & NEATNESS	GOOD Yes ☐ No ☐	COMBUSTIBLE WASTE REMOVED ON SCHEDULE (PROMPTLY) Yes ☐ No ☐
			COMBUSTIBLE DUST, LINT OR OIL DEPOSITS ON CEILINGS, BEAMS OR MACHINES Yes ☐ No ☐
	ELECT. EQUIP.	DEFECTS NOTED Yes ☐ No ☐	DESCRIBE AREAS NEEDING ATTENTION INCLUDING YARD:
		SAFETY CANS USED Yes ☐ No ☐	
	FLAM. LIQUIDS	EXCESSIVE IN MFG AREAS Yes ☐ No ☐	DRAINAGE OBSTRUCTED Yes ☐ No ☐
			VENT FANS ON Yes ☐ No ☐
	SMOKING REGULATIONS	LOCATIONS WHERE VIOLATIONS NOTED	
	CUTTING & WELDING	PERMITS ISSUED FOR ALL C&W OPERATIONS Yes ☐ No ☐	LISTED PRECAUTIONS TAKEN Yes ☐ No ☐
	STORAGE	WELL ARRANGED Yes ☐ No ☐	AISLES CLEAR Yes ☐ No ☐
		ADEQUATE SPACE BELOW SPRINKLERS Yes ☐ No ☐	CLEAR OF LAMPS, HEATERS Yes ☐ No ☐

DOORS AT CUT OFF WALLS	CONDITION	OBSTRUCTED Yes ☐ No ☐	BLOCKED OPEN Yes ☐ No ☐
Sprinkler Alarms	TESTED Yes ☐ No ☐		OPERATION SATISFACTORY (IF "NO" - COMMENT BELOW) Yes ☐ No ☐

OTHER ITEMS

INSPECTED BY:		DATE
REVIEWED BY:	TITLE	DATE

FIGURE 1.2. Back of Factory Mutual inspection blank.

Management should establish a clear line of authority among guards. Even when only two guards are present, one should be designated leader. Supervision of guards furnished by outside firms should be through the designated representatives of the company providing the guard service. That company should be given, in writing, specific details of the service they are expected to provide.

Management should see that guards are provided with suitable communications equipment. When watchclocks or time recording systems are used to assure performance of patrols, any irregularities in records should be investigated promptly.

PERIODIC FIRESAFETY INSPECTIONS

Periodic, recorded inspections of the fire protection equipment and fire systems are essential. The facility should also be checked for fire hazards. The fire prevention manager is responsible for scheduling the inspections and ensuring that they are made properly. Inspections are usually made at weekly intervals.

All sprinkler and fire protection water system control valves should receive attention. Closed valves have caused many industrial fire disasters. The inspector should personally examine, at every inspection, each and every control valve on piping supplying water to the fire protection systems. Water pressure, and the condition of dry-pipe valves, fire pumps, sprinkler alarms, and fire doors should also be checked and recorded. The inspection usually also includes noting fire hazards including defects in housekeeping.

A sample inspection form suitable for general industrial use is shown in figures 1.1 and 1.2. Similar forms can be adapted for use in properties of other types.

THE PLANT EMERGENCY ORGANIZATION

Every property should have an organization to deal with fires or related emergencies, which occasionally happen in spite of the best efforts to prevent them. Proper action in the first few minutes can make the difference between a minor incident and a disaster.

The makeup of the emergency organization (or fire brigade) will depend on individual conditions — the size and fire hazards of the property, the type of fire protective equipment, and the availability of the public fire department.

The organization, in its simplest form, would consist of a chief and two or three assistants.

In a typical sprinklered manufacturing plant there should be, for each operating shift,

1. A person in charge.
2. A team or squad for each major section.
3. An individual assigned to sprinkler control.

Other special assignments will usually be made depending on the size and type of the plant.

The person in charge, who usually has the title of fire chief, should be trained in fire fighting and be thoroughly familiar with the plant processes, fire hazards, and fire protective equipment. (See Chapter 3, Plant Emergency Training and Organization, for a more detailed discussion of the plant emergency organization.).

BIBLIOGRAPHY

NFPA CODES, STANDARDS, AND RECOMMENDED PRACTICES

Reference to the following NFPA Codes, Standards, and Recommended Practices will provide further information on good industrial risk management practices discussed in this chapter. (See the latest *NFPA Codes and Standards Catalog* for availability of current editions of the following documents.)

NFPA 1, *Fire Prevention Code*.
NFPA 26, *Supervision of Valves Controlling Water Supplies for Fire Protection*.
NFPA 27, *Private Fire Brigades*.
NFPA 601, *Guard Services in Fire Loss Prevention*.
NFPA 601A, *Guard Operations in Fire Loss Prevention*.

ADDITIONAL READING

Bennett, G., "Investigating the Ins and Outs of Fire Prevention," *Control and Instrumentation*, vol. 12, no. 2 (February 1980), p. 59.
Brannigan, Francis L., *Building Construction for the Fire Service*, 2nd ed., National Fire Protection Association, Quincy, MA, 1983.
Cooper, D., "Moves Towards Greater Self Regulation," *Fire Engineers Journal*, 40 (120), December 1980, pp. 33-34.
Fire Inspection Management Guidelines, National Fire Protection Association, Quincy, MA, 1982.
Hubitsky, J., "Preventing Fires on Job Sites," *Fire Command*, vol. 45, no. 7 (July 1978), pp. 34-35.
Industrial Fire Protection, IFSTA, Stillwater, OK, 1982.
Introduction to Fire Protection (a training package), National Fire Protection Association, Quincy, MA, 1982.
McKinnon, G. P., ed. *Fire Protection Handbook*, Fifteenth Edition, National Fire Protection Association, Quincy, MA, 1981.

Planner, R. G., *Fire Loss Control*, Marcel Dekker, Inc., New York, NY, 1979. (Available from NFPA)

Rutsein, R. and M. B. J. Clarke, "The Probability of Fire in Different Sectors of Industry," *Fire Surveyor*, vol. 8, no. 1 (1979), pp. 20-23.

"Self-Fire Inspection: A New Concept in Fire Safety," *Fire Chief Magazine*, vol. 24, no. 5 (May 1980), pp. 34-35.

Stephens, H. F., "Fire Strategy for Management," *Fire Prevention*, (133), November 1979, pp. 29-30.

Thor, J., G. Sedin, "Principles for Risk Evaluation and Expected Cost to Benefit of Different Fire Protective Measures in Industrial Buildings," *Fire Safety Journal*, vol 2., no. 3 (March 1980), pp. 153-166.

Tuck, C. A., Jr., ed., *NFPA Inspection Manual*, National Fire Protection Association, Boston, MA, 1982.

Waters, D., "The Fire Protection of Plant and Equipment," *Fire*, 70 (867), 1977, pp. 185-186, 189.

2

Life Safety in Industrial Occupancies

Ron Coté

The potential for loss of life due to fire in an industrial occupancy is directly related to the fire hazard risk of the manufacturing processes performed in the occupancy. Fire records show that a majority of the industrial fires that result in multiple deaths are the result of flash fires in highly combustible contents; or explosions involving combustible dusts, flammable liquids, or gases.

Although annual industrial fires constitute a high percentage of the total national fire loss (from a property standpoint), such fires have not, as a general rule, resulted in extensive loss of life. A number of favorable operating features common to industrial occupancies contribute to this, and continued emphasis on good fire protection and prevention and day-to-day attention to industrial safety and training programs can help to continue this trend.

If, during the initial planning of buildings, more consideration were given to the life safety of the future occupants, safer buildings would be designed. When a building is designed, whether it is an industrial structure or not, many factors deserve consideration in the

---

Ron Coté is Life Safety Specialist on the NFPA staff.

designer's attempt to produce an economical structure that best satisfies the needs of the owner. One of the most important is the life safety of the building's occupants from fire.

Many times buildings are in advanced stages of design before local firesafety codes are consulted. For the most part, local firesafety codes set forth minimum requirements that do not deal with all types of situations. Therefore, to assure that a building will be constructed with an acceptable level of life safety, it is necessary to include firesafety features during the initial design stages. All too often building owners take it for granted that architects and designers automatically include such features.

Other important factors to be considered during the planning of a building are: (1) the people who will occupy the building and their activities, (2) the type and severity of any fire that might occur in the building, and (3) the inherent design features of the building that could affect life safety, both positively and negatively, should a fire occur.

PEOPLE FACTORS

The design of means of egress is largely governed by human characteristics. Important human characteristics that need to be considered for safety of occupants from fire and similar hazards are as follows:

1. Physical and psychological/physiological characteristics
2. Age
3. Agility
4. Decision-making capabilities
5. Awareness
6. Training
7. Special knowledge and beliefs

Not included in the list of characteristics (since it is not a characteristic but the result of a situation) is people control. This is a factor that becomes important in many occupancies, including certain industries where the hazards of quick spreading fire and explosion are prevalent. It has been demonstrated that evacuation during fire emergencies will be orderly and rapid when the people involved have been exposed to disciplinary control and training. There will be little likelihood of panic, and people so trained will be able to deal with the unexpected.

Most occupants and employees of industrial buildings are ambulatory and of an age and agility that enables them to be fully capable of quick response to fire situations, as well as capable of rapid exit once properly alerted. To capitalize on this employee capability, many industrial plants include life safety measures in their emergency pre-planning. A well thought out preplan provides a valuable tool for helping to prevent loss of life. The provisions of an emergency preplan for alerting employees should include identifying and posting exit routes, special arrangements for assisting handicapped persons in the work force, establishing group assembly areas outside the building, and procedures for determining if all employees have safely exited. Responsibilities are usually established in the preplan to ensure that the tasks required to facilitate safe exiting of the building are accomplished. The preplan should be routinely evaluated by simulated fire exercises and fire drills. It is only through such drills that weaknesses in the preplan can be recognized and the plan modified.

Influence of Psychological and Physiological Factors

Despite the generally good physical and mental capabilities of most occupants of industrial properties, psychological and physiological factors must be considered in planning exit facilities. People cannot always be expected to behave logically during the stress of fire conditions. Panic is contagious, and the danger is greater in large groups. Fear, rather than actual fire danger, is the main factor in panic. Fatal panics have occurred where there was no fire, but people thought there was. On the other hand, because people have had confidence in a building and its exits, there have been orderly evacuations without panic, even though actual danger was present. As long as people can keep moving toward a recognized place of safety, there is little danger of panic, but any stoppage of movement is conducive to panic. Once panic starts, the exits may be quickly blocked.

TYPE AND SEVERITY OF FIRE

Anticipating the type and severity of the fires which may occur in an industrial plant is important because such predictions influence decisions on how to manage both the fire and the exposed occupants. The fire hazards inherent in each of the manufacturing processes involved in a production line must be carefully considered, particularly those operations that could be the source of flash fires and explosions. The fire characteristics of the materials being processed, such as their ignition temperatures, calorific values, rates of burning, smoke developed, and susceptibility to ignition, should also be studied carefully. The fire characteristics of materials that make up the structure must be evaluated before attempting to predict the kind and severity of any fire or explosion that may occur.

Information on equating fire loading (lbs of combustibles per sq ft of floor space) to fire severity in terms of time of expected duration is available and should be consulted during the evaluation. However, fire loading is only one of the parameters to be considered when predicting the kind and severity of a fire. All of the characteristics previously mentioned should be studied, and there are variables within some of those characteristics, e.g., variation of ignition temperature with size of specimen, size and shape of specimen, effect on rate of burning, etc.

Since the type of ignition source can relate to the kind of fire anticipated, a study of fire ignition sequences, e.g., the heat sources, the kindling fuels, and the events (human actions or natural acts that can get an ignition source and fuel together to start a fire), is appropriate. It is also appropriate to study other fires in similiar occupancies. Such reviews often reveal situations that some building designers do not believe can happen; the expression, "there is nothing to burn here," has often been shattered by history.

BUILDING DESIGN FEATURES

A fundamental of good industrial building design is to provide a structure that will best suit its intended purpose. Before any design is started, the designer should spend time determining the clients' needs, and, more specifically, determining the kinds of activities to be performed in the building. The building must provide for the most efficient utilization of the space available for the specific type of production processes involved. It must also provide an environment that will result in its efficient use by the people who will occupy it. Therefore, it should be recognized that certain buildings have inherent design features which should be considered when life safety protection for the building is studied.

Although life safety experience in industry has been relatively good, a major problem may be emerging with the trend to construct larger industrial plants housing hazardous operations. The modern industrial building has compounded life safety exposure to employees from fire. Compared to industrial buildings of the early twentieth century, the modern industrial complex places a larger number of employees in a more complex and increasingly hazardous environment. The introduction of new building materials with heightened fire hazards has increased the need for additional life safety measures to help ensure employee safety from fire. These trends have increased the need for industrial management to concentrate on life safety principles, not only during the design stage, but also during day-to-day plant operations.

Influence of Fire Protection Equipment

It is questionable practice to rely solely on fire extinguishment to the neglect of exits because of the possibility of both human and mechanical failures, and because loss of life may occur before fire fighting facilities can be effective. Under no conditions can manual fire fighting, however valuable, be accepted as an excuse for not providing proper exits.

When a complete system of automatic sprinklers is installed, it is

LIFE SAFETY IN INDUSTRIAL OCCUPANCIES

sufficiently reliable to have a major influence on life safety. Originally developed for industrial property protection, the automatic sprinkler has been largely responsible for the excellent life safety record of industrial occupancies. That this record has been recognized by fire protection engineers and fire authorities is evidenced by the widespread adoption of automatic sprinkler systems specifically designed for life safety protection in buildings with extensive life safety exposure. Automatic sprinkler protection in industrial occupancies has been a principal factor in life safety through control of fire spread. Limitation of fire size by sprinklers provides sufficient time for safe evacuation of personnel exposed to fire. The contribution of the automatic sprinkler to industrial life safety can only be fully evaluated when it is recognized that most industrial plants have a wide range of fire risks due to the variety of processes and product related hazards present.

NFPA *LIFE SAFETY CODE*® REQUIREMENTS

To a great extent, NFPA *101*®, hereafter referred to in this chapter as the NFPA *Life Safety Code*, was developed from a review of past catastrophic events as well as from considerable research and applied engineering judgment. The result is an approach to building design and operation which, if properly applied, reduces the potential for loss of life from fire. The NFPA *Life Safety Code* is concerned only with life safety. This is the difference between the NFPA *Life Safety Code* and fire protection provisions in building codes which concern themselves with the preservation of property in addition to life.

Industrial buildings of modern design have not, as yet, accumulated a major loss of life experience from fire. When fully incorporated, the measures in the NFPA *Life Safety Code* are sufficient to ensure against major loss of life from industrial plant fires.

In order to properly arrange the exit facilities of an industrial occupancy, the plant life safety risk should be fully evaluated. From this evaluation, the exit facilities and protection for employees from the effects of fire can be properly designed to ensure the necessary degree of employee life safety.

The Risk to Life Safety

To properly design exit facilities for an industrial plant, the occupany must first be classified for relative degree of fire and life safety risk. Operations involving low or ordinary hazard materials, processes, or contents are classified as general industrial occupancies for life safety plan-

---

®Registered trademark of the National Fire Protection Association, Inc.

ning purposes. Where high hazard materials, processes, or contents are housed in the building, the occupancy is classified as a high hazard industrial occupancy. Examples of a general industrial occupancy include electronic and metal fabrication operations, textile mills, automobile assembly operations, steel mills, and clothing manufacturing operations. Incidental high hazard operations, such as a paint spray booth or flammable liquid storage room in a low or ordinary hazard occupancy, do not classify the entire building as a high hazard occupancy. Examples of high hazard occupancies include paint and chemical plants, explosives manufacturing plants, grain or other combustible dust handling operations, plus any operation involving extensive quantities of flammable or hazardous materials.

Occupancy classification is necessary because the same life safety risk does not occur in all industrial occupancies. Once so classified, the design for life safety can be made using the basic design principles set forth in the NFPA *Life Safety Code*.

Some low and ordinary hazard industrial operations involve the use of extensive machinery or equipment which occupies a majority of the available floor space. In such instances the entire building need not be evaluated for exit purposes, since there are few people in the building. In a special purpose building, exits need only be provided for the actual number of persons in the structure.

MEANS OF EGRESS DESIGN REQUIREMENTS

Requirements in the NFPA *Life Safety Code* for arrangement of means of egress in industrial occupancies include many of the features required in any structure. The travel distance to an exit in a general industrial occupancy is 100 ft (30.5 m), except where the building is completely protected by an automatic sprinkler system. With automatic sprinkler protection, travel distance may be increased to 150 ft (45.7 m).

In large industrial buildings, the 150-ft (45.7 m) travel distance may not be possible without major building renovations, especially in large structures. In such cases, exit facilities consisting of exit tunnels, overhead passageways, or travel through fire walls utilizing horizontal exits will provide the necessary safeguards. In unusual situations additional precautions may permit increased travel distance of up to 400 ft (122 m). In such cases the contents should be limited to low or ordinary hazard in a general industrial or special purpose occupancy. Additional provisions required to make use of the increase in allowable travel distance should include, as a minimum, all the following:

1. Only one-story buildings should be considered.
2. Interior finish should be limited to Class A or B.
3. Full emergency lighting should be provided in the building.
4. An automatic sprinkler or other automatic fire extinguishing system

LIFE SAFETY IN INDUSTRIAL OCCUPANCIES

should be installed and fully supervised for malfunctions, closed valves, and water flow or operational alarm.
5. Smoke and heat venting or some other engineered means to limit spread of fire and smoke is necessary. The system design should ensure that occupants will not be overtaken by heat or smoke within 6 ft (1.8 m) of floor level before reaching the exits.

In high hazard industrial occupancies the maximum travel distance to an exit is reduced to 75 ft (22.8 m). The shorter travel distance ensures that employees can reach the exit rapidly when exposed to fire in hazardous materials or processes. A common path of travel of 50 ft (15 m) to a point from which one can then diverge to two separate exits in a general industrial occupancy is allowed. There is no allowable common path of travel in a high hazard industrial occupancy. Regardless of size, every high hazard occupancy is required to have at least two separate and remote exits.

Illumination of the Means of Egress

Emergency lighting has become very important in industrial occupancies, particularly with the increasing popularity of windowless structures which encompass large floor areas. Equally important is illumination of the means of egress. Where a building is only occupied in daylight hours and the means of egress is fully illuminated by means of skylights, windows, or other means of natural lighting, the normally required electrical illumination may be waived. In special purpose industrial occupancies where routine human habitation is not the case, emergency lighting may also be omitted.

PROTECTION FOR LIFE SAFETY IN INDUSTRIAL OCCUPANCIES

As a general rule, vertical openings should be fully enclosed. The enclosure may be omitted from vertical openings which are not used for exit purposes in existing buildings with low or ordinary hazard contents protected by an automatic sprinkler system. Another exception to the requirement for enclosure of openings is in specially designed industrial buildings housing operations, processes, or equipment requiring openings between floors. Should this be the case, each floor connected by the openings should be provided with exits, such as enclosed stairways, which are fully protected from obstruction by fire or smoke in the interconnecting floors.

Due to the size and complexity of most industrial structures, a fire alarm system is necessary to insure prompt and effective action. The fire alarm system should alert responsible persons in a continuously manned location so that positive steps to start fire fighting, employee evacuation, shutdown of hazardous processes, and other actions needed to limit the fire and life safety hazard are promptly taken. In high hazard occupancies, the fire alarm system should also immediately sound an alarm to notify employees to evacuate the building.

High hazard industrial occupancies present a unique fire control problem and a severe life safety hazard. High hazard industrial occupancies, operations, or processes should have automatic extinguishing systems or other equally effective protection, such as explosion venting or suppression, to minimize the life safety hazard. The protection system should allow occupants to escape before being exposed to a fire or explosion.

In general industrial occupancies with a total capacity of fewer than 100 persons, or fewer than 25 persons normally above or below the street level, the fire alarm system may be waived.

EXIT DRILLS

During a fire or an emergency in an industrial plant, a two-fold problem can arise: (1) the need to move essential personnel, such as the plant emergency organization, to locations where they are needed, and (2) the need to evacuate occupants quickly and efficiently.

Providing a means of egress is not enough to assure personnel safety. Exit drills are essential in order to obtain effective use of exits. They are required and commonly practiced in educational, health care, business and some residential occupancies. Drills should be conducted in industrial occupancies where life hazard from fire is high. In all cases some form of exit drill organization designed to ensure that someone will direct evacuating occupants and to avoid confusion between fire fighting and evacuation is desirable. Necessary functions include checking exits, selecting evacuation routes, controlling traffic, searching for stragglers, checking for occupants after they have exited the fire area, and controlling their return to the building when it is safe. A most important decision is determining the time when it is necessary to evacuate; in case of doubt, the building should always be evacuated.

Responsibility for planning exit or evacuation drills is generally assigned to the plant fire risk manager and staff. Plans should be discussed with both general and direct management to assure their understanding and cooperation. If there is no fire risk manager, the plant manager may assume the responsibility, or assign it to a member of the staff.

All employees should be advised of the evacuation signal and the exit route that they should follow. They should be instructed to shut off equipment immediately upon hearing the signal and report to a predetermined assembly point. In large facilities, primary and alternate routes should be established, and all employees should be trained in the use of either route.

When employees are assembled, the direct managers of each area should account for all personnel under their supervision, paying particular attention to the whereabouts of any handicapped persons in the area. (Pre-assigning mobile persons to accompany handicapped persons along the exit route in an emergency would be a helpful arrangement.) If one or more employees are not accounted for, this fact should be reported to the fire risk manager so that search and rescue efforts can be initiated. Only trained search and rescue personnel should be permitted to re-enter an evacuated area.

After each exit drill a meeting of the responsible managers should be held to evaluate the success of the drill and to work out details that might have been faulty or misunderstood.

The timing of drills will depend, to some extent, on the nature of the operation. As a general rule, drills conducted a few minutes before the lunch break have been found to minimize loss of time and production. The frequency of drills should be determined by the degree of hazard involved in the operation and by the complexity of shutdown and evacuation procedures.

If a facility does not maintain a security organization with responsibility for daily inspection of emergency exits and designated evacuation routes, one employee in each area should be given the assignment. Maintenance of doors, panic fire exit hardware, exit lights, etc., should be given high priority to assure that repairs will be made without delay.

BIBLIOGRAPHY

NFPA CODES, STANDARDS, AND RECOMMENDED PRACTICES

Reference to the following NFPA Codes, Standards, and Recommended Practices will provide further information on the safeguards for life safety in industrial occupancies discussed in this chapter. (See the latest *NFPA Codes and Standards Catalog* for availability of current editions of the following documents.)

NFPA 13, *Installation of Sprinkler Systems.*
NFPA 13A, *Care and Maintenance of Sprinkler Systems.*
NFPA 101, *Life Safety Code.*

ADDITIONAL READING

Automatic Sprinkler Systems Handbook, National Fire Protection Association, Quincy, MA, 1983.
Beddows, N. A., "Safety and Health Criteria in Plant Layout," *Chemical Engineering*, vol. 83, no. 22 (1976), pp. 133-136.
Canter, David, ed., *Fires and Human Behavior*, John Wiley & Sons, New York, NY, 1980.

"The Fire Resistance of Industrial Buildings," *Fire Protection*, June 1980, pp. 24-25.

"Growing Industrialization Calls for Renewed Emphasis on Fire Safety," *Fire News* (Alberta, Canada), vol. 4, no. 3 (September 1980), pp. 4-5.

Guide to NFPA National Building Firesafety Standards, National Fire Protection Association, Quincy, MA, 1983.

Life Safety Code Handbook (1981), National Fire Protection Association, Quincy, MA, 1981.

Mattoon, B. P., "Controlling the Human Factor in Industrial Fires," *Supervisory Management*, vol. 25, no. 6 (1978), pp. 2-8.

McKinnon, G. P., ed., *Fire Protection Handbook*, Fifteenth Edition, National Fire Protection Association, Quincy, MA, 1981.

3

Plant Emergency Organization and Training

John T. Higgins

Most industrial properties have an organization designed to deal with the fires and related emergencies that occasionally occur despite efforts to prevent them. It has been proven that proper action in the first few minutes of an emergency can make the difference between a minor incident and a disaster. A plant emergency organization can make that difference. The size and makeup of a plant emergency organization will depend on the plant's size, the hazards present, the type of fire equipment available, and the proximity and capability of the local public fire department. In addition, Subpart L of Title 29 of the Code of Federal Regulations, known as "OSHA Subpart L," introduces additional criteria for creating an emergency organization.

    It is assumed that the facility is fully protected by a system of automatic sprinklers meeting the requirements of NFPA 13, *Installation of Sprinkler Systems*. For plants not so protected, management should consider more training in manual fire fighting by plant personnel than is recommended here.

John T. Higgins is Manager of the Midland Plant Safety and Security Department of the Dow Corning Corp., Midland, MI.

Chapter 3

OPTIONS FOR HANDLING A FIRE EMERGENCY

Beginning in September 1980, new requirements for employee safety during fire emergencies and for employee fire brigades went into effect as Title 29 of the Code of Federal Regulations (Subpart L), published by the Occupational Safety and Health Administration (OSHA). The aim of the new requirements is to provide performance-oriented standards that give the employer a choice between five options for fire response.

OSHA Subpart L contains the following main sections:

1. Miscellaneous definitions	1910. 35(i & j)
	1910. 37(n)
2. Emergency action plan	1910. 38(a)
3. Fire prevention plant	1910. 38(b)
4. Fire brigade requirements	1910.156
5. Fire equipment requirements	1910.158–163
6. Fire detection systems	1910.164
7. Employee alarm systems	1910.165

In this chapter, however, only emergency action plans (1910.38[a]) and fire brigade requirements (1910.156) will be discussed.

A plant owner, usually in consultation with the person responsible for plant fire protection, decides how personnel will handle a fire emergency. Figure 3.1 is a decision tree, based on the OSHA guidelines outlined in 1910.156, which shows the various options that may be followed to fight a fire. Basically, five options are available, ranging from total evacuation, to a group of employees organized, trained, and equipped to fight interior structural fires (i.e., the structural fire brigade).

The first option requires that, upon hearing the alarm, all employees immediately evacuate the fire area to a safe location. Evacuation routes and procedures must be communicated to all employees. In this situation the employer does not need to train employees in how to extinguish fires.

If the second option is chosen, then every employee must be trained in the use of fire extinguishers. When a fire occurs, employees in the fire area pick up extinguishers and attempt to put out the fire. If the fire and resultant smoke are beyond control with extinguishers, the alarm must be sounded, and all employees must be evacuated to a safe location.

The third option is to assign designated employees who are trained to use fire extinguishers. In case of a fire, this group of employees responds with their extinguishers while all other employees evacuate or are assigned to emergency duties. If the fire is beyond control with extinguishers, the alarm is sounded and all employees must evacuate.

The fourth option is to establish a fire brigade to fight *incipient* fires only. This means that employees will be organized, trained, and equipped to fight fires and to assist with evacuation. Incipient, or the beginning stage of a fire, is the type of fire which can be controlled or extinguished using portable fire extinguishers, a small hose system 5/8 in.

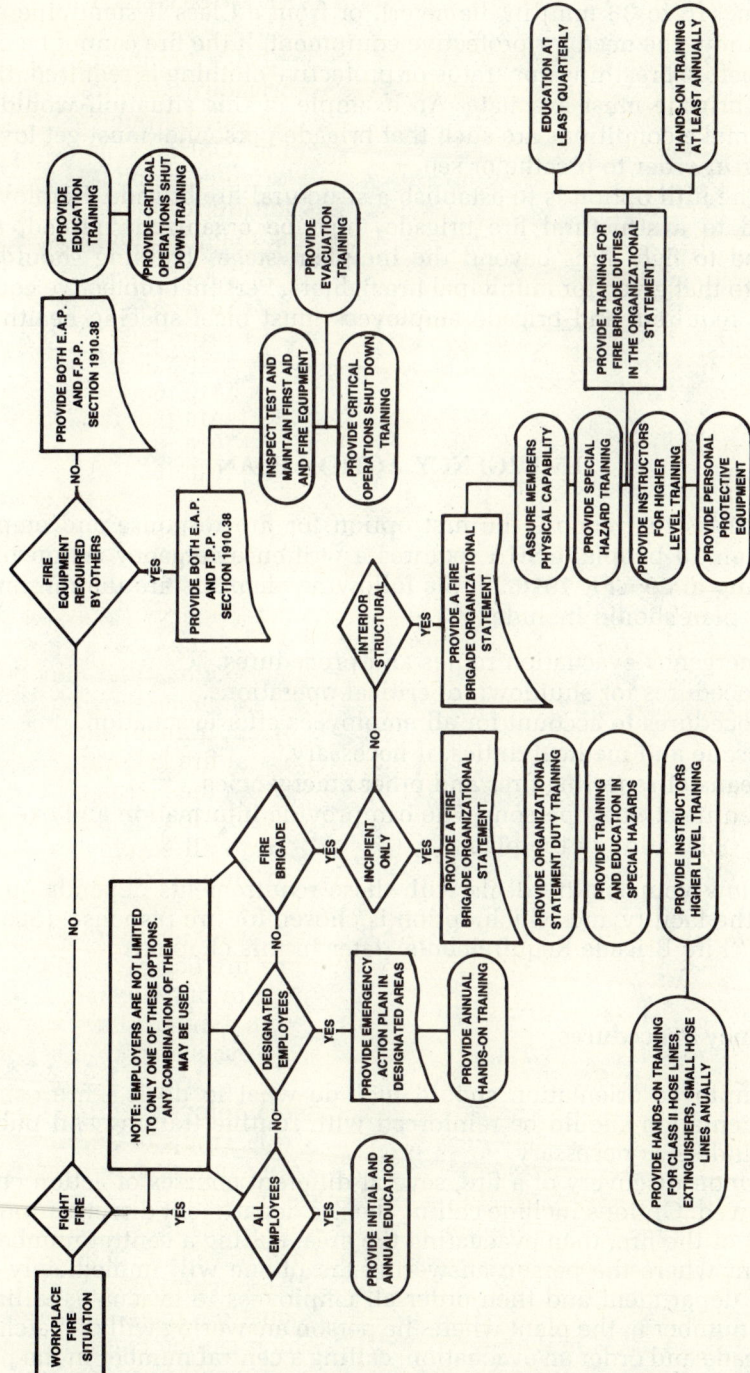

FIGURE 3.1. Decision tree showing the five options available to fight a fire.

to 1½ in. (16 to 38 mm) in diameter], or from a Class II standpipe system, without the need for protective equipment. If the fire cannot be controlled before breathing apparatus or protective clothing is required, then the fire brigade must evacuate. An example of this situation would be where smoke conditions are such that brigade personnel must get low to the floor in order to breathe or see.

The fifth option is to establish a structural fire brigade. Employees assigned to a structural fire brigade must be organized, trained, and equipped to fight fires beyond the incipient stage. Training should be similar to that given for municipal fire fighters. Personal protective equipment is required, and brigade employees must meet specific health requirements.

EMERGENCY ACTION PLAN

For facilities that choose the first option for fire response and employ more than 10 people, OSHA requires a written emergency action plan. According to 29 CFR 1910.38, the following elements are the minimum that the plan should include:

> Emergency evacuation routes and procedures.
> Procedures for shutdown of critical operations.
> Procedures to account for all employees after evacuation.
> Rescue and medical duties, if necessary.
> Means of reporting fires and other emergencies.
> Identification of persons who can provide information and explanation of the plan.

How your facility deals with these requirements depends on the size of the facility and which option is chosen for fire response. (See the section "Fire Brigade Requirements" later in this chapter.)

Emergency Procedures

New employee orientation should include what to do if a fire occurs. This orientation should be reinforced with routine training and publicity reminders as necessary.

Upon discovery of a fire, several different courses of action could be followed. Options include calling the fire department directly from the vicinity of the fire, then evacuating the area; calling a central number in the plant where the person answering the phone will immediately call the fire department and then order all employees to evacuate; calling a central number in the plant where the person answering will dispatch the fire brigade and order an evacuation; calling a central number in the plant where the person answering will call either the fire department or fire bri-

PLANT EMERGENCY ORGANIZATION AND TRAINING

gade while the original caller tries to suppress the fire with a fire extinguisher or a standpipe fire hose; or a combination of these actions.

Every employee should be directed to report a fire immediately upon discovery, then to evacuate or, if trained, to extinguish the fire with an extinguisher or hose. If employees are provided with an internal number to call when fire is discovered: 1.) management is notified that a problem exists and, 2.) someone notifies everyone else in the plant of the problem. If the plant is too small to assign someone to answer an emergency telephone, then the call should go directly to the fire department. In this case, a method must also be established to notify all other employees at the same time, perhaps by a page call using the telephone, a separate voice system for alarm purposes, or activation of a bell or horn alarm. The overall plan and training should specify what actions employees should take when they hear the alarm. If the initial action is not to evacuate, then a unique signal needs to be sounded to evacuate, if circumstances dictate this action.

If a central emergency number is used, the person answering the telephone must be trained to call either the fire department or fire brigade, to notify employees according to the pre-established plan, and to notify management so that the emergency is properly managed. Management will decide when to evacuate employees and when to call for more help. This chain of command must be established as part of the emergency action plan.

Evacuation Plans

The evacuation plan is the most important part of the emergency action plan since it addresses the safety of employees. This plan must be carefully prepared, and thoroughly understood by *all* employees. It should be reviewed annually with all employees and a practice evacuation should be held.

If the decision is made to evacuate, then all employees must evacuate. A plan must be in place on how to do this and where to go. Each area of the plant must have an evacuation plan from that area of the plant. The plan should be written and posted on the employee bulletin board. It should include the conditions when the employee should evacuate, a description of the evacuation signal that the employee will hear, a map showing the route the employee should take including an alternate route in case the primary route is blocked, the location where the employee should report, and the person to whom the employee should report. The evacuation gathering point may be outside or inside, as long as it is in an area isolated from smoke and safe for employees.

The plan must designate one person and at least two alternates to ascertain that everyone in the area, including visitors, have evacuated. It should name the person who is to meet the employees at the designated evacuation gathering point and account for everyone. This person will then report to the fire chief, fire brigade leader, or designated manager the

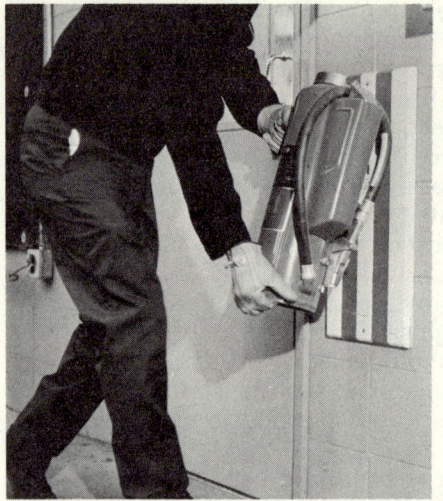

FIGURE 3.2. Extinguishers should be mounted on wall brackets for quick removal during an emergency. (Ansul)

FIGURE 3.3. Extinguishers should be carried by the handle, not by the valve stem or hose. (Ansul)

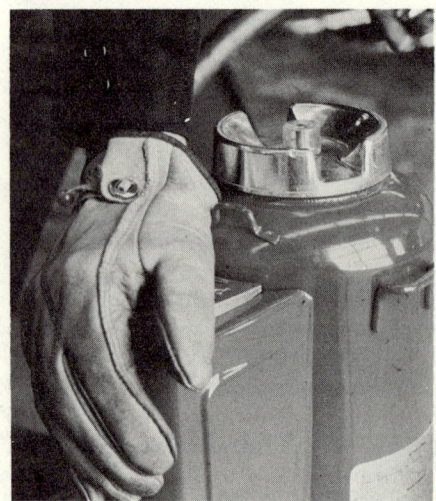

FIGURE 3.4. To release the propellant of this multipurpose dry chemical extinguisher, push the lever on the small canister after the hose has been removed. (Ansul)

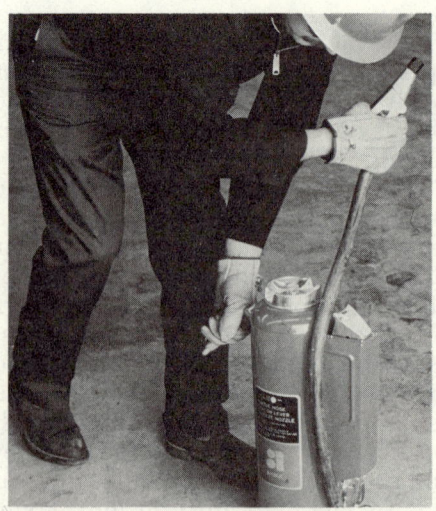

FIGURE 3.5. Release the extinguishing agent and aim it at the base of the fire. (Ansul)

PLANT EMERGENCY ORGANIZATION AND TRAINING

FIGURE 3.6. Apply two or more extinguishers to the fire at the same time if it cannot be extinguished by one person with one extinguisher. (Ansul)

status of all employees — who is safe, who is missing, and where they may be. This person should remain with the evacuated personnel and act as the communication link between the emergency forces at the emergency scene and the evacuated employees. If the situation changes, this person should lead the employees to a safer location and keep them informed of the fire fighting progress.

Fire Extinguisher Training

For facilities that choose the second or third options, OSHA regulations require that every employee (in the case of option 2) and only designated employees (in the case of option 3) be trained in the use of fire extinguishers and instructed to evacuate if they do not quickly extinguish the fire. The training must be done annually and can be a movie or classroom talk on how extinguishers work. It is far superior, however, to have employees actually practice using the extinguishers on a small fire, if local law allows this kind of training. If it does, the local fire department or fire extinguisher supplier will usually be happy to provide this training. Allowing employees to practice using extinguishers on a small fire will give them confidence to deal with a real fire in their area. The standard fire attack procedures common to virtually all extinguishers currently in use are shown in Figures 3.2 through 3.6. Extinguisher training is also necessary when the fourth or fifth options are chosen: if 1½-in. (38 mm) standpipe fire hose or a small hose system is available, those choosing option 4 must also include training in the use of this equipment in their program.

It is recommended that employees be instructed to: first, report the fire; second, attempt to extinguish it with a fire extinguisher; and, if that fails, third, evacuate the area. Employees should also be instructed to send someone to meet the responding forces.

Plant Emergency Plan

If options 1, 2, or 3 are chosen, someone at the plant must be designated to be in charge of fire protection; otherwise, the program will not work. That person's responsibility will include assuring proper testing and maintenance of all fixed fire suppression equipment such as sprinkler and Halon systems, fire extinguisher training as needed, and training in the emergency action plan. If there is an incipient or structural fire brigade, or if the plant is fairly large, a more detailed plan is probably needed. The complexity of the plan depends on the individual situation.

It is recommended that the plant emergency plan be divided into two or three phases, depending on whether there is a fire brigade. Phase 1 covers the initial report of the fire and the actions taken immediately thereafter. This should include procedures for: calling the fire department or fire brigade, use of fire extinguishers by employees in the fire area, and evacuation. A management employee should respond to the fire scene and decide if further action is necessary. If the fire is extinguished quickly, then this person will supervise cleanup, decide when employees can return to work, and initiate the fire investigation.

Phase 2 covers fire fighting by the fire brigade if the plant has one. Management's representative would assure that all employees in the affected area are evacuated safely and that the fire is quickly handled by the available brigade personnel. The management representative may want to notify other management team members during the emergency. When the fire is out, the procedure is as in Phase 1. The fixed fire suppression system may have been activated, so action must also be taken to return this system to service as quickly as possible.

The last phase of the plan, Phase 3, should be activated if the fire cannot be quickly controlled by the fire brigade and the fire department must be called to assist in extinguishment. At this time a management team should be appointed to deal with all problems associated with the emergency. This team should establish a headquarters and maintain constant communication with their respresentative at the scene of the emergency if the locations are different. Figure 3.7 shows a sample organization chart for a medium size manufacturing plant. Note all of the

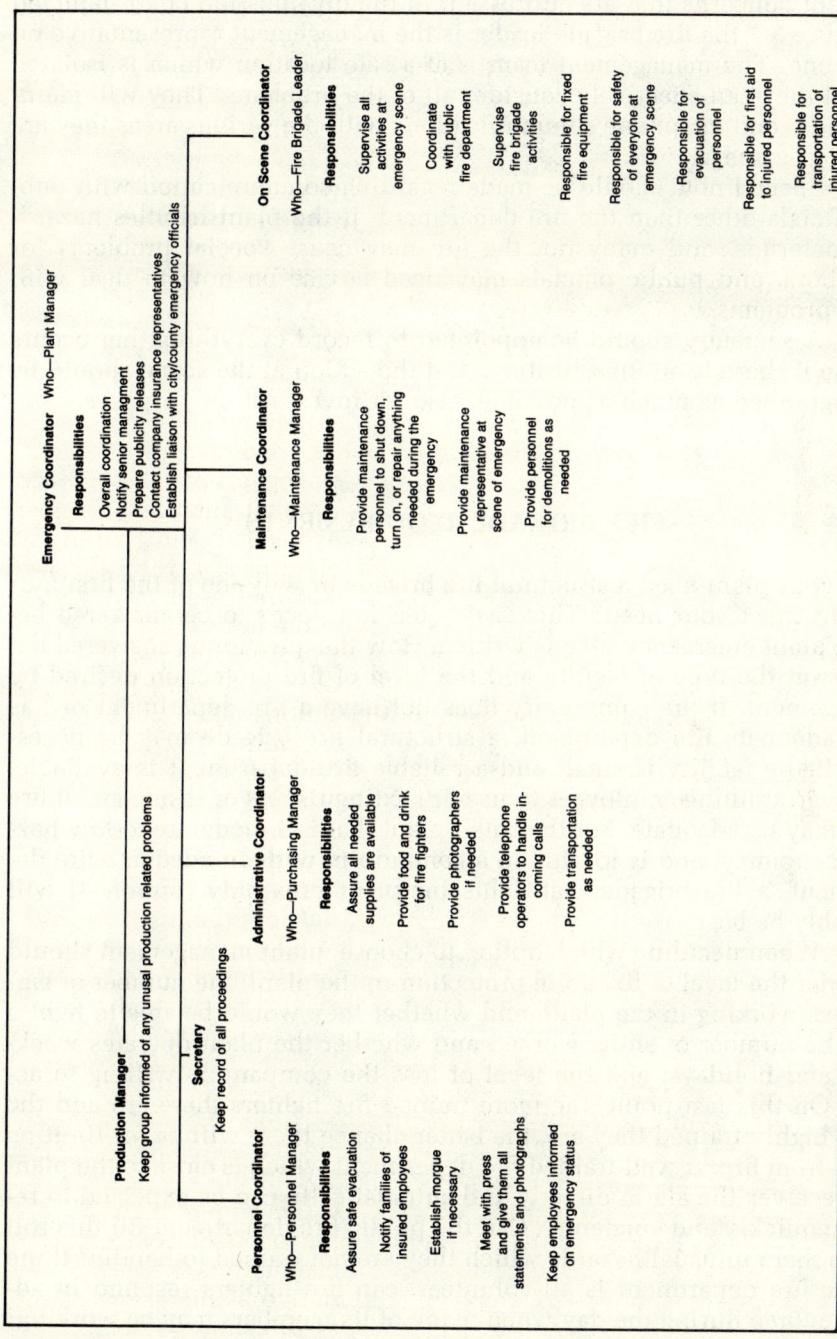

FIGURE 3.7. Sample organization chart for a medium sized manufacturing plant.

different concerns that are addressed. In the organization chart depicted in Figure 3.7 the fire brigade leader is the management representative on the scene. The management team is at a safe location which is isolated so that they can effectively consider all of the problems. They will maintain radio and telephone communications with the various areas they are trying to manage.

Special note should be made regarding communication with public officials other than the fire department. If the plant handles hazardous materials, and many do, the fire may cause special problems for neighbors, and public officials may need advice on how to deal with these problems.

A secretary should be appointed to record everything that occurs for use if there is an investigation, and the action at the scene should be photographed as much as possible, also for investigation purposes.

FIRE BRIGADE REQUIREMENTS

Does your plant need a structural fire brigade or will one of the first four options meet your need? This basic question needs to be answered before a plant emergency plan is written. How the question is answered depends on the type of facility and the level of fire protection desired by management. If the community does not have a fire department or has an inadequate fire department, a structural fire brigade may be necessary. If the facility is small and a reliable fire department is available, however, training employees to use fire extinguishers or inside small fire hose may be adequate. For the small plant that is a moderate-to-low hazard occupancy, and is located in a community with an adequate fire department, a fire brigade that fights incipient fires only (option 4) will probably be best.

When deciding which option to choose, plant management should consider the level of fixed fire protection in the plant; the number of employees working in the plant, and whether they would be able to fight a fire; the number of shifts worked and whether the plant operates weekends and holidays; and the level of loss the company is willing to accept. On this last point, the more trained fire fighters there are and the more highly trained they are, the better chance there will be for limiting a loss from fire. A well-trained fire department which is close to the plant and receives the alarm directly and automatically can be expected to respond quickly and efficiently. Can the public fire department fill this roll or are there unusual hazards which they are not trained to handle? If the public fire department is all-volunteer, can fire fighters respond in adequate force during the day when many of its members may be working? Whether a fire brigade is necessary, and the level of its training, are decisions which should be made after this kind of analysis is done.

Fire Brigade Organization Statement

If the decision is made to have an incipient stage or structural fire brigade, a written fire brigade organizational statement is required by OSHA regulations. This statement must explain the purpose of the fire brigade (incipient or structural); the organization for each shift, including officers and number of members; the training which members will receive, including content and time spent; and where the employees who form the brigade will come from within the plant. In addition, for structural fire brigades, OSHA requires that all brigade personnel be screened for medical problems that may be dangerous under fire conditions and that protective equipment for head, face, eyes, lungs, body, hands, legs, and feet be provided. Figure 3.8 shows an organizational statement using a plant security department as the fire brigade, although many other possible sources for staffing the brigade exist.

Selection of the Fire Brigade

The fire brigade chief and the officers selected should have some knowledge of fire fighting techniques. More importantly, they must have leadership ability and motivation, for it is their enthusiasm that will determine how successfully the brigade will work together as a team. If it is necessary to choose leaders who do not have fire fighting experience, such leaders should attend one of the previously mentioned fire schools prior to assuming their fire responsibilities.

The fire brigade chief is responsible for the selection and appointment of qualified brigade members; for adequate brigade coverage on all shifts (including vacations, downtime, and shutdown); for supervision of fire fighting activities and all brigade activities; for evaluation, inspection, maintenance, and replacement of brigade equipment; for brigade training (or supervision of the training officer); and for protection during scheduled and nonscheduled automatic fire protection system impairment.

The brigade members should either be volunteers from plant employees or persons hired with the understanding that some of their responsibilities will be with the fire brigade. They should be able to leave their work stations and allowed to remain at the emergency scene until their fire fighting responsibilities are completed. Employees selected for the brigade should ideally have some knowledge of fire fighting, plant systems, and plant layout. Each person serving on a structural fire brigade must be physically able to fight fires. This means that each of the structural brigade members must have an annual physical exam to assure, from a health standpoint, that they are able to continue fighting fires.

Fire brigade policies should require that each member maintain a specified level of proficiency. This policy may require members to at-

ORGANIZATION STATEMENT
PLANT FIRE BRIGADE

The Security Department shall perform the duties of a fire brigade at this site.

The organization structure shall be as follows:

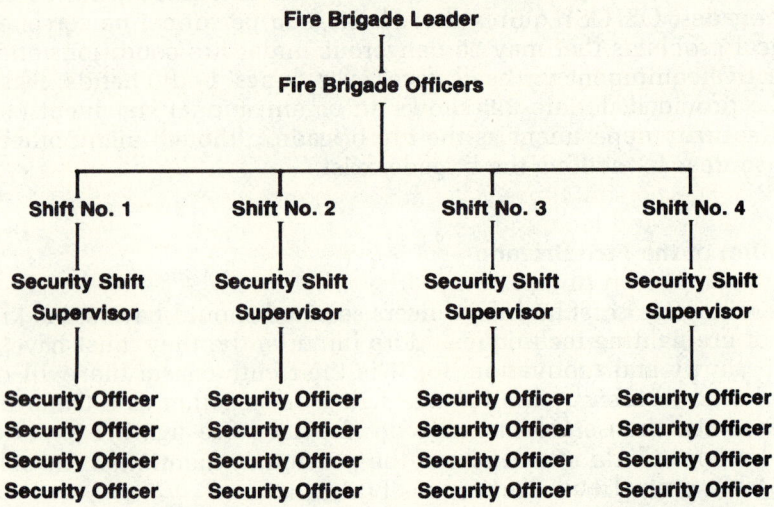

In the absence of the fire brigade leader, the senior fire brigade officer present will assume command. The Security Shift Supervisor on duty during an emergency is in charge of communication but the senior fire brigade officer is in charge of on scene operations until the arrival of the fire brigabe leader. The fire brigade leader is responsible for training and direction of fire fighting activities.

Training will be provided in the use of self contained breathing apparatus, personal protective equipment including coats, hats, gloves, and boots, and the use of fire trucks, and fire hose, including actual application of water and foam in three training sessions anually. Each session will consist of a minimum of two hours of training. When employees are assigned to the brigade they will reveive eight hours of initial training and eight hours of training annually thereafter.

The organization will consist of a minimum of a fire brigade leader, four fire brigade officers, four security shift supervisors, and 16 security officers. At lease one fire brigade officer and four security officers will be at the scene of any fire. A security shift supervisor will handle the alerting of personnel and all other communications.

The fire brigade is expected to report to any fire on the site and provide extinguishment using water, dry chemical, or various foams. Fires will usually be small but can involve structural fires as well. Back up is provided by the city fire department.

FIGURE 3.8. Organization statement.

tend training sessions and pass an annual exam. A probationary period for each new member may also be a requirement for brigade status.

Fire Brigade Training

Several different methods for training a fire brigade are available and a combination of these should meet most needs. There are many excellent fire schools in the United States that will train fire brigades. Some of these are listed in OSHA Subpart L and others advertise in national magazines. Also, NFPA's *Fire Almanac* contains a list of state fire training facilities. Training manuals are available, some with student workbooks and classroom materials, that can be used by the brigade leader in the training program. The local fire department may be willing to help train the brigade and should, in any case, be included in overall planning and training. The person chosen as fire brigade leader must know fire fighting principles and techniques.

The nature of the fire brigade training is determined by the type of fire brigade chosen (incipient or structural) and what type of equipment is available. Some of the information that should be included in a training program is described below.

Any program should begin with a discussion of the basics. Discuss what fire is, including the fire triangle and tetrahedron, the chemistry of fire, the toxicity of smoke, and the types of fire hazards encountered in the plant. Be sure to include any flammable and combustible liquids that are stored or used and discuss how they should be protected. Describe storage arrangements, including high-piled stock. Identify any unusual fire hazards such as rolled paper; baled rags, cotton, or fiber; peroxides; and oil under pressure. Review the types of construction found in the plant and the problems associated with fighting fire in each. For example, is there one very large open floor manufacturing facility, many open structures, or several temporary structures such as trailers? Each of these structure types requires different fire fighting techniques.

The next phase of fire brigade training should address the types of fire fighting equipment available for use at the site. This training should include how each piece of equipment works and the techniques used to maximize the equipment's fire fighting potential. Discuss the fire extinguishers available, what type (class) of fire each is best-suited for, the limitations of each, and any situations where it is dangerous to use an extinguisher. The use of standpipe hoses and the location of these hoses should be reviewed. If the fire brigade is a structural one, also discuss the use of larger hose and fire hydrants, how fire pumps work (if there are any), and the use of any fire apparatus on site. The use of personal protective equipment, including self-contained breathing apparatus, coats, boots, hats, and gloves, should be discussed. How, where, and when to use this equipment should be demonstrated.

Phase three of the training program is actual practice in the use of available equipment. Part of this practice should include proper re-

Chapter 3

FIGURE 3.9. Many types of industrial trucks can be used to carry fire brigade equipment to the scene of a fire. (L. L. Bean, Inc.)

sponse techniques and how to operate each piece of equipment available to the brigade, while another part of the practice should be to use the equipment on actual fires, if they can be ignited in the area. Be sure to get permission from whatever local authorities must be contacted, including state environmental authorities and the local fire department. The local fire department may even be willing to help with this phase of training. If actual fire fighting cannot be provided locally, one of the fire schools mentioned earlier should be considered because practice in fighting actual fires is a very important part of the training program, particularly for a structural fire brigade.

The last phase of training should cover the automatic fire suppression equipment available on the premises, including how the equipment works, how it is maintained, how it helps fire fighting, and what actions the brigade will take to use this equipment during a fire (how long to let it run and monitoring the control valve). In some plants the brigade has some responsibilities for inspection and/or maintenance of the equipment. This is helpful to the brigade since the members will have an opportunity to learn the location of the equipment and how it works. It will also serve as a method to get the brigade into the plant so they know their way around the areas where they do not normally work.

A training schedule must be established and followed. Plant management must be committed to the brigade and should allow the brigade members to be available for training. Training must be well planned to get the most out of the time available. Training should be at least quarterly for structual fire brigades (monthly would be even better) and at least annually for incipient stage fire brigades.

A final note on brigade training: since, hopefully, it will not be necessary to use the brigade very often, it is important to keep their spirits high. Using them to maintain equipment helps, putting on fire displays at least annually helps, but when management recognizes the efforts of these dedicated people, morale is boosted. There are many ways to recognize members, including an annual dinner, a special badge to wear on work clothing, stories in the plant newspaper, or any other method of recognition which will establish pride in being part of this important function.

Prefire Plan

Brigade officers should develop a prefire plan for all areas of the plant. This plan should include the level of local fire department response, including when they will be called. If you do not include the local fire department as part of the training program already discussed, you should invite them into the plant for a tour and a review of your prefire plans, including their part in the plans. Be sure the fire department is familiar with your fixed fire suppression equipment and the importance of keeping it in operation as part of the prefire plan.

It is important that brigade and fire department officers maintain a working relationship with each other so that they can work together smoothly in case of a fire. Brigade officers should remember two things when dealing with the local fire department. First, the law usually mandates that the local fire chief is in complete charge when called to a fire. Second, the brigade officer knows more about the plant and its unique problems than the fire chief probably does. Therefore, it is very important that both groups work together with a mutual respect for and an understanding of each other's responsibilities.

SUMMARY

In this chapter, the various options available for fire response have been explained. An emergency plan must be prepared and communicated to employees. It can be as simple as calling the fire department and then evacuating the facility, or it can be as sophisticated as establishing a structural fire brigade and detailed plant emergency actions. However, there are many options between these two extremes. The main thrust of the plan should be employee safety with a second objective of preserving company assets by limiting damage as much as possible.

BIBLIOGRAPHY

NFPA CODES, STANDARDS, AND RECOMMENDED PRACTICES

Reference to the following NFPA Codes, Standards, and Recommended Practices will provide further information about fire brigades. (See the latest *NFPA Standards and Codes Catalog* for availability of current editions of the following documents.)

NFPA 1, *Fire Prevention Code.*
NFPA 10, *Portable Fire Extinguishers.*
NFPA 13, *Installation of Sprinkler Systems.*
NFPA 12A, *Care and Maintenance of Sprinkler Systems.*
NFPA 14, *Standpipe and Hose Systems.*
NFPA 27, *Private Fire Brigades.*
NFPA 30, *Flammable and Combustible Liquids Code.*
NFPA 101, *Life Safety Code.*

ADDITIONAL READING

Bugbee, Percy, *Principles of Fire Protection*, National Fire Protection Association, Quincy, MA, 1978.

Chinnock, J. H. J., "Pre-Planning for Safety in Industry," *Fire Engineers Journal*, 36(104) (1976), pp. 25-27.

Corey, D. C., "Plant Brigades Need Fire Exposure," *Fire Command*, vol. 46, no. 7 (July 1979), pp. 18, 20.

"FPA Guide to Industrial Fire Safety," *Fire Prevention*, (113), 1976, pp. 21-28.

Gold, David T., *Fire Brigade Training Manual* and *Instructor's Guide*, National Fire Protection Association, Quincy, MA, 1982.

Introduction to Fire Protection (training course for employees involved in facility safety programs), National Fire Protection Association, Quincy, MA, 1982.

Jenaway, W. F., "Training the Fire-Fighting Brigade," *Plant Engineering*, vol. 31, no. 23 (1977), pp. 165-167.

Katzel, Jeanine, "Planning for Plant Emergencies," *Plant Engineering*, vol. 36, no. 1 (1983), pp. 34-41.

Mattoon, B. P., "Controlling the Human Factor in Industrial Fires," *Supervisory Manager*, vol. 23, no. 6 (1978), pp. 2-8.

McKinnon, Gordon P., ed., *Fire Protection Handbook*, Fifteenth Edition, National Fire Protection Association, Quincy, MA, 1981.

Miller, J. S., "Is Your Plant Fire Brigade Familiar with Subpart L?", *National Safety News*, vol. 123, no. 6, (June 1981), pp. 33-35.

National Safety Council, *Accident Prevention Manual*, National Safety Council, Chicago, IL, 1981.

Peige, John D., ed., IFSTA 200, *Essentials of Fire Fighting*, Oklahoma State University, Stillwater, OK, 1977.

"Prefire Planning," *Record*, vol. 55, no. 4(1970), pp. 3,5, 7-9.

Scott, A., "Fire Safety: Industry Must Look to Itself," *Fire Engineers Journal*, vol. 40, no. 120 (December 1980), pp. 35-37.

U.S. Department of Labor, 29 CFR Part 1910.156, *Subpart L*, 1980.

Part Two

FIRE HAZARDS IN MAJOR INDUSTRIES

4

Electric Generating Plants

John Luley, P.E.

Electricity is produced in electric generating plants from several energy sources. These sources include fossil fuel, nuclear fuel, and water (hydroelectric). While this chapter concerns only those electric generating plants that produce electricity from coal or oil, many of the situations and recommendations contained herein may be applicable to generating plants that use other energy sources.

American consumers used more than 2 trillion kWh (kilowatt-hours) of electricity in 1982. More than one-third of this power was used in residential circuits and appliances. Ninety-four percent of the balance was used by industrial and commercial consumers. The electric utility industry employed approximately 390,000 people to operate and maintain the electric systems that supplied that power. Gross revenue derived from sales of electricity in that year exceeded $121 billion.

In 1982, 69.5 percent of the electricity produced in this country came from coal- or oil-fueled generating units. These plants are found throughout the United States and are generally located near a body of water because of their operational requirement for large quantities of

---

John Luley is Loss Control Administrator for the Potomac Electric Power Company of Washington, DC.

cooling water. Each unit normally consists of a boiler and a turbine-generator combination. Beyond these basic similarities, individual generating plants tend to differ widely due to factors such as local climate, unit size, selected unit operating conditions, and number of units per plant. Likewise, fire protection design for electric generating plants is not uniform. However, certain representative hazards can be identified and some of the protection methods used to counter these hazards will be discussed in this chapter.

Some factors to be considered in the fire protection design of new coal or oil-fueled generating plants are:

1. The fuel itself is the most obvious fire hazard; its purpose at the plant is for controlled burning.
2. The process of producing electricity creates large heat loads in the generator. This heat must be constantly removed, and many generators are cooled with hydrogen — a flammable gas.
3. Rotating machinery, such as a turbine-generator, requires lubrication. Combustible lubricating oil may be found in large quantities in an electric generating plant.
4. The size of the equipment used in a generating plant combined with operational and maintenance requirements for access to the equipment generally results in large undivided spaces. In many cases the entire building constitutes a single undivided occupancy.
5. Economic, ecological, and social restraints have resulted in the location of many new plants in rural areas. Plants may, therefore, need to be self-sufficient for fire protection purposes.
6. The necessity to protect our environment from the effects of large scale boiler exhaust emissions has dictated new and large pollution control equipment including precipitators, baghouses, and scrubbers. Since each of these serves to trap undesirable materials exiting the boiler, they can also accumulate unburned fuel residues.

RAW MATERIALS

Coal

Chemically, coal consists largely of carbon, with varying amounts of hydrogen, oxygen, nitrogen, sulfur, and mineral ash. The physical form of coal varies from anthracite, which is a clean, dense, hard coal with low volatile content, to lignite, which is a soft, brownish coal in which the original woody structure is still apparent.

All coals may undergo spontaneous heating. In general, the lower the grade of coal the greater the probability of that happening. Freshly mined coal absorbs oxygen more rapidly and is more likely to heat spontaneously than coal that has been out of the mine for some time. Factors such as high moisture content, small particle size, and high sulfur compound content all tend to increase the spontaneous heating probability.

Transfer and storage of coal should involve as little handling as possible to prevent breaking the coal into small particles. Roll packing (compacting layers by "rolling" over them) storage piles into tight, thin layers helps prevent spontaneous heating because it excludes oxygen and helps the storage pile shed water instead of absorbing it. Coal storage piles should be located remote from important structures.

Oil

Fuel oil is composed of various compounds of hydrocarbons. There are several grades of fuel oil, and these are basically classified according to their degree of refinement. *Crude oil* is unrefined and contains a spectrum of components from the lightest, most volatile compounds to heavy, waxy compounds. The specific composition of crude oil can rarely be predicted from batch to batch. Most crude oils give off more vapor than refined fuel oils due to their volatile light fractions. Their flash points may be below ambient temperature. Crude oil should be handled as a Class I flammable liquid. (See Chapter 37 for a complete discussion of "Handling and Storage of Flammable Liquids.") *Number 6 fuel oil* is a processed, heavy distillate with a flash point above 140°F (60°C). Therefore, Number 6 fuel oil is a Class III combustible liquid. *Numbers 4 and 5 fuel oils* are also heavy distillates, but with flash points at or above 130°F (54.4°C). *Numbers 1 and 2 fuel oils* are middle distillates with flash points above 100°F (37.8°C) and below 165°F (73.9°C) and 190°F (87.8°C), respectively.

Under most conditions, Numbers 1 and 2 fuel oils can be stored, atomized, and pumped at ambient temperatures. The other grades generally require heating or agitation, or a combination thereof, to keep them fluid.

Since the quantities of fuel oil required are generally sizable, large aboveground storage tanks are necessary. There are basically two types of nonpressurized or atmospheric tanks used for fuel oil storage — cone-roof tanks and floating-roof tanks. Cone-roof tanks require venting during filling or removal operations. They also contain a vapor space between the liquid surface and the fixed roof, which, under certain circumstances, may contain a flammable or explosive mixture. A floating-roof tank employs a bouyant roof which floats on the surface of the liquid and, therefore, does not contain a vapor space above the liquid surface. There are three general types of floating roofs: (1) pan, (2) single-deck pontoon, and (3) double-deck pontoon. The pan type is lowest in cost, but is subject to submergence and can be damaged by wind when the tank is full. The single-deck pontoon is like the pan type, except it has built-in compartments that make the roof less likely to submerge. The double-deck pontoon roof has a number of compartments that promote better floating characteristics and can take larger snow and water loads. Another type of storage tank, the internal floating roof tank (also known as a "hard hat floater"), has a floating roof contained in a cone-roof tank. This type of tank has gained popularity because it provides more protection than an internal floating roof tank. It requires less maintenance, has less evaporation loss, and preserves the purity of the fuel oil better.

Large fuel oil storage tanks should be remotely located from important structures. Since hydrocarbon vapors are heavier than air, tankage should be located downgrade from the plant if possible. Prevailing winds should be taken into consideration to minimize the possibility of released vapors drifting to an ignition source. Tanks should be diked in accordance with NFPA 30, *Flammable and Combustible Liquids Code*. Since some fuel oils are susceptible to ejection from the tank through slopover, frothover, or boilover, even properly designed dikes may be defeated.

THE PRODUCTION PROCESS

An electric generating plant converts the potential chemical energy of the fuel into electrical energy by means of a mechanical process. The fuel is conveyed from the storage area to the generating plant where it is prepared for introduction into the boiler. Coal is pulverized to a powder and then entrained in a stream of air. Oil is heated, if necessary, and pressurized. The fuel is then burned in the boiler to generate steam. The steam produced may be at pressures up to 5,000 psi (34 474 kPa) and at a temperature of approximately 1,000°F (538°C). The steam drives a turbine which, in turn, drives the generator. The turbine-generator combination is usually either a cross-compound or tandem-compound arrangement in which turbine sections operate at various pressures: high, medium, and low. Generators normally operate at either 1,800 or 3,600 rpm. They produce electricity at voltages which are characteristic of the specific design of the machine. The output voltage is raised to transmission voltage (69 to 500 kV) by a main power transformer for transmission to consumer locations.

The electric utility industry, in an effort to produce electricity ever more efficiently, has developed several efficiency enhancing additions to the above process which are common or standard in most modern power plants. For instance, nearly all modern turbines incorporate a condensor to return the steam to liquid form after it has passed through the turbine. This increases efficiency by creating a vacuum at the end of the path of steam travel which increases the pressure differential available for use in the turbines. Another method of increasing unit efficiency is by preheating the combustion air so that it will absorb, and waste, as little heat as possible in its travel through the boiler and up the stack. The air preheater in a steam generating unit reclaims heat from the flue gas and adds this heat to the air required for combustion.

Another concern of the industry, and society as a whole, is pol-

ELECTRIC GENERATING PLANTS

lution control. Pollution control requirements have resulted in additional equipment being located in the boiler flue gas train. These include precipitators, bag-house filters, and scrubbers. Some types of bag filters are combustible and should be avoided. Scrubbers often have a combustible liner. All flue gas train equipment may be subject to fuel carryover from incomplete combustion in the boiler.

This entire production process is usually controlled from a central control-room complex. Often a process computer will be used. The control-room complex will generally include the control room, a cable spreading room, a relay room, a battery room, and a computer room. (See Chapter 40 for a complete discussion of the fire hazards associated with Computer Centers.) The control-room complex is critical to the operation of the generating plant.

THE FIRE HAZARDS

The first step in the production process includes receiving and storing the fuel, recovering it from storage, preparing it for combustion, and then moving it to and into the boiler. The raw coal which is delivered to the plant may contain foreign substances such as scrap iron, wood shavings, rags, and rocks which can interrupt coal feed, damage or jam equipment, or become a source of ignition within a pulverizer. This foreign material should be removed prior to coal crushing. Coal is moved by conveyors. A fire in the conveyor system can be especially serious since it is often a running fire (carried on a moving belt). After removal from storage, coal is generally broken into pebble-sized pieces and transferred to a "bunker" inside the plant. At this point, the coal presents a potential hazard from spontaneous ignition and also from off-gassing. The hazard is significantly increased if the coal must be retained in the bunker for any length of time. The coal passes from the bunker to a pulverizer where it is ground to a very fine consistency and then entrained in an air stream to be carried to the boiler. The failure of any of the piping carrying the coal dust suspension between the pulverizer and burner can release an explosive cloud of coal dust or result in a "blowtorch" effect if ignited quickly. Erosion of the piping, especially at bends, is the most common reason for piping failure. Of primary concern is the design of burner piping (and the operation of burner piping systems) so that transport air velocities which prevent the settling of fuel are maintained. Fuel settling is dangerous because a furnace explosion may result when airflow is increased in a burner pipe in which fuel has settled.

Oil is generally carried from storage to the boiler by fixed piping. The hazards here are the pressure and temperature of the oil. The heavier oil distillates require heating to become pumpable. Usually they are heated to between 130°F and 180°F (54.4 and 82.2°C) and pumped into the plant at about 100 psi (690 kPa) pressure. Depending upon the atomizing method used, the fuel oil pressure can be raised to between 150 and

300 psi (1034 and 2069 kPa), for air or steam atomizing, or around 1,000 psi (6895 kPa), for mechanical atomizing. Rupture of an oil line at these high pressures can result in an extensive spill or an atomized spray if the failure is only a crack.

The major hazard associated with a boiler-furnace is a fire box implosion or explosion. An uneven or low fuel flow at the burner can result in loss of flame in a portion of the boiler. Continued introduction of fuel eventually builds an explosive "cloud" which often finds a source of ignition, either from other burner flames or from the hot boiler surfaces. Loss of fuel luminescence or changes in airflow may cause furnace implosions which can be extremely damaging. (See Chapter 42 for a more detailed explanation of the fire hazards associated with boilers.)

Carryover of unburned fuel particles may result from poor or inadequate fuel/air ratio control, malfunction of a fuel admission valve, or the loss of a forced draft fan. This carryover presents a hazard to equipment such as air preheaters, electrostatic precipitators, and baghouse filters in the flue gas train. Flue gas scrubbers, which are located in the flue gas train, generally operate wet and are not normally affected by carryover. However, they usually have combustible linings which are a hazard whenever the scrubber is not in operation, or if the water spray system fails during operation.

Steam-driven turbine-generators are constructed to very close tolerances and balances because they operate at high rotational speeds. To keep them running smoothly, substantial lubrication systems are necessary. Also, many turbine-generators are controlled hydraulically. Because of the large quantities of oil involved in both turbine and generator, the extensive piping systems and high pressures necessary to deliver this oil to bearings, hydraulic mechanisms, coolers, etc., and the various auxiliary equipment used for instrumentation, the lubricating and hydraulic oils must be considered significant hazards.

Once the steam has passed through the turbine, it is cooled in the condensor. This requires large volumes of water and, occasionally, cooling towers. Cooling towers can be constructed almost entirely of noncombustible materials, eliminating any fire hazard. All too often, however, for economic or other reasons, wood or plastic materials which present significant hazard are used. This is especially true in the fill area. Fill, hangers, drift eliminators and compartment walls are often made of combustible material while the balance of the structure is noncombustible.

Generators create, as a by-product of generating electricity, large amounts of heat. They must be continuously cooled while in operation. Many generators are cooled by hydrogen. A very low energy input will be sufficient to ignite a mixture of hydrogen and air. If not controlled carefully, hydrogen can become a serious fire hazard.

As the electricity is produced, it must be converted to a suitable voltage for transmission. This is usually done by a large power transformer located adjacent to the turbine building. Several other station service and auxiliary transformers are also located adjacent to the turbine building. Generally, these transformers are mineral oil insulated. The combination of large amounts of combustible oil contained in a mechanism handling very large amounts of energy can be disastrous in the event of a transformer failure.

The prime hazard in the control-room complex is the cable insulation. Large numbers of cables are required to monitor instrumentation and control equipment. These cables enter the control complex from all over the plant. Usually they enter through a cable spreading room where they are sorted and rearranged before continuing up into the control room. The control room is constantly manned and presents little hazard. The fire hazard is not much greater in the cable room, but this area is generally not manned constantly, and access may be difficult in the event of a fire. Although the hazard in these areas is not as significant as it is in many other areas, the critical nature of these areas makes them extremely important.

Finally, most modern generating plants are situated in remote locations. This remoteness usually increases the fire hazard level because the public protection which must be relied on is both small and far away. Response times by a fire department can often be 20 minutes or more. Even when the fire department arrives, equipment which is perfectly adequate for the area protected is not sufficient in type or in numbers to combat a fire of the magnitude possible in these plants.

THE SAFEGUARDS

Protection of an electric generating plant against fire should consist of multiple "lines of defense" which include:

1. Protection against fire through good fire prevention design. This is the best method since no damage or interruption of service is incurred.
2. Early detection of a fire. Early detection allows plant personnel to respond to a fire rapidly and limit the extent of a loss.
3. Fire protection. Fixed and portable, automatic or manual, for suppressing the detected fire.
4. Physical separation. To prevent the extension of a fire beyond its area of origin.

The following will attempt to outline these "lines of defense" as they apply to the fire hazards associated with electric generating plants.

Building Layout and Construction

In the design of a new power plant, careful study should be made of the materials of construction and the physical layout of the buildings and the equipment contained therein. Fire-resistive or superior noncombustible construction should be utilized throughout the structure. Particular attention should be given to the selection of wall paneling or siding. Protected steel walls, plastic paneling, and certain insulated sandwich-wall paneling present peculiar structural fire fighting problems in electric generating plants where walls are undivided and extend upwards of 100 ft (30.5 m). Structural steel may require fire-resistive treatment in areas where flammable liquid storage and handling may present a hazard.

The building layout demands particular attention. High vertical undivided walls without windows for "borrowed" light, limited wall openings at lower elevations (doors and windows), "catwalks" which utilize open steel grated floors (as opposed to large solid floor areas), and the lack of vertical floor separation from the top to the bottom of the boiler room demand that consideration be given to providing adequate means of access and egress throughout the station for the operators and fire fighting personnel. Two means of egress should be available from each of the various elevations throughout the boiler room, and properly enclosed stair towers (at least one) should be provided for safe evacuation of personnel from upper elevations should the station become involved in a fire. Access should be provided to roofs so that effective fire control operations can be conducted. Also, attention must be directed to the problem of smoke and heat removal under fire conditions which could affect the safety of station personnel and, conceivably, the structural strength of roof members. The use of windows or roof monitors should be considered to ensure smoke and heat venting of the structure rather than relying entirely upon motor-operated fans which, under fire conditions, could fail when power supplies are affected or motor thermal protection operates.

Where there is no wall separation between the turbine room and the boiler room area of the plant, the use of draft curtains 3 to 6 ft (0.91 to 1.83 m) in depth may prevent extension of damage by limiting the release of large volumes of smoke and heat into the upper elevations of the boiler room.

When considering the physical arrangment and location of power plant equipment and structures, the bulk storage of hazardous materials should be located outside the structure in detached buildings or compartmented at grade elevation in fire-resistive rooms. Fuel oil pump rooms and rooms for storing and handling lubricating oils and other flammable liquids should be enclosed in fire-resistive construction with proper ventilation and drainage, and all electrical equipment should meet the requirements of Article 500 of the *National Electrical Code.*® Maintenance supplies of paints, lacquers, thinners, and hydraulic oils should be located outside or in similar fire-resistive rooms. Attention is directed to the problem of adequate emergency venting capacity for flammable-liquid storage tanks. In addition to normal fill and discharge venting equipment on such tanks, emergency venting is necessary to prevent collapse. Coal bunker rooms should be segregated from the remainder of the plant by suitable fire resistant walls so as to contain or stop the

---

®Registered trademark of the National Fire Protection Association, Inc.

ELECTRIC GENERATING PLANTS

FIGURE 4.1. Aerial view of generating station. The lower building in the foreground is the turbine building—totally open for its entire length; the taller sections behind are the boiler houses. They are over 200 ft (61 m) high with no vertical separations. Note the lack of separation between the boiler house and turbine building. The cooling towers are shown in the background. (Potomac Electric Power Co.)

extension of a fire from the bunker areas into the boiler house. Such rooms should have adequate ventilation including explosion venting, and all reliefs, vents, rupture diaphragms, etc., should be of adequate size and piped immediately to the outside.

Water Supply

Fire protection systems for a generating plant require a large volume of water. Intake and discharge canals, cooling tower basins, ponds, or tanks may be used as the source. Duplication of sources is recommended to permit shutdown and maintenance.

The method of determining the adequacy of a water supply to protect an electric generating plant is generally the same method used for other industrial applications. An adequate supply is that quantity of water required to satisfy the largest sprinkler demand which can occur in combination with manual fire hose streams.

Each potential fire area must be evaluated to determine its greatest demand for water. Where isolation of sprinkler systems exists, either through fire barriers or wide separation, the largest demand can be considered to be that which is required by the largest single system. If, however, such spacing does not exist, and the provision of fire walls is not feasible, the largest demand must be figured as that which is required by the largest number of contiguous sprinkler systems likely to operate simultaneously. This is based on the theory that a fire originating near the juncture of two sprinkler systems will require both systems to activate for proper containment. Generally, such dual sprinkler system demands arise on lengthy coal-handling systems or on large combustible-fill cooling towers. Consideration must also be given to multiple effects from a single occurrence, such as several bearing failures in the event of a major vibration of the turbine-generator.

With large water requirements for individual systems, it is usually found expedient to avoid simultaneous deluge system operations, unless the water supply and distribution systems are of unusually large proportions.

If this supply is to be furnished from a limited stored source, a quantity sufficient to provide for two hours of operations is usually considered adequate. It is usually considered necessary to provide multiple fire pump installations sized to provide the required demand with the largest single pump out of service.

Water supplies and power used to run pump motors, fire detection systems, a release system, etc., should be dependable. It is advisable to use electric motor-driven pumps for normal operation (preferably supplied from an outside power source) and provide a standby diesel engine-driven pump of proper capacity in the event electric power is unavailable during an emergency.

Fire protection water distribution should be via fire-service mains that are separate from station-service supplies. Desirable design layout should allow for maximum flexibility and reliable water supply distribution to hydrants, hose connections, and automatic protection systems through adequate control valving.

Fire Protection Safeguards — Coal

Since fires involving coal can be very difficult to cope with and no extinguishing method is entirely satisfactory, fire prevention and rapid detection are of the utmost importance.

All possible sources of ignition should be eliminated by prohibiting open flames, smoking, etc.; by selecting and installing electrical equipment in accordance with Article 500 of the *National Electrical Code*; by effectively grounding and bonding belts, pulleys, etc.; and by enforcing a rigid maintenance schedule for the equipment (such as motors, conveyor belts, etc.) used in the coal-handling system. These controls are important from both a gas and dust ignition standpoint.

Precautions should be taken to avoid the formation and accumu-

FIGURE 4.2. *Coal conveyor. (Potomac Electric Power Co.)*

lation of coal dust. Continuous dust suppression systems should be employed at the car dumper, receiving hopper, discharge of receiving hopper feeders, inlets and outlets of crushers, emergency reclaim hopper feeders, and any other transfer points where dusting is a problem. Design features such as the elimination of as many horizontal surfaces as possible, smooth interior wall surfaces, and a continuously operating dust collector system can be incorporated to minimize any accumulation of dust. Frequent removal of dust with a vacuum cleaning apparatus is also recommended. (Compressed air should never be employed.)

From the viewpoint of spontaneous ignition, bunkers should be designed to eliminate pocketing of coal. Dead storage of coal in bunkers should be avoided, and foreign materials in coal should be controlled.

The rapid detection of a coal fire is very important. Consideration should be given to providing means of alarm and actuation of automatic extinguishing systems for the essential coal conveyors, tunnels containing coal conveyors, coal "tripper" conveyor rooms, and buildings containing coal-crushing equipment. The detectors available are of various types: fixed temperature, rate compensation, combination fixed temperature and rate-of-rise, and continuous line-type sensors.

Automatic and manual release devices should be provided for all extinguishing systems, and should also automatically shut down the conveyor belts upon system trip and actuate all alarm components.

Because of the possibility of formation of toxic gases, self-contained breathing apparatus for emergency use by personnel is essential.

Fixed fire extinguishing systems should be provided for the essential coal conveyors, tunnels containing coal conveyors, coal "tripper" conveyor rooms, and buildings containing coal-crushing equipment.

Automatically actuated water-type sprinkler systems provide the maximum extinguishing capability for coal fires. Automatic water spray deluge systems are ideally suited to conveyor protection. When used in conjunction with sensitive detection, a water spray system can minimize the spread of fire from one conveyor to another. An automatic sprinkler system should be provided for enclosed buildings and for those locations where an excess flow of water might have detrimental effects.

Deluge or alarm valves for coal-conveyor systems will frequently be housed in separate buildings located throughout the coal yard. Some of these buildings may be remote and unattended. Since the power supply to these buildings could be lost, freeze protection, and provisions for a building "low temperature" alarm annunciating in the control room should be considered.

Any rapid increase in coal pulverizer mill outlet-air temperature may indicate a fire condition that demands attention. Mills and exhausters should not be shut down nor should doors or access ports be opened until all evidence of fire has disappeared. Details concerning the operation of pulverizers, including provisions for inerting, are contained in NFPA 85E, *Pulverized Coal-Fired Multiple Burner Boiler-Furnaces,* and NFPA 85F, *Pulverized Fuel Systems.*

Duct or burner pipeline fires and explosions generally result from excessive air temperatures, from improper control of air velocities that allows coal to drop out of suspension, or when lines are not properly purged. When proper air flows are reestablished, coal/air mixtures pass through the explosive range, and a damaging explosion may occur in the presence of hot spots or other sources of ignition. When shutting down, operators should be particularly careful to completely purge lines. The application of hammer blows to clear coal piping and ducts should be avoided, since coal may be put into suspension creating a potentially explosive condition. If a fire is detected in coal lines or ducts, it may be necessary to resort to water spray extinguishment. The introduction of water into coal pipes will require cleanup of equipment to avoid plugging of certain restrictive zones such as coal-pipe distributors, etc., and to prevent further difficulties.

Effective fire prevention requires vigilance in all areas where fuels are transported and utilized. A small leak in an undesirable location could create a serious potential for a difficult fire. Coal burners can "coke-up," solid fuels can erode away metal of pipelines and ducts, vibration of equipment and movement resulting from expansion and contraction due to temperature changes can cause failures of gaskets and flexible connectors or induce stresses to cause pipeline failures, all of which would lead to the release of fuel in proximity to the hot surfaces of the boiler. Visual inspection of equipment can keep dangerous situations from developing, and automatic detection devices can detect potential or actual fires in such locations.

FIGURE 4.3. *Coal conveyor after fire. (Potomac Electric Power Co.)*

Fire Protection Safeguards — Oil

Care should be exercised in the selection of design, location, and material used for the construction of fuel-oil handling facilities.

Piping: A leak-free piping system is important to the safe handling of liquid fuels. Design, fabrication, assembly, and testing of the system should conform to all applicable fire and safety codes and regulations. It is recommended that fuel oil piping be of all-welded construction with a minimum of connections, and these should be flanged rather than screwed. However, piping and fittings of 2-in. (50.8 mm) nominal size and smaller may be welded, flanged, or screwed, if they are maintained in a leak-free condition. Pressure- and thermal-relief valves should be installed on piping runs exceeding 100 ft (30.5 m), if the piping system can be blocked between valves. Relief-valve discharge lines should be connected to the tank return line, or should be piped to a collection vessel operating at atmospheric pressure. Refer to NFPA 85D, *Fuel Oil-Fired Multiple Burner Boiler-Furnaces*, for more information on fuel oil systems for boiler firing.

Heat tracing and insulating of piping is usually required to transfer heavy liquid fuels. Heat tracing can be accomplished by attaching a small steam line of steel or copper to the oil piping, by electric-resistance heating, or by electric-induction heating of the line.

Heaters: Heavy liquid fuels require heating to maintain pumpability and firing atomization. There are several acceptable types of storage tank

Chapter 4

FIGURE 4.4 Fuel-oil pump. (Potomac Electric Power Co.)

heaters. The most commonly used heaters are: (1) the bundle type inserted into the side of the tank; (2) heater coils located near the bottom of the storage tank; and (3) circulation-type heaters located outside of the tank. All oil heaters should be equipped with thermal-limiting devices to prevent overheating the oil. All external flow-through-type oil heaters should also be equipped with a flow switch to stop the heating process if circulation is interrupted.

Pumps and filters: Centrifugal or rotary pumps can be used for liquid fuel transfer. Pumps should be selected according to the product to be pumped. When fuels such as crude oil and numbers 1 and 2 fuel oils are to be pumped, it is recommended that double-seal pumps be used. A curb and closed-drain system should be provided for pump-seal leakage. In extremely cold conditions, pump-case heating may be required to maintain fuel oil flow.

Strainers should be installed at the pump suction to protect the pump. If fuel oil filters are utilized, they should be placed in the discharge line of the pump and should be of a type that is easily cleaned.

All pumps, filters, and fuel-oil handling equipment should be located outside of tank dikes. When possible, pumps should be located to take advantage of prevailing winds to minimize heat exposure, allowing the pumps to be used for transferring contents of a burning tank to other storage facilities.

Electrical: The *National Electrical Code* classifies and defines electrical requirements for flammable and combustible liquid installations. Additional information on electrical area classification may be found in NFPA 31, *Installation of Oil Burning Equipment*, or API RP 500 A, *Classification of Areas for Electrical Installations in Petroleum Refineries*.

Ventilation: Ventilation should be provided for pump houses and any enclosed area where hydrocarbon vapors may accumulate.

In locations where flammable vapors may be present, precautions should be taken to prevent ignition by eliminating or controlling sources of ignition. Sources of ignition may include open flames, lightning, smoking, cutting and welding, hot surfaces, frictional heat, sparks (static, electrical, and mechanical), spontaneous heating, chemical reactions, and radiant heat.

All fuel-oil handling facilities should be maintained in good repair at all times. Personnel should be trained in the operation, maintenance, and emergency procedures for all fuel-oil handling equipment. Periodic tests should be made to check the pressure and thermal-limiting devices.

Automatic and/or manually operated fixed fire protection systems should be considered for major liquid fuel exposures. The types of systems include foam, sprinklers, water spray, carbon dioxide, dry chemical, and halon. Each exposure must be evaluated to determine the type of fixed system that would provide the most effective protection. An important consideration in designing a protection system is determination of the potential extent of an area which may be exposed by an oil fire from a pressurized oil source.

Portable fire extinguishers should be located throughout the handling facilities in accordance with the guidelines of NFPA 10, *Portable Fire Extinguishers*. Where fire protection water is readily available near fuel-oil handling facilities, a ready-connected fire hose should be provided, with a spray/straight-stream nozzle. Fire drills and coordination with local fire authorities should be considered as a part of emergency planning.

Boilers

Various boiler surfaces, steam pipes, and valve bonnets under normal operating conditions may have exposed surfaces at temperatures exceeding 500°F (260°C). They present a ready ignition source for flammables should they spray onto, or come in contact with, such surfaces. Fuel oil and hydraulic oils escaping from ruptured piping or fittings under pressures up to 2,000 psi (13 790 kPa) can form a fine mist which can carry a considerable distance, and, if ignited by contact with a hot surface, spark, or open flame, will result in a torch-like fire that is very intense and may involve a large area. Fuel lines and lubricating and hydraulic oil lines should be run in as direct a manner as possible, while avoiding hot surfaces. Where the latter is not practical, metal baffles or shields can be utilized to deflect oil away from such parts should lines accidently rupture.

Where pressurized oil lines are located close to boilers, steam pipes, and high-temperature air ducts, the lagging (insulation) of hot surfaces should be covered with oil-impervious fire-retardant paints or sheet metal to prevent it from acting as a wick which will eventually permit the oil to penetrate to the hot surface, vaporize, and ignite. Whenever possible, flanges, unions, and screwed fittings should be avoided in areas where hot surfaces are present. Oil piping systems in such areas should be of all-welded steel construction. In areas proximate to the burners of the boiler, consideration should be given to the use of solid flooring or platforms provided with curbs and drains in order to prevent burning liquids from cascading downward through the structure.

There is considerable activity in the industry toward automating boiler operation and scanning furnace conditions to prevent furnace explosions. Systems designed to control the composition of furnace atmospheres to prevent fuel explosions also are being developed. There are many techniques and systems being investigated and utilized at this time. Each system should be carefully reviewed with regard to protection of surrounding areas and the training of operators.

It is impossible to foresee each emergency that might arise in the operation of a power plant. It is important, however, to evaluate potential hazards and give adequate instruction and training to operators so that they can handle the more likely emergency operating conditions.

The importance of operator training and retraining cannot be emphasized enough. A worker's "failure" is usually involved in losses reported; however, no records are available regarding the many potentially serious fires or explosions that were prevented because the operator knew what to do.

Turbine-Generators

Since hot valve bonnets, steam pipes, steam chests, etc., operating at or above 700°F (371°C) provide ready ignition sources for lubricating or hydraulic oils that may come in contact with them, attempts should be made wherever feasible to install oil lines remote from such high-temperature parts. High-pressure oil lines should be run within another pipe so that the outer pipe will contain the oil from a high-pressure leak. Where it is impracticable to secure adequate separation between hot steam lines or parts and oil pipes, the oil lines might be encased within a metal duct or baffles might be installed to deflect oil away from hot surfaces that might cause ignition. Covering the steam-pipe lagging (insulation) with a paint that is oil impervious, water impervious, and fire retardant, or encasing the lagging in sheet metal, will also assist in preventing the ignition of oil. In order to lessen the chance of oil and hydrogen piping becoming dislodged or broken, all such items, including oil- and hydrogen-instrument piping, should be of welded steel construction. Cast iron fittings and piping should be strictly avoided. Thermometer and thermocouple wells should also be welded. Care should be taken with the manner in which such piping is supported so that chances of pipe failure due to vibration are minimized.

Drainage systems should be trapped to prevent the spread of an oil fire from one area to another.

Trays used to support cables in the turbine-generator area may act as chutes to convey and spread burning oil. Such cable runs should either be totally enclosed in an oil-tight duct or laid in open "grating" type trays. Important control cables should be so located and protected as to preclude exposure to, or involvement in, a fire. In areas where high ambient temperatures prevail, cable insulation that is suited to such an environment should be employed to preclude insulation deterioration which could result in a fire.

It may be very important to the overall protection of a station to be able to shut down a turbine-generator and its auxiliaries as rapidly as possible, in the event of a major incident, through the use of an emergency shutdown panel or station located in an accessible and tenable area. Consideration should be given to the inclusion of such items as stop buttons for the main and standby oil pumps (both lubricating and seal oil), and the hydrogen dump valve, along with other required primary controls. The control room is, of course, the logical place for such an emergency panel — provided the control room has been properly designed from a firesafety viewpoint. In cases where the control room is not so designed, it might be more desirable to locate the emergency controls elsewhere. Likewise, other controls, valves, etc., within the plant which might be helpful in isolating a fire involving either the oil system or hydrogen system should also be at readily accessible locations.

In addition to the auxiliary pump for the lubricating system, it is also advisable to provide an auxiliary seal oil pump so that, in the event of failure of the pump normally in service, the hydrogen seals will be automatically maintained by the second pump. Both standby pumps should be supplied from dependable power sources.

A means should be provided for removing hydrogen from the oil to ensure that it will not find its way into the lubricating system and oil reservoir. This can best be accomplished by good system design on the part of the manufacturer and by including an oil seal in the loop to prevent the passage of hydrogen into the reservoir. As an additional precaution, hinged explosion-relief covers should be provided on the reservoir.

A fire resistant synthetic base fluid is commonly used for hydraulically operated steam valves and should be employed where possible and practicable. Consideration should always be given to the use of fire resistant fluids in lieu of conventional oil, which is a prime fire hazard of the turbine-generator area.

Fire prevention vigilance should be observed during the overhaul of a machine. Caution should be used when handling oil so that it is not spilled onto steam pipes and other apparatus which, upon start-up of the unit, may ignite the oil as the temperature of the unit rises during normal operation. Caution should be observed in using cutting and welding apparatus as this, too, might serve as an ignition source for exposed oil or oily equipment and apparatus. Rags, tarpaulins, temporary cribbing, and combustible scaffolding are among the items which could provide fuel for a fire during overhaul. Combustibles should be reduced to a bare minimum and, wherever possible, noncombustible materials should be used. Good day-to-day housekeeping should be employed at all times, and fire fighting equipment should be readily available.

Specific instructions covering both normal and emergency operating procedures should be prepared and frequently reviewed with personnel.

After reducing the possibilities of fire by the means heretofore mentioned, it is still necessary to protect the turbine-generator area with adequate and proper fire fighting equipment.

In view of the major proportions a fire involving a turbine-generator can rapidly attain and the limited number of personnel that may be available to fight such a fire, serious consideration should be given to the installation of a fixed-pipe extinguishing system. Such a system should be designed and installed so that all possible fire areas in the turbine-generator zone from turbine floor to basement are covered. This should include the front standard of the turbine, bearings, oil lines, turbine oil reservoir, hydrogen detraining tank, hydraulic valves, and any other oil piping or tanks (hydraulic, cooling, and lubricating) that may be in the vicinity. Depending on location of equipment, one system or several separate systems may be preferred. The portion of the system provided for basement coverage should be so designed and installed that ample floor coverage will be achieved. The basement floor should be equipped with drainage ditches or curbs in order to confine oil and water (if a water spray system is used) so that fire will not spread to other equipment or units. The drainage system should be adequate, and the power supplies to sump pumps should be dependable.

Because of the usually abundant supply of water and its effectiveness on oil fires, a water spray (deluge) system is very desirable. Care should be taken to prevent the inadvertent spraying of water on hot turbine parts, and an attempt should be made to aim spray nozzles away from such exposed parts. If an *automatic* water spray (deluge) system is used, inadvertent discharge of water on hot parts (due to possible false operation) can be avoided by using a combination open-nozzle and closed-nozzle system. In areas where there is a possibility of spraying water on hot turbine parts, a closed-type, temperature-sensitive nozzle can be employed. Where there is no danger of striking hot parts with water spray, conventional open-type spray nozzles can be used. A preaction sprinkler and/or spray system can also be used to advantage where there is danger of water striking hot metal parts.

Another manner in which accidental discharge of water spray onto hot turbine parts might be avoided is to use the fire detection system for fire detection only (rather than for deluge valve operation) and activate the system manually from a remote station upon receiving a fire alarm signal. This scheme involves the human element, and the decision has

ELECTRIC GENERATING PLANTS

FIGURE 4.5. Turbine-generated building. (Potomac Electric Power Co.)

to be made by someone as to whether or not the extinguishing system should be tripped.

Where water spray systems are employed, electrical equipment in the area of coverage should be of the waterproof type. In the case of a new plant, electrical auxiliaries can be designed as such; in older plants, waterproofing can be achieved by providing splash hoods, gasketing, etc.

In cases where the water supply might be questionable or where the drainage system might not be adequate to handle the volume of water and oil involved, a fixed foam system or a combination fixed foam and water spray system may be used to conserve water and to reduce drainage requirements. Since foam is a conductor of electricity, it should not be employed in areas where electrical equipment is installed unless such equipment is suitably waterproofed or shut down prior to discharge. Also, precautions concerning hot parts and waterproofing, as previously outlined, should be observed.

A low-pressure (multi-ton) carbon dioxide system may also be considered for fixed protection of the turbine-generator area. This system would present no drainage problems. However, the supply of carbon dioxide is limited (as compared with water), and, since the turbine-generator and its auxiliaries are generally located in an open area, difficulty might be encountered in maintaining sufficient concentration for a long enough period to achieve extinguishment. Although it is nontoxic, carbon dioxide can cause suffocation due to oxygen deficiency. Thus, personnel should be clear of the entire area during and for some time after

FIGURE 4.6. *Low-pressure (multi-ton) carbon dioxide storage tank. (Potomac Electric Power Co.)*

the operation of the system, unless equipped with self-contained breathing apparatus and life lines.

It may be desirable to supplement a major fixed extinguishing system with relatively small, piped dry chemical systems for spot protection of certain inaccessible areas, such as bearing locations, for control of a minor fire. In any case, adequate access openings should be provided in the turbine lagging to permit the application of portable extinguishing equipment to potentially hazardous areas.

Regardless of the type of system used (manual or automatic; water spray, foam, or carbon dioxide), an accessible auxiliary manual release should be provided. Since personnel may have to abandon the plant, an assumption based on past experience, it may be desirable to locate this auxiliary trip near a likely plant exit.

Temperature settings of thermostats (and closed-type nozzles, where used) should be in accord with ambient temperatures found in the various areas protected. In some cases, three or four different temperature settings might be required.

It is of the utmost importance that extinguishing systems be carefully and thoroughly tested and inspected at regular intervals to make sure that they are in proper operating condition. In the case of a new installation, it is advisable to actually discharge the system, taking the necessary precautions of covering up equipment which might be damaged or unnecessarily wetted. All items, including thermostats, release mechanisms, tamper (monitor) switches, the supervisory system, relays, remote releases, etc., should be tested.

In addition to fixed fire extinguishing systems, an adequate number of suitable hand-portable and wheeled extinguishers and fire hoses equipped with spray nozzles should be provided. This equipment could be used to supplement the fixed systems should the fire spread be-

ELECTRIC GENERATING PLANTS

FIGURE 4.7. Hydrogen bulk storage facility. (Potomac Electric Power Co.)

yond the confines of those systems. These hand-portable or wheeled extinguishers should be of the carbon dioxide, dry chemical, or foam type. An adequate number of hose stations equipped with approved fire hose spray nozzles should be strategically located for general area coverage. In the event it becomes necessary to abandon the plant (or access cannot be gained to inside protection), outdoor hydrants with enough hose (perhaps on carts) to reach the turbine-generator area should be provided. In cases where low-pressure carbon dioxide, foam, or dry chemical systems have been provided for fixed system protection, hose lines from such systems may be provided for minor fires or mop-up operations. For areas where foam might be used safely, foam generators or eductors might be considered for introducing foam into fire hose lines.

Because of smoke and heat conditions, self-contained breathing apparatus may be required for performing special or unusual tasks.

As in the case of fixed pipe systems, all fire extinguishers, hose line equipment, masks, suits, etc., should be regularly inspected and tested.

Good judgment should be used in locating the various hydrogen-system components and related equipment so that hydrogen leakage will not become pocketed and to avoid exposure of vital equipment to fire. The supply of hydrogen, whether in high-pressure cylinders, bulk, or generated on the site, should be located outdoors. Where high-pressure cylinders are employed, handling should be limited by providing storage platforms at a convenient loading and unloading height and cylinders (CO_2 and hydrogen) should be securely racked and chained to preclude upsetting. Care should be taken in selecting a suitable location for the installation. It should not be susceptible to damage from vehicles and the like, and, in the event of a hydrogen incident, vital equipment should be far enough away so that it will not become involved.

Emergency controls for generator venting should be installed outdoors, if practicable, or at least at accessible locations. (This is in addition to any indoor control facilities.) Insofar as possible, every effort should be made to locate other hydrogen facilities out-of-doors where leakage will not result in explosive concentrations. The construction of semi-outdoor power plants may make it more practical to do this.

Where it is necessary for hydrogen facilities to be located inside major plant buildings, the arrangement of these facilities should be such that long runs of hydrogen supply and instrument piping are avoided. Hydrogen pipe runs should be routed with caution. Hydrogen piping should be kept away from traffic areas and any other area that could expose the pipe to potential damage. Pipe runs from hydrogen installations located outside the plant should preferably be run below grade with a vented and guarded pipe system.

The control room in the station should not be needlessly exposed to the hydrogen hazard by piping hydrogen to control room instrument boards; rather, instrument piping should be as short as possible and should terminate at panels close to the generator, using a transmitter system between the panel and the control room.

Where hydrogen is delivered to power stations on tube trailers and connected to the exterior manifold, care should be taken to be sure that the vehicle trailer is chocked and grounded before the manifold is coupled. Care should be taken to prevent the discharge of gas should the tube trailer be removed without being disconnected from the manifold. Strict administrative control or some type of mechanical interlocking should be considered.

Enclosures into which hydrogen may leak should be adequately vented to safe locations. This should include such items as control panels and the different housings on the generator, such as the collector end housing, turning gear housing, and the generator lead enclosure. Care should be taken in selecting safe locations above the roof for the termination of vents, including the generator vent. It may be convenient to install a common or master vent pipe to which all vented areas of a particular unit could be connected. Where venting is impracticable, the possibility of ventilating areas with explosion-proof fans should be considered. However, the hazard of introducing electrical equipment into a possibly explosive atmosphere should be kept in mind. Electric motors and other possible ignition sources used in the ventilating system should be installed in a known-to-be-safe area or, otherwise, should be installed in accordance with Article 500 of the *National Electrical Code.*

Valves for isolating equipment (such as the control cabinet) in the event of a hydrogen fire should be provided at readily accessible locations. It is much safer to turn off the supply of hydrogen to the affected equipment and allow the fire to burn itself out than to attempt extinguishment. Extinguishing a gas fire could ultimately result in an explosion much more damaging than the original fire.

Since the hydrogen equipment and related facilities are adjacent to and closely associated with the lubricating oil and seal-oil systems, the protection provided for the latter should be extended to include protection against hydrogen fires (hydrogen protection is only one part of overall turbine-generator protection). Protection may take the form of deluge systems (fixed water spray) which serve to cover the turbine-generator area, or a suitable substitute, such as piped carbon dioxide, dry chemical, or foam systems. (If a foam system is used, due respect must be given to any energized equipment in the area.) Suitable hand-portable equipment, such as dry chemical and carbon dioxide extinguishers and hose lines with spray nozzles attached, should also be provided in adequate quantity.

Cooling Towers

Much has been done in recent years to eliminate combustible components in cooling tower construction. The frame and shell may be constructed of reinforced concrete or corrosion-resistant metal. The fan deck, distribution deck, louvers, and walls may be constructed of reinforced concrete, cement asbestos, or an impregnated fire-retardant-treated plywood or lumber. However, impregnated lumber used as a fill cannot maintain its fire-retardant characteristics over several years due to the effects of leaching.

Fill material of galvanized steel or cement asbestos is noncombustible. Some PVC (polyvinyl chloride) fill has satisfactorily undergone fire tests which indicate that, with severe fire exposure, a self-propagating fire cannot be established.

Sprinkler protection should be provided for cooling towers with significant amounts of combustible materials. On very large towers having a fill section height exceeding 37 ft (11.3 m), two or more horizontal levels of sprinkler protection are needed.

In large combustible towers, vertical fire partition walls are desirable if it is necessary to limit the water-demand requirements for the protection system. Without partition walls, the water supply must be capable of meeting the system demand of two adjacent sprinkler systems plus simultaneous hose stream use. Providing partition walls of reinforced concrete, double cement-asbestos board, or double ½-in. (12.7 mm) treated plywood (or equivalent) could reduce the water demand to one system, plus simultaneous hose stream use.

Careful engineering is essential in choosing the type and location of fire detection used to actuate the deluge systems because of varying draft conditions and tower designs. Since a tower with combustible fill is vulnerable during construction, the protection system should be in service before fill is installed.

Chapter 4

Transformers

Protection against faults is necessary for transformers. This may be accomplished by various combinations of shielding, grounding, lightning arresters, interrupting devices, and relays.

Overhead shielding creates zones of protection in which damage to equipment, as a result of atmospheric electrical discharge or direct lightning strokes, is minimized. Such shielding may be afforded by grounded overhead wires (static wires), grounded masts, steel towers, or nearby grounded metallic structures (see NFPA 78, *Lightning Protection Code*). The extent of the shielding considered necessary will depend upon the equipment involved and system conditions; it should be such that all important electrical equipment will be included within the zone of protection. In addition, it is generally considered advisable to extend the shielding so that incoming and outgoing lines are protected for a substantial distance beyond the area, the other alternative being to protect the full length of the lines by shielding.

An effective grounding system should be provided so that excessive flows or vagrant currents caused by artificial or natural disturbances that might cause damage will be safely dissipated. All equipment such as transformers, breakers, lightning arresters, etc., should be connected to an effective common grounding system by conductors of adequate size. Metallic underground systems should be provided to serve as electrodes. These may consist of buried plates, grids, driven ground rods, or some combination of these. All such electrodes should be made of a corrosion-resistant material.

Provision should be made to safely discharge surges due to lightning and other causes. For this purpose, lightning arresters, surge protection capacitors, reactors, or spill gaps of adequate rating may be used. The need for such protective equipment will vary depending upon the type and importance of the equipment as well as the severity of the exposure to electrical disturbances.

Interrupting devices arranged to isolate circuits or equipment automatically should be installed as a protection against surges and to afford electrical fault protection. Such protective equipment includes oil and air circuit breakers and vacuum-tube or gas-filled interrupters, with the associated automatic actuating devices, and fuses. Dry types of equipment should be chosen where it is likely that oil will present a hazard that cannot readily be segregated.

The selection and extent of such protection will depend upon system conditions, the equipment involved, and the degree of protection required. In critical or highly important situations, the establishment of several protective zones may be desirable, and additional relaying systems for backup protection may be justified.

There should be a well-developed program for the testing and maintenance of station electrical equipment, based on the manufacturer's recommendations and adjusted to local experience.

Tests on the oil in transformers and other oil-filled apparatus, on

ELECTRIC GENERATING PLANTS

FIGURE 4.8. Main power transformer. Note the water spray system. (Potomac Electric Power Co.)

the resistance of insulation in transformers, and on bushings and bus work will frequently reveal conditions that could later cause a failure with an accompanying fire.

With regular periodic maintenance, worn or faulty equipment will often be discovered in time to prevent failure and the possibility of a fire. Many destructive oil fires are the result of a ruptured transformer or broken oil line. Periodic cleaning of drainage systems will facilitate the removal of burning oil and also the water from deluge systems or fire hoses. Fire prevention is one of the by-products of proper maintenance.

Fixed systems for the protection of major oil-filled electrical equipment may either be manually operated or automatic. However, with reduction of manpower and the trend toward automated stations, automatic fixed extinguishing systems should be seriously considered.

Where water is present in sufficient quantity at suitable pressures, water spray (deluge) systems should be given first consideration. Generally, the spraying of bushings (except base) and lightning arresters should be avoided to preclude flashover. (Refer to NFPA 15, *Water Spray Fixed Systems.*)

If water is unavailable, a dry chemical system should be considered as a suitable alternate, especially if the transformer is in an enclosure. Dry chemical is very effective for extinguishing oil fires, but due to its limited capacity, is generally a second choice to water. The system should be designed to operate successfully against adverse winds. In addition, operation should be automatic to effect extinguishment before metal parts become hot enough to cause a reflash of the oil, perhaps extending the fire beyond the capability of the dry chemical in the system.

Carbon dioxide fixed systems are usually of questionable value out-of-doors (due to effect of winds).

Automatic actuation of fire systems may be by fixed-temperature or rate-of-rise thermostats, flame or products-of-combustion detectors, by relays, or by combinations of these devices. The types chosen will depend upon the equipment protected and the type of fire protection system being used.

When water spray systems protect equipment remotely supervised or where excessive use of water could be a problem, consideration should be given to a fixed fire protection system that automatically shuts off when the fire is extinguished, or after a predetermined time interval, and then returns automatically to operational mode.

Supplemental protection for a transformer should consist of hand and, in some cases, wheeled dry chemical extinguishers. The dry chemical units are superior on flammable liquid fires and are, therefore, recommended for protection of major oil-filled equipment. Although foam is effective on flammable and combustible liquid fires, it is a conductor of electricity and should, therefore, not be used on or near energized equipment.

Because water in the form of a solid stream is conductive and ineffective on oil fires, only nonadjustable spray nozzles or nozzles which cannot be adjusted to the solid-stream position should be used in these areas. It should be noted that, where very tall equipment is installed, extinguishment with hose lines (or extinguishers) may be very difficult.

All fire fighting equipment, including hand and wheeled extinguishers, hydrants, hose, fire pumps, tanks, automatic systems, etc., should be properly maintained, inspected, and tested to ensure that they will function properly when needed.

Control Room Complex

A fire in the central control room of an automated modern plant or in its computer or relay room can have catastrophic consequences. An apparently small or insignificant fire in electronic data processing equipment, or in control or power wiring, can cause extensive shutdowns and be extremely costly since many of these systems are individually designed and have intricate wiring systems and special component parts. Repair to these systems requires involved calibration, testing, and assembly techniques. The design and layout of the control room should, therefore, be considered in conjunction with the design of related fire protection and/or detection systems, as warranted by the importance of maintaining continuity of station operation.

Electronic data processing and process control equipment should be isolated in a separate room, and attention should be given to fire protection for cable spaces beneath raised floors. Where cables enter the data processing console section they should be carefully sealed. These rooms should be of noncombustible construction, and attention should be given

ELECTRIC GENERATING PLANTS

FIGURE 4.9. Partial view of modern control room. (Potomac Electric Power Co.)

to air conditioning to ensure that the rooms are pressurized to preclude entry of dust and, in the event of a fire, entry of smoke. Make-up air supplies for these rooms should be filtered and taken from outside the building rather than from boiler or turbine rooms. Combustibles, paper records, and data sheets should not be left out in the open, and should be removed from these rooms or kept in metal record or filing cabinets.

BIBLIOGRAPHY

REFERENCES

API RP 500A, *Classification of Areas for Electrical Installations in Petroleum Refineries*, American Petroleum Institute, Washington, DC.

Edison Electric Institute, *Edison Electric Institute Statistical Yearbook of the Electrical Utility Industry for 1982*, Washington, DC, 1983.

Factory Mutual Engineering Corporation, "Electric Generating Stations," *Loss v Prevention Data* 5-15, April 1978, Factory Mutual System, Norwood, MA, pp. 1-25.

F.M. Eng. Corp., "Powerhouses, Generators, and Switchgear," *Handbook of Industrial Loss Prevention*, Second Edition, 1967, pp. 27-1 to 27-8.

McKinnon, G. P., ed., "Electrical Power Sources," *Fire Protection Handbook*, Fifteenth Edition, National Fire Protection Association, Quincy, MA, 1981, pp. 8-2, 8-9.

"Power Engineering," *Power Plant Primer*, Barrington, IL.

Richardson, Donald O., "Fire Hazards and Protection in Utility Stations Today," *Generation Planbook 1978*, McGraw-Hill, Inc., 1978, pp. 103-114.

NFPA CODES, STANDARDS, AND RECOMMENDED PRACTICES

Reference to the following NFPA Codes, Standards, and Recommended Practices will provide further information on the safeguards for electric generating plants discussed in this chapter. (See the latest *NFPA Codes and Standards Catalog* for availability of current editions of the following documents.)

NFPA 10, *Portable Fire Extinguishers.*
NFPA 11, *Foam Extinguishing Systems.*
NFPA 11A, *Medium and High Expansion Foam Systems.*
NFPA 12, *Carbon Dioxide Extinguishing Systems.*
NFPA 13, *Installation of Sprinkler Systems.*
NFPA 13A, *Care and Maintenance of Sprinkler Systems.*
NFPA 14, *Standpipe and Hose Systems.*
NFPA 15, *Water Spray Fixed Systems.*
NFPA 16, *Foam-Water Sprinkler Systems and Spray Systems.*
NFPA 17, *Dry Chemical Extinguishing Systems.*
NFPA 20, *Centrifugal Fire Pumps.*
NFPA 22, *Water Tanks for Private Fire Protection.*
NFPA 30, *Flammable and Combustible Liquids Code.*
NFPA 31, *Installation of Oil Burning Equipment.*
NFPA 37, *Stationary Combustion Engines and Gas Turbines.*
NFPA 50A, *Gaseous Hydrogen Systems.*
NFPA 51, *Oxygen-Fuel Gas Systems for Welding and Cutting.*
NFPA 54, *National Fuel Gas Code.*
NFPA 70, *National Electrical Code.*
NFPA 71, *Central Station Signaling Systems.*
NFPA 72A, *Local Protective Signaling Systems.*
NFPA 72D, *Proprietary Protective Signaling Systems.*
NFPA 75, *Protection of Electronic Computer/Data Processing Equipment.*
NFPA 78, *Lightning Protection Code.*
NFPA 80, *Fire Doors and Windows.*
NFPA 85D, *Fuel Oil-Fired Multiple Burner Boiler-Furnaces.*
NFPA 85E, *Pulverized Coal-Fired Multiple Burner Boiler-Furnaces.*
NFPA 85F, *Pulverized Fuel Systems, Installation and Operation.*

NFPA 85G, *Prevention of Furnace Implosions.*
NFPA 101, *Life Safety Code.*
NFPA 120, *Coal Preparation Plants.*
NFPA 214, *Water Cooling Towers.*
NFPA 220, *Standard Types of Building Construction.*
NFPA 1961, *Fire Hose.*
NFPA 1962, *Care, Maintenance and Use of Fire Hose.*
NFPA 1963, *Screw Threads and Gaskets for Fire Hose Connections.*

ADDITIONAL READING

Averill, E. L., "Fire Protection Methods for PSO Electric Generating Stations," IEEE Power Engineers Society Summer Meeting, 1977, Mexico City, Mexico.

"Fire Protection for Power Stations," *Electrical Review* (London), vol. 200, no. 13 (1977), pp. 20-21.

Hathaway, L. R., "Coal Handling Fire Protection," in *Coal Technology '82, the Annual Coal Utilization Exhibition and Conference* (4th), December 7-9, 1981, vol. 2: Storage and Handling, American Gas Association, pp. 164-168.

Hollister, R. H., C. L. Moore, P. Voytik, "Gas Vapor and Fire Resistant Transformers," in *Proceedings of the 14th Electrical/Electronics Insulation Conference, Boston, MA,* 1979, IEEE Publication no. 79CH1510-7-EI, pp. 239-242.

Pearce, H. A., "Fire Resistant Fluids for Industrial Transformers," *Fire Technology*, vol. 14, no. 2 (1978), pp. 159-165.

Power Plant Engineers Guide, Frank D. Graham, ed., Bobbs Merrill, Chicago, IL, 1983.

Sawyer, R. G., "Preplanning Electrical Generating Stations — Parts I and II, *Fire Command*, vol. 45, nos. 10 and 11 (1978), pp. 24-25, 26-27.

5

Nuclear Power Plants

D. A. Decker

This chapter describes the processes and identifies the fire hazards associated with nuclear fueled electric power generating stations. It is restricted to the hazards that occur from the time nuclear fuel arrives at the plant site until electricity is generated. Hazards associated with outdoor power transformers, switchyards, and off-site power distribution are not discussed because they are similar to the hazards associated with conventional fossil and hydroelectric power generating stations.

Like other electric power generating stations, the primary product of nuclear stations is electricity. It is used by industrial, commercial, institutional, and residential complexes as an increasingly important power source. The plant itself consumes a considerable quantity of electricity for powering equipment, control devices, and auxiliary heating. Other products are generally considered waste. They include spent fuel, excess heat, and radioactive wastes. Spent fuel is stored on site until it is shipped to reprocessing plants. Most excess heat is released into the atmosphere through cooling towers or into water bodies, although an increasing effort is being devoted to the use of excess heat. Projects that

---

D. A. Decker, P.E., is a Safety and Fire Protection Engineer and a Loss Prevention Consultant based in Centre Hall, Pennsylvania.

involve heating water bodies in order to encourage aquatic growth appear to be a possible solution to the disposal of excess heat in certain cases. Radioactive wastes are stored on site until they are shipped to a disposal site for burial and radioactive decay.

Currently there are approximately 130 commercial nuclear power generating stations under construction or in operation in the United States. The operating nuclear plants produce about 12 percent of our electricity. Present estimates project an increase to 30 percent by 1995.

As is common at fossil-fuel fired generating stations, each generator and its associated steam supply system is referred to as a unit. Nuclear plants may have one or more units, and each unit has its own nuclear reactor and generator. Some auxiliary facilities may be shared between units. Electrical generating capacity has grown to more than 1,200 MW (megawatts) per unit. Such plants take over ten years to build and cost more than $500 per kW (kilowatt).

Nuclear power plants are staffed twenty-four hours per day throughout the year. The number of employees on site varies from fewer than ten during night shifts to more than a few hundred during maintenance and refueling operations. Supplying operating materials and support services involves hundreds of additional personnel. During peak construction periods a few thousand people may be on site. Because the nuclear power industry is becoming an increasingly vital source of employment, plant loss or operation interruption may have a substantial economic impact on the community.

The fire loss record of nuclear power plants has been very good due to low fire loads, fixed protection of specific hazards, and management's concern with fire protection.

The most significant loss occurred at the Browns Ferry Nuclear Power Plant of the Tennessee Valley Authority on March 22, 1975.* The plant, located near Decatur, Alabama, consists of three boiling water reactor (BWR) units, each designed to produce 1,097 MW of electric power. At the time of the fire, Units 1 and 2 were operating; Unit 3 was under construction. A candle being used to check a polyurethane foam cable penetration seal ignited the foam. The fire penetrated the wall between the cable spreading area and the Unit 1 reactor building through the seal, moving into the cable spreading room. Major damage occurred to the high concentration of electrical cables in the area. The damage resulted in control power loss to much of the installed equipment. Sufficient equipment remained functional to shut down the plant safely, and no radiation release in excess of normal operating limits occurred.

The fire was most significant because it generated a public awareness of the need for adequate fire defense in nuclear plants. Numerous public hearings, studies, and recommendations followed the event. It caused regulatory agencies, standards organizations, and the public to de-

---

*A detailed analysis of the Browns Ferry fire appears in the July 1976 issue of *Fire Journal*.

vote a concentrated effort toward ensuring that a fire at a nuclear plant would not jeopardize nuclear safety.

Many of the fire hazards found in other types of electric generating stations or large industrial plants are similar to those in nuclear power plants. (See Chapter 4, "Electric Generating Plants," for a thorough discussion of the fire hazards associated with electric generating plants.) These include combustible lubricants, combustible shipping containers, filtering materials, organic insulations, and flammable gases. Although the hazards may be typical of those found in other facilities, with respect to combustibility, their total hazard, when present in a nuclear power plant, can be significantly greater. Fire and explosion hazards in nuclear plants must be identified and evaluated with respect to the hazard's influence on the safe shutdown capability of the plant and the release of radiation in excess of acceptable limits. Plant designs are required by regulatory agencies to have safeguards that ensure safe shutdown under various natural phenomenal events—including fire. Certain plant provisions, such as shield walls and ventilation systems, are necessary for the protection of personnel and the public from excessive radiation. These safeguards and provisions are often redundant in order to ensure their availability.

The hazards common to nuclear power plants reach beyond the typical property and life loss potential to include the loss of necessary nuclear safety functions with its associated threat to public safety.

RAW MATERIALS

Nuclear fuel is delivered to the plant by special truck or rail transport vehicles. The fuel usually consists of uranium (U^{233} and U^{235}) and plutonium (Pu^{239}) oxides and is not considered a fire hazard. The fuel is generally encased in metal fuel rods or elements, and the elements are usually arranged in groups or fuel bundles. The bundles are generally enclosed in a polyethylene bag inside a metal shipping canister which is shipped in an outer shipping crate of combustible construction. The overall length of the shipping crate is approximately 12 ft (3.7 m). The bags and outer shipping crate present an ordinary combustible fire hazard and should therefore be promptly and properly discarded.

New fuel elements do not present a radiation hazard prior to their use in the reactor. They are stored in special new fuel storage vaults or pools. The storage is especially designed and arranged to prevent formation of a critical mass.* Accidental formation of a critical mass must be avoided by using, storing, and handling the fuel in accordance with the plant design.

In addition to the fuel for the nuclear reactor, numerous supplies,

---

*The minimum mass the fissile material must have in order to maintain a spontaneous fission chain reaction. For pure U^{235} it is computed to be about 20 lbs (9 kg).

many of which are combustible, are used at a nuclear power plant. These include disposable clothing, filter media such as charcoal and polystyrene resins, lubrication oil and grease, diesel fuel for emergency generators, flammable gases for blanketing, and cleaning materials, as well as shipping containers for many noncombustible items. The receiving, and main storage of these materials should be completely separate from the operating areas. Necessary supplies required for operation may be stored in closed metal cabinets or specially protected storage rooms in the area of use. Combustible shipping containers must be disposed of promptly after use.

THE PRODUCTION PROCESS

The primary process in a nuclear power plant is the generation of steam with the heat from fission of the nuclear fuel in the reactor vessel. The steam is used to power a turbine which in turn powers a generator to produce electricity. Many other processes are also present, including heat recovery, air purification, waste processing, and power transformation.

Most nuclear power generating stations in the United States use a *pressurized water reactor* (PWR) cycle or a *boiling water reactor* (BWR) cycle for generating steam. Two other steam generation cycles have also been considered for power plant use. They are the *high temperature gas reactor* (HTGR) cycle and the *liquefied metal fast breeder reactor* (LMFBR) cycle. PWRs, BWRs, and HTGRs are in commercial operation or under construction at the present time in the United States. The LMFBR is currently in the test and development stages, but it is predicted by some to be the plant of the future because of its ability to produce more fuel than it uses.

Pressurized Water Reactor Cycle

The pressurized water nuclear power reactor cycle contains a closed loop of pressurized water which removes the heat energy from the reactor core and transfers that energy to a second water system in order to generate steam therein. The steam drives a turbine generator set which produces electric power. (See Figure 5.1.)

The reactor system consists of a reactor vessel which contains the nuclear fuel necessary for the generation of heat energy, a steam generator in which the heat energy is used to generate steam, a circulating pump which circulates the coolant, and a pressurizer which maintains and controls system pressure. The coolant circulation pumps are powered by 4,000- to 8,000-hp electric motors. Each motor contains several hundred gallons of combustible lubrication oil. The reactor coolant is demineralized water. The reactor system is assembled with a series of coolant loops which radiate from the reactor vessel. A single reactor has only one

NUCLEAR POWER PLANTS

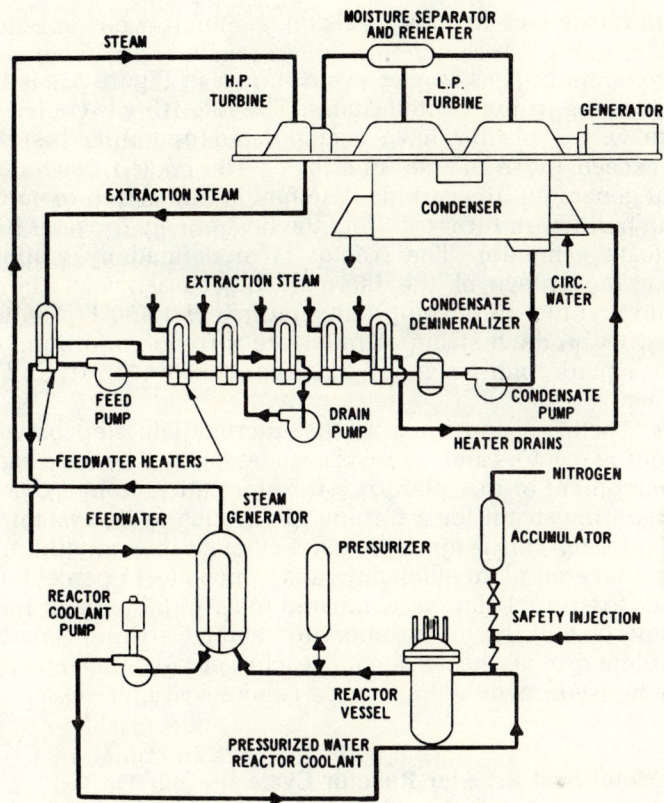

FIGURE 5.1. System schematic of pressurized water reactor cycle.

pressurizer regardless of the number of loops. One loop is provided for each steam generator required, and there are either one or two pumps per loop, depending on the manufacturer.

Boiling Water Reactor Cycle

The boiling water reactor cycle is shown in Figure 5.2. It is a direct cycle steam generating system in which the water is boiled by the fissioning core in the reactor vessel. The wet steam passes through moisture separators and steam dryers, leaves the reactor vessel and goes to the turbine. After passing through a suitable heat cycle, the condensate is pumped through demineralizers, feedwater heaters, and then back to the reactor. Since the steam for driving the generator is radioactive, radiation hazards must be considered in the turbine building of a BWR plant.

High Temperature Gas Reactor Cycle

The high temperature gas reactor cycle shown in Figure 5.3 is the newest type operating in the United States. The English gas-cooled Magnox and advanced gas reactor have accumulated operating histories that match or exceed those of the American water-cooled reactors. In this design heat generated in a carbide-base fuel is transferred to compressed helium; the helium, in turn, transfers the heat energy to water in a once-through steam generator. The reactor is moderated by graphite. The design takes advantage of the thermal characteristics of the helium, carbide, and graphite to develop temperatures of 1,450°F (788°C) in the coolant and to produce steam at moderate turbine conditions of 2,500 pounds per square inch gauge (17 238 kPa), 955°F (513°C) and 1,000°F (538°C) reheat.

This reactor, a converter, is the intermediate step between uranium fissioning reactors and future breeder reactors. The next logical step in the development of this plant is a transformation from a closed-cycle system generating steam for a turbine to an open-cycle system working with a gas turbine. This is apt to be more efficient than any thermal cycle in use today. Overall plant efficiency runs from 39 to 41 percent, the same as a modern fossil fuel station, compared to 31 to 33 percent for a water reactor plant. One of the most important characteristics of this design is that a complete primary heat system—reactor, circulators, and steam generators—is housed inside a prestressed concrete reactor vessel.

Liquefied Metal Fast Breeder Reactor Cycle

The liquefied metal fast breeder reactor concept is not currently used for commercial electric power generation in the United States. Figure 5.4 shows the basic heat transport cycles for the experimental Clinch River Breeder Reactor Plant. A system of three identical piped circuits transports heat from the reactor through both primary and intermediate sodium loops to steam generator modules which produce steam for the turbine. Sodium is transported through the primary loop using 33,500 gallons per minute (127 m^3/min) pumps at a maximum pressure of 175 psig (1207 kPa). Guard vessels surround the primary components in order to contain sodium in the event of a leak in the coolant boundary.

Each circuit has an intermediate heat exchanger which transports the heat from the radioactive primary sodium system to the nonradioactive intermediate sodium loop. The sodium in the intermediate loop is transported through the steam generator using 29,500-gpm (112 m^3/min) pumps and produces steam for the turbine. Leakage, if it were to occur, would be from the intermediate loop to the primary loop because a higher pressure is maintained on the intermediate loop. Turbine steam temperatures are approximately 900°F (482°C) where the steam enters the first stage.

Each of the basic nuclear steam supply processes discussed de-

NUCLEAR POWER PLANTS

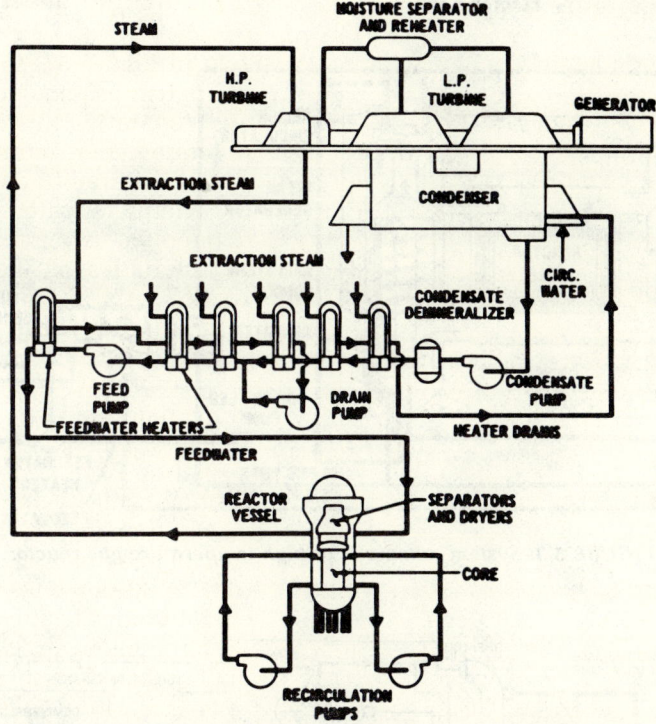

FIGURE 5.2. System schematic of boiling water reactor cycle.

pends on numerous support systems for the safe operation of the cycle. The support systems have different degrees of importance relative to the safe operation of the plant. Some are required for safe shutdown, while others are only needed for ease of plant operation. The system names vary among manufacturers and designs. Likewise, the systems critical to safe plant shutdown also vary. Plant designers and operators must clearly define the systems, structures, and components needed for safe operation in order to license nuclear power plants. The following essential fuctions must be provided by support systems:

1. Reactor shutdown
2. Cooling water supply
3. Electric power supply to essential equipment
4. Air filtration and purification
5. Radiation containment
6. Radioactive waste processing
7. New and spent fuel handling
8. Combustible gas control.

Support systems normally have redundant systems to ensure that their essential functions can be performed. In some cases, designs are ar-

Chapter 5

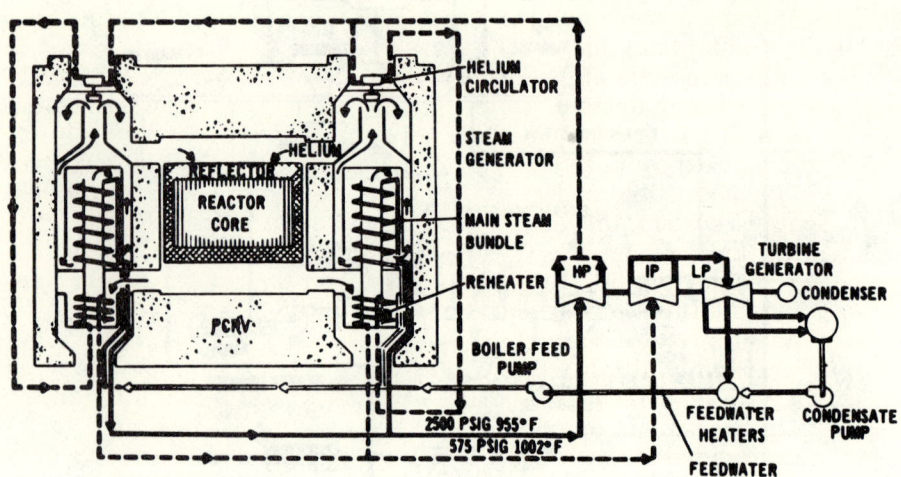

FIGURE 5.3. System schematic of high temperature gas reactor.

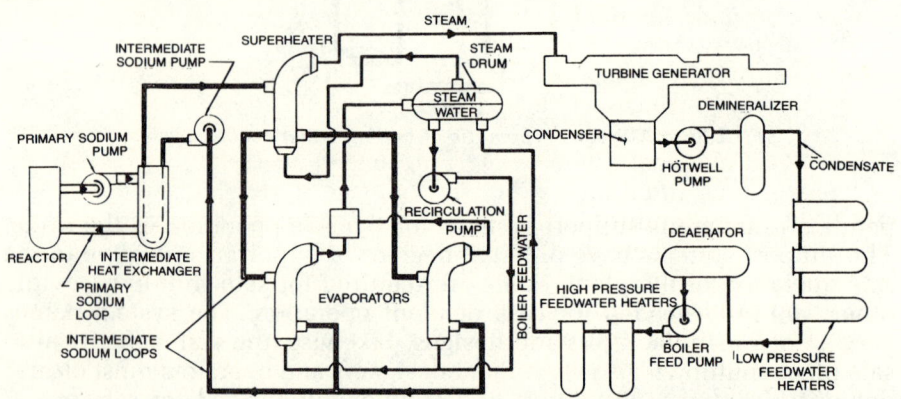

FIGURE 5.4. System schematic of liquefied metal fast breeder reactor cycle.

ranged to provide the essential functions by a means other than a redundant system. Although most of the support systems, like the steam supply cycles, present essentially no fire hazard, the extreme importance of safety necessitates that they be protected from the hazards of fire exposure.

Safe shutdown requires that electrical power be available to certain equipment, and control circuits be functional. It also requires a supply of cooling water and control of cycle pressure. Some designs also require that various chemical addition systems be functional. Due to design variations, the functions necessary for safe shutdown must be analyzed on an individual plant basis.

Cooling water for nuclear power plants is supplied by numerous methods. Some plants use large natural water bodies while others use large storage reservoirs and cooling towers. Pumps are used to circulate and deliver water to demand points. Essential safety pumps are supplied with electrical power from emergency bus bars and are separated in order to prevent exposure to common fire hazards. Pump house construction is noncombustible or fire-resistive. Where cooling towers are used in the supply system, they can be large hyperbolic towers or horizontal mechanical draft towers. Towers in newer designs are noncombustible with noncombustible fill. Tower construction is similar to that for towers at large fossil-fuel power plants and industrial plants.

The supply of electrical power to essential equipment is accomplished through multiple off-site power feeds and redundant on-site emergency diesel-driven generators. The diesels are large and usually have tanks that contain one day's supply of fuel near the engine. On-site storage is provided for the diesel fuel supply. Storage arrangements vary but include atmospheric tanks with dikes, underground tanks, and large tanks within buildings.

Due to the possible presence of airborne radiation, the building air-handling systems for nuclear power plants are of special design. Many systems are redundant. Some areas, such as the control room, must be manned during in-plant nuclear accidents as well as off-site accidents, such as toxic gas release. The systems are provided with complex filter plenums that house roughing filters, high efficiency particulate air filters (HEPA), and charcoal filters. Figures 5.5 and 5.6 show the arrangement of a hopper-type and drawer-type plenum, respectively. These filters present the last opportunity to remove radiation prior to recirculation or release to the atmosphere through the plant vent. All discharges must pass by radiation monitors. See Figure 5.7 for a simple system diagram.

The primary method used to prevent radiation release is containment. The buildings, equipment, and piping systems that contain radiation are designed to retain the contents under expected upsets. Shield walls and barriers are provided to absorb radiation.

All liquid waste is collected and processed through filters and decay tanks in a manner similar to that used in the air-handling system. Contaminated solid waste is collected and compacted. Some contaminated solid waste is then drummed and encapsulated in organic resins. Solid waste could include disposable clothing, disposable filters, shipping containers, equipment components, etc.

New and spent fuel is handled in a separate fuel-handling area. Special water-filled storage pools and transfer canals are provided for the fuel. The water in pools must be maintained cool and filtered; therefore, special systems circulate and treat the pool water.

During the radiation of water in the steam generation cycle of the PWR and BWR, some diassociation of the water molecules occurs. This results in the release of hydrogen gas. Therefore, provisions are included

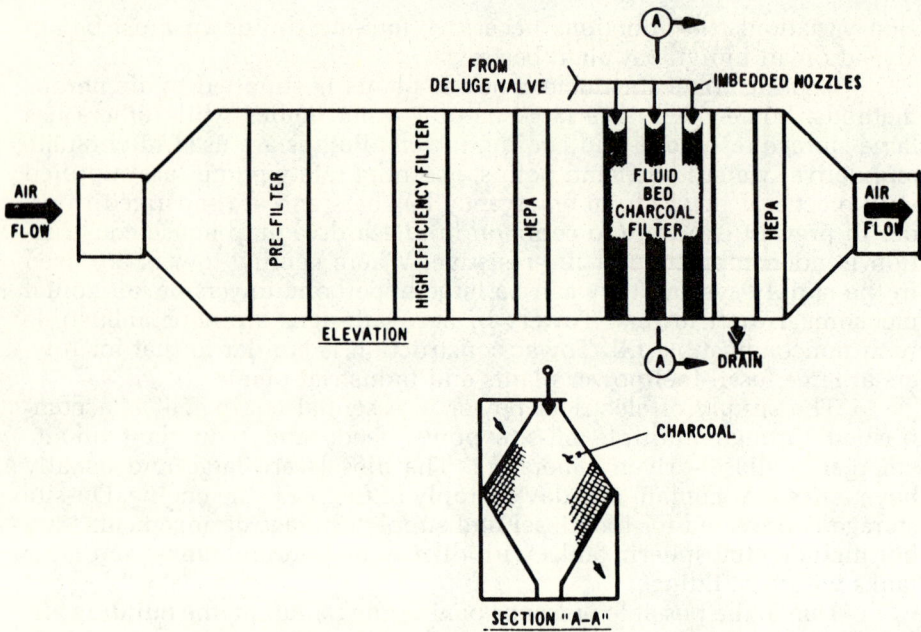

FIGURE 5.5. *Fluidized-bed-type charcoal filter plenum.*

in the plant design to dilute and recombine the hydrogen. Some designs collect the gas in decay tanks prior to recombining. Although the hydrogen concentrations are normally maintained well below the flammable range, it is a hazard that must be safeguarded against.

THE FIRE HAZARDS

The fire load of nuclear power plants is low compared to that of fossil-fuel power plants or many industrial plants. Common combustibles and hazardous materials present in nuclear plants are listed in Table 5.1.* The quantities and locations of these combustibles vary among nuclear power plants. Identification of these combustibles and their associated fire characteristics gives only a partial indication of the fire hazard. The bearing that these will have on nuclear safety must also be considered in defining the total fire hazard. Nuclear safety factors include the maintenance of safe shutdown capability as well as the prevention of radiation release in excess of acceptable limits. The nuclear safety system's concrete construction, and its inherent resistance to fire damage, is another

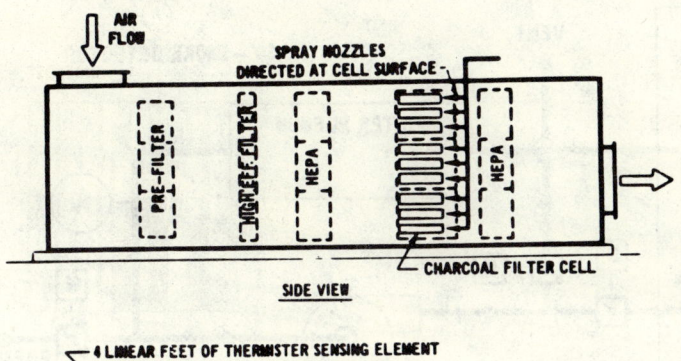

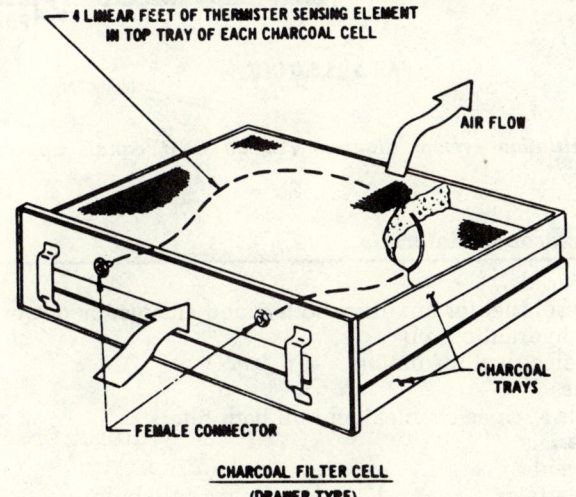

FIGURE 5.6. *Drawer-type charcoal filter plenum.*

element in the total fire hazard potential.*

The fire hazards related to nuclear power plants include the following:

Oil fire hazards associated with large reactor coolant pump motors;
Oil fire hazard involving emergency turbine-driven feedwater pumps;
Diesel fuel fire hazard at diesel-driven generators;
Fire hazard involving charcoal in filter plenums;
Fire hazard associated with electrical cable insulation;

---

*Taken from National Nuclear Risks Insurance Pools and Associations, *International Guidelines for the Fire Protection of Nuclear Power Plants* (Zurich: Swiss Pool for the Insurance of Atomic Risks, 1974).

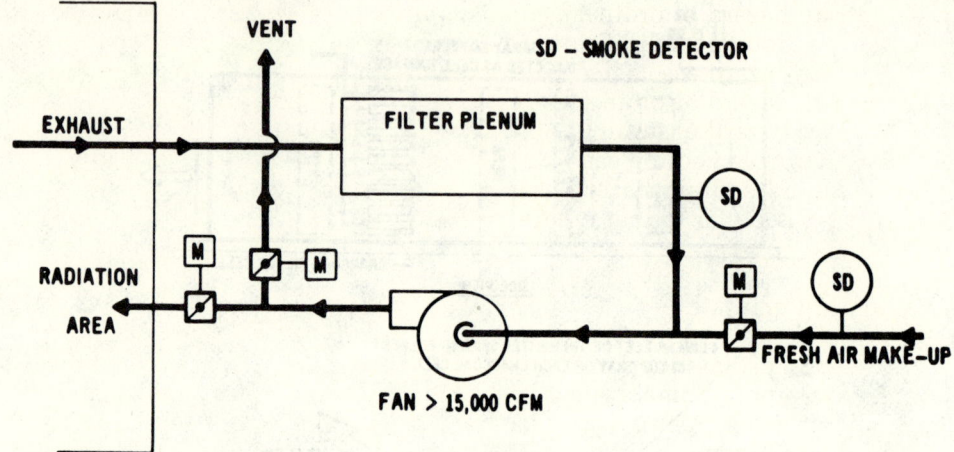

FIGURE 5.7. *Ventilation system diagram (15,000 CFM equals approximately 425 m^3/min).*

TABLE 5.1. *Hazardous Materials*

Combustible materials
1. Fuel oil or diesel fuel for auxiliary boilers and emergency power units
2. Lubricants or hydraulic fluid
3. Thermal and electrical insulation
4. Interior finishes
5. Filter materials such as charcoal and oil bath filters
6. Cleaning materials
7. Paint and solvents
8. Packaging materials
9. Neutron shield (if organic)

Explosive and flammable gases
1. Hydrogen for generator cooling
2. Hydrogen or methane for coolest conditioning of gas-cooled reactors
3. Propane or other fuel gases
4. Oxygen or hydrogen by radiolysis in the core and hydrogen for improved recombination
5. Welding and cutting gases

Radioactive substances
1. Sealed sources such as radiated and/or plutonium-containing fuel elements, unradiated control code, neutron sources
2. Unsealed sources such as ion exchanger fillings and filter cartridges

Acids and alkalis
1. Used in water treatment (boric acid used in PWR for regulating long-term reactivity)

Fire hazard of ordinary combustible wastes awaiting compaction and solidification in organic resins;
Fire hazard associated with flammable off gases;
Fire hazard of protective coatings;
Fire hazard of turbine lube oil and hydrogen seal oil;
Hydrogen cooling gas hazard in generator buildings;
Liquid sodium hazard associated with the LMFBR, a hazard peculiar to that type of nuclear plant.

Coolant Pump Motors

Each main coolant loop in PWR and BWR designs has a large coolant circulation pump. Some designs have as many as four loops and therefore have four such pumps with one reactor. The motors are over 5,000 hp, and many designs approach 9,000 hp. They have hydraulic bearing lift oil and lubrication oil systems. Oil quantities vary among designs but range between 50 gal (189 L) and 500 gal (1893 L). The oil is contained in a welded steel piping system. Newer designs use guard enclosures to collect any leaks. Since these motors are located adjacent to the reactor in most designs, a minor oil fire could do extensive damage to control circuits in the area. These areas are quite inaccessible during operation due to high radiation levels. Care must be taken to ensure that redundant safety circuits are not exposed to the same pumps. Curbs and drains are needed to isolate and control oil spread. The pressure-containing capability of the pumps and associated piping is needed for safe shutdown. Thus, the pump and pipe supports must be able to withstand anticipated maximum fire exposure without failure.

Turbine-driven Pumps

A supply of cooling water is necessary for safe shutdown in water-cooled reactors; therefore, many designs incorporate turbine-driven pumps as backup to electric pumps. Such turbines have high steam temperatures, and some designs have extensive oiling systems. A misty oil leak could be ignited by hot steam pipe surfaces. The hazard is dependent on the size and operating conditions of the oiling system, the steam temperatures, and the degree of isolation from redundant pumps and control circuits.

Diesel-driven Generators

Diesel fuel is handled in closed metal tanks and piping systems at emergency generator locations. Special drainage systems and curbs are used to prevent the spread of leaks. Since the diesel fuel fire hazard cannot be removed, care is taken to maintain separation of redundant diesel generators. The actual fire exposure, or the products of combustion, cannot be allowed to affect redundant diesels. Therefore, diesel generators are generally located in separate rooms with separate combustion air intakes.

Charcoal Filter Plenums

Charcoal filters are the last filters in most nuclear plant ventilation system design and serve to collect airborne radiation prior to release of the air. The fire hazard associated with the charcoal is quite low because the charcoal is densely packed between perforated metal enclosures. (See Figures 5.5 and 5.6.) Ignition is extremely difficult under such conditions. Some have theorized that ignition could occur from the heat of decay from trapped radioactive substances. The spontaneous ignition temperature of charcoal is approximately 450°F (232°C). A more important hazard consideration relates to deabsorption of radiation which occurs near 300°F (149°C) and which could result from fire exposure. Such fire exposure could be from a fire involving oil and dust accumulations in ducts, fire in areas being exhausted, or fire adjacent to the filter plenums.

Electrical Cable Insulation

In nuclear power plants the fire hazard associated with electrical cable insulation and jackets is similar to that of other occupancies which use cable trays to support huge numbers of power and control cables. However, in nuclear plants the added hazard associated with loss of redundancy exists. Cable insulation and jackets are organic compounds in most cases and therefore will burn under the proper circumstances. More recent designs have used fire retardant materials for insulation and jackets. Such materials reduce the hazard. The hazard related to loss of redundancy involves the potential of fire fueled by cable insulation and jackets and, more importantly, the hazard of exposure to fires fueled by stationary or transient combustibles. Since redundant safety circuits cannot be sacrificed, the redundant circuits cannot be subjected to simultaneous loss from any fire. Complete isolation of the circuits from each other, and design to prevent exposure to common fire hazards are a necessity.

Combustible Wastes

The collection, compaction, and solidification of solid wastes is a housekeeping hazard common to other industries in many respects. The possible radioactive contamination of these materials in a nuclear plant increases the hazard beyond that of their combustibility. Burning of combustible wastes could expose safety systems and contaminate the plant atmosphere. Wastes must be promptly removed from the point of generation to the waste collection and processing area. Compaction and solidification, likewise, must be promptly performed. Monomers and solvents for the solidification process must be handled in a safe fashion. Ignition sources must be eliminated. Electrical equipment may need to be suitable for Class I, Division 1 or 2 hazardous locations if solidification involves flammable gases or vapors.

Flammable Off Gases

Hydrogen off gas exhibits the flammable hazards associated with the gas and also is radioactive. Therefore, it must often be held in decay tanks prior to recombination with oxygen or venting to the atmosphere. Constant monitoring of the off-gas systems for oxygen is a necessary safeguard. The system piping and equipment design must be extra heavy to help protect against failures. Electrical equipment must be suitable for Class I, Group B hazardous locations (as defined in Article 500 of NFPA 70, *National Electrical Code*) where leaks are possible. Constant high-rate ventilation systems must be provided for decay tank and equipment areas. Where the hydrogen is vented, safeguards to ensure an adequate dilution stream of air are needed. The hydrogen off gas presents one of the most severe hazards in nuclear plants and therefore requires constant attention.

Protective Coatings

Nuclear shield walls, structural elements, and equipment are provided with various protective coatings which are made of organic compounds. Flame spread ratings are normally below 50 for fuel contribution, flame spread, and smoke development. These coatings are applied in various thicknesses. In most cases after they have cured, they present little fire hazard. Caution is needed if the application is done in operating plants.

Lubrication and Seal Oil

The hazards associated with turbine lubrication oil and hydrogen seal oil are similar to those at conventional fossil-fuel or hydroelectric generating stations. Steam temperatures may be somewhat lower at nuclear plants so the ignition potential may be less. Steam in BWR plants is radioactive in the turbine building. As a result, access to the turbine for manual fire attack is somewhat restricted, and fire protection water could become contaminated.

Chapter 5

Hydrogen Cooling Gas

The hydrogen gas used for generator cooling presents hazards similar to those at fossil-fuel plants. Auxiliary boilers and transformers also present hazards similar to those at fossil-fuel or hydroelectric plants. (See Chapter 4, "Electric Generating Plants," for a thorough discussion of the fire hazards associated with electric generating plants.)

The hazard of liquid sodium and a liquid metal fire is present in the LMFBR plant. Sodium melts at 207.5°F (97.5°C), and the metal or its vapor may ignite spontaneously in air at temperatures above 239°F (115°C). The resulting smoke, chiefly sodium oxide, can be a serious respiratory hazard. Burning sodium can react violently with many materials. Liquid sodium in contact with concrete spalls the concrete. The main requirement for extinguishing a sodium fire is the addition of an excess of an inert material to exclude air and cool the burning material to below the ignition temperature. Met-L-X (a proprietary combustible metal extinguishing agent), dry calcium carbonate, dry sodium chloride, dry sand, and dry graphite may be used as extinguishing agents.

As is apparent from the hazards previously described, the identification and evaluation of hazards in nuclear plants are complicated by nuclear safety considerations. Regulatory agencies have therefore required that a detailed documental fire hazards analysis be performed on all nuclear power plants. Various systematic approaches can be used to identify and evaluate the impact of each hazard on nuclear safety. It is important to recognize the complexity of the problem, and the need for such a systematic approach, in planning the plant fire defense. The analysis must be performed by a team of knowledgeable engineers and scientists under the direction of a qualified fire protection engineer. It must include a documented record of all combustibles, their fire characteristics, the redundant safety systems being exposed, and the safeguards. The analysis must consider all parts of the plant and exposures. As changes occur in construction or operation, the analysis must be reviewed and updated, and the safeguard revisions must be identified if needed. The need for the complete documentation of the analysis exists due to the constant personnel changes which occur over the life of the plant.

THE SAFEGUARDS

The safeguards for nuclear power plants are based on the concept of defense-in-depth which strives to achieve an adequate balance of fire prevention, fire detection, and fire suppression techniques with an inherent resistance of essential plant functions to fire damage. The fire defense is a total fire management program including construction safeguards, fire detection and protection safeguards, and life safety provisions.

The importance of the safeguards' availability necessitates the establishment of a quality assurance and control program which verifies

NUCLEAR POWER PLANTS

and documents the fire defense designs and operating procedures, and periodically audits the program.

Construction Considerations

Proper plant layout is a very important fire protection consideration in nuclear power plant design. Nuclear plants as shown in Figures 5.8, 5.9, 5.10, and 5.11 are generally composed of a group of fire resistant or noncombustible buildings separated by 3-hour fire-rated barrier walls. These buildings usually include a reactor containment building, control building, fuel-handling building, auxiliary building, waste management building, diesel generator buildings, turbine building, auxiliary boiler building, and pump buildings. In addition, warehouses and service buildings and all elevators are enclosed in 1- or 2-hour fire-rated shafts as indicated in Figures 5.9 and 5.10. The separation of hazards and exposures into separate fire-rated cubicles is common practice. Extensive vertical and horizontal runs of cables, pipes, or ducts are usually set in fire-rated chases. High hazard equipment such as turbine oil reservoirs should be located in separate fire-rated rooms. Redundant systems should be separated by fire-rated barriers or provided with spatial separation commensurate with the area fire hazard. Control rooms and areas below the operating floor of turbine buildings for multiunit plants should be separated by 3-hour fire-rated walls.

Hydrogen or other gas manifolds should be located outdoors in well-ventilated protected areas. Crushed stone or concrete areas free of combustible vegetation should be provided, and the areas should be fenced. Likewise, large oil-filled transformers should be located outdoors. Hydrogen vent gas collection systems including compressors, decay tanks, and recombiners should be isolated from other combustibles in well-ventilated fire-rated enclosures. Indoor transformers should be limited to inert gas or fire-resistive oil-filled types. Outdoor oil-filled transformers should be separated from important plant buildings, or any walls within 50 ft (15.2 m) should have a minimum of a 2-hour fire rating. Good drainage should be provided for the rapid removal of oil to a safe location; and fire walls should be provided between transformers extending to the top of the bushings if adequate spatial separation between the transformers does not exist. The new fuel storage should be in a 2-hour fire-rated room or vault, and the radioactive waste drumming area should be separated from the radioactive waste storage area. It should be noted that all fire separations discussed previously are considered minimum and should be upgraded if the hazard evaluation so indicates.

The effectiveness of fire barriers in nuclear power plants, as in other plants, depends on the establishment and maintenance of the barrier as a fire-resistive membrane that will not allow propagation of fire or fire gases through the barrier. Since it is often necessary to penetrate these barriers with pipes, cables, and ducts, effective penetration seals must be provided. These often must meet seismic and radiation expo-

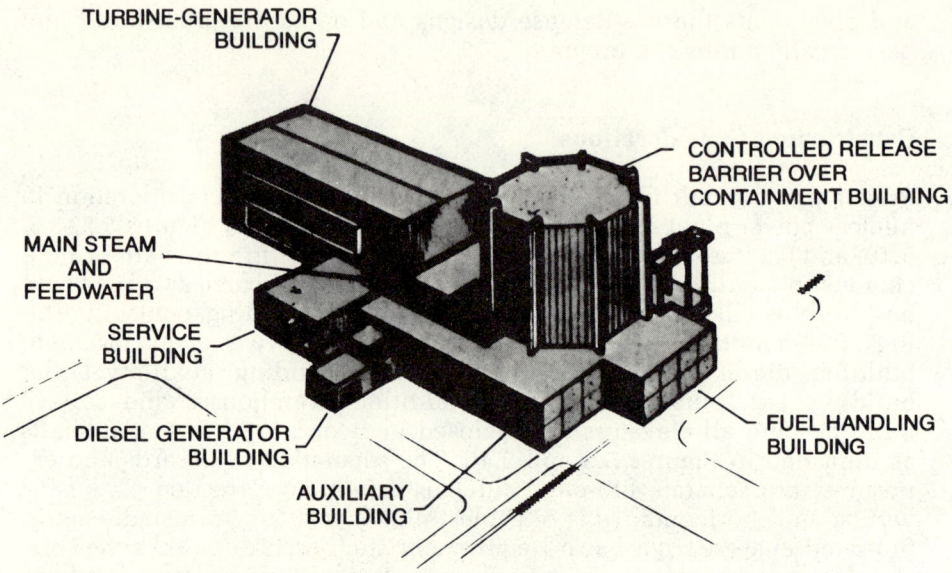

FIGURE 5.8. *Typical single unit nuclear plant arrangement.*

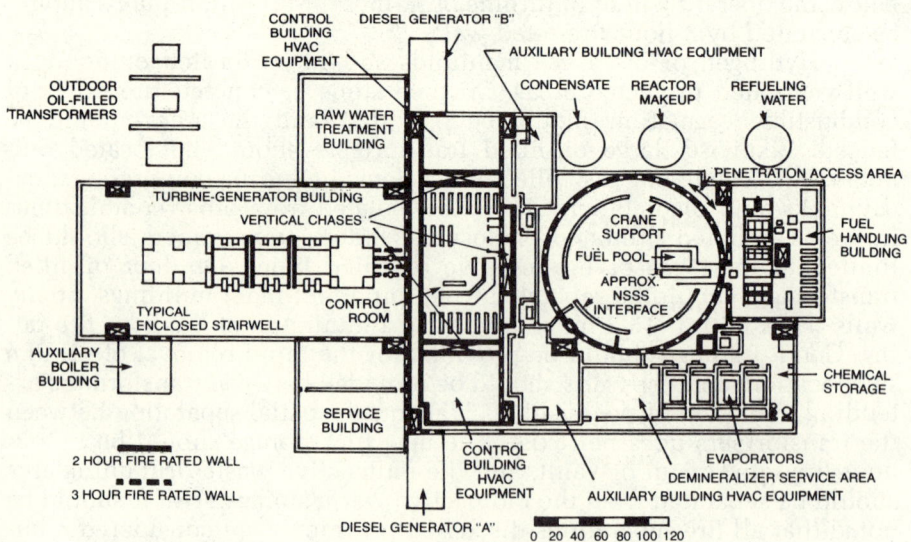

FIGURE 5.9. *Typical plan of control room floor elevation.*

sure requirements, in addition to the fire rating. Materials that have been used include high and low density cellular concrete, silicon foams, neoprene multicable transit devices, and Flamemasti® cover mineral fiber and cement asbestos board. Fire dampers must be installed in ducts un-

NUCLEAR POWER PLANTS

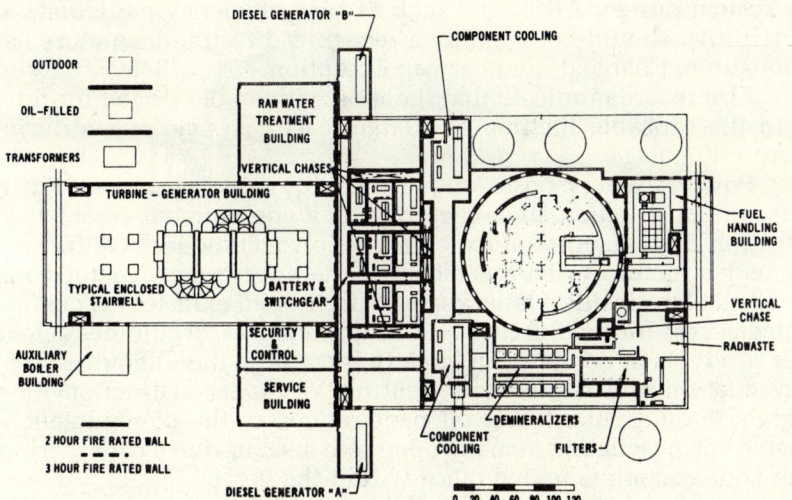

FIGURE 5.10. *Typical plan of grade floor.*

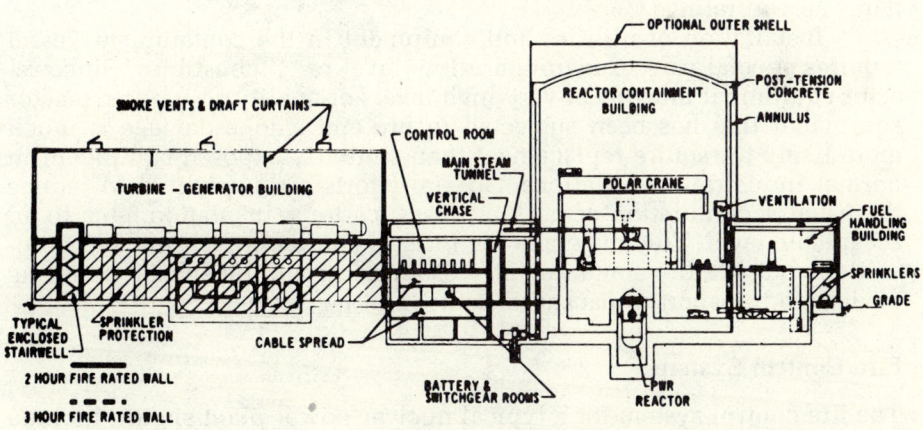

FIGURE 5.11. *Section view of typical single unit nuclear power plant.*

less such dampers would jeopardize nuclear safety. In some cases the duct may require a fire-rated enclosure to ensure its effectiveness.

Another important construction consideration involves ventilation system design. All filters, such as high efficiency particulate air filters (HEPA), should be listed by a recognized testing laboratory as noncombustible. Charcoal filters are an exception.

Fire records indicate that the most vulnerable period for fire damage in the probable lifetime of a large reactor system exists during the construction stage.

Power reactors pose unique fire protection problems during the course of construction since they require a containment vessel as a final protection system. And, since construction techniques require the containment structure to be erected first, the subsequent reactor construction takes place inside a large vessel with limited exits for evacuation and limited access for fire fighting. In addition, the vessel confines smoke and other products of combustion, greatly increasing the difficulties inherent to evacuation and manual fire fighting. A good construction firesafety program should ensure that all penetrations of the containment vessel suitable for evacuation remain open and usable during the period that other construction is taking place within the vessel.

Insofar as access for fire fighters is limited, and the escape of smoke is not facilitated as in normal outdoor construction, it is imperative that all combustible materials incident to construction be severely limited. Metal formwork, scaffolding, platforms, stairways, etc., are preferable to wood. As a minimum, the use of wood in extensive quantities is limited to those appropriately treated to reduce the combustibility and flame spread ratings.

Installation of utilities and equipment in the containment vessel requires special care to maintain a low level of combustibles. Since reactor equipment must meet very high levels of quality assurance, reactor equipment that has been subjected to fire and smoke damage is much more likely to require replacement than similarly exposed equipment in normal industrial installations. Special efforts are warranted to reduce the usual accumulation of packing cases, cartons, insulation, etc., to an acceptable level. This may take the form of conducting all uncrating operations outside the containment vessel and providing special handling devices to transport unpackaged items into the vessel.

Fire Control System

The fire control system for a typical nuclear power plant should include a water supply and distribution system, a fire detection and alarm system, gaseous extinguishing systems, as well as fixed water spray and automatic sprinkler systems. Types of fixed detection and extinguishing systems used in typical plant designs are summarized in Table 5.2.*

---

*This table appears as Table 10.1.2 in NFPA 803, *Fire Protection for Light Water Nuclear Power Plants*.

TABLE 5.2. Fixed Extinguishing Systems

Area or Hazard To Be Protected	Water Systems			Liquefied Compressed Gases			Foam Systems			Dry Chemical
	Automatic Sprinklers	Pre-Action Sprinklers	Deluge-Water Spray	Halon 1301	Halon 1211	Carbon Dioxide	Mech.	Foam Water Systems	High Expansion	
Auxiliary Oil or Gas Boiler Room	X	X					X (oil)	X (oil)		X
Battery Rooms	X	X		X	X	X				
Cable Penetration Rooms and Spreading Rooms	X (note 1)	X (note 1)	X (note 1)	X (note 2)	X (note 2)	X (note 2)				
Cable Tunnels, Shafts, Chases, Cable Tray Run Concentrations	X (note 1)	X (note 1)	X (note 1)	X (note 2)	X (note 2)	X (note 2)				X (multi-purpose only)
Combustible (Charcoal) Filters			X							
Computer Rooms (including under floor space)	X	X		X		X (note 4)				
Control Room		X		X						
Diesel Fire Pump Room	X									
Diesel Generator Rooms (note 3)	X	X		X	X	X	X	X	X	X

TABLE 5.2. Fixed Extinguishing Systems (Continued)

Area or Hazard To Be Protected	Water Systems			Liquefied Compressed Gases			Foam Systems			Dry Chemical
	Automatic Sprinklers	Pre-Action Sprinklers	Deluge-Water Spray	Halon 1301	Halon 1211	Carbon Dioxide	Mech.	Foam Water Systems	High Expansion	
Diesel Fuel Day Tank Rooms	X	X	X	X	X	X	X	X	X	X
Diesel Fuel Storage Tank, If Not Buried	X		X	X	X	X	X	X	X	X
Filter Rooms and Plenums	X	X	X							
Flammable Liquid Storage Areas	X		X	X	X	X	X	X	X	X
Fuel Oils Storage Tank	X		X				X	X	X	X
Hydrogen Seal Oil Units			X							X
Instrument Rack Rooms				X	X	X				
Laboratories	X	X		X	X	X				
Oil Lines & Reservoirs at Steam Turbine Driven Equipment (if more than 50 gal) (190 L)	X	X	X	X (note 5)	X (note 5)	X (note 5)				X
Reactor Coolant/Recirculation Pumps	X	X	X			X	X	X		X

TABLE 5.2. Fixed Extinguishing Systems (Continued)

Area or Hazard To Be Protected	Water Systems			Liquefied Compressed Gases			Foam Systems			Dry Chemical
	Automatic Sprinklers	Pre-Action Sprinklers	Deluge-Water Spray	Halon 1301	Halon 1211	Carbon Dioxide	Mech.	Foam Water Systems	High Expansion	
Record Storage Rooms	X			X (note 6)	X (note 6)	X (note 6)				
Relay Rooms/Cabinets	X	X		X	X	X				
Switchgear Rooms		X		X	X	X				
Transformer Outdoor (if combustible oil filled)			X							
Transformer Indoor (if combustible oil filled)		X	X	X	X	X				X
Turbine Building Beneath Operating Floor Where Oil Can Spread	X	X					X			X
Turbine Generator Governor Housing (if combustible fluid)		X	X	X (note 5)	X (note 5)	X (note 5)		X		X
Turbine Generator Bearing (seals)		X	X	X (note 5)	X (note 5)	X (note 5)	X			X
Turbine Generator Lube Oil Conditioning or System Room	X	X	X	X	X	X		X	X	X

TABLE 5.2. Fixed Extinguishing Systems (Continued)

Area or Hazard To Be Protected	Water Systems			Liquefied Compressed Gases			Foam Systems			Dry Chemical
	Automatic Sprinklers	Pre-Action Sprinklers	Deluge-Water Spray	Halon 1301	Halon 1211	Carbon Dioxide	Mech.	Foam Water Systems	High Expansion	
Turbine Generator Lube Oil Storage Rooms	X		X	X	X	X	X	X	X	X
Steam Valves (if combustible hydraulic fluid)			X			X				X
Staging, Storage and Warehousing Areas	X	X							X	
Truck and Railroad Bays (other areas of combustible occupancy)	X	X								
Cooling Towers (combustible)			X							

Note 1. Systems shall be designed so that water is directed into every tray. Where closed head sprinkler or thermal detection systems are used, means shall be provided for prompt actuation.
Note 2. Ceiling sprinklers may be required in addition.
Note 3. Release of extinguishing agent shall not be prevented if equipment is operating.
Note 4. Under floor only.
Note 5. Design concentrations shall be maintained during the entire coast down period.
Note 6. Combustible storage should be in metal cabinets.

Water supply and distribution systems: The fire protection water supply and distribution system should be capable of providing water to the point of highest demand (with one side of the loop out of service) at an adequate pressure and in sufficient volume to supply the largest fixed fire protection system plus three or four 250-gpm (950 L/min) hose streams. Ideally, the supply is from two 300,000-gal (1136 m^3) water storage tanks, a cooling tower basin, or large and reliable natural water bodies. A typical system, as shown in Figure 5.12, would include two 2,500-gpm (9.5 m^3) or three 1,500-gpm (5.7 m^3) fire protection water supply pumps, as well as a jockey pump and a pressure maintenance tank. One of the main pumps would normally be electrically driven and the "back-up" pump(s) would normally be diesel engine driven. The electric fire pump normally starts automatically upon pressure drop, and the diesel pump would start automatically at a lower pressure (diesel pumps are usually shut off manually). Provisions are often made to test the fire pumps using an approved flow meter located in the pump house.

The fire protection water distribution system usually includes a complete exterior 12-in. (0.30 m) underground yard hydrant loop as well as an interior distribution piping system, as shown in Figure 5.13. Hydrants with individual shut-off valves are provided at approximately 250- to 300-ft (76.7 to 91.4 m) intervals along the loop and approximately 50 ft (15.2 m) from the buildings. Each hydrant normally has a hose house equipped with 250 ft (76.2 m) of 2½-in. (63.5 mm) fire hose, spray nozzles, and miscellaneous tools and hose couplings. The interior distribution headers, as shown in Figure 5.14, supply hose cabinets or continuous flow hose reels and the various water-type fixed fire extinguishing systems. More recent designs make greater use of separate supplies from the exterior underground system for interior fixed systems and standpipes. Hose reels are often equipped with approximately 75 ft (22.9 m) of 1½-in. (38 mm) noncollapsible hard rubber "suction-type" hose and a combination fog nozzle. Hose cabinets have the same amount of single jacket collapsible hose.

The interior and exterior distribution system piping is provided with adequate sectionalizing valves to prevent the removal of fire protection water service from more than five components (fixed systems, hydrants, or hose reels) at the same time.

Fire detection and alarm systems: The fire detection and alarm system must be designed to provide rapid and reliable detection and location of fires in their early stages. Detection is normally provided in high hazard areas such as cable spreading areas, critical containment vessel areas, and electrical rooms. Detection devices for fixed fire protection systems are discussed under the various fixed system descriptions. All fire detection devices are connected to a central monitoring unit normally located within the main control room. The monitoring unit includes an annunciator, a standby power supply, and complete trouble supervisory capability. It also serves to monitor the fire pumping system as well as the fixed fire protection system and annunciate "fire" and "trouble" condi-

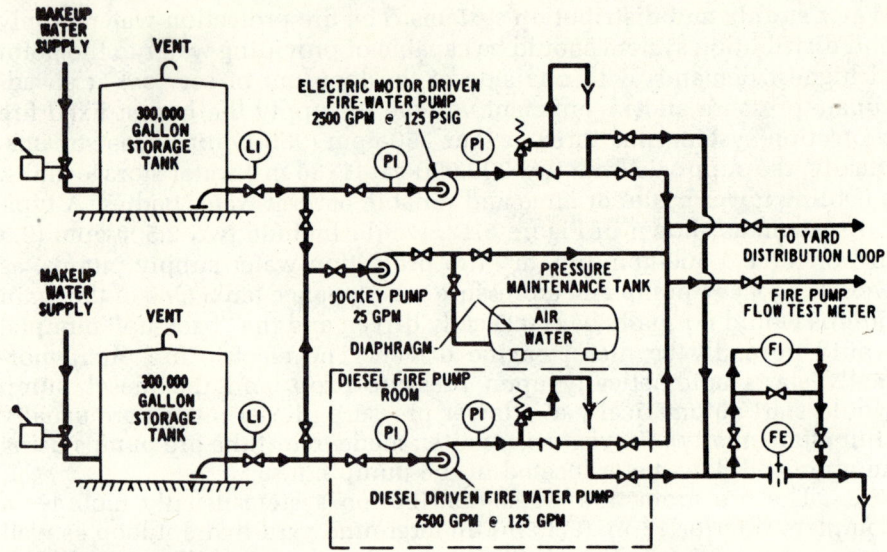

FIGURE 5.12. Fire protection water supply system (1 gal = 3.785 L; 1 psig = 6.895 kPa).

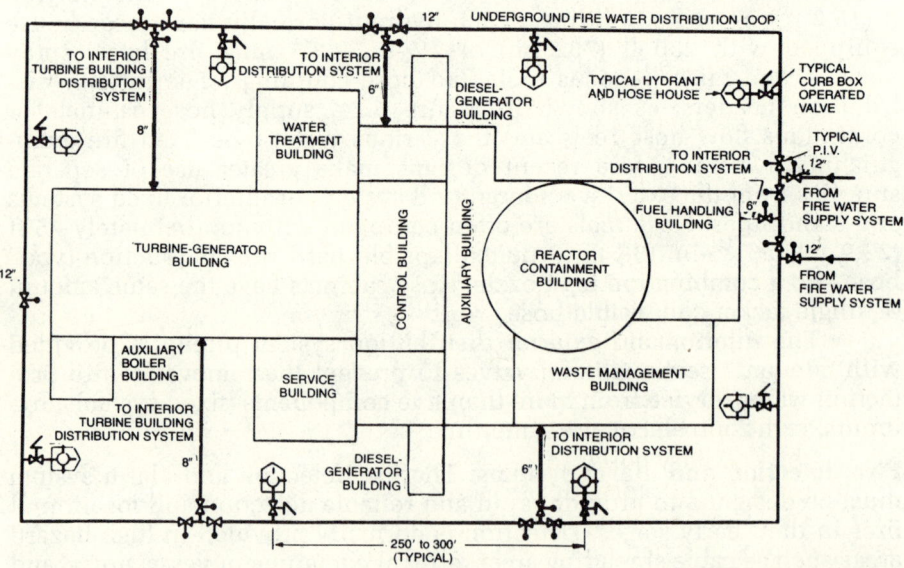

FIGURE 5.13. Yard fire protection water distribution system (1 ft = 0.3 m; 1 in. = 25.4 mm).

tions for these systems. Recent systems are of a loop design and use a computer or multiplexing technique to monitor the loop. A schematic diagram of a typical system is shown in Figure 5.15. Strategically located manual pull stations are also provided.

NUCLEAR POWER PLANTS

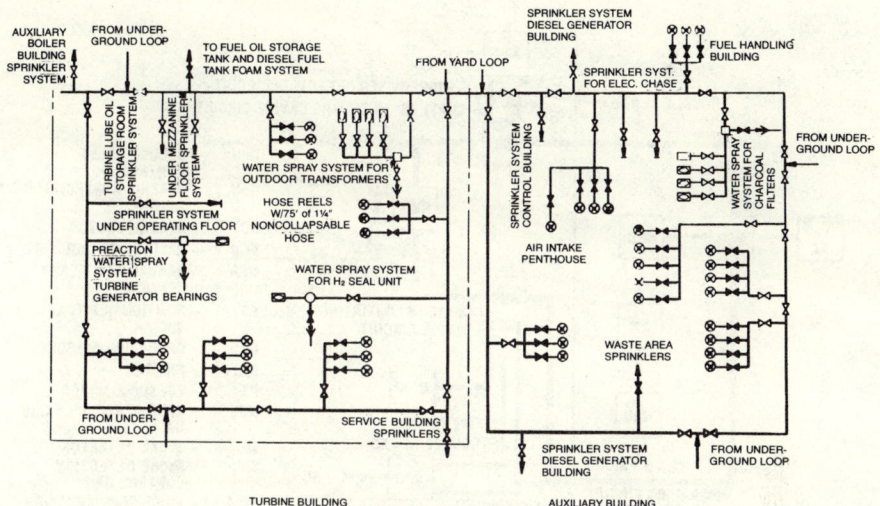

FIGURE 5.14. Interior fire protection water distribution system.

Gaseous extinguishing systems: Gaseous extinguishing agents such as carbon dioxide or bromotrifluoromethane (Halon 1301) are often used in fixed fire extinguishing systems for the cable spreading areas, possibly including areas inside the containment building where a high concentration of cables exists, and for the raised control room floor section, termination cabinets, and computer equipment. These systems are automatically activated by ionization or optical-type smoke detectors. In addition, each system is provided with a manual release capability. They provide an extinguishing concentration of gas and maintain it for an adequate "soaking" period.

Fixed extinguishing systems: Fixed water extinguishing systems include automatic sprinkler systems and water spray systems supplied from the interior distribution system as shown in Figure 5.14. Automatic wet-pipe sprinkler systems are normally installed in the turbine buildings below the operating floor, the service building, areas with combustible contents within the control complex and auxiliary buildings, diesel generator rooms, the auxiliary boiler room, and the diesel fire pump room. The piping network for all sprinkler systems, except the turbine building systems below the operating floor, is usually sized in accordance with the sprinkler piping schedule for the type of hazard being protected as given in NFPA 13, *Sprinkler Systems*. The sprinkler systems below the turbine building operating floor are hydraulically calculated for 0.20 gpm/sq ft [8.1 (L/min)/m^2] water density assuming the most remote 10,000-sq ft (929 m^2) area is flowing. Each system includes an approved shut-off valve, alarm valve, shut-off valve position switch, and a water flow pressure switch. Most sprinklers are 165°F (73.9°C) or 212°F (100°C) temperature-rated except in high temperature areas, such as the auxiliary boiler

95

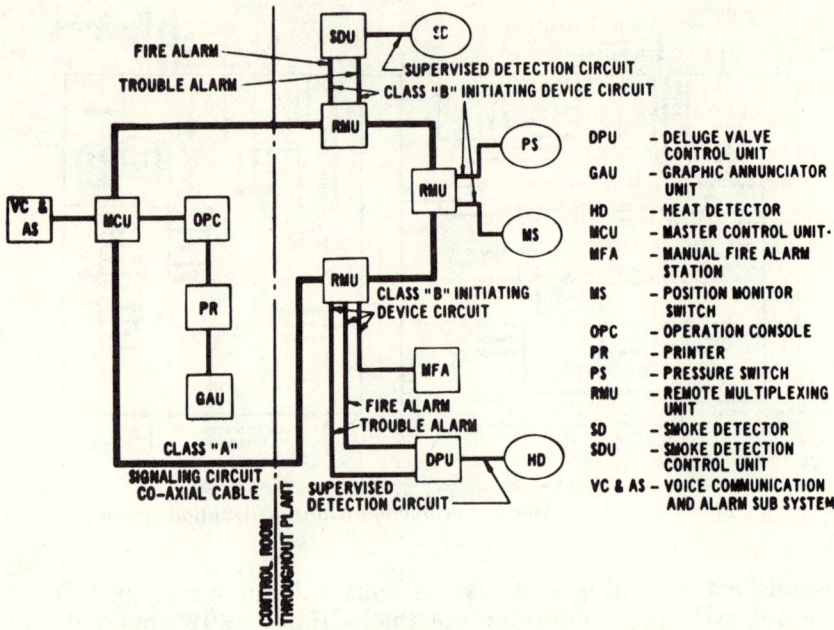

FIGURE 5.15. Fire alarm monitoring and annunciating system.

room. These systems are designed to extinguish fires and control their spread beyond the point of origin. Each provides a fire alarm signal upon the fusing of a sprinkler, and a trouble alarm signal upon the closing of a controlling shut-off valve.

Hydraulically designed water spray systems, utilizing directional solid-cone or fog-type spray nozzles controlled by an approved deluge valve, offer the best protection for all large outside oil-filled transformers, all steam turbine bearings and oil lines, hydrogen seal oil units, charcoal filters, and hydraulic oil areas not using fire-resistive hydraulic fluid. These systems are designed to wash down oil leakage, cool equipment, protect equipment exposed to fires, and extinguish fires.

Transformer and oil equipment water spray systems are automatically activated by heat detectors, as well as manually activated at strategically located pull stations. Recent charcoal filter systems have a thermistor heat detector strip attached to the charcoal surface which alarms at a bed temperature several hundred degrees lower than the spontaneous ignition temperature and below the de-absorbing temperature of the charcoal. The charcoal water spray system can be operated manually if the temperature cannot be controlled by other means. Directional nozzles protecting the bearings of steam turbine driven equipment are of a closed frangible bulb type. The piping supplying these closed nozzles is normally supervised with air pressure so that no water can reach the bearings without the operation of a heat detector and the opening of a nozzle. Each water spray system is designed to efficiently apply water at a rate between 0.25 gpm/sq ft and 0.50 gpm/sq ft [10.2 and 20.4 (L/min)/m^2]. A controlling shut-off valve with a position monitor switch, a deluge valve, and strainer are incorporated into each system. A fire alarm is initiated upon deluge valve operation, and a trouble alarm is sounded when the controlling shut-off valve is closed.

Portable extinguishers: Portable fire extinguishers of the water, carbon dioxide, and dry chemical types are provided for use on small incipient fires. These extinguishers are located and selected in accordance with NFPA 10, *Portable Fire Extinguishers*. Containment vessel extinguishers are located outside the vessel and limited to carbon dioxide and dry chemical types.

Each fire protection system at a nuclear energy plant is designed to assure that operation or failure of any portion of it will not produce an unsafe condition. All areas where fire protection water could be released by system operation or a system failure are provided with floor drains, and critical equipment is supported above the normal floor elevation to prevent damage due to flooding. Special attention is given to sizing and routing of fire water lines in order to minimize exposure of electrical equipment. Often a redundant safety system, such as a charcoal filter system, is protected by its own independent fire protection system. Failure of the fire protection system of one safety system must not affect the operation of another safety system. These fire protection systems are often designed so that two fire detection devices must detect fire in order to initiate automatic operation of the system. Manually operated fire protection systems protecting safety systems often use a dual action release device to intiate operation. In special cases, when the hazard analysis so indicates, seismic designed piping must be employed. Failure mode and effects analysis, done point by point, is normally conducted on the fire protection system including all of its major components. "No Effect" is used in the analysis to mean that the mode of failure would not affect plant safety or prevent fire protection system operation as designed.

Plant Fire Brigade

A trained and equipped plant fire brigade is required by Nuclear Regulatory Commission regulations at all nuclear power plants. It is responsible for providing an adequate manual fire fighting capability for all plant safety functions. The brigades must have at least five members on each shift, and the leader plus two of the members must be trained and knowledgeable of plant safety related systems to understand the effects of fire and fire extinguishing agents on the safe shutdown capability.

Fire brigades are equipped with complete personal protective equipment such as turnout coats, boots, emergency communications equipment, portable lights, portable ventilation equipment and self-contained breathing apparatus.

The training includes classroom instruction followed by fire fighting practice and plant fire drills. The amount and frequency of the training is clearly established by the regulations.

Life Safety Provisions

Life safety considerations at nuclear power plants are similar to those at other large industrial complexes with the added need to consider radiation. Provisions must be included in the building design to allow for safe egress of occupants and safe access for manual fire attack. At least two exits should be available to all occupied areas. Exit sizing must be adequate to handle the increased labor force required during periods of maintenance and refueling. The two exits must be as remote from each other as possible. Most areas of nuclear power plants are considered special purpose industrial occupancies for determining exiting requirements; exceptions to this would include office areas and supply storage warehouses.

Since control areas must be inhabitable during periods of toxic gas releases, fires, etc., special attention must be given to provisions for emergency breathing air supplies and ventilation in these areas. Self-contained breathing apparatus is needed for manual fire fighting. Fire fighters must also be monitored for radiation exposure, and their exposure must be limited to safe dose rates.

BIBLIOGRAPHY

REFERENCES

Decker, Dick A., "Nuclear Power Plant Fire Protection." Presented at Second International Fire Protection Engineering Institute, University of Maryland, College Park, May 1975.

Federal Register vol. 45, no. 225(Wed. Nov. 19, 1980) Rules and Regulations Nuclear Regulatory Commission 10 CFR Part 50. Branch Technical Position (BTP - APCSB 9.5-1) and Appendix A to same. Final Rule.

Generic Requirements for Light Water Nuclear Power Plant Fire Protection, Final Draft RI, ANSI N 18.10, May 15, 1977.

Hawauer, Stephen H., chairman, Recommendations Related to Browns Ferry Fire, NUREG-0050, U.S. Nuclear Regulatory Commission, Washington, DC, 1976.

National Nuclear Risks Insurance Pools and Associations, International Guidelines for the Fire Protection of Nuclear Power Plants, Swiss Pool for the Insurance of Atomic Risks, Zurich, 1974.

Nuclear Energy Liability and Property Insurance Association, Basic Fire Protection Requirements for Nuclear Power Plants, Hartford, CT, 1970.

Regulatory Guide, 1.120, Fire Protection Guidelines for Nuclear Power Plants, U.S. Nuclear Regulatory Commission, Revision, Washington, DC, April 1977.

NFPA CODES, STANDARDS, AND RECOMMENDED PRACTICES

Reference to the following NFPA Codes, Standards, and Recommended Practices will provide further information on the safeguards for nuclear energy plants discussed in this chapter. (See the latest *NFPA Codes and Standards Catalog* for availability of current editions of the following documents.)

NFPA 12, *Carbon Dioxide Extinguishing Systems*.
NFPA 12A, *Halon 1301 Fire Extinguishing Systems*.
NFPA 12B, *Halon 1211 Fire Extinguishing Systems*.
NFPA 13, *Installation of Sprinkler Systems*.
NFPA 13A, *Care and Maintenance of Sprinkler Systems*.
NFPA 15, *Water Spray Fixed Systems*.
NFPA 30, *Flammable and Combustible Liquids Code*.
NFPA 72D, *Proprietary Protective Signaling Systems*.
NFPA 101, *Life Safety Code*.
NFPA 220, *Standard Types of Building Construction*.
NFPA 251, *Fire Tests of Building Construction and Materials*.
NFPA 255, *Tests of Surface Burning Characteristics of Building Materials*.
NFPA 801, *Facilities Handling Radioactive Materials*.
NFPA 802, *Fire Protection Practice for Nuclear Research Reactors*.
NFPA 803, *Fire Protection for Light Water Nuclear Power Plants*.

ADDITIONAL READING

Ackroyd, G. C., P. J. Lake, "Fire Protection in Nuclear Power Stations," *Nuclear Engineering International*, vol. 23, no. 276 (1978), pp. 27-29.

Alvares, N. J., A. K. Hasegawa, "Fire Hazard Analysis for Fusion Energy Experiment," *Fire Safety Journal*, vol. 2, no. 3 (March 1980), pp. 191-211.

Barnes, R. D., J. D. Behn, "Fire Protection for Nuclear Safety Systems," *Professional Safety*, vol. 26, no. 2 (February 1981), pp. 26-32.

Carrou, Al, "New Fire and Security Rules Change USA Nuclear Power Plant Emergency Plans," *Nuclear Engineering International*, vol. 23, no. 271 (1978), pp. 43-46.

"Do the Fire Hazards Multiply at Nuclear Power Plants?," *Record*, vol. 54, no. 6 (1977), pp. 3-11.

Herman, L. P., "Nuclear Fire Protection: A Fourth Dimension," *Fire Journal*, vol. 74, no. 4 (July 1980), pp. 51-55.

International *Guidelines for Fire Protection of Nuclear Power Plants*, Mutual Atomic Energy Reinsurance Pool, P.O. Box 688, Norwood, MA, 02062.

Joint ANI-MAERP, *Basic Guidelines for Fire Protection of Nuclear Power Plants*, Mutual Atomic Energy Reinsurance Pool, P.O. Box 688, Norwood, MA, 02062.

Keyer, W., "Nuclear Plant Fires; Planning for the Next Threat," *Firehouse Magazine*, vol. 4, no. 7 (July 1979), pp. 33, 52.

Kowaller, S. I., "Meeting Standards for Fire-Resistive Construction at Nuclear Power Plants," *Power NY*, vol. 122, no. 2 (1978), pp. 69-71.

McKinnon, G. P., ed., *Fire Protection Handbook*, Fifteenth Edition, National Fire Protection Association, Quincy, MA, 1981, Section 9, Chapter 17.

Purington, Robert G., and Wade Patterson, *Handling Radiation Emergencies*, National Fire Protection Association, Quincy, MA, 1977.

Rittenhouse, R. C., "Fire: Detection and Prevention at Power Plants," *Power Engineering*, February 1981, pp. 42-50.

Townley, J. P., "Fighting Fire in Nuclear Power Plants, *Fire Chief*, vol. 24, no. 10 (October 1980), pp. 35-37.

6

Bulk Grain Handling

Max R. Spencer

The grain industry encompasses the movement, storage and processing of grains such as wheat, corn, oats, sorghums and oilseeds such as soybeans and sunflower seeds. It can be separated into two broad categories — grain elevator operations, and millers or processors. Grain elevators primarily accumulate, store and ship whole grains and oilseeds. Their end product is virtually identical to their raw material with only minimal drying, cleaning and fumigation to preserve and meet product quality standards. Grain millers and processors, on the other hand, take whole grains as their raw material and process these into products such as: animal feeds, cereals, flour, starch, vegetable oils, corn sugars, brewery products, etc. The discussion which follows is primarily aimed at the grain elevator segment of the industry and that portion of grain milling and processing which handles and stores whole grains on a bulk commodity basis. Additional chapters in this book related to grain milling applications include: "Vegetable and Animal Oil Processing," Chapter 7; "Lumber Kilns and Agricultural Dehydrators and Driers," Chapter 33;

Max R. Spencer is Vice President, Engineering, Continental Grain Company, New York, NY.

TABLE 6.1. Production of U.S. Grains and Oilseeds¹ in Bushels*

	Corn	Wheat	Soybeans	Others	Total (1,000 bu)
1950	3,131,009	1,026,755	287,010	2,105,195	6,549,969
1955	3,184,836	938,159	371,276	2,548,501	7,042,772
1960	4,352,668	1,363,443	558,778	2,317,196	8,592,085
1965	4,084,342	1,315,613	845,608	2,227,321	8,472,884
1970	4,151,938	1,351,588	1,127,100	2,272,590	8,903,216
1975	5,828,961	2,122,459	1,547,383	2,092,432	11,591,235
1980	6,644,841	2,374,306	1,792,062	1,767,121	12,578,330

*1 bushel = 0.3523 m³

"Food Processing," Chapter 22; "Grinding and Milling Operations," Chapter 34; and "Solvent Extraction," Chapter 32.

The growth of the grain industry in the United States has paralleled the phenomenal growth in production, consumption, and export of grains and oilseed products. This growth has been the result of many factors — the most important of which is the ideal temperate climate of the United States to produce these agricultural commodities, combined with the mechanization of farming, the development of high-yielding hybrids, the free market system, and uniquely efficient storage and transportation systems. Table 6.1 shows dramatically the rate of growth in harvested grain and oil seed crops.[1]

RAW MATERIALS

Grain and oilseeds consist primarily of starch or carbohydrates, protein, fiber, and various vegetable oils, all capable of burning under very specific conditions. In their raw, whole-kernel states, these commodities are stable and not readily subject to combustion if protected from moisture, insects, and fungi. So-called spontaneous combustion is a product of microbiological spoilage created by fungi combined with certain levels of moisture and temperature[2]. The incidence of this source of ignition of fires and explosions in grain elevators is rare, although it has been known to occur. Good grain industry practice is to store grains at moistures and temperature levels which will not only prevent spontaneous combustion, but will keep the grain from deteriorating in quality. Where grain is stored for long duration in large storage vessels, temperature can be monitored by installation of temperature detection systems.

As size reduction of the grain kernel proceeds, the susceptibility to fires and explosions increases dramatically.[3] The frequent handling of grain from farm to consumer progressively creates more broken kernels, hence more dust. This reduction in size also occurs as an integral part of the milling and processing segment of the industry.

Grain is essentially handled as a bulk commodity. The economies of scale have dictated larger and more automated handling units. Only a small fraction of the total harvest is bagged, consisting primarily of seed stocks and small amounts for export to a few emerging nations which are not equipped to discharge bulk vessels.

STORAGE

Bulk storage facilities can consist of upright concrete silos, wooden bins, steel silos, steel tanks, conventional warehouse structures referred to as flat storage, or even outdoor temporary piles during harvest periods. The function of the specific elevator, i.e., whether it is for long-term storage, short-term holding, or an adjunct to a processing or milling plant, determines the configuration and type of storage the designer may choose. These facilities are often categorized as country elevators, inland terminals, transfer stations, truck-to-rail terminals, river stations, or export terminals. Many have a multipurpose function and can be included in more than one definition.

Size of storage structure is determined by the nature of an individual facility's operation. Individual bins may range from 1,000 bushels (35.4 m^3) or smaller to 500,000 bushels (17 700 m^3) or larger.

HANDLING

The primary process of a grain handler, other than storage and quality preservation mentioned above, is that of conveying grain horizontally and vertically into and out of the storage facility. Conveyors common to most bulk handling industries are used in this process. The discussion below shows how these conveyors are used in the grain industry and highlights modified designs tailored to grain industry needs.

Belt Conveyors

The most common conveyor employed to move grain horizontally from point to point is the troughed-belt conveyor. By virtue of the wide choices of speeds and belt widths available, any desired volume can be accommodated in this manner. The fire hazards associated with these conveyors are the combustible rubber construction material of the belting and the tendency of dust to be liberated from the grain moving on open belts. The fire hazard can be minimized by the use of flame resistant belting and good hot work permit procedures. (See Chapter 24, "Welding and Cutting.") The dust hazard can be mitigated by enclosures, aspiration, belt speed control or a combination of these.

Chain Conveyors

Alternatives to the belt conveyor include the en masse, drag or chain conveyors which are totally encased in a housing that prevents the escape of dust. These are usually of more limited capacity and convey at reduced linear speeds with much deeper grain depths. Normally, only the loading point or discharge point needs to be aspirated. Higher energy use than belt conveyors and installation costs must be considered.

Screw Conveyors

The helical screw conveyor is a standard for low volume conveying for short distances. The grain industry generally uses screw conveyors only for specialized purposes because of the damage they cause to whole grain.

Pneumatic Conveyors

The milling end of the industry uses pneumatic conveying systems extensively. These are especially effective for confinement and movement of finely ground commodities in a complex processing operation requiring multipoint pickup or distribution. There is some concern about static electricity build-up and discharge in ungrounded systems.

Bucket Elevators

The common principle used in grain elevator design is that of elevating the commodity to the highest optimum point, then permitting the material to flow by gravitational force down through the various garners, weighing scales, cleaners, and finally through spouts into the storage compartment (see Figure 6.1). This not only makes efficient use of the energy required to move the mass, but eliminates re-elevating and rehandling with all the inherent physical damage to the kernel. Each subsequent rehandling contributes to reducing the quality of the grain and generation of additional quantities of fine particles and dust. The bucket elevator (or "leg" as it is referred to in the industry) is the primary equipment used to gain these elevations. Inclined belts have been used in several large export terminals, but are impractical for most applications.

The bucket elevator is not only the workhorse of the industry, but is also the single most hazardous piece of equipment from an explosion standpoint. The dust concentration within a bucket elevator is likely above the lower explosive limit during normal operations. This, combined with the pumping action of the buckets moving in a confined space and the high amount of mechanical energy inherent to its operation, have led to its identification as the principal ignition source in explosions.[4,5]

BULK GRAIN HANDLING

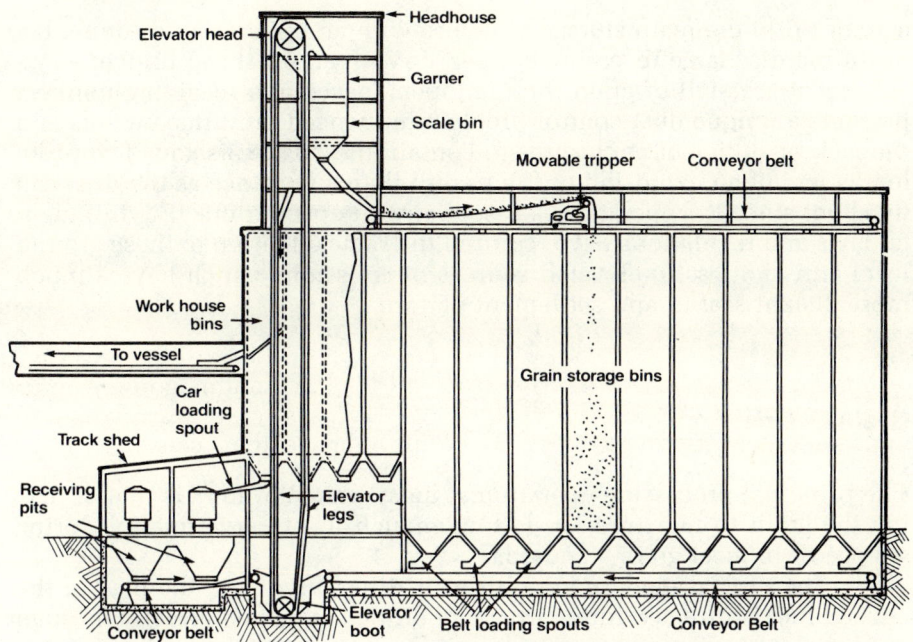

FIGURE 6.1. Diagrammatic section view of a terminal-type grain elevator.

Spouting and Lining

Grain and oilseeds are very abrasive commodities that can rapidly erode steel conveying spouts used to channel the flow through elevators. This creates an almost constant demand for maintenance to eliminate leaks and dust emissions. While patching and repair are adequate temporary measures, they seldom fully restore a spout. An alternative has been the wide use of abrasion resistant liners which can be totally replaced without disruption of the outer spout. Materials most commonly used include abrasion resistant alloy steel plates, high-density synthetic plastics, and certain vitreous ceramics. These can usually be formed or molded to the contour of the spouts and bolted into place without the need for welding or other heating devices. Although the high density plastics are not easily ignited, they can burn and must be removed or protected whenever welding or cutting is conducted on the spout.

Receiving and Shipping

The first and last adjuncts to a grain storage elevator are the machinery and structures needed to unload or load the grain from and into a carrier. Whether a railroad or truck receiving system is used, these operations are at ground level, partly or completely in the open, and usually

Chapter 6

connected to the main storage structure with an underground tunnel beneath the discharge receiving hopper, or with an overhead bridge.

The free-fall of grain through open spaces into receiving hoppers presents a unique dust control problem influenced by surface winds and the lack of sufficient enclosures to contain the dust emission. The problem is less of an explosion or fire hazard than a nuisance as the dust can fly about since the operation is essentially an open air one. Environmental laws and regulations have required the same attention to these ground level emissions as to elevated sources of emissions which have a much more distant scatter and settlement pattern.

Grain Drying

The principal processing operation at most grain elevators is that of drying the grain to moisture levels low enough to preserve quality during storage or to meet grain standards.

The typical modern grain drier is direct-fired, i.e., the heat of the burned fuel is directed into a stream of air that is passed directly through the moist grain. The fuels are principally natural gas, fuel oil, or vaporized liquid propane. Although driers are designed to minimize their fire hazards, inadequate maintenance can result in ignition of the grain being dried. The most frequent initiator of these fires is improper cleaning of drier screens, racks, or columns. The resulting "hang ups" and overdried extraneous material readily ignites when burner temperature gets too hot or small embers are entrained in the plenum air.

Ideally, a drier should be housed in a structure completely removed from the storage facility. The grain is fed to the drier from an elevated source, and returned from the drier to storage in a second separate conveyance for distribution into silos. Locating a direct-fired drier away from the storage unit itself and good operating procedures minimize the risk of more serious fires or explosions within the silo structure. In cases where a drier serves solely to heat grain without cooling, the hot grain is often cooled in the storage silos by high pressure fans forcing ambient air up through the grain mass until air-temperature equilibrium is reached.

The grain temperature seldom exceeds 150°F (65.5°C) in a drying operation avoiding heat damage to the kernels.[2] As the wet grain is heated, the evaporating water at the surface of the kernels has a cooling effect that prevents the kernels from overheating. The heated air leaving the drier is in a near saturated state and some coarse dust and fine particles escape with it. (See Chapter 33, "Lumber Kilns and Agricultural Dehydrators and Dryers.")

Grain Cleaning

The second most important processing activity in a grain handling facility is usually a screening, cleaning, or scalping system that removes extraneous material from the grain, such as bits of stalks, stems, seed pods, husks, corn cobs, weed seeds, or fine broken grain particles. These materials not only affect the quality of the grain, but are also more prone to ignition than the grain itself by virtue of their extremely dry state and high fiber content. The grain is cleaned by passing it over vibrating or gyratory motion screening devices or stationary gravity screens for simple size separation. A positive air aspiration system is used to remove dust generated by the grain movement within the device.

Grinding and Cracking

Some grain facilities serve specialty industries which require grain to be cracked or ground. This entails the use of hammermills or grinders to accomplish the size reduction. These are common sources of dust explosions, particularly in feed milling operations. The hammermill is frequently used to grind corn and other feed grains for use in rations. Care must be taken to exclude foreign objects from entering the grinding mechanisms, especially stones and metallic objects. (See Chapter 34, "Grinding and Milling," for further information on grinding processes.)

Dust Control

Supplementing the conveying, elevating, drying, screening, and storage activity is the dust control process. Each point of grain handling can produce suspended dust nearby. A complete dust control system at an elevator will consist of a combination of passive dust control methods such as the use of enclosures and reduced grain handling speeds. Active dust control provided by mechanical dust collection systems is most effective in reducing fugitive emissions.

The dust collector most commonly employed is the "baghouse" or fabric filter wherein the dust-laden air is passed through filter media and exits virtually 99.95 percent dust-free. The dust is recovered in the filter housing, stored in a remote bin, and sold as a by-product of grain for use by the animal feed industry. Such systems are complex processing operations with self-cleaning devices for the bags, automatic discharge mechanisms, continuous conveying to disposal points or storage tanks, and separate load-out systems.

The cyclone collector, which was in common use prior to clean air laws, served as a quasi-separator for dust aspirating systems, but was only 75 to 80 percent efficient overall. For fine particles, efficiency is only

50 to 60 percent and merely concentrates the collected dust at a central point for discharge into the atmosphere. The cyclone is still often used as a preskimmer in dust systems ahead of filters to remove and recover the very large particles for reentry into the grain stream. The fine dust particles only are then collected in the filters for disposal or for readmission into the grain if circumstances and practice permit.

Dust collection systems themselves can sometimes increase the fire hazard because they concentrate the dust in specific pieces of equipment. For this reason, dust filters should be located outside and should be equipped with deflagration vents to minimize possible damage should an explosion occur.

THE FIRE HAZARD

Although dust explosions are the overriding hazard in the grain industry, there are several other fire hazards unique to the grain industry. These are primarily the grain drier, the grain itself, and the use of propane.

The grain drier's most significant fire hazard is that of igniting the grain or extraneous material. The large amount of heat needed — in some driers more than 20,000,000 Btu hr (21.12 million kJ/hr) — combined with the large size — up to 80 ft high (24.4 m) — requires that control of this hazard be focused on good operating procedures and mechanical/electronic safeties such as temperature sensors, UV flame detectors, and fuel train safety devices.

In addition to the preventive measures used to reduce the possibility of a grain drier fire, each elevator must consider an emergency plan for dealing with a fire. The size and unique design of a grain drier makes fighting the fire with hose streams difficult. Probably the most effective consideration in fighting a drier fire is moving the burning grain to a safe area away from the elevator where water can be selectively applied.

The hazard of grain itself igniting must also be addressed from a preventive, as well as a fire fighting, standpoint. Preventive measures are basically to keep open flame (welding, torch cutting, cigarettes) away from the grain. Once grain ignites, it becomes quite difficult to extinguish, and, if the extinguishing method creates a dust cloud, an explosion is likely. Application of large amounts of water in a storage bin will not only deteriorate the grain, but in the case of soybeans, could cause enough swelling of the soybeans to rupture the bin itself. Elevator management should consider meeting with local fire departments to review emergency procedures, should a grain fire occur.

Another potential fire hazard in grain elevator operations is the use of propane. Whether used for grain drying or comfort heating, propane presents a serious fire hazard because of the below grade tunnels normally present in grain elevators and because the heavy rail or truck traffic can cause deterioration in underground propane pipes. If a break oc-

BULK GRAIN HANDLING

TABLE 6.2. *Summary of Grain Handling Facility Explosion*

Year	# Of Grain Handling[4] Facility Explosions	# Of Deaths[6]	# Of Injuries[6]
1969	15	4	13
1970	21	1	14
1971	17	4	14
1972	14	7	23
1973	22	2	10
1974	25	13	37
1975	9	4	19
1976	28	22	82
1977	31	65	84
1978	20	7	47
1979	29	N/A	N/A
1980	45	N/A	N/A

curs, the liquid propane vapors can seep into the below grade tunnels until discovered or ignited. Elevator management should insist that propane lines be installed according to NFPA codes and maintained regularly. (See Chapter 39, "Liquid Petroleum Gases at Industrial Plants.")

THE EXPLOSION HAZARD

Dust explosions certainly must be considered the number one hazard in the grain industry. They have been present in the industry probably since its inception and there are recorded instances as far back as 1785.[3] More recently, in the three-year period between 1976 and 1978, the number of serious explosions peaked and resulted in 94 deaths, as indicated in Table 6.2.

As a result of this increase in grain dust explosion incidents and fatalities, the National Grain and Feed Association established a Research Council to increase the industry's understanding of known causes of explosions and to conduct research into the basic factors contributing to grain dust explosions. Their research is an ongoing effort which promises to result in an improved understanding of the causes and development of preventive measures necessary to reduce the incidents of dust explosions.

Elements of a Dust Explosion

The elements of a grain dust fire or explosion are almost axiomatic; namely, that to be initiated and sustained there must be fuel, oxygen and an ignition source. To have an explosion, a fourth element, confinement,

TABLE 6.3. *Lower Explosive Limits of Grain Dusts*[3]

Dusts	g/m^3
Corn Cob	45
Corn Starch	40
Alfalfa	100
Malt Barley	55
Peanut Hull	45
Rice	50
Soybean Flour	60
Sugar	45
Wheat, Flour	50
Wheat, Starch	45
Cottonseed Meal	55

which confines the rapidly expanding heated gases of combustion within a constraining enclosure until the pressures exceed the ultimate strength of the enclosure, is necessary.

The precise circumstance under which an explosion of a grain dust will occur is a complex combination of dust particle size, concentration in the air (gram/cubic meter), the energy of the ignition source, and less easily determined factors, such as the moisture content of the dust (or percent relative humidity of the air) and the actual composition of the dust. Dust from each agricultural commodity has its own explosion characteristics. While researchers have not agreed precisely on the limits of the various characteristics of a particular dust, their conclusions are generally within an acceptably narrow range. The table below is generally accepted as being a good approximation of lower explosive limits. Others have reported lower explosive limits as low as 20 g/m^3 and as high as 100 g/m^3.

The fuel: Dust particles emanating from various emission points within a grain elevator are of varying composition and of a wide range of sizes. It is generally agreed by researchers that particle sizes below the 100μ (micron) range constitute the greatest hazard.[8] As indicated in Table 6.4, a considerable portion of dust within the elevator environment is smaller than 100μ. Larger particles not only tend to settle out rapidly (see Table 6.5), but have a lower ratio of surface area to mass available for combustion. Approximations indicate a concentration of 20 g/m^3 (near the lowest required for an explosion) would be nearly impossible to see through at a path length of one meter.[4]

While it seems improbable that such a dense cloud would exist within the ambient space of an elevator structure where personnel are present, such concentrations have been measured within the confines of bucket elevators and may also occur in conveyor housings, bins being loaded, silos, dust collecting systems, and connecting spout work.

The mechanism of an explosion depends upon the immediate heat release of a burning particle to ignite and support the burning of adja-

TABLE 6.4. Size Distribution[9]

Dust size (μ)	Dump pit (mostly beeswings)		Belt loading (mostly starch dust)		Main elevator (60:40 mixture; beeswings: starch)		Beans dust		Mesh	Inches	mm
	Retained on %	% cum	Retained on %	% cum	Retained on %	% cum	Retained on %	% cum			
+150	94.8	94.8	—	—	56.0	56.0	16.0	16.0	+100	—	—
150–100	3.7	98.5	—	—	11.3	67.3	12.1	28.1	100	0.0059	0.150
100–74	1.1	99.6	—	—	7.0	74.0	13.4	41.5	150	0.0041	0.104
74–38	0.4	100.0	—	—	6.0	80.0	9.2	50.7	200	0.0029	0.074
38–21	—	—	31	31	6.0	86.0	16.3	67.0	450	0.0017	0.043
21–16	—	—	28	59	5.0	91.0	16.0	83.0	630	—	—
16–8	—	—	22	81	4.0	95.0	11.0	94.0	937	—	—
8–6	—	—	10	91	3.0	98.0	4.0	98.0	1875	—	—
6–4	—	—	3	94	2.0	100.0	2.0	100.0	2300	—	—
4–2	—	—	2	96	—	—	—	—	4500	—	—
2–1	—	—	3	99	—	—	—	—	6250	—	—
–1	—	—	1	100	—	—	—	—	12500	—	—
Total	100.0	—	100.0	—	100.0	—	100.0	—	—	—	—

TABLE 6.5. *Settling Rates for Various Sized Particles*[9]

Size (μ)	Rate of Fall			
	in./min	in./hr	m/min	m/hr
100	320.0	—	8.128	
50	160.0	—	4.064	
10	7.0	—	0.178	
5	1.8	—	0.045	
1	—	5.0	—	0.127
0.5	—	1.4	—	0.035
0.1	—	0.05	—	0.001
smaller	—	~0	—	~0
—	Brownian motion			

Note: Specific gravity = 1.0

cent particles.[3] As this rapid spread of flame proceeds from particle to particle, pressure waves and thermal expansion of the air can create an intense shock sufficiently strong to rupture the typical reinforced concrete structure. In studies performed by the US Bureau of Mines, the maximum pressures for corn dust are greater than 100 psig (690 kPa). Concrete structures found in elevators can usually withstand no more than 25 psig (172 kPa).

Suspended dust is not the only fuel to be concerned with. Dust accumulation on floors, walls, and equipment may become suspended if disturbed by vibration, fires, or small explosions. If this accumulated dust is suspended in sufficient concentration, the resulting explosive dust concentration can become ignited and progress into an explosion. This resuspended dust can encompass large volumes and propagate minor explosions through an entire elevator.

Oxygen: The oxygen concentration required for a dust explosion is that found in the atmosphere. Using inert gases in equipment to reduce the oxygen concentration enough to prevent explosions has been suggested, but has limited application due to the large size and volume of grain handling equipment. Processing equipment such as hammermills may be more likely candidates for inerting.

Ignition sources: The second most important element of a grain dust explosion is ignition of the suspended dust cloud by a source of energy of sufficient intensity and duration. One ignition source that has been identified in a large percentage of known instances is improper use of welding and cutting equipment. (See Tables 6.6 and 6.7.)

Other identified ignition sources include heat caused by the fric-

TABLE 6.6. *Probable Location of Primary Explosion (1958–1978)*[6]

Location	No. Of Facilities	Percent Of Facilities
Unknown	107	42.8
Bucket elevator	58	23.2
Hammer mills, roller mills or other grinding equipment	17	6.8
Storage bins or tanks	13	5.2
Headhouse	9	3.6
Adjacent or attached feed mill	8	3.2
Basement	4	1.6
Processing equipment	3	1.2
Dust collector	3	1.2
Tunnel	2	0.8
Distributor heads	2	0.8
Passenger elevator or manlift shaft	2	0.8
Grain drier	2	0.8
Outside and adjacent to facility	2	0.8
Pellet collector	2	0.8
Conveyor system	2	0.8
Receiving pit	2	0.8
Other handling equipment	2	0.8
Processing plant	1	0.4
Down spout	1	0.4
Corn tester	1	0.4
Feed room	1	0.4
Sampler	1	0.4
Storage room	1	0.4
Boiler or feed mill	1	0.4
Electrical switch	1	0.4
Auger conveyor	1	0.4
Electric panel	1	0.4
Sample size	250	100.0

tional energy of mechanical equipment such as bucket elevators, bearings and belt drives. Heat or arcing caused by the failure of electrical equipment, such as lighting, motors and wiring, have also been identified as ignition sources. Miscellaneous ignition sources include open flames from matches or smoking, space heaters, lightning, and internal combustion engines on vehicles such as industrial trucks.

Welding and cutting is a particular concern in the grain industry because many elevators do not have full-time maintenance personnel. This results in contractors performing welding operations within the grain elevators. These contractors may be unfamiliar with the fire and explosion potential of grain dust and need to be monitored by elevator management.

Inasmuch as the majority of known locations of explosions in grain elevators stems from the bucket elevator, it would follow that this single

TABLE 6.7. *Probable Ignition Sources (1958–1978)*[6]

Source	No. Of Facilities	Percent Of Facilities
Unknown	103	41.2
Welding or cutting	43	17.2
Electrical failure	10	4.0
Tramp metal	10	4.0
Fire other than welding or cutting	10	4.0
Unidentified foreign objects	9	3.6
Friction from choked leg	8	3.2
Overheated bearings	7	2.8
Unidentified spark	7	2.8
Friction sparks	7	2.8
Lightning	6	2.4
Extension cords caught in legs	4	1.6
Faulty motors	4	1.6
Static electricity	3	1.2
Fire from friction of slipping belt in leg	3	1.2
Leaking flammable vapor	3	1.2
Smoldering grain or meal handled	2	0.8
Smoking material	2	0.8
Lighted firecracker	1	0.4
Volatile chemical escaped from soybean processing	1	0.4
Fire from cob pile outside facility	1	0.4
Heating system	1	0.4
Pocket of gas in bin ignited	1	0.4
Extinguishing fire	1	0.4
Leak in gas pipe ignited	1	0.4
Electric control panel exploded	1	0.4
Slipping conveyor belt	1	0.4
Sample size	250	100.0

piece of operational equipment presents the most serious ignition hazard to the grain handler.[4,5]. This conveyance produces ignition energy in a number of ways. Overloading or stalling of the belt generates intense frictional heat on the revolving drive pulley. This has been known to burn the belting to the point of failure, allowing the severed and flaming pieces to drop within its housing. Failure of belt splices could lead to the same results.

Misalignment of belts can cause the leg casing to become hot enough to ignite combustible materials such as dust or lubricants. The misaligned belt itself probably does not get hot enough to ignite while it is moving, since it has a chance to cool as it moves, but may ignite after the equipment is turned off and the belt is stationary next to the hot casing.

Another potential ignition source associated with bucket elevators is that of overheating bearings. Some older conveyor designs have tail or head pulley bearings located inside of the casings. If these bearings overheat, they can provide sufficient heat for ignition of static or suspended dust. Even bearings located outside of casings can ignite layered dust which in turn can be drawn into the leg casing by the dust collection aspiration or the "air pumping" action of the leg itself.[5]

Extraneous foreign material such as scrap metal, tools, wood, stones, or pieces of concrete are also a concern within a bucket elevator. They may or may not produce sparks with sufficient energy to ignite dust, but certainly can result in plugged spouts, damage to the belting or deformation of the elevating cups. When this occurs, the chances of belts stalling or continuous frictional rubbing is increased.

Nonconductive belting moving over the pulleys in bucket elevators can create significant static electricity on the buckets. Research seems to indicate that sufficient energy to initiate a dust explosion is not released from this static discharge, but prudent practice is to reduce this static accumulation by using electrostatically conductive belting and good grounding practices.[10]

The energy inherent in electrical systems is extremely high, but can be minimized as an ignition source by strict adherence to the provisions of NFPA 70 for Class II, Group G atmospheres. This applies with equal importance to portable electrical devices, lighting, low voltage control circuitry, extension drop-lights, and communications equipment.

Confinement: As with all dust explosions (except detonations), the pressures generated after ignition of a grain dust explosion increase until the fuel or oxygen is consumed or until the explosion is vented. If there is no confinement (i.e., unlimted venting), explosion pressures are minimal and should more properly be called a flash fire. On the other hand, as confinement increases, the explosion pressures can build up to experimentally observed levels over 100 psi (690 kPa). Grain elevator structures and equipment cannot withstand pressures anywhere near these levels, so considerable damage will occur, unless these pressures can be vented. Due to structural considerations, most existing grain elevators would be nearly impossible to retrofit with adequate venting, but newly designed elevators are adaptable to increased venting and the resulting reduction in explosion confinement.

THE SAFEGUARDS

Building Design

NFPA 61B, *Grain Elevators and Bulk Handling Facilities*, has been prepared as a standard to assist designers of grain handling facilities, devoting considerable attention to the physical structure. NFPA 61B is not intended to apply to existing facilities which may have been constructed prior to its publication (1980). The following discussion summarizes the current industry thinking concerning the design of new facilities and does not necessarily reflect the view of the author.

The modern design approach is to construct upright storage silos of reinforced concrete or steel, with a high contiguous structure known as a headhouse to elevate the grain high enough to pass through necessary scales, samplers, garners, cleaners and distributors by gravity into the storage bins. Structures should be of noncombustible materials. Enclosures, such as silos, tanks, bucket elevator housings, and ancillary structures, should be designed to relieve explosion pressure waves as much as possible if such an incident happens. This is not always physically possible because of the very nature of the configuration of the enclosure. Current research (1983) is being conducted to determine if traditional gaseous explosion venting ratios of 1 sq ft/50 cu ft ($0.1 \text{ m}^2/1.4 \text{ m}^3$) volume are appropriate to grain elevator dust explosions.

The use of doors, windows, multiple explosion relief panels, and light gauge structural coverings on steel structures is a practical approach to an acceptable design. Stairwells and elevator shafts, where they are enclosed, should be protected with approved fire doors. Fire walls of approved design can be provided to separate the grain handling function from adjoining grain processing operations, such as flour milling, preparation for oil extraction, feed milling, or grinding. Where tunnels, basements, or other underground structures can be avoided, the alternate use of ground level or aboveground structures can give opportunity for maximum openings to the atmosphere. Where sufficient land area is available, the placement of structures for weighing, cleaning, and other operational functions can be remote from silo storage and interconnected with inclined belt systems.

All surfaces, bin bottoms, and spout inclinations should be designed to be self-cleaning when the grain or grain products flow. Structural ledges, beams, or other horizontal surfaces should be designed, where practicable, with an inclination of a minimum of 60 degrees from the horizontal to prevent gradual buildup of static dust. Where possible, design should accommodate the flushing down of accumulated dust on vertical walls and overhead structures with water.

No direct interconnecting openings should be constructed between silos or storage bins. Other methods of controlling the displaced air of filling and emptying should be provided. Where possible, the placement of bucket elevators, in open air, apart from the structure should be considered. Additionally, office buildings, employee welfare rooms, maintenance shops, inspection labs and control rooms should be located remote from concrete structures when constructing new elevators.

Mechanical Design

Many devices are available to detect, warn, and control faulty operation of mechanical equipment or their components. Many of these systems were neither available nor sufficiently developed to be applicable in

Class II, Group G dust atmospheres until recent years. Among them are hot bearing sensors, speed indicators, alignment devices, level-sensing gauges, slowdown detectors, overflow alarms, and pressure gauges, all of which can serve to indicate or react to malfunctioning units. Particularly important among these are the speed indicator and alignment devices for monitoring bucket elevators. If properly designed, installed and monitored, these devices can prevent ignition from stalled and rubbing belting.

A combination of mesh screens, grating, scalpers, or magnets will prevent large extraneous material and metal from entering into the grain handling machinery. Depending on individual elevators, permanent or electro-magnets, large mesh screening, grizzlies, and specific gravity separators may be appropriate. Their use is particularly important just prior to the point where the grain enters hammermills or other types of grinders.

Dust Control

The need for adequate dust control has been established in earlier discussions. Dust control in elevators should not be limited to installation of dust collection systems, but should be a systematic approach to dust control. This systematic approach should include designs which limit the amount of dust liberated from the grain, mechanical dust collection at grain transfer points where dust occurs, containment of dust to selected points where mechanical dust collection can be effective, reduction of dust concentrations within equipment, maintenance of dust collection equipment, and manual housekeeping to complement mechanical dust control.

The design of dust collection systems must provide sufficient capture velocity at the point of emission, particularly at conveyor loading and discharge points. Design must also provide sufficient air velocity within the ducts to prevent settlement of the particles and subsequent plugging. Blast gates and fresh air inlet dampers are necessary to balance the air flow to all points served by the system so that starving of some remote emission points is avoided. Operators must understand the dust collection systems, so that they can monitor and maintain their effectiveness on a daily basis.

If a baghouse or fabric filter is used, the porosity of the filter media must be maintained to avoid diminishing the air flow. High humidity and fine dust particles often form a cake on the bags and reduce the air passage. In such instances, replacement or laundering of the bags is necessary. (See Chapter 45, "Air Moving Equipment," NFPA 91, *Blower, Exhaust Systems for Dust, Stock, Vapor Removal*; and NFPA 66, *Pneumatic Conveying of Agricultural Dusts*, for further information on the application and design of air moving systems.)

Containment of dust to selected points not only makes the mechanical dust collection system more effective, but also limits the amount of manual housekeeping effort needed to maintain the lowest feasible level of accumulated dust.

Typical methods to reduce dust liberation from grain include reducing the distance and velocity grain is allowed to free fall onto conveyors or from spouting (choke feeding, where possible, is the best approach to limit grain velocity). Reduction of conveyor belt speeds and adequate belt tension can both be effective dust control methods where horizontal belt conveyors are used. Well-designed transition points which do not have abrupt changes in direction can also limit the amount of dust liberated from the grain. Additives such as water, soybean oil, and mineral oil have been proposed as an additional method to control dust. Research is being conducted to determine whether these additives deteriorate grain quality for the end user and whether grain so treated can meet the grading standards of the USDA.

Regardless of the facility design or amount of mechanical dust collection, manual housekeeping efforts are important in grain elevator facilities. Although safe levels of accumulated dust are impossible to determine, housekeeping programs at each facility should be designed to keep the dust accumulation at the lowest feasible level. This housekeeping program should focus on not only grain dust, but any type of combustibles to reduce the possibility of their providing an ignition source for a dust explosion.

Electrical Design

Electrical wiring and equipment in grain dust environments should be installed in accordance with Articles 500 and 502 of the *National Electrical Code (NEC®)*. These articles need to be thoroughly understood by those installing or maintaining electrical equipment in a grain elevator facility. Individual paragraphs in these articles cover equipment requirements for maximum surface temperature, transformers and capacitors, surge protection, wiring methods, switches/breakers, etc.; motors; ventilating piping; utilization equipment; lighting equipment; flexible cords; receptacles and plugs; signal and communication systems; live parts; and grounding. No attempt is made here to paraphrase these articles. The reader is encouraged to review the most current issue of NFPA 70, *National Electrical Code*.

Away from the technical requirements of the NEC, the development of economical and miniaturized electronic components for use in Class II, Group G atmospheres has made the operation and control of grain elevators more sophisticated and less labor intensive. Remote control of the entire facility from a centralized control center is now a reality in new installations. This technological advance has dictated the need for redundant detection of malfunctions since people are no longer in an operating area to observe the minute by minute activity.

The more common safeguards include electrical interlocking to simultaneously shut down all activities in a sequenced fashion when one unit of a sequentially connected operation fails. Annunciator panels can display the exact cause of shutdown to the operator. Ammeters and load indicators can depict the exact weight being carried by conveying equipment. Spring-return pneumatic devices can activate shutdown conditions even in a total power failure. Fully electronic scales now safely transmit millivolt differentials detected by a load cell or strain gauge to a remote amplifier for direct conversion into information on weight.

Maintenance

A preventive maintenance program is essential not only to ensure trouble-free operation and minimize breakdowns, but in grain elevator facilities it is also a critical safeguard for eliminating ignition sources. This preventive maintenance program should include lubrication of bearings, checking of belt splices, replacement of bent or missing bucket elevator cups, and the early remedy of leaking, worn-out spouting.

Dust filters should be checked frequently for plugging, excessive pressure drop, and worn or poorly sealed rotary air locks. Duct work in the aspiration system should be free of plug-ups and bent ducts, and blast gates should be opened or closed consistent with system balancing. Grain traps should be emptied frequently and floor sweep openings kept closed except when in use.

Electrical junction boxes should be kept closed and lighting fixture protective globes kept in place. Whenever maintenance work on the electrical systems is performed, qualified personnel should ensure that the power is disconnected or that hazardous dust concentrations are not present.

Welding and any other hot work should be rigidly controlled by facility management. The use of a written "hot work permit" is generally the best system to ensure that hot work is not performed in the presence of explosive dust concentrations. This is particularly important with contractors whose employees may not be cognizant of the dust explosion hazard.

Improved maintenance diagnostic equipment can also be effective in monitoring equipment which could become an ignition source for an explosion. This diagnostic equipment includes bearing temperature monitors, bearing vibration monitors and infrared analysis of electrical and mechanical equipment.

Pesticides

Pesticides and fumigants present minimal hazard from a fire standpoint if properly used. The only two concerns are the flash point of the pesticide being used and the toxic products of combustion to which fire fighters could be exposed.

Liquid fumigants and pesticides as applied to grain normally have flash points well in excess of 100°F (37.8°C). However, some undiluted residual type pesticides have flash points near this 100°F (37.8°C) level. Whenever possible, these concentrates should be stored away from the elevator and only the dilute mixtures utilized within the elevator.

Solid phosphine based fumigants also have a potential fire hazard. Self-ignition of phosphine gas can take place at 1.79% volumetric concentration. This concentration is more than 10 times the level that manufacturers recommend for effectiveness. However, phosphine fumigant storage canisters and poorly distrubuted fumigants can achieve these ignition concentrations. Fumigant storage canisters should always be opened away from other flammables and dust hazard environments, and only qualified or certified personnel should be allowed to apply phosphine.

Fire Control Systems

Grain elevators have few applications where traditional automatic sprinkler systems would provide much protection. The nature of the explosion hazard is such that, even if sprinklers are installed, the lines would likely be ruptured by the shock waves. Additionally, explosions are normally initiated so rapidly that fusible links would have insufficient time to react. Combine this ineffectiveness with the lack of combustibles present and the catastrophic damage which could result from water leaks seeping into grain storage bins and it becomes clear that general purpose automatic sprinkler systems would do little to mitigate damage from an explosion hazard. There are, however, several specific applications where sprinklers, fixed water spray nozzles and explosion suppression systems can be effective.

Wooden elevators which have a significant fire hazard potential, as well as the explosion hazard, are the major exception where automatic sprinkler installation can be justified. Additional specific applications where sprinklers or fixed water spray nozzles can have some benefit are within the drive pulley enclosure of bucket elevators; at remote, elevated belt gallery structures where manual fire fighting would be impossible; and within some grain driers in which, because of design, it would be impossible to use hose streams effectively.

Explosion suppression offers the grain industry a possible tool to reduce the number of explosions. Explosion suppression technology (rapid detection of incipient explosion and immediate introduction of suppressing agent) has been used effectively in other industries for some time. Recent research efforts directed at applying the technology to the grain industry have resulted in the development of self-contained explosion suppression devices which are suitable for installation on bucket elevators.[11] Although a number of bucket elevators have had these suppressors installed, it is too soon to evaluate their long range effectiveness and ability to survive in the elevator environment.

One additional fire control method which needs consideration is

FIGURE 6.2. Elevator explosion scene.

that of manual fire fighting. A number of explosions have been reported to occur after hose streams were used to fight fires within an elevator.[13] Application of hose streams within an elevator where the possibility of a dust explosion exists must be a well thought out endeavor. Any application of high pressure water or use of fire extinguishers which disperses accumulated dust must be avoided to prevent an airborne dust concentration which could be ignited.

SUMMARY

Much of the above discussion is a reiteration of concepts and principles which have been known by the grain handling industry for a number of years or recently learned through its research efforts. The devices, systems, and machinery now available are much more extensive than in earlier years. There are an estimated 10,000 grain elevators in the United States ranging from crossroad country stations to newly constructed export facilities. Many of these newer facilities have incorporated the devices mentioned above. In spite of this, grain dust explosions still occur and, in the majority of instances, the exact location of the primary explosion and ignition source are not precisely known. However, this much is known: the initial explosion has the characteristic of transmitting pressure waves, vibration, and air movement throughout the facility which could result in other deposited dust being thrown into suspension. This additional dust might have been lodged on the ledges, beams, walls, floors, and inaccessible places and is exposed to the progressing flame front. The result is a series of additional explosions known as secondary explosions.[13]

In summarizing the potential hazards associated with the handling and storage of grain, one must remember that its dust is a powerful fuel. Prudent consideration of its inherent dangers requires respect, care, and adherence to building codes and regulations in the design and construction of each facility. Once the facility is built, each day requires positive management appreciation of the inherent dangers, and action to ensure that all those working in the facility understand the basics of proper operations, maintenance, safety, housekeeping, and training.

Investigations are continuing into the causes of grain dust explosions in hope of more clearly defining the practices and methods needed to reduce the hazard. It would appear that the most fruitful field for concentration is the reduction in the potential of bucket elevators as ignition sources. This will require a better understanding of the methods to reduce dust concentration within the enclosure and better methods for detecting the impending mechanical failures which could lead to ignition.

If it is assumed that some dust will aways be present within the grain elevator facility, the elimination of the explosion can only be accomplished by controlling the dust and removing its ignition sources.

BIBLIOGRAPHY

REFERENCES CITED

1. Source: Statistical Reporting Service, US Department of Agriculture.
2. Christensen, C. M. (Editor), *Storage of Cereal Grains and Their Products*, American Association of Cereal Chemists, St. Paul, MN, 1982.
3. Palmer, K. N., *Dust Explosions and Fires*, Chapman and Hall, London, 1973.
4. Source: *Prevention of Grain Elevator and Mill Explosions*, National Materials Advisory Board, National Academy of Science, Publication #NMAB 367-2, National Academy Press, Washington, DC, 1982.
5. Anderson, Robert, *Proceedings of International Symposium on Grain Dust Explosions*; Session III; Grain Elevator and Processing Society, Minneapolis, MN, 1977.
6. Source: *Prevention of Grain Elevator Explosions — An Achievable Goal*, US Department of Agriculture, 1979.
7. Jacobsen, M., Nagy, J., Cooper, A. R., and Ball, F. J., *Explosibility of Agricultural Dust*, US Bureau of Mines, R1 5753, 1961.
8. Lilienfeld, Pedro, *Special Report on Dust Explosibility*, GCA Technology Division, GCA-TR-78-17-6, EPA Contract No. 68-01-4143 Technical Service Area 1, Task Order No. 24, March 27, 1978.
9. Matkovic, Dr. Ing. Mijo, *Dust Composition, Concentration and Its Effects*, Pro-

ceedings of International Symposium on Grain Dust Explosions, Grain Elevator and Processing Society, Minneapolis, 1977.
10. Dahn, J., *Electrostatic Grounding Characteristics of Grain Facilities*, Fire & Explosion Research Council, Washington, DC, 1982.
11. Gilles, J., *Explosion Venting and Suppression of Bucket Elevators*, National Grain and Feed Association, Fire and Explosion Research Council, Washington, DC 1980.
12. Source: *Guidelines for Investigation of Grain Dust Explosion*, National Materials Advisory Board, National Academy of Science, Publication #NMAB 367-4, 1983.
13. Tamanini, F., *Dust Explosion Progagation in Simulated Grain Conveyor Galleries*, National Grain and Feed Association, Fire and Explosion Research Council, Washington, DC, 1983.

NFPA CODES, STANDARDS, AND RECOMMENDED PRACTICES

Reference to the following NFPA Codes, Standards, and Recommended Practices will provide further information on the safeguards for grain and grain milling discussed in this chapter. (See the latest *NFPA Codes and Standards Catalog* for availability of current editions of the following documents.)

NFPA 10, *Portable Fire Extinguishers.*
NFPA 13, *Installation of Sprinkler Systems.*
NFPA 14, *Standpipe and Hose Systems.*
NFPA 15, *Water Spray Fixed Systems.*
NFPA 61B, *Grain Elevators and Bulk Grain Handling Facilities.*
NFPA 61C, *Prevention of Fire and Dust Explosions in Feed Mills.*
NFPA 61D, *Prevention of Fire and Dust Explosions in the Milling of Agricultural Commodities.*
NFPA 69, *Explosion Prevention Systems.*
NFPA 70, *National Electrical Code.*
NFPA 77, *Recommended Practice on Static Electricity.* NFPA 78, *Lightning Protection Code.*
NFPA 91, *Blower, Exhaust Systems for Dust, Stock, Vapor Removal.*
NFPA 493, *Intrinsically Safe Apparatus.*

ADDITIONAL READING

Bluhm, D. D., *Grain Elevator Explosions: A University View*, Iowa State University, 1978.
Cross, Jean and Farrer, Donald, *Dust Explosions*, Plenum Publishers, 1982.
Dust Control For Grain Elevators, NGFA, Washington, DC, 1981.
Faber, J., "Grain Explosions," *The Dock and Harbours Authority*, 59(690), 1978, pp. 7-8.

Frank, T. E., "Fire and Explosion Control in Bag Filter Dust Collection Systems," *Fire Journal*, vol. 75, no. 2 (March 1981), pp. 73, 75-80, 94.

Jaffee, H. M., "Grain Elevator Protection: What's Being Done Today?" *Fire Journal*, vol. 74, no. 3 (May 1980), pp. 131-132.

McKinnon, G. P., ed., *Fire Protection Handbook*, Fifteenth Edition, National Fire Protection Association, Quincy, MA, 1981.

"Moving Fire: Fire Hazards of Belt Conveyors," *Record*, vol. 54, no. 6 (1977), pp. 18-21.

NGFA, *Fire & Explosion Research Council Research Reports* For complete up-to-date listing of Research Reports available from NGFA, contact: Technical Director, National Grain & Feed Association, PO Box 28328, Washington, DC 20005.

A Practical Guide to Elevator Design, NGFA, Washington, DC, 1979.

Proceedings of the International Symposium on Grain Dust, Division of Continuing Education, Kansas State University, Manhattan, KS, 1979.

Weatherhead, D., "Intrinsic Safety," *Measure Control*, vol. 10, no. 9 (1977), pp. 341-349.

Williams, G. M., "Quantitative Method for the Analysis of Electrostatic Hazards and Risks," Annual Industry Applications Society Meeting, Oct. 2-6, 1977, pp. 1058-1064.

7

Vegetable and Animal Oil Processing

Ernest Holt

The United States is a major industrial processor of vegetable and animal oils. The major portion of these processed oils are used in edible foods and soap products which are used daily by consumers.

The industry employs many thousands of persons to make such products as margarine, shortening, salad and cooking oils, fatty acids, and glycerine.

From a firesafety and loss standpoint, the hazards of the industry are not severe and, in general, primarily involve the equipment and materials used as adjuncts to the main processing of the edible fats and oils and soaps. Examples of the materials and equipment that pose hazards are the generation and use of hydrogen to "harden" the "softer" fats and oils, the use of high temperature heat transfer boilers (generally involving an organic heat-transfer fluid or a similar material which is quite flammable); and the use of high temperatures on the products themselves (as in the case of fatty acid distillation and fractionation). Also, high temperature and high pressure splitting of the fats and oils into fatty

---

Ernest Holt of West Yarmouth, Massachusetts, formerly Process Engineering Manager, Lever Brothers Co., New York, NY, wrote this chapter. W. M. Neuner, Processing Engineering Manager, and M. J. Oricchio, Chief Process Engineer for Lever Brothers Co., NY, revised it for the second edition.

acids and glycerine, and the spray drying of soaps or other detergents from a wet slurry into a dry powder form causing buildups of the partially dried powder in the heat zone can occasionally result in fires.

These hazards must be prevented by proper design and materials of construction of both buildings and equipment; the proper location of these items; the proper safety devices for both plant and personnel; and the periodic safety and maintenance check of the equipment itself, particularly where the corrosive effects of fatty acids at high temperatures are involved.

RAW MATERIALS

The indoor and outdoor storage of oils, fats, and fatty acids, although presenting no exceptional fire hazards, do require specific attention to certain safety regulations. Generally, oils and fats may be stored and handled in steel equipment. Fatty acids must be stored in Type 2S aluminum, polyester-fiber glass or Types 304 or 316 stainless steel containers for the same service, because of the corrosive effect the fatty acids have on steel. All tanks should be covered, vented, and have easily accessible manway openings in their covers for gaging, sampling, and general observation. Large outdoor storage tanks should have gasketed and bolted manholes in their sides about 6 in. (152 mm) from the bottom for cleaning, for repairing heating coils within the tanks (the fats and oils tend to solidify in storage), and for general inspection purposes.

Once the tanks have been in use, care must be taken to completely empty the tanks prior to any welding repair on the tanks themselves; pipelines must be thoroughly steamed out and drained prior to any welding.

A particular caution that must be observed is that the steam supplied to the heating coil in the bottom of each tank, used to melt the tank contents prior to pumping, must enter the top of the tank and pass downward within the tank to the coil (see Figure 7.1). This ensures liquidation of the fatty material within the tank, at least in one area, from the top to the bottom. It provides an open channel through which the material being heated at the bottom of the tank can expand upward within the tank, thereby avoiding any excessive buildup of pressure at the bottom of the tank. This is imperative because the periodic steaming out of pipelines connected to the tank adds steam condensate to the fat, which subsequently settles to the bottom either as free water or as an emulsion. If the tank coil has not been properly installed, it can become superheated to the point of instantaneously converting the trapped water to steam, in which case the tank can explode. Severe explosions of this type have occurred.

On one tank farm, an explosion occurred in the dead of winter, rupturing and tipping over a 6-million-pound (2.7 million kg) tank of fat, which in turn dislodged other tanks and piping, with the net effect that

VEGETABLE AND ANIMAL OIL PROCESSING

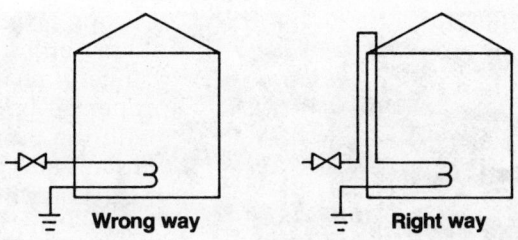

FIGURE 7.1. *Heating coil piping of fat and oil storage tank.*

many millions of pounds (1 million pounds equals approximately 454 metric tons) of fat were poured over the area, in some places to a depth of 8 ft (2.4 m). This in turn resulted in a temporary shutdown of the main line of a railroad running adjacent to the plant property, grave financial losses due to the cleanup and repair, and the loss and degradation of raw materials. Fortunately, no fire resulted, nor were there injuries or loss of life, although the potential was there.

EDIBLE OIL PROCESSING

The first step in the manufacture of shortenings, margarines, and salad and cooking oils generally consists of refining and bleaching. However, some crude oils, such as soybean oil, contain significant quantities of phosphatides which, if removed prior to refining, can yield a lecithin emulsifying agent that is of considerable value. This lecithin is sometimes removed prior to refining by a degumming process consisting of batch reaction with water, followed by centrifugal removal.

Refining

Refining removes free fatty acids from crude oils by reacting them with a slight excess of sodium hydroxide solution to form soap. The soap is then removed from the now-refined oil by either settling or centrifuging. Large modern plants usually employ centrifuging or some modification of it. Steam or miscella refining may also be used in some cases. Excellent generalized information on these and other up-to-date oil-processing techniques may be found in Reference 1.

Bleaching

Bleaching treats the refined oils (often first dried under vacuum in vacuum bleachers) with about 0.5 to 4.5 percent by weight of bleaching

FIGURE 7.2. *View of second floor of fatty acids plant showing product storage tanks (in background), weigh tanks, and instrument panel board (at left) on second floor. Auxiliary fatty acid and water tanks are shown in foreground. (Lever Brothers Co.)*

earths[2], activated earths, and/or activated carbon. The vacuum bleachers can be either batch or continuous, the former having an agitator and steam heating coils, and the latter a series of heated trays. In general, the vacuum used will vary from 27 to 28.5 in. (0.68 to 0.72 m), and the bleaching temperatures from about 194 to 230°F (90° to 110°C), with actual bleaching time in the bleacher after earth addition of about 15 min. Following bleaching, the "spent" earth has to be filtered from the bleached oil. This can be done either using the conventional plate and frame filter press or the more modern semiautomatic filters[3]. The spent filter cake, after washing and steaming (or, in some cases, solvent extraction) to remove as much residual oil as possible, must then be disposed of.

Disposing of the spent filter cake is the only operation in degumming, refining, and bleaching that may present an exceptional fire hazard. The more highly unsaturated an oil is, the more susceptible the spent earth is to spontaneous ignition, particularly when it is piled in large heaps; therefore, the spent earth should be removed from the filter presses as soon as possible and taken to a dump for disposal and burial.

It should not be stored in the plant in large quantities. This earth should be handled in steel pans, hoppers, and trucks; and all storage hoppers equipped with automatic carbon dioxide extinguishing systems. If a flammable solvent is used for the oil extraction from spent earth, all the required safeguards for storage and handling of the solvent itself, its use in the filter presses, and subsequent recovery for reuse must be observed and the spent earth treated accordingly. See Chapter 32, "Solvent Extraction," for information on the extraction process.

Hydrogenation

After bleaching, the oils must be "hardened" by the process of hydrogenation. In this process, the unsaturated stocks, such as soybean and cottonseed oil, have hydrogen molecules added to their double bonds to make them more saturated and "harder," thereby giving them higher melting points. To follow the progress of the hydrogenation, and to know when the required "hardness" has been obtained, a simple test called "checking the refractive index (R.I.)" is periodically run throughout the process. This test closely correlates with the iodine value (I.V.), which in turn tells the degree of "hardness" or saturation of the fat or oil.

In order for hydrogen to react with the double bonds, certain conditions must be met:

1. Catalyst is required. This is generally in the form of finely divided nickel and may be purchased quenched in oil or fat to protect it from deterioration by the atmosphere.
2. The material being hydrogenated must be at a temperature from about 248 to 356°F (120 to 180°C).
3. The hydrogen should have a purity of at least 99.5 percent and must be under pressures in the hydrogenation vessel of from 15 to 90 psig (103.4 to 620.5 kPa).

Finally, agitation within the vessel must be vigorous, both to keep the catalyst in suspension, and also to keep whipping the hydrogen into the oil.

Although continuous processes have been developed for hydrogenation, they are generally used only for partial hydrogenation, i.e., only down to an iodine value of 65 to 70. When they can be used properly, however, continuous units are less expensive to build and operate.[12]

Batch hydrogenation usually employs a cylindrical steel vessel which will hold about 40,000 lbs (18 140 kg) of fat or oil charge at 248 to 356°F (120 to 180°C), with the top liquid level about one-third to one-quarter from the top of the tank. Generally, the tank height is about twice its diameter. The vessel should have a central agitator with three turbine-type blades — the top one close enough to the oil surface to whip the gas into the oil. It must have adequate heating and cooling coils, a vent to atmosphere well above the roof line, a single-stage vacuum ejector, and a suitable hydrogen distribution sparge at the bottom of the tank. It should be designed for 300 lbs (2068 kPa) pressure.

During any initial use of the vessel, i.e., whenever the empty vessel contains air, the unit should first be evacuated via the single-stage ejector, then slowly filled with the correct preweighed amount of oil or fat to be hydrogenated along with the required amount of catalyst. During filling, the agitator should be in operation. The exhaust line to the ejector system should then be closed, the ejector shut off, and hydrogen permitted to enter the vessel via the hydrogen compressor and sparge at the bottom of the vessel. If hydrogen is normally stored under high pressure, the compressor would not, of course, be required.

As some hydrogen pressure builds up within the hydrogenator, the vent line to the atmosphere should be cracked open long enough to bleed off any residual air-hydrogen mixture. This vent may then be closed, heat turned on the vessel, and the hydrogen pressure brought to the required amount, where it should be maintained automatically. At about 248°F (120°C), hydrogenation will start.

With high purity hydrogen, good catalyst, and clean dry oil, "dead-end" hydrogenation may usually be used, i.e., no hydrogen vented, or circulated, during hydrogenation. This is generally preferred. But if the hydrogenation process should slow too much, or stop, cracking the vent line to atmosphere briefly may correct this condition.

When the batch is finished, it is cooled to below 212°F (100°C) and filtered through a plate and frame filter press using the hydrogen pressure within the hydrogenator to force the oil through the press. When the tank empties, some hydrogen will pass through the press into the room. The operator must be prepared for this and quickly close the line between the bottom of the hydrogenator and the filter press. By this time, hydrogen into the vessel should have been shut off.

The hydrogenator should then be ready to receive another charge of oil, but this time without the need for evacuation and venting.

Some of the filtered catalyst may be recycled in subsequent batches. Sometimes the presses are steamed before catalyst recovery, but they are *never* air blown.

There are, quite obviously, certain fire hazards throughout this entire process.

Hydrogen has an extremely wide flammable range (4 to 75 percent by volume of hydrogen in air) — the two limits between which it is susceptible to ignition. Therefore, the following precautions should be given very careful consideration: hydrogenation should take place in a separate building, although this is not mandatory in many areas. In any event, the building should be of fire-resistive or noncombustible construction and have nothing but "explosion-proof" electrical wiring, lighting, motors, outlets, etc., suitable for Class I, Division 1, hazardous locations. All hydrogen lines should be conspicuously identified, and no repairs or welding of any sort should be done while the plant is in operation. Tools should be of the nonsparking variety. Since hydrogen is the lightest element known and has only $1/16$ the weight of air, it rises very rapidly in

VEGETABLE AND ANIMAL OIL PROCESSING

the atmosphere—it does not have any tendency to settle or accumulate in "dead" areas. For this reason, the ceiling, or roof, of a hydrogenation plant should be high, well-ventilated, and gas-monitored. A complete automatic sprinkler system should be installed. When hydrogen vessels or lines have to be worked on, they must be thoroughly steamed beforehand, with particular attention paid to high spots in the lines to make sure they have been purged properly.

If the hydrogenation plant is a part of another building, a fire wall should separate it from the remainder of the plant, and all exits and entrances should have adequate fire doors.

Finally, all personnel working in or passing through that part of the plant where hydrogenation takes place should be carefully warned of the hazards which may be present. The operators should wear rubber-soled shoes.

Manufacturing Hydrogen

In the manufacture and storage of hydrogen, all of the precautions observed in the hydrogenation process must be taken, plus several more. The manufacturing facilities should definitely be in a separate building well removed from the remainder of the plant. The same applies to storage tanks, which, in addition, should be in a fenced-in area.

There are many ways of making hydrogen. Where inexpensive electric power is available, it can be manufactured by the electrolysis of water. It can also be made by the oxidation of finely divided metals with steam at high temperatures, as in the steam-iron process:

$$3Fe + 4H_2O \rightarrow 4H_2 + FeO + Fe_2O_3$$

Yet another process, the Bosch process, makes hydrogen from steam and coke at 1832°F (1000°C):

$$C + H_2O + \text{heat} \rightarrow H_2 + CO$$

The mixture of hydrogen and carbon monoxide is then blended with an excess of steam at 932°F (500°C) and passed over a catalyst of metal oxides such as iron and chromium oxide to form a mixture of hydrogen and carbon dioxide:

$$H_2 + CO + H_2O \rightarrow 2H_2 + CO_2$$

The carbon dioxide may then be separated from the hydrogen by passing the mixture through cold water under high pressure, where the carbon dioxide is dissolved in the water and the hydrogen may be collected as a free gas. This is possible only because carbon dioxide is 100 times more soluble than hydrogen under these conditions.

Another process frequently used is stream reforming with natural gas and a catalyst in a direct-fired heater to yield CO_2 and H_2. After cleaning by passage through molecular sieves, hydrogen with less than 10 ppm impurities is available.

There are several other commercial processes for making hydrogen, but the four previously described are those most common. For further details, see Reference 6.

TABLE 7.1. *Commercial Deodorization Conditions*

Absolute pressure	1–6 mm Hg
Deodorization temperature	410–525°F
	(210–274°C)
Holding time at elevated temperature:	
Batch type	3–8 hrs
Continuous and semicontinuous types	15–120 min
Stripping steam (wt % of oil):	
Batch type	5–15%
Continuous and semicontinuous types	1–5%
Product free fatty acid (wt % of oil):	
Feed (including steam refining)	0.05–6%
Deodorized oil	0.02–0.05%

To fully guard against the fire and explosion hazards of any of these processes, one must rely on the professional expertise of the plant suppliers and designers for full protection, depending on the process one selects for the making and storage of hydrogen.

Following hydrogenation, the stocks are often post-bleached with a small amount of earth and/or carbon prior to deodorization. This is basically a repetition of the earlier bleaching.

Deodorization

The next major step in the treatment of the oils and fats is generally deodorization. However, this is sometimes preceded by the blending of various stocks to meet the requirements of a given finished product. Occasionally, this may be done prior to hydrogenation, and blending may also occur after deodorization. Hence, the plant should be designed with sufficient versatility to permit these options.

There are three types of deodorization: batch, semicontinuous, and continuous.

Batch is the oldest method of deodorization and still gives very satisfactory results. It has the disadvantage of large space requirements, more duplication of vacuum equipment, the greater use of steam (since deodorization is essentially a steam "stripping" process done to remove free fatty acids and other flavor- and odor-bearing materials from the oils), and more time to completion. Table 7.1 shows some of these operating variables.

The semicontinuous method gives equally good, if not better, results, but it may be slightly more expensive. The continuous method is generally employed only where long runs of the same stock are required. The three methods, with their relative advantages and disadvantages, are all completely explained and discussed in the texts given in References 7 and 13.

Following deodorization, the oil or fat is refiltered but this time either through cartridge filters or "closed" presses dressed with a fine grade of filter paper. These filters are blown with nitrogen, and the finished oil also is stored in tanks under a blanket of nitrogen.

Heat Transfer Systems

There are no particular special fire hazards in the deodorizing process itself, but the deodorizing temperature of 410 to 525°F (210 to 274°C) almost invariably employs a high heat-transfer media. The most frequently used media is an organic eutectic mixture of diphenyl and diphenyl oxide, a nontoxic fluid, which, upon ignition (flash point of 255°F, 124°C), burns with a hot smoky flame. Boilers for heat transfer systems employing diphenyl-diphenyl oxide mixtures are direct fired, generally using either gas or oil. There is always some danger of a leak developing between the firebox and boiler which could result in fire or explosion when the heat transfer media enters the firebox. Although this is not a frequent occurrence, it has occurred and adequate precautions should be taken.

The diphenyl-diphenyl oxide mixture does not vaporize at atmospheric pressure until it reaches a temperature of 496°F (258°C). At 15 psig (103 kPa), the temperature of the vapors is 559°F (293°C). The boilers are seldom operated above this pressure because, in many areas, an operator's license is not required to operate a boiler which does not exceed 15 psig (103 kPa). Therefore, many boilers can be operated by personnel who also have other functions to perform. These boilers can also be automated.

Other features of the frequently used diphenyl-diphenyl oxide transfer media should be mentioned. The smallest leak of liquid or vapor (which has the easily detected characteristic odor of geraniums) can contaminate an oil or fat, necessitating, at the very least, redeodorization of the stock. The heat transfer media, which physically resembles a thin oil when hot, can find the minutest leak, and the circulation of hot liquid by a pump for heating purposes is avoided whenever possible because it is so difficult to keep the pump glands tight. For this reason, vapors are used for heating rather than circulated liquid. Thus, by placing the heat transfer boiler at least one floor below the equipment to be heated, the vapors can rise into the heat exchanger, and the condensed vapors can return to the boiler by gravity in a completely enclosed system.

The boiler itself should always be in a separate fire resistive or noncombustible room or building with adequate sprinkler protection and fire doors. If above grade, the room should be diked and scuppered to the ground out-of-doors. A good protection feature is to have carbon dioxide piped into the boiler firebox with the valve controlling the CO_2 discharge located outside the room. Similarly, the storage tank for the heat transfer media should be located below the boiler for complete drainage of the boiler by gravity, and the tank's control valve should also be located outside the room for easy access. The tank must have steam heating coils, a vent to atmosphere, and a pump for charging the boiler. The heating coils are required because the diphenyl-diphenyl oxide mixture solidifies at 53.5°F (11.9°C).

The heat transfer boiler itself must have a relief valve properly set and vented to atmosphere in an area where no personnel are normally present, to avoid someone being burned with hot condensate should the valve blow. A steel rooftop or similar area would make a good choice.

The piping for the heat transfer system should have at least a single-stage vacuum ejector at its highest point in order to remove most of the air from the system prior to heating. Otherwise the pressure within the boiler will not correspond with the vapor pressure curves for the diphenyl-diphenyl oxide mixture (supplied by the manufacturer), and the required temperatures cannot be met within the required pressure range.

Because of the low surface tensions and viscosities associated with some high temperature heat transfer media, special care in construction of joints and connections is required. Welded construction is recommended wherever possible. Rolling of tubes is satisfactory if expansion is cared for and if scale, rust, etc., is removed beforehand.

Screwed joints are satisfactory for pipe of ¾ in. (19.0 mm) and smaller. Perfectly cut threads are necessary to prevent leaks. Flanged joints welded to pipes and joined with soft metal gaskets are satisfactory.

Piping construction should follow the best high temperature practice. All steel valves fitted with extra deep stuffing boxes are satisfactory. Only the best grades of high temperature packing should be used. Ordinary composition "superheat" gaskets and packing materials, which are composed mainly of asbestos, may be initially tight but do not stand up very long under service above 550 or 600°F (288 or 315°C).

For liquid-level gages, a flat type using a mica window instead of glass is recommended. Ordinary round gage glasses are not satisfactory due to leakage around the glass and breakage due to high temperature.

Before start-up of a heat-transfer system, it should be boiled out with water to which about 5 lbs (0.23 kg) of 50° Baume' sodium hydroxide has been added. It should then be drained, rewashed with fresh water, flushed, and blown free of water with air. It should then be evacuated for a day or two to remove all residual moisture.

Following this, ammonia gas at about 15 psig (103.4 kg) should be added to the system. While under this pressure of ammonia gas, all parts of the system should be checked externally with a swab of glass wool wet with concentrated hydrochloric acid. Any leaks will be apparent immediately by the formation of a white cloud of ammonium chloride vapor in the area of the leak. Needless to say, all leaks found should be tightened or corrected before testing is continued.

When tight, the system should be vented of ammonia, evacuated, and charged with the heat-transfer media. The system may then be slowly heated, with repeated evacuation as necessary, until the full temperature required is attained *at the correct pressure*, as shown by the

VEGETABLE AND ANIMAL OIL PROCESSING

FIGURE 7.3. *View of gravity return, gas-fired boiler showing safety valves, four explosion discs, gage glass, and storage tank underneath steel walkway (at left). (Lever Brothers Co.)*

manufacturer's temperature versus pressure chart. When this step is attained, with no leaks apparent in the system and no sound of boiling, which might be due to residual water present, the unit should be ready for operation. If all these precautions are followed, no serious hazards from the operation of the unit should be expected.

The remainder of the oil processing, handling, and ultimate finished product storage operations present no unusual fire hazards. They involve winterization, blending, chilling, and tempering techniques all described in some detail in Reference 1. Briefly, winterization is controlled chilling of the blended oils to permit removal, generally by filtration, of the more solid fats from the more liquid oils. It is done so that the desired end products — salad and cooking oils — will remain as clear liquids at room temperature, or even below, in the home kitchen. The techniques involved are given in Reference 8.

The manufacture of shortenings and margarines from deodorized oils is quite similar, except that the latter employs more additives in the form of water, milk solids, emulsifiers, lecithin, color, and flavor additives, vitamins, salt, and preservatives. These are all dealt with in Reference 9, including the blending, chilling, and tempering required.

For final processing, the salad and cooking oils are bottled in glass, the shortenings in cans, and the margarines in cartoned and foil-wrapped quarter-pound sticks or in plastic tubs for the softer varieties. All are packed in cases and subsequently stored and handled on pallets in temperate, or chilled, storage, depending on the need.

Warehouses should of course be adequately sprinklered, with care taken that the pallets are not stacked so high as to make sprinkler coverage less than fully efficient. (Good general storage practices are covered in Chapter 48, "Industrial Storage Practices.")

SOAPMAKING

Soapmaking in the United States normally requires the use of about 65 to 80 percent of inedible tallow and 20 to 35 percent of coconut oil in the finished product as fatty material.

The three main processes generally used for manufacturing soap are:

1. *Kettle saponification* of the fats and oils, or modifications thereof, with in-tank separation of the soap and lyes.
2. *Continuous saponification* of the fats and oils, and subsequent separation of the soap and lyes, both done in centrifuges.
3. *Continuous fat splitting* and subsequent continuous distillation and saponification of the free fatty acids.

Kettle saponification: the oldest of these processes and requires much space, time (7 to 10 days to make a kettle of "neat" soap), and considerable expertise in manufacture. In this process, the blend of tallow and coconut oil is added to a kettle that has a cone-shaped bottom and an open steam sparge near the base of the cone and a swing pipe part near the top of the tank wall. The oil charge is then boiled with open steam, while sufficient caustic solution is added to saponify the entire charge. Salt is added during the process to maintain an open "grain" on the soap. This, and all subsequent processing, requires skill on the part of the operator. When thoroughly saponified, the charge is allowed to settle at least overnight, during which time it separates into three layers: a lower layer consisting of glycerine, strong lye, and salt; a middle layer of "dirty" soap with some lye and glycerine called "nigres"; and an upper layer of open-grained white soap. The lower layer is then drawn off for glycerine recovery, the middle layer is reworked separately to recover more soap and glycerine, and the top layer then goes through a series of washings and settlings to remove glycerine and salt, finally resulting in a finished soap of about 30 percent moisture (63 percent total fatty matter), and about 0.02 to 0.05 percent excess caustic. This soap is dried (usually in continuous vacuum driers) down to about 10 percent moisture and the dried soap is passed through a pelletizer. The soap pellets are then mixed batchwise with any desired additives, such as perfumes, glycerine, etc., extruded into bars, cut into cakes, stamped, wrapped, and packaged for storage and shipment.

Continuous Soap Making: There are various procedures for continuous saponification. The processes are essentially the same as the kettle method in that neutral oil is saponified with caustic soda and the glycerine is separated from the soap using salt (or brine).

In one system the blended tallow and coconut oil, along with the caustic soda, are fed to a circulating loop of saponified material. The loop includes a small reactor with a high speed agitator. The complete system is maintained under a slight pressure to increase the reaction rate. A portion of the circulating stream is continuously withdrawn, mixed with brine and washed in a centrifuge to remove the glycerine. The soap is then fed to the circulating loop of another small agitated reactor column where small amounts of water and caustic are added to properly adjust the grain of the soap for finishing. Finishing is achieved by passing the soap through centrifuges which separate the feed into finished soap of about 30 percent moisture and nigre (mixture of water with some soap and small concentrations of lye and glycerine which is recycled). The continuous saponification reduces the time required for soapmaking from days to hours, dramatically reduces the stocks in process, and diminishes the level of expertise required by the operators as compared to the kettle saponification. The continuous system does, however, add considerable burden on mechanical maintenance and requires expertise in the operation of the centrifuges.

Continuous fat splitting: fatty acid distillation with some fractionation and continuous saponification (but without centrifuge) replaces the kettles. This method has at least two advantages over the previous methods in that lower grade fatty raw materials may be used because of the fatty acid distillation and fractionation steps, and glycerine recovery is substantially simplified since vast amounts of salt do not have to be removed and handled during processing, as little salt is needed during the saponification step. Continuous fat splitting along with glycerine recovery is described more fully in References 10 and 11.

The storage and handling of fats and oils, the protection of processing equipment, and the safety precautions in using heat-transfer media in soap manufacturing are the same as previously outlined for edible oil processing. Two additional fire hazards should, however, be emphasized. These hazards are common to the soap drying process and the fat splitting process.

Soap Drying

In any spray drying of soap or detergent, as in making soap "beads" for clothes washing purposes, there is always some chance of fire occurring in the spray tower itself or in the air exhaust ducts from the tower. The same can apply to the spray drying of synthetic detergents. Whenever the soap or detergent being sprayed impinges on and adheres to the walls of the spray tower, there is always the chance that buildup of material will become extremely dry and certain materials which are pyrophoric will auto-oxidize and ultimately smolder and ignite. To a lesser degree, this could also happen in the exhaust air ducts. Careful design and operation of the spray tower is required to avoid hazards of ignition.

Generally, the spray tower walls should be vertical, but if any tapering of the tower is to be done, the unit should be slightly larger in diameter at the bottom than it is at the top, never the reverse, since this would only add to the buildup of soap or detergent on the tower walls.

The tower should be internally well-lighted by 12-in. (0.30 m) diameter light and sight glasses at eye level on each floor through which the tower projects. Preferably there should be at least three combination light and sight glasses and three easy-open, man-sized cleanout doors at each floor level. Scrapers with handles 10 or 12 ft (3.0 or 3.6 m) long should be available at each floor level for general tower cleaning and/or knocking down any charred powder observed before it has an opportunity to ignite. Good fire extinguishing capacity can be provided by 2½-in. (63.5 mm) standpipe hose lines available at each floor level. With good visibility through ports, easy access through large doors and with scrapers and standpipe hose lines at the ready, all the basic elements are available for coping with possible fires within the towers. Similarly, the exhaust ducts should have accessible ports and standpipe hose lines within reach.

Deluge systems: an additional fire protection means is a deluge system, which, when activated, will spray high volumes of water through the tower to both quench the system and extinguish fires. Deluge systems which operate in conjunction with furnace safety controls are typically actuated by the tower operator in response to alarm conditions such as high temperature or high pressure.

Fat Splitting

The hazard with the fat splitting process of soapmaking is related to the high pressure-high temperature fat splitter and the fatty acids still and fractionating equipment. Today, this equipment is generally placed outdoors, which has some obvious fire protection and fire fighting advantages.

Fatty acids are much more susceptible to fire than their parent fats and oils, basically because of their molecular structure. The danger of fire

VEGETABLE AND ANIMAL OIL PROCESSING

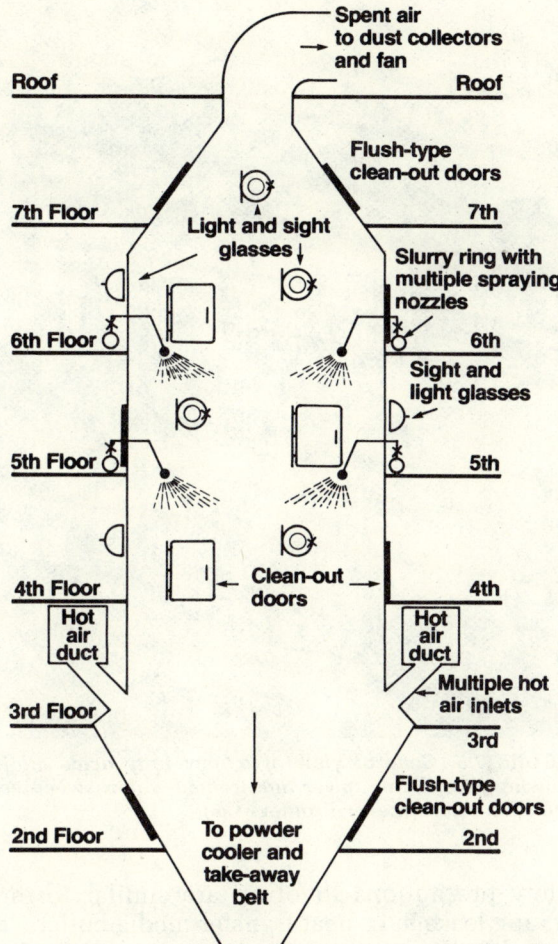

FIGURE 7.4. Soap/detergent spray drying tower.

is contingent upon leakage of the high-temperature equipment which can unobtrusively permit fatty acids to soak into the heavy insulation surrounding the equipment and there build up even higher temperatures, due to auto-oxidation, up to the point of self-ignition. Therefore, it is imperative to:

1. Check new equipment prior to insulating it for any possible leaks.
2. Train all operators to report any "wet" insulation at once, so that the plant may be shut down and the leak found and corrected at the first opportunity.
3. At least every two years check all high-temperature fatty acid equipment for leakage and repair and recheck any leakage found.

Chapter 7

FIGURE 7.5. General view of a new fatty acids addition showing fat-splitting tower and treated-water, sweet-water, and skim tanks. (Lever Brothers Co.)

The safety precautions involved are similar to those previously given for checking leakage in heat-transfer media boilers, except that the areas being tested for leaks should be warm enough to assure that there could be no solidified fatty acids plugging an otherwise potential leak. This can be done by warming the metal to be tested with a steam hose just prior to the actual test.

Ignoring these safety checks and procedures can be disastrous. In several instances fatty acid plants have been completely destroyed by fire caused by leaky fractionating equipment.

It is important to remember the corrosive hazards of fatty acids, and utilize the outlined precautions. Fatty acids are highly corrosive to most metals. In general, all high-temperature fatty acid equipment should be fabricated throughout of Type 316L stainless steel. Types 304, 306, and 307 will all be corroded by high-temperature fatty acids and should not be used. Some of these lower-grade stainless steels have been inadvertently and mistakenly used in place of Type 316L.

Fat splitters operate at 500°F (260°C) and 660 psig (4551 kPa). They are heated by 800 psig (5516 kPa) steam.

Although fatty acids stills and fractionating columns operate under high vacuums, down to 4 mm Hg absolute pressure, they are heated by heat-transfer media vapors which may be as hot as 560°F (293°C), and the fatty acids within the units may reach temperatures as high as 460°F (237°C), as in the case of pitch.

THE SAFEGUARDS

Fire-resistive or noncombustible construction should be used for vegetable and animal oil processing plants, but must be used for the structure or area housing the heat-transfer boiler or vaporizer. Sprinkler protection throughout the plant is another excellent safeguard, but, again, is absolutely essential in structures housing the heat-transfer boiler or vaporizer, and in the areas where processing equipment supplied by the heat-transfer system is located. The latter is particularly important if the operating temperature of the heat-transfer media is above its flash point and liquid phase heating is used (not the common practice in the industry).

Storage areas for the finished product should be separated from the remainder of the plant by fire walls of approved design with openings protected by fire doors. Particular caution should be observed so that materials in storage, usually in a palletized arrangement, do not exceed the effectiveness of the sprinkler protection (see Chapter 48).

BIBLIOGRAPHY

REFERENCES CITED

1. *Journal of the American Oil Chemists Society*, World Conference on Oilseed and Vegetable Oil Processing Technology, Amsterdam, March 1-5, 1976, "Processing Vegetable Fats and Oils," June 1976 issue, Session III.
2. Siddiqui, M. K. Hasnuddin, *Bleaching Earths*, Pergamon Press, NY, 1968, pp. 1-18.
3. Walker, Lewis, McAdams, and Gilliland, "Filtration," *Principles of Chemical Engineering*, Third Edition, McGraw-Hill, NY, 1937, pp. 323-365.
4. Same as (1). Papers by Bauman, Patterson, and Goebel.
5. Same as (1). Papers by Neumunz and Coenen.
6. Quagliano, *Chemistry*, Second edition, Prentice-Hall, Englewood Cliffs, NJ, 1963, pp. 129-151.
7. Same as (1). Paper by Zehnder, Calvin T., Chemetron Corp., Louisville, KY.
8. Same as (1). Papers by Neumunz and Kreulen.
9. Same as (1). Papers by Haighton and Gander.

10. Ladyn, Henry W., "Fat Splitting and Continuous Soapmaking," *Chemical Engineering*, Aug. 17, 1964.
11. Meinhold, Ted F. and Peisker, Edward A., "Soap Made Directly From Fatty Acids at a Rate of 1,900 Bars Per Minute," *Chemical Processing*, Jan. 1965, pp. 56-59.
12. Dravo Continuous Hydrogenation Process, Dravo Chemical Plants Division, 1 Oliver Plaza, Pittsburgh, PA., F.O. 5289.
13. *Journal of the American Oil Chemists Society*, Nov. 1977, pp. 528-532.

NFPA CODES, STANDARDS, AND RECOMMENDED PRACTICES

Reference to the following NFPA Codes, Standards, and Recommended Practices will provide further information on the safeguards for vegetable and animal oil processing discussed in this chapter. (See the latest *NFPA Codes and Standards Catalog* for availability of current editions of the following documents.)

NFPA 12, *Carbon Dioxide Extinguishing Systems*.
NFPA 13, *Installation of Sprinkler Systems*.
NFPA 14, *Standpipe and Hose Systems*.
NFPA 36, *Solvent Extraction Plants*.
NFPA 70, *National Electrical Code*.
NFPA 231, *Indoor General Storage*.

ADDITIONAL READING

Chemical Technology, An Encyclopedic Treatment, "Edible Oils and Fats and Animal Food Products," vol. 8, Barnes & Noble Books - Imports, Totowa, NJ, 1975.
Factory Mutual Engineering Corporation, "Heat Transfer by Organic Fluids," *Loss Prevention Data 7-99*, August 1975, Factory Mutual System, Norwood, MA.
McKinnon, G. P., ed., *Fire Protection Handbook*, Fifteenth Edition, National Fire Protection Association, Quincy, MA, 1981.

8

Paints and Coatings Manufacturing

Melvin V. Harris

Paint and coating operations are a very important and highly complex part of American industry. They utilize the largest number and widest variety of raw materials of any segment of the chemical industry. A single plant can utilize as many as several hundred ingredients in the manufacturing of thousands of finished products. These finished products will range from items for mass marketing of a whole line of products to a custom-made product to fulfill a specific function for one customer.

Paints and coatings not only add color to our lives, they do us a great economic service. They protect and maintain from weather, corrosion, chemicals, and other environmental factors the many surfaces on which they are applied such as our homes, public buildings, factories, equipment, machinery, and vehicles.

Paints and coatings generally fall into three general end use categories: (1) architectural coatings, (2) product coatings, and (3) special purpose coatings.

Architectural coatings: Paints and coatings that are stock types formu-

---

Melvin V. Harris is Vice President of Engineering for Fire Safety International, Arlington, VA.

lated for normal environmental conditions and generally applied on new and existing residential, commercial, institutional, and industrial structures are known as architectural coatings. Products in this category include such items as interior and exterior latex and oil base house paints, stains, and clear coatings.

Product coatings: When coatings are specifically formulated for original equipment manufacturers to meet specific conditions of application and performance and are applied to such a product as part of the manufacturing process, they are called product coatings. This category includes coatings for furniture, metal containers, appliances, automobiles, and hundreds of other manufactured items. Product coatings are often referred to within the industry as "industrial coatings."

Special purpose coatings: These coatings may also be stock types but are different from general architectural coatings because they are formulated for special application and/or special environmental conditions, such as extreme temperatures or corrosive atmospheres. They are often used for heavy duty maintenance work. Marine coatings are an example of special purpose coatings.

The paint and coating manufacturing industry is comprised of approximately 1,500 manufacturing facilities. These tend to be located near population centers due to the expense of transporting paints long distances. In fact, five states, California, New Jersey, New York, Illinois, and Ohio contain more than 40 percent of these sites, with each state having over 100 plants. Since paint manufacturing is not a labor intensive industry, the manufacturing facilities tend to be small operations with 60 percent of the facilities employing fewer than 20 production workers. The total number of employees involved in the production of paint and coatings is approximately 43,000. The paint and coating industry produces, annually, more than 950 million gallons of product with an estimated value of over 4 billion dollars.

There are two types of coatings manufacturing: solvent-based and waterborne. Within the industry, 25 percent of the plants produce only solvent-based products, and in another 20 percent of the plants, solvent-based products comprise at least 90 percent of the production. Additionally, 51 percent of the plants produce a mixture of both types; 4 percent produce only waterborne products. Waterborne manufacturing involves little or no flammable liquid usage, and therefore the fire hazard is low in plants devoted entirely to this type of production.

While today's trend is toward waterborne coatings facilities, the production of solvent-based products is considered hazardous due to the large quantities of flammable and combustible liquids involved and because of the use of heated reactors for making resins and other specialized products. Available loss information, though limited, indicates that the industry does not experience as many fires as would be expected with the materials involved. Unfortunately, when a major fire occurs, it can have far reaching effects and could have a severe impact on the economy of a city or area that was adjacent to that plant.

PAINTS AND COATINGS MANUFACTURING

The main hazards in coatings plants are those associated with flammable liquid handling and storage. (See Chapter 37, "Handling and Storage of Flammable and Combustible Liquids.") When such liquids are mixed, transferred, and dispensed, flammable vapors may be released providing a ready source of fuel. Since some of the liquids are also prolific generators of static electricity, the potential of an ignition source is ever present and flash fires are extremely easy to initiate.

The uncontrolled release of flammable and combustible liquids that can spill from floor to floor in multistory buildings, flow under fire doors to other sections of the building, and flow from one building to another is another potentially severe hazard common to any flammable liquid operation.

Reactors are used in the manufacturing of resins, latex, and other special products. These reactors operate under varying pressures and temperatures. Reactions can become exothermic, and the reactors must be designed to control reactor conditions. Due to the complexity of resin manufacturing and the limited number of coatings plants that manufacture resins, the processing, storage, and fire protection of resin facilities are not covered in this chapter. See Chapter 31, "Chemical Processing Equipment," for further information on the hazards associated with reactor vessels.

RAW MATERIALS

Paints and coatings consist basically of four broad types of raw materials: resins, pigments, solvents, and additives. The quantity and variety of these raw materials varies widely with the size of the plant and the type of products produced.

Resins

The film-forming binders used in paint manufacture are resins and vary widely from naturally occurring resins, such as shellac, to very complex synthetic polymers, such as polyvinyl chloride. Four families of the most commonly used resins are alkyds, acrylics, vinyls, and epoxies. Resins may be either solvent-borne or waterborne types. Solvent-borne resins are those which are dissolved in organic solvents such as hydrocarbons, alcohols, or oxygenated solvents. Because of the flammability characteristics of these solvents, resins dissolved in them can present a storage and handling fire hazard. Waterborne resins are those resins in which most of the solvent is water. Because these resins generally contain a small amount of organic coupling solvent, they often exhibit a flash point when tested in closed-cup flash point instruments; however, studies[1] have shown that waterborne coatings do not present a potential fire hazard despite the flash point characteristics exhibited in the closed-cup test.

Pigments

The materials used in coatings to impart color, opacity, body, corrosion resistance, flatting, and certain functional properties are known as pigments. The four basic types of pigments are: (1) prime white pigments such as titanium dioxide and zinc oxide, (2) colored organic and inorganic pigments such as pthalocyanine blues and greens and lead chromate yellows and oranges, (3) filler and extender pigments such as talc, calcium carbonate, chinaclay and silica, and (4) metallic powders such as zinc dust, aluminum and bronze powders. The majority of pigments are provided in dry powder form while some pigments, which are used in large quantities, are provided in a water slurry form. The hazard associated with this type of material ranges from the nuisance dust problems of low density pigments, such as talc and titanium dioxide, to the possible explosion and fire hazard of the dry metallic powders such as aluminum and zinc dust.

Solvents

The prime function of a solvent in a paint or coating is to dissolve or reduce the resin or film-former, thereby providing the coating with a consistency suitable for the desired type of application. The paint industry is a large consumer of organic solvents, but the consumption of the solvents is steadily declining and will continue to decline as new environmental regulations force the use of more waterborne and high solid-type coatings. The major organic solvents used are: aliphatic hydrocarbons, such as mineral spirits and VM&P naphtha; aromatic hydrocarbons, such as toluene and xylene; ketones, such as MIBK, MEK, and acetone; esters such as ethyl acetate and butyl acetate; alcohols, such as butanol and isopropanol; and glycols, such as ethylene glycol and propylene glycol.

Additives

In addition to the resins, pigments, and solvents, the industry consumes a wide variety of chemical additives. These include anionic and nonionic surfactants and dispersing agents, driers, cellulosic thickeners, preservatives, mildewcides, antiskinning agents and antifoaming agents.

Nitrocellulose

Nitrocellulose resins are used extensively in the manufacture of coatings for furniture and automobile refinishing. These resins exhibit several unique properties that may make them desirable, but they also present some handling and storage hazards. Nitrocellulose (low nitrated cotton fibers) is shipped and utilized in the coatings plant in a variety of forms.

The primary method of shipping is in 55-gal (208 L) drums with the nitrocellulose wetted with alcohol, normally at the 30 percent level. Alcohol wet nitrocellulose looks similar to a fluffy white cotton substance although its appearance varies depending on whether it is shipped in the dense or fibrous state. The level of alcohol is carefully controlled to ensure safe transfer of the material.

Another form of nitrocellulose is made by replacing the alcohol with water. Nitrocellulose wet with water is classified as a flammable solid and is known as water-wet nitrocellulose.

A third form of nitrocellulose utilized in this industry is nitrocellulose wetted with a plasticizer, normally at the 18 percent level. This material is shipped in a chip form in fiber drums with plastic liners. In recent years, the use of this product has become more widespread.

Oils

The paint and coatings industry does use significant amounts of such naturally occurring oils such as linseed oil, soybean oil, tall oil, and castor oil in manufacture of paints and coatings, but their use in the last few years has been declining and will continue to decline in the future.

Raw Material Storage

The large variety of raw materials used in the coatings industry is so great that it would be difficult to describe how each is stored and moved to the production area of the plant. Raw materials arrive at a typical plant by railroad tank car, tank truck, semitruck, or by smaller trucks. Although the liquids are usually transferred immediately to the company's tank farm, occasionally a railroad tank car or tank truck will be utilized temporarily as a supplemental storage tank.

Most bulk flammable and combustible liquids are stored in tank farms outside of the plant proper or in underground tanks; smaller amounts are stored in 55-gal (208 L) drums or 550-gal (2.08 m^3) portable tanks either inside or outside the building. The liquids are usually pumped into the production area but occasionally gravity flow is utilized.

Highly viscous materials usually require heated storage to allow the liquids to flow sufficiently to be pumped and mixed. Usually such tanks are located in heated tank rooms so that heating of the individual tanks is not necessary.

Flammable or combustible liquid tanks are sometimes located inside buildings where code restrictions will not allow tank farms or where plant expansion requires the tanks to be enclosed due to the lack of adequate land for the plant. Such installations present special fire hazards. See Chapter 37, "Handling and Storage of Flammable and Combustible

Liquids," for hazards associated with the storing of flammable and combustible liquids.

Raw materials in flake or powder form are usually shipped and stored in paper or plastic bags. At times, certain materials will be shipped in fiber drums. Quite often this type of material is stored in unheated portions of the plant or in small simple buildings outside of the main plant. Occasionally bulk storage of dry materials may utilize silos with air being used to move the material from storage to processing.

Storage of Nitrocellulose

Nitrocellulose drums require special consideration. They are usually stored under cover in the yard away from the plant or in a building that is specifically designated for nitrocellulose storage only. Decomposition is retarded by storage in a cool place away from any source of heat.

Nitrocellulose drums are handled by industrial lift trucks with special clamps that fasten to the rim of the drum or that clamp around the drum itself. Inside the building the drums are usually handled by manual methods. Extreme care must be exercised when handling these drums. Nitrocellulose in drums is easily ignited by friction, and drums should never be skidded on the floor. The friction of a wheel of a hand cart rubbing on a drum has been known to ignite the contents of a drum. The number of nitrocellulose drums in the manufacturing area should be limited to that required for a single work shift and they must be kept closed until used.

Nitrocellulose is never to be stored more than two drums high; one high being the recommended practice.

Raw Material Fire Hazards

The storage of flammable and combustible liquids presents the potential for large fire losses. The provisions of normal safety features, however, can make such storage relatively safe. NFPA 30, *Flammable and Combustible Liquids Code*, covers the requirements for both the tank and container storage found in industry in general, including the coatings industry. NFPA 35, *Organic Coatings Manufacture*, deals with the coatings industry specifically and provides information about the protection from the hazards involved in both the production and storage operations typically found in the coatings plant. Both of these standards should be carefully reviewed when dealing with coatings industry fire hazards.

Two common problems with the tank storage of flammable and combustible liquids is the lack of and/or inadequacy of emergency and breather venting. Improper venting of either type can have a catastrophic effect.

Breather venting is essential to all tanks, whether buried underground or located aboveground. The function of breather venting is to allow the filling or emptying of the tank without causing a rupture from filling the tank faster than the air can be exhausted, or a collapse by emptying a tank faster than the air can displace the liquid. The breather vent for tanks with Class I flammable liquids should always be equipped with flame arrestors to prevent a fire from igniting the vapors and being pulled into the tank and causing an explosion. Breather vents must always be extended to the outside so that the vapors will not be discharged inside a building.

A conservation vent, a type of breather vent, can be installed when it is desirable to prevent the tank contents from being constantly exposed to the air. These vents are equipped with poppet-type valves that open only when the contents are either being replenished or withdrawn.

Emergency venting is a very vital part of tank fire protection. If a fire exposes a tank containing a flammable or combustible liquid, the heat will cause the vapor pressure to increase. If pressure relief is not provided, the tank will catastrophically fail in a condition known as a Boiling Liquid Expanding Vapor Explosion (BLEVE). This BLEVE is a very spectacular and destructive phenomenon even in containers as small as a 55-gal (208 L) drum. (Detailed requirements for emergency venting for storage containers is found in NFPA 30.)

Self-closing valves are a positive method to prevent the uncontrolled flow of the liquids in bulk dispensing operations. This type of valve should be used wherever the control of the liquid flow is by manual means. Pump switches with momentary contacts will also achieve this control. Self-closing valves can also be installed in conjunction with the pump switch so that the single operation of the valve will provide total control.

Gravity flow of bulk solvents from storage tanks is generally discouraged; however, it can be done safely by the use of fail-safe, remote-actuated valves so that the uncontrolled flow of liquid can be prevented. Fusible link valves are also used to prevent liquid flow when a tank or dispensing point is exposed to a fire.

Aboveground tanks located outside of buildings need to be protected to prevent the accidental discharge of their contents from presenting a hazard to the plant, adjacent property, or to waterways. This protection usually consists of dikes but can also be achieved by remote impounding.

Tanks must have sufficient clearance between themselves and buildings on the same property, to the property line, and the nearest public way. The required distances depend on the liquid involved, its stability, the tank size, the emergency venting provided, and the protection provided to both the tank and the exposures.

When unloading bulk quantities of flammable liquids from railroad tank cars and tank trucks, the same precautions must be taken to prevent the uncontrolled flow of flammable liquids as with tanks. All bulk unloading must be under the direct supervision and control of plant employees. The unloading process requires constant attendance.

The tank truck motor must be stopped and all pumping be done with plant pumps and not the pump on the tank truck. This eliminates the hazard of the running truck motor being an ignition source. Some spectacular coatings plant fires have occurred when this precaution was ignored.

Phthalic anhydride and maleic anhydrides are sometimes stored in a heated molten state and are pumped directly into reactors for processing. In a molten state, these anhydrides are a flammable liquid and appropriate precautions must be taken.

Phthalic and maleic anhydrides are also used in a dry flake form. By themselves they are not a high hazard material but, if exposed to fire, they melt at a relatively low temperature — 268°F (131°C) and 127°F (52.8°C) respectively — and then become a flammable liquid. This burning liquid can spread a fire just like a burning solvent.

Organic peroxides are sometimes used in coatings plants although usually in limited quantities. Since these materials are considered unstable chemicals, special care in handling and storage is required. Cool, isolated storage is required, with the quantities allowed inside the plant limited to only what is needed for a single shift.

Alcohol wet nitrocellulose exhibits good stability at room temperatures but, with increased temperatures, the rate of decomposition increases rapidly. Once the nitrocellulose has been incorporated in the final lacquer finish, the danger of decomposition is removed. Nitrocellulose molecules themselves contain sufficient oxygen to burn without outside air. Detailed information on proper safety practices that must be used in handling and storing this material should be obtained from the nitrocellulose manufacturer.

Fire Protection for Raw Materials Storage

The basic fire protection system for combustible raw materials storage in the coatings industry is automatic sprinklers, and all areas of the plant should be protected. A hydraulically designed sprinkler system ensures that a specific water density will be available at the most hydraulically remote portion of the system. This will ensure an adequate water supply to control the fire.

NFPA 30 gives guidance design densities, operating areas, sprinkler spacing, sprinkler temperature rating, hose stream demands and water supply duration for hydraulically designed systems for various types of flammable and combustible liquid storage configurations. NFPA 231, *Indoor General Storage*, and NFPA 231C, *Rack Storage of Materials*, give information on safeguards, including sprinkler design densities, for storage of materials other than flammable liquids found in coatings plants.

Particular attention should be given to the protection arrangements for nitrocellulose in storage. Nitrocellulose itself contains sufficient oxygen to sustain combustion, and, once ignited, heat release is extremely rapid. Large quantities of water are needed to put out the fire, and deluge sprinkler systems are the preferred type of protection. Extinguishing agents that act to smother the fire, such as carbon dioxide, or to break the combustion chain reaction, such as a halon, and both ordinary and multi-purpose dry chemical are not effective on burning nitrocellulose; however, all are effective on nitrocellulose solution fires.

Although basic fire protection to tank farms is through separation, automatic fire protection systems can provide a higher degree of basic protection or more specific protection from adjacent exposures. This protection can consist of water spray systems and foam systems. Where tanks are located inside of buildings, carbon dioxide total flooding systems are also used for tank protection.

Carbon dioxide and nitrogen are sometimes used inside of tanks when conservation type breather vents are used. This not only reduces the fire hazard to the contents but also prevents contamination of the contents from the air that would normally be in the tanks.

THE PRODUCTION PROCESS

Paint manufacturing is a batch process with batch sizes ranging from 25 to 6,000 gal (0.09 to 22.7 m^3), and it involves simple mixing and dispersing techniques. All paint manufacture involves six basic steps:

1. Pigment dispersion
2. Mixing of raw materials and intermediates
3. Thinning and tinting
4. Quality control testing and adjusting
5. Filtering
6. Filling into shipping containers

Mixing and Dispersion

In many cases, pigment dispersion and mixing of raw materials can be done in one step using high speed dispersion equipment. This consists of a circular tank into which is inserted a vertical rotating shaft powered by an electrical motor. The end of the shaft is equipped with various dispersion heads or blades which both disperse the pigment agglomerates into very fine particles and thoroughly mix all of the materials in the tank.

As indicated earlier, paints can be classified as solvent-borne or waterborne products. Some differences in production of these products does exist. With solvent-borne paints, pigments are usually dispersed in a combination of the appropriate resin and organic solvent, whereas, with waterborne paints (such as latex house paints), a water and surfactant mixture is used.

Where "hard-to-disperse pigments" are called for, or an extremely fine dispersion is necessary because of product requirements, other types

Chapter 8

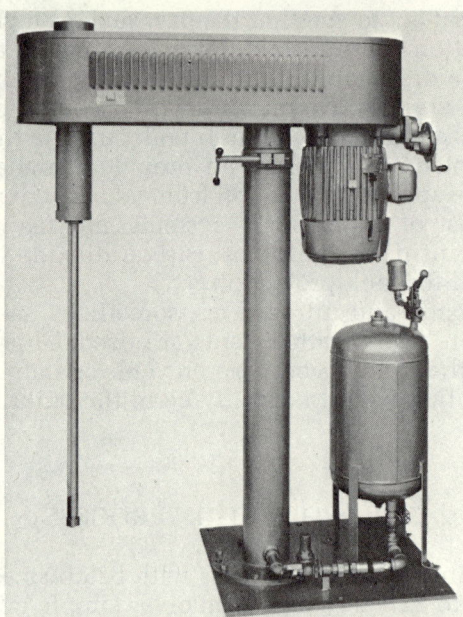

FIGURE 8.1. Typical high-speed disperser. (Hockmeyer Equipment Corp.)

of dispersion equipment, such as sand mills, pebble mills, or steel ball mills, may be used to disperse the pigment in the resin. Of these types of dispersion equipment, the most commonly used is the sand mill. The sand mill has the advantage of being a continuous process disperser, whereas steel ball mills and pebble mills are batch type. As its name implies, this mill uses sand as the grinding media, but the type of media used in this piece of equipment is not limited to sand. Glass beads, steel shot, ceramic, and zirconium oxide are also used. The mill consists of a shaft which is fitted with equally spaced discs and rotated in a vertical cylindrical housing filled with sand. The premixed dispersion paste is pumped into the mill from the bottom of the housing and is forced upward and out the top of the mill housing. A circular screen which fits around the top of the mill allows the paste to flow out but retains the sand. Although no longer in general use, roller mills are occasionally used to grind the pigments into a very fine dispersion.

Thinning and Tinting

After the desired pigment dispersion is attained, the pigment paste is then pumped or gravity fed into a thin-down tank where it is thinned with additional resin and solvent to reach the proper balance of pigment, resin, and solvent that is indicated by the formula. The batch is

PAINTS AND COATINGS MANUFACTURING

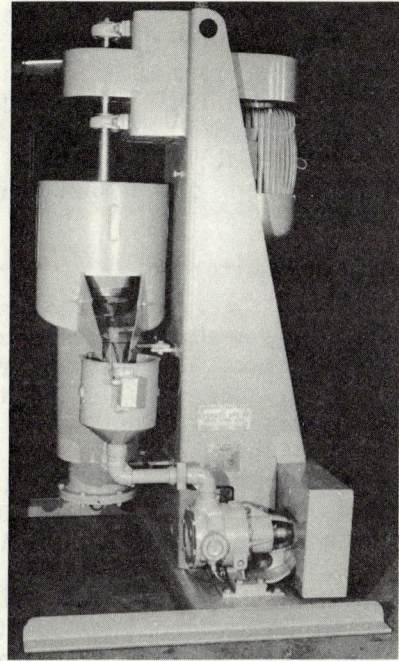

FIGURE 8.2. Typical sand mill. (Epworth Manufacturing Co., Inc.)

then tinted to match the desired color and a sample taken to the quality assurance laboratory for testing.

Quality Control Testing and Adjustment

Tests are made in quality control laboratories to verify that the batch meets the desired product specifications for viscosity, weight per gallon, gloss, color drying, and application properties. The batch may then be adjusted by the additions of resin, solvent, or additives if necessary.

Laboratories are usually equipped with spray booths and other types of application equipment so that the finished product can be applied to actual material to see if its characteristics meet the design criteria.

Filtering and Filling

The final steps in manufacturing are filtering and filling the product into the desired shipping containers. The filtering process may range from simply passing the batch of material through a nylon mesh bag to remove the oversized pigment agglomerates and foreign material, to running the entire batch through a pressurized filtering apparatus that will filter out particles down to 5 microns in size. Most architectural paints are filled and sold in small containers: half-pints (236 mL), pints (473 mL), quarts (946 mL), gallons (3.78 L), and 5-gal (19 L) pails. Industrial coatings are usually filled in 5-gal (19 L) pails, 30- and 55-gal (113 and 208 L) drums, portable tanks, or in bulk in tank trucks.

Aerosol Manufacturing

A popular method of applying small quantities of coatings is by the use of pressurized aerosol containers. The coatings product is put into a small container and then pressurized by the injection of a propellent. This propellent is usually propane or isobutane and is an integral part of the final product as well as acting as the propellent.

The charging is done either in a separate building or a specially equipped room inside of the plant. Conveyors are used to transport the containers from the filling area to the charging room and on to the packaging line.

After the containers are charged, they pass through a warm water bath as a check for leaks in the container. (See Chapter 35, "Aerosol Filling and Storage," for hazards associated with aerosols.)

Open Varnish Cooking

Although largely phased out due to environmental regulations, the cooking of varnish in large open vessels over open flame burners may be found. Such operations are always carried out in separate buildings.

Finished Products Storage

When the finished coatings product has passed through the filling line and is in its container, all containers of 1 gal (3.78 L) size or less are put into cartons and placed on pallets. Sometimes pallets are automatically loaded by the use of palletizers. Five-gal (19 L) pails are stacked one on top of the other on pallets. Sometimes they are encapsulated in plastic film.

Sometimes conveyors will carry the containers from the filling machines to other parts of the plant for packaging.

The filled pallets are moved by industrial lift truck into the finished goods warehouse and distribution centers. The storage configuration is either palletized on the floor or in rack storage. Loaded pallets may be stored on top of each other to about 12 ft (3.6 m) in height. Rack storage is usually less than 25 ft (7.6 m) high although special warehouses with automated inventory storage may be as high as 80 ft (24.4 m).

The size of finished products warehouses will vary with the plant. Warehouses of from 50,000 to 100,000 sq ft (4645 to 9290 m^2) are considered typical.

Fifty-five (208 L) and 30-gal (113.5 L) drums are filtered and filled directly from large thin-down tanks or agitated storage tanks. On pallets, they are then moved into the warehouse and, like the cartons, are stored either on the floor or in racks.

Bulk quantities of finished products are pumped directly into tank trucks for direct shipment.

Protection for Finished Products Storage

The various protection considerations outlined under the section on protection for raw materials storage apply to finished products storage as well. The same type of containers, storage configurations, etc., are used.

THE FIRE HAZARDS

Experience in the coatings industry has shown that nitrocellulose lacquer manufacturing and aerosol charging operations present the most severe fire hazard. Coatings manufacture using Class I — flash points below 100°F (37.8°C) — solvents and varnish cooking present the next level of fire hazard followed by resin manufacturing and coatings manufacturing using Class II — flash points at or above 100°F (37.8°C) — solvents. Manufacturing of waterborne coatings presents the lowest fire hazard.

There are two broad classifications of fire hazards that have to be guarded against while processing coatings products. The first is the accidental ignition of flammable vapors from the material involved in the normal processing operations, and the second is the uncontrolled release of flammable and combustible liquids within the plant so that the potential for ignition can exist. When manufacturing waterborne coatings these two problems do not exist and the relative hazard is substantially reduced.

In pigment dispersing operations flammable liquids are often exposed to the air either on a steady basis or during the charging and/or emptying phases. Controlling ignition sources must be constant and is very critical.

Static Electricity

The most common ignition source of coatings plant fires is static sparks. Since flammable liquids are pumped, agitated, mixed, filtered, and otherwise kept in motion during the manufacturing process, the generation of static electricity is a constant problem.

Generally, all equipment found in a coatings plant can be grounded, thereby bleeding off static charges. An exception is pebble mills which cannot have the generated charges bled off due to the insulating qualities of the ceramic linings in the mills. Typically, after the dispersion has been completed and the mill emptied, wash solvent is frequently put into the mill and the mill operated to wash out the residue. This mill action generates static electricity prolifically.

After the mill is washed, the hatch is removed and the charging chute or hose is inserted for the purpose of adding materials for the next batch. If the chute or hose is of conductive material and is grounded, quite often the static charge inside the mill will spark across the gap and a flash fire can occur. An inert gas such as nitrogen should be used to reduce the oxygen inside the mill to no more than 8 percent in order to minimize the likelihood of ignition.

Chemical Heating

With some coatings processes utilizing reactors, ingredients must be added in a certain sequence to prevent a chemical reaction from occurring that may generate sufficient heat to ignite the ever-present vapors. Solutions into which nitrocellulose is introduced can be especially prone to this phenomenon.

Reactor Hazards

Some of the points that must be considered are: proper quantity and type of ingredients, adequate cooling capacity for cooling any exothermic reaction, and adequate rupture disc capacity with a means for conveying any rupture disc discharge to a safe location. (See Chapter 31 for further information on reactor vessels.)

Maintenance and Housekeeping

Maintenance and housekeeping are a vital part of any coatings plant operation. Leaking valves and fittings can release flammable liquids and vapors. Key items such as valves, pumps, and automatic controls should always operate as intended. Explosion-proof electrical equipment must be kept constantly maintained or it will not function as designed and may present an ignition hazard.

Paints and coatings can be readily spilled at several stages of the production process. If not promptly cleaned up the residues will solidify, making cleaning difficult, and, in the event of a fire, they can add

PAINTS AND COATINGS MANUFACTURING

fuel to the fire. If spills are cleaned up promptly, augmented by a weekly general cleaning, coatings plants can be kept clean.

Handling Nitrocellulose

In a solvent solution, alcohol-wet nitrocellulose poses the same hazards as any other Class I flammable solvent mixture; however, during the dissolving process, nitrocellulose poses some unique problems. For example, it is sensitive to caustic and other alkaline materials such as amines. In direct contact with strong alkalis, nitrocellulose interacts with the evolution of heat causing ignition of the material. Improper tank cleaning involving alkalines has resulted in several fires.

Extreme care must be taken in following the required procedures when dissolving nitrocellulose. Improper procedures have resulted in many fires.

Nitrocellulose requires special housekeeping attention. All spilled residue must be swept up immediately with a natural bristle broom, deposited in a covered metal container, and wet with water. Each day the container must be emptied. Disposal must be done by burning after drying. Nitrocellulose will dry rapidly when exposed to air and even the friction from a shoe on the floor can cause ignition.

Any rags used to wipe up nitrocellulose are to be disposed of in the same way as the nitrocellulose fibers.

Areas that process nitrocellulose will, over a period of time, have a buildup of fine nitrocellulose dust on horizontal surfaces, such as ledges, roof bar joists, beam flanges, equipment, etc. This dust should be periodically cleaned by hosing down the area with water or by the use of an explosion-proof vacuum cleaner that utilizes a water chamber to deposit the dust particles. If this cleaning is not done, any flash fire and related explosion may cause the particles to become suspended in the air and a rapid flash fire can develop throughout the room.

Solidified dissolved nitrocellulose residue on equipment, floors, etc., is highly combustible and, once ignited, will burn rapidly and is very difficult to extinguish.

THE SAFEGUARDS

The following safeguards apply primarily to solvent-based manufacturing plants where flammable and combustible liquids are utilized.

It is desirable to house coatings manufacturing operations in buildings that do not have other types of operations in them. Topographical features must be considered so that an uncontrolled liquid flow will not endanger other parts of the plant or other buildings. Drainage or ramps and sills may be required to prevent such flow.

All portions of the plant, such as offices, laboratories, general offices, raw materials storage, finished goods storage, and processing areas should be separated from each other by 2-hour fire walls with openings protected by approved fire doors.

NFPA 30, *Flammable and Combustible Liquids Code*, has information on the minimum distances that processing vessels should be located from property lines and public ways.

Fire department access is important and it is desirable to provide such access on all sides of the plant.

Buildings should be either of fire-resistive construction or noncombustible construction without basements. In multistory buildings, stairways should be enclosed with 2-hour enclosures with openings protected by approved fire doors.

Where Class I liquids are handled, electrical and heating equipment suitable for Class I, Division 1 hazards are required. (NFPA 35, *Manufacture of Organic Coatings*, and NFPA 497, *Class I Hazardous Locations for Electrical Installations in Chemical Plants*, give guidance on locations that are considered hazardous in evaluating the need for explosion-proof equipment in process areas.) Heating can also be provided by indirect methods such as hot water, steam, or hot air.

Ventilation

Ventilation is one of the most important safeguards in coatings plants. If flammable vapors can be diluted, removed, or both, ignition cannot occur. Mechanical ventilation in process areas should be a minimum of 1 cu ft per min per sq ft [0.30 $(m^3/min)/m^2$] of solid floor area. Exhaust pickups should be at floor level and should be located so that the flow of air will be across the entire area in order that "dead spots" will be eliminated.

Local ventilation at the point of vapor liberation can effectively remove the vapors at their source. This eliminates both the fire hazard from the vapors as well as the potential health hazard exposure to employees. Such local ventilation can be utilized for up to 75 percent of the recommended ventilation of 1 cu ft per min per sq ft [0.30 $(m^3/min)/m^2$], thereby reducing the amount of general dilution ventilation required as well as the heating required in the winter time.

All low areas, such as pits, require local ventilation due to vapors of flammable and combustible liquids being heavier than air. These vapors will collect in such low areas.

Explosion Venting

Explosion venting is highly desirable where Class I liquids, unstable liquids, or finely divided flammable solids are processed. This can be achieved by open-air construction, lightweight wall or roof panels, or windows of the explosion-relief-type. NFPA 68, *Guide for Explosion Venting*, can be used for a guide in providing appropriate venting.

Smoke and Heat Venting

Smoke and heat venting is also desirable. Flammable liquid spill fires have a very rapid heat release, often accompanied by a fireball. By providing heat vents, the heat is allowed to escape and will not open an excessive number of sprinklers. Such vents will also limit the amount of damage to the building and contents. Another important factor is that, after venting, the smoke and heat buildup within the building is sufficiently reduced to allow fire fighters to move to the seat of the fire and effectively suppress it. NFPA 204, *Guide for Smoke and Heat Venting*, is a good source of information for this type of venting.

Containment of Liquid Spills and Vapors

Openings where pipes, tanks, etc., protrude through floors need seals that can prevent migration of either liquids or vapors to the area below where an ignition source may exist. Some equipment may extend through floors in such a way that the space between the equipment and floor cannot be made vapor-proof. Barriers are needed around the equipment to prevent liquid spills from running to the floor below. Self-closing lids at charging ports for mills or tanks also help to keep vapor migration to a minimum. Covers on tanks will also help to reduce the emission of vapors into the room when general area dilution ventilation is used.

Preventing the accidental flow of flammable liquids under doors from one part of a plant to another is also important. This can be prevented by noncombustible, liquid-tight, raised sills or ramps of at least 4 in. (102 mm) in height at all door openings. Drains may be used in place of sills or ramps to capture the liquid and remove it to a safe place outside of the building.

Aerosol Safeguards

Aerosol manufacturing requires special considerations (See Chapter 35). The charging room, whether located inside of a building or in a separate building, must be designed so that an explosion will limit the damage to the room or building itself. Adequate explosion venting is essential.

High velocity ventilation of one complete change of air per minute is required. Combustible gas detection should be installed to detect any leaks when flammable propellants are used. A commonly used approach is to have an audible alarm sounded when a gas concentration of 25 percent of the lower explosive limit of the propellant is reached. At the same time the ventilation system is increased to two complete air changes per minute. If the gas concentration increases to 50 percent of the lower explosive limit, the charging equipment is automatically shut down and a fail-safe solenoid valve is activated to shut off the supply of propellant to the charging area.

Static Electricity Control

Static electricity may be generated in most coatings plant operations, and its control is one of the most important fire prevention measures. It is easily achieved by proper bonding and grounding procedures.

"Bonding" is the practice of providing an electrical connection between various items of equipment, piping, tanks, etc., so that any static electricity charges will be equalized between the various items. "Grounding" is providing an electrical connection to earth so that static electricity charges in the bonding system can be harmlessly bled off to ground.

Bonding is a very simple and effective system, and it works. The only problem is that it depends on the human element to see that bonding clamps are on all portable tanks and equipment. Clamps of the point-type, and not the commonly used alligator clamps, are required. The points can "bite" through any coatings residue on the equipment and make a good contact with the metal.

Equipment involving the storage and handling of Class I liquids must be bonded and grounded. This includes tank cars or trucks delivering raw materials through to the final disposition of the finished product. Where nonconductive piping, hoses and flexible connections are used, a separate bonding arrangement must be provided to have electrical continuity.

A driven rod to earth is the best ground. Water piping may have high resistance connections or sections of plastic that are nonconductive. (Some water companies deliberately isolate the piping electrically at the water meter to prevent the water system from being used as an electrical conductor.)

Static generation can also be reduced by eliminating the free fall of bulk flammable liquids into mixers or containers. Fill pipes should go to the bottom of the container or be diverted to the side of the container to prevent splashing. The velocity of liquid in piping should be below 15 fps (4.6 m/sec) in order to reduce static generation.

Maintaining a relative humidity above 30 percent in processing areas will also reduce the rate of static generation.

Certain commercial additives can be added to solvents which reduce the static generation rate of the liquid by increasing its conductivity. These are used successfully by some companies.

Introducing polar solvents to a coating batch first, before other solvents are added, helps to reduce generation of static electricity. Polar solvents have a lower resistivity than nonpolar solvents. Also, it has been found that the addition of five percent by volume of highly conductive solvents to solvents with very low conductivity will increase the conductivity of the mixtures to a point where static electricity will not present a problem in properly grounded systems. Caution must be exercised so that the agent will not cause any problems with formulations.

Filters, because of their large surface area, can produce as much as 200 times more electrostatic charge than the same piping system without filtration. This hazard is controlled by installing filters far enough upstream of discharge points to provide a 30-second liquid relaxation time prior to discharge.

For further information on the proper techniques of bonding and grounding and information on the generation and control of static electricity and test equipment or procedures, refer to NFPA 77, *Static Electricity*, and *Generation and Control of Static Electricity*, the latter published by the National Paint and Coatings Association.[2]

Struck Spark Control

Nonsparking tools of a nonferrous material, such as bronze or berylium, should be used around nitrocellulose and flammable liquids. Wrenches of nonferrous material are required to remove the retaining ring holding the cover on nitrocellulose drums in place. Forks and scoops used to scoop nitrocellulose out of a barrel must be of a similar material.

Covers on mixing tanks should also be of nonferrous material.

Spray Booths

Any spray booths should be installed and protected in accordance with NFPA 33, *Spray Application Using Flammable and Combustible Materials*. (See Chapter 25, "Spray Finishing and Powder Coating," of this book for information on the safeguards for spray finishing.)

Materials Handling

Where Class I liquids are processed, materials handling equipment should either be by manual means or with the appropriate classification of industrial fork lift truck. NFPA 35 specifies which classification of fork lift truck is allowed in the various portions of a coatings plant. (See Chapter 46, "Materials Handling Systems," for further information on the classification and use of industrial trucks.)

Mill Inerting and Operation

Ball mills, which utilize steel balls, will bleed off a static charge through the body of the mill. However, porcelain-lined pebble mills will insulate the static charge. With a static charge present inside the mill, any introduction of a grounded object, such as a conductive hose or a grounding probe, will likely cause a static discharge to occur with a resulting flash fire. For this reason, only nonconductive hose should be used, and the pebble mill should be adequately inerted. Some companies use both methods.

The introduction of an inert gas, such as nitrogen, into a ball or pebble mill will safeguard against a flash fire occurring due to static. Care must be exercised so that enough inert gas is injected to adequately displace the oxygen throughout the entire mill. When using an inerting agent, it is necessary to reduce the oxygen content to a maximum of 8 percent by volume to prevent ignition. One company has now developed an automatic system to monitor and maintain the correct percentage of inert gas and has adapted it specifically to the coatings industry.[3] This system is available for both fixed and portable installations and has proven to be highly effective.

Other hazards associated with ball and pebble mill operations are loss of a gasket or a seal, or the manhole cover on the mill coming loose during operation. Spill of material from the revolving mill will then occur.

Periodically, some coatings plants will charge a mill and let the operation continue with the plant unoccupied. This is usually at night and on weekends. There are several safeguards which could be employed whenever ball or pebble mill operations are taking place, and when the department is unoccupied. A hydrocarbon detection system, tied into local and central station supervision alarms, is a recognized means of detecting a large spill. In addition, a guard service, with frequent inspection of the operating area, is also desirable. Ideally, mill operations should be conducted only when personnel are working in the area.

Fire Control Systems

Automatic sprinklers are the basic protection system in processing areas of coatings plants. Particular care must be given to be sure that sprinklers are installed under tanks, equipment, mezzanines, and other objects that will obstruct the sprinkler discharge pattern. A key point to remember is that flammable liquids will flow under such objects, and if they ignite, sprinkler discharge must be able to reach all areas.

Special extinguishing systems are sometimes used to provide additional protection to tanks and special hazards. These systems would include foam, carbon dioxide, and dry chemical. The particular agent used would depend on the materials involved. Carbon dioxide is most commonly used over open tanks or mixers since the agent does not contaminate the product in process.

It is essential that waterflow signals from sprinkler systems, signals to indicate that special protection systems have been activated, manual fire alarm system signals, and any other fire detection or protection systems should go either to a constantly monitored place at the plant, a central station supervisory service, or directly to the fire department. In this way, immediate notification is given to the fire department of any fire emergency. Local signals should be given as well as those to alert the plant fire brigade.

Signals from these protective systems can be tied into flammable liquid pumps so that, in the event of a fire, the pumps will automatically be cut off.

Combustible gas detection systems are sometimes installed in processing areas to provide early detection of gas concentrations. Signals from such systems can be tied in to sound an alarm, to increase the ventilation rate, and to shut down flammable liquid pumps or other process equipment. The location of the detection heads is critical and must be done with the assistance of the detection equipment manufacturer.

Life Safety Considerations

The exiting requirements for general industry must be provided throughout the plant. In those areas of the plant where flammable liquids are handled, exiting must be provided as specified for high hazard occupancies. This extra hazard area will require a minimum of two well-separated exits with doors swinging in the direction of exit travel and equipped with panic hardware. The exit path should be a maximum of 75 ft (23 m). Exit details are outlined in NFPA *101, Life Safety Code*.

Emergency exit drills should be conducted at least once a year so that all employees know emergency procedures.

Fire Prevention

Fire prevention activities in a coatings plant can play a significant role in reducing the likelihood of a fire. In-house inspections will reveal unsafe conditions or acts that can be corrected. Most insurance companies have self-inspection forms to assist in this activity.

The National Paint and Coatings Association has developed an in-plant visual system that can be utilized to apprise the plant employees of health, flammability and reactivity hazards associated with the materials in use within the plant. The system can be used as a label attached to raw material packages or as a placard fastened to a wall, tank, pallet or pipe. In addition, it can be printed on the batch card that follows the product through the various manufacturing phases.

The symbol for this system is a square with four horizontal colored areas. The top section has a blue background with the health hazard rating, the second section has a red background with the flammability hazard rating, the third section has a yellow background with the reactivity rating, while the bottom section has a white background and indicates the personal protection to be utilized while handling the material.

Each hazard is identified by a numerical designation ranging from 0 to 4. A rating of 0 indicates a very low degree of hazard while a rating of 4 represents a very high degree of hazard. An alphabetical designation of "A, B, C..." advises the worker of the appropriate personal protective equipment to be worn while handling that particular material during the manufacturing process.

Many coatings companies are now utilizing this system to rapidly and concisely communicate the relative degree of hazard of the materials that an employee is working with. Complete information on how it is utilized can be obtained from the National Paint and Coatings Association.[4]

Smoking should be limited to offices, lunch rooms and other specifically designated areas and not allowed in laboratories, storage areas, and process areas.

Specific procedures for the control of welding and cutting within the plant must be strictly enforced (see Chapter 24, "Welding and Cutting"). Repair work that requires welding should be scheduled for nights or weekends when there is no production within the plant. Cutting and welding permit programs should be established.

It is highly desirable for the local fire department to visit the plant and preplan their suppression activities in the event of a fire. This will allow them to become acquainted with the materials and processes in the plant so they will know what to expect during a fire emergency. Their fire fighting efforts will be more efficient with such preplanning.

BIBLIOGRAPHY

REFERENCES CITED

1. National Paint and Coatings Association, "Evaluation of the Fire Hazard of Water-Borne Coatings," *Scientific Circular 804*, Washington, DC, 1977.
2. ———, "Generation and Control of Static Electricity," *Scientific Circular 803, Revision 2*, , Washington, DC, 1983.
3. "Inerting for Safety in the Paint and Coating Industry," Neutronics, Inc., King of Prussia, PA.
4. National Paint and Coatings Association, "Hazardous Materials Identification System," *Raw Materials Rating Manual*, Washington, DC.

REFERENCES

American Conference of Governmental Industrial Hygienists, *Industrial Venti-*

lation, A Manual of Recommended Practice, 17th Edition, Committee on Industrial Ventilation, Lansing, MI, 1982.

Banov, Abel, *Paints Coatings Handbook for Contractors, Architects, Builders and Engineers*, Structures Publishing Company, Farmington, MI, 1973

Biddle, James G., *Getting Down to Earth . . . A Manual on Earth-Testing for the Practical Man*, James T. Biddle Co., Plymouth Meeting, PA, 1970.

Buford, R.R., "The Use of AFFF in Sprinkler Systems," *Fire Technology*, National Fire Protection Association, Boston, MA, 1976.

Crouse-Hinds Company, *Crouse-Hinds Code Digest, Articles 500-503 and 510-517, 1978 National Electrical Code*, Syracuse, NY, 1977.

Definitions Committee, *Paint/Coatings Dictionary*, Federation of Societies for Coatings Technology, Philadelphia, PA, 1978.

DuPont Company, *Static Protection for Flammable Materials*, Wilmington, DE, 1972.

———, *Nitrocellulose Properties and Uses*, Wilmington, DE, 1976.

Factory Mutual Engineering Corporation, *Loss Prevention Data*, vols. 1-5, Factory Mutual System, Norwood, MA, 1979.

Factory Mutual Research Corporation, Fitzgerald, Paul M. and Young, John R., "The Feasibility of Using 'Light Water' Brand AFFF in a Closed-Head Sprinkler System for Protection Against Flammable Liquid Spill Fires," *FMRC Serial No. 22352*, The 3M Company, St. Paul, MI, January 1975.

———, Fitzgerald, P. M., Newman, R. M., and Young, J. R., "Fire Protection of Drum Storage Using 'Light Water' Brand AFFF in a Closed-Head Sprinkler System," *FMRC Serial No. 22464*, Allendale Mutual Insurance Company, Johnston, R.I., Factory Insurance Association, Hartford, Conn., The 3M Company, St. Paul, MI, March 1975.

———, Tavares, Raymond, "Investigation of Solvent-Based Coatings in 55-Gal Drum Test Apparatus," FMRC J.I. 0G3R1.RA 070(A), National Paint and Coatings Association, Washington, DC, August 1982.

Fuller, Wayne R., *Understanding Paint*, The American Paint Journal Company, St. Louis, MO, 1965.

Haase, Heinz, *Electrostatic Hazards, Their Evaluation and Control*, Verlag Chemie-Weinheim, New York, NY, 1977.

Johnsen, Montfort A., Dorland, Wayne E., Dorland, Eleonore Kanar, *The Aerosol Handbook*, Wayne E. Dorland Co., Caldwell, NJ, 1972.

Jones, Charles L., *Safety in Lacquer Plants*, Hercules Powder Company, Wilmington, DE, 1946.

Martens, Charles R., ed., *Technology of Paints, Varnishes and Lacquers*, Robert E. Krieger Publishing Company, Huntington, NY, 1974.

National Paint and Coatings Association, "The Electrical Code Problem in the Paint and Coatings Plant," *Safety Health Bulletin No. 36*, Washington, DC, 1978.

———, "Floor Cleaning in Coatings Plants," *Safety Health Bulletin No. 37*, Washington, DC, 1979.

Rase, Howard F., *Chemical Reactor Design for Process Plants, Volume One: Principles and Techniques*, John Wiley & Sons, Inc., New York, NY, 1977.

Sanders, Paul A., *Handbook of Aerosol Technology*, Second Edition, Van Nostrand-Reinhold Co., New York, NY, 1979.

Schneberger, Gerald L., *Understanding Paint and Painting Processes*, Second Edition, Hitchcock Publishing Co., Wheaton, IL.

NFPA CODES, STANDARDS, AND RECOMMENDED PRACTICES

Reference to the following NFPA Codes, Standards, and Recommended Practices will provide further information on the safeguards for paints and coatings manufacturing discussed in this chapter. (See the latest *NFPA Codes and Standards Catalog* for availability of current editions of the following documents.)

NFPA 12, *Carbon Dioxide Extinguishing Systems.*
NFPA 12A, *Halon 1301 Fire Extinguishing Systems.*
NFPA 12B, *Halon 1211 Fire Extinguishing Systems.*
NFPA 13, *Installation of Sprinkler Systems.*
NFPA 15, *Water Spray Fixed Systems.*
NFPA 16, *Deluge Foam-Water Sprinkler and Spray Systems.*
NFPA 17, *Dry Chemical Extinguishing Systems.*
NFPA 30, *Flammable and Combustible Liquids Code.*
NFPA 35, *Manufacture of Organic Coatings.*
NFPA 68, *Explosion Venting Guide.*
NFPA 77, *Recommended Practice on Static Electricity.*
NFPA 101, *Life Safety Code.*
NFPA 231, *Indoor General Storage.*
NFPA 231C, *Rack Storage of Materials.*
NFPA 497, *Class I Hazardous Locations for Electrical Installations in Chemical Plants.*

ADDITIONAL READING

Coatings Technology, M. T. Gillies, ed., Noyes Press, Park Ridge, NJ, 1978.
Fawcett, Howard H., Wood, William S., *Safety and Accident Prevention in Chemical Operations*, John Wiley & Sons, Inc., New York, NY, 1965.
Federation of Societies for Coatings Technology, *Federation Series on Coatings Technology* (units 1-25), Philadelphia, PA.
"Fire Risks in Paint Manufacturing," *Fire Prevention*, October 1981, pp. 26-29.
Harrington, J. L. and R. B. Hopkinson, *Rack Storage Protection*, Worcester Polytechnical Institute, Worcester, MA, 1977.
Henry, Martin F., ed., *Flammable and Combustible Liquids Code Handbook*, First Edition, National Fire Protection Association, Quincy, MA, 1981.
Magison, Ernest C., *Electrical Instruments in Hazardous Locations*, Third Edition, The Instrument Society of America, Research Triangle Park, NC, 1978.

McKinnon, G. P., ed., *Fire Protection Handbook*, Fifteenth Edition, National Fire Protection Association, Quincy, MA, 1981.

Mellan, Ibert, *Industrial Solvents Handbook*, Noyes Data Corporation, Park Ridge, NJ, 1977.

National Paint and Coatings Association, *Abstract Review*, Washington, DC.

———, "Flammability of Paint Study," *Scientific Circular 801*, Washington, DC, 1974.

Paint and Powder Coatings, Predicasts, Cleveland, OH, 1981.

Payne, Henry Fleming, "Oils, Resins, Varnishes, and Polymers," *Organic Coating Technology*, vol. I, John Wiley & Sons, Inc., New York, NY, 1954.

———, "Pigments and Pigmented Coatings," *Organic Coating Technology*, vol. I, John Wiley & Sons, Inc., New York, NY, 1954.

Pigment Handbook, (3 vols.), Temple C. Patton, ed., John Wiley & Sons, Inc. NY, 1973.

Pilborough, L, *Inspection of Chemical Plants*, Gulf Publishing Co., Houston, TX, 1977.

Schram, Peter, ed., *The National Electrical Code Handbook*, Third Edition, National Fire Protection Association, Quincy, MA, 1983.

Talbot, G., "Static Electricity," *Fire*, 70(871), pp. 397-398.

Vervalin, Charles H., ed., *Fire Protection Manual for Hydrocarbon Processing Plants*, Gulf Publishing Company, Houston, TX, 1973.

9

Rubber Products

E. N. Proudfoot

The rubber products industry manufactures products from both natural and synthetic rubber. Rubber was first noted by Westerners in the West Indies when early explorers noticed the natives using natural rubber from trees. It was later discovered that rubber trees prosper anywhere within 10 degrees of the equator. Brazil, Africa, and the Far East now produce large amounts of the product, and rubber plantations have been established by many larger rubber manufacturers to ensure proper quantities and types of natural rubber.

Synthetic rubber, made basically from petroleum products, was developed prior to World War II but was not manufactured on a production basis until the normal natural rubber supplies in the Far East were cut off in 1941. The development and production of synthetic rubber has continued to the point where today many types are necessary to obtain characteristics needed for better rubber products. Approximately 75 percent of the rubber used today is synthetic.

The major use of rubber is for *tires* and *tubes*, including those for automobiles, trucks, airplanes, farm equipment, motorcycles, and heavy

---

E. N. Proudfoot is Manager of Fire Protection for The Goodyear Tire & Rubber Company of Akron, Ohio.

off-the-road construction equipment. Another large use of rubber is in *fabricated rubber products* which include molded custom tire parts, shoe products, rubber clothing, rubber flooring, sponge and foam rubber goods, and a host of other miscellaneous items. Other categories include *rubber hose* and *belting*, both making up a substantial part of the rubber market, and also *rubber* and *plastic footwear*.

The 1975 U.S. Census of Manufacturers showed that the value of the rubber products shipped was approximately $11 billion; in 1981 this figure moved up to approximately $18 billion.

The rubber industry in the U.S. employs about 243,000 people, down from 265,000 reported in 1978. At one time tire and tube production was centered in Akron, Ohio, where most corporate offices still remain; but since World War II, most new plants have been constructed in locations across the country.

There are many different rubber products, and it would be prohibitive to discuss all of them in this chapter. Since many of the mixing processes involved are similar for a large number of rubber products, the author has chosen to follow the tire building cycle as representative of the operations used; the hazards common to most rubber plant operations will be given attention in the course of the discussion.

Most ingredients used in the manufacture of rubber products are combustible; therefore, it is necessary that fixed fire protection be included in the construction of plants. Although early plants used wood in their building construction, the more recent plant construction is of noncombustible type, using unprotected steel and masonry. Some of the older international rubber plants are known to be without sprinkler protection throughout, and this is where the most serious fire losses have occurred.

The principal hazards involve storing natural and synthetic rubber and various compounds including sulfur, oil, and hydrocarbon solvents; mixing of solvents and rubber; mixing of rubber; spreading cement on rubber; spreading rubber on fabric; dipping rubber fabric in cement; paint spraying of tires; and storing tires. From the raw material to the finished product there is a continuity of combustibles, and modern tire plants usually have several hundred thousand square feet (100,000 sq ft equals 9290 m^2) in one fire area. The storage of tires is usually limited to about 120,000 sq ft (11 148 m^2) per fire area because of their combustibility and piling height.

RAW MATERIALS

For natural rubber products, the rubber tree is tapped and latex is collected. The latex is taken to a collection point where it is coagulated and run through a mill to produce sheet rubber. After baling, it is shipped to the rubber factory.

Synthetic rubber is produced by reacting petroleum products such

RUBBER PRODUCTS

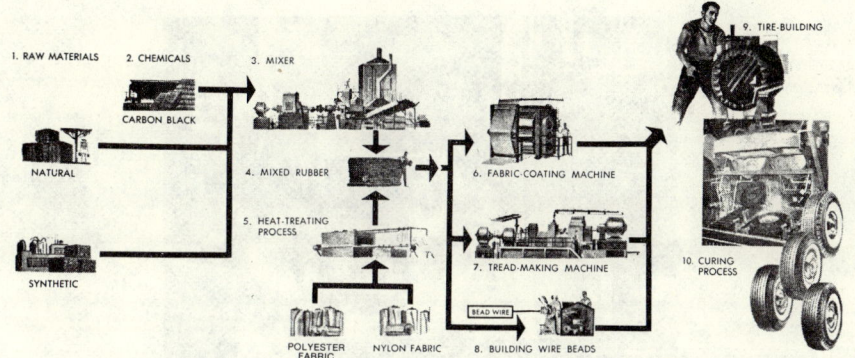

FIGURE 9.1. Steps in tire manufacturing. (The Goodyear Tire & Rubber Co.)

as butadiene and styrene under heat and pressure conditions. A slurry of crumb is produced which is then washed, dried, and baled for marketing. The storage and handling of the petroleum products through the reaction stage subject the synthetic rubber plant to the usual chemical industry hazards. Beyond this, the drying, while in crumb form, represents a hazard of a Class A combustible material which needs special attention.

Other principal raw materials are carbon black (stored in silos), sulfur, woven fabric in 60-in.-wide beams (1.5 m), hydrocarbon solvents, and oils.

THE PRODUCTION PROCESS

The manufacture of rubber tires (as shown in Figure 9.1) essentially involves ten phases:

1. Mixing the rubber
2. Heat treating of fabric
3. Calendering rubber fabric
4. Rubber extrusion
5. Cement mixing
6. Bead building
7. Building tires
8. Curing tires
9. Final finishing
10. Storage.

Chapter 9

FIGURE 9.2. *Feed conveyor to banbury mixer. (The Goodyear Tire & Rubber Co.)*

FIGURE 9.3. *Banbury mixer at feed level. (Farrel Company Div., USM Corporation)*

Mixing the Rubber

The banbury mixer is used throughout the rubber industry for breaking down various types of natural and synthetic rubber and for mixing rubber with other compounding materials. The rubber and some compounds are front loaded into the banbury using a conveyor (see Figures 9.2 and 9.3), while carbon black and oils are usually charged by direct connections to the chute. All are weighed for accuracy and quality control. The mixing action causes the temperature to rise to a desired level, approximately 200 to 300°F (93.3 to 149°C), and then the rubber is

RUBBER PRODUCTS

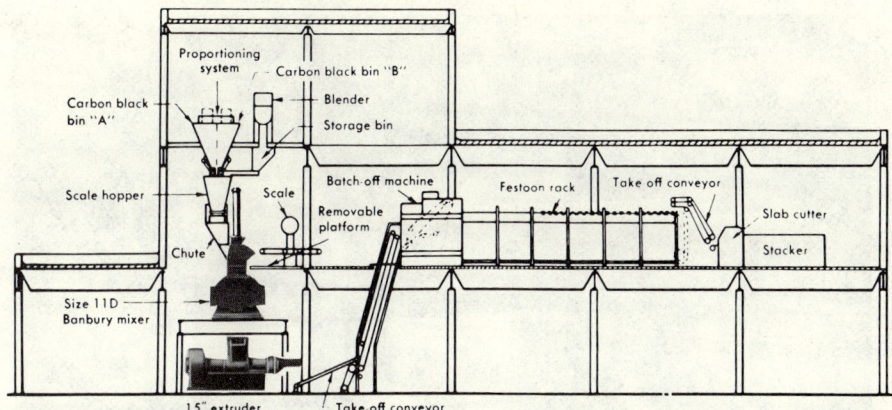

FIGURE 9.4. *Banbury mixer and process equipment. (Farrel Company Div., USM Corporation)*

dumped onto a mill where it is extruded or pelletized for further processing. (See Figure 9.4.)

The heat is created by two rotors revolving in opposite directions and at different speeds in an enclosed space of the banbury. The rotors are designed with two or four wings which work eccentrically to the center of the roll, causing the breakup of the materials within. Sufficient heat from this action may be created to develop higher-than-desired temperatures within the banbury; thus, the rotor and banbury body have features which provide cooling water circulation.

The rubber is usually delivered by truck or conveyor from the banbury area to warm-up mills in sheet form (see Figure 9.5). The mill rolls rotate at two different speeds to create friction, thus warming up the rubber. The strip rubber is then transferred by conveyor to calenders and extruders.

Heat Treating of Fabric

The fabric, which is to be filled by the rubber in the calendering phase, is first processed through a heat-tension conditioner so as to fix the fabric at its greatest resiliency and strength. For adherence purposes later in the process, the fabric, in approximately 60-in. widths (1.5 m), is dipped in a latex solution prior to its subjection to heat and tension. The route of the fabric through the conditioner is several hundred yards long (100 yards equals 91.4 m), including the oven sections. There is a fire hazard in the vacuuming-off of solids following the dipping operation. The solids are deposited in an exhaust system designed to collect them. A second hazard exists in the vaporizing of oils from the fabric in the oven section; these oils are deposited throughout and have a tendency to penetrate all surfaces including the insulation. Both the solids fallout and

Chapter 9

FIGURE 9.5. Warm-up mill. (The Goodyear Tire & Rubber Co.)

oil penetration in insulation are subject to spontaneous combustion. (For a comprehensive discussion of the hazards associated with ovens, see Chapter 27, "Heat Processing Equipment.")

Calendering Rubber Fabric

Various kinds of fabric (nylon, polyester, rayon, wire, fiberglass, carbon, etc.) are furnished to this area in approximately 60-in.-wide (1.5 m) beams to have rubber impregnated on each side to a prescribed thickness. The calender is a three-roll or four-roll unit with the rubber applied between two of the rolls on one side and fabric entered on the opposite. (See Figure 9.6.) It takes two sets of three-roll calenders to apply rubber to both sides while a single four-roll calender can apply rubber to both sides. The speed of the fabric can be up to 100 yd/min (91.4 m/min) between roll changes, each roll having up to 1,000 yd (914.4 m). A festoon is utilized to store fabric while a roll of fabric is being changed, thus allowing the calender to continue running at slower speeds.

Rubber Extrusion

Previously, it was normal to furnish heated rubber from a mill for the rubber entrusion; however, more recently, cold rubber is fed to the extruder. Tread stock is the principal component extruded, although other products such as white sidewall stock, tubes, and plain rubber stock are also extruded. Some extruders are known as duplex units because they have two screws (not necessarily of the same size) which allow extruding of the tread and sidewalls as one unit; however, each has its own particular rubber compounding. A duplex extruder used in the rubber industry

RUBBER PRODUCTS

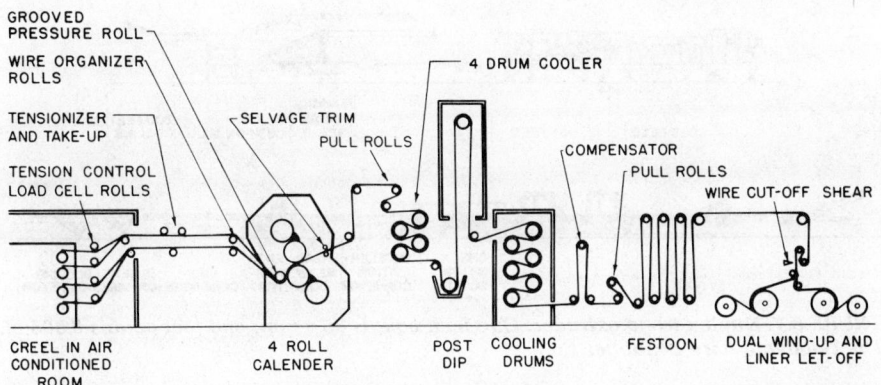

FIGURE 9.6. *Flow diagram of calendar-train process. (The Goodyear Tire & Rubber Co.)*

could be an 8 × 10 in. (200 × 250 mm) or a 10 × 10 in. (250 × 250 mm), etc.; the dimension is the screw size. (See Figure 9.7.)

Immediately following the extruder is a cementing operation where a coat of flammable cement is applied by the roll-coat method. The tread stock is then conditioned in a water solution and cut to proper lengths.

Cement Mixing

Rubber is dissolved in petroleum solvents to make a mix of approximately 90 percent solvent and 10 percent rubber to be applied to extruded stock. Several mixers and hold tanks are usual in the typical rubber factory. Since many solvents are low flash point materials, this area presents fire and explosion hazards. (For detailed information see Chapter 39, "Handling and Storage of Flammable Liquids," and NFPA 30, *Flammable and Combustible Liquids Code*.) Spray paints to be applied to green and finished tires are also prepared in this area, and they are transferred to the area of use either by portable tanks or by a circulating pipe system.

Bead Building

Bead building involves several strands of wire that are wound into a circle the approximate diameter of a tire wheel. The wire goes through a die to allow rubber encasement, cement application, and wrapping with a fabric. Beads are then transported by truck to the tire building area.

Chapter 9

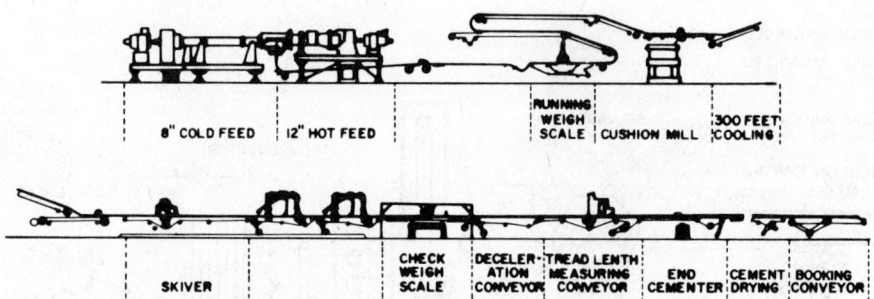

FIGURE 9.7. *Duplex tread extruder. One inch equals 25.4 mm; one foot equals 0.305 m. (The Goodyear Tire & Rubber Co.)*

Building Tires

Rubberized fabric is brought to the tire machine from the calendering operation, having been cut in particular angles and widths on a bias cutter machine. The first layer of rubber fabric is placed around a rotating cylindrical drum the approximate bead diameter of a tire. Other plies of fabric are placed on the drum, and the beads are set in each side followed by additional fabric (which is available from the ply stock) to turn up and around the bead. Small wheels known as stitchers automatically press the fabric firmly around the bead in a manner to eliminate air pockets and to ensure good adhesion. The number of ply stocks placed on a drum in the building operation depends on the type and ply rating of the tire. Some additional fabric is added to the sidewall encompassing the beads before the components are placed under the tread section, the latter which, depending on the type of tire, may be radial or bias. The last section to go on is the tread sidewall rubber stock which is precut to length. After another series of stitching, the metal drum on which the tire is built is collapsed allowing the tire to be removed. A low flash point solvent is often used to make the rubber stock tacky. Truck and larger tires require greater usage of the solvent and it is usually applied with a hand swab.

This briefly describes the passenger tire building cycle. At this stage a bias belted tire (called a "green tire") resembles a barrel with open ends; however, the radial "green tire" is constructed so that it looks more similar to the final product. (See Figure 9.8, bias type and Figure 9.9, radial type.) Other larger tires including those 11 ft (3.35 m) or more in height and weighing several tons (one short ton equals 907.2 kg) are much more intricate, involving many layers of fabric and beads.

Following the building stage, green tires are taken to a paint spraying area where the inside is sprayed with a mold release agent and the outside is sprayed with a material containing carbon black. (See Chapter 25, "Spray Finishing and Powder Coating," for a thorough treatment of this hazard.)

RUBBER PRODUCTS

FIGURE 9.8. Bias-type green tires on conveyor before curing. (The Goodyear Tire & Rubber Co.)

FIGURE 9.9. Radial-type green tires before curing. (The Goodyear Tire & Rubber Co.)

Curing Tires

Passenger tires are cured in a press, two at a time, using steam or some other inert heating medium (See Figure 9.10). Recently designed presses function automatically; a worker sets the green tires in particular spots in front of the units, and the cured tires are deposited onto a conveyor on the opposite side. Curing time for passenger tires is 15 to 20 min compared to several hours for large off-the-road types. Fire hazards are minor in the curing process, the major concern being the combustibility of the tires.

Chapter 9

FIGURE 9.10. Radial tires awaiting the automatic curing process. (The Goodyear Tire & Rubber Co.)

Final Finishing

Following the curing cycle, tires are conveyed to the finishing area where excess rubber produced by venting holes is removed. This is done by automatic rotating cutters for passenger tires, and by hand using special knives for larger tires. White sidewall grinding is also performed in this area using abrasive wheels. Exhaust ducts carry the dust away to various kinds of dustcollecting equipment. (A comprehensive discussion of air moving equipment can be found in Chapter 45.) In more recent years, advanced, complex balancing equipment and force variation machines to determine out-of-roundness have been located in this area. Inspection of the tires is the other task performed in the finishing area, and this is done physically by trained personnel.

Storage

Tires are conveyed from the final finish area to storage by belt conveyors, trucks, tow conveyors, etc., where they are placed directly in trucks or railroad cars or stored in various arrangements within the warehouse. Tires may be stored on their tread or side, in fixed or portable racks with wood pallet bases or on all metal racks. Although at one time 20 ft (6.09 m) was the usual maximum height of storage, manufacturers have increased it in the last few years to at least the 24-ft (7.32 m) level. There is a difference in the fire hazard depending on whether one uses on-tread or on-side storage, the former requiring the greater protection. The fire area of finished tire storage is usually confined to a maximum of 120,000 sq ft (11 148 m^2) of space, determined originally by insurance authorities because of the economic value attributed to the area, and thus the potential loss. (See Chapter 48 for a detailed discussion of industrial storage practices.)

THE FIRE HAZARDS

The use of flammable liquids, paint spraying, roll coating, and the combustibility of rubber in process add up to substantial fire hazards in the rubber manufacturing operation. The heat created in the mixing operation, static electricity generated by rubber moving over rolls, buffing of rubber, and other known ignition sources such as welding, burning, and electrical and spontaneous combustion are the substantial causes of fires in the industry.

The delivered oils and solvents are the first hazards to unload at the tire plant and usually these are placed in aboveground tanks at least 100 ft (30.5 m) from the principal buildings. Other miscellaneous oils and solvents are purchased in drums which are stored in a separate inside room. Some plants also use propane as a truck fuel in addition to a backup fuel for boilers and process heating equipment. (See NFPA 30, *Flammable and Combustible Liquids Code*, or NFPA 58, *Liquefied Petroleum Gases*, for specific guidelines on the handling of these fluids or gases.) Carbon black, not a substantial fire hazard, is sometimes unloaded from railroad cars using an underground conveyor, then an overhead conveyor is used from the storage bin to the use point. (See Figure 9.11.) Since these conveyors utilize rubber belts, a fire can develop from a friction point involving the rubber and the carbon black causing extensive damage and interrupting a necessary function of the production cycle.

Raw Material/Rubber Storage

Baled natural rubber may be shipped in various kinds of containers including wood cartons, and synthetic rubber can be shipped in either cardboard or wood crates. (See Figure 9.12.) When these materials are palletized and stored to a level of 12 to 14 ft (3.65 to 4.27 m), they offer a significant fire hazard, because splintered wood and cardboard are easily ignited. In recent years, some synthetic rubber has been shipped in returnable aluminum containers (see Figure 9.13), thus reducing the number of Class A combustibles in the raw material storage area and improving the housekeeping condition. (For more detailed information see

Chapter 9

FIGURE 9.11. Carbon black storage and elevated conveyor to transport the material to small hopper in the banbury area. (The Goodyear Tire & Rubber Co.)

Chapter 48, "Industrial Storage Practices," and NFPA 231D, *Storage of Rubber Tires*.)

Banbury Mixing

The banbury is used to mix the rubber and compounds, raising the temperature to a predetermined level where the batch is discharged to the next step. The discharge action is determined by the temperature of the compounded rubber. If there were a malfunction of the temperature-indicating equipment, the rubber would be allowed to continue mixing, creating higher temperatures to the point where the added oils begin to vaporize. Should the temperature of the compound reach the autoigni-

RUBBER PRODUCTS

FIGURE 9.12. Natural rubber in wood crates. (The Goodyear Tire & Rubber Co.)

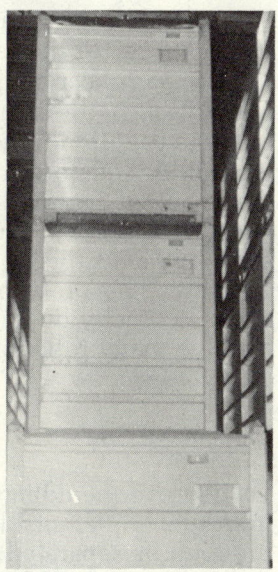

FIGURE 9.13. Synthetic rubber in aluminum containers. (The Goodyear Tire & Rubber Co.)

tion temperature of the oils or rubber, the vapors would ignite on exposure to oxygen; this usually occurring after the compound is discharged manually. One way to overcome this problem is to install two temperature-indicating devices, in effect providing a backup system to ensure temperature indication. At times the compounds in a banbury do not mix adequately in the usual period of time, thereby creating some doubt as to the expected temperature of the mix. The dual temperature method gives the operator a positive means of temperature determination.

Exhaust ventilation is necessary for the banbury operation to remove the carbon black and compound dust and to remove the vapors created by the heated mixture of rubber. Dust collectors are provided to trap the compounds of dust, and these units can be located in the building or on the roof. However, if a fire occurs in the banbury, it can easily spread to the exhaust duct and dust collector.

From the use of oils and compounds, an oil-soaked residue accumulates on the exterior of the banbury mixer, and other associated equipment, creating a fire hazard. Although a fire in this area is not expected to spread beyond the control of 10 sprinklers in an adequately designed system, it does create an interruption of production and some loss of materials and minor equipment. Periodic cleaning of the banbury equipment with a noncombustible solution will prevent this unnecessary fire hazard.

Calendering

The calendering operation does not offer unusual fire hazards in the adding of rubber to the fabric unless there is reason to add adhesive qualities to the rubber by dipping it into a rubber cement solution containing a low flash point flammable liquid. If the dipping is a part of the process, it is usually followed by a rapid drying method using a heated oven section. Adequate ventilation is a necessity to exhaust the vapors and to keep the concentration of vapors below 25 percent of the *lower flammable limit* (LFL). Usually combustible gas analyzers are provided with sufficient detectors to assure that the vapors are being exhausted adequately. The problem area in this dipping process is the creation of static electricity due to the rapid speed of the rubberized fabrics passing over the many rolls involved.

Cement Mixing

To improve the adherence of the rubber components a cement is made from rubber, flammable liquids, and certain other compounds. The mixing room is separated from other rubber manufacturing operations. The main quantity of flammable liquids is stored in exterior tanks and pumped to the mixing area by pipelines using automatic shut-off valves at the dispensing point. This area may include 8 to 10 mixing vessels of the 50- to 100-gal (190 to 380 L) size, plus some holding tanks of up to 400 gal (1514 L). In addition to the piped flammable liquids, a number of 55-gal (208 L) drums will be used and stored in this mixing area. (For the storage and handling of combustible liquids see Chapter 37 and NFPA 30, *Flammable and Combustible Liquids Code*.)

Because of the potential release of a flammable liquid or a fire at the opening of a mixer, and the possibility of having all of the room involved, this area is separated from the plant by a 4-hour fire wall. Static electricity and friction sparks developed from tools or the process are the principal sources of ignition in this area. The control of static electricity is improved by the provision of 50 percent or greater humidity in this mixing area, the colder months of winter being the primary concern.

Tread Cementing

The extruded tread is warm as it enters the cement roll-coating process, and this accelerates the vaporizing of the flammable liquid following its exit from this small enclosure. The container of cement in the enclosure will hold 3 to 5 gal (11.4 to 19 L) and is returned to its source whether it be a portable tank or a circulating pipeline. The exhaust ventilation system is electrically interlocked so as to have the ventilation in operation to reduce vapors to 25 percent or less of the LFL before the extruder conveyor line can operate. Static electricity and friction sparks are the principal sources of ignition in this area, and when fire occurs, the entire enclosure becomes involved immediately, and the fire will easily spread to the exhaust duct.

Bead Manufacturing and Building of Tires

These processes use only small quantities of flammable liquid or cement; therefore, these materials do not represent a substantial fire hazard. The quantities of rubber stock present on some of the floor space do represent large Btu (one Btu equals 1.056 kJ) content; however, due to the extensive one-floor layout, the areas are not confined and are available from all sides for fire fighting purposes. Following the completion of the green tire, it is taken to the paint spraying area where the concentration of tires can create an additional fire hazard, especially with the barrel-like form of the bias tire at this stage. A fire in the paint spraying operation could spread to the storage area; in fact, the shape of the tire itself acts as a shield to the water discharge of the overhead sprinkler system.

Curing and Finishing Area

Since inert heating mediums are used in the curing of tires, little fire hazard exists in these areas except for the tires themselves, which are few when compared to the square footage of the area involved in this process and the rubber conveyor belts which carry them away. The grinding of white sidewall tires to expose the white surface produces rubber dust

FIGURE 9.14. Palletized portable rack, on-side storage arrangement.

which is exhausted through a duct and collected by usual methods. Some operations prefer the use of water separators which reduce the fire hazard at the collection point; this, however, does not eliminate the problem of periodic fires in the exhaust ducts due to erratic accumulations. Hot grindings from an abrasive wheel or spontaneous combustion from accumulations in the duct can cause a fire which usually extends to the entire length of the duct. If a water-separator-type collector is not provided, this fire can also spread to the roof.

Tire Storage

Rubber tire storage is rated as a very high hazard because of the Btu (kJ) input and the obstructed airspaces that are inherent. While rubber tires are more difficult to ignite than the usual Class A combustibles, such as paper, cardboard, etc., once ignited, the spread of fire and smoke can be rapid. The use of industrial trucks, electrical equipment, and heating sources represents the greatest ignition potential in the storage area. Incendiarism is also a suspected cause in a number of warehouse fires, apparently due to the knowledge that great loss is a potential in rubber tire storage.

The higher the storage of rubber tires, the greater the fire hazard and the more difficult a fire is to control by sprinklers at the ceiling level. When tires are stored on their side in pallets (see Figure 9.14) with minimal support between levels, the tires tend to fall into the fire as they burn, and this helps to confine the spread of the fire. If the portable racks are more stable, the fire will spread at a faster rate and usually will involve a greater area. One fire test in 1970 involving tires stored on their treads in portable racks to the 17½-ft (5.33 m) level resulted in the reali-

RUBBER PRODUCTS

FIGURE 9.15. *Open portable rack, on-tread storage arrangement.*

zation that tires stored by this method (see Figure 9.15) may produce a greater fire hazard than those stored on their sides.

Welding and Cutting

This ignition source presents a severe hazard in the rubber manufacturing plant due to the constant presence of flammable liquids, combustible oils, and rubber materials. Because of the concentration of heavy machinery in many locations in the plant, daily welding and cutting is a usual practice. (The welding and cutting guidelines in NFPA 51B, *Cutting and Welding Processes*, should be followed to keep this ignition source controlled.)

THE SAFEGUARDS

Construction

Noncombustible construction should be used throughout the rubber plant with the roof conforming to the assemblies, and with installation strong enough to withstand the strongest wind having occurred in the past 100 years. In the usual rubber plant there is a significant amount of finished goods stored. These areas are separated from the manufacturing area by a 4-hour fire wall, with any openings protected by 3-hour fire doors. If the plant size is 500,000 sq ft (46 452 m^2) or more, it would be prudent to install another fire wall located between the raw materials/mixer area and general manufacturing. A wall at this location cuts off the storage area of raw materials which represents the greatest hazard at this end of the plant.

Any cement mixing room should be cut off from the other buildings area by a 4-hour fire wall or, preferably, be located 50 ft (15.3 m) from the major plant. If located in the plant, any openings in the walls common to the plant are to have 4-in. (102 mm) curbs, the floor is to be depressed equally, or grating and drains are to be placed at the doorways. Drains from this room or scuppers should be provided to carry the sprinkler water away. Explosion relief equal to 1 sq ft (929 cm^2) of relief to 50 cu ft (1.42 m^3) of volume should also be provided for the outside wall(s) with common interior walls constructed to withstand any explosion relieved through the exterior wall. (See NFPA 69, *Explosion Prevention Systems*, for more specific guidelines.)

If a flammable liquids storage room is necessary in the plant, its construction should be according to NFPA 30, *Flammable and Combustible Liquids Code*.

Fire Control Systems

Water supplies: A separate water supply is necessary for fire protection, and the basic quantity and pressure are determined by the greatest hazard in the plant, usually the height of finished-tire storage. If there is no finished-tire storage, the basic quantity may be determined by raw material storage or a deluge system protecting the cement mixing operation. For example, if there is 20-ft (6.1 m) onside tire storage, according to NFPA 231D, *Storage of Rubber Tires*, 0.60 gpm/sq ft [24.43 (L/min)/m^2] of water should be applied to the most remote 3,000 sq ft (279 m^2), assuming 286°F (141°C) sprinklers are used. The quantity is 1,800 gpm (6814 L/min) plus about 200 gpm (757 L/min) for imbalances in the hydraulically calculated area. After adding 750 gpm (2839 L/min) for hose streams, the total quantity of water expected to control a fire in a tire storage area is 2,750 gpm (10 410 L/min). This quantity can be supplied by a 2,000-gpm (7571 L/min) pump using its 150 percent factor, and the total quantity of water needed for the 3-hour reserve according to NFPA 231D is approximately 500,000 gal (1893 m^3). Figure 9.16 shows two 2,000-gpm (7571 L/min), 125-psi (862 kPa) horizontal centrifugal fire pumps — one driven by a diesel engine and the other driven by an electric motor. The redundant number of pumps and reservoirs for a rubber plant depends on the fire insurance carrier with respect to the value being insured.

If there is no major tire storage, and the greatest hazard is the raw material area with storage 13 to 14 ft (4 to 4.3 m) high, the density would be 0.35 [14.25 (L/min)/m^2] applied to the most remote 3,000 sq ft (279 m^2) or about 1,900 gpm (7192 L/min) including 750 gpm (2839 L/min) for hose streams. For a 2-hour water supply, as given by NFPA 13, "Ordinary Hazard Group III" (the category in which tire manufacturing falls), one would need approximately 250,000 gal (946 m^3). If tire storage is a

RUBBER PRODUCTS

FIGURE 9.16. Industrial fire pump installation. (Industrial Risk Insurers)

part of the rubber manufacturing operation, it is unlikely that a typical gravity tank would be of benefit because of the size and height that would be needed to meet the same demand of a pumping supply as given above. Generally, it is unusual to find a city water supply that will give 2,000 gpm (7571 L/min) or greater at 125 lbs (862 kPa) pressure; therefore, one cannot depend on a gravity tank or the city water supply to provide water needs at a tire manufacturing facility unless the quantity and pressure are adequate for the hazard.

Sprinkler protection: Sprinkler protection should be provided throughout the plant. Generally, the usual manufacturing area can be protected by an ordinary hazard schedule system as defined in NFPA 13, *Sprinkler Systems*. Exceptions to the ordinary hazard schedule are as follows:

1. For 13 to 14 ft (4 to 4.3 m) of raw material storage, the area should be hydraulically calculated for 0.35 gpm/sq ft [14.25 (L/min)/m^2] of floor area for the most remote 3,000 sq ft (279 m^2). For storage of 10 ft (3.05 m) or less, use the extra hazard schedule.
2. The cement mixing area should be hydraulically calculated for 0.35 gpm/sq ft [14.25 (L/min)/m^2] of floor area assuming all of the sprinklers are operating. The system can be a closed sprinkler or open deluge type.
3. If flammable liquids are stored in an inside storage room or an individual warehouse, an extra hazard schedule should be used; however, if there are multiple heights of containers anticipated, see the chapter on container storage in NFPA 30, *Flammable and Combustible Liquids Code*, for adequate protection.
4. In paint spray booths and post dipping of fabric, provide an extra hazard schedule of sprinklers even though there may be another type of fixed protection provided. Also provide a special indicating valve up-

stream of this protection in a location where it can be turned off easily without having the entire sprinkler system out of service.
5. Sprinkler protection for tire storage can be determined from the following chart reprinted from NFPA 231D, *Storage of Rubber Tires*. (See Figure 9.17.)

In addition to the aforementioned areas, it would be of benefit to reference some other equipment areas where special protection is warranted to reduce losses or to give commentary on some actual experiences.

Where there are enclosed-type carbon black conveyors, or a similar type, to handle coal, one should consider sprinklers to protect the rubber conveyor belt and, therefore, the structure. Without sprinklers on an ordinary schedule basis, a fire could go undetected until the equipment shuts down or until it is observed by personnel, and there are few employees in these locations.

There are three areas associated with the banbury mixing operation where special sprinkler protection should be considered:

1. Because of the special problem of rubber overheating on occasion and igniting on discharge from the banbury, a recessed sprinkler nozzle may be placed in the discharge chute controlled by a manual indicating valve near the operator's control panel. The main point here is to provide cooling water as soon as possible on the rubber at the point of discharge. It is also good to have a 1-in. (25.4 mm) preconnected water hose in the area in addition to backup of typical 1½-in. (38.1 mm) fire hose to allow quick water application.
2. A fire in the banbury may spread to the exhaust ventilation duct and, if not stopped in this area, is likely to spread to the dust collectors. By providing sprinklers on an ordinary schedule basis at 12½-ft (3.8 m) intervals for horizontal runs and at the turns in elevated runs, the fire is likely to be extinguished. Special indicating valves should be placed convenient to the operator but away from the duct work so as to close the valve immediately following the extinguishment of the fire.
3. The carbon black dust collectors should be protected as described in the NFPA's *Fire Protection Handbook*.

In addition to the sprinkler protection indicated for the cement mixing room, it is advisable to provide automatic carbon dioxide or other noncontaminating extinguishing mediums to extinguish the flash fires which may occur at the opening of the individual mixers and the exhaust ducts from the mixers. An automatic extinguishing system which applies the medium to all areas simultaneously is preferred, thereby preventing a flash fire from spreading to adjacent mixers.

It is also usual to provide the undertread cementer (roll-coating operation) with automatic carbon dioxide protection or other noncontaminating extinguishing mediums. When this automatic system operates, it

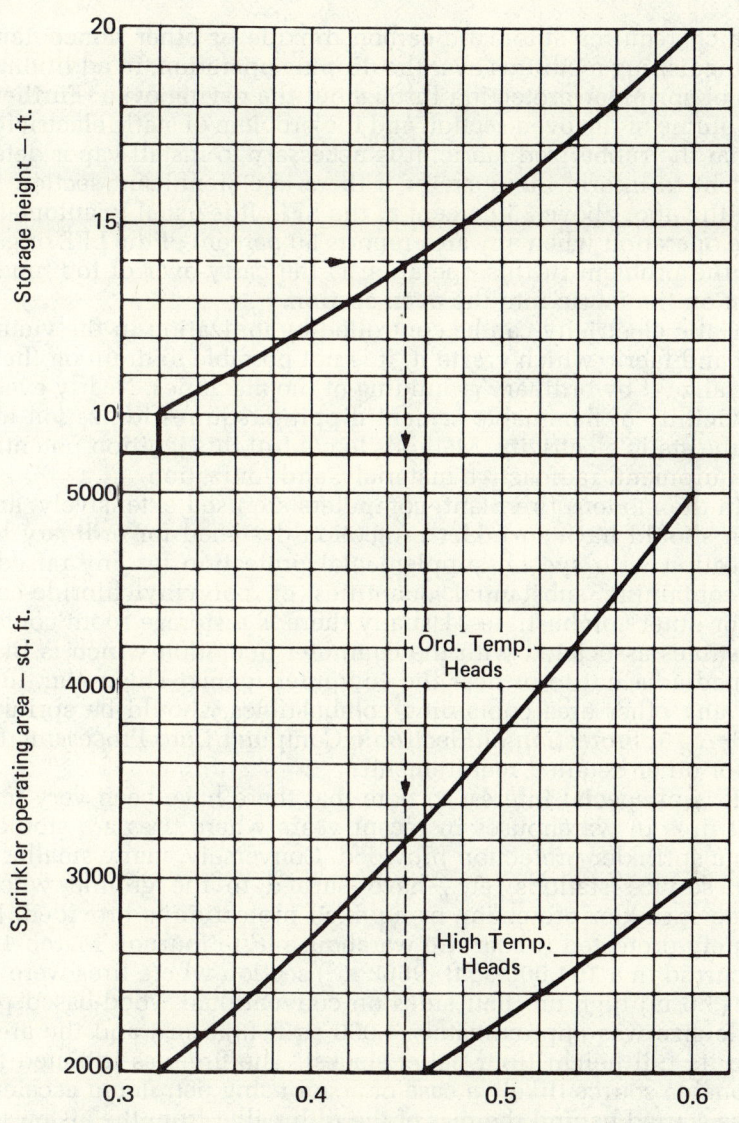

FIGURE 9.17. Sprinkler system design curves for palletized storage and fixed rack storage with pallets. To use curves, enter at storage height; read density then down to sprinkler operating area; 3,200 sq ft for ordinary heads, 2,000 sq ft for high temperature heads. [1 ft = 0.3 m; 1 sq ft = 0.93 m²; 1 gpm/sq ft = 40.746 (L/min)/m²]

should stop the flammable cement flow, shut down the conveyor, stop the exhaust fan, and close the damper in the exhaust duct.

Post dipping of rubberized fabric is a flammable cement application which requires automatic carbon dioxide or other noncontaminating extinguishing mediums over the dipping operation, in addition to the backup of sprinkler protection throughout the drying oven. Further, due to the volume of the oven section and the problem of static electricity created from the rubberized fabric, it is necessary to install vapor detection equipment to inform the operator if there is a significant section in the oven with vapor above 25 percent of the LFL. It is usual to automatically stop the operation when any area reports 50 percent of the LFL or greater. Part of the problem in this operation is the carry-over of too much dip solution on the fabric into the oven section.

Static electricity can be controlled by ionization in the vicinity of the roll and fabric which create it. It is not possible to drain off the static or neutralize it by ordinary grounding of the machines. Nearly every roll in the vicinity of flammable cement dipping requires ionization to neutralize the static electricity. Methods to control this ignition potential are X-ray equipment, radioactive materials, and ionization.

In the modern tire plant, computers are used extensively, and the location should have sprinkler protection designed for ordinary hazard classification with special supplemental protection for any raised floor space containing substantial quantities of polyvinylchloride-covered cables or other combustibles. Usually there is a storage room containing combustibles associated with the computer operation which is likely to be located adjacent to or near the computer rooms. This room, in addition to any other area containing combustibles, should be sprinklered. See NFPA 75, *Protection of Electronic Computer/Data Processing Equipment*, for other detailed requirements.

It is of special interest to note that there have been very few significant fires in warehouses in recent years where tires are stored with adequate sprinkler protection provided. Conversely, many smaller warehouses, service stations, etc., have burned to the ground where inadequate sprinkler protection or no fixed protection is provided. In one adequately protected rubber tire warehouse (*Fire Journal*, March 1974) a fire occurred in a 104,000-sq ft (9662 m^2) section where tires were stored 18½-ft (5.6 m) high on their sides on conventional wood-based pallets. The pile size was approximately 4,000 sq ft (372 m^2) and the area was used to its full height (four-pallet stacks). The fire was initiated from a questionable source (likely a case of arson using petroleum accelerators) and was started behind the rear of the fifth pallet from the aisleway near a fire wall. Also, the location was one most remote from the sprinkler system, thus creating a real test case of whether or not the sprinkler system could control the fire. The word control is used appropriately since it is unlikely that sprinklers will extinguish a rubber-tire fire due to the void spaces that retain air and shed the water. Seven sprinklers operated to control the fire and continued to operate for 3 hrs while the volunteer fire department with the aid of industrial trucks removed the tires in the area to expose the fire. As stated in the appendix to NFPA 231D, *Storage of Rubber Tires*, it is mandatory to allow the sprinklers to continue their operation while the overhauling is done.

The density of the protection for this tire storage was 0.48 gpm/sq ft [19.5 (L/min)/m^2] for the most remote 2,400 sq ft (223 m^2). This design results from tests that were conducted in 1955. However, after the fire it was calculated that the first sprinkler which operated produced approximately 0.85 gpm/sq ft [34.6 (L/min)/m^2] of floor area and, even with seven sprinklers and three 1½-in. (38.1 mm) hose streams, the density continued at a high level: 0.74 gpm/sq ft [30.1 (L/min)/m^2].

The total loss occurring in this fire was approximately $80,000, 62 percent of which was for tire cleanup and about 10 percent for building damage.

Fire Alarms

Fire alarms in the form of public address systems are common in the rubber factory today. The system doubles as a routine communications network using a special signal to identify a fire alarm. To initiate an alarm, wall telephone sets are installed according to NFPA 72C, *Remote Station Signaling Systems*, throughout the factory.

Fire Brigades

Fire brigades are composed of 16 to 24 employees using a small inside fire truck to convey necessary equipment to the fire. In some plants the area is so great that it is preferred to have two functioning fire brigades, each responsible for approximately one-half of the plant. If the first brigade is in need of temporary help before the arrival of the public fire department, it would initiate a second alarm for the other fire brigade. The reliability of the automatic sprinkler systems installed in the plants is the major factor in confining fire losses, often making it unnecessary for the public fire department to respond. The areas most respected in a tire plant by fire fighters are the high-piled storage of tires and the areas where flammable liquids are stored and used. (See Chapter 3, "Plant Emergency Training and Organization.")

BIBLIOGRAPHY

NFPA CODES, STANDARDS, AND RECOMMENDED PRACTICES

Reference to the following NFPA Codes, Standards, and Recommended Practices will provide further information on the safeguards for the manufacturing operations discussed in this chapter. (See the latest *NFPA Codes and Standards Catalog* for availability of current editions of the following documents.)

NFPA 13, *Installation of Sprinkler Systems.*

NFPA 13A, *Care and Maintenance of Sprinkler Systems.*
NFPA 30, *Flammable and Combustible Liquids Code.*
NFPA 51B, *Fire Prevention in Use of Cutting and Welding Processes.*
NFPA 58, *Liquefied Petroleum Gases.*
NFPA 69, *Explosion Prevention Systems.*
NFPA 72C, *Remote Station Protective Signaling Systems.*
NFPA 75, *Protection of Electronic Computer/Data Processing Equipment.*
NFPA 231, *Indoor General Storage.*
NFPA 231C, *Rack Storage of Materials.*
NFPA 231D, *Storage of Rubber Tires.*

ADDITIONAL READING

Cotton, P. E. and Newman, R. M., "Fire Tests of Palletized and Racked Tire Storage," FMRC Serial No. 19037, July 1970, Factory Mutual Research Corporation, Norwood, MA.

"Designing to Limit Loss in High-Rise Rack Warehouses," Kemper Group Report, vol. 10, no. 3 (September 1981, pp. 2-9.)

Goring, G., "Sprinkler Protection of Storage Risks," *Fire Protection*, vol. 8, no. 2 (June 1981), pp. 20-25.

Harrington, J. C. and R. B. Hopkinson, *Rack Storage Protection*, Worcester Polytechnical Institute, Worcester, MA, 1977.

McKinnon, G. P., ed., *Fire Protection Handbook*, Fifteenth Edition, National Fire Protection Association, Quincy, MA, 1981.

Morton, Maurice, *Rubber Technology*, (2nd ed), Krieger Publishing, Co. Inc., Melbourne, FL 1981.

Nash, P., "A Non-Electric Resetting Zoned Sprinkler System for High Racked Storages," Fire Prevention Science and Technology, (18), 1978, pp. 14-20.

"Prevent Cutting/Welding Fires," *The Minnesota Fire Chief*, vol. 16, no. 5 (May/June 1980), pp. 16, 48, 54.

Proudfoot, E. N., "Sprinklers Control High-Piled Tire Warehouse Fire," *Fire Journal*, vol. 68, no. 2, March 1974, pp. 70-72.

Rubber Tire Manufacturing, (IES Committee Reports Series), Illumination Engineering Society of North America, NY, 1969.

U.S. Department of Commerce, *1981 Annual Survey of Manufacturers, Value of Product Shipments*, M82(AS)-2, Bureau of the Census, Washington, DC, May 1983, pp. 16-17.

Young, R. A. and P. Nash, "The Fire Protection of Modern High Bay Storages," *Fire Prevention Science and Technology*, (18), 1977, pp. 4-13.

10

Mining

William H. Pomroy

The mining of coal, metallic and nonmetallic mineral ores, sand and gravel, and stone is a significant component of our national economy. Our modern industrial society depends heavily on mined raw materials for fuel, manufactured goods, chemicals, and construction materials. The relative abundance and low cost of our mineral resources contribute materially to our high standard of living. Mineral production also directly impacts national security, as many minerals are considered critical and strategic for defense purposes.

The United States is the world's largest mineral producer. Mineral production and employment are summarized in Table 10.1.

Fires and explosions pose an everpresent threat to the safety of miners and to the productive capacity of the nation's mines. Historically, some of our most tragic industrial disasters resulted from fires and explosions in underground mines. More recently, advances in mining technology and safety have cut losses to a tiny fraction of their former levels. But fires and explosions continue to occur, demanding constant vigilance to minimize losses.

William H. Pomroy is Group Supervisor, Mine Fire Safety Group, Twin Cities Research Center, US Department of the Interior, Bureau of Mines.

TABLE 10.1. U.S. Mineral Production and Employment

	Production (000,000's tons*)	Value ($000,000)	Employment (000's)
Coal	780.9	27,331	243
Metallic ores	674.0	9,799	94
Nonmetallic ores	846.4	11,426	47
Sand and gravel	694.0	1,956	40
Stone	875.3	3,388	88
Total	3,874.6	53,900	512

Sources: U.S. Department of the Interior, U.S. Department of Labor and U.S. Department of Energy

*one million short tons equals 907 200 metric tons.

This chapter discusses the mining methods and equipment, fire and explosion hazards, and safeguards involved in the extraction of coal, metal and nonmetal ores, sand and gravel, and stone. The processing, milling, and refining of mined products are not covered.

METHODS AND EQUIPMENT

Mining can be broadly grouped into two categories: surface mining and underground mining. Surface mining requires the removal of the overlying dirt and rock strata (overburden) before the desired minerals are excavated. Underground mining permits the selective extraction of desired minerals with minimal disturbance to overlying strata.

Since surface mining requires the removal of all overburden before the desired mineral is mined, surface mining is feasible only if the amount of overburden relative to the amount of recoverable mineral (stripping ratio) does not exceed a predetermined economic limit. Where deposits are too deeply emplaced, the amount of overburden becomes excessive, necessitating extraction by underground methods.

The economics of large scale production and mechanization strongly favor surface mining. Over 95 percent of all mine production is accomplished by surface methods.[33] Other mining methods such as "in situ" mining and "deep seabed" mining have been demonstrated on a research scale and in limited commercial applications. However, production by these methods is negligible compared to conventional surface and underground methods.

Surface Mining Methods

Numerous surface mining methods have evolved in response to varying surface and subsurface geological conditions, and mining equipment innovations.

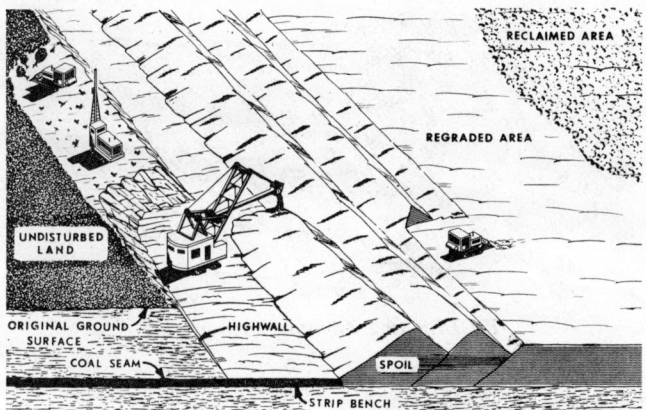
FIGURE 10.1. Area strip mining.

Methods of surface coal mining, or strip mining, include area strip and various forms of contour mining.[3,5] Area strip mining is practiced where the terrain is relatively flat and the coal seams are of moderate thickness—3 to 30 ft (0.91 to 9.1 m). It begins with the removal of overburden from above the coal seam in a roughly linear strip, usually 100 to 170 ft (30 to 52 m) wide and up to several miles (1 mile equals 1.6 km) long. Overburden depths of 60 to 120 ft (18 to 37 m) are common. As coal is uncovered, it is loaded into trucks for haulage to storage or processing facilities. When an entire strip has been mined, the process is repeated, with the overburden above the adjoining strip of coal cast onto spoil piles in the previously mined cut (see Figure 10.1). The rock overburden and coal are generally blasted prior to removal. Huge stripping machines are most often used to remove overburden. Where large stripping machines are not used, the overburden is removed by smaller draglines or shovels which load the material into haulage trucks. For coal seams between 50 and 200 ft (15 to 61 m) thick, specialized mining systems more closely resembling metal and nonmetal open pit methods have been developed.

Contour strip mining is most common where flat-lying coal seams occur in rolling or mountainous terrain. In contour mining, an initial cut is established along a hillside at the point where the coal seam is exposed, or outcrops. Successive cuts are then made into the hillside until an economic stripping limit is reached. Mining then proceeds laterally along the hillside, following the outcrop (see Figure 10.2). Numerous variations of this basic technique have been developed in response to recent regulations aimed at controlling the adverse environmental impact of contour strip mining. Included are the slope reduction method, parallel fill, box cut method, head-of-hollow fill, block cut, and mountain top removal. Contour mining follows the same drill-blast-strip-load-haul cycle as area strip mining, but the equipment employed is generally much smaller.

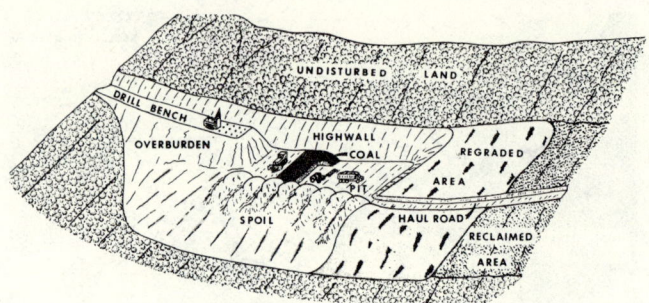

FIGURE 10.2. Contour strip mining.

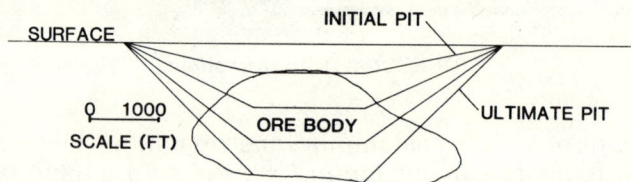

FIGURE 10.3. Open pit development.

Most surface metal and nonmetal mining is by the open pit or open cast method.[24] The overburden is first stripped to expose the top of the orebody, then stripping and mining proceed concurrently, the pit growing ever larger and/or deeper, until an economic pit limit is reached (see Figure 10.3). Pit geometries vary due to the irregular shape and orientation of orebodies, but pits of ½ to several miles (1 mile equals 1.6 km) across and 100 to over 1000 ft (0.03 to 0.30 km) deep are common (see Figure 10.4). Both overburden and ore must be drilled and blasted prior to removal. Excavation is generally by shovel or loader. Truck haulage is most common, but pit railroads and belt conveyors are sometimes utilized.

Specialized mining systems have been developed for certain metal and nonmetal ore deposits. Examples are dragline stripping and hydraulicking of phosphates, and dredging of alluvial beach sands, gold and tin.

Sand and gravel are generally mined with smaller equipment from shallow pits and small surface excavations or with draglines and dredges from natural or artificial ponds. Drilling and blasting is seldom required. After washing and sizing, sand and gravel are frequently loaded into over-the-road trucks for transport directly to the end user.

Crushed stone and dimension stone are mined from rock quarries. Stone for crushed aggregate is generally mined by conventional open pit methods. Dimension stone is removed in large blocks with the use of wire saws, channeling machines, or similar means.

Other than maintenance facilities, few fixed structures are directly or indirectly involved in the surface mining process. Rather, surface mining is performed out-of-doors by specialized mobile and portable mining

FIGURE 10.4. Open pit mine.

equipment. The major equipment types are described in detail in the following sections.[18]

Blasthole drills: Blasthole drills vary in size and design to accommodate varying mining conditions, rock types, production rates, blasting practices, and other factors. The smallest drills, generally used for small quarry blasting, are pneumatically operated, mounted on lightweight mobile carriages, and produce holes up to about 2 in. (51 mm) in diameter. The largest drills varying mining conditions, rock types, production rates, blasting practices, and other factors. The smallest drills, generally used for small quarry blasting, are pneumatically operated, mounted on lightweight mobile carriages, and produce holes up to about 2 in. (51 mm) in diameter. The largest drills are enclosed and self-contained, diesel or (more commonly) electrically powered, and capable of drilling holes over 20 in. (0.51 m) in diameter (see Figure 10.5).

Stripping machines: Stripping shovels, draglines, and bucketwheel excavators are used to strip overburden from above relatively shallow — up to 150 ft (46 m) deep — flat-lying deposits, usually coal (see Figure 10.6). Stripping shovels and bucketwheel excavators operate from the bottom of the cut while draglines operate from atop the highwall. The machines are generally quite large, moving up to 6,500 cu yd (4970 m^3) of material per hour over spans of up to 400 ft (122 m). Shovel and dragline bucket capacities range to 180 and 220 cu yd (138 and 168 m^3) respectively. The size of the main machinery deck can exceed 5,000 sq ft (46 m^2) and dragline booms over 300 ft (91 m) are common. Unlike shovels and draglines which operate on a dig-swing-dump-swing cycle, bucketwheel excavators operate continuously. The rotating bucketwheel feeds belt convey-

Chapter 10

FIGURE 10.5. Blasthole drill.

ors which extend the length of the machine and deposit overburden directly onto the spoil piles. Bucketwheel excavators can span 300 to 400 ft (91 to 122 m) from the bucketwheel to the end of the discharge conveyor. Stripping shovels, draglines, and bucketwheel excavators are electrically powered, with the electricity supplied through heavy cables from nearby power substations.

Loading shovels: Loading shovels are used to load blasted overburden, coal, and ore into trucks or rail cars for transport from the working face (see Figure 10.7) Shovels are either diesel or electrically powered and bucket capacities range to over 40 cu yd (31 m^3).

Haulage trucks: Haulage trucks are the primary means of materials haulage in surface mining (see Figure 10.8). Haulage trucks generally represent the single largest component of a mine mobile equipment fleet. Off-highway trucks are most common; however, on-highway trucks are frequently used at contour strip coal mines and sand and gravel operations. Off-highway trucks range from 35 to 350 tons (32 to 318 metric tons) capacity with trucks in the 50 to 170 ton (45 to 154 metric ton) range the most common. Motive power is supplied by diesel engines ranging from 500 to 3,500 horsepower. Drive systems are either mechanical or electrical. Trucks under 85 ton (77 metric tons) capacity employ mechanical transmissions. Trucks in the 85 to 120 ton (77 to 109 metric tons) capacity range are available with either mechanical or electric drive. Most trucks over 120 ton (109 metric tons) capacity are electric

FIGURE 10.6. Stripping shovel.

FIGURE 10.7. Mining shovel.

drive. In electric drive vehicles the diesel engine powers an alternator which produces electricity to drive large wheel motors.

Wheeled loaders: Wheeled loaders, or front-end loaders, are rubber-tired, diesel powered, highly mobile machines which can perform a variety of functions (see Figure 10.9). They are most commonly used for loading overburden, coal, or ore at a working face, stockpile reclaiming, general cleanup and utility duties, and load-haul-dump applications where they act as both a loading and haulage means. Most wheeled loaders are mechanically driven; however, electric drive vehicles are available in the larger size ranges. Bucket capacities range to 36 cu yd (27.5 m^3); how-

Chapter 10

FIGURE 10.8. Haulage truck.

FIGURE 10.9. Wheeled loader.

ever, machines in the 8 to 12 cu yd (6 to 9 m^3) range are more common. Engines range to 1,400 horsepower.

Dozers: Dozers are diesel-powered wheeled or (more commonly) tracked machines equipped with a large blade for pushing material (see Figure 10.10). Primary applications include reclamation (leveling coal mine spoil piles is typical) and general cleanup.

A wide variety of miscellaneous equipment is required at most mines to support the actual mining process. Included are scrapers, explosives haulers, graders, water trucks, pumps, power substations, generators, crushers, conveyors, cranes, backhoes, and maintenance and utility vehicles.

MINING

FIGURE 10.10. Tracked dozer.

Underground Mining Methods

Access to underground workings from the surface is by shaft, slope, or adit. A shaft is a vertical or nearly vertical opening, usually equipped with hoisting facilities for transporting personnel, supplies, and/or mined products. A slope (also incline or decline), is an inclined opening which may be designed to accommodate a belt conveyor for materials haulage, a hoist for personnel and supplies, or diesel haulage. An adit mine (or drift mine) is entered through a horizontal opening in a hillside. An individual mine may employ any or all of these means of access, depending on the depth and orientation of the coal seam or orebody and the requirements of the mine production system.

Where a single, uniform, relatively flat-lying coal seam or orebody is worked, the entire mine is generally confined to a single working level. Where multiple, irregularly shaped, and/or steeply inclined coal seams or orebodies are worked, mining may be spread over many levels. Large mines may comprise hundreds of miles of drifts or entries (horizontal openings in metal or coal mines, respectively) on many levels interconnected with shafts, slopes, raises (vertical openings between levels), and ramps (inclined openings between levels).

The predominant constraint influencing the selection of an underground mining method is the degree to which the opening formed when coal or ore is extracted must be supported to avoid unwanted failure of the roof and/or walls. Factors related to this constraint are the physical character, size, shape, orientation, and depth of the coal seam or orebody and the characteristics of the overlying strata. In addition, planned

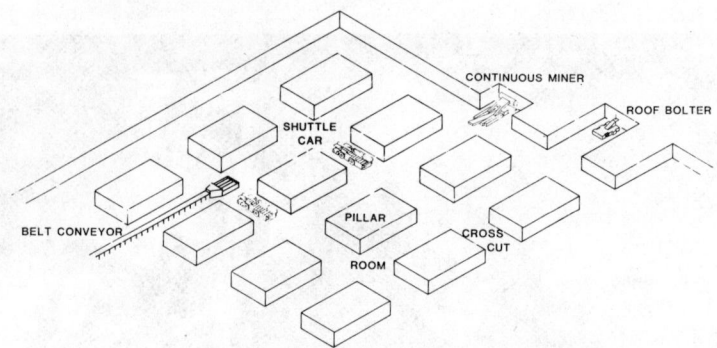

FIGURE 10.11. Room and pillar coal mining.

production capacity, and the presence of water, methane, high rock temperatures, radon, or other environmental factors may be important considerations.

Since nearly infinite variability exists among these factors from deposit to deposit and even within a single property, no two mines share precisely the same operating practices. However, the broad spectrum of underground mining methods currently in use can be effectively subdivided into three generic classifications. Included are methods producing openings which are naturally supporting (or requiring minimal artificial support), methods which provide for substantial artificial support of the roof, and caving methods where failure of the roof is integral to the mining process.

With few exceptions, coal mining systems are profoundly different from noncoal mining systems, and they are treated separately in the following sections. Where similarities do exist, they are limited to the softer, bedded nonmetallics such as potash, salt, and trona which are amenable to coal mining systems and equipment.

Underground coal mining:[3] The two basic coal mining methods are room and pillar, producing openings which are naturally supporting; and longwall, which requires substantial artificial support. Room and pillar mining, as its name implies, involves the extraction of coal in a regular pattern of rooms separated by pillars of in-place coal to support the roof (see Figure 10.11). The size and shape of the rooms and pillars depend on the stability of the coal and roof, the thickness of the seam, which is generally 3 to 7 ft (0.9 to 2.1 m), and rock pressures. Room widths of 15 to 25 ft (4.6 to 7.6 m) and pillar dimensions of 30 by 30 ft (9.1 by 9.1 m) to 40 by 80 ft (12.1 by 24.3 m) are typical.

The objective of a room and pillar mining plan is to extract as much coal as possible without risking roof collapse or pillar failure. Generally, about 40 to 60 percent of the available coal is mined, with pillars accounting for the remainder. Where conditions permit, pillar mining is practiced. The coordinated mining of pillars and controlled caving of the roof can result in overall recovery of 85 percent of the available coal without jeopardizing the safety of miners.

Two systems are presently used for room and pillar coal mining. In conventional mining, the coal is extracted in a sequence of operations, with specialized equipment required to execute each step. Using a machine resembling a large chain saw, a slot is cut in the coal face the entire width of the room along the floor, center, or roof line. Next, a drill is used to bore holes in the coal face for explosives. The coal is broken by the explosives and gathered by a loading machine which feeds a self-propelled shuttle car or it may be loaded and hauled by a self-propelled scoop. In either case, the coal is discharged onto a conveyor belt or into rail cars for transport to the surface or to hoisting facilities.

In continuous mining, a single machine, called a continuous miner, is used to mechanically cut the coal from the face and load the broken coal into a shuttle car.

After each drill/blast/load cycle in conventional mining or each machine advance in continuous mining, the mining equipment is withdrawn and the roof is secured by a machine called a roof bolter. Holes are drilled into the roof and long bolts inserted and anchored to strengthen the roof span and prevent roof failure.

Longwall mining involves the extraction of coal in a nearly continuous operation using a specialized and integrated mining and roof support system (see Figure 10.12). Using standard room and pillar techniques, longwall panels 300 to 600 ft (91.4 to 183 m) wide by 3,000 to 6,000 ft (914 to 1828 m) long are prepared. Each panel is mined in linear slices at full seam height by a mechanical cutting machine or plow that is drawn back and forth across the 300 to 600 ft (91.4 to 182.8 m) face. As the coal is cut, it is forced onto a chain conveyor which extends the entire length of the face. The roof immediately adjacent to the face is supported by large, self-advancing hydraulic jack units (chocks or shields). As the roof support line advances with the face, the roof behind it is permitted to cave.

Shortwall mining, a modification of longwall mining, involves the use of a continuous miner, under the support of shocks or shields, to extract coal along a 150 to 300 ft (45.7 to 91.4 m) face. Shuttle cars or a flexible conveyor are used to transport coal from the miner to a belt conveyor in an adjacent entry.

Most coal mining equipment is electrically powered. Undercutting machines, drills, loading machines, continuous miners, shuttle cars, scoops, and roof bolters are generally supplied power through a trailing cable. Belt conveyors are generally wired directly to a permanent or semi-permanent power center, as is longwall equipment. Self-propelled rail mounted equipment is generally supplied power from an overhead trolley wire. Battery powered scoops, personnel transports, and utility vehicles are also common. Some mobile haulage equipment, notably scoops, load-haul-dump (LHD) vehicles and rail locomotives are diesel powered. Though diesel usage is growing, it is not yet extensive in underground coal mining.

Chapter 10

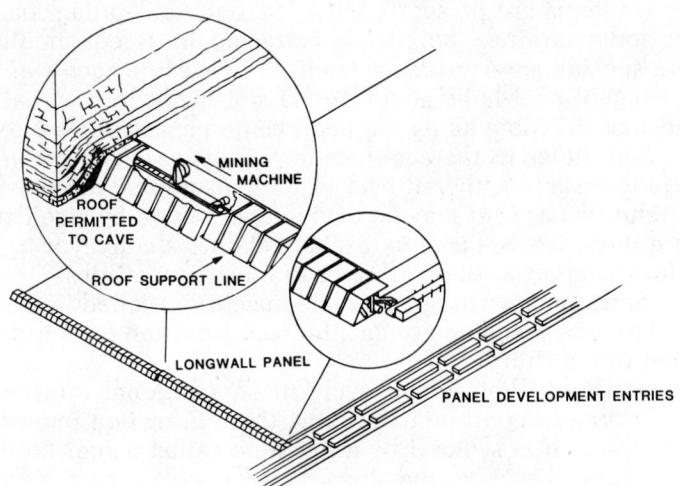

FIGURE 10.12. Longwall coal mining.

Numerous ancillary operations must be coordinated with the mining process. The main haulage network, whether rail or belt, must function smoothly to prevent materials handling bottlenecks from interfering with production. The roof in active areas is constantly monitored for failures or signs of impending failure. In heavily traveled areas, such as main haulageways and important ventilation routes, heavy timbers or steel arches are often installed to help control the roof. Water supply, power distribution, and ventilation systems require periodic attention, and equipment and facilities inspection, maintenance and repair are continuous. Dust control also receives high priority, both to protect miner health and to prevent explosive conditions.

Underground metal and nonmetal mining:[4,12] Underground metal and nonmetal mining methods are more diverse than underground coal mining methods because there is greater variability in the physical character, size, shape, and orientation of metal and nonmetal orebodies than most coal seams. Room and pillar and sublevel stoping methods are used where minimal artificial roof support is required in the extraction areas. Shrinkage stoping, cut-and-fill, and square set methods are used where the roof and walls in the extraction areas require substantial support. The two most common caving methods are sublevel caving and block caving.

The room and pillar method is used for relatively flat-lying deposits of moderate thickness — up to about 90 ft (27.4 m) — with a competent roof. The other methods are used for deposits which are vertical or steeply inclined, or deposits which extend hundreds or thousands of feet (1,000 ft equals 305 m) in all directions. Simplified descriptions of these methods are provided in the following sections. In actual practice, however, adaptions to these basic methods are necessary to suit local conditions.

Room and pillar mining of metal and nonmetal ores (with the exception of the softer, bedded nonmetallics) is generally characterized by larger openings than typical room and pillar coal operations. Rooms of 30 to 50 ft (9.1 to 15.2 m) wide or more and heights greater than 15 ft (4.6 m) are common, permitting use of larger, more productive mining equipment. The ore is first drilled for blasting using diesel or electrically powered multi-boom "drill jumbos" equipped with pneumatic or hydraulic rotary or rotary percussive rock drills. Following blasting, the roof is scaled (rock loosened by blasting is removed) and bolted. Diesel powered LHDs or wheeled loaders (often identical to units used in surface mines) then load the broken ore, or "muck," into diesel haulage trucks. The ore is transported by truck from the working face to an underground crusher.

Sublevel stoping, used in vertical or nearly vertical orebodies of highly competent rock, requires considerable premine development to prepare the orebody for production mining. The development work consists of driving a network of drifts, raises, and draw points to facilitate access, mining, ventilation, and retrieval of broken ore. Development work generally follows a conventional drill/blast/load/scale/bolt cycle, with advances of 5 to 15 ft (1.5 to 4.6 m) per cycle typical. Smaller equipment, such as jacklegs and stopers (portable, manually operated pneumatic rotary percussive rock drills) and overshot loaders (small, pneumatic, rail, crawler or rubber tire mounted loading machines), are used for most development work; however, diesel drill jumbos and LHDs are often used in openings greater than 8 ft by 8 ft (2.4 by 2.4 m) cross section. Mechanical boring machines are frequently used for driving long raises.

Sublevel drifts along the longitudinal centerline of the orebody are driven in a vertical column between the mine's main levels (see Figure 10.13). On the main levels, or haulage drifts, draw points perpendicular to the sublevel drifts are driven into the orebody. A complete radial pattern of blastholes or parallel rows of vertical blastholes are then drilled at regular intervals along the entire length of each sublevel drift. Each ring or row of holes is individually loaded with explosives and blasted. Every blast results in a vertical slice of ore falling to the bottom of the open cavity, or stope, where it is loaded out through the draw points. If LHDs are used to retrieve ore from the draw points, they also tram the ore to a central ore pass system which connects all the main levels to an underground crusher. If overshot loaders or slushers (a scraper blade suspended on a cable between a hoist and a pulley which gathers broken ore as it is dragged over a muck pile) are used in the drawpoints, other means for ore haulage between the draw points and ore pass, usually rail cars, must be provided.

If the ore or surrounding country rock are too weak or rock pressures too great, open stopes are not feasible. Such orebodies require ar-

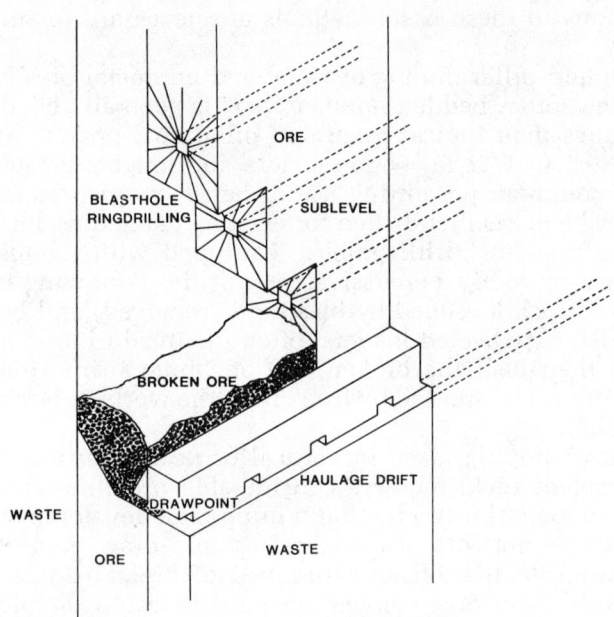

FIGURE 10.13. Sublevel stoping.

tificial support of the stopes during mining. Shrinkage stoping, cut and fill, and square set stoping are the most common methods employed to extract such orebodies.

Shrinkage stoping requires very little preproduction development. Draw points are driven into the orebody at regular intervals along the main levels and raises are driven in the ore between levels (see Figure 10.14). The ore is extracted in horizontal slices, starting at the bottom of the stope and advancing upward. The broken ore in the stope provides a work platform for miners and supports the walls of the stopes. Since blasted ore occupies a greater volume than in-place ore, a portion of the broken ore must be drawn off after each slice is mined to maintain a suitable working height between the top of the broken ore and the in-place ore above (back). The process of mining horizontal slices and drawing excess ore is repeated until the upper limit of the stope is reached. The stope is then emptied of ore and refilled with waste rock if needed. Ore may be loaded out from the drawpoints by LHDs, overshot loaders, or slushers.

Preproduction development for cut and fill mining is similar to shrinkage stoping except that only one or two draw points are required (see Figure 10.15). Mining is similar as well, with the orebody mined in horizontal slices from bottom to top. However, the stope is emptied of broken ore after each slice (cut), and the stope filled with waste material before mining resumes. The fill provides a work platform for miners and

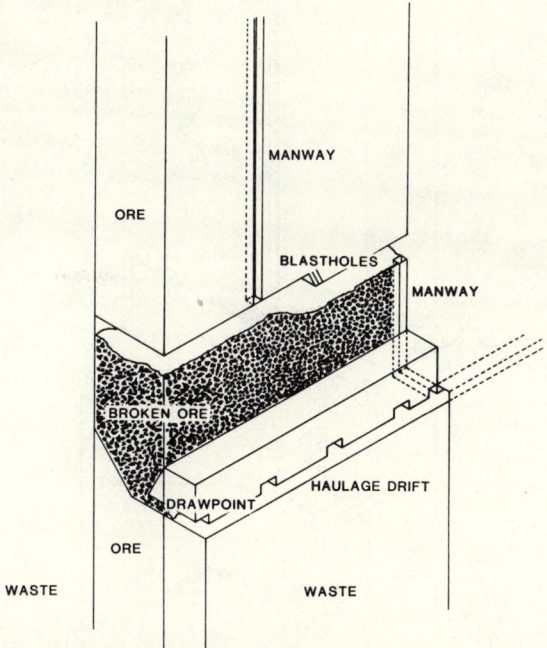

FIGURE 10.14. *Shrinkage stoping.*

supports the stope walls. Waste rock from mine development work or quarry stone is sometimes used as fill; however, hydraulically placed sand fill is more common. The sand fill, generally a waste product from the mine's concentrator, is slurried with water and piped directly into the stope. The drained fill has excellent support properties and provides a smooth work surface. If the fill is mixed with cement, its support properties are enhanced and the work surface is harder and more durable. Ore is removed from the stope through timber or steel ore chutes imbedded in the fill. Additional stope accessways (manways, ladderways) can be constructed in a similar manner. In smaller stopes, slushers are commonly used to load out the ore and jackleg or stoper drills are used to drill holes for explosives and roof bolts. In larger stopes, LHDs are commonly used to load and haul the ore to the ore chutes and multiboom pneumatic or diesel hydraulic drill jumbos are used for drilling blastholes. The ore is generally hauled in rail cars from the draw points to the ore pass on the main level.

Preproduction development and mining in square set stopes is identical to cut and fill mining. Support in the mined-out portion of the stope, however, is provided by an elaborate network of timber sets (see Figure 10.16). Each timber set consists of a vertical post and two horizontal members at mutual right angles. Successive sets are interlocked to form a complete cellular timber support structure which conforms to the irregular shape of the stope. The space between the timbers normally is

Chapter 10

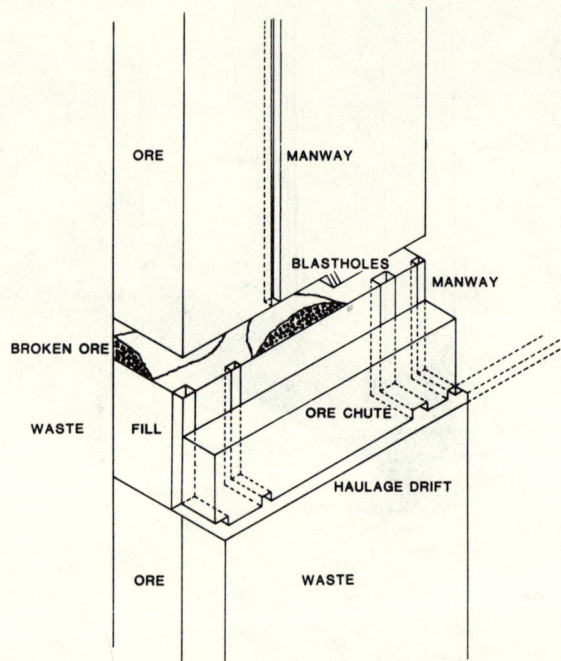

FIGURE 10.15. Cut and fill mining.

filled with waste rock, leaving only limited openings which serve as drifts, manways and ore chutes. Slushers are generally used for ore handling in the stope and haulage on the main levels is generally in rail cars. Extraction of the ore is an entirely manual operation and handling the timbers and erecting the supports are laborious and costly operations. Hence, this mining method is suitable only for small bodies of high grade ore. Although square set mining was common in years past, it is seldom practiced today.

    Caving systems are characterized by a high degree of mechanization, high production capacity, and low production costs. In sublevel caving, preproduction development is similar to sublevel stoping, with sublevel drifts driven in the ore at 30 to 50 ft (9 to 15 m) intervals (see Figure 10.17). Draw points on the main levels are not required, however, as the ore is retrieved through the sublevel drifts. Fan shaped blasthole patterns are drilled upward at regular intervals along the entire length of each sublevel drift. The drill fans are blasted individually, with the broken ore caving into the sublevel drifts. LHDs are used to load the ore and tram it to the ore pass system, which is accessible from the sublevels. The rock overlying and adjoining the orebody is also permitted to cave, and some dilution of the broken ore with waste rock is inevitable. When this dilution reaches an economic limit at a given muck pile, the next drill fan is blasted.

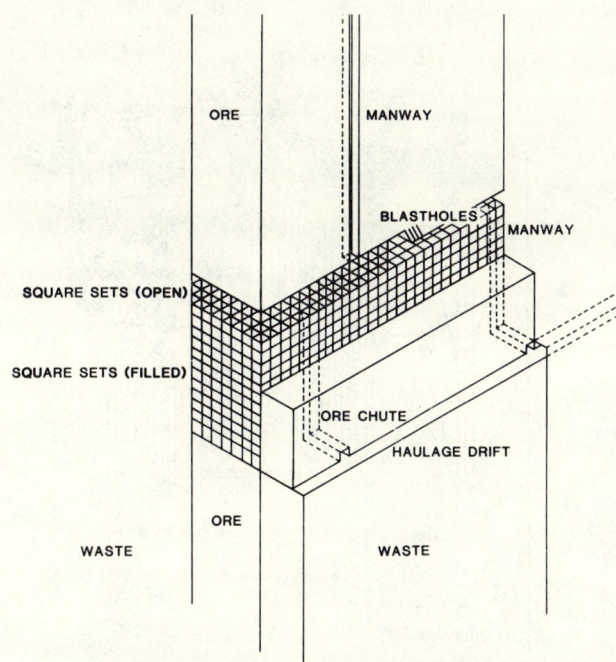

FIGURE 10.16. Square set mining.

Block caving has the lowest production cost of all underground methods. Preproduction development is extensive, however, and the method is suitable only for very large deposits. The deposit must have enough horizontal area to cave freely, and under ideal conditions, the ore should part easily from the sidewalls. Block caving in flat-lying deposits is feasible only if the thickness of the orebody is sufficient to warrant the cost of mine development. Minimum thickness is usually about 100 ft (30.5 m), but mining cost is the final determining factor. Wide veins, thick beds, or massive deposits of homogeneous ore are generally the most desirable for block cave applications.

The principle of block cave mining is the natural fracturing and caving of ore which has been undercut over a large area. A minimum of drilling and blasting during the ore extraction phase is required, because the ore fractures itself (a result of gravity forces acting on the undercut rock masses).

Preproduction development is in five steps (see Figure 10.18). First, a network of haulageways is developed beneath the ore horizon. Next, finger raises are driven up to a set of slusher drifts. From the slusher drifts, a second set of finger raises is driven, gradually widening to form funnels at the undercut level. Finally, the entire ore block is undercut.

The weight of the overlying ore and overburden creates enormous

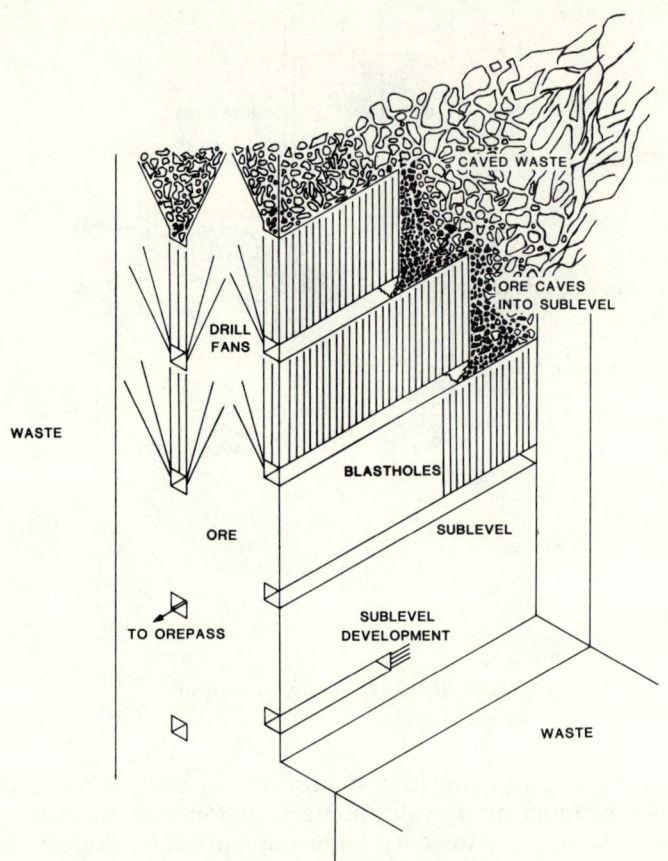

FIGURE 10.17. *Sublevel caving.*

stresses in the ore at the bottom of the block, causing it to fracture and cave into the funnels. Slushers are used to load ore from the upper fingers into the lower fingers leading to the haulage level. Haulage is generally by train.

One common variant of this system involves the substitution of drawpoints and LHDs for the upper fingers and slushers. In this system, LHDs load and tram ore from the drawpoints to ore passes leading to the haulage level.

As with underground coal mining, the ore extraction process is supported by a vast array of ancillary operations. Included are mining, mine development, and haulage equipment maintenance, which at most mines is performed in large, underground shops. Mines relying on diesel equipment also must provide facilities for storage and/or transfer of diesel fuel. In addition, power distribution, ventilation, hoisting, roof con-

MINING

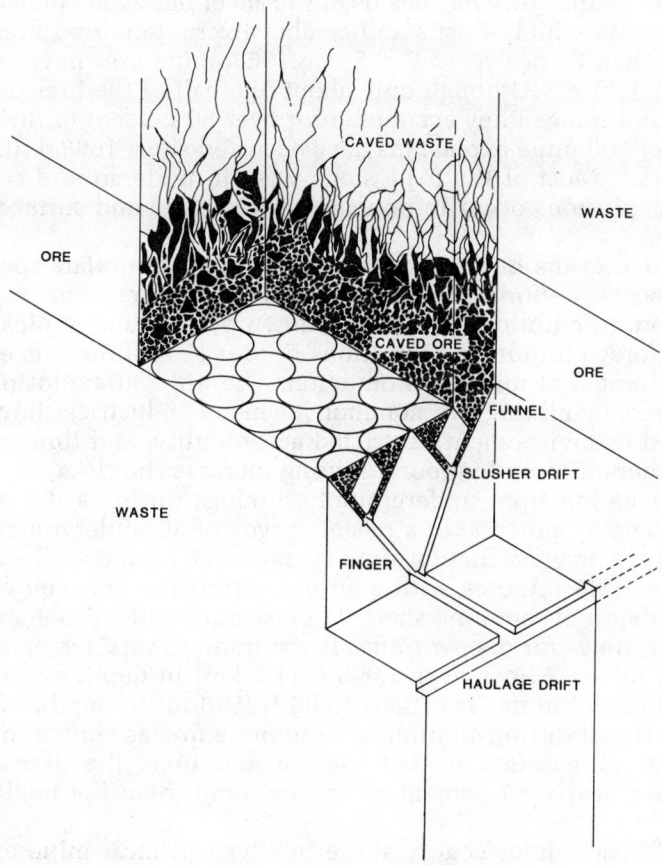

FIGURE 10.18. Block caving.

trol, and dewatering equipment require frequent inspection, maintenance and repair.

THE FIRE AND EXPLOSION HAZARDS

During the 1960s and 1970s, about 50 fires per year in underground and surface mines, pits, and quarries were reported to federal mine safety authorities. Nonreportable fires, those lasting less than 30 minutes and not resulting in an injury, were estimated at another 200 per year. In the mid-1970s, a steady decline in the fire frequency was observed, probably due to the adoption of improved mining and safety technology (roof bolts for roof support, AC powered face equipment in coal mines, etc.). This downward trend, however, has begun to level off, as is characteristic of frequency data which must stabilize above zero. Reported fires now average less than 20 per year.[2,19,20] Since 1968, mine fires have resulted in over 150 fatalities. Although only about one half of the fires occurred in underground mines, they accounted for over 80 percent of the fatalities.

Over 100 mine explosions have occurred since 1968, killing nearly 200 miners.[28] Most of the explosions were in underground coal mines, but fatal explosions occur in noncoal underground and surface mines as well.

Added to the human cost are the millions of dollars spent on rescue and recovery efforts, equipment repair and replacement, and mine rehabilitation. In addition, mines shut down by fires and explosions were forced to forego hundreds of millions of tons (1 million tons equals 907 200 metric tons) of mineral production. The fire and explosion hazards encountered in mining are common to many industries; however, the harsh mining environment, restricted accessibility, and limited fresh air supply inherent to underground mining increase the risks.

Evacuation from underground workings during a fire emergency can be a lengthy process. In a recent survey of 50 underground noncoal mines, emergency evacuation time averaged 27 minutes. The range was 5 minutes to 85 minutes with a strong correlation between evacuation time and depth of the mine shaft.[30] Compounding the problem of longer evacuation times for deeper mines is the trend toward larger workforces in deeper mines. Mines over 3,600 ft (1.1 km) in depth average double the workforce of mines less than 1,800 ft (548 m) in depth. Mine ventilation is critical during an underground mine fire, as contaminated air is the primary life safety hazard. Smoke and toxic fire gases have accounted for nearly 80 percent of underground mine fire fatalities since 1945.[30]

Methane and/or coal dust are involved in most mine explosions; however, natural gas, bottle gas, acetylene, and liquid hydrocarbon vapors are involved in many surface explosions. Practically all mine explosions start with a methane ignition. However, if the explosion propagates more than a few hundred feet, coal dust will be the most important source of fuel. In underground coal mines, methane explosions outnumber coal dust explosions about 6 to 1.[22] Along with methane explosions are methane ignitions (explosions which do not result in fatalities, injuries, or damage to mine workings or equipment). Methane ignitions outnumber methane explosions about 7 to 1.[22]

In surface coal mines, about two-thirds of the ignitions and explosions involve coal dust and one-third methane. In underground coal mines, 95 percent involve methane.

The high frequency of methane explosions and ignitions can be attributed to the low energy required to ignite a methane-air mixture. As little as 0.3 millijoule of electrical energy can ignite methane under the proper conditions (equivalent to $\frac{1}{50}$ of the static electricity accumulated by the average person walking across a carpeted floor on a dry day). Methane can be ignited by a miner's pick striking a rock, a buffing wheel,

TABLE 10.2. *Ignition Sources of Mine Explosions and Methane Ignitions (1970–1977)*

	Coal Mines			Noncoal Mines
	Underground		Surface	
	Explosions	Ignitions		
Friction	32	285	0	0
Electric arc	7	13	10	2
Welding	3	24	18	6
Smoking	6	2	3	4
Open flame	0	0	1	10
Battery	2	1	0	1
Lightning	0	2	1	0
Explosives	4	8	2	0
Safety lamp	2	0	0	0
Other	0	0	5	0
Total	56	335	40	23

Source: Mine Safety and Health Administration.

or nails in a miner's boot. A lighted cigarette might not ignite methane, as the temperature of burning tobacco is too low; but burning cigarette paper and the sparks from a match or cigarette lighter will readily ignite methane. Frictional sparks generated by mining machines impacting rock or inclusions in the coal face are a frequent cause of methane ignitions and explosions. Sparks generated by contact of aluminum and rusty iron are even more highly incendive. Research has shown that sparks sufficient to ignite methane can be generated by a rail car wheel (rusty iron) running over a candy wrapper (aluminum foil).[22]

The LEL (lower explosive limit) for a suspension of coal dust in air is 0.05 oz per cu ft (50 g/m^3). (The presence of methane in the atmosphere, however, produces a linear reduction in the LEL for coal dust).

The minumum electrical energy to ignite a coal dust cloud is from 70 to 200 times greater than that required to ignite methane (depending on the type of coal). High volatile bituminous coal dust in the absence of methane can be ignited by frictional sparks and explosives, but the energy required is much higher than that for methane. Coal dust explosions are generally triggered by methane ignitions which cause settled coal dust to become suspended in air and also provide the necessary ignition source.

The destructive potential of ignitions and explosions results from the high temperature flame fronts and static and dynamic pressure produced. Flame temperatures of 3,600°F (1649°C) and static pressures in excess of 40 psig (276 kPa) are possible. Dynamic, or "wind," pressure is directional, and depends upon the air velocity produced by the explosion which, in turn, depends upon the static pressure. A static pressure of 5 psig (35 kPa), for example, produces a maximum air velocity of 280 ft/sec (85.3 m/sec) which results in a dynamic pressure of 90 psf (4309

TABLE 10.3. *Mine Fire Ignition Sources (percentages)*

	Underground Mines		Surface Mines	
	Coal	Noncoal	Coal	Noncoal
Electrical	43.7	18.5	19.0	16.0
Friction	7.8	2.1	9.5	2.0
Spontaneous combustion	10.9	2.1	—	—
Explosives	0.8	—	—	—
Hot surface (engines)	—	14.4	38.1	32.0
Welding and cutting	8.5	12.4	9.5	18.0
Other or unknown	24.7	47.4	23.8	34.0

Source: U.S. Bureau of Mines.

Pa). A dynamic pressure of this magnitude applied to a mine rail car with a cross-sectional area of 20 sq ft (1.86 m^2) for a period of 2 seconds results in an applied impulse force of over 3,600 lb sec (1633 kg sec), more than adequate to propel the car an appreciable distance.

The major fire and explosion hazards are discussed in detail in the following sections.

Ignition sources of recent mine explosions and methane ignitions are shown in Table 10.2.[22] The ignition sources, equipment, and burning substances involved in mine fires since 1968 are listed in Tables 10.3, 10.4, and 10.5, respectively.[2,20]

Electrical

The largest category of fires is comprised of those of electrical origin. This is especially true in underground coal mines where nearly half of all fires are electrical. During the 1960s and 1970s, about 40 percent of underground coal mine electrical fires involved face equipment powered through a trailing cable, and about half of those fires were caused by faults in the cable. Face equipment fires have been significantly reduced in recent years by the conversion from DC to AC power. Fires on such equipment are now relatively rare.

Mobile equipment accounted for 37 percent of the electrical fires in underground noncoal mines.[2] Over half of these fires involved electrical faults on diesel vehicles which ignited insulation and/or combustible liquids. Faults in high voltage power cables igniting nearby timber accounted for 17 percent of the underground noncoal mine electrical fires.

The majority of surface mine electrical fires occurred in permanent or semi-permanent power substations. The remainder occurred in fixed structures, stripping machines, and conveyor belt drives.

About 5 percent of underground coal mine ignitions and explosions result from electric arcs.[22] In view of the extensive use of electrical equipment in underground coal mine face areas and the low electrical en-

TABLE 10.4. *Equipment Involved in Mine Fires (percentages)*

	Underground Mines		Surface Mines	
	Coal	Noncoal	Coal	Noncoal
Stationary electrical equipment	27.5	21.6	23.9	18.0
Mobile equipment	35.5	39.1	42.7	32.0
None or unknown	31.2	25.4	6.8	13.5
Other equipment	5.8	13.9	26.6	36.5

Source: U.S. Bureau of Mines.

ergies required to ignite methane, this figure is surprisingly low. (See Chapter 47 for a more detailed discussion of the hazards associated with electrical installations.)

Diesel Equipment

Except in underground coal mines, more fires orginate on diesel equipment than from any other source. About 60 percent of underground noncoal diesel equipment fires involve LHD vehicles, and the trend established since 1970 is toward increasing numbers of LHD fires. Combustible liquids (hydraulic fluids, diesel fuel, lubricants) constitute the greatest hazard; however, wiring, hoses, tires, and combustible refuse are also contributors. Wheeled loaders, haulage trucks, and utility vehicles account for the remaining diesel equipment fires.

Fires onboard large, surface-mining mobile equipment are a serious safety hazard to operators and the cause of substantial property damage and mine production loss (see Figure 10.19). Three factors are responsible, in large part, for the frequency and severity of such fires. First, this equipment is designed for high production with large turbocharged diesel engines and extensive high-temperature, high-pressure hydraulics. Second, the equipment must operate in an extremely harsh enviornment with extremes in temperature, constant vibration, shock loads, dust and rocks thrown by wheels, and generally poor housekeeping. Finally, normal duty cycles can be quite demanding with operations typically running around the clock. Each factor alone creates potential hazards, and when combined, the hazards are compounded. (The hazards of fluid power systems are detailed in Chapter 43.)

The large size of this equipment further increases the firesafety problem by reducing the operator's view of fire hazard areas and making emergency egress especially difficult. A typical fire involves leaking hydraulic or fuel lines that spray a heated mist of highly combustible liquid onto an ignition source such as a hot exhaust manifold or turbocharger.

Chapter 10

FIGURE 10.19. Haulage truck fire.

Welding and Cutting

Welding and cutting operations are the third leading cause of fires in mines. Fires result from direct flame impinging on combustible material, combustible material contacting the metal surface which is heated in the welding or cutting operation, and hot slag, sparks, and discarded welding rod contacting combustible material. Welding and cutting operations are the primary cause of shaft fires in both coal and noncoal underground mines. The potential seriousness of shaft fires cannot be overemphasized because fresh air to underground workings and safe mine evacuation may be impaired by a fire in a shaft.

Welding and cutting operations are the third leading cause of underground coal mine explosions and methane ignitions and the second leading overall cause of mine explosions and ignitions. In underground coal mines, the trend since 1970 has been toward increasing numbers of explosions and ignitions caused by welding and cutting operations. (See Chapter 24 for a detailed discussion of the hazards associated with welding and cutting.)

Spontaneous Combustion

Exothermic oxidation reactions occur in both coal and metal sulfide ores. The chemical mechanisms are distinctly different, but neither is well understood. When the heat generated by these reactions is not dissipated, the temperature of the rock mass or pile increases. If sufficient temperatures are reached, rapid combustion of coal, sulfide minerals, and other combustibles can result.[23] Although spontaneous combustion fires occur relatively infrequently, they are generally quite disruptive to mine operations and are most difficult to extinguish. Spontaneous combustion fires may be localized or they may affect large areas. They often occur in remote, abandoned sectors of the mine where access for fire fighting is difficult. Half of underground coal mine fires lasting more than 24 hours and 57 percent of noncoal fires lasting more than 24 hours are caused by spontaneous combustion.

Friction

Heat generated by friction between conveyor belt and conveyor drives or idlers had been a leading cause of fires in underground coal mines until about 1970 when requirements for sequence and slippage switches became effective. Other frictional heat sources include parking brakes on mobile equipment which are engaged during vehicle operation and overheated rubber tires.

Frictional sparking between cutting bits and rock surfaces (continuous miners, roof bolters, undercutters, face drills, longwall equipment) causes over 80 percent of the ignitions and explosions in underground coal mines. Other sources of frictional sparking are drill steel striking machine frames, sandstone impacting sandstone or other hard rock during a roof failure, and mobile vehicle brake linings.[22]

Timber

Historically, timber was the most common material used for temporary and permanent support of underground openings. Large amounts of timber were used in extraction areas, haulageways, shafts, slopes, and anywhere else roof control problems were experienced. The extensive use of timber resulted in a large number of timber fires. Prior to 1967, over half of all metal and nonmetal mine fires involved timber. In more recent years, the use of roof bolts to provide support in underground openings has significantly reduced the need for timber underground. But older mines still contain large amounts of timber from previous years, and timber is still used today at most mines, though in lesser quantities.

About one fire in ten in underground coal mines and one fire in five in underground noncoal mines now involve timber. However, timber fires tend to constitute a greater overall hazard than these figures indicate because timber fires spread more rapidly, burn longer, and are more difficult to extinguish than fires involving most other extraneous combustible materials. In underground noncoal mines, over three-quarters of the fires lasting longer than 24 hours involve burning wood.

Fires involving discrete pieces of equipment generally burn themselves out when the available combustible materials are consumed. Thus, even when fire fighting attempts are unsuccessful, such fires are gener-

TABLE 10.5. *Burning Substances Involved in Mine Fires (percentages)*

	Underground Mines		Surface Mines	
	Coal	Noncoal	Coal	Noncoal
Electrical insulation	25.2	20.2	23.8	20.0
Timber	8.9	22.2	—	—
Coal/combustible ore	28.8	4.0	—	—
Rubber	16.0	14.6	14.3	6.0
Explosives	0.4	1.0	—	—
Flammable/combustible liquids	16.0	26.8	60.9	73.4
Other or unknown	4.7	11.2	1.0	0.6

Source: U.S. Bureau of Mines.

ally of short duration. However, in areas where timber is used extensively for support, fuel sufficient for fires of several months' duration is readily available. When such fires occur in abandoned or caved areas or where extensive roof failures occur as a result of the burning of support timber, fire fighting is further complicated.

Coal

The ultimate objective of underground coal mine fire protection efforts is to prevent the ignition of the coal. As noted above, equipment fires are generally limited by the amount of combustible material present. However, once the coal is ignited, the entire mine represents the available fuel supply. Over one-third of fires which involve coal as a burning substance last longer than 24 hours, with the average duration being 6 days.[19] Failure to extinguish a fire while it is just developing results in a 50 percent chance that more than 8 hours will be required to bring it under control and a 5 percent chance that part or all of the mine will be sealed.[21]

Flammable and Combustible Liquids

In noncoal underground mines and surface mines, flammable and combustible liquids are involved in more fires than any other combustible material. Most of these fires occur on mobile equipment where high pressure hydraulic systems represent the primary hazard.

In underground noncoal mines, the storage of flammable and combustible liquids is an important concern. Many mines store large quantities — up to 30,000 gallons (114 m^3) — of flammable and combustible liquids underground. The potential seriousness of a fire in an underground flammable and combustible liquid storage area has prompted extreme care in the design of storage areas and the implementation and strict enforcement of safe operating procedures. As a result, few fires have been reported in such areas. (See Chapter 37 for a more detailed discussion of the hazards associated with the handling and storage of flammable and combustible liquids.)

THE SAFEGUARDS

Fire and explosion safety in the mining industry is based on the general principles of fire and explosion prevention, early fire detection and warning, fire suppression, limiting fire propagation, and providing for miner safety during a fire emergency. Implementation of these general principles is discussed in the following sections.

Fire Prevention

Fire prevention practices fall into three categories: limitations on fuel sources, limitations on ignition sources, and limitations on fuel source and ignition source contact.

Limitations on fuel sources: Where the frequency and/or severity of fires involving certain combustible materials is sufficient, and where suitable and less hazardous substitutes for those materials are available, the less hazardous materials are preferred, or in some cases, mandated by law. Examples include fire resistant hydraulic fluids, conveyor belting, hydraulic hoses, and ventilation tubing, all of which are required under certain conditions in underground mines.[37] Even more basic is good housekeeping to prevent unsafe accumulations of trash, oily rags, coal dust, and other combustible materials.

Limitations on ignition sources: Sources of fire which are not essential to the mining process can be banned altogether. Examples are prohibitions on open fires and on smoking in underground coal mines and certain areas of underground noncoal and surface mines. Means to prevent the unwanted buildup of heat, such as slippage and sequence switches on conveyors and thermal cut-outs on electric motors, are also common.

Limitations on fuel source and ignition source contact: Many combustible materials and potential ignition sources are essential to mining operations. Firesafety in these instances depends on preventing ignition source and fuel source contact. Precautions observed during welding and cutting operations are typical. When welding cannot be performed in fire safe enclosures, areas are wet down and, if practical, nearby combustibles are covered with fire resistant materials or relocated. Fire extinguishers must be readily available and a fire watch posted for as long as necessary to guard against smoldering fires. On mobile equipment, the hydraulic fluid, fuel, and lubricant lines can be rerouted away from hot surfaces, electrical equipment, and other possible ignition sources. Spray shields can also be installed to deflect sprays of combustible liquid from a broken fluid line away from a potential ignition source. Areas with a high loading of combustible materials such as timber storage areas, explosives magazines, flammable and combustible liquid storage areas, and shops both underground and on surface, can be designed to minimize possible ignition sources.

Explosion Prevention

Explosion prevention practices fall into the same three catagories as fire prevention: limitations on fuel sources, limitations on ignition sources, and limitations on fuel source and ignition source contact.

Limitations of fuel sources: Coal dust and methane are the primary fuels involved in underground coal mine explosions. Methane may also be present in noncoal mines.[31]

Methane is treated in two ways. Most commonly, methane is diluted and exhausted from the mine by ventilation air. Careful control and constant monitoring of the ventilation system are required to ensure that explosive pockets or layers of methane do not form. In coal mines where ventilation alone is not entirely satisfactory, or where ventilation volumes must be increased to very inefficient levels to dilute the methane, "degassification" in advance of mining may be practiced. Typically, wells are driven into the coal seam from the surface or holes are bored from underground to drain methane from virgin coal. Drainage is coordinated with mining so that degassification is completed before the coal is worked. Under favorable conditions, draingage of nearly 50 percent of the methane present is possible.[10]

Although every attempt is made to minimize the generation of coal dust in mining processes, the tiny amount of dust needed to propagate a coal dust explosion — a layer 0.005 in. (0.13 mm) thick, if suspended in air, will propagate an explosion, less if methane is present — is unavoidable. The hazard of coal dust can be controlled, however, by mixing the dust with an inert material.[22] When coal dust is mixed with an inert material such as pulverized limestone, dolomite, or gypsum (rock dust), in a 35 percent coal to 65 percent inert ratio for mine sized dust (up to 850 micron particles) or a 20 percent coal to 80 percent inert ratio for float dust (average particle size of 20 microns, return airways only), the resulting mxitures will not propagate an explosion. Float coal dust, so designated because it is so fine that mine ventilation currents transport it long distances before it settles out, is more reactive and easily dispersed than mine sized dust, hence the higher incombustible content required to arrest propagation. The incombustible content must also be increased in proportion to the amount of methane present. Rock dust is applied to all horizontal and vertical surfaces within 40 ft (12.2 m) of the working face and trickled into ventilation streams periodically as needed to maintain a minimum safe incombustible content in settled float coal dust.

Limitations on ignition sources: Every effort is made to eliminate extraneous ignition sources from the mining environment. Smoking is prohibited in coal mines and electrical equipment operating where methane may be present must be permissible or intrinsically safe. Permissible electrical equipment is equipment which has been tested by the federal government and, due to design, construction and installation, will not cause a mine explosion or mine fire. Use of explosion-proof enclosures, specially designed plugs and receptacles, automatic circuit-interrupting devices, and voltage limitations are typical. Intrinsically safe equipment is equipment which has been tested by the federal government and is incapable of releasing sufficient electrical or thermal energy under normal or abnormal conditions to ignite an explosive gas mixture. Explosives and diesel powered equipment must also be permissible. Because such a high proportion of methane ignitions and explosions result from frictional sparking between cutting bits and rock surfaces (continuous miners, roof bolters, longwall machines), considerable research effort has been directed toward reducing the frictional sparking problem. Bit shape, speed, attack angle, and composition have all been related to ignition potential, and control of these parameters can reduce ignitions.[9,29] Water sprays which impinge upon the back side of cutting bits have been shown in research experiments to be nearly 100 percent effective in eliminating longwall ignitions; however, this technique has not been demonstrated on operating equipment in a mine setting.[1]

Limitations on fuel source and ignition source contact: Because of geologic nonuniformity, varying mining rates and atmospheric pressure, and other factors, the liberation of methane into mine ventilation air from the workings is not uniform with time. Also, incomplete mixing of methane with ventilation air may occur, with explosive pockets or layers forming even though the average methane concentration is below the LEL. Thus, dilution and drainage of methane cannot be relied upon to always produce nonexplosive atomspheres. Methane checks are made frequently with hand-held and machine-mounted methanometers. If the methane concentration reaches 1 percent (20 percent of the LEL), all potential ignition sources (electrical equipment, welding apparatus, etc.) are shut down. Specialized ventilation equipment, such as spray fans, may be used to prevent pockets of methane from forming at the face.

Explosion hazards which may exist in surface mines, such as accumulations of natural gas or bottle gas from gas line leaks, acetylene from oxy-fuel cutting apparatus, and liquid hydrocarbon vapors can be avoided by proper inspection and maintenance of facilities and equipment and strict enforcement of safe work procedures.

Chapter 10

Early Fire Detection and Warning

The elapsed time between the onset of a fire and its detection is critical because fires tend to grow in size and intensity with time. Early fire detection and warning permit the initiation of a mine's emergency plan (evacuation, fire fighting, etc.) while the fire is still small, or ideally, while it is still in the incipient stage. Over 70 percent of fires detected within 15 minutes cause little or no damage to the mine.[20] Underground miners represent fire detection in its most basic form. Fire bossing (the systematic inspection of active and inactive mine workings for fire) is a more organized approach to manual early fire detection and warning. Advanced fire detection and warning systems utilizing sensitive heat, flame, smoke, and gas analyzers provide the most rapid and reliable indication of fire.[7,8,14,32] Thermal fire detection systems are commonly installed over conveyor belts, belt drives and take-ups, and other unattended equipment. Thermal detection systems can also be installed on attended equipment where the outbreak of fire may not be immediately apparent to the operator, providing additional time to respond.

In localized, high hazard areas such as flammable and combustible liquid storage areas, refueling areas, and shops, faster acting fire detection devices may be appropriate. Optical flame detectors, sensing either ultraviolet or infrared radiation emitted by a fire, have been used successfully in these areas.

Smoke and/or gas detection is the most cost-effective approach to providing large-area (or even whole-mine) fire detection coverage. Detection instruments sampling the mine ventilation air streams are capable of discriminating increases in submicron particulates (smoke) as little as 1 mg/m^3 and trace increases in combustion gases. Carbon monoxide (CO), for example, can be reliably detected at levels as low as 2 parts per million (ppm). Two techniques are available for measuring smoke and gas levels: fully electronic systems, and pneumatic "tube bundle" systems. In a fully electronic system, individual detection instruments are placed at various key locations underground and electronically linked by hardwired telemetry to a control unit.[7,35,36] Systems may consist of any number of detection instruments, with one or more detectors located at each monitoring point (see Figure 10.20).

A pneumatic fire detection system consists of centralized analytic instruments which determine the composition of gases drawn by a vacuum pump through a series of individual sampling tubes terminating at remote locations.[11,17,33]

The number of sampling points required depends on the desired system response time, ventilation airflows and velocities, types and characteristics of detection instruments, and extent of the mine workings. Mathematical models have been developed which relate these factors and provide guidance in siting sampling points.[16]

Once a fire has been detected and responsible mine officials notified, a fire warning signal must be communicated to each miner. In coal mines, the most common means of fire warning are shut-down of elec-

FIGURE 10.20. Fire detection instruments in an underground mine.

tric power and/or notification by telephone and messengers. In noncoal mines, shut-down of electric power is generally ineffective because so little equipment is electrically powered. Telephones and messengers are used on a limited basis, but most miners are too remote from telephones and too widely scattered for either of these means to provide a timely warning. Stench warning is the most common method of emergency communication in noncoal underground mines.[25]

In a typical stench system, ethyl mercaptan, a highly odoriferous organic compound, is injected into the compressed and/or ventilation air supply of the mine. The liquid is quickly vaporized, and the stench is carried in the airstreams to the working areas underground. Miners, upon smelling the stench, respond in accordance with the mine's emergency plan.

Fire Suppression

The most common types of fire suppression equipment used in underground coal mines are hand portable extinguishers (generally multipurpose dry chemical), water hoselines, rock dust (applied manually or from a rock dusting machine), and foam generators. Water lines for fire suppression and a supply of fire hose are required in underground coal mine conveyor belt, and track haulage entries and deluge-type water spray systems, sprinkler systems, dry chemical systems, or foam generators are required at conveyor belt drives and for certain other unattended electrical equipment. The hydraulic fluid in attended coal mine equipment must be fire resistant or the equipment must be provided with a fire

Chapter 10

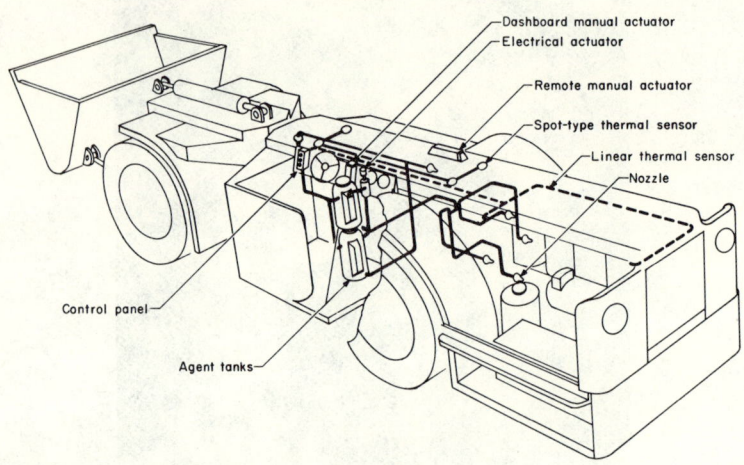

FIGURE 10.21. Fire suppression system for underground LHD vehicle.

suppression system. Fire fighting equipment, including hand portable extinguishers, rock dust, hoselines, and portable water or dry chemical cars, are required in each coal mine working section and at various other fire hazard areas such as electrical installations, oil storage areas, and areas where welding, cutting, and soldering are performed.

Fire suppression equipment in noncoal underground mines is generally limited to hand portable extinguishers and water hoselines. Fire suppression systems, either manual or automatically actuated, are becoming more common, however, for such applications as mobile equipment, combustible liquid storage areas, conveyor belt drives, and electrical installations[15,39] (see Figures 10.21 and 10.22).

Where large areas of coal or timber are burning and fire fighting is complicated by extensive roof falls, ventilation uncertainties, and accumulations of explosive gas, the only practical alternatives are inerting (with nitrogen, carbon dioxide, or the combustion products of an inert gas generator), flooding, and/or sealing parts or all of the mine.[21]

All underground miners are trained to use basic fire fighting equipment; however, fires which grow to an advanced stage are fought only by highly trained and specially equipped fire fighting teams. In some cases mine sealing or inerting may be the only practicle recourse.

Mobile equipment is the primary fire hazard in surface mines. Hence, most surface mining fire suppression equipment is specifically designed for maximum effectiveness on typical mobile equipment fires. One hand portable extinguisher is required on each piece of mobile equipment; however, larger machines are often equipped with manual and/or automatically actuated suppression systems. These systems range from four nozzle dry chemical units for small loaders and dozers, to

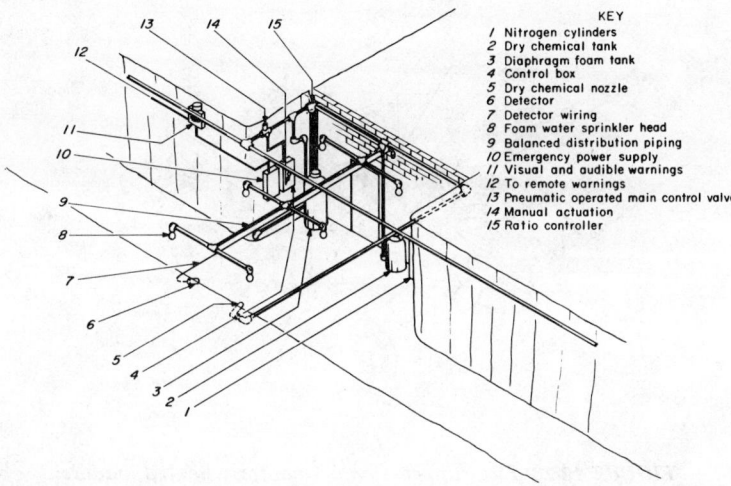

FIGURE 10.22. *Fire suppression system for underground diesel refueling area.*

twelve nozzle fully automatic dry chemical systems for large haulage trucks, to integrated dry chemical and Halon 1301 systems for stripping machines, loading shovels, and blasthole drills[13,26,27,38] (see Figures 10.23 and 10.24). Sprinkler systems using water containing a suitable anti-freeze additive and automatic sprinkler heads have also been used successfully on stripping machines.

At many mines, the water trucks that are used for dust control are equipped with pumps and hoses for fire fighting and the pickup trucks used by roving foremen are equipped with large dry chemical hosereel units.

Limiting Fire Propagation

Means for confining or limiting the rate of propagation of an underground mine fire can help ensure a safer mine evacuation and lessen the hazards of fire fighting. In underground coal mines, oil and grease must be stored in closed, fireproof containers and the storage areas must be of fireproof construction. Transformer stations, battery-charging stations, substations, shops, and other installations must be in fireproof areas or housed in fireproof structures. Unattended electrical equipment must be mounted on a noncombustible surface and separated from the coal or protected by a fire suppression system. Materials for building bulkheads and seals, including wood, brattice cloth, saws, nails, hammers, plaster or cement, and rock dust, must be readily available to each working sec-

Chapter 10

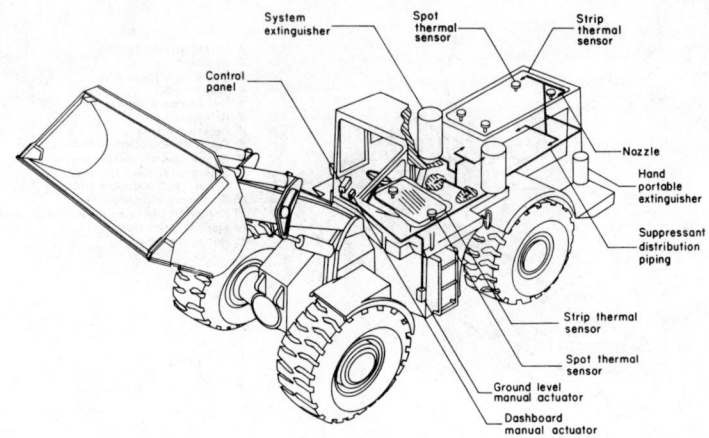

FIGURE 10.23. *Fire suppression system for wheeled loader.*

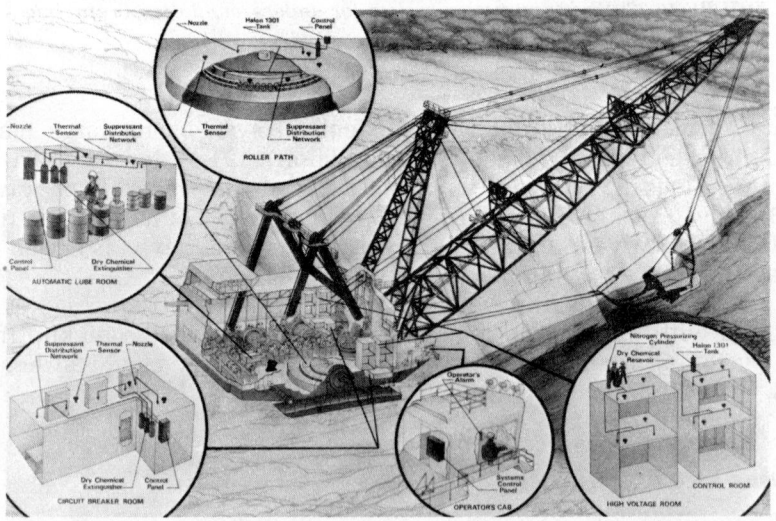

FIGURE 10.24. *Fire suppression system for walking dragline.*

tion. In underground noncoal mines, oil, grease, and diesel fuel must be stored in tightly sealed containers in fire resistant areas at safe distances from explosives magazines, electrical installations, and shaft stations. Ventilation control doors and fire doors are required in certain areas to prevent the spread of fire, smoke, and toxic gas.[40]

The relationship between the amount of combustible material present, fire size, and the ventilation air velocity can also affect propagation. For example, research has demonstrated that in a horizontal drift of 8-ft by 8-ft (2.4 by 2.4 m) cross section with timber support accounting for about 40 percent of the surface area of the drift (exclusive of the floor), and a fire size of 6000 kW, fires will propagate if the ventilation velocity is less than or equal to 200 fpm (107 m/min) but will not propagate if the ventilation velocity is increased to 250 to 300 fpm (76 to 91 m/min). Higher timber loading results in propagation at higher ventilation velocities.[35]

Providing for Miner Safety During a Fire Emergency

The primary concern during an underground fire emergency is the safety of underground personnel. Smoke and toxic fire gases represent the greatest threat; hence, the mine's ventilation system and its relationship to the various working areas, potential fires, primary and secondary escape routes, refuge chambers, etc., are critically important. Detailed emergency preplanning which considers these factors is essential. The use of computer models which simulate mine ventilation systems can be very helpful in developing and optimizing emergency plans and designating escape routes. Advanced models which further simulate the effects of a mine fire on the ventilation system (heat induced bouyant forces causing throttling, reversals, etc.) and can track the distribution and concentration of contaminants are particularly valuable.[6] To ensure smooth implementation of the emergency plan, miners are provided comprehensive training and annual retraining. Fire drills, complete with the activation of the mine's warning system and the total evacuation of all mine personnel, are run frequently to reinforce the training and identify weaknesses in the emergency plan.

Although every attempt is made to avoid such situations, evacuating miners are sometimes required to travel through areas contaminated by smoke and toxic gas. Therefore, appropriate respiratory protection for all miners is required. In coal mines, self-contained breathing apparatus supplying a minimum of one hour of oxygen must be provided. In noncoal mines, one-hour rated self-rescue devices which convert carbon monoxide to carbon dioxide must be provided.

BIBLIOGRAPHY

REFERENCES CITED

1. Agbde, R. O., K. L. Whitehead, R. L. Mundell, and R. D. Saltsman, *Frictional Ignition Suppression by the Use of Cutter Drum Mounted Sprays*. Bituminous Coal Research Inc., BuMines Contract Final Report J0395040, 1982.
2. Baker, R. M., J. Nagy, and L. B. McDonald, *An Annotated Bibliography of Metal*

Chapter 10

  and Nonmetal Mine Fire Reports, U.S. Bureau of Mines Contract Final Report J0295035, 1980.
3. Cassidy, S. M., ed., *Elements of Practical Coal Mining*, Society of Mining Engineers of the American Institute of Mining, Metallurigical and Petroleum Engineers, Inc., Port City Press, Baltimore, MD, 1973.
4. Cummins, A. B. and I. A. Given, ed., *Mining Enginering Handbook*, Society of Mining Engineers of the American Institute of Mining Metallurgical and Petroleum Engineers, Inc., Port City Press, Baltimore, MD, 1973.
5. *Economic Engineering Analysis of U.S. Surface Coal Mines and Effective Land Reclamation*, BuMines Contract Final Report S0241049. Skelly and Loy, Engineers-Consultants, 1975.
6. Edwards, J. C. and R. E. Greuer, *Real-Time Calculation of Product-of-Combustion Spread in a Multilevel Mine*, U.S. Bureau of Mines Information Circular 8901, 1982.
7. Griffin, R. E., *In-Mine Evaluation of Underground Fire and Smoke Detectors*, U.S. Bureau of Mines Information Circular 8808, 1979.
8. Kacmar, R. M., *Reliability of Computerized Mine-Monitoring Systems*, U.S. Bureau of Mines Information Circular 8882, 1982.
9. Hanson, B. D., *Cutting Parameters Affecting the Ignition Potential of Conical Bits*, BuMines RI 8820, 1983.
10. Hartman, H. L., J. M. Mutmansky, and Y. J. Wang, *Mine Ventilation and Air Conditioning*, John Wiley & Sons, New York, NY, 1982.
11. Hertzberg, M., and C. D. Litton, *Multipoint Detection of Products of Combustion with Tube Bundles*, BuMines RI 8171, 1976.
12. Hustrulid, W. A., *Underground Mining Methods Handbook*, Society of Mining Engineers of the American Institute of Mining, Metallurgical and Petroleum Engineers, Port City Press, Baltimore, MD, 1982.
13. Johnson, G. A. and D. R. Forshey, *Automatic Fire Protection Systems for Large Haulage Vehicles*, U. S. Bureau of Mines Information Circular 8683, 1975.
14. Johnson, G. A. and D. R. Forshey, *In-Mine Fire Tests of Mine Shaft Fire and Smoke Protection Systems*, U.S. Bureau of Mines Information Circular 8788, 1978.
15. Johnson, G. A., *Automatic Fire Protection for Mobile Underground Mining Equipment*, U.S. Bureau of Mines Information Circular 8954, 1983.
16. Litton, C. D., *Guidelines for Siting Product-of-Combustion Fire Sensors in Underground Mines*, U.S. Bureau of Mines Information Circular 8919, 1983.
17. Litton, C. D., *Design Criteria for Rapid Response Pneumatic Monitoring Systems*, U.S. Bureau of Mines Information Circular 8912, 1983.
18. Martin, J. W., T. J. Martin, T. P. Bennett, and K. M. Martin, *Surface Mining Equipment*, Martin Consultants, Inc. Golden, CO, 1982.
19. McDonald, L. B., and R. M. Baker, *An Annotated Bibliography of Coal Mine Fire Reports*, U.S. Bureau of Mines Contract Final Report J0275008, 1979.
20. McDonald, L. B. and W. H. Pomroy, *A Statistical Analysis of Coal Mine Fire Incidents in the United States from 1950 to 1977*, U. S. Bureau of Mines Information Circular 8830, 1980.
21. Mitchell, D. W. and F. A. Burns, *Interpreting the State of a Mine Fire*, U.S. Mine Safety and Health Administration Informational Report 103, 1979.

22. Nagy, J., *The Explosion Hazard in Mining*, U.S. Mine Safety and Health Administration Informational Report 1119, 1981.
23. Ninteman, D. J., *Spontaneous Oxidation and Combustion of Sulfide Ores in Underground Mines*, U.S. Bureau of Mines Information Circular 8775, 1978.
24. Pfleider, E. P., ed., *Surface Mining*, Society of Mining Engineers of the American Institute of Mining, Metallurgical and Petroleum Engineers, Maple Press, York, PA, 1968.
25. Pomroy, W. H. and T. L. Muldoon, *A New Stench Gas Fire Warning System*, Mines Accident Prevention Association of Ontario, North Bay, Ontario, Canada, 1983.
26. Pomroy, W. H., N. Goodwin and F. Lynch, *Economic Analysis of Surface Mining Mobile Equipment Fire Protection Systems*, U. S. Bureau of Mines Report of Investigations 8698, 1982.
27. Pomroy, W. H. and K. L. Bickel, *Automatic Fire Protection Systems for Surface Mining Equipment*, U.S. Bureau of Mines Information Circular 8832, 1980.
28. Richmond, J. K., G. C. Price, M. J. Sapko and E. M. Kawenski, *Historical Summary of Coal Mine Explosions in the United States 1959-81*, U.S. Bureau of Mines Information Circular 8909, 1983.
29. Roepke, W. W., and B. D. Hanson, *Bit Ignition Potential with Worn Carbide Tips*, BuMines TPR 121, 1983.
30. Stevens, R. B., *Mine Shaft Fire and Smoke Protection System*, FMC Corp. BuMines Contract Final Report H0242016, 1975.
31. Thimons, E. E., R. P. Vinson, and F. N. Kissell, *Forecasting Methane Hazards in Metal and Nonmetal Mines*, U.S. Bureau of Mines Report of Investigations 8392, 1979.
32. Welsh, J. H., *Computerized, Remote Monitoring Systems for Underground Coal Mines—Fires and Explosive Atmospheres*, BuMines IC 8875, 1982.
33. Thomas, E. C., *A Pneumatic Sampling Fire Detection System in an Underground Haulageway*, IEEE Trans. on Industry Applications. v. IA-19, No. 3, May/June 1983.
34. *Mineral Commodity Summaries 1983*, U.S. Bureau of Mines, 1983.
35. *Underground Metal and Nonmetal Mine Fire Protection*, U.S. Bureau of Mines Information Circular 8865, 1981.
36. *Metal Mine Fire Protection Research*, U.S. Brueau of Mines Information Circular 8752, 1977.
37. *Coal Mine Fire and Explosion Prevention*, U.S. Bureau of Mines Information Circular 8768, 1978.
38. *Dragline Fire Protection*, U.S. Bureau of Mines Technology News 106, 1981.
39. *Automatic Fire Protection Systems for Underground Fueling Areas*, U.S. Bureau of Mines Technology News 160, 1982.
40. *Improved Fire Doors for Noncoal Underground Mines*, U.S. Bureau of Mines Technology News 188, 1983.
41. *Mines Injuries and Worktime*, Quarterly, U.S. Mines Safety and Health Administration, 1983.
42. *Weekly Coal Production*, Energy Information Administration, U.S. Department of Energy, 1984.

NFPA CODES, STANDARDS, AND RECOMMENDED PRACTICES

Reference to the following NFPA Codes, Standards, and Recommended Practices will provide further information on the safeguards for mining operations discussed in this chapter. (See the latest *NFPA Codes and Standards Catalog* for availability of current editions of the following documents.)

NFPA 10, *Portable Fire Extinguishers.*
NFPA 11A, *Medium and High Expansion Foam Systems.*
NFPA 12A, *Halon 1301 Fire Extinguishing Systems.*
NFPA 12B, *Halon 1211 Fire Extinguishing Systems.*
NFPA 13, *Installation of Sprinkler Systems.*
NFPA 17, *Dry Chemical Extinguishing Systems.*
NFPA 30, *Flammable and Combustible Liquids Code.*
NFPA 51, *Oxygen-Fuel Gas Systems for Welding and Cutting.*
NFPA 51B, *Fire Prevention in Use of Cutting and Welding Processes.*
NFPA 70, *National Electrical Code.*
NFPA 120, *Coal Preparation Plants.*
NFPA 121, *Fire Protection for Mobile Surface Mining Equipment.*

ADDITIONAL READING

Cashdollar, K. L., J. M. Singer, C. K. Lee, et. al., "Smoke Characteristics of Coal-Lined Tunnel Fires," *Fire and Materials*, vol. 5, no. 2 (1981), pp. 47-51.

Cybulski, W., *Coal Dust Explosions and Their Suppression* (translated from Polish), National Technical Information Service, Springfield, VA, 1981.

Chamberlin, R. E., "Where's the Fire?", *Coal Mining and Processing*, vol. 17 (December 1980), pp. 56-58.

Factory Mutual Research, *Analysis of Full Scale Fire Tests in a Simulated Mine Gallery*, U.S. Bureau of Mines Contract Reports H0202037 & H0252085, 1982.

Hall, Christopher, *Mine Ventilation Engineering*, Society of Mining Engineers, Dearborn, MI, 1980.

Hartman, Howard L., et al., ed., by Jan M. Mutmansky and Y. J. Wang, *Mine Ventilation and Air Conditioning*, John Wiley & Sons, Inc., New York, 1982.

Holtzberg, J. T., "Carbon Monoxide Detectors Revolutionize Fire Protection," *Coal Mining and Processing*, vol. 18, no. 4 (April 1981), pp. 118-120.

Lazzara, C. P., *Spontaneous Combustion Susceptibility of U.S. Coals*, U.S. Bureau of Mines Report of Investigations, 8474, 1980.

Lazzara, C. P., *Fire Resistance Test Method for Conveyor Belts*, U.S. Bureau of Mines Report of Investigation 8521, 1981.

Lee, C. K., P. A. Croce, F. S. Newman, "An Experimental Investigation of the Fire

Hazards Associated with Timber Sets in Mines: Main Volume," Factory Mutual Corp., Norwood, MA, 1981.

Lehmann, E. J., *Mine Safety, Par, Fires and Explosions (A Bibliography with Abstracts)*, National Technical Information Service, Springfield, VA, 1978.

Mining Information Services, *Mining Methods and Equipment*, McGraw Hill, New York, 1980.

Richmond, J. K., L. I. Liesman, and A. E. Bruszak, "A Physical Description of Coal Mine Explosions, Part II," 17th International Symposium on Combustion, Combustion Institute, Pittsburgh, PA, 1979.

Sapko, M. J., et. al., *Fire Resistance Test Method for Conveyor Belts*, Bureau of Mines, U.S. Dept. of the Interior, Investigative Report 8521, U.S. Government Printing Office, 1981.

Stevens, R. B., *A Guide to the Selection of Mine Shaft Fire and Smoke Protection Systems*, FMC Corp, Engineered Systems Division, Santa Clara, CA. BuMines OFR-73(2) - 78, 1976.

Title 30, *Code of Federal Regulations*, U.S. Government Printing Office, 1984.

Treuhaft, M. B., B. C. Dial, *Study of Air Compressor Hazards on Underground and Surface Mines*, Southwest Research Institute, San Antonio, TX, 1981.

11

Aluminum and Nonferrous Metals Processing

Peter K. Schontag and Richard E. Caines

This chapter is devoted primarily to the aluminum and aluminum alloy industry, the largest segment of the nonferrous metal industry by a significant margin. Many of the fire hazards common to the aluminum industry are also found in the other nonferrous metal industries, although the aluminum industry does have some unique hazards.

Aluminum is a nonmagnetic and nonsparking lightweight metal. Pure aluminum is soft and easily malleable; thus, it is used mostly in an alloy form with small amounts of copper, silicon, magnesium, manganese, or other elements added for hardness. Primary uses, in order of importance, are containers and packaging, building and construction, transportation, electrical, consumer products, machinery and equipment, and exports. United States production of aluminum has declined in recent years due to reduced product demand. In 1982 the U.S. imported 243 million pounds (110 thousands metric tons) of aluminum. Due to high energy costs, recovery of scrap aluminum has become an important supply source. In 1982, primary aluminum production in the United States

---

Peter K. Schontag is a fire protection engineer and consultant with Rolf Jensen & Associates. Richard E. Caines is "Corporate Manager, Property Conservation and Insurance" with Kaiser Aluminum & Chemical Corporation.

was 6.5 million short tons (5.9 million metric tons) with scrap recovery adding another 2 million short tons (1.8 million tons).[1]

Electric power producing or regulating equipment is the primary source of fire hazards in the aluminum industry due to the heavy use of electric power. Hydraulic systems, conveyors, and the use of oil in rolling operations are the other major hazards. (A comprehensive discussion of fluid power and materials handling systems can be found in Chapters 43 and 46, respectively.) Explosions in molten metal have also been a major source of concern, with considerable research being done to determine the cause and best methods of prevention. In 1981 the Aluminum Association issued new guidelines for handling molten aluminum. These guidelines describe the results of extensive research which attempted to determine the causes of molten metal explosions. While the exact conditions that result in molten aluminum explosions are not fully understood, despite years of study by the aluminum industry, much is known that can be applied to minimize the hazards. The production of aluminum powder has some of the hazards associated with the magnesium industry.

RAW MATERIALS

Small amounts of aluminum are found in almost all parts of the earth, although it is not always in metallic form. The economics of current processes limit recovery of aluminum to earth having an aluminum content of 45 percent or more. The basic material for aluminum is bauxite, a red mud or clay, the richest deposits of which are found in a wide band extending on both sides of the equator. The bauxite is usually mined by the open-pit process utilizing draglines, front end loaders, and power shovels. Caustic and lime are used in the refining process, and pitch is used in smelting.

The United States, while the largest aluminum-producing country in the world, has to rely on imports for approximately 90 percent of its bauxite. Aluminum is one of the most plentiful minerals found in the earth, but most deposits in the United States are found in various types of clay and are uneconomical to refine by presently known methods. Most bauxite used in the United States is imported from Jamaica, New Guinea, Surinam, Haiti, and the Dominican Republic. The domestically mined bauxite is found primarily in Arkansas, with minor deposits in Alabama and Georgia.

THE PRODUCTION PROCESS

Once the mined bauxite has been loaded onto trucks or trains, it is transported to a central location where it is processed through a crusher and washing screens to remove clay, silica, and other waste materials. The bauxite is then dried in rotary kilns and stored in large buildings or silos until it is refined or loaded onto ships for transport to distant refining facilities. An increasingly large amount of bauxite is being refined in the country of origin before being imported by the United States. Refined bauxite in the form of alumina comes primarily from Australia, Jamaica, and Surinam.

Refining of Bauxite into Alumina

The dried bauxite contains other elements besides the alumina (aluminum oxide), and these must be removed before the alumina ore can be smelted. A combination of chemical, mechanical, and thermal forces are used to separate iron oxide, silica, titanium dioxide, and other impurities from the ore. Alumina plants are primarily located in Arkansas, Louisiana, and Texas. These plants often produce their own power through the burning of natural gas or fuel oil and are faced with increasing energy costs.

The most common process used to refine bauxite to alumina is the *Bayer Process*, which produces 1 lb of alumina from 2 lbs of bauxite. The bauxite is mixed with a strong solution of caustic soda, which is then pumped into large pressure tanks. Steam is applied to create heat and pressure. This causes the alumina in the bauxite to dissolve into the caustic soda, forming a sodium aluminate solution. The other materials in the bauxite are insoluble and remain as solids. The resulting hot wet mixture is passed through a series of pressure-reducing tanks and cloth-filter presses. The solids, which are commonly known as "red mud," are held back by the filters, and the liquid containing the dissolved alumina passes on through to the next stage. At most plants the red mud is discarded, but some plants further refine it to remove trapped alumina. The dissolved alumina flows through a cooling tower into large tanks called precipitators. As the solution continues to cool, large amounts of crystalline alumina hydrate are added to hasten the separation process known as *seeding*. This seeding speeds the precipitation of the alumina in the settling tanks. The alumina is then filtered from the solution and washed to remove any remaining chemicals and dried in kilns at 1,800°F (982°C). This drying at high temperatures removes any remaining water, and the pure alumina is then ready for the smelting process. Until the alumina is shipped, it is stored in large silos. At this point it is noncombustible and resembles white sugar or salt.

Smelting Alumina to Aluminum (Reduction)

The process by which alumina (aluminum and oxygen) is broken down into its two components takes place in an electric furnace called a *pot*. These pots are deep, rectangular, steel shells lined with a carbon com-

Chapter 11

FIGURE 11.1. A series of reduction cells (pot line). (Kaiser Aluminum & Chemical Corp.)

pound which serves as the cathode. The alumina is dissolved in a bath of molten cryolite (sodium aluminum fluoride), and carbon anodes are suspended in the molten bath so that a high-amperage, low-voltage direct current can pass through the bath to the cathode. In the smelting process, pitch is mixed with calcined carbon to form anodes and cathodes for the electrical current. This electrical force separates the dissolved alumina, with the molten aluminum falling to the bottom of the pot and the oxygen combining with the carbon anode to be released as carbon dioxide. This process is continuous with more alumina being added to the bath to replace that consumed. This smelting of aluminum is usually referred to as reduction. The electric current maintains the cryolite bath in a molten condition, and anodes are replaced as they are spent. The molten aluminum is siphoned from the pot, and this process continues until the pot reaches the end of its life, which is sometimes more than two years. The large amounts of direct current needed for this process require very careful control, and major transformers are needed to reduce high-voltage alternating current to low-voltage direct current. The molten aluminum is usually cast or alloyed at this point to conserve energy that would otherwise be needed to remelt the aluminum at a later time.

Finishing

Much of the aluminum produced is sold in ingot form to companies for manufacturing or finishing into various products. A large segment of the

FIGURE 11.2. Pouring molten aluminum into 1,000-lb (454 kg) molds. (Kaiser Aluminum & Chemical Corp.)

aluminum industry is also involved in this type of manufacturing. Each ingot is passed through a series of presses which consist of sets of large rollers. They take the ingot and squeeze it into longer and thinner sections until it becomes sheet, plate, or, if continued, aluminum foil. Another segment of the industry uses aluminum in smaller ingots which are referred to as *logs* or *pigs* due to their shape. These are heated and squeezed through dies to form extruded products. Further extrusion produces rods which are used to make electrical cable after they are drawn through additional dies to reduce them to wire sizes. Ingots are also melted and used to cast various products. In this type of operation the aluminum is almost always alloyed with other metals to produce the desired metal characteristics.

THE FIRE HAZARDS

The fire and explosion hazards associated with the aluminum industry are common to many other industries except for the risk of a molten metal explosion and the heavy concentration of electrical hazards. As with all industrial and manufacturing operations, the aluminum industry is subject to common fire hazards such as welding and cutting and flammable liquids handling and storage. This section of the chapter outlines the hazards that are commonly encountered, with details given only on the few hazards that are unique to the aluminum industry.

Mining

Almost all of the open-pit mining equipment utilized has large hydraulic systems using flammable or combustible hydraulic fluid. Some of the larger pieces of equipment also pose a major hazard with regard to the fuel system.

Chapter 11

FIGURE 11.3. Huge ingot of aluminum alloy is lifted by crane from soaking pit to be moved to hot rolling line. (Kaiser Aluminum & Chemical Corp.)

Driers

The driers or kilns used both in the drying of bauxite and the drying of alumina operate at very high temperatures and burn either natural gas or fuel oil. Combustion controls on this type of equipment are very difficult to design and require special maintenance effort in order to be reliable from a safety standpoint while remaining practical enough to allow the kilns to function.

Transportation

Bauxite and alumina are moved by truck, train, and ship; however, much of the material is also moved throughout each of the various stages of the process by conveyors. The fire hazards associated with conveyors are covered in "Materials Handling Systems," Chapter 46 of this book. Certain conveyors are critical to the alumina production process, and contingency plans should be developed so that production can be maintained even if the conveyor is lost. If this is not possible, then protection should be installed.

Power Generation

Many alumina plants and a certain number of smelters produce their own power. Power generation presents its own set of fire hazards which are covered in Chapter 4, "Electric Generating Plants."

Electrical Control Equipment

Aluminum smelters must take the alternating current supplied to or generated by the plant and convert it to direct current. Large rectifying transformers and sophisticated electrical control equipment are used to accomplish this and regulate power to the pots. The hazards inherent in oil-filled transformers and large amounts of combustible power cable are common to this part of the operation. The fire hazards associated with electrical control equipment are discussed in Chapter 47, "Electrical Installations for Industrial Locations."

Molten Metal

Molten metal explosions are common in most metal-working industries, and the aluminum industry is no exception. It is not known if the explosions are the result of trapped water flashing to steam or if they are caused by a reaction involving the metal itself. While not a true fire or flammable/combustible gas explosion, these explosions are violent and present considerable life and property damage loss potential.

Rolling Operations

A major portion of finished aluminum passes through various rolling processes with the metal in either the hot or cold state. These operations are known as *hot rolling* and *cold rolling*. Large amounts of combustible oils are used in these operations and, while the oil used in most hot rolling is water soluble, the oil used in cold rolling is usually very similar to kerosene in ignition and burning characteristics.

Hydraulic Systems

Most finishing operations involve the movement of heavy ingots or coils requiring hydraulic equipment. Hydraulic systems that utilize combustible hydraulic fluid need protection. (A comprehensive discussion of hydraulic equipment can be found in Chapter 43, "Fluid Power Systems".)

Chapter 11

FIGURE 11.4. Light-gage aluminum alloy plate is rolled to finish thickness on 144-in. (3.65 m)-wide plate mill. (Kaiser Aluminum & Chemical Corp.)

Scalping

Rough ingots are often scalped or planed to smooth the surfaces and remove burrs. This produces shavings and fines which can produce a fire and explosion hazard in dust collection systems.

Related Processes

The smelting process requires some auxiliary operations to produce cathodes and anodes. These present fire hazards separate from the main aluminum production steps.

Carbon and pitch mixing: Anodes and cathodes are manufactured from an almost inert carbon which is mixed with a combustible pitch. This is usually done in mixing vessels heated by steam or hot oil. Hot-oil heating of pitch-mixing vessels presents a fire hazard. The storage and handling of pitch also present fire hazards, along with an explosion hazard if solid pitch instead of liquid pitch is used.

Anode baking: The pitch and carbon mixture is used to form anodes which are baked in furnaces to harden them. These furnaces are, in reality, pits in the ground into which the "green" anodes are stacked. The top of the pit is covered, and an oil or gas flame is introduced. The heat from this firing, combined with the burning pitch volatiles, bakes the an-

ALUMINUM AND NONFERROUS METALS PROCESSING

FIGURE 11.5. A scalped ingot starts down the hot line, the first step in rolling ingots into plate, sheet, or foil. (Kaiser Aluminum & Chemical Corp.)

odes. Since combustible volatiles are given off by this process, the fire hazard in the exhaust duct is a serious one. It also exposes the air-handling system used for control of pollution products. Control of this baking process must be carefully maintained at all times to reduce the chance of a fire.

Powder

Aluminum powder is used in many applications, and its manufacture requires special precautions. Intense fires and explosions can occur in aluminum powder.

Magnesium

Among the metals that are alloyed with aluminum is magnesium. The fire hazards of magnesium, while fairly common knowledge, are not a major hazard since the aluminum industry uses fairly large pieces, and these do not present a severe hazard. Storage and handling practices should be designed to keep this hazard at a minimum.

Chapter 11

THE SAFEGUARDS

Construction

The predominant type of construction used in the aluminum industry is noncombustible which normally consists of aluminum or steel on steel frame. The following areas need special consideration:

Conveyors: Enclosed conveyors should be constructed of noncombustible material since conbustible construction would only add to the fuel provided by the conveyor belt itself if a fire occurs.

Rectifier buildings: Older rectifier buildings were normally of fire resistant construction; however, it is common practice today to house the rectifiers outside in the switchyard with the transformers. This leaves only the small control center building, which should be of noncombustible construction. Vertical cable runs, cable passes through floors, and cable openings into control panels should all be fire-stopped with a "flame-mastic-type" material. Horizontal cable trays should be covered with noncombustible covers whenever they are stacked above each other. Design considerations should be taken to prevent large airflows in passage ways containing cable trays or runs. Electrical switchgear yards and electrical transformers should be arranged so there is plenty of room for maintenance and good separation between major oil-filled equipment. (See Chapter 47, "Electrical Installations.")

Smelter potlines: The buildings containing the reduction pots are of noncombustible construction. The primary hazard in these buildings is the failure of a potlining, i.e., allowing molten metal to spill out onto the floor of the building. This is called a *tap out*. Design of the building, with special attention to utilities, should take into consideration the exposure this molten metal presents, and all utilities must be arranged so that they are not damaged by the molten metal.

Processing or finishing mills: Larger buildings in this type of plant normally have steel deck roofs, and all components of the composition covering below the roof felts themselves should be noncombustible. Many noncombustible adhesives, vapor barriers, and insulations are available for this purpose. Operations within the plant or mill that utilize large amounts of hydraulic or lubricating/cooling oils should be designed so that the room or basement housing the filters and tanks is cut off from the rest of the plant with fire resistant construction and rated fire doors. Warehouses containing storage with combustible packaging or wrapping should also be separated from the main production buildings by a sufficient distance or by fire resistant construction.

FIGURE 11.6. This 130-in. (3.3 m) cold mill can process coils weighing up to 40,000 lbs (18.1 metric tons), 120 in. (3 m) in width, and up to 0.250 in. (6.35 mm) in gage. (Kaiser Aluminum & Chemical Corp.)

Fire Control Systems

It is common practice to provide only the minimum complement of fire extinguishers in large sections of aluminum industry plants because fire hazards appear to be at a minimum. Where common hazards such as combustible construction, storage or use of flammable liquids, large hydraulic or oil systems using combustible fluid, and other similar hazards exist, sprinkler protection, as recommended by NFPA 13, *Sprinkler Systems*, should be provided. Areas requiring additional protection or special considerations are discussed below.

Mining: Under normal conditions, fire protection systems are not needed in the open-pit operations used by the industry. Occasionally, dry chemical extinguishing systems are installed on high-value dragline equipment.

Transportation: Fire protection for conveyors is a controversial area with many insurance interests recommending that the conveyors be sprinklered. Since most conveyor fires are due to careless cutting and welding (See Chapter 24), control of this type of maintenance operation greatly reduces the need for fixed protection systems. Often dry standpipes are installed in areas subject to freezing and wet standpipes in other areas where the climate allows. These standpipes provide a medium level of protection. The other common cause of conveyor fires is the result of friction due to the drive pulley slipping under a heavy load. Installation of "limited slip" switches along with a few sprinklers at these points in the conveyor system will prevent or minimize the hazard of fire. Conveyors that are critical to production and cannot be bypassed may need fixed extinguishing systems. (See Chapter 46, "Materials Handling Systems," for further information.)

Power generation: Determination of adequate fire protection for fire generating plants usually will require substantial analysis. (See Chapter 4, "Electric Generating Plants.")

Electrical control equipment: Generally, companies in the aluminum industry have not installed fire protection on major oil-filled transformers. Rather, they have relied on rigorous maintenance of the transformers, including annual testing of the dielectric strength of the oil, to prevent fires from occurring. This must be maintained even during production cutbacks, however. New installations should include spatial or physical separation of the transformers as well as dikes filled with crushed stone to cool and retain any oil that might be spilled. Training of the plant fire brigade (see Chapter 3, "Plant Emergency Training and Organization") in the use of hose lines to cool adjacent transformers which may be electrically energized is also advised. Cables are usually not protected by sprinklers except where a "flamemastic-type" material is not used or metal covers cannot provide adequate protection. When computers are used to control processes, they should be protected to prevent major business interruption. (See Chapter 1, "Industrial Fire Risk Management"; also, NFPA 27, *Private Fire Brigades*.)

Molten metal: Prevention of molten metal explosions is still under study by the aluminum industry. Results of the latest research and tests are available from the Aluminum Association.*

Rolling operations: Sprinkler systems should be installed over all rolling operations, regardless of whether they are *hot rolling* or *cold rolling*. In addition, special protection must be provided for cold-rolling mills due to the light oil used for cooling and lubrication. Carbon dioxide (CO_2) extinguishing systems should be installed using the "Local Application Requirements" of NFPA 12, *Carbon Dioxide Extinguishing Systems*. Both primary and reserve supplies are required. Hydraulic systems, fume exhaust fans, and other related electrical circuits should be interlocked with the CO_2 system so that all related equipment will be shut off if a

---

*Contact Occupational Safety, Health and Loss Prevention Committee of the Aluminum Association, 818 Connecticut Ave., NW, Washington, DC 20006.

fire occurs. The only exception is that certain high-speed mills need a sequenced shutdown to prevent damage to the machinery and aluminum in process. Whenever hydraulic equipment is not in an area protected by sprinklers, all hydraulic lines should be shielded from possible ignition sources. An emergency shutoff switch should be located at least 75 ft (22.9 m) from the equipment and wired to shut down all hydraulic pumps and accumulators.

Scalping: Scalping or planing operations produce shavings and fines which are usually removed by air-handling exhaust systems to cyclones or similar recovery collectors. These should be designed and protected as outlined in NFPA 65, *Processing and Finishing of Aluminum.*

Related processes: The following areas related to the smelting process need additional fire protection consideration:

> Carbon and pitch mixing: While the carbon used in anode and cathode manufacturing is almost inert due to the combustibles having been driven off by the calcining process, the pitch used does present a hazard. Liquid pitch should be treated as a combustible fluid, especially when heated. Solid pitch must be handled carefully to prevent dust explosions, and electrical equipment should be dust-tight and designed to meet NFPA 70, *National Electrical Code.* Pitch mixers that are heated with a combustible oil should have sprinkler protection over this hazard, especially if the oil is heated above the flash point.
>
> Anode baking: Air-handling equipment, such as above-ground air ducts, fans, and pollution control bag houses, requires special water spray protection since the buildup of condensed pitch vapor will occasionally catch fire in the anode baking process. Water spray protection for fans must be automatically actuated by heat sensors to assure fan integrity during fires in the duct system. The use of air pollution equipment on carbon bake furnace exhaust systems is quite new, and each system needs its own detailed analysis in order to determine the fire protection required.

Powder: Protection for aluminum powder manufacturing operations should be designed and installed in accordance with NFPA 651, *Manufacture of Aluminum and Magnesium Powder.*

Magnesium: When magnesium is used as an alloy with aluminum, NFPA 48, *Magnesium Storage and Processing,* should be consulted to determine the precautions and protection needed.

Chapter 11

Life Safety Provisions

Other than the risks normally found in manufacturing operations, most of the safety and life hazards in the aluminum industry are unrelated to fire hazards. Records of many agencies have shown that few aluminum industry injuries or deaths are caused by fire. Special consideration is needed, however, for areas where molten metal explosions are possible and for basement areas where flammable liquid filtering or pumping equipment is located, especially when it is below ground level.

BIBLIOGRAPHY

REFERENCES CITED

1. The Aluminum Association, Inc., *Aluminum Statistical Review 1982.*
2. The Aluminum Association, Inc., *Guidelines for Handling Molten Aluminum,* Sept., 1980.

NFPA CODES, STANDARDS, AND RECOMMENDED PRACTICES

Reference to the following NFPA Codes, Standards, and Recommended Practices will provide further information on the safeguards for aluminum processing discussed in this chapter. (See the latest *NFPA Codes and Standards Catalog* for availability of current editions of the following documents.)

NFPA 10, *Portable Fire Extinguishers.*
NFPA 12, *Carbon Dioxide Extinguishing Systems.*
NFPA 13, *Installation of Sprinkler Systems.*
NFPA 15, *Water Spray Fixed Systems.*
NFPA 27, *Private Fire Brigades.*
NFPA 30, *Flammable and Combustible Liquids Code.*
NFPA 48, *Magnesium Storage, Handling, and Processing.*
NFPA 65, *Processing and Finishing of Aluminum.*
NFPA 70, *National Electrical Code.*
NFPA 75, *Protection of Electronic Computer/Data Processing Equipment.*
NFPA 101, *Life Safety Code.*
NFPA 481, *Titanium, Production, Processing Handling and Storage.*
NFPA 482, *Zirconium, Production, Processing, Handling and Storage.*
NFPA 651, *Manufacture of Aluminum or Magnesium Powder.*

Additional Reading

"Aluminum, The Magic Metal," *National Geographic*, Aug, 1978.

The Aluminum Association, Inc., *The Story of Aluminum*, 1968.

Chemical Technology, An Encyclopedic Treatment, Metals and Ores, vol. 3, Barnes and Noble Books, Imports, Totowa, NJ, 1970.

McKinnon, G. P., ed., *Fire Protection Handbook*, Fifteenth Edition, National Fire Protection Association, Quincy, MA, 1981.

12

Wood Products

NFPA Industrial Fire Protection Section
Pulp & Paper — Wood Products Industry Committee

The wood products industry in the United States is a major industry with processing facilities located in all sections of the country where wood is available. The south and northwest, however, have the greatest concentration of plants because those areas contain the largest stands of softwood trees, the major raw material for the industry.

For the purposes of this chapter, the wood products manufacturing industry is considered to include (1) sawmills, which reduce rough timber to standard-sized lumber, a process that generates much waste stock, and (2) the manufacture of plywood and particleboards, the latter product using as a principal raw material the waste generated in sawmill operations. It does not include the process of converting wood into finished products, such as furniture and other woodworking activities. The converting processes, as they apply to wood, are discussed in Chapter 13, "Furniture Manufacturing."

The principal fire hazards associated with manufacturing of wood products are the storage of logs in large piles that make fire control difficult once ignition occurs; the generation of large quantities of waste ma-

---

This Chapter is based on contributions made by members of the Forest Porducts Industry Committee, a standing committee of the FNPA's Industrial Fire Protectioin Section.

Chapter 12

FIGURE 12.1. Aerial view of sawmill (includes log storage, lumber storage, and waterfront).

terial, including fine dusts, in the process of reducing the raw materials to finished products; the use of heat-producing equipment, such as dryers and presses for conditioning the plywood veneers and particleboards; and the storage of finished products (sized lumber and plywood and particleboard sheets).

Each of the three basic segments of the industry, i.e., sawmills, plywood manufacturing, and particleboard manufacturing, will be discussed as separate entities in the following pages.

SAWMILLS

A sawmill's principal function is to transform logs into finished lumber of varying sizes. The basic areas and processes associated with sawmill operations are (1) the log storage yard, (2) the sawmill building itself, which can include a variety of lumber sizing and dimensioning operations, (3) sorting the cut and dimensioned lumber, and (4) drying, finishing, and storing the finished lumber.

Raw Materials

The type of wood processed at a sawmill is usually dictated by the supply that is available within a reasonable shipping distance from the plant. Logs of varying lengths, depending on the needs of the mill, are delivered either by truck or rail car and are moved into the log yard or log pond by conveyor, crane, or industrial trucks to be stored in stacked (high cone-shaped) piles or in ranked (evenly arranged) piles, the latter usually 20 to 50 ft (6.1 to 15.2 m) long but often up to 500 ft (152.4 m). Ranked piles are usually referred to as "cold decks" or "log decks," to differentiate them from log pond storage. They are most commonly found at sawmills, while huge stacked piles are more a feature of pulpwood operations associated with paper mills (see Chapter 15).

Log decks should be arranged with adequate aisles and fire lanes to permit ready access to all areas of the log yard by mobile fire fighting equipment (NFPA 46, *Storage of Forest Products*, provides guidance on fire protection and prevention for log yards where the logs are piled in "cold," or "log", decks.) Where the logs are piled to an approximate normal height of 10 to 15 ft (3.05 to 4.6 m), main thoroughfares through the yard should be at least 60 ft (18.3 m) wide. With a growing trend toward medium- and high-decking arrangements — 40 ft (12.2 m) or more — good separation becomes increasingly important as a safeguard against radiant heat from the face of one high-piled deck seriously exposing adjoining decks. Areas with high decking should be separated from adjacent piles by 100-ft (30.5 m) cleared distances.

Log storage areas should be separated from major plant buildings by considerable clear space. The distance should be based upon the severity of the exposure. The area, height, occupancy, construction, and protection of the exposing structure and the type of piling and height of adjacent piles all should determine what constitutes a safe distance. One principal manufacturer of wood products has established that distance as a minimum of 100 ft (30.5 m).

The danger from forest, brush, and grass fires spreading to the log yard is everpresent in areas where sawmills are located. Adequate clear space around perimeters free of all combustible vegetation is the best defense. At least 20 ft (6.1 m) clear space for grass exposures and 100 ft (30.5 m) for light brush exposures are recommended, with greater distances required where there are forested lands exposing the mill.

Another source of fire danger is sparks from refuse burners associated with sawmill operations, boiler stacks, locomotives, and vehicle exhaust. Diesel locomotives entering the yard should be equipped with approved devices, such as spark arrestors, to prevent escape of glowing carbon particles from the exhaust stacks. Industrial trucking equipment used to move logs should be refueled and serviced in areas isolated from the storage area (see Chapter 46, "Materials Handling Systems," for further information on safe practices to observe with industrial trucks).

Fire control systems: A yard fire hydrant system connected to an ample water supply is the basic protection for a log yard. The system should have a minimum capacity supplying at least 1,000 gpm (3785 L/min); larger capacities would be required for more severe situations. Fire protection procedures developed by one wood products manufacturer anticipates that flows of up to 3,000 to 4,000 gpm (11 356 to 15 141 L/min) at high pressure would be required to supply effective hose streams.

It is good practice to have all areas of log yards within 200 to 300 ft (61 to 91 m) of hydrant protection. In large yards this may present problems, as hydrants become a serious maintenance problem if they do not have substantial protection from physical damage. A minimum layout envisions an underground loop around the perimeter of the log yard area with hydrants spaced between 250 and 300 ft (76 to 91 m) apart. A hydrant house at each hydrant in the yard is recommended by NFPA 46, *Storage of Forest Products*; however, some plant installations rely on a few strategically located hydrant houses in the log yard if prompt public fire department or plant fire brigade response can be anticipated and the public department's equipment is compatible with the yard system.

Summary: No two log storage yards are alike; land contours, site configurations, and locations of buildings greatly influence the layout of the area. But if there is an underlying principle in good yard storage arrangement, it is to avoid large, concentrated single decks that mass the log inventory into a single fuel source. Loss experience has shown that it is nearly impossible to extinguish an established log deck fire using hose streams alone. Even the aid of chemical "bomber" planes may be ineffective in putting out fires; thus, it must be anticipated that once fire takes hold in a huge log deck, effort must be directed at saving logs in nearby piles, either by wetting them down or physically removing them from danger.

The Sawmill Building

Logs are moved from the log storage yard by mechanical moving equipment to conveyors, usually of the drag chain or rubber belt type, that carry the logs to the sawmill building. Once a log is in the mill, the initial steps are to cut it to a workable length and then to remove its bark by either a mechanical or hydraulic debarker. After debarking, the log is placed onto the carriage and passed through the head rig, made up either of a band saw or a circular saw. Band saws are favored for two reasons: (1) they make a smaller kerf (size of cut), which increases productivity, and (2) they are capable of sawing in both directions (one cut up and one back), which also increases productivity.

After the log has passed through the main saw, it is fed through either a set of gang saws or a resaw. The purpose of the gang saws and resaws is to further reduce the size of the "cant" (a piece of wood 8 in. by 10 in. (0.20 to 0.25 m) in diameter or larger). The lumber is further

WOOD PRODUCTS

FIGURE 12.2. Abrasive mechanical debarker.

improved by an edger which squares-up the board and removes any additional bark which might still be attached.

All the boards are then passed through a trim saw where they are trimmed to within 1 in. (25.4 mm) of their desired length. The trim saw is also used to cut out defects and knots. The boards then pass from the mill proper to a sorting operation.

Many variations of the above operation may be encountered. Additional processes, such as manufacturing of moldings, gutter stock, and other special shapes, are often done in conjunction with the basic sawing operations.

Sawmill buildings are usually one-story with a basement area, the latter housing a mechanical conveyor system that carries away debris and waste material (sawdust, bark, waste wood, etc.) that drops to the conveyor through channeled openings from the sawing floor above. The waste material is then carried to a "hogger" which is essentially a sizing mechanism that reduces all the wood to a common, smaller size to ease further processing. Clean sawdust is collected for shipment or conveyed to nearby paper or particleboard mills for use as raw materials in those production processes. Waste bark, shavings, and poor quality sawdust are used as hog fuel to fire plant boilers.

Fire hazards: An extremely large number of sawmill buildings are of wood construction, which, coupled with the combustibility of the lumber itself and the waste material generated in the sawing process, represents an extremely high fire load. Other hazards include the use of combustible hydraulic fluids in some process machinery and the presence of temporary heating devices such as salamander-type heaters and barrels with fireplace-sized logs. Sparks from welding tools, faulty electrical equipment, and smoking are also major causes of fires in sawmill buildings.

It is obvious that good housekeeping practices are very important because of the large amount of loose bark, waste wood, and sawdust generated in the sawing operations. The underfloor areas in the vicinity of waste conveyors are particularly prone to accumulations of waste that escape from the collecting and conveying system. (See Chapter 50, "Industrial Housekeeping Practices.")

Safeguards: All areas within the sawmill building should be protected by automatic sprinklers. Because mills are usually located in remote areas where quick response from paid or volunteer fire departments is unlikely, reliance on automatic protection serviced by good water supplies (as required for log yard protection) is the best first line of fire defense. In addition, standpipe hose stations with 1½-in. (38.1 mm) hose lines and portable fire extinguishers located throughout the millyard are essential for attacking incipient fires.

Lumber Sorters

From the sawmill operations, the sized lumber goes to a "green chain" or a sorter where the lumber is sorted and stacked according to length, width, and thickness. The green chain, an operation that is now seen less and less because of its high labor intensity, is basically a long table about 10 ft (3.04 m) wide on which the lumber is laid out as it emerges from the saws. Employees on each side of the table manually sort the different-sized lumber and put it in bins or on carts to await further processing.

Most prevalent in the industry today are mechanical sorters of varying types. In one commonly found arrangement the lumber is sorted in a tower structure at the feed end of the sorter which is essentially a series of "trays" stacked one on top of another. As the lumber is lifted toward the top of the sorting tower, various sizes of lumber are diverted into predetermined trays. A tray may have its diverting mechanism set to accept lumber having particular dimensions, and the sorter operator can adjust these mechanisms as necessary. Once on a tray, the lumber is moved over tray rollers toward the discharge end of the sorter where a swinging arm conveyor is often used to unload the individual trays. Other types of sorters may have bins in which the sorted lumber is deposited.

Generally, sorters have roofs to protect them from the weather, an arrangement that requires sprinkler protection at the ceiling level to protect the sorter and its load of lumber. The sorting operation often creates

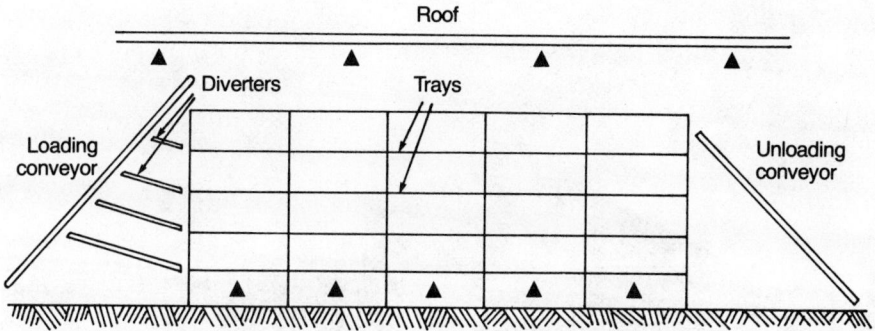

FIGURE 12.3. *A simplified representation of an elevated sorter showing the location of the diverting mechanisms, the trays that hold the varying lengths and grades of lumber, and the areas requiring sprinklers.*

a latticework of horizontal and vertical flue spaces around the lumber lying in the trays. Fire can easily work its way up through the sorter, feeding on the loose lumber in the sorter; however, the loss record is favorable due to the lower hazard of the green lumber being processed.

Particular attention must be paid to the area at the bottom of the sorter. Depending upon the type of sorter, loose pieces of wood can drop from the trays and accumulate at the underside if housekeeping is not pursued vigorously. Automatic sprinklers installed under the lowest tray in the sorter are effective in protecting against fire originating in accumulations of waste.

A simplified representation of the different components of the sorter and the location of sprinkler protection are shown in Figure 12.3.

Drying

After sorting, the lumber is prepared for drying to predetermined moisture contents. Drying is accomplished either in kilns or by stacking the lumber in the yard for air drying. (See Chapter 33, "Lumber Kilns, Agricultural Dehydrators and Dryers.") To reduce the drying time and to assure uniformity in the drying operation, small boards are placed between each horizontal layer of dimensional lumber. Lumber stacked in this manner is often termed "stickered" or "stuck" lumber.

The stickered storage arrangement is not without hazard. The separation between each layer of lumber creates horizontal flues through which fire can burrow making extinguishment extremely difficult. In fact, where stickered lumber is stored in high piled quantities it presents a very severe hazard that requires large hose stream equipment and powerful water supplies for fire control. Portable turrets, deluge sets, and monitor towers are good large stream devices for the type of protection

Chapter 12

FIGURE 12.4. "Stickered" lumber adjacent to kiln cooling sheds.

required and are capable of delivering in excess of 1,000 gpm (3785 L/min). In large yards where the hazard is severe many of these devices may be operated simultaneously.

Good fire lanes permitting access to individual stickered piles and wide separation distances that minimize exposure hazards are points of concern. NFPA 46, *Storage of Forest Products*, and NFPA 80A, *Protection from Exposure Fires*, are sources for guidance in providing fire and exposure protection where lumber is stored in yards.

After the lumber is dried in the kilns, it is cooled in cooling sheds. Usually this is done for the softwood species only, as it is dried at much higher temperatures than hardwood. When the lumber is removed from the cooling sheds it is in the final stage of production.

Lumber Finishing

From the cooling shed the seasoned lumber goes to the planer building where the boards are trimmed by high-speed rotating knives which shave the boards down to desired size, such as the common 2 by 4s (50 by 100 mm) and 2 by 6s (50 by 150 mm). The boards are again passed through a trim saw, which is used to cut out defects, such as knots, split ends, or curls, and to cut the boards to precise lengths. The final stage in the operation is sorting the lumber by grade, stacking it, and preparing it for shipment or storage.

It is in the planing mill that pneumatic stock conveying systems may be found. They are used to convey the fine shavings and dust away from the planes and sanders. (See Chapter 45, "Air Moving Equipment," for a discussion of the hazards associated with air moving equipment.)

It is essential that good housekeeping practices are observed to keep dust collections and wood debris at a minimum. As with other buildings in the sawmill operation, the planing mill should be fully equipped with sprinklers and sufficient hand extinguishers and standpipe hose stations to attack incipient fires.

Storage of Finished Lumber

The finished lumber may be stored outdoors or in sheds. The lumber may either be banded in even-ended or random length bundles or stored in unbanded or stickered piles. Fire protection requirements for outside storage of lumber are governed to a large degree by the type of storage. Yards holding only flat piled green lumber, for example, require only the minimum type of protection, which is generally the same degree of protection given to log storage yards. High piles of lumber stickered for air drying, however, present a very severe hazard that will require large stream equipment and greatly expanded water supplies for fire control.

Large stream equipment, such as portable turrets and deluge sets, require supplies of 750 to 1,000 gpm (2839 to 3785 L/min) for each appliance. Monitor towers may require supplies in excess of 1,000 gpm (3785 L/min). It is particularly important that no part of the storage area is more than 50 ft (15.2 m) distant in any direction from access by motorized fire fighting equipment and that wide fire lanes provide the needed access. NFPA 46, *Storage of Forest Products*, gives guidance on good arrangement of outdoor lumber storage and the protection that it requires.

Shed storage requires good sprinkler protection. One manufacturer of wood products requires varying sprinkler densities governed by the storage configuration. Stickered lumber, for example, requires a density of 0.45 gpm per sq ft [18.3 (L/min)/m^2] while banded, even-ended lumber requires only 0.20 gpm per sq ft [8.1 (L/min)/m^2]. In either case, though, the storage pile is limited to 20 ft (6.1 m) in height.

COMPOSITE PANEL PLANTS

Two important classes of wood products are made from the waste resulting from the sawing of logs into dimension lumber. These products are particleboard, which is medium density fiberboard (MDF), and hardboard, which is oriented strand board (OSB). The processes for producing them are much the same. Basically the products are either wood flakes or wood fibers bonded by a resin and pressed into boards.

Raw Materials

Chips, shavings, and sawdust, the principal raw materials for particleboard and hardboard, are delivered to the plant by truck, rail, or conveyed from adjacent woodworking processes. They are usually stored in open-bay buildings. Free-fall dumping of incoming materials and retrieval by front-end loaders can put wood dust into suspension and create an explosion hazard. Since this is an inherent hazard, protective, rather than preventive, measures are indicated. First, the raw material, when possible, should be prescreened to remove small particles or "fines" that could be potentially explosive later in the production process. The fines may be stored in bins or silos for future use, which might be fuel for dryer equipment if the plant has a burner that can handle wood waste. Removal of the fines at this point also helps to minimize the accumulation of dust through the plant.

Front-end loaders used for retrieval and transport of the raw materials in the storage area should be of a type suitable for use in dusty locations where the danger of explosions may be present. Most common machines are diesel-powered units that require safeguards with fuel and exhaust systems that prevent the machines themselves from becoming ignition sources. NFPA 505, *Powered Industrial Trucks*, contains guidance on the selection and use of various types of industrial trucks.

The Production Process

Briefly outlined, the production process starts with a grinding operation to mill the particles to desired size. After grinding, the particles are dried and screened to remove fines and return oversized particles for further grinding. The particles are then dried and mixed with a resin, such as urea-formaldehyde or phenol-formaldehyde, and a preservative and then formed into boards or panels.

Grinding: The first step in making particleboards is to grind or mill the chips, shavings, panel trim, etc., to the desired size in hammer mills, knife hogs, flakers, etc. (machines that are similar to those described in Chapter 34). But before the materials enter the grinders and mills, they should pass through separating equipment (magnets, air dropouts, and chip washers) to remove "tramp metal" or other foreign objects which might cause sparks or machine failure. Where production is heavily dependent on these grinding and milling machines and the raw material is fairly dry, explosion suppression or spark detection/suppression systems offer good protection. Activation of each system should shut down the equipment, stop material from entering and leaving, and activate deluge sprinkler systems at key collecting or transfer points in the stock-conveying system downstream from the grinders.

Wood fiber refiners operating on green material are not as hazard-

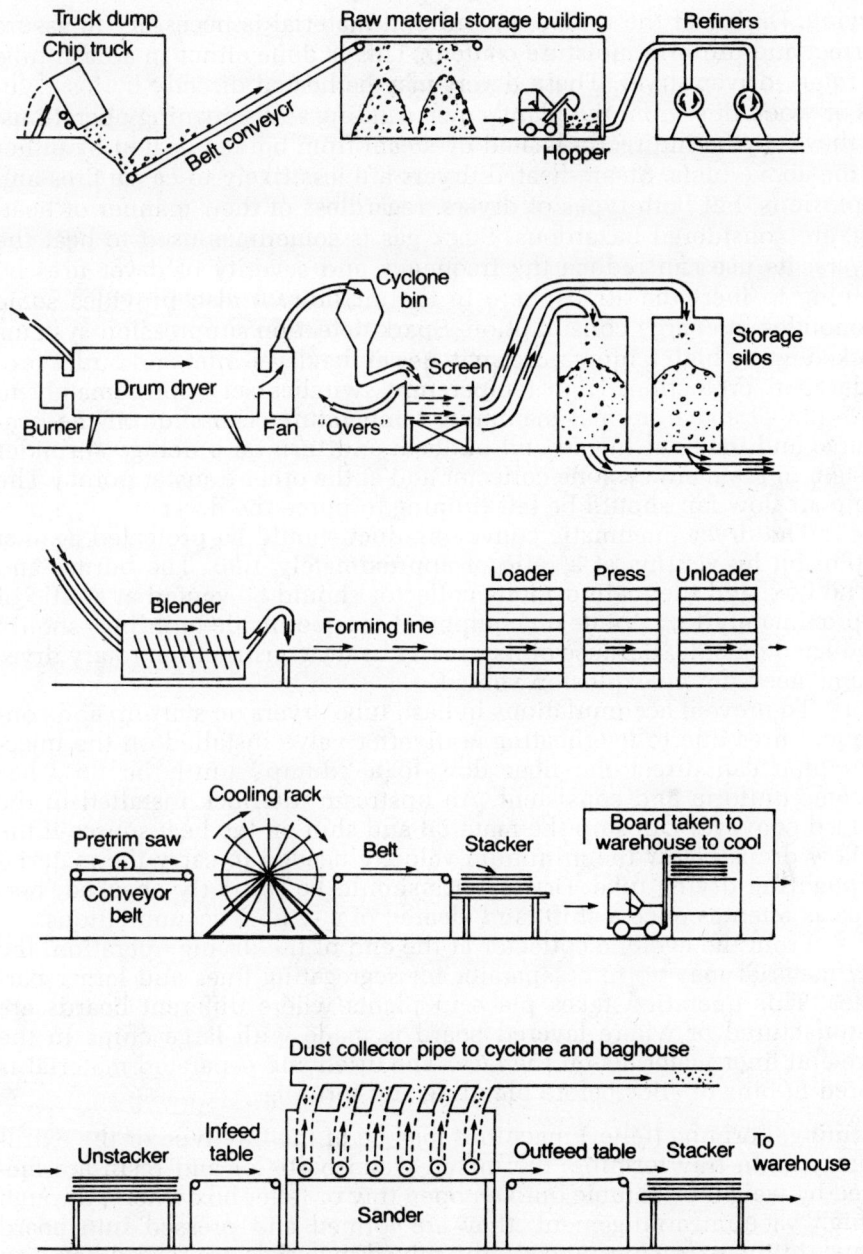

FIGURE 12.5. Flow diagram of a basic particleboard production process.

ous due to the high moisture content of the material being processed, and no special protection seems to be warranted.

Drying: Drying of the milled and ground material is necessary to assure correct and uniform moisture content. This is done either in a flash tube or rotary-driven dryer. These dryers may be heated directly by fossil fuels or wood dust from the plant's own sanding and screening operations, or they may be indirectly heated by steam from boilers heated by either of the above fuels. Steam-heated dryers are less likely to cause fires and explosions, but both types of dryers, regardless of their manner of heating, are considered hazardous. Stack gas is sometimes used to heat the dryers. Its use can reduce the frequency and severity of dryer fires by helping to inert the atmosphere in the air flow. It also provides some economies in energy consumption. Spark detection/suppression systems backed up by high temperature switches at the dryer inlet and outlet provide good protection. High temperature switches set approximately to 50°F (28°C) above normal maximum temperatures can shut off the heat source and material infeed and outfeed, and turn on a deluge sprinkler system in the main cyclone collector and at the other transfer points. The main air flow fan should be left running to purge the dryer.

The dryer pneumatic conveying duct should be protected against explosion by venting at a ratio of approximately 1:30. The burner and blend box, and the main cyclone collector should be vented at a ratio of approximately 1:35. Vents on equipment located inside buildings should be ducted outside. Because of its usually heavy construction, rotary dryer drums need not be explosion vented.

To prevent accumulations in flash tube dryers on start-up and consequent fires due to overheating, a diverter valve installed on the injection pipe can direct the fiber flow to a "dump" until the flow has become uniform and consistent. An upstream interlock installed in the in-feed conveyor can stop the material and shut off the heat source if the airflow drops below the minimum velocity needed to carry the material through the drying tube. Dryer tubes should be regularly checked, perhaps as often as once a shift, and cleared of any fiber accumulations.

From the cyclone collector at the end of the drying operation, the raw material may go to a separator for segregating fines and larger particles. This operation takes place in plants where different boards are manufactured or where layered board is made with large chips in the core and finer materials on the face. The dried and separated material is stored in bins or silos before blending and forming.

Forming: Forming (felter) machines may be of the tray type or the extrusion type. In tray forming, the mixed wood particles and resin are metered by weight or volume onto an open tray or felter box where, through a high vacuum arrangement, they are formed and pressed into board shape. In the extrusion process, the mixed materials are forced between heated platens. Thick boards made by the extrusion process may be produced with hollow cores by placing rods in the aperture ahead of the platens.

The properties of the finished particleboard will depend upon the size and orientation of the particles and the method used to produce the board. The tray-type method tends to orient the particles with long dimension approximately parallel to the face of the board, while extrusion leaves the particles randomly oriented.

Hardboard and medium density fiberboard are often produced by much the same method. The difference between these boards and particleboards or strand boards is that they are made of wood fiber instead of flakes. The fiber is produced by heating and pressurizing wood chips to 1,000 psi (6895 kPa) in a steam chamber. When the pressure is abruptly relieved, the chips puff into a brown fiber. This fiber is dried and then subjected to the forming processes similar to those for particleboard.

The forming boxes for composite panels present some hazards. Particleboard and OSB having a low percentage of dust are not particularly susceptible to explosion at this point in the process; however, a fire hazard does exist, and automatic sprinklers or manual waterspray nozzles provide good protection at the forming boxes.

Fiberboard and hardboard, however, have a high concentration of light, fine fibers which can easily form explosive concentrations in air. Forming boxes should be protected by an explosion suppression system. Protection should extend to the suction boxes below the felter, and the suppression system should stop the pneumatic feed to, and the conveyor from, the forming boxes and stop the fans for the suction boxes and the scalping system.

A pneumatic conveying system to the felters and from the suction boxes may also require suppression systems or venting, and a fire protection system similar to that for particleboard forming boxes provides good backup protection.

Pressing: From the forming boxes, composite panels go to the hot presses where they are further consolidated by pressure. Under the heat, the resin liquefies, flowing around the wood particles or fibers. As the resin is thermosetting, it soon solidifies to form a solid sheet of wood and plastic.

The hot presses use fluid power systems to provide the required force. Such systems often use petroleum-based fluids, which, if permitted to escape, present a serious hazard. Ruptures in the fluid power lines can dump large quantities of flammable liquids into the press pit, and leaks can spray them over considerable distances. Because the hot press itself is an ignition source, such leaks and ruptures may result in serious fire. (See Chapter 43, "Fluid Power Systems.")

It is imperative, then, that hot presses be adequately protected. Easily accessible shutoff valves on the hydraulic fluid reservoirs of the presses can minimize flows in emergencies. Fire suppression systems ideally include automatic sprinklers, deluge systems, or both, for press pits, press hoods, and ventilating fans. The discharge nozzles on the systems should be located in normally hard-to-reach places, such as between the press and hydraulic cylinders. Foam generators for blanketing oil fires in press pits are another good protection measure.

Ovens and humidifiers: After pressing, the boards are usually humidified in oven-like enclosures. Fires occurring in oven-humidifiers have proven difficult to extinguish, chiefly because of the equipment design. Opening the equipment to get at the fire delivers more oxygen and increases the severity. The usual precaution is to install sprinklers inside the equipment. Special consideration should be given to designing the automatic sprinkler system to match the configuration of the enclosure and the usual arrangement of the material being processed. As high temperatures and corrosive conditions in ovens and humidifiers can affect the reliability of sprinkler systems, it is necessary to have a good maintenance program.

Sanders: In the final production stage, the boards are cooled and sanded as part of the finishing operation. The large belt sanders and the dust-collecting systems associated with them are particularly susceptible to fire and explosion hazards. Hot bearings and static sparks are two common sources of ignition of dust residues at the sanders. Bearing temperature monitors interlocked with controls are frequently installed to shut off the machines if the temperature becomes critical. Low static sandpaper belts are available, and they are useful in reducing the amount of static electricity generated, but the entire sander system should be grounded or bonded as insurance against static buildup.

The sanders should have regular periodic inspections to ensure that they are properly lubricated and the belts are in good condition. Whenever they produce any unusual vibrations, they should be immediately shut down for necessary repairs.

The dust-collecting systems include cyclones, baghouses, condensers, cloth screen arrestors, centrifuges, and similar devices to separate solid material from the air which transports it (see Chapter 45, Air Moving Equipment). The air-dust mixture is potentially explosive, and it is usually within the collecting system that explosions originate. The separating equipment is best located outside of buildings and equipped with backdraft dampers to keep explosions from flashing back to the sanding equipment.

All dust-producing equipment must be dust-tight or connected to adequate exhaust systems to be effective. Collecting systems for sanders are usually vacuum type with the fan located beyond, or on the "clean side" of, the separator. The air flow capacity of the system should be sufficient to keep the dust concentration at a low level, thus helping to prevent an explosive mixture from forming in the conveying ductwork.

Spark detection/suppression systems in the main duct located as close as possible to the sander can be arranged to shut down the sander in an emergency and to stop conveying systems to and from the machine as well as to activate fire extinguishing equipment in the collector downstream; however, it is good practice to keep the main airflow fan operating in an emergency in order to purge the system. If steam is available, a manual backup smothering system should be considered for the dust removal system.

For further information on dust collecting systems see Chapter 45,

"Air Moving Equipment." NFPA 91, *Blower and Exhaust Systems*, and NFPA 664, *Prevention of Dust Explosions in Wood Processing and Woodworking Facilities*, contain detailed requirements for installation of exhaust systems and collecting equipment and information on hazards involved in the production of finely divided wood particles, their removal from the point of operation, and their subsequent disposal.

Conveying Systems

Conveying systems adaptable to composite panels plants are of six major types. These are low pressure pneumatic systems, high pressure pneumatic systems, screw conveyors, flight conveyors, bucket elevators, and belt conveyors. Because they move combustible, and potentially explosive, materials, they may be the mechanisms for rapid fire spread throughout an entire plant.

Some mention of the pneumatic systems has already been made; however, they should be considered once again. Low pressure systems have relatively high airflow capacities and large diameter ducts. These systems are primarily used for conveying finely divided materials or dusts which can easily reach explosive concentrations. The dust-collecting systems found at sanders are an example.

High pressure pneumatic systems have relatively small airflow capacities and small diameter ducts of heavy construction. Such systems can move larger-sized particles than can the low pressure systems, and they carry dusts in high concentrations, usually above the upper explosive limit. However, when the dust is discharged into the collector, it will pass through the explosive range; thus, explosion protection for the collector is needed. High pressure ducts are strong enough that venting is not considered to be necessary. Spark detection/suppression equipment can also be utilized on these systems.

Screw conveyors moving fine, dry material horizontally or on a slight incline are prone to explosion. This is because the material does not fill the duct in which it is flowing. There is enough air in the system to create an explosive mixture. By cutting out one revolution of the screw helix, a plug can be created to block air movement. But this is not complete protection, and a baffle should be added where the helix is removed to block the top portion of the conveyor housing. This will help to prevent flashback, but if a fire or explosion does occur, the conveyor must be stopped to prevent smoldering material from moving farther downstream. In addition, the conveyor housing should be properly vented, especially when the conveyor is directly connected to a possible ignition source such as a dryer or grinder.

Flight or drag link conveyors consist essentially of an endless series of baffles or scrapers which pull the material along. If they are enclosed, they present hazards similar to those of screw conveyors, but especially in the flight return space. Here, there will be air to mix with the residual dust carried in by the conveyor. Enclosed flight conveyors should be vented and equipped with an automatic infeed cutoff if conditions indicate they are necessary.

Bucket conveyors (elevators) are inherently hazardous because they produce a column of dust which can readily reach explosive concentrations. They are not practical for particleboard plants. If they are used, they should be installed outside of buildings and the entire top should be designed as an explosion vent. Elevators more than 30 ft (9.1 m) long should have additional venting. Belt slippage is a leading cause of fire and explosion in elevators. For this reason, the tail-spool shaft should be equipped with a motion sensor that will automatically stop the drive motor when there is an abnormal slowdown of the belt.

Belt conveyors are not often used for conveying dusts and small particles. If they are used, they normally are enclosed and thus present the same explosion hazards as other enclosed conveyor systems. These enclosures should be of explosion-relieving construction. Dusts should be fed to belt conveyors through a choke to prevent the formation of dust clouds.

All mechanical conveyor systems should be equipped not only with the proper vents and chokes, but also with the means of diverting burning material. Diverter gates can shunt the material to a "fire dump," or the conveyor can be reversed to allow the material to be removed at a safe location. Sensors or detectors can be connected to the gates or drive mechanism to automatically divert the burning material. Pneumatic systems require that the interception be accomplished at the infeed or outfeed points of the system.

The Safeguards

Buildings housing composite panels plants ideally are of fire-resistive, heavy timber, or noncombustible construction. Lightweight, unprotected, steel-type construction is not encouraged for general use, although it may be useful in special cases such as miscellaneous storage and attached shops. Structural steel used in buildings without automatic sprinklers where there is enough combustible material to present the danger of a serious fire should be encased in a protective ("fire-resistive") covering.

Good separation between major plant buildings and each step of the manufacturing process is essential to prevent the spread of fire from one section of the plant to another. This can be done by fire walls, eliminating unnecessary openings between sections or floors, and providing good protection at the openings. A clear space between buildings without yard storage is also a good separation measure.

Those portions of the plant where large amounts of dust are present should be provided with explosion venting. NFPA 68, *Explosion Venting*, contains extensive recommendations on ways to provide good venting for dust explosion hazards.

Fire control systems: Complete automatic sprinkler protection for the entire plant is the basic fire protection system to protect against the hazards associated with high fuel loadings, such as those represented by the raw stock storage and the manufacturing process. Particular attention must be given to protection of system components from physical damage in those areas where dust in suspension creates unusual hazardous conditions, such as the raw materials storage building. A good practice is to "shield" risers by structural building components, provide four-way sway bracing for risers, feed mains, and cross mains, and connect the larger sizes of pipe — 2 in. (50.8 mm) and above — by welds or welded flanged fittings. These measures offer protection to the system from damage by explosive pressures.

Special extinguishing systems, such as deluge sprinkler systems, offer good protection for special hazards such as at hot presses and forming boxes as previously mentioned. Strong water supplies are needed to supply the sprinklers, special systems, and yard hydrants. One major manufacturer of wood products requires a minimum flow of 1,500 to 2,000 gpm (5678 to 7571 L/min) at about 100 psi (690 kPa) for at least 3 hours. A good primary supply would be automatic fire pumps taking suction from ground reservoirs or ground storage tanks supplemented by elevated tanks or reservoirs or a connection to a public water supply.

NFPA 13, *Installation of Sprinkler Systems*, and NFPA 24, *Installation of Private Fire Service Mains and Their Appurtenances*, give guidance on the basic requirements for sprinkler systems and systems of underground yard piping and hydrants.

Housekeeping: As in any plant or structure where dusts are present, careful housekeeping is the primary defense against fire and explosion. Good housekeeping begins with building construction, repair, or remodeling. Wherever possible, ledges should be sloped 60 degrees from horizontal to prevent dust accumulation. Beam flanges and webs should be enclosed and all horizontal surfaces eliminated where possible. If horizontal surfaces are unavoidable, they should be cleaned on a cycle that prevents buildup of dust. NFPA 664 suggests that cleaning done by vacuum sweeping apparatus minimizes the scattering of dust, and that blowing down the dust by compressed air is prohibited since it would pose the danger of throwing up clouds of dusts. Some operators, however, have found that very frequent cleaning using compressed air, particularly when normal operations are shut down, does not create large dust clouds and is a practice they can follow with confidence if strict precautions are observed.

Good housekeeping applies not only to the prevention of dust buildup, but also to the maintenance of equipment. Milling and grinding equipment need regular inspection to prevent bearing failure, loosening of parts, or internal breakage, all of which might cause sparks or sufficient heat buildup to ignite the wood dust. Fluid power lines, hoses, and fittings on hot presses also require frequent inspection.

Summary

Composite panels plants are inherently susceptible to fires and explosions which may result in injury or death to employees and extensive property damage. The hazards can be reduced by one or more preventive or containment measures. Among these are:

1. Prevention of dust buildup by good housekeeping.
2. Minimizing ignition sources by eliminating spark-producing mechanisms and static electricity.
3. Venting buildings, conveyors, and equipment, or use of explosion suppression systems.
4. Isolation of burning materials by use of airlocks, diverters, and flashback dampers, etc.
5. Installing spark detection/suppression and other fire suppression systems.
6. Inerting air-dust mixtures with stack gases.

These may be considered as preventive measures which should be used where applicable to reduce the hazard. They cannot be completely relied upon to prevent fire or explosion, but will help to minimize the consequences. The extent to which any of them should be utilized will depend upon the degree of hazard, personnel exposure, proximity of adjacent structures, cost, and feasibility.

PLYWOOD PLANTS

Plywood, an "engineered" forest product, has a wide range of uses, especially in the construction field. An added economic benefit is that it helps to conserve timber resources for it produces more usable board feet from a given log than could be obtained by sawing into dimension lumber.

Plywood's usefulness comes from its many built-in properties, of which strength and rigidity are perhaps the most important. The strength and rigidity come from the cross-laminated construction. Plywood panels have the wood grain running at right angles in each adjacent lamination. This provides resistance to flexural bending and deflection. And because wood tends to shrink only minimally lengthwise to the grain, plywood panels have greater dimensional stability than do sawed boards. These proper ties make plywood excellent for sheathing wood-framed structures.

Raw Materials and Processes

Plywood is made by peeling thin sheets, or veneers, from debarked logs. Peeler logs are generally about 8 ft- 9 in. long (2.65 m). They are de-

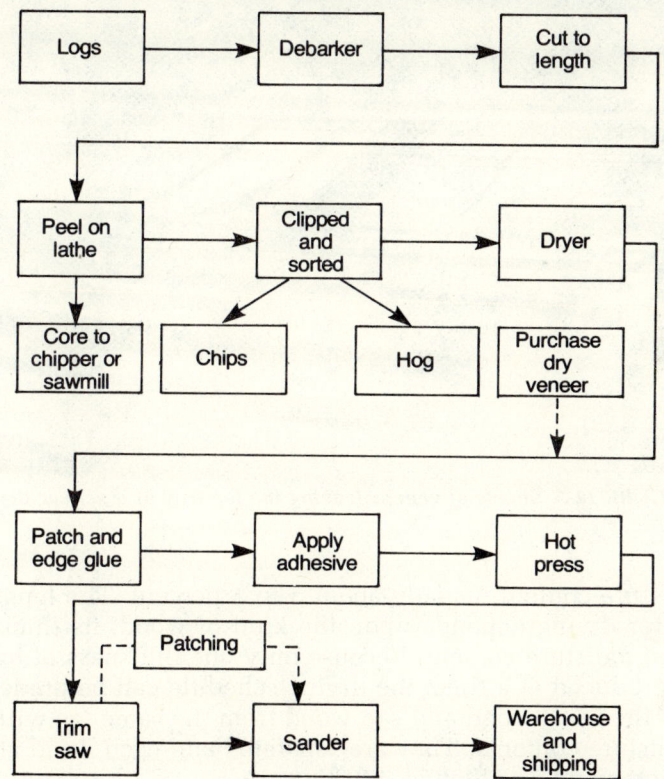

FIGURE 12.6. Flow diagram of a plywood production process.

barked either mechanically or hydraulically. The hydraulic process uses a high-pressure jet of water to remove the bark. After being debarked, the logs usually go to a steam vault to soften the wood for the veneering process.

Veneers, or plys, are produced by rotating the log against a sharp cutting knife on a lathe-like machine that extends the full length of the log. This produces a continuous sheet $\frac{1}{16}$ to $\frac{1}{4}$ in. (1.59 to 6.4 mm) thick and up to 200 ft (61 m) long depending upon the thickness of the ply and the size of the log. This strip is known as green veneer, which is cut to a uniform width by knife clippers. The usual width is either 24 or 48 in. (0.61 or 1.22 m), though other widths may be produced. During the clipping process, unusable veneer is eliminated.

Clipped veneer then goes to a dryer where it is brought to a uni-

FIGURE 12.7. Sheets of veneer leaving the top tray of a veneer dryer.

form moisture content, usually about 3 to 5 percent. The length of time required for drying depends upon the kind of wood, its thickness, and its original moisture content. Because only one thickness of one kind of wood is produced at a time, the drying schedule can be predetermined. However, the heart wood and sap wood from the same log will have different moisture contents. They are separated and each dried at temperatures which may be as high as 400°F (204°C).

Once it has been dried, the veneer is graded in accordance with appearance, for this will determine its final use. The finest veneer is graded N for natural finish. The other grades are A, B, C, and D in descending order. N grade plywood is 100 percent heart wood free of visible defects and is used as the face ply for decorative panelling, cabinet work, and furniture. D grade plywood is generally used for inner and back plies of panels.

Veneer strips 24 in. (0.61 m) wide, or narrower, may be edge-glued to form sheets 48 in. (1.22 m) wide. This is done by planing the edges smooth and applying glue. The pieces are then fed through a dielectric heater to set the glue.

After the veneers have been graded and sorted, they are assembled in the desired arrangement at a glue spreader. The core veneer is glued on both sides by passing it through rollers. It is then laid on one of the outer veneers, and the second is laid on top of it. If the panel is to

FIGURE 12.8. Automatic plywood press during unloading.

FIGURE 12.9. Plywood finishing line.

be of more than three plys, the filler plies are added before the face ply is laid on.

Once it has been glued and laid up, the panel is conveyed to a hot press where it is subjected to pressures up to 200 psi (1379 kPa) and temperatures as high as 300°F (149°C) for as long as eight minutes. This process cures the glue and assures a firm bond between the plies. Different types of glue or adhesive are used for interior panels and those intended for exposure to the weather, and varying heat and pressures are required for proper curing.

From the press, the panels go to a storage pile, known as the hot stack, for at least 4 hours to allow for complete setting of the adhesive. They then are trimmed to exact lengths and widths and, if necessary, sanded to produce the desired finish. If inspection indicates a need for repair, the panels are patched or plugged with small pieces of veneer and resanded. Finished panels are grade marked and identified as to interior or exterior type by a green or red line, respectively, on the panel end and then sent to storage.

The process flow is indicated in Figure 12.6.

The Fire Hazards

Plywood plants do not exhibit the same level of hazard as particleboard plants. This is due principally to the absence of the dust hazard in the early stages of preparing the veneers for assembly into finished plys. (Assembly of 4- by 8-ft [1.2 by 2.4 m] particleboards involves grinding, separating, and drying finely divided wood particles, all processes conducive to creating dusty conditions.) The two manufacturing processes do, however, have similar dust hazards in the finishing departments where the plywood sheets and particleboards are given a finish which uses sanders and the attendant hazard of generating a considerable amount of fine dust. And, too, as with sawmills, plywood plants need log storage yards and rotary saws for cutting logs to proper length. Though plywood plants use whole logs as raw material, they do produce chips, flakes, and sawdust which are the raw materials for particleboards; consequently, plywood and particleboard plants may be located within relatively short distances from each other or may even share certain common facilities, such as the finishing operation where both types of product receive the same attention.

Equipment such as hot presses, sanding machines and dryers are all potential ignition sources in plywood plants. In order to minimize these hazards, plywood plants should have the same careful attention given to prevention and containment measures as do particleboard plants and sawmills.

The Safeguards

Safety precautions for plywood plants should extend from log yards to storage areas for finished panels. As in sawmills, log yards should have adequate aisles and fire lanes depending upon the size of the log decks. Yards should be situated in cleared areas and at least 100 ft (30.4 m) from the nearest building. They should have fire control systems as outlined in the first section of this chapter.

Housekeeping: Inside the plant, the same housekeeping measures as outlined in the section on particleboard plants should be followed (see Chapter 50, "Industrial Housekekeping Practices"). Dust should be held to a minimum by periodic cleanup, preferably by vacuuming to prevent dust dispersion and turbulence. Process equipment should be regularly inspected and maintained to eliminate any potential spark generation or heat buildup.

Processing: Any dusts or particles resulting from the peeling, or veneering, process should be removed from the area by negative-pressure pneumatic systems (see Chapter 45, "Air Moving Equipment"). Trimmings from the clipping process should be transported from the area by a suitable mechanical or pneumatic conveyor so there will be no hazardous buildup of combustible materials in the area.

Veneer driers require special attention because the temperatures reached within them approach 392°F (200°C), the nominal heat of ignition for wood. They should be frequently inspected, and, if necessary, cleaned to remove any accumulation of dust and fibers which might ignite. An automatic deluge sprinkler system is a good protective feature for the interior of dryers. Activation of the deluge system should automatically shut down the fans, burners, and drive machinery. Manual deluge trips should be provided at each end of the dryer. Further protection can be provided by hose stations, supplied by a separate system. Each station should have at least 75 ft (22.9 m) of hose and be located to permit reaching all parts of the dryer.

The high frequency dielectric heaters used in the edge glueing of veneers do not create high temperatures of sufficient duration to present a fire hazard. However, the hot presses for consolidating and partial glue curing do present hazards. As the temperature required is below the ignition temperature for wood, the hazards here are primarily those resulting from leaks or ruptures in the fluid power lines of the presses. Petroleum-based fluids leaking or flowing from broken lines may be ignited by some immediate or distant ignition source. Press pits should be protected as in particleboard plants.

Trims saws and sanders produce dust which should be removed by a vacuum system and transported to separators and storage bins.

All dust and particle conveying equipment should be constructed and vented as is suggested in the section on particleboard plants. Bag houses and separators should be located outside structures, properly vented, and equipped with backdraft dampers to prevent flashback in case of explosion.

As specified for particleboard plants, all structures housing plywood operations should be of fire-resistive, heavy timber, or noncombustible construction. They should be equipped with automatic sprinkler systems with good, strong water supplies.

BIBLIOGRAPHY

REFERENCES

McGraw-Hill Encyclopedia of Science and Technology, McGraw-Hill, New York, NY, 1977.

McKinnon, G. P., ed., *Fire Protection Handbook*, Fifteenth Edition, National Fire Protection Association, Quincy, MA, 1981.

National Particleboard Association, *Dust Explosion Hazards*, Silver Spring, MD, 1977.

NFPA CODES, STANDARDS, AND RECOMMENDED PRACTICES

Reference to the following NFPA Codes, Standards, and Recommended Practices will provide further information on the safeguards for wood products manufacturing discussed in this chapter. (See the latest *NFPA Codes and Standards Catalog* for availability of current editions of the following documents.)

NFPA 10, *Portable Fire Extinguishers.*

NFPA 13, *Installation of Sprinkler Systems.*

NFPA 14, *Standpipe and Hose Systems.*

NFPA 16, *Deluge Foam-Water Sprinkler and Spray Systems.*

NFPA 46, *Storage of Forest Products.*

NFPA 68, *Explosion Venting Guide.*

NFPA 69, *Explosion Prevention Systems.*

NFPA 70, *National Electrical Code.*

NFPA 80A, *Protection from Exposure Fires.*

NFPA 91, *Blower and Exhaust Systems for Dust, Stock, and Vapor Removal.*

NFPA 101, *Life Safety Code.*

NFPA 505, *Powered Industrial Trucks.*

NFPA 664, *Prevention of Fires and Explosions in Wood Processing and Woodworking Facilities.*

Additional Reading

Best, Richard, "Storage Collapse Kills Five," *Fire Command*, vol. 47, no. 11 (November 1980), pp. 20-21, 24.

Dust Explosion and Fires: A Manual. National Particleboard Association, Silver Spring, MD, 1977.

Goring, G., "Sprinkler Protection of Storage Risks," *Fire Protection*, vol. 8, no. 2 (June 1981), pp. 20-25.

"Wood and Wood Products," Fire Safety Aspects of Polymeric Materials, National Materials Advisory Board, National Academy of Sciences, Technomic Publishing, Westport, CT, 1977.

Wood, Paper, Textiles, Plastics and Photographic Materials, vol. 6, *Chemical Technology: An Encyclopedic Treatment*, Barnes and Noble Books-Imports, Totowa, NJ, 1973.

13

Furniture Manufacturing

Edwin P. Bounous

The household furniture manufacturing industry consists mainly of upholstery, bedroom, dining room, occasional pieces, wall units, and decorative items.

Annual factory shipments for the industry presently amount to approximately 9 billion dollars per year with a projected growth rate of approximately 5 percent per year. More than 5,000 companies engage in furniture manufacturing; over half of these have fewer than twenty employees. The larger manufacturers employ between 800 and 1,000 people. Upholstery can be produced and transported very efficiently by small regional plants with a minimum investment through the utilization of purchased components. Including manufacturers of components, the household furniture industry gives employment to approximately 300,000 people.

In recent years, large loss fires occurring in the manufacture of wood products, upholstery, paper, and other related goods have cost an estimated 20 million dollars per year. Except for the newest facilities,

---

Edwin P. Bounous is retired from Drexel Industries, Drexel, NC. This chapter was reviewed by Mr. Lewis Webb, Senior Regional Engineer with Philadelphia Manufacturers Mutual Insurance Company, Greensboro, NC.

many of these older properties have characteristics that make fire control by sprinkler systems and other preventive measures difficult. These are highly combustible structures and, in the case of furniture, are filled with wood and/or fabric-covered finished and semifinished goods as well as other raw materials; ceilings are high and many places have gabled roofs; finishing room exhaust, sawdust, and shaving removal systems make for high draft conditions throughout most plants.

The two principal problem areas in furniture manufacturing are *spray finishing* and *dust collection (See Chapters 25 and 45)*. Most finishing room fires start from maintenance workers' torches, saw or grinder sparks, or overheated motors. Other contributing factors to fire include dirt, rags used in finishing operations, accumulation of overspray, and poor ventilation which leads to a dangerous accumulation of solvent vapors.

Even with the best dust removal systems, woodworking operations remain dusty. The threat of explosion and fire is always present. Primarily, dust fires start at the wood hog where an accumulation of "tramp" metal in the hog trap can become red hot as the crusher knives hammer the metal as well as wood blocks, rip saw edgings, and other waste materials. From the wood hog, fire can spread to cyclones, to the dust house, and often out onto the building roof. Dust fires are hard to control since fires in dust can burrow and smolder even when wet by sprinkler spray. (See Chapter 34 for a discussion of characteristics of and protection for dust explosions.)

RAW MATERIALS

In solid furniture the principal ingredient is lumber, and the most extensively used species are mahogany, walnut, hard maple, cherry, oak, pecan, and white pine. Veneered furniture uses the same species, but logs are sliced and peeled to obtain thin flitches for laminating to tops and panels. Laminates may have a core of hardwood, particle board, hardboard, etc. Generally, posts, bands, rails, and pedestals on furniture match the tops and ends but are solid.

Some furniture is made with plastic components and subassemblies. Most of these parts, such as door assemblies, mirror frames, bed headboards, carved drawer fronts, and posts, have a high labor content. Monomers that have found favor in furniture are urethane, polyester, and styrene. After a base coat has been applied, the finishing process for plastic parts is compatible with that for wood parts.

More and more simulated wood grain materials are being employed on wall and furniture panel material in the form of high pressure laminate, and vinyl prints, or printed directly onto substrate by special cylinders.

Sofas and parlor chairs are built around wooden frames with spring units attached to form seat and back. Urethane encapsulated with

FURNITURE MANUFACTURING

fiber fill as additional insulation and padding to hard-framing materials is frequently added. Sewn covers are applied to padded surfaces as a final operation.

Generally, all case goods, dining room chairs, tables, and some upholstered pieces are packed in fiberboard cartons of 150 to 200 psi (1034 to 1379 kP) bursting strength for storage and transportation.

Lumber is stored in multiple hacks to facilitate handling by forklift trucks. Generally, furniture companies will store a 90-day supply of lumber to allow for air seasoning as well as for protection against seasonal harvesting cycles due to climatic conditions. Dry kilns and dry storage areas are generally sprinklered, but open areas where raw lumber is stored have only fire hydrant protection.

THE PRODUCTION PROCESS

The various departments in a furniture manufacturing complex are:

- Lumber yard, dry kilns, and dry lumber storage
- Rough mill
- Veneer and/or gluing
- Finish machining
- Machine sanding
- Subassembly and final assembly
- Repair, bleach, and finishing
- Rub and pack
- Storage and shipping

After lumber has been dried and conditioned to about 7 percent moisture content, it is moved into the rough mill where long planks are converted to rough blanks from which legs, pedestals, door and drawer parts, posts, rails, dowels, turnings, etc., are later converted to finished parts in the machining department. In the rough mill, cross-cut saws and rip saws as well as planing, moulding, and gluing equipment are located. More often than not, in order to obtain desired sizes, smaller pieces of wood must be glued together under pressure using highly water-resistant adhesives.

The Machine Room

To get specified appearance and provide means of assembly and subassembly, the rough blanks are machined at various stations in the machine room. Parts are stored and transported between stations on hand trucks except where ultrahigh production prevails and belts or dead roller conveyors are used (mainly in rough mill departments). A woodworking machine room layout has shapers, routers, bandsaws, carvers, turning lathes, boring equipment tenoners, variety saws and panel siz-

Chapter 13

FIGURE 13.1. *The outside roof installation of a pneumatic system used to convey wood dust to storage areas and to the boiler room for use as fuel. Note the neat, stickered piles of lumber air-drying in the foreground and set well back from the building itself.*

ers, mitre and bore machines, dovetailers, mortisers, and other special equipment. By the time parts have traveled through the rough mill and the finish machine departments, approximately 50 percent of the material originally purchased has been discarded by defecting at cut-off saws, ripsaws, planing stations, yard and kiln degrade, etc. About 35 percent of this material is conveyorized to the wood hog in the form of blocks and ripsaw edgings to be converted to wood splinters and chips for use in burning in order to generate process steam and heat or in manufacturing particle board, hardboard, or paper. After machining, whether in flat form or irregular shape, all parts are sanded very carefully. Without sanding a good finish cannot be achieved.

Assembly and Inspection

Subassembly and final assembly connection and joining is accomplished by glue, screws, nails (regular and automatic), mortise and tenon, dowels, tongue and groove, and dovetails in case goods; and by hog rings, staples, tacks, etc., in upholstery.

Prior to finishing, case goods are inspected "in-the-white." "Touch-up" and repairs are then made.

FIGURE 13.2. A shaping machine, one of many operations found in a furniture plant machine room. Note the suction inlet on the bench for carrying away shavings and wood dust.

Finishing Operations

Spray finishing followed by hot-air curing of finished furniture is much the same as in other industries that involve coating and curing operations (see Chapter 8). Furniture finishes can be simple combinations of stain-filler-glaze, sealer coats, and top coats. However, the better finishes entail more than twenty steps when bleaching, distressing, padding, application of decorative steps, and sanding between coats are added to the process. Nitrocellulose coatings, wash coats, and top coats continue to be industry favorites because these materials allow notably easy refinishing of damage sustained in manufacturing, in transit, and in storage warehouses. Urethane, polyester, acrylic, and other synthetic finishes have many good features, but repairing is not one of them; consequently, they have yet to gain extensive use. Water soluble finishes are presently in the development stage.

The principal flammable solvents used in the finishing process and their approximate flash points are: acetone, 4°F (-15.5°C); methyl ethyl ketone, 16°F (-8.9°C); naphtha (VM&P), 50°F (10°C); xylene, 81°F (27.2°C); and toluene, 40°F (4.4°C).

In modern furniture plants the finishing process can employ as many as twenty spray booths in a conveyorized operation using 20,000 to 40,000 sq ft (1858 to 3716 m^2) of floor space.

FIGURE 13.3. A sanding machine for applying a smooth finish to furniture legs prior to finishing. The sanding operation is particularly hazardous because of the fineness of the dust created. Visible at upper right are the inlets for suction hoods located over the sanding machine.

Ovens for curing the finished pieces operate at about 250°F (121°C), but even with the best high temperature, high air velocity — 4,000 fpm (1219 m/min) maximum — and conveyor movement through the curing ovens, most multicoat finishes require overnight storage to be absolutely certain that all the solvents have been released prior to rubbing.

Doors and heavily carved legs, crowns, chair backs, etc., are generally polished with oil and pumice, worked with the softest of steel wool; tops and other flat surfaces are machine rubbed with oil, pumice, and very fine cut down paper.

Finished Furniture Storage

Except for upholstery, which is often "shrink-wrapped" and stored in metal racks prior to shipment, case goods are packed, stacked in the warehouse, and shipped in corrugated cartons. Most furniture shipped east of the Mississippi River goes by truck; case goods bound for the west coast travel by rail. (See Chapter 48, "Industrial Storage Practices," for information on the various storage methods that are encountered and the safeguards that can be applied to them.)

THE FIRE HAZARDS

Furniture is manufactured in a hostile environment: in densely populated factories (the industry needs one employee per $25,000 in annual sales), and in structures involving processes that do not always tend toward best practices in fire protection.

Ironically, many fires are inadvertently triggered by those best trained and equipped to fight and put out fires. However, the worst fires in the industry often occur when outside contracted people are on the job — personnel untrained in fire prevention and fire fighting. An 8 million dollar factory and equipment fire that occurred near Asheville, NC in September 1974 started in the finishing room due to poor welding practices. (See Chapter 24, "Welding and Cutting.")

As stated previously, the two most hazardous areas of fire origin in a furniture factory are spray finishing and dust systems. Other than "total fires," the most costly fires are finishing, rubbing, and finished goods storage fires. Great losses here are due to a combination of fire, smoke, and water damage because cartons, wood finishes, decorative hardware, packing materials, and upholstery supplies are severely damaged by water and smoke. Lumber yard fires are also very damaging and hard to control.

Yard Storage

In the lumber yard, when materials are delivered by primary manufacturers who convert trees from stumpage to logs to boards, the pieces are individually inspected and built into hacks for drying and handling. Inspection is either done manually or with the aid of an elaborate mechanical conveyor which eliminates most of the physical effort. Lumber hacks are about 14 ft (4.3 m) long, 6 to 8 ft (1.8 to 2.4 m) wide, and vary in height from 4 to 6 ft (1.2 to 1.8 m). Each layer of boards making up the hack is separated by spacers known as *kiln sticks*. These separators are placed about every 12 in. (0.305 m) along the length of the boards and are ¾ in. thick (19 mm). The sticks are placed in line vertically (one over another) in order to minimize degradation such as bending and drooping during the seasoning and drying process in the yard, in kilns, and in cooling sheds. More important, the spacing allows for good air circulation which is needed for moisture removal at various stages of drying and storage.

In the yards, racks are piled by forklift trucks back to back and up to 16 ft (4.9 m) high. Lengths and widths of aisles for forklift maneuvering depend on the terrain and the type of equipment used.

Among the main fire hazards around areas of lumber concentration are lightning, discarded cigarettes, and boiler room sparks in dry weather. Also, portable heaters during winter, and cleanup fires. Bonfires and barrel fires for warmth also pose hazards.

The honeycomb construction of these lumber piles makes it vir-

FIGURE 13.4. A 30-spindle carving machine. This operation generates much wood dust as evidenced by the accumulation on the floor under the machine. Constant attention to cleanliness is important, particularly at installations where pneumatic removal equipment is difficult to install.

tually impossible to extinguish a fire once it is underway. What adds to the fire risk in a lumber yard is the fact that these stations are virtually unattended except for 8 hrs each day, Monday through Friday.

Milling, Machining, and Sanding

In the rough milling, veneering, finishing, and sanding sections, waste materials that range from edgings to fine dust are removed by conveyors, hand trucks, and pneumatic systems. The larger parts, such as slivers, edgings, and blocks are conveyed to the wood hog by belt conveyor for reduction into particles small and light enough to be moved to boiler room fuel silos or storage bins. Magnetized pulleys for conveyor systems that carry wood refuse to hogs are a recent development that have dramatically reduced the danger of fire caused by tramp metal entering the hogs from the conveyors. Trucks take waste material from machines not serviced by the conveyor to the hog chute which directs material into the hog's pulverizer. Sweepings are a great source of trouble because metal tools, etc., can often be picked up with the debris dropped into the chute to become a source of friction sparks in the pulverizer.

Smaller chips, shavings, and wood dust are carried from their points of origin by air moving equipment connected to cyclones, bag fil-

FURNITURE MANUFACTURING

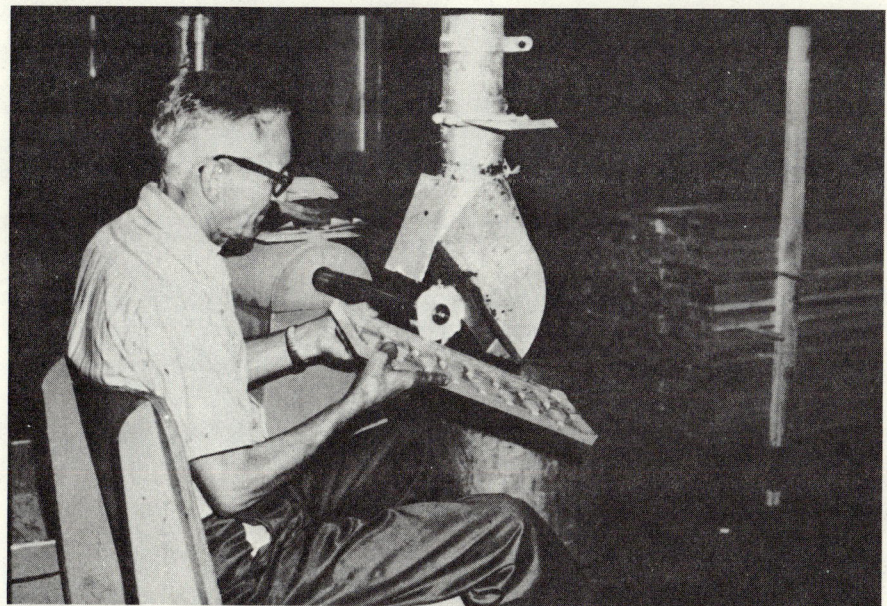

FIGURE 13.5. A single-spindle carver with a well-arranged overhead hood for removing wood dust. Despite mechanization, many woodworking operations continue to rely on manual skill.

ter systems, or equivalent equipment for separation of filtration. In many installations about 80 percent of the air from the pneumatic system is returned to the plant, while solid waste is blown to the fuel silos or sold for use in pulp or particle board manufacturing. Re-injection of the pneumatic system's air into the plant atmosphere can help to cut down on the air makeup requirement for the system and reduce plant heating fuel needs; however, fire and explosion hazards may be introduced by the practice, particularly from transient flammable solvent vapors that may enter the pneumatic duct system for moving wood waste. Chapter 45, "Air Moving Equipment," contains information on good practices for installing air moving equipment and safeguards that should be observed in their operation. NFPA 664, *Woodworking and Flour Manufacturing Plants*, and NFPA 91, *Blower and Exhaust Systems for Dust Spark Vapor Removal*, also contain guidance for the installation and protection of air moving systems for wood waste in furniture plants.

Finishing Operations

The principal hazard of the spray application of finishers involves the hazard of flammable and combustible liquids and their vapors or mist, and combustible residues in spray booths. Properly constructed booths with adequate mechanical ventilation are the ideal way to discharge va-

Chapter 13

FIGURE 13.6. A conveyor carrying unfinished furniture to a spray booth seen at left. This arrangement is typical of modern furniture plants which often have as many as 20 spray booths in the finishing department.

por to a safe location in order to reduce the possibility of explosion. Likewise, booths are a good way to control overspray residue, much of which is not only combustible but also subject to spontaneous ignition. The elimination of all sources of ignition in the spraying area and constant supervision of the overall finishing process are essential to a safely conducted operation. Chapter 25, "Spray Finishing and Powder Coating," gives detailed information on spray booths and spraying equipment and the safeguards that must be observed with them. NFPA 33, *Spray Application Using Flammable and Combustible Materials*, outlines practical requirements associated with good practices in operations involving flammable coatings.

Mechanical ventilation of the spraying area is important to remove solvent vapors and to control overspray. Generally, ventilation provides an average velocity of not less than 100 fpm (30.5 m/min) over the open face of the booth, sufficient to contain overspray to the booth interior. In a good installation, each spray booth exhausts about 25,000 cfm (708 m^3/min). For example, in a spraying area containing 20 booths, about 500,000 cfm (14 158 m^3/min) of air is exhausted. The make-up air required to replace the exhausted air would represent a considerable expense for heating fresh air taken from the outside. It also means that air is moving at a significant velocity throughout the plant; often doors, windows, staircases, and other openings can be like wind tunnels with inrushing air until the exhaust system is turned off. These conditions can help a fire to spread and become more intensive. See Chapter 45 of this text and NFPA 91 for further information on ventilating spray finishing areas.

FURNITURE MANUFACTURING

The curing ovens associated with the finishing process are essentially Class A ovens as identified by NFPA 86A, *Ovens and Furnaces Design, Location and Equipment*. These are enclosures that operate at approximately atmospheric pressure and in which there is an explosion or fire hazard from flammable volatiles and combustible residues. They are also continuous-type ovens in that the finished furniture passes through the oven on a conveyor. The oven exhaust system may go directly to outdoors for removal of the vapors or it may be permitted to exhaust directly into the immediate area. In the latter instance, the area needs balanced ventilation, as with spray booths, to bring fresh make-up air from outside the building to exhaust the vapor-laden air to the outdoors. Chapter 27, "Heat Processing Equipment," contains further information on Class A ovens.

THE SAFEGUARDS

Older furniture factories are most difficult to make safe against fire. Some started as sawmills or sash and door plants. They are generally of combustible construction, located in out-of-the-way places, and over the years have been enlarged as needed with little planning except to provide more space.

In the last 20 years, newly constructed woodworking facilities, as in many other industries, have been primarily single-story structures of noncombustible construction with exposed steel framing and steel roof decks. Frequently they have large open areas that offer the production advantages of more efficient work flow patterns, materials handling, and inventory control. In view of the open area concept for manufacturing, it is desirable, wherever practicable, to have the spraying operations confined to a detached building, particularly if they are quite extensive, or at least separated from the remainder of the manufacturing operation and finished furniture storage areas by fire walls with protection at openings. If conveyor systems pierce the walls, it is sometimes difficult to arrange automatic fire doors to protect the openings. Chapter 46, "Materials Handling Systems," contains suggestions for protection of conveyor openings in fire walls.

Fire Control Systems

Complete automatic sprinkler coverage is the best fire protection for furniture plants. However, woodworking properties, such as furniture plants, involve a variety of factors that could work against effective control of fire by sprinklers. These include the combustibility of contents, obstruction to water distribution, high ceilings, open-sided structures, and, mostly in the case of older structures, unprotected floor openings and steeply pitched roofs. NFPA 13, *Installation of Sprinkler Systems*, recognizes that there may be adverse circumstances influencing protection and, in an appendix to that standard, suggests that upward adjustments in the sizes of piping supplying sprinklers may be advisable in anticipation of the likelihood that more sprinklers may operate in a fire in view of the adverse hazard conditions. The same general recommendations also single out areas for paint spraying and upholstering and similar hazardous processes that require extra hazard protection (larger pipe sizes and closer sprinkler spacing) as defined in NFPA 13.

Automatic water-spray extinguishing systems have proved to be effective protection for dust handling systems. Flame radiation detectors mounted inside ducts sense glowing objects and send signals to a control device which actuates a valve controlling water-spray nozzles located downstream of the detectors.

It follows that extremely good water supplies are necessary in view of the combustibility of contents of furniture plants and the presence of extensive yard and shed storage of lumber. Piles of lumber "stickered" for air drying can present a severe hazard. The degree of protection that may be needed can vary from plant to plant, but a basic recommendation is that a yard system of mains and hydrants should be capable of supplying at least 1,000 gpm (3785 L/min), enough to sustain four 2½-in. (63.5 mm) hose streams simultaneously. If conditions warrant, expanded supplies and larger fire stream appliances may be needed for effective fire control. Good unobstructed fire lanes are needed in the yards so that fire equipment can approach the lumber piles. Chapter 12, "Wood Products Manufacturing," contains further information on the hazards and protection requirements associated with yard and shed storage of lumber. NFPA 46, *Storage of Forest Products*, and NFPA 24, *Installation of Private Fire Service Mains and Their Appurtenances*, also give guidance on protection for lumber in storage and on yard systems supplying sprinklers, hydrants, monitor nozzles, etc.

BIBLIOGRAPHY

NFPA Codes, Standards, and Recommended Practices

Reference to the following NFPA Codes, Standards, and Recommended Practices will provide further information on the safeguards for furniture manufacturing discussed in this chapter. (See the latest *NFPA Codes and Standards Catalog* for availability of current editions of the following documents.)

NFPA 10, *Portable Fire Extinguishers.*
NFPA 13, *Installation of Sprinkler Systems.*
NFPA 24, *Installation of Private Fire Service Mains and Their Appurtenances.*
NFPA 27, *Private Fire Brigades.*
NFPA 30, *Flammable and Combustible Liquids Code.*

NFPA 33, *Spray Application Using Flammable and Combustible Materials.*
NFPA 46, *Recommended Safe Practice for Storage of Forest Products.*
NFPA 51B, *Fire Prevention in Use of Cutting and Welding Processes.*
NFPA 63, *Dust Explosions in Industrial Plants.*
NFPA 68, *Explosion Venting Guide.*
NFPA 69, *Explosion Prevention Systems.*
NFPA 70, *National Electrical Code.*
NFPA 86A, *Ovens and Furnaces, Design, Location and Equipment.*
NFPA 91, *Blower and Exhaust Systems, Dust, Stock, and Vapor Removal or Conveying.*
NFPA 101, *Life Safety Code.*
NFPA 231, *Indoor General Storage.*
NFPA 260A, *Standard Methods of Tests and Classification System for Cigarette Ignition Resistance of Components of Upholstered Furniture.*
NFPA 664, *Prevention of Fires and Explosions in Wood Processing and Woodworking Facilities.*

Additional Reading

Buehrer, P., "Check List for the Prevention of Fires Arising from Welding and Allied Processes," *Welding in the World*, vol. 14, no. 5-6 (1976), pp. 122-125.
Deacon, F. C., "Designing Fire Protection to Limit Monetary Loss," SFPE Technology Report No. 80-2, Society of Fire Protection Engineers, Boston, MA, 1980.
"Fires in Furniture Factories," *Fire Protection*, March 1980, pp. 34-35.
McKinnon, G. P., ed., *Fire Protection Handbook*, Fifteenth Edition, National Fire Protection Association, Quincy, MA, 1981.
"Moving Fire: Fire Hazards of Belt Conveyors," *Record*, vol. 54, no. 6 (1977), pp. 18-21.
"Prevent Cutting/Welding Fires," *The Minnesota Fire Chief*, vol. 16, no. 5 (May/June 1980), pp. 16, 48, 54.
Richards, D., "Fire Risks in the Furniture Industry," *Fire Engineers Journal*, vol. 38, no. 10 (1978), pp. 23-24.
Trinks, W., *Industrial Furnaces*, (vols. 1 and 2), John Wiley and Sons, NY, 1967.

14

Pulp and Paper Processing

Jack C. Castleberry and Peter A. Smith

Paper is manufactured from renewable resources such as wood. Most large primary paper mills are equipped to receive wood in its natural form which is then reduced to pulp, and further reformed into an engineered fiber structure called paper.

The common uses of paper products in daily life are well known. Wood pulp is processed to form a wide variety of products ranging from soft absorbent tissue to hard waterproof boards, and also modified into man-made fibers such as rayon and plastics. Major paper products include newsprint, magazines, tissue, bags, cartons, milk containers, corrugated containers, wrapping paper, books, and writing paper. The conversion of pulp and paper into various end products such as milk containers can occur at or near the mill site. Normally, however, processing at the mill ends with the winding, slitting, or coating of paper. Rolled and sheet paper or pulp is then shipped to converting or processing plants located nearer the marketplace. (A comprehensive discussion of paper products can be found in Chapter 15.)

Peter A. Smith is Manager of Property Insurance and Conservation for the Corporate Insurance Department of the International Paper Company, New York, New York. Jack C. Castleberry is Fire Protection Coordinator in Process Technology of the International Paper Company, Mobile, Alabama.

Overall, the pulp and paper industry employs approximately 704,000 persons in the United States. Paper manufacturing accounts for an annual gross product of approximately $48.2 billion.[1] Pulp and paper mills are located near major natural sources of wood. Mixed hardwood-softwood forests are located in the eastern United States, softwood forests are found on the west coast, and major pine forests are located in the southern United States. The primary pulp and paper mills are concentrated in these areas as well as throughout Canada.

A paper mill processes and stores large quantities of combustible material in the form of logs, wood chips, and paper. Usually located in remote areas, most large paper mills rely on their own utility services including electric power, steam, and water, as well as fire fighting capabilities. Large chemical and heat recovery boilers with unique explosion potential, high speed paper machines, and automated process control systems all represent areas where, without the proper safeguards, production losses could be catastrophic. The loss record of the pulp and paper industry[2] indicates that the major hazard areas are:

1. Log and chip piles
2. Black liquor recovery boilers
3. Paper machines
4. Roll paper storage warehouses
5. Process control centers and wiring.

Other areas of special concern include power and waste fuel boilers, turbogenerators, large wood cooling towers, material handling systems, and chlorine bleaching processes.

The critical factors governing fire loss control at a pulp and paper mill are:

1. Fire service water supply and distribution
2. Automatic sprinkler protection
3. Supervision and maintenance of equipment
4. Preplanning procedures and fire brigade organization and training
5. Building construction, separation, and cutoffs.

RAW MATERIALS

Wood

Although paper can be made from rags, bagasse, straw, and other fibrous material, the most common material used is wood, a natural and renewable resource. The basic ingredients of wood are:

1. Cellulose, a crystalline linear polymer of glucose
2. Hemicellulose, composed of various other sugar-type polymers
3. Lignin, an aromatic amorphous polymer that serves as the "cement" which holds the fibers of wood together.

Other extractives include tannins, oils, gums, and resins.

The fire hazard properties of wood are well documented. Wood is a typical Class A combustible solid, and red oak serves as the standard measure for flame spread ratings. Normally, wood must be heated to a temperature well above 500°F (260°C) to result in flaming combustion.[3] Ease of ignition and rate of fire growth depend greatly on physical conditions such as size, shape, moisture content, contamination, and storage configuration.

Spontaneous heating and ignition can occur under certain conditions when heat cannot be readily dissipated. Such heating can occur in a large wood chip pile left standing for a long period of time. The heat of combustion of wood varies from 7,000 to 9,000 Btu/lb (16 282 to 20 934 kJ/kg) depending on its carbon and hydrogen content.[4] Cellulosic materials, which make up 62 to 80 percent of the wood, contain approximately 7,500 Btu/lb (17 445 kJ/kg). Lignin, which makes up 18 to 30 percent, contains approximately 13,000 Btu/lb (29 068 kJ/kg).

Chemicals

A variety of chemicals and chemical processes are used in the manufacture of pulp and paper. These processes involve the delignification of wood. Chemicals are used to dissolve the lignin in the wood to form pulp, and to bleach pulp to the desired brightness.

Pulping chemicals: These include sodium hydroxide, sodium carbonate, sodium sulfide, sodium sulfate, calcium carbonate, sodium sulfite, and sulfur dioxide. Chemical recovery boilers used to recover these chemicals after their use present a special hazard. If water accidentally comes in contact with the hot chemical smelt produced by the boiler, a damaging explosion can occur. Such explosions appear to be physical in nature rather than chemical.

Bleaching chemicals: These include chlorine, sodium chlorate, sodium chloride, methanol, chlorine dioxide, sulphuric acid, caustic soda, and oxygen. Bleaching chemicals such as chlorine, although noncombustible, will promote burning of combustible materials and can form explosive mixtures with flammable gases. Methanol is a flammable liquid. Chlorine dioxide is a water solution and is not hazardous; however, when subjected to below freezing temperatures, crystals are formed which, when heated, can be explosive.[5]

Fire and spills involving either pulping or bleaching chemicals can produce toxic fumes. Full protective clothing, including self-contained breathing apparatus, is recommended.

Fuel

Energy on the order of tens of millions of Btu (kJ/kg) per ton of paper is required by a paper mill.[6] The paper industry requires energy for electric power, steam, and hot water. Traditionally, all or most of the required energy is produced on site. Sources of fuel include (1) wood, oil, gas, and coal which are burned in conventional power or bark boilers, and (2) organic wood residuals, such as lignin from the chemical pulping process, which must be burned in large specialized recovery boilers. (A comprehensive discussion of conventional power boilers can be found in Chapter 42, "Boiler Furnaces.") The storage, handling, and burning of wood, oil, gas, and coal (in the form of combustible solids or dusts, flammable or combustible liquids and gases) all present well-known hazards and require appropriate safeguards. (See the appropriate NFPA standards listed at the end of this chapter for specific recommendations and guidelines.)

Special handling and safeguards are required for the burning of organic residuals which are derived from the pulping process in the form of a solution known as *black liquor*. These safeguards are designed to prevent explosions in the recovery boiler where black liquor is burned.

Handling and Storage

Wood, the major raw material in the pulp and paper process, is received directly from the forest in the form of trees, logs, or chips, or as residuals from other woodworking operations. Shipments are made by barge, rail, and truck. Logs are generally cut to 4-ft (1.2 m) lengths for ease of handling. (See Figure 14.1.) Logs and chips are stored in the woodyard in large piles awaiting processing. The size, age, and moisture content of log and chip piles determine the degree of fire hazard. A well-established log or chip pile fire is an extreme challenge requiring enormous quantities of water.

THE PRODUCTION PROCESS

The three basic processes for making pulp are (1) mechanical, (2) chemical, and (3) semichemical.

Mechanical: This process is frequently called the ground wood process because the fibers of wood are torn out by a grinding stone. Today, the preferred method involves the shredding and grinding of chips between metal shearing disks in a *refiner*. The major advantage of this process is the high yield of pulp produced. Newsprint is generally made from this type of pulp.

Chemical: This process is the most popular method of obtaining pulp. In chemical pulping the lignin that holds the wood together is degraded and

PULP AND PAPER PROCESSING

FIGURE 14.1. *Pulpwood railcars deliver cut logs to the mill woodyard. (International Paper Company)*

dissolved by various chemical reagents at elevated temperatures and pressures in a digester. The two main chemical pulping processes are kraft and sulfite. The kraft process uses caustic soda and sodium sulfide. The sulfite process uses sodium sulfite and sulfur dioxide. In the kraft process, the more prevalent of the two, black liquor containing lignin and spent chemicals from the digester is burned in the recovery boiler. The chemicals are recovered in the furnace smelt bed and recycled while process steam is being generated. Chemical pulp usually yields strong paper.

Semichemical: This is a combination process whereby wood is treated with chemicals and then ground, resulting in a high yield and a strong paper.

The Kraft Process

The kraft production process provides the best illustration of a typical pulp and paper mill operation. Figure 14.2 shows an overall process flow diagram. (1) Logs are removed from storage and conveyed to the barking drum. Bark is removed by revolving the logs in the drum, causing the logs to repeatedly strike one another. The flume is a common method of conveying logs from the woodyard to the debarking process. The bark is collected and later burned as fuel in a boiler. (2) Debarked logs are transferred to a chipper which consists of a large rotating disc fitted with a series of knife blades which reduce the log to chips. The wood chips are screened to approximately one-quarter inch (6.4 mm) and transferred by continuous belt conveyors to the chip pile for holding or the chip bin for transfer to the digester. The continuous belt conveyors can be run underground or overhead. (3) Chips are charged into a digester where they are "cooked" under pressure and steam heated in a solution of sodium hydroxide and sodium sulfide known as *white liquor*. (See Figure 14.3.) The white liquor degrades and dissolves most of the lignin which holds the wood together.

After cooking, the chips are expelled into a blow tank where a drop in pressure breaks the chips into pulp. The pulp is put through a washer where the spent cooking liquor containing the dissolved lignin is separated and sent to the recovery process in the form of weak black liquor, and the pulp is sent on for further refining.[4] The chemical recovery cycle starts with weak black liquor being concentrated in multiple-effect evaporators which increase the solids concentration from 15 percent to approximately 40 to 50 percent. Prior to being burned in the recovery boiler, black liquor is further concentrated in a direct contact evaporator whereby hot furnace flue gases in contact with the liquor increase the concentration to about 63 percent. Many new recovery units replace the direct contact evaporator with a concentrator because of emission control problems created by direct contact of the liquor with flue gas. Proper control of liquor concentration is most important. Firing weak liquor solutions could very well result in serious smelt bed explosions. After concentration, salt cake (sodium sulfate) is added to make up chemical losses in the system. The black liquor is then piped to the recovery boiler and fired through "guns" into the furnace where lignin and other organic materials burn while the inorganics (sodium salts) form a molten ash or smelt in the bed of the furnace and the sodium sulfate is reduced to sodium sulfide. It is the smelt which can react violently to contact with water.

Steam is generated as the organics in the black liquor are burned in the recovery boiler significantly reducing the mill requirements for outside sources of energy. (See Figure 14.4.) Combustion gases exit the recovery boiler through a direct contact evaporator, induced draft fan, electrostatic precipitator, and then through the stack. The smelt, composed mostly of sodium sulfide and sodium carbonate, is tapped from the furnace into a dissolving tank where, under the proper conditions, water is added to form a solution referred to as *green liquor*. The green liquor is clarified and converted to white liquor by treatment with slaked lime. Calcium carbonate is precipitated as lime mud and heated in a lime kiln to regenerate quick lime which is slaked and reused. The white liquor is then recycled through the digester. [5] Pulp from the digester is processed through various screens, beaters, and refiners in preparing the wood fibers for the papermaking stage.

PULP AND PAPER PROCESSING

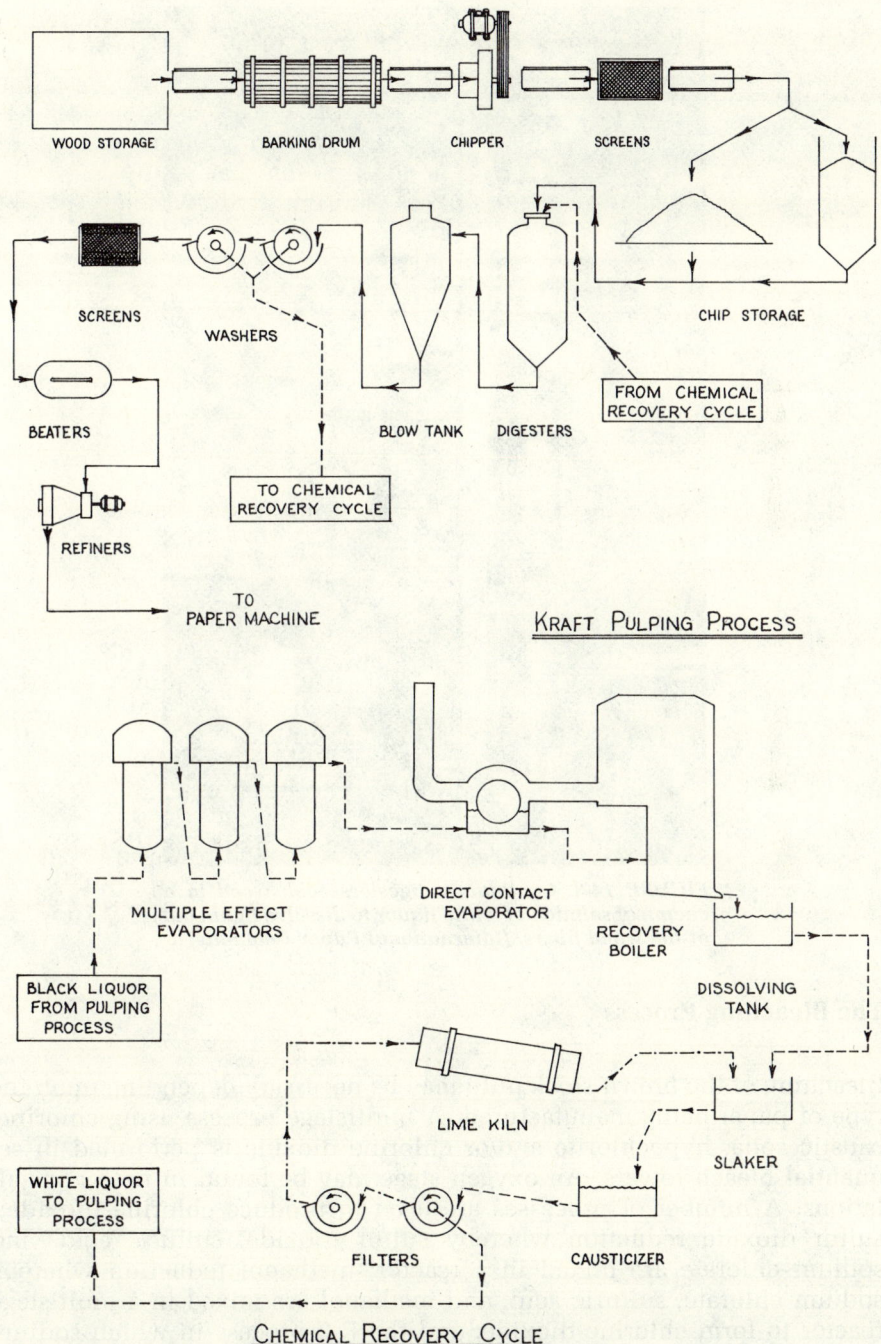

FIGURE 14.2. Kraft process flow diagram. — Kraft pulping process (top) and chemical recovery cycle (bottom). (International Paper Company)

Chapter 14

FIGURE 14.3. Continuous digesters cook wood in a chemical solution of white liquor to dissolve lignin out of the wood fibers. (International Paper Company)

The Bleaching Process

Bleaching of the *brown stock* pulp may be necessary depending upon the type of paper being manufactured. A multistage process using chlorine, caustic soda, hypochlorite and/or chlorine dioxide is performed in sequential bleach towers. An oxygen stage may be found in newer installations. A number of processes are used to produce chlorine dioxide:[7] sulfur dioxide reduction whereby sulfur dioxide, sulfuric acid, and sodium chlorate are mixed in a reactor; methanol reduction whereby sodium chlorate, sulfuric acid, and methanol are mixed in a multistage reactor to form chlorine dioxide; and the R-2 process in which sodium chlorate and sodium chloride are mixed in a reactor along with sulfuric acid and yield chlorine and chlorine dioxide.

FIGURE 14.4. *The black liquor recovery boiler and supporting structure towers above the pulp mill. (International Paper Company)*

The Paper Machine

Refined or bleached pulp is finally transferred to the paper machine where the wood fibers are formed into paper. A modern paper machine can measure as much as 400 in. (10.2 m) in width and hundreds of feet in length (100 ft is approximately 30.48 meters), representing a substantial capital investment. Sheets of paper are formed on the machine when a suspension of pulp fibers in a 99 percent water solution is introduced from the head box onto a moving mesh of fine wire called a *fourdrinier*. The water drains through the mesh by gravity and vacuum pumps, leaving a sheet of paper. The sheet is picked up at the end of the fourdrinier by a moving wool or synthetic felt at the press section.

The press section presses the wet paper between large rolls to squeeze out additional water, and at the end of the press section the paper is roughly 60 percent water and 40 percent fiber. The paper machine, up to this point, is referred to as the "wet end."

The paper now passes through a large series of steam-heated dryer rolls. The dryer section is enclosed by a noncombustible hood which can extend down into the basement area below the machine and encloses the drive gear train on the "back side" of the machine. Within the upper hood area is a plenum chamber running the full length of the hood or hood section. Exhaust ducts from the plenum chamber are run through economizers which recover the waste heat for additional heating capacity in cold weather. From the dryer section, final surface smoothing and

Chapter 14

FIGURE 14.5. The dryer sections of the paper machine are enclosed by large noncombustible hoods. (International Paper Company)

compressing are performed on a series of rolls known as a *calender stack*. After calendering, paper is slit and wound on rolls for storage or shipment. (See Figure 14.5.)

Lubrication of paper machine bearings is accomplished from an oiling system known as a *Bowser system*. A central oil reservoir is located under the machine in the basement area from which oil is pumped to the machine bearings.

Paper machines can be driven by electric motors or steam turbines. Speeds range up to 3,000 ft/min (914 m/min) (about 35 mph, or 55 kph).

A paper break occurring within the machine can send large quantities of paper (*broke*) into the basement (*broke pit*) located below the machine. Conveyors are used to transfer broke to a hydropulper for reprocessing. The operation of a modern paper machine can be monitored and controlled by computer.

Finishing

Some grades, such as fine or white paper, require additional finishing.

Sizing: Sizing modifies surface characteristics for improved printing quality. Chemicals, such as starch, are added by large-diameter rolls usually located after the first dryer section of the paper machine.

Coating: Coating enhances the paper in various ways — for example, making it smoother or adding a protective coating — by applying chemicals to the paper surface. Polyethylene-coated milk container stock is finished on large extruding machines where a film of molten polyethylene is applied to both sides of bleached paper board. Polyethylene beads are forced through electrically heated dies by an extruder screw. The barrel of the screw machine is mounted on a large retractable base which is moved across the paper face. Direct contact gas burners preheat the paper sheet just prior to coating. Hot gases are exhausted through a hood and duct system.

Supercalendering: Supercalendering adds a very high gloss to publication paper by compressing the sheet between heavy steel rolls mounted in a vertical *stack*.

Roll Paper Storage

Paper is normally wound on rolls for storage and shipping. Roll weights and sizes vary; the largest may exceed 2 tons (1814 kg), measuring 6 ft (1.8 m) in diameter. The common categories of paper include linerboard, mediums and krafts of various grades, tissue, newsprint, printing and writing paper, polycoated board, and pulp. Most of the heavier grades of paper, such as linerboard and kraft, are wound with steel bands at both ends of the roll. Printing and writing papers are wrapped with heavy plain paper. Pulp and polycoated paper may or may not be provided with banding. Tissue paper is normally not banded.

Rolls are frequently stored in a vertical position for the most efficient use of warehouse floor space; however, on-side storage is not unusual and, at some mills, rack or "reel" storage is used (especially where large-diameter rolls are being held temporarily for further cutting or processing). Such large rolls may be stored horizontally three high — approximately 20 ft (6.1 m) -in a heavy steel framework. Vertical storage can be handled by clamp truck to an average height of approximately 20 to 25 ft (6.1 to 7.6 m). Roll storage is generally arranged in tightly stacked rows with every two rows separated by a narrow inventory aisle — approximately 18 in. (0.46 m). Main aisles are 12 ft (3.7 m) or more in width. (A comprehensive discussion of general storage practices can be found in Chapter 48, "Industrial Storage Practices.")

Many mills handle rolls with an overhead crane requiring high ceiling clearances. Such *high bay* warehouses can be as much as 60 ft (18.3 m) in height with a clear space (from the tops of rolls to the ceiling) of over 30 ft (9.1 m). (See Figure 14.6.)

THE FIRE HAZARDS

Common Hazards

The common hazards associated with heating, light, power, smoking, and welding are present throughout the pulp and paper mill. Proper installation and maintenance of heating and electrical equipment are essential fire prevention features including provisions for adequate combustion safeguards, overcurrent protection, and grounding. Of special concern are the extent and location of electrical power and control cables with combustible insulation. Significant concentrations of cables having varying degrees of combustibility are found in the power house, pulp mill, paper machine building, and below process control and computer rooms.

Good housekeeping is important in these areas, as well as around and beneath the paper machine, and in all other areas where wood, chips, sawdust, paper, paper dust, oil mists, or other combustible materials can collect. Because of these combustibles, any welding done in the mill presents a major hazard, especially in the woodyard and preparation buildings, near the paper machine, and in the roll paper warehouse. Designated welding areas and a supervised hot-work permit system are essential. (A comprehensive discussion of safe welding practices can be found in Chapter 24, "Welding and Cutting.")

Raw Material Storage

Log piles: These piles are arranged in *stacked* or *ranked* order in the woodyard. Stacked logs are arranged at random in cone-shaped piles. Ranked logs are divided in an even manner. Stacked piles are generally larger in size and height and present the greater fire exposure. Dried-out wood, refuse, and poor housekeeping, combined with uncontrolled ignition sources such as welding, smoking, open burning, or poorly maintained material handling equipment, present considerable fire hazards in the woodyard.

Chip piles: These piles are maintained by bulldozers and built up into large piles by traveling-belt or pneumatic conveyors. (See Figure 14.7.) Moisture content of the chips seriously affects the fire hazard, and control of ignition sources is essential.

Log and chip piles are particularly susceptible to fire during pe-

PULP AND PAPER PROCESSING

FIGURE 14.6. A high bay roll paper warehouse with overhead crane. (International Paper Company)

riods of extended dry weather. If a fire is established in a log or chip pile, control becomes extremely difficult especially during high winds. The size of the pile then becomes the controlling factor in the extent of the loss.

Wood Preparation

Debarkers and chippers: These machines produce large quantities of combustible residue in the form of bark and wood dust. Welding is a major cause of fire in and around the equipment along with friction and electrical defects. (For further information on the handling of combustible dusts, see Chapter 45.)

Conveyors: These conveyors employ continuous rubber belts to move the chips to storage silos or to and from chip piles. Conveyors can be enclosed and run overhead or in tunnels; this creates problems for manual fire attack should a fire occur. (A comprehensive discussion of materials handling systems can be found in Chapter 46.)

FIGURE 14.7. *Chip piles are fed from large overhead belt conveyors. (International Paper Company)*

Pulp Mill Group

The bleaching process: This process involves the use of oxidizing agents such as chlorine and chlorine dioxide in solution. Depending on the process, chemicals used to produce chlorine dioxide can be toxic or flammable. Major hazards involve the shipping, handling, and storage of flammables or toxic gases such as methanol and chlorine. (See Chapter 37, "Handling and Storage of Flammable and Combustible Liquids," for a detailed discussion of the handling of flammable liquids.)

Process and control: Electrical cables are found in significant concentrations in the pulp mill, primarily between motor control centers and process control rooms. Large modern process control rooms can handle the operations of the entire pulp mill including the pulping and bleaching process.

Grouped electrical cables are sometimes concentrated in raised floor areas or rack rooms below the control room. Cable fires in these areas can shut down the entire mill for extended periods. Electrical cable and cable tray installations in and around the pulp mill are subject to corrosive atmospheres, and chemical and pulp spills increase the fire exposure to the cables. (See Chapter 47, "Electrical Installations for Industrial Locations," for a detailed discussion of electrical installations.)

PULP AND PAPER PROCESSING

The Powerhouse Group

The recovery boiler: This boiler presents the greatest hazard in the powerhouse. A violent reaction may result if water comes in contact with hot smelt. Shock waves from such a reaction cause severe damage to the recovery boiler and result in extensive interruptions to production. Water could accidentally reach the smelt bed of the furnace from leaks in the pressure components or tubes of the boiler, from the firing of a low solids concentration or "weak" black liquor, or from the misuse of external water sources in the vicinity of the boiler. Auxiliary gas- or oil-fired burners, which are used in the recovery boiler for preheating and temperature control, introduce an additional fuel explosion hazard associated with power boilers. Fires can also occur in combustible residues in the flue gas circuit of the recovery boiler affecting the direct contact evaporator, induced draft fan, and precipitator.

Power boilers: These boilers are used to meet additional steam requirements of the mill. Power boilers are designed to burn gas, oil, coal, or bark or a combination of these fuels. The hazards of fuel explosions and fires in the storage and handling of combustible dusts, combustible or flammable liquids and gases are associated with the operation of these boilers. (For a discussion on the safe operation of boilers, see Chapter 42, "Boiler Furnaces.")

Turbogenerators: These generators use steam to generate all or part of the electrical power requirements of the mill. Ranging in size between approximately 5,000 and 75,000 kW, these units are equipped with central lubrication and hydraulic systems using combustible oil. Oil spraying from leaking pressure piping could be ignited by hot steam pipes in the area. Oil leaking into the basement presents a hazard of severe fire exposure to the turbine from below. Larger generators may also be equipped with hydrogen cooling units. During the long deceleration periods required to safely bring the turbine to a complete halt, the lubrication system cannot be shut down without the threat of major bearing damage. (See Chapter 4, "Electric Generating Plants," for more information on electric generators.)

Process and control: These cables are found in large concentrations in the powerhouse, predominantly in the areas around and below the control room.

Combustible materials exposing cables present a serious fire loss potential. Fire originating in or below a modern centralized control room can shut down the entire powerhouse and close down the mill for extended periods of time.

Cooling towers: These are usually constructed of combustible wood fill with exposed dry surfaces even under full operating conditions. Fires can result from defective fan drives or by welding. In some cases, waste heat cannot be redirected, so the cooling tower can be essential to uninterrupted production.

The lime kiln: This kiln can be gas- or oil-fired and presents fuel explosion hazards similar to the power boiler. Noncondensible gases, such as hydrogen sulfide originating around the mill in various processes such as in the digester, are normally collected and piped to the lime kiln for incineration.

Paper Machine and Finishing Buildings

The paper machine: This machine is the major fire hazard, particularly the machine hood, ducts, economizer, broke area, and oil lubrication system. The process is "wet" to the first dryer section of the paper machine, and thereafter contains combustible paper, wool or synthetic felts, paper scraps, lint, and oily deposits which can quickly accumulate on the machine, hood, and duct surfaces. Large quantities of broke can accumulate in the basement area below the machine.

Sources of ignition within the machine or hood include overheated bearings, friction, defective lighting or electrical equipment, static electricity, and spontaneous combustion of residues from steam pipes or other hot surfaces. Other common sources of ignition are improper smoking and welding in the area.

Clamp trucks: These trucks are used to move finished rolls into the warehouse, and they present fire hazards associated with fuel and hot engines. (See Chapter 46, "Materials Handling Systems," for more information on industrial trucks.)

Solvents: Solvents of a flammable or combustible nature such as kerosene may be used to clean calender and other rolls and equipment.

Extrusion machines: These machines combine the hazards of combustible paper, polyethylene, and lint with the ignition sources of electrically heated dies and gas-fired flame impingement preheaters.

Supercalenders: Supercalenders are large frameworks housing stacks of polishing rolls piled one on top of the other through which paper sheets pass in a vertical direction. A large hood is sometimes used to cover the top end of the stack.

Roll Paper Storage

The fire loading in a roll paper warehouse can be extremely high, ranging up to 500 lbs (227 kg) of combustibles per sq ft of floor area or greater. Within minutes, a fire in vertical roll storage can develop temperatures dangerous to exposed structural steel. The severity of a roll paper storage fire is attributed to the rapid unwinding or *exfoliation* of the paper as the fire burns through layer after layer. Fire tests indicate that the rate of exfoliation increases when the flue spaces between adjacent rolls are increased.[8] Fire severity also depends on storage height, with the greatest exposure to be found in high-piled warehouses which have a large clearance between the top of stock and the ceiling. This condition promotes the rapid upward acceleration of hot fire gases.[9]

The type of paper is also a significant fire factor since paper type determines the speed at which a fire will develop. For example, lighter papers such as tissue will burn faster than heavier papers such as linerboard.[12] The fire spread across the surface of roll paper storage is reduced when a lighter grade of paper such as newsprint is totally wrapped with a heavy grade outer wrapper.

Outside storage of paper creates a severe exposure to nearby buildings or equipment. Damaged rolls and loose sheets torn by wind can cause ignitable debris to accumulate in the area. Yard storage of unprotected combustible material also presents an attractive target for vandals. (A comprehensive discussion of storage practices can be found in Chapter 48, "Industrial Storage Practices.")

THE SAFEGUARDS

Construction

Pulp and paper mill construction is generally heavy noncombustible or heavy timber. Modern mills may make use of laminated beam construction. The powerhouse group, pulp mill group and paper machine, finishing and storage groups are usually well separated.

To prevent the spread of fire throughout interconnected areas, fire walls with adequate fire resistance ratings and stability are needed to cut off the warehouse, finishing building, paper machine building, and any other exposed buildings containing high fire loading or hazardous operations. Electrical centers and process control rooms should also be cut off. All openings in fire walls need protection by appropriate and approved fire door assemblies. Wall and floor penetrations around electrical cable trays and piping should be tightly sealed with a noncombustible material.

General Fire Control Systems

Water supplies: Water supplies for fire service must be reliable and capable of delivering the highest total mill demand including sprinklers, hose, and high expansion foam if provided. Primary sources of fire service water are taken from large reliable mill service reservoirs or systems. Back-up sources can be taken from rivers, ponds, or tanks where suction can be reliably maintained. The high sprinkler demands of the warehouses generally require two or more automatic fire pumps with diesel, electrical, or steam drives. To maintain the highest order of reliability, mill service should not be taken from the fire system. The fire service system pressure should be maintained by low volume, high pressure "jockey" pumps.

Yard systems: Yard systems supplying water to sprinklers, and the yard hydrants, must be reliable and of adequate capacity to supply the required total water demand. This usually requires looped mains of 10 in. (254 mm) or larger provided with section control valves to minimize impairments resulting from main breaks. Major areas of the mill requiring adequate hydrant or monitor nozzle coverage include the paper machine, finishing and warehouse building group, the power plant group, the pulp mill group, wood storage and handling areas and miscellaneous shops, storerooms, and office facilities. Hose houses equipped with 2½-in. (63.5 mm) and 1½-in. (38.1 mm) hose and associated nozzles and equipment are required at each hydrant. Many paper mills are providing a modern fully equipped fire truck as a supplement to or replacement for hose houses at each hydrant.

Fire brigades: Brigades are required for adequate mill fire protection especially in remote areas where outside protection is minimal. Adequate training programs and drills must be conducted on a regular, scheduled basis. Preplanned fire tactics are important, particularly in areas around the woodyard, bleach plant, paper machine, and roll storage buildings. Adequate communication and control is essential, and a reliable internal alarm system must be established. Sufficient equipment such as self-contained breathing apparatus and emergency lighting must be available. Fire trucks are advisable for large mills. This will allow a reduction in the number of hose houses and equipment, and less equipment maintenance will be required. Where public fire departments provide service at the mill, preplanning and coordination with the local fire service is strongly recommended.

Sprinkler systems: Sprinkler systems are needed to protect buildings with combustible construction or contents. Sprinkler protection is generally not required over the "wet end" of the paper machine unless the machine building has a combustible roof. Sprinkler protection is needed over the dry end of the paper machine on the operating floor, in enclosed offices, and in control rooms or other enclosures located on the operating floor which have combustible construction or partitions, are used for storage of combustible material, or are subject to accumulations of broke.

Special sprinkler protection is required for high-piled stock and rack storage areas located in the finishing building or warehouse in accordance with the applicable storage standards. Fires in waste paper storage can become deep-seated and be accompanied by large quantities of smoke. Means for smoke removal such as vents or eave line windows should be available for use during manual fire fighting operations.

Preventing or minimizing impairments to water supplies and sprinkler systems is essential. This can be accomplished by electrical supervision of systems and valves, guards, inspections, sealing or lock-

PULP AND PAPER PROCESSING

FIGURE 14.8. *A modern integrated pulp and paper mill; background, woodyard; right center, pulp mill group with continuous digesters; center, powerhouse group with recovery, power boilers, and turbogenerator building; left center, paper machine and finishing building; upper left, roll paper storage building. (International Paper Company)*

ing of valves, or a combination of these methods. Restoration of impaired fire protection systems and equipment requires priority attention.

Special Fire Control Systems

Woodyard protection: Protection for log and chip piles is best provided by maintaining small piles well separated by a clear space of 100 ft (30.5 m) or more, with all areas of the piles covered by yard hydrants provided with strong water supplies. Alarm communication stations are needed throughout the woodyard for immediate notification of the fire department or mill fire brigade. Large log piles may need special 2-in. (50.8 mm) monitor nozzles and towers for adequate high volume water stream coverage. Immediate fire alarm and water application are essential. Fires becoming buried deep within the pile require enormous quantities of water and must be separated from the remainder of the pile. This requires considerable manpower and equipment when piles are excessive in size.

Wood debarker and conveyor equipment: This equipment may need special protection when accessibility to manual hose stream attack is difficult. Overhead, enclosed, or tunnel chip conveyors are best protected by automatic sprinklers. Large debarking drums can be protected with a ring of spray nozzles provided with a manual control valve.

Bleach plant: Chlorine dioxide generating equipment and reactors may need special explosion relief venting, as well as flow and high temperature limit controls depending on the process. Manufacturers' safety control specifications for equipment must be followed and applied. Storage and handling of flammable liquids must follow the applicable flammable liquid codes and standards. Self-contained breathing apparatus and protective equipment must be available for emergency crews in the event of leaks or spills of toxic materials. (A comprehensive discussion of the handling of flammable liquids can be found in Chapter 37.)

Recovery boiler: Safeguards, preplanning, and training regarding emergency shutdown procedures (ESP) are essential to prevent or minimize dangerous smelt water reactions or auxiliary fuel explosions. The Black Liquor Recovery Boiler Advisory Committee (BLRBAC), with membership from the paper industry, insurance industry, and recovery boiler manufacturers, has developed various safety standards applicable to recovery boilers.[10] Major safeguards include monitoring of proper concentrations of black liquor solids being burned, provisions for rapid draining of the boiler in the event of a pressure part failure, and combustion controls for the auxiliary fuel burners. Fire control for the flue gas circuit is provided by automatic steam flooding in the direct contact evaporator, induced draft fan, and precipitator. Due to the complexity of the recovery boiler and the special hazards associated with its operation, proper operator training is considered a major factor in reducing loss. The American Paper Institute (API) has developed various programs and procedures to aid in the proper training of recovery boiler personnel. [11]

Power boilers: Boilers firing gas, oil, coal, or wood require conventional combustion safeguards and fire and explosion control equipment as outlined by the appropriate codes and standards.

Turbogenerator: Oil reservoirs and associated piping can be protected by automatic sprinklers. Running high pressure oil lines inside of return lines helps to contain oil leaks and prevent oil from spraying hot steam pipes or surfaces.

Process control: Process control rooms and associated rack rooms or underfloor spaces containing concentrations of electric cables can be protected by approved automatic Halon or carbon dioxide extinguishing systems when these areas are enclosed or sealed off. Automatic sprinklers provide highly effective protection, especially for concentrations of important, closely grouped power or control cables or cable trays that cannot be cut off. Where accidental discharge of water could cause damage, preaction sprinklers should be used. Approved cable coatings or adequate separation of cable trays can limit fire spread where cables are not exposed by other combustibles. Good housekeeping is essential, and covers on cable trays help prevent the buildup of combustible debris in the trays.

Cooling towers: Cooling towers with combustible construction or fill need automatic sprinkler protection in accordance with the appropriate codes and standards. Protection is especially important for cooling towers that are required to maintain production.

Lime kilns: Lime kilns are primarly gas- or oil-fired and require conventional combustion safeguards. Combustion controls can be interlocked with safety shutoff valves controlling noncondensible gas burning where provided in the kiln.

Paper machine protection: Protection here is best provided by automatic sprinklers located within the hood and under any obstructing paper machine walkway. In addition, sprinklers are needed in the basement (broke pits) below the dryer sections, over lubricating or hydraulic oil tanks, pumps, and filters, inside exhaust plenums and ducts, and inside economizers. Appropriate high temperature ratings should be used on sprinkler heads located within the hoods and ducts, and where the ducts or economizer are subject to freezing temperatures, dry-pipe or deluge systems may be required. Manual fire fighting equipment and techniques in and around the paper machine area are an important supplement to fixed protection. Small hose stations are needed to cover both sides of the machine on the operation floor and in the basement. One- and one-half-inch (38.1 mm) hose equipped with combination spray nozzles is recommended, and, to avoid stress on steam-heated dryer rolls, sprays rather than solid streams should be used. The most efficient manual fire control can be achieved through the organization and training of machine crew fire teams able to react immediately to a fire, that is, prior to the arrival of the mill fire brigade. Machine maintenance, frequent inspection, cleaning of machine hoods and exhaust systems, and continuous cleanup of broke accumulations are all essential in preventing costly paper machine fires from occurring.

Extruding and coating machine protection: Protection is provided by building sprinklers, small hose, and large portable or wheeled carbon dioxide extinguishers located in the area. Where polyethelene extruders using electrically heated dies and direct gas flame preheating experience a high frequency of small fires at the equipment and in the ducts, special extinguishing systems may be required. Manual steam flooding of extruder exhaust ducts has been used effectively where an adequate and reliable source of steam is available.

Roll paper storage protection: Requirements depend greatly on piling height, storage configuration, use of steel banding, and building construction and height. Tight stacking or *nesting* of rolls, the use of steel bands, and on-side storage reduce the hazard by preventing or controlling the rapid *exfoliation* of rolls in a fire situation.

Specially designed high-density sprinkler systems are required to adequately protect vertical storage of high-piled rolls, particularly in buildings with exposed structural steel construction.

An especially challenging fire exposure occurs in high-piled *high bay* warehouses where the clearance from the top of the rolls to the roof can be 30 ft (9.14 m) or more. Fire tests indicate that high-density sprinkler systems with conventional sprinkler heads can be overwhelmed by strong upward fire plume velocities created in these high bays. Special large drop sprinkler heads may be required depending on the building height.[13] (See NFPA 231F, *Storage of Roll Paper*, for protection recommendations and guidelines.)

When outside storage of paper is unavoidable, it should be limited to the minimum height and area possible and separated by adequate distance from all important buildings and equipment. Where adequate separation distances are impossible, exposure protection such as outside sprinklers should be installed. To guard against vandalism, protection measures such as guards, fencing, and lighting should be employed where applicable.

Life Safety

Preplanning, training, evacuation procedures, drills, and exit facilities are the most important life safety considerations in the event of fire or other emergency. Special considerations are needed in areas subject to toxic spills or explosions such as the bleach plant or recovery boiler area. Notification and evacuation of all but emergency personnel and adequate training of operators or other emergency personnel in the proper shutdown procedures are essential. Inspection and audits of safety equipment and procedures will ensure a constant state of readiness.

BIBLIOGRAPHY

REFERENCES CITED

1. Personal communication with the American Paper Institute (API), New York, NY.
2. National Fire Protection Association, "Paper Mills," *Occupancy Fire Record*, FR64-1, Boston, MA, 1964.
3. Thompson, Norman J., *Fire Behavior and Sprinklers*, National Fire Protection Association, Boston, MA, 1964, p. 3.
4. Clark, Robert T., "Energy," *Pulp and Papermaking Technology*, Edited by Esther G. Dorfman, International Paper Company, Corporate Research and Development Division, Tuxedo Park, NY, 1976, p. 107.
5. Factory Insurance Association (now Industrial Risk Insurers), *Recommended Good Practices for the Protection of Pulp and Paper Mills*, Hartford, Chicago, and San Francisco, 1966, pp. 38-42.
6. Clark, "Energy."
7. FIA, *Recommended Good Practices for the Protection of Pulp and Paper Mills*.
8. Rhodes, J. M., *Fire Hazards of Roll-Paper Storage*, Factory Mutual System, Nor-

wood, MA, 1963 (address at the TAPPI Fall Conference, New Orleans, La, October 30, 1963).
9. Troup, Emile W. M., "The High-Challenge Fireball: A Trade Mark of the Seventies," *Fire Technology*, vol. 6, no. 3, August 1970, pp. 214-215.
10. Factory Mutual Engineering Corporation, "Black-Liquor Recovery Boilers," *Loss Prevention Data* 6-21, April 1980, Factory Mutual System, Norwood, MA, p. 9.
11. Personal communication with the American Paper Institute.
12. Chicarello, P. J. and Troup, J. M. A., "Technical Report, Fire Tests in Roll Paper Storage, Phase II," prepared for: American Paper Insitute, Inc., February 1981, Factory Mutual Research, Norwood, MA, pp. 9-12.
13. Chicarello, P. J and Troup, J. M. A., "Technical Report, Fire Tests in Roll Paper Storage, Phase II," pp. 9-10.

NFPA CODES, STANDARDS, AND RECOMMENDED PRACTICES

Reference to the following NFPA Codes, Standards, and Recommended Practices will provide further information on the safeguards for paper mills discussed in this chapter. (See the latest *NFPA Codes and Standards Catalog* for availability of current editions of the following documents.)

NFPA 10, *Portable Fire Extinguishers.*
NFPA 11, *Foam Extinguishing Systems.*
NFPA 11A, *Medium and High Expansion Foam Systems and Combined Agent Systems.*
NFPA 12, *Carbon Dioxide Extinguishing Systems.*
NFPA 12A, *Halon 1301 Fire Extinguishing Systems.*
NFPA 13, *Installation of Sprinkler Systems.*
NFPA 13A, *Care and Maintenance of Sprinkler Systems.*
NFPA 13E, *Fire Department Operations in Properties Protected by Sprinkler and Standpipe Systems.*
NFPA 14, *Standpipe and Hose Systems.*
NFPA 15, *Water Spray Fixed Systems.*
NFPA 17, *Dry Chemical Extinguishing Systems.*
NFPA 20, *Centrifugal Fire Pumps.*
NFPA 21, *Operation and Maintenance of Steam Fire Pumps.*
NFPA 26, *Supervision of Water Supply Valves.*
NFPA 27, *Private Fire Brigades.*
NFPA 30, *Flammable and Combustible Liquids Code.*
NFPA 43A, *Storage of Liquid and Solid Oxidizing Materials.*
NFPA 43C, *Storage of Gaseous Oxidizing Materials.*
NFPA 46, *Storage of Forest Products.*
NFPA 49, *Hazardous Chemicals Data.*
NFPA 50, *Bulk Oxygen Systems at Consumer Sites.*
NFPA 51B, *Fire Prevention in Use of Cutting and Welding Processes* .

NFPA 54, *National Fuel Gas Code.*
NFPA 58, *Storage and Handling of Liquefied Petroleum Gases.*
NFPA 70, *National Electrical Code.*
NFPA 71, *Central Station Signaling Systems.*
NFPA 72A, *Local Protective Signaling Systems.*
NFPA 72B, *Auxiliary Protective Signaling Systems.*
NFPA 72C, *Remote Station Protective Signaling Systems.*
NFPA 72D, *Proprietary Protective Signaling Systems.*
NFPA 72E, *Automatic Fire Detectors.*
NFPA 75, *Protection of Electronic Computer/Data Processing Equipment.*
NFPA 80, *Fire Doors and Windows.*
NFPA 80A, *Protection from Exposure Fires.*
NFPA 85A, *Prevention of Furnace Explosions in Fuel Oil- and Natural Gas-Fired Single Burner Boiler-Furnaces.*
NFPA 85B, *Prevention of Furnace Explosions in Natural Gas-Fired Multiple Burner Boiler-Furnaces.*
NFPA 85D, *Prevention of Furnace Explosions in Fuel Oil-Fired Multiple Burner Boiler-Furnaces.*
NFPA 85F, *Pulverized Fuel Systems, Installation and Operation .*
NFPA 101, *Life Safety Code.*
NFPA 204M, *Smoke and Heat Venting.*
NFPA 214, *Water-Cooling Towers.*
NFPA 220, *Standard Types of Building Construction.*
NFPA 231, *Indoor General Storage.*
NFPA 231C, *Rack Storage of Materials.*
NFPA 231F, *Storage of Roll Paper.*
NFPA 491M, *Manual of Hazardous Chemical Reactions.*
NFPA 505, *Powered Industrial Trucks.*
NFPA 601A, *Guard Operations In Fire Loss Prevention.*
NFPA 664, *Prevention of Fires and Explosions in Wood Processing and Woodworking Facilities.*
NFPA 1961, *Fire Hose.*
NFPA 1962, *Care, Maintenance and Use of Fire Hose.*

ADDITIONAL READING

Casey, James P., *Pulp and Paper*, John Wiley and Sons, New York, NY, 1983.
Factory Mutual Engineering Corporation, "Black-Liquor Recovery Boilers," *Loss Prevention Data* 6-21, Factory Mutual System, Norwood, MA.
F.M. Eng. Corp., "Black Liquor Recovery Boilers Fire Protection For Direct-Contact Evaporators And Associated Equipment," *Loss Prevention Data* 6-23.
F.M. Eng. Corp., "Cables And Bus Bars," *Loss Prevention Data* 5-31.
F.M. Eng. Corp., "Electronic Computer Systems," *Loss Prevention Data* 5-32.

F.M. Eng. Corp., "Fire Protection For Belt Conveyors," *Loss Prevention Data* 7-11.

F.M. Eng. Corp., "Paper Machines And Economizers," *Loss Prevention Data* 7-4.

F.M. Eng. Corp., "Steam Turbine-Driven Generator Units," *Loss Prevention Data* 5-12.

Hall, Keith F., "Wood Pulp," *Scientific American*, vol. 230, no. 4, April 1974, pp. 52-62.

Halpern, M. G., *Paper Manufacture*, Noyes Data Corp., Park Ridge, NJ, 1975.

Hollands, E. E., "Fire Protection of Grouped Electrical Cables," *Pulp and Paper Magazine of Canada*, vol. 78, no. 6 (1977), pp. 49-51.

Kline, James E., *Paper and Paperboard Manufacturing and Converting Fundamentals*, Miller Freeman Publications, Inc., San Francisco, CA, 1982.

Leech, D., "Close Up: Paper and Board Industry, Part 2," *Fire Surveyor*, vol. 10, no. 6 (December 1981), pp. 13-21.

McKinnon, G. P., ed., *Fire Protection Handbook*, Fifteenth Edition, National Fire Protection Association, Quincy, MA, 1981.

Osborn, Howard, ed., *Paper Finishing*, International Publications Service, New York, NY, 1972.

"The Pulp and Paper Industry," *The Sentinel*, (4th Quarter 1981), pp. 3-7.

"Pulp and Paper Mills," Industrial Risk Insurers, Hartford, CT, 1983.

Riley, R. G., "Planning to Avoid Disaster - 2. Fire Protection Equipment," *Pulp and Paper*, vol. 51, no. 4 (1977), pp. 98-101.

Riley, R. G., "Planning to Avoid Disaster - 3. Importance of Housekeeping," *Pulp and Paper*, vol. 51, no. 5 (1977), pp. 122-125.

Smith, Kenneth E., ed., *Pulping Processes: Mill Operations, Technology, and Practices*, Miller Freeman Publications, Inc., San Francisco, CA, 1981.

Steam, Its Generation and Use, 38th ed., The Babcock and Wilcox Company, New York, NY.

15

Paper Products Manufacturing

NFPA Industrial Fire Protection Section
Pulp and Paper — Wood Products Industry Committee

This chapter discusses the processes, fire hazards, and protection principles associated with converting paper into finished products. The manufacturing of cardboard shipping containers, milk cartons, and disposable diapers have been selected as representative products. Throughout each of the three types of manufacturing there runs a common thread of rolled combustible stock as the raw material and converting operations in which a web stock is passed through various machines to produce finished products. The fire hazards are also similar; they involve large quantities of combustible materials in storage and the creation of combustible waste material (paper dust and scraps). The printing phases of some of the converting processes can involve the hazards of flammable coatings, solvents and inks. None of the operations involves the actual manufacturing of the paper stock.

This chapter is based on contributions made by members of the Forest Products Industry Committee, a standing committee of the NFPA's Industrial Fire Protection Section.

SHIPPING CONTAINERS

The shipping container industry's principal activity is to take roll paper stock, usually a kraft paper, and produce corrugated cardboard that is then converted into shipping containers of various sizes and shapes. Corrugated paper, however, is no longer used just for boxes; die-cut store and window display cards, printed in multicolors, are an increasingly large part of the industry's output. In summary, the fire hazards with shipping containers involve the storage of large amounts of rolled paper, and the waste paper generated in the converting process.

Raw Materials

The average shipping container plant receives rolled paper stock in widths up to 87 in. (2.21 m) either by truck, rail, or water transport. Normally the rolls are stored on end, usually three high, to a height of about 22 ft (6.7 m) depending on the width of each roll. In older plants, where ceiling clearances do not permit vertical storage, the rolls are stored horizontally. Forklift trucks, roll-clamp trucks, or overhead hoists are used in handling the rolls in the warehouse. A six-week supply of paper, the exact amount dictated by the production rate for finished containers, is normally an adequate inventory.

There is a trend in the industry to build roll warehouses with much larger floor areas so that rolls may be stored two-high rather than three-high, thus avoiding the complications associated with high-piled paper storage. See Chapter 14, "Pulp and Paper Mills," and Chapter 48, "Industrial Storage Practices," for further information on rolled paper storage.

The Production Process

Most corrugated box plants are built to service the packaging requirements of a cross-section of industry. They require a rather broad selection of machinery as many different processes are involved in making corrugated shipping containers. First, rolls of paper must be converted into corrugated board. This is done on a corrugating and combining line. The corrugated board blanks are then processed through converting and finishing machines or through specialty machines depending on the type of end product required.

The corrugating and combining line contains the most important equipment in a container plant. It is a combination of various machines placed in line to produce scored and trimmed, single- or double-faced corrugated board blanks from rolls of paper. The blanks are then converted into corrugated shipping containers and other specialty die-cut items. The basic components in line are a mill roll stand, a single facer, a preheater for liner stock, a preconditioner for the corrugating medium,

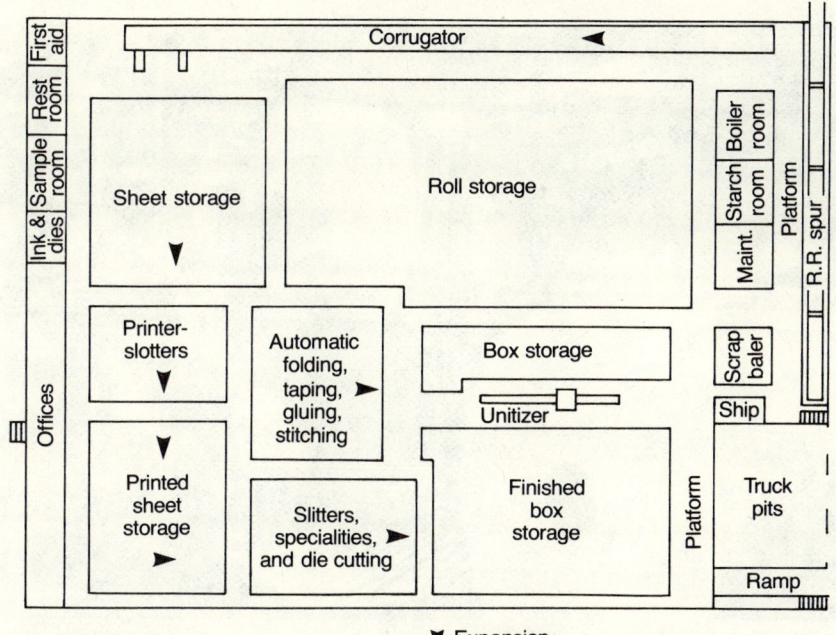

FIGURE 15.1. *Corrugating plant layout with a U-Shap (double right angle turn) production flow pattern (S & S Corrugated Paper Machinery Co., Inc.)*

an overhead bridge conveyor, a glue machine, a double facing machine, a slitter-creaser, a continuous position cutoff, and equipment to collect cut sheets of stock for delivery to the converting operations.

In a typical corrugating line, paper rolls are delivered to the mill roll stand by forklift truck, mechanical conveyor, crane, or floor dolly. The first operation in manufacturing corrugated paper is making "single-face" paper which is a web of paper bonded to a corrugated or "fluted" web to form a two-ply flexible sheet. But before the two webs enter the single facer unit, one from each side, the corrugating medium web goes through a preconditioner that adds moisture to the stock while the liner web goes through a preheater that drives off moisture in the stock. The two different webs must be treated differently; the liner dried to control the quality of the bond to the corrugating medium which is substantially moistened to give it plasticity so that it will form good corrugations. Both preheater and preconditioner have 24-in. to 36-in. (0.61 to 0.91 m) steam-heated drums that operate at steam pressure up to 200 psi (1379 kPa).

The single facer unit performs two basic steps. First it forms the flutes in the corrugated medium by pressing the web against a corrugating roll that has a fluted surface. The fluter rolls are finished to very close tolerances since the flutes must be extremely uniform in height and contour for subsequent operations. The second operation is applying the adhesive (normally a starch paste or silicate-based glue) to the tips of the

Chapter 15

FIGURE 15.2. Paper roll stand for quart-size rotary press in a carton manufacturing plant.

flutes on the corrugating medium and joining the lining paper to the medium forming a continuous sheet of single-faced corrugated stock.

Bridges and conveyors: From the single-facer machine the stock goes to the glue machine, the first step in the double-facing operation. As the single-face paper emerges from the single facer, it is fed into an inclined conveyor which carries the paper upward and away from the single facer, dropping it down onto a lower speed horizontal belt conveyor. Since the lower conveyor operates at a speed substantially below the inclined conveyor, the paper "fanfolds" into loops permitting the lower conveyor to store a quantity of single-face paper to feed the double-facer when the single-facer is slowed down for adjustments or paper changes. This also permits the single facer to operate at normal speed if the double facer is slowed down or stopped. Holding the stock on the conveyor also provides extra time for drying and bonding for a more permanent adhesion; however, the amount of stock normally held on the conveyor is not considered to represent an excessive fire load.

Double facer: Emerging from the conveyor, the single-faced paper again goes through a preheating unit to condition the stock for application of glue to the tips of the flutes in an adhesive unit positioned directly before the double facer. Here the second liner is glued to the corrugated medium to produce double-walled cardboard familiar to us as the base stock for cardboard containers. Once through the double facer, the cardboard

PAPER PRODUCTS MANUFACTURING

FIGURE 15.3. Checking paper on a corrugating machine

web is cut to lengths in rotary cutoff machines, and the cut blanks are automatically stacked into neat piles that are transferred to trucks, skids, dollies, or conveyors to await processing in the converting operation.

Converting and finishing: Corrugated sheets delivered from the corrugator are set aside for a few hours for conditioning and drying. They are then ready for conversion to box blanks and, finally, for jointing. Converting blanks to cartons requires printer-slotters of various kinds and sizes, semiautomatic and automatic folding machines, and taping, stitching, or gluing machines. Boards used for interior packing and partition pieces require partition-slotters or other specialty machines. Special types of cartons may require flap cutting or surface-coating machinery. Die-cutting and printing units are required for display products of unusual shape.

In recent years, combined finishing operations that involve one-step combination printer-folder-gluers have found increasing use. They take the board directly from the corrugating line and deliver finished box blanks into counted piles.

The converting process, because of the great variety of end products, does not lend itself to a totally standardized operation. But whatever combination of machines may be found in the converting section, they include the basic steps of cutting, creasing, scoring, and trimming a paper product; printing legends on box blanks or die-cut sheets; and gluing, taping, or stitching box blanks together.

Tying and bundling machines are placed at the ends of the finishing machines, such as automatic folder-tapers, folder-gluers, and stitchers. They tie piles of boxes in small lots of 15, 20, or 30 for easy loading and shipping. Strapping machines combine many such tied piles into a large bundle, usually from about 300 to 800 boxes, and the piles are usually placed on skids for handling.

Sheet plants: A variation in shipping container manufacturing is the sheet plant that handles small lot orders. They are usually within daily trucking distances of the sheet supplier so large storage areas for blanks are usually not necessary. On the other hand, some sheet plants have built up a substantial business in so-called stock sizes of boxes that require storage areas for finished box blanks waiting shipment.

A modestly equipped sheet plant would consist of a slitter-scorer, a rotary slotting machine, and a semiautomatic taping machine. No space would be needed for the corrugating line, boiler room for process steam, and adhesive storage.

The Hazards

The principal fire hazards associated with paper container manufacturing are the storage of the raw stock and finished products, and the waste paper generated in corrugating and converting operations. Flammable inks and solvents are sometimes associated with the printer-slotter phase of the converting process; however, water-base inks are the rule.

Paper storage: Both the rolled paper raw material and the finished boxes represent the hazard of high fire loads in storage; more so if they are in high-piled configurations. High-piled rolled paper, particularly if the rolls have differing diameters, lends itself to rapid fire spread through the flues formed by the rolls. The problem is compounded if the rolls are not restrained from "exfoliating," or unwinding or peeling from the roll to expose more unburned fuel. Heat radiated and reradiated from one stack to another within the flue space adds to the intensity of the fire.

Finished box blanks on skids may be stored in a rack storage arrangement or in the more simple method of pallet load piled one on another. In the latter arrangement the height of the piles is limited by stability of the piles.

Idle pallets: Piles of idle pallets are a severe hazard. After pallets are used for a short time they can dry out and edges become frayed and splintered, a condition that can lead to easy ignition from a small source. A 6-ft (1.83 m) storage height is considered a maximum to keep pallet fires within limits manageable by ordinary sprinkler protection.

Paper waste: Scrap paper in varying amounts is produced at various points in the production process. Unless it is promptly removed, it usually collects on the floor around the machines to the point where it becomes a major housekeeping problem (see Chapter 50, "Industrial House-

PAPER PRODUCTS MANUFACTURING

FIGURE 15.4. View inside a typical corrugated box plant. The feed end of a corrugating machine is at the right.

keeping Practices"). Almost every modern container plant is equipped with pneumatic systems for conveying the pieces of waste from their points of origin to the collection point where the scrap is usually baled for sale. At some plants the scrap is brought to the balers in bulk carts from localized collection points.

The basic components of the pneumatic system are the pickup devices (pickup tubes, sweep boots, automatic hoppers located at the printer-slotters, slitters, partition machines, and other scrap-producing operations), the duct work, a suction blower or chopper blower, a cyclone separator, and an accumulator or garner (see Chapter 45, "Air Moving Equipment," for further information on blower systems).

The accumulator, or storage chamber for the waste, located below the cyclone and before the balers, should be a point of concern. Corrugated scrap is accumulated here in a tangled mass and often does not flow readily. It has a pronounced tendency to bridge or hang up. Large volumes of waste also tend to become packed by their own weight and are troublesome to break apart. This can create a fire hazard that can be minimized by installing sprinklers in the bin and providing access doors that can be used to fight fires in the bin.

Balers for scrap waste ideally are located conveniently close to both the major source of waste and the shipping dock, and should be cut off from the remaining portions of the plant by substantial fire walls with a fire resistance rating of at least 1 hour. Ample storage space should be

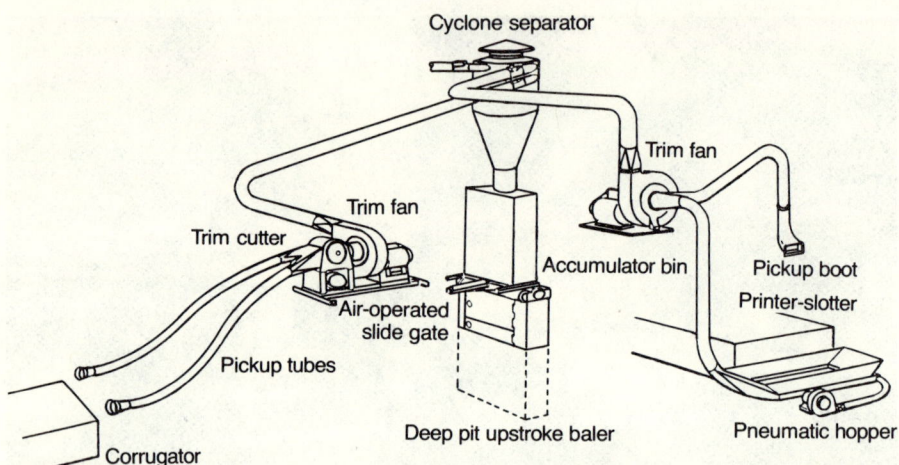

FIGURE 15.5. *Typical baling system connected to upstroke baler. (S & S Corrugated Paper Machinery Co., Inc.)*

allowed for storing bales waiting for shipment. Balers are either of the vertical type or continuous horizontal type. The deep pit, upstroke, chain-type that makes a 900-lb (408 kg) bale is the most commonly found vertical baler, and its charging chamber holds enough scrap to make a bale in a single compression. Horizontal balers can make bales up to 1,000 lbs (453.6 kg) in weight in an automatic horizontal compressing operation that produces bales on a demand basis; consequently, it is ideally suited for use with pneumatic waste-collection systems since large waste storage chambers are not needed, thus eliminating that hazard.

Baled waste is often stored one to three bales high on its side. Fire tends to burrow into solid piles of bales, and the bales can become mushy and difficult to handle when wet by sprinkler discharge and hose stream. In fact, the integrity of the bales slowly disappears as they are wetted, not only making it difficult to remove the bales in a fire, but also presenting the danger of the piles collapsing and causing structural damage to the building or injury to fire fighters.

Inks and solvents: The hazards of printing operations in container plants are similar in nature to those found in the printing industry in general. The raw materials are primarily paper, water-based or solvent-based inks, solvents, and thinners. Some printer-slitter machines have the capability of printing in four colors. Flexography, a process that uses flexible rubber plates and fast-drying solvent- or water-based inks, is the name of the particular printing process associated with printing on corrugated board. Anything that can go through a web-fed arrangement can be printed by the flexographic process. The machine that does the printing is often identified as a flexographic printer-slotter folder-gluer. Fast-drying solvent inks and the thinners used to clean rolls present the hazards of flam-

PAPER PRODUCTS MANUFACTURING

FIGURE 15.6. An hydraulic downstroke (vertical) baler. (Economy Baler Company)

mable liquids. Safety cans are a must and good practices must be observed in storage of flammable inks and solvents (see Chapter 37, "Handling and Storage of Flammable and Combustible Liquids").

The Safeguards

Ideally, corrugated container plants should be of fire-resistive construction, while heavy timber construction, with its ability to retain strength under severe fire exposure, is the next best. Least desirable, but most often used, is light duty (open truss) exposed steel, a construction type that is particularly vulnerable to the fire damage in the roll storage area if sprinkler protection is not adequate.

Of principal concern is the separation of the roll paper storage area from the production area and the latter from the finished products area by substantial fire walls with good protection at door openings. One major container producer requires a minimum of 3-hour fire resistance for

FIGURE 15.7. *A continuous horizontal-type baler. (The American Baler Company)*

such separations. Isolating the baling operation as well as the boiler room or maintenance area by fire walls of lesser resistance is also good practice.

Fire control systems: A complete automatic sprinkler system is the best primary protection for a container plant. The water supply should be strong and reliable since it must satisfy design requirements for sprinklers located at the ceiling level, sprinklers in rack storage (if any), standpipe and hose systems for indoor use, and yard hydrants. The sprinkler density discharges must be calculated carefully for the particular type of storage arrangements that are used. Full-scale rolled paper fire tests were conducted in 1982-1983 and sprinkler density requirements are reflected in NFPA 231F, *Storage of Roll Paper.* If rack storage is used for finished cartons, the location of sprinklers in the racks and the discharged densities should be appropriate for the particular rack configuration that is used. Chapter 48, "Industrial Storage Practices," discusses sprinkler water supply requirements for different storage configurations as described in NFPA 231, *Indoor General Storage,* and NFPA 231C, *Rack Storage of Materials.* These two standards contain requirements for sprinklers in storage occupancies that are supplementary to the requirements of NFPA

PAPER PRODUCTS MANUFACTURING

FIGURE 15.8. Flexo folding and printing machine in a corrugating box plant.

13, *Installation of Sprinkler Systems,* which gives criteria for the basic sprinkler installation.

A full complement of portable fire extinguishers, suitable for the types of hazards that may be found in the plant, and standpipe hose stations throughout the storage and manufacturing areas are the other fire protection provisions necessary for a good basic protection program. NFPA 10, *Portable Fire Extinguishers,* and NFPA 14, *Standpipe and Hose Systems,* are the sources for guidance when installing these auxiliary protection measures.

MILK CARTON MANUFACTURING

Polycoated cartons are manufactured for storage of milk, juices, fresh fruits, and a variety of other liquids and frozen goods. Since milk cartons are the largest single product produced, this industry is commonly referred to as "milk carton manufacturing."

Raw Materials

Bleached polycoated paper is received in rolls and stored in the raw materials warehouse, usually in vertical columns 18 to 20 ft (5.5 to 6 m) high. The rolls may vary in size, but they average approximately 25 in. (0.64 m) in width, 60 in. (1.52 m) in diameter, and weigh approximately 2,000 lbs (901 kg) each. The rolls are normally held together by one or two metal bands or wrapped in paper, and they are primarily handled by clamp-type lift trucks. Other raw materials include printing inks and press cleaning solvents, which are usually stored in 5- and 55-gal (18.9 and 208.2 L) drums (most often the inks and solvents are flammable, or "red label," liquids), and cardboard cartons for shipping finished goods (corrugated shipping containers) stored flat on pallets.

The Production Process

Rolls of paper are brought by lift trucks from the roll stock warehouse and positioned at the printer to be fed in a continuous web through a multicolor printer, a gas-fired drier, and finally through a cutter-creaser machine which cuts and creases the web into individual flat cartons in a manner similar to the cutting and creasing operations found in a corrugated shipping container plant. The cartons are then placed on pallets or stored for up to two days to allow the inks to cure.

The final production steps take place when the carton flats are brought from the temporary storage area to the sealing department. Here the flats are fed through a gas-flame sealer with a folder section. The flame melts the polycoating at the edge of the flat and bonds the carton together, forming a tube (the top and bottom are left open). The cartons then go through an automatic counter and are packed in corrugated shipping containers.

After the cartons are packaged, they are shipped immediately or taken to the finished goods warehouse for storage while awaiting shipment. Handling is by forklift truck and storage of finished goods is either bulk-type storage, on pallets, or on pallets in racks.

The Fire Hazards

The fire hazards in a milk carton plant are similar to those in corrugated container plants. They also involve the storage of large quantities of paper products and flammable liquids and the use of these materials in manufacturing. But unlike corrugated container plants, the amount of waste material generated in forming the cartons is not appreciable; thus, a waste removal system is not required.

Storage in the roll stock warehouse consists of polycoated rolled paper, usually to 20 ft (6 m) in height. It is not definitely known how the hazard of polycoated paper compares with uncoated paper; however, based on fire experience and since no formal fire tests have been made, there is no reason to assume that polycoating increases (or decreases) the relative fire hazard of rolled paper. Fire sources in the storage area include common faults like lift truck fuel leaks that are apt to ignite from

PAPER PRODUCTS MANUFACTURING

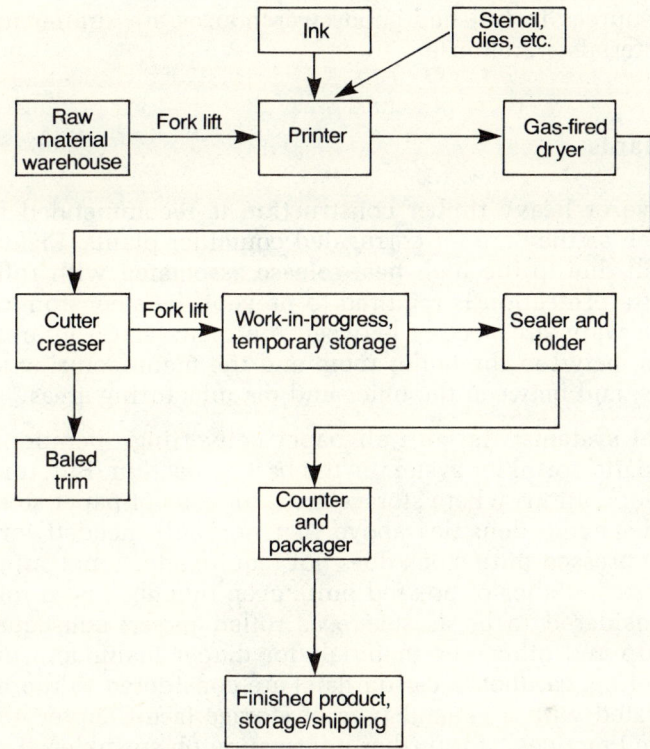

FIGURE 15.9. Flow diagram of milk carton manufacturing process.

faulty ignition systems or hot mufflers, and human failures such as poor smoking habits and poor housekeeping practices. Large quantities of flammable inks and press cleaning solvents are needed in the printing operation. These inks and solvents are best stored in a flammable liquids storage room, and the hazard is typical of any flammable liquids storage area (see Chapter 37, "Handling and Storage of Flammable and Combustible Liquids").

In the manufacturing area, fire hazards include the use of the flammable inks and solvents at the printers, gas-fired driers, and in-process storage of partially finished milk cartons. Safe operating procedures such as limiting the amount of flammable liquids at each printer to one shift's supply, keeping lids on unused ink cans, maintaining good housekeeping, adhering to "No Smoking" regulations, and using cutting and welding procedures properly should be followed to limit the hazard.

The finished goods warehouse will typically contain finished products in corrugated shipping containers, stored either on pallets, in bulk piles, or on pallets in racks. Storage is commonly approximately 20 ft (6 m) high. The finished goods warehouse may also contain corrugated cartons (stored flat) on pallets to be used to package the finished milk car-

327

tons. Fire sources in finished goods warehouses are similar to those in the raw materials warehouse.

The Safeguards

Fire-resistive or heavy timber construction is recommended for carton plants, much as they are for corrugated container plants. Light steel can be used, but due to the high heat release associated with rolled paper storage, extra protection is required to prevent the steel from collapsing in fire. Fire walls are needed between the manufacturing area and the warehouses, between the boiler room and the manufacturing and warehouse areas, and between the office and manufacturing areas.

Fire control systems: As with all paper converting operations, a complete automatic sprinkler system is the best protection. But, unlike other converting operations where storage areas for rolls of paper stock require sprinkler discharge densities above that normally needed for stock in storage, the pressed pulp stock does not require additional supplies. The burning characteristics of pressed pulp, even though it is in rolled form, are not considered to be as severe as rolled paper; consequently, the pressed pulp and other raw materials for diaper manufacturing (fabric, plastic sheeting, cardboard carton flats) are considered to represent hazards associated with a general type of storage (see Chapter 48, "Industrial Storage Practices," for further information on sprinkler discharge requirements for industrial storage occupancies).

DISPOSABLE DIAPER MANUFACTURING

Disposable diaper manufacturing is limited to mechanical converting and packaging processes of various paper and plastic materials into a finished product.

Raw Materials

The basic materials for diaper manufacturing consist of absorbent paper wadding composed of pressed pulp, plastic, fabric, packaging cartons, and assorted fastening devices. As in corrugated container and carton manufacturing, paper, as well as the plastic and fabric sheeting, are delivered to the converting plant as roll stock for warehousing in vertical or horizontal piles until ready for use. The same materials handling equipment found in box and carton plants will be found in warehouses at the converting plant.

The Production Process

Modern diaper manufacturing involves high speed converting machines which automatically sandwich an absorbent layer of paper wadding between another plastic sheet (to retain liquid) and an inner moisture penetration fabric. Additional equipment may include folding, fastener attachment, and packing machines. The rolls of pressed pulp, plastic, and fabric are mounted on reels at the beginning of the converting machines, and the stock is webbed through the machines for combining into the finished product.

The Fire Hazards

Inherent hazards of normal converting operations are limited to accumulations of paper dust at the converting machines which, if not quickly and efficiently removed, can contribute to the potential for fire. Historically, hot bearings and belt friction in the first stage of the converting machine can cause ignition of the fine paper dust-lubricating oil mix routinely found in these units.

A pneumatic dust removal system with sweep units located strategically at the converters is the best method of controlling the dust hazard. The dust is accumulated in collectors equipped with filter units (see Chapter 45, "Air Moving Equipment").

The Safeguards

The fire protection systems and equipment that are found in corrugated box and carton plants are equally acceptable in converting plants. Of particular importance is the protection given to the dust collecting system. The equipment should be designed to withstand anticipated explosion pressures, allowance being made for explosion relief vents. The equipment should be constructed of steel or enclosed in steel and located outside the building. To avoid collapse of the equipment, which could result in an explosive dust cloud, equipment should be well-supported on steel, masonry, or concrete. Equipment and discharge ducts should be well-separated from combustible construction and unprotected openings in the buildings. Collectors which must be located indoors and cannot be constructed of sufficient strength to withstand anticipated explosion pressures can be located near outside walls to facilitate explosion relief venting. Automatic sprinklers are recommended within the dust collecting enclosures.

Chapter 15

BIBLIOGRAPHY

REFERENCES

Corrugated Box Manufacturers' Handbook, Third Edition, S & S Corrugated Paper Machinery Co., Inc.

Factory Mutual Engineering Corporation, "Corrugated Paper Machines," *Loss Prevention Data* 12-24, July 1967, Factory Mutual System, Norwood, MA.

NFPA CODES, STANDARDS, AND RECOMMENDED PRACTICES

Reference to the following NFPA Codes, Standards, and Recommended Practices will provide further information on the safeguards for paper products manufacturing discussed in this chapter. (See the latest *NFPA Codes and Standards Catalog* for availability of current editions of the following documents.)

NFPA 10, *Portable Fire Extinguishers.*
NFPA 13, *Installation of Sprinkler Systems.*
NFPA 14, *Standpipe and Hose Systems.*
NFPA 30, *Flammable and Combustible Liquids Code.*
NFPA 68, *Explosion Venting Guide.*
NFPA 69, *Explosion Prevention Systems.*
NFPA 70, *National Electrical Code.*
NFPA 77, *Static Electricity.*
NFPA 91, *Blower and Exhaust Systems, Dust, Stock, and Vapor Removal or Conveying.*
NFPA 101, *Life Safety Code.*
NFPA 231, *Indoor General Storage.*
NFPA 231C, *Rack Storage of Materials.*
NFPA 231F, *Storage of Roll Paper.*

ADDITIONAL READING

"Designing to Limit Loss in High-Rise Rack Warehouses," Kemper Group Report, vol. 10, no. 3 (September 1981), pp. 2-9.

Harrington, J. C. and R. B. Hopkinson, *Rack Storage Protection*, Worcester Polytechnical Institute, Worcester, MA, 1977.

King, P. W. and J. Magid, *Industrial Hazard and Safety Handbook*, Butterworths Publishers, Inc., Woburn, MA, 1979.

McKinnon, G. P., ed., *Fire Protection Handbook*, Fifteenth Edition, National Fire Protection Association, Quincy, MA, 1981.

"Moving Fire: Fire Hazards of Belt Conveyors," *Record*, vol. 54, no. 6 (1977), pp. 18-21.

Talbot, G., "Static Electricity," *Fire*, 70 (871), 1978, pp. 397-398.

16

Printing and Publishing

John J. Schillo

The printing and publishing industry represents about 3 percent of the Gross National Product (GNP). Advancements in electronics, the paper industry, and the introduction of new and exotic inks have resulted in a forecast by the U.S. Department of Commerce that the printing industry will grow at an annual rate of more than 8 percent. By the year 2000, the annual dollar volume has been projected to $103 billion.

Of the 38,000 commercial printers in the United States, over 80 percent employ fewer than 20 persons, and this segment of the industry accounts for more than 25 percent of the total industry receipts. Those employing 20 to 100 persons account for 50 percent of the total industry receipts.

The industry is fragmented into many different processes, each with its own particular and unique problems and hazards. The end result is characters or designs laid on paper, metal, cloth, or synthetic material, arrived at by basically four methods: lithography (55 percent),

---

John J. Schillo is Technical Services Specialist, Loss Control Department for the Utica National Insurance Group, of Utica, NY; and Consulting Safety Engineer for the Printing Industries of America, Inc., of Arlington, VA.

Chapter 16

FIGURE 16.1. Rolls of paper for use on web presses.

letterpress (20 percent), gravure (15 percent), screen printing and a few minor processes (10 percent).

RAW MATERIALS

Paper

The raw materials of the printing and publishing industry are primarily paper, ink, solvents, and thinners. Paper is received in roll or sheet form depending on the kind and size of the printing establishment and the type of printing press to be used. Other materials, such as cloth and metal, or synthetic materials, such as cellophane, are printed from rolls or bolts. Paper in sheet form is received strapped onto stackable skids or in wrapped bundles.

Ink

Ink is received in quantities ranging from tank car shipments to 1-pt (473 mL) cans. Ink manufacturing is now a more than half billion dollar industry in the United States alone, a departure from the early days when soot or lampblack was mixed with glue or oil according to each printer's secret formula. Very little color was used until the discovery of coal tar dyes in the middle of the nineteenth century. Now inks are formulated in literally thousands of colors. Specialty inks are very popular: among them are magnetic inks for bank coding and nontoxic inks for food packaging. Jet printing uses inks that break down into miniscule droplets that are electrostatically charged and directed by computers to image 150,000 characters per sec. Solventless inks that are dried by ultraviolet light represent another fast-growing field.

The drying of inks is important because a printed piece cannot be handled or used until the liquid or plastic ink film has solidified and dried. Printing inks dry by absorption, selective absorption, oxidation, polymerization, evaporation, and precipitation, or a combination of two or more of these mechanisms.

New systems for drying or curing inks have been developed to eliminate pollution caused by the evolution of solvents and other pollutants associated with ink drying. New infrared units using inks of the low solvent type have been developed for drying inks on sheet-fed presses. Also, water- and alcohol-soluble coatings are being used to overcoat wet inks immediately after printing. The coatings dry rapidly, keeping the inks from scuffing or marking while they dry.

Solvents

Various solvents and thinners are ordered in quantities ranging from tank car lots to 5-gal (19 L) cans. Storage and handling of the basic liquid raw supplies present certain problems (see Chapter 37, "Handling and Storage of Flammable and Combustible Liquids").

THE PRODUCTION PROCESS

In general, the term printing is used to describe any method of producing copies of designs or symbols by transferring pigment to paper from a prepared printing surface. Although most people generally think of printing as ink on paper, it is not limited to any particular material, or to ink. The embossing process, for instance, uses no ink at all, and all shapes and sizes of metals, wood, and solid or thin plastics are common receivers of printed messages.

Four main techniques, called printing processes, are used by printers to reproduce graphic images. These processes are: letterpress printing, intaglio printing, screen printing, and lithography. They differ according to the method used to apply ink to the printing surface.

The *Letterpress* process is also referred to by other terms, such as flexography, and relief printing. Whatever the label, letterpress is a method transferring an image using a raised printing surface (see Figure 16.2).

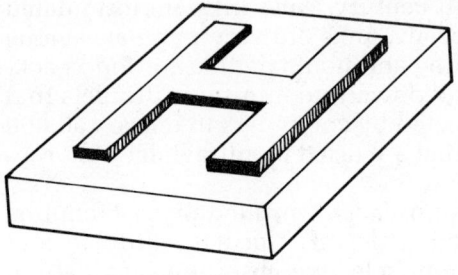

FIGURE 16.2. Printing from a raised surface is called letterpress printing.

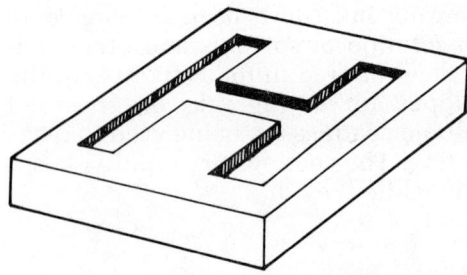

FIGURE 16.3. Printing from a lowered surface is called intaglio, or gravure printing.

Intaglio printing is the reverse of relief printing. In intaglio printing an image is transferred from a sunken printing surface (see Figure 16.3). The term intaglio refers to such subprocesses as etching, engraving, collotype, and gravure.

Screen printing is a method of transferring an image by allowing ink to pass through an opening or stencil (see Figure 16.4). The screen process is sometimes called silk screen printing, mitography, or serigraphy.

Lithography is a relatively new printing process. In lithography, an image is transferred from a flat printing surface using chemistry. Certain areas of the printing plate are chemically treated to accept ink while other areas are left untreated so they will repel ink. The inked image is transferred to a blanket cylinder which, in turn, contacts the paper and transfers the image to the paper (see Figure 16.5). The lithographic process is sometimes called planography, offset lithography, or simply offset printing.

Letterpress

In a letterpress operation, the paper or material being printed comes in direct contact with the inked type. Generally, letterpress equipment is much older than offset equipment. Today, much of the older hand-fed

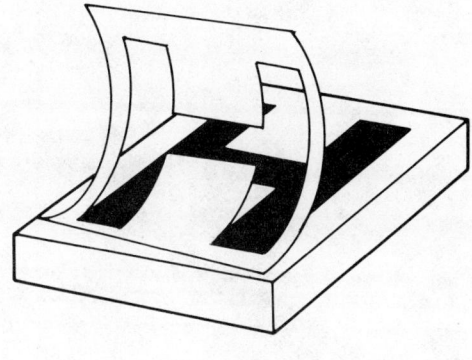

FIGURE 16.4. *Printing by means of a stencil is called screen printing.*

FIGURE 16.5. *Printing from a flat surface is called lithography, or offset printing.*

letterpress equipment is found in basement or garage shops. Replacement parts are difficult to find because many of the original manufacturers have disappeared.

Letterpresses include platen presses, flatbed cylinder presses, sheet- or roll-fed (web) rotary presses and direct wraparound presses. Each printing method has its own make-ready process. The type used for the letterpress operation is called hot type. A small job, e.g., an order of business cards or envelopes, might be set from individually cast type which has been hand-set for the particular job. In the letterpress process (and for photomechanical plate-making), type is locked up in a *chase* (a heavy rectangular steel frame). Lockup is done on a large table called a *stone*, so named because it originally had a stone top; today the table top is steel. The form does not take up the entire area inside the chase, and the empty spaces are filled with *furniture* (wood or metal blocks). In addition to the furniture, *quoins* (steel wedge-shaped devices) are placed between the furniture on two adjacent sides of the type. The quoins are tightened slightly, and the type is checked for levelness. Once all the parts are made level, the quoins are tightened to hold the type securely in place. The type is then proofed and put directly on a flatbead press. Plates for rotary cylinder presses are cast in a half-cylinder form from paper-mache molds called stereotypes. The stereotype operation is fast becoming extinct.

Many newspapers are still printed by the letterpress process, and

Chapter 16

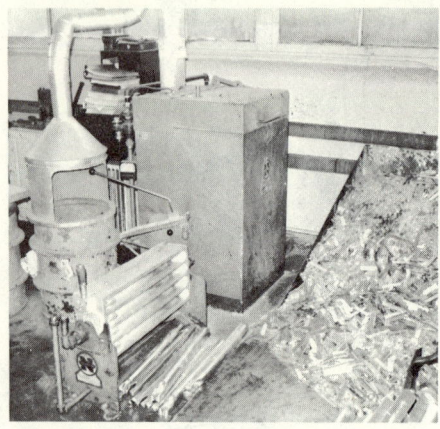

FIGURE 16.6. Metal type from linotype operation can be remelted to make new "pigs."

FIGURE 16.7. Vertical letterpress—material printed comes in direct contact with type.

Many newspapers are still printed by the letterpress process, and presses are being converted to use wraparound plates of zinc, magnesium, copper, or photopolymer. Some steel-backed plates are mounted directly on magnetic cylinders.

Rapid technological changes are manifest in all phases of the graphic arts industry. One example is the Cameron belt press which is a letterpress operation that prints complete books in one pass through the press. This eliminates the practice of printing large signatures which must be folded, stored or stacked, and later collated into the final book. On the Cameron press, the web is fed at one end and the bound book is delivered in boxes at the other end.

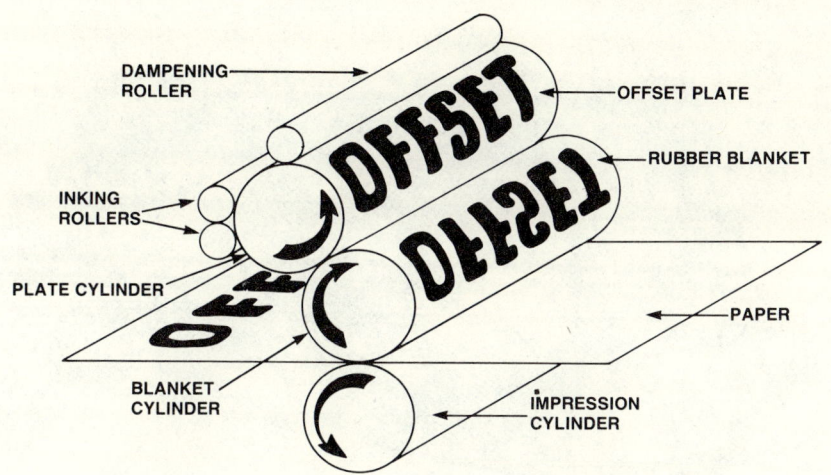

FIGURE 16.8. The major parts of an offset lithographic press.

Lithography

Lithography, printing from a flat surface, works because oil and water do not mix. It has advanced from its original state in which images were drawn on a flat stone with a greasy marker into one of today's most advanced printing processes. The lithographic printing process works as follows: First, a greasy image is placed on a thin, flat, metal printing surface called a printing plate (usually done photochemically). Water is then applied to the plate. The water covers the nonimage portion of the plate, but does not cover the greasy image area because oil and water do not mix. Next, an oil-based ink is applied to the plate. The ink adheres to the greasy image area, but it does not adhere to the wet nonimage areas of the plate because oil and water do not mix. Finally, paper is pressed against the surface of the plate, and the inked image is transferred from the plate to the paper.

Since the image can wear away rapidly when the paper rubs against it during the printing process (especially true when the plates are used on high-speed presses), the image is "offset" (transferred) to a rubber "blanket" before it contacts the paper (see Figure 16.8).

While phototypesetting can be used to some degree in both the letterpress and gravure operations, it is most commonly associated with lithography, or offset printing. Here, advances are also being made very rapidly. It is common today for copy to be transmitted by laser beam. Photocomposition lends itself to great speed and accuracy. Copy and pictorials are color separated and photographically transferred to a metal plate.

Sheet-fed offset presses can be found in all sizes: from the small duplicators found in offices to presses large enough to accommodate

Chapter 16

FIGURE 16.9. High-speed sheet-fed offset press. (Heidelberg Eastern)

FIGURE 16.10. Delivery end of a web offset press.

sheets of paper 6 ft (1.8 m) wide. Many of the new sheet-fed presses are called perfectors; both sides of the paper are printed at the same time. Now the trend is toward offset webs. This combines the versatility of the offset process with the web press: a high speed operation with a minimum of set-up time.

Gravure

Whereas the letterpress process uses a raised or relief surface and the offset process uses a flat surface, gravure uses a sunken or depressed sur-

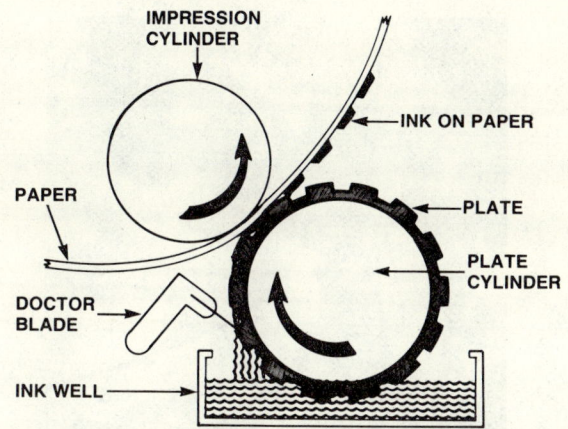

FIGURE 16.11. *The operating principle of a gravure press.*

face for the image. The image area consists of cells or wells etched into copper cylinders or wraparound plates; the cylinder or plate surface represents the nonprinting areas. The plate cylinder rotates in a bath of ink. The excess is wiped off the surface by a flexible steel plate or doctor blade. The ink remaining in the thousands of recessed cells forms the image by direct transfer to the paper as it passes between the plate cylinder and the impression cylinder (see Figure 16.11).

Gravure printing is considered to be excellent for reproducing pictures, but the high plate-making expense usually limits its use to runs of 50,000 or more impressions.

Most of the gravure presses are webs; however, they can also be sheet-fed.

Screen Printing

Formerly known as silk screen printing, this method employs a porous stencil of fine silk, nylon, dacron, or even stainless steel mounted on a frame. A stencil is produced on the screen either manually or photomechanically; the nonprinting areas are protected by the stencil. Printing is done on a simple press by feeding paper under the screen, applying ink with a paint-like consistency to the screen, and spreading and forcing the ink through the fine mesh openings with a rubber squeegee.

Thermography

A process called thermography creates special effects in printing stationery, invitations, greeting cards, and paper decorations. A raised surface of printing resembling genuine engraving is formed without using costly

FIGURE 16.12. Perfect binder — pockets hold books to be dipped in glue.

engraving dies. Special nondrying inks are used in conventional printing, either by letterpress or by offset, and the wet inks are dusted with a powdered compound. After the special powder is removed from the nonprinting areas by suction, the sheet passes under a heater which fuses the ink and powdered compound. The compound swells and raises to produce an engraved effect. The drying process is usually accomplished in an electric furnace.

Flexography

A form of rotary web letterpress, flexography uses flexible rubber plates and fast-drying solvent or water-based inks. The rubber plates are mounted to the printing cylinder with adhesives. Almost anything that can go through a web press can be printed by the flexographic process, including decorated toilet tissue, bags, corrugated board, and materials such as foil, cellophane, polyethylene, or other plastic films. It is well suited to printing large areas of solid color. Inks can be overlaid to obtain high gloss and special effects.

Bindery

A bindery is an integral part of the graphic arts industry. It can be part of the printing shop or a separate job shop. The binding equipment includes cutters, shears, stitchers, collators, folders, and gluing machines known as *perfect binders*. Perfect binders use hot glue. Thermostats ensure that they are performing adequately.

FIGURE 16.13. Modern typesetting equipment. (Compugraphic Corp.)

Finished Material Storage

Finished material from the printing plant or the bindery is stored on skids and usually covered with plastic to prevent damage from dust or sprinkler leakage.
    Clamp or suction-jaw industrial lift trucks and overhead traveling cranes are used for handling paper rolls. Lift trucks are well adapted to on-end piling. Traveling cranes are adapted to both on-end and on-side piling.
    On-end storage is more varied and more widely used than on-side storage. One advantage is its adaptability to industrial lift trucks. Another is the accessibility of individual rolls, particularly important for mixed storages.

Values

Proper values are difficult to obtain, but a rule of thumb for the graphic arts industry would be approximately $23,000 worth of contents equipment for every employee. This would vary, of course, by the number of shifts worked and how modern the equipment is.
    Values can be very high for stored material, especially finished products. Finished items create added value; often this new value is greatly in excess of what the printer or binder has invested in the job. For example, the printer or binder will assemble sales kits and store them for delivery at a certain date. Their involvement might only be a small percentage of the total value of the kit. Another example is a printer who completes a large order of return addressed envelopes for a customer. The

delivery of the envelopes could be spread out over a long period of time. Insurance falling in the category of care, custody and control, or bailee coverage is a very grey area for the printer/binder, and must be thoroughly explored.

The replacement cost of all graphic arts equipment must be given serious consideration. A very large percentage of the equipment purchased today is foreign made and, while most distributors carry parts and supervise erection and major repair, there are bound to be parts delays. Only relatively small pieces of equipment can be purchased off the shelf. Most equipment is built to customer specifications or modified to the individual's requirements. Constant technological developments make it difficult to replace, in kind, a press or piece of bindery equipment only three years old.

THE FIRE HAZARDS

Letterpress

The speed of the press, especially in web-fed letterpress operations, produces an ink mist which floats and then adheres to ceilings and walls. This fuzzy black coating on walls and fixtures can be easily observed. Because of the oil content of the ink, this presents a fire hazard if allowed to accumulate. Many letterpresses are old and, consequently, leak oil and grease, presenting both a housekeeping problem and a fire hazard.

Lithography

Some high-speed offset operations use driers to obtain the proper humidity content of the paper. These are usually gas-fired, but can be oil or electric. In most offset operations, a wetting agent is added to the water. The primary wetting agent is isopropyl alcohol which has a flash point of approximately 53°F (11.7°C). (A comprehensive discussion of flammable liquids can be found in Chapter 37, "Handling and Storage of Flammable and Combustible Liquids.")

On most sheet-fed offset operations, a powder is sprayed over each sheet as it comes to the delivery end of the press. This is, in essence, a drying process. The powder is very fine with a cornstarch base and a high tendency to float. Accumulations can be readily noted on overhead fixtures. This powder has explosive qualities greater than coal dust. All accumulations should be removed periodically by a vacuum cleaner — not blowing. Cleaner bags should be emptied immediately.

The offset operation involves much wiping and cleaning with solvents or washes. The ideal wash solvent will cut the ink and dry very quickly. Suppliers will generally try to formulate such washes regardless of flash point. Data sheets should be checked and flash points verified.

PRINTING AND PUBLISHING

FIGURE 16.14. *A fire hazard exists when wiping rags are not stored in covered metal containers.*

All operations that have paper in movement have the capability to develop static electricity, thus providing an ignition source.

Gravure

The plates or cylinders of a gravure press are etched with acids. The inks used in gravure printing are thinned with solvents such as naphtha or toluene. Most presses are completely enclosed in order to contain the fumes which can be cleaned and reprocessed.

Screen Printing

While some inks or paints now being used in the screening process are water-based, most have an oil base and use highly volatile thinners. The washing solvents used in cleaning the screens are also highly volatile. Because of the heavy consistency of the ink or paint used, it is common to have drying ovens connected to the press.

Solvents, Thinners, and Washes

Many commonly used solvents, thinners, and washes are purchased locally from chemical or petroleum supply houses (sometimes the corner gas station; lead-free gas is an excellent wash). Particular attention should be paid to the washes used on the presses. Pressmen prefer a fast-drying wash (highly volatile by nature). Safety cans are a must, and data sheets should be checked for flash points.

Chapter 16

FIGURE 16.15. On-end roll storage with access aisle.

Rollstock

Indoor storage of roll paper presents serious fire hazards aggravated by materials handling methods and economic pressures that increase piling heights and reduce aisle widths. Where clamp-jaw equipment is used, jaw clearance space of at least 4 in. (0.1 m) is usual.

In any storage area where heat can be radiated and re-radiated from one stack to another within spaces, intense fires can develop. Fire grows rapidly in the flue-like spaces formed from storing differing diameter rolls on-end in columns. Such storage occurs in bag manufacturing, general printing, and similar converter operations. (A comprehensive discussion of storage can be found in Chapter 48, "Industrial Storage Practices," and Chapter 15, "Paper Products Manufacturing.") The maximum challenge to sprinkler protection is presented by rolls stored on-end as separated columns under a roof supported by uninsulated structural steel. Fire spread and development is unusually severe and rapid and can quickly raise exposed building steel to temperatures at which it fails structurally. Steel failure can rupture sprinkler piping and deprive roll paper of fire protection at a time of maximum need.

Fire spreading up the side of a paper column quickly burns through the outer ply; paper then unwinds and peels away from rolls. Peeled-away material greatly and quickly increases the burning surfaces. Rolls continually shed outer layers wet by sprinkler discharge, thereby exposing dry paper. Peeled material spreads fire by contact with adjacent columns.

Initial fire development and intensity differ only slightly for different papers. Hard, thin, and laminated papers burn somewhat faster than soft, thick, and coated papers. The differences are not important to fire protection in on-end storage because the extremely high temperatures developed so quickly are common to all papers. The important concern is the method of storage.

Although rolls stored on-side avoid dangerous peeling, rolls stored in this manner may be nested between rolls of a lower tier. A fire here, which is well shielded from fire fighting efforts, can involve a large portion of the storage and become quite intense in the vertical flues created between roll ends.

Paper is sometimes supported horizontally on racks by rods which run axially through the rolls. Such an arrangement, because of the separation between rolls, has the same general fire characteristics as separated vertical stacks.

Paper in sheet form that is normally received strapped to skids can be stacked and presents no unusual problems.

THE SAFEGUARDS

Strict supervision of storage areas and thorough training of personnel in good fire prevention practices are essential (See Chapters 48 and 50 for a complete discussion of Industrial Storage and Housekeeping Practices). Usual fire causes should be given particular attention. These are cutting and welding, smoking, and electrical equipment. (A comprehensive discussion of welding and cutting can be found in Chapter 24; electrical installations are covered in Chapter 47.) It is especially important that industrial trucks be kept in first-class condition, and that all refueling be done outside of storage areas.

Rolls should be handled carefully to prevent pieces of paper from being torn loose and left hanging. Such loose pieces should be trimmed off or taped tightly to rolls. Waste paper or broken rolls should not be stored with roll paper.

Where all rolls in a sector of storage have the same diameter, adjacent stacks can be placed in contact or nearly so. In addition to space economy, such an arrangement has important fire protection advantages. Peeling is prevented and air supply is restricted. If there are fewer than 4 in. (0.1 m) between stacks, fire growth is slow and intensity is relatively low. Close stacking in both directions requires extra care and effort by equipment operators, particularly where clamp-jaw equipment is used (some rolls cannot be handled by suction). All stacks of differing diameter rolls are considered to be separated by more than 4 in. (0.1 m).

Paper dust or other combustible deposits should not be allowed to accumulate on tops of rolls or on ceilings, structural members, or piping. Fire can flash across such deposits, opening sprinklers to the full extent of the accumulations. Particular attention should be given to floor drains and scuppers. A small piece of scrap paper can effectively plug a floor

Chapter 16

FIGURE 16.16. Paper dust should not be allowed to accumulate on equipment.

drain, causing a flood and thereby ruining an enormous amount of paper after the sprinkler system has saved it from fire.

In the event of a fire, access aisles are of great help. A width of at least 8 ft (2.4 m) is needed. Distances between aisles in one direction should not exceed 50 ft (15.2 m). Absorbent papers such as tissue stock and newsprint, to a lesser extent, swell when wet. Consequently, a 2-ft (0.61 m) expansion space is needed between storage and building walls. Storage should be kept at least 3 ft (0.91 m) below sprinkler deflectors to ensure effective water distribution.

Housekeeping is by far the biggest problem in the bindery. Every piece of paper is cut, trimmed, or punched, and every operation generates waste that must be removed. Cyclone or suction systems are often used to remove the waste. Humidity is a critical factor since any movement of paper generates static electricity, thereby creating an explosive atmosphere with regard to the constant presence of paper dust.

The larger binderies have automated baling machines to handle the waste. They resemble hay balers. Valuable baled waste is usually stored inside until it is picked up for recycling. This is a very critical nonsmoking area and, if not sprinklered, fire hoses should be readily available.

Adequate fire extinguisher protection is imperative. In gravure, screen, flexography, and heat-set-offset operations, carbon dioxide extinguishers are preferred because of the cleanup. No water-type extinguisher should ever be allowed near a press room because all fires involve solvents or washes, and the numerous electric motors present a severe electric shock hazard.

BIBLIOGRAPHY

REFERENCES

Factory Mutual Engineering Corporation, "Roll Paper Storage," *Loss Prevention Data* 8-21, November 1983, Factory Mutual System, Norwood, MA.

NFPA CODES, STANDARDS, AND RECOMMENDED PRACTICES

Reference to the following NFPA Codes, Standards, and Recommended Practices will provide further information on the safeguards for printing and binding operations discussed in this chapter. (See the latest *NFPA Codes and Standards Catalog* for availability of current editions of the following documents.)

NFPA 12, *Carbon Dioxide Extinguishing Systems.*
NFPA 13, *Installation of Sprinkler Systems.*
NFPA 14, *Standpipe and Hose Systems.*
NFPA 30, *Flammable and Combustible Liquids Code.*
NFPA 69, *Explosion Prevention Systems.*
NFPA 77, *Recommended Practice on Static Electricity.*
NFPA 101, *Life Safety Code.*
NFPA 231F, *Storage of Roll Paper.*

ADDITIONAL READING

Adams, Michael and David D. Faux, *Printing Technology*, 2nd ed., Wadsworth Publishing Co., Belmont, CA, 1982.
Cox, Robert, *Printing Processes*, Van Nostrand Reinhold Co., NY.
Grice, A. M., "Fire Hazards Peculiar to the Printing Trade," *Fire Engineers Journal*, September 1980, pp. 29-31.
Lodge, J., "Fire Risks in the Printing Industry," *Fire Surveyor*, vol. 10, no. 1 (February, 1981), pp. 27-37.
McKinnon, G. P., ed., *Fire Protection Handbook*, Fifteenth Edition, National Fire Protection Association, Quincy, MA, 1981.

Prevention and Control of Fire in the Printing Industry, British Printing Industries Federation and Fire Protection Association, 1976.

Printing and Graphics Industry, Unipub, Xerox Publishing, NY.

Russell, B., "Good Housekeeping - Its Value and Importance," *Fire Surveyor*, vol. 8., no. 5 (October 1979), pp. 25-28.

17

Textile Manufacturing

Bruce Gray and E. E. Smith

Products of the textile industry are used for wearing apparel, household goods, house furnishings, carpeting, automotive upholstery, and for other industrial and domestic uses that are not quite so apparent. A major industry in the United States, in 1982 textile mills produced goods with an estimated sale value of $42 billion or about 2 percent of all manufacturing industry sales. In that same year, 750,000 were people employed in textile manufacturing with another 1.2 million in apparel and related jobs.

The basic fire hazards in textile manufacturing are congestion of machinery and stock, the ease with which the stock is ignited, and the ease with which fire can spread due to the relatively open flow of material between individual processes. The leading causes of fire are electrical equipment failures, sparks from foreign material in the stock, and heat from friction. Table 17.1 shows a breakdown of fires in textile plants of one company over a nine-year period (1975-1983).

Bruce Gray is Manager of the Service Center of West Point Pepperell, Langdale, Alabama.
E. E. Smith is involved with Special Projects at the Service Center.

TABLE 17.1. *Processes Involved in Textile Mill Fires in a Nine Year Period, 1975 to 1983**

	Number Of Fires	Percent Of Total
Opening through slubbers	689	9.34
Spinning through twisting	228	3.09
Slashing through drawing-in	20	0.27
Weaving	4,054	54.96
Cloth room operation	77	1.04
Warehousing	37	0.50
Waste machines	1,515	20.54
Ovens, tenters, dryers	499	6.77
Other	257	3.49
	7,376	100.00

*Based on forty-five plants in one company.

RAW MATERIALS

Raw materials used in the textile industry are varied. In a plant making 100 percent cotton yarn or fabrics, for example, the basic raw material is bales of cotton. However, plants producing blends of synthetic and natural fibers rely on a mix of raw materials based upon the type of end product produced. The production processes involved in converting raw stock to finished goods in the textile industry differ in their requirements because of the nature of the raw material used. There are, however, certain common denominators of fire hazard, such as the storage and opening of bales, the cleaning of raw stock, and weaving and finishing of fabrics, which are affected by the inherent combustible characteristics of the raw material. In a general way, a discussion of the basic processes of one segment of the textile industry has application to the industry as a whole; thus, the discussion in the major portions of this chapter is centered on the cotton industry as representative of the textile industry as a whole. However, the increased use of synthetics may change this in the next few years.

Cotton Stock

Bales of cotton are of either the domestic, compressed, or imported types. The domestic type is less lightly packed than the other two; consequently, domestic bales are more adversely affected by fire and water damage. The bales are delivered to the plant either by truck or box car. Normally, the bales are stored in piles of 500 or more, stacked to a height of up to 20 ft (6.1 m) (see Figure 17.1). Sometimes a smoldering or "fire packed" bale which is brought to the storage area undetected, or fire from another source, can flash over the surfaces of a loosely packed bale. The problem can take on severe proportions, particularly when the ties hold-

TEXTILE MANUFACTURING

FIGURE 17.1. *Bales of cotton in warehouse storage. Fire can spread rapidly over the loose fibers in the bales where bands have been broken and the burlap covering ripped away. Contrast this poor storage arrangement with the neat piles of bales of synthetic fibers shown in Figure 17.2. (West Point Pepperell)*

ing the bales together break, the bales open, their contents spill out, and the stacks fall.

Bales of synthetic fibers in the form of cut filaments (see Figure 17.2) are not as easy to ignite as cotton fibers. However, they can burn readily when exposed to heat from burning combustible wrappings or cartons. Wool fibers and other raw materials that are handled as bales may not exhibit the same susceptibility to ignition and ease of burning as cotton; however, they collectively present fire problems similar to those for bales of cotton.

Most handling of baled raw materials at the plant site (from unloading of box cars and trucks to storage facilities and through to placing the bales at the first machine in the manufacturing process) is by lift trucks. (See Chapter 46 for hazards associated with industrial lift trucks.)

THE PRODUCTION PROCESS

In a cotton weaving and spinning mill the basic processes (in the order in which they generally will be encountered) are: opening of bales, picking, carding, combing, drawing, roving, spinning, weaving, and finishing. Each of these processes has fire hazards distinctive to the conditions and machinery involved.

When bales of cotton enter the production process from the storage facility they are usually delivered to the opener room. Before the bales enter the opener room, the most common practice is to remove the ties and buckles from the bales using a tie cutter. When it is not practi-

FIGURE 17.2. Bales of synthetic fibers in warehouse storage. (West Point Pepperell)

cal to remove all ties in an outside area, the practice is to remove all but two, and those must be accounted for when they are removed in the opener room to make sure that they do not get into the processing machinery to become a source of mechanical damage or friction sparks.

Opener Rooms

Opener rooms should be located at grade level and cut off from the remainder of the plant by fire walls of at least 1 hour fire resistance. Ideally, there should be enough openings in the walls, protected by suitable rated fire doors, for access to the room for manual fire fighting operations as well as for quick removal of undamaged cotton bales from the room in a fire emergency if they can be moved to safety without exposing personnel to undue danger.

When the bales are brought into the opener room, a good arrangement is to lay them down in piles of twelve bales each with a 5-ft (1.52 m) space between the individual piles and an equal distance between opened bales being fed into the machinery and the machinery itself. In no event should there be more than four twelve-bale lots in each opening production line, and each opening line should be separated from the adjoining line by a 20-ft (6.10 m) space. Lines painted on the floor help to keep the storage of bales within prescribed limits and the spaces around the machinery clear.

If the plant operations require unopened cotton bales in the opener room (not in the opening lines), a good arrangement is to separate the unopened bales from the open bales on the opening lines by at least 25 ft (7.62 m). Another arrangement is to cut off the storage area from the processing area by a 1-hour rated fire partition with automatic closing fire doors at the necessary entrances to the storage area.

The machines found in an opener room go by various names: blending feeders, vertical openers, horizontal openers and cleaners, Centrif-Air machines, and gyrators. Some of these machines are slow moving while others operate more rapidly, but their general purpose is the same — to open up the bales and clean the stock through a process of beating, tearing, and shaking the cotton fibers — removing foreign material, such as motes (small particles), leaves and dirt.

Blending feeders: These are among the first machines into which the fiber is introduced. Their speed is slow and they do little more than slightly loosen the stock and blend the various grades of cotton fibers from the different bales on the opening lines (see Figure 17.3). Fully enclosed blending feeders offer several desirable features, including antifriction bearings that are not as prone to becoming choked by fibers as are the bearings on older blending feeders, exhaust systems for removing dust and fine lint, and, on some machines, an automatic device for proper loading of hoppers. (It is common practice to limit loading hoppers to two-thirds of their capacities; this keeps the risk of ignition through friction heat caused by choked bearings to a minimum.)

Vertical openers: These are the first machines in the manufacturing process which have rapidly rotating parts. The cotton is pneumatically conveyed upward through the machine, where arms on a rotating, vertically mounted steel beater throw the cotton outward against adjustable steel grid bars surrounding the beater. Foreign matter in the cotton passes through these grids and is collected and removed. Beaters operate at about 660 rpm; speeds faster than 800 rpm are not advisable.

Horizontal cleaners: These machines are usually fed by a lattice apron and a fluted feed roll. The beater is mounted horizontally, giving either an upward or downward stroke to the cotton held between the feed rolls. Surrounding the beater are a number of grid bars that remove fly, dirt, or motes. The average horizontal cleaner speed is about 650 rpm, although it is safer at around 550 rpm. The number of fires appears to increase rapidly with faster operating speeds.

Centrif-air machines: The cotton enters one end of the machine and is pneumatically carried through it. The beater is mounted horizontally; this is one of the few opener-cleaning machines having nonferrous beaters (or beaters protected by brass sheathing) in order to cut down the danger of sparks generated by pieces of metal striking the beaters. There is, however, a tendency for cotton to wind around the beater shaft at the discharge end of the machine. Thus causes a "lap" (a layering of the fibers) and provides a source of fuel for fire ignited by friction heat.

Gyrators: These machines further mix and clean the cotton before it reaches the picker hopper. They are slow moving and do not create much

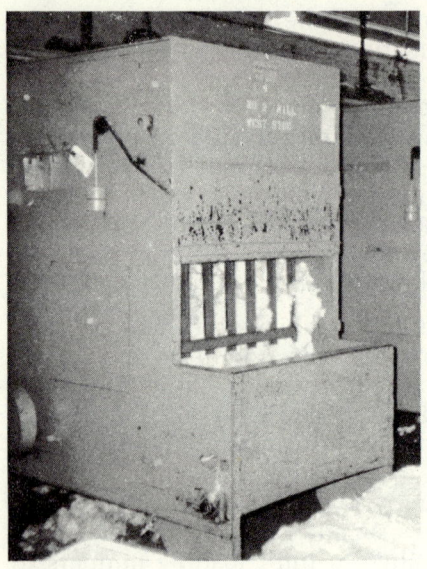

FIGURE 17.3. *A blending feeder, one of the first machines in the stock cleaning and opening process. (West Point Pepperell)*

additional fire hazard. When fire does occur, the usual cause is friction from cotton wrapped around the bearings. Gyrators are an excellent trap for heavy foreign material that may have passed through the opener line. Such material will be found at the base of the gyrator by removing the top plate at the lower end of the incline.

Magnetic separators: The purpose of magnetic separators is to remove ferrous metallic foreign material ("tramp metal") that could cause mechanical damage and friction sparks from the raw stock. The separators are installed in conveyor lines and ducts at a point where the ferrous particles can be extracted before they enter the machinery. Separators can be of two distinct types: permanent magnet units, found in most modern installations, and electromagnets. Electromagnets require a reliable source of electric current for operation. Figure 17.4 shows a permanent-type magnetic separator installed in a conveyor duct between an opening line and a condenser, which is the entry point of the cotton to the picker room.

Opener Room Layouts

Arrangement of the equipment in the opener room is important. No one arrangement will meet all demands; the nature of the equipment used in a particular location, whether it is for blending feeders, horizontal and vertical openers, dusters, etc., dictates the most efficient arrangement. Some general observations on good installations are: (1) vertical openers, where used, are placed first in the opener line; (2) if only two types of

TEXTILE MANUFACTURING

FIGURE 17.4. *A magnetic separator (arrow) in a pneumatic conveyor duct that carries cotton fibers from the opener line to a condenser. The separator stops "tramp metal" from entering machinery further along in the process to cause mechanical damage and fires from friction sparks. (West Point Pepperell)*

openers are used, such as a Centrif-Air machine and a horizontal cleaner, the Centrif-Air unit is first in line; and (3) where horizontal cleaners and vertical openers are used in combination, it is preferable to place the horizontal cleaner ahead of the vertical opener. A well thought out arrangement of opener room equipment allows sufficient aisle space around the machines and good subdivision of the bales as they wait to enter the machinery.

The Picker Room

From the opening room the stock is usually conveyed pneumatically through one or more systems of ducts to the picker room where more cleaning and alignment of fibers is done in the picking machines by a series of beaters, screens, and rollers. At the end of the process the fibers come out as a flat sheet of stock which is made up into a roll or lap (see Figure 17.5). In the more modern plants this stock is fed directly into the next process without forming a lap.

Picker machines: A picker unit usually consists of a condenser, a rake or belt conveyor, and one, two, or three beater sections. In older picker rooms, the picker unit was supplied by an open-feed hopper. Modern in-

FIGURE 17.5. *The delivery end of a picker machine showing a lap being formed under pressure to ensure density of stock. (West Point Pepperell)*

stallations consist of a reserve chamber and enclosed cleaners and feeders. The stock from the opening process is deposited in the reserve chambers as a lightly felted mass by the condenser in the pneumatic conveyor duct. The condensor is essentially a revolving screen that allows the air current to pass through but traps the air-blown cotton fibers and deposits them in the reserve chamber below. (Cotton in the reserve chamber can act as a dampener on fire that could enter from the blower — causing it to smolder rather than burn rapidly. The result is that less heat is produced, extinguishment is easier, and less cotton is lost.)

Fire prevention: Most fires in conventional opener-picker operations are traceable to the presence of tramp metal in raw or waste stock. The magnetic separators provide an automatic means of extracting metal from the conveyors and ducts. Fires that do occur in the conventional preparation process, however, usually originate in the opener equipment and are quickly carried to the picker area. The amount of fire damage in the preparatory area is considerably affected by the extent of fire spread within the picker room itself. When several opener lines have been connected together by pneumatic conveyor ducts, the amount of fire damage is increased in almost direct proportion to the amount of stock handled by a single conveyor system. Preferably, stock from each opener line should be conveyed through an independent duct and picker distribution system.

Good fire prevention practices in the opener-picker operations are aimed at keeping the machinery in top condition through cleanliness and maintenance. Definite maintenance and cleaning schedules will go a long

TEXTILE MANUFACTURING

FIGURE 17.6. Cotton fiber accumulations on the floor around machines are fuel for fires that can disrupt an entire processing line. Note the excessive lint on the floor under the lap at the delivery end of the picker machine. (West Point Pepperell)

way toward keeping risk at a minimum. Overhauling and repairing one machine every two weeks and keeping accurate records of such work is another routine well worth observing. After the repairs are made all surplus nuts, bolts, screws, and other metallic pieces must be removed from both inside and outside the machines; or they will become tramp metal and add to the hazard.

Constant vigilance is required to keep loose cotton from under and around machinery and to eliminate leakage of cotton from machines. Chokes or stoppages in opener lines and conveyor systems are particularly hazardous (see Figure 17.6). The same is true for foreign material entering machinery where it could cause damage and sparks.

The Card Room

In the conventional production process the picker lap is transferred by hand, wheeled rack, or conveyor from the picking room to the card room for further cleaning and straightening of fibers. The card machines may also be fed by a lap directly from a picker or from an automated line through a card feeding unit, which is usually referred to as a chute feed system.

Card machines: In the first part of a card machine (usually referred to as a "card") the fibers are separated into a web of fine stock that goes through a series of rollers that forms or "doffs" the web into either a rope-like or a flat web-like product, depending on whether the end product will be woven or nonwoven. The cylinder speeds are from 165 rpm (low speed cards) to over 300 rpm (high speed cards) with the latter type the most prevalent. Ventilation systems are needed with high speed cards to draw lint and fly away from the cylinders and flats and from around the feed and doffer rolls.

As in opener and picker operations, card fires can be caused by sparks from tramp metal, excessive amounts of stock caused by the improper laying of laps or lumps, thick areas in the lap, and electrical defects. Fires in low speed cards, which normally do not have ventilation systems, can spread over lint deposits and involve all cards in the room. Fires are less frequent in high speed cards because their ventilation systems keep them relatively clean and can pull an ignited fiber away from the card to the filter assembly. Sliver and lint accumulations on conveyor belts located below in-line cards can spread fire from one card to another, interrupting the production of an entire line.

Automated Processes

The preparatory process, encompassing opening, picking, and carding operations, is increasingly accomplished by automated systems which are totally enclosed from the time stock is placed in the opener until the sliver (a light web of fibers gathered into a tape shape) is "doffed," or removed, from the carding machines. Although the speed and amount of stock being processed is greatly increased by automated systems (and there are fewer operators around to watch for fires), automated systems are not necessarily more hazardous. Firesafety measures can be incorporated into the processing equipment, and automated mills can be as safe as those employing conventional processes.

Conveyor duct fires in automated systems must be detected quickly and the process fans shut down to prevent fire spread to other machinery, particularly to machinery containing large quantities of combustible fibers, i.e., automixers, flock feeders, and card condensers. Damage to one unit of an automated processing line will mean shutdown of all equipment in that automated processing line, and the resulting production shutdown will generally be more serious than the shutdown of a single unit in a conventional processing line.

Most fires developing in automated opener equipment will be transmitted through ducts, either to another machine or through the exhaust system to the filter bank that controls dust collection. Unless there is quick detection and subsequent application of extinguishing agents, such a fire is likely to cause equipment damage and prolonged shutdown. If, for example, a fire develops in the opener area (where large quantities of cotton await loading on the opener) the heat developed there could be carried through the enclosed system to open the sprinklers over the carding machinery at the far end of the line, and the resultant water damage could be high.

TEXTILE MANUFACTURING

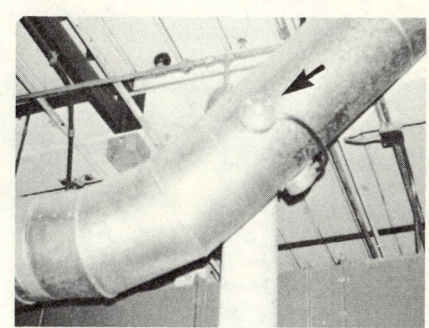

FIGURE 17.7. *Flame detector (arrow) shown mounted in a conveyor duct of a chute feed card system to shut down fans in the event fire enters the stock conveying system. (West Point Pepperell)*

Protection for Automated Systems

Protection for automated systems, aside from observing the same restraints on storage of unopened bales and lay-down at openers (no more than twelve bales at each opener), centers on the detection and control of fires in conveyor ducts and within unit enclosures. Smoke or flame detectors set up to shut down fans should be mounted in the ducts of process equipment as shown in Figure 17.7. (Flame detectors have proven more satisfactory in cotton systems, since dust may cause false alarms with smoke detectors.) Machines containing large concentrations of combustible fiber can be protected by automatic dry chemical extinguishing systems with discharge nozzles located within the machine enclosures or in the downstream ducts. The systems should be activated by a flame or smoke detector so that the dry chemical will reach downstream fiber buildup quickly. Complete extinguishment cannot be expected; thus, it is essential that an alarm is sounded so mill personnel can remove smoldering cotton from the machines involved.

Combing, Drawing, Roving, and Spinning

From the card machines stock normally goes to the combing process or to a drawing frame, depending on the requirements of the end product. Combing and drawing, as well as roving and spinning, are all operations which further align the fibers, make the yarn more uniform in diameter and weight, and increase its strength.

Combing: The object of a cotton comber is similar to that of a carding machine, but it is more exact and removes all of the short fibers under a predetermined length. Several laps are fed into the comber where they are broken up into strips of fibers of uniform length. Some waste is generated in the combing process.

Drawing: Draw frames essentially consist of three or four pairs of drawing rolls placed one ahead of another. Cans of card sliver are set in back of the frames, and several slivers are fed into the frame simultaneously

FIGURE 17.8. *Two types of spinning frames, a ring type (left) and open-end type (right). "Lap-ups" or bunching of fibers between moving rolls and ignition of the jammed stock by friction sparks are a hazard associated with fast-moving spinning frames. (West Point Pepperell)*

to emerge as a single sliver. The drawing action, designed to cause the fibers to lie parallel and to make the weight of the sliver uniform along its length, is produced by having each pair of rolls run faster than the preceding pair.

Roving: Cans of sliver from the drawing frames are taken to slubber machines for further processing. The slubber draws down the sliver by means of drawing rolls to a size called "roving," puts in enough twist to give the yarn the required strength, and winds the roving upon a bobbin.

Spinning: Spinning is the reducing of the roving to the final weight required in the finished yarn. There are two types of cotton spinning machines in general use: the ring-type spinning frame and the open-type spinning frame (Figure 17.8).

In any of the processes listed above there is a danger of fire as a result of ignition by friction between moving rolls and the jammed stock "lap-up" or bunching of fiber between rolls, and it is most likely to occur on the faster rotating machinery, i.e., the spinning frames. Electrical arcs from worn wiring insulation or from faults in electrically driven overhead cleaning equipment can also be a source of fires.

Warp Preparation

Spooling, winding, quilling, warping, and slashing are all processes involved in preparing yarns for weaving. The yarn is transferred from bob-

FIGURE 17.9. A slasher in which warp is dipped in a sizing solution and dried on a series of heated cylinders. (West Point Pepperell)

bins to spools or to large packages known as "cheeses," quills, or beams in preparation for use as warp or filling yarn in the weaving process.

At the slasher the warp is run through a sizing solution consisting of starch and special additives and then dried on heated cylinders. This process strengthens and protects the yarn during the weaving operation (see Figure 17.9). Dust explosions can occur in the starch used in the sizing solution if it is improperly handled in the dry state, particularly if a silo is involved.

Blending Fibers

If the desired product is a blend of cotton and synthetic, the mixing may take place in the opener room, or both stocks may follow the full process separately and be blended at some other point in the process. Two processes may be used to obtain synthetic yarn, both beginning with a synthetic material in the form of filaments. The most direct method of making synthetic yarn requires two additional steps beyond the preparatory processes: reeling and throwing. Reeling consists of twisting several filaments together to form strands; throwing consists of twisting several strands together to form yarn.

A second method that is similar to the conventional processing of cotton described earlier may be used. This process produces fabrics that are a blend of synthetic and natural fibers. The synthetic filaments are first cut into short lengths called staple. When blended with cotton, synthetic staple lengths are usually within ¼ in. (6.35 mm) of the length of cotton staple. The equipment used to process this length of staple is the same as for cotton; however, changes may be made in operating speeds.

Chapter 17

FIGURE 17.10. An older type of cam loom with alley takeup. Vibrations in looms can cause electrical breakdowns, a major cause of weave room fires. (West Point Pepperell)

Synthetic products involving the use of longer staple length synthetics may be processed on a modified worsted or woolen system. Cleaning operations may be omitted for 100 percent synthetic fiber.

Synthetics are commonly blended with cotton either in the opener room or on a drawing frame in which several card slivers, or loosely coiled tapes of stock, are fed into the machine to emerge as a single sliver. Several variations may be used when blending occurs in the opener room. The cotton may be introduced directly into the synthetic blending line without precleaning; or it may be opened, blended, and cleaned separately before it is introduced into the blending line with a synthetic fiber. When blending occurs on a drawing frame, both cotton and the synthetic fiber go through opening, blending, picking, and carding operations separately. They are blended by combining synthetic card sliver with cotton sliver on a drawing frame. Due to the flexibility of modern textile equipment, different types of synthetics as well as cotton may be processed interchangeably in the same machinery with equipment modifications. Therefore, all equipment should be protected as outlined earlier.

Due to the toxic nature of the fumes omitted, fire fighting personnel are more likely to require protective equipment if a fire involves synthetics.

Weaving Operations

The weaving operation is the culmination of all the preceding processes. The mechanisms of a loom are too well known to need minute description. In order to produce a fabric in a loom, the warp yarns (those which run the length of the cloth) must bind in the filling (the yarn which runs across the cloth). The movements of the warp yarns are produced in the loom by a cam-operated harness that raises or lowers the necessary ends of the warp as the shuttle passes back and forth with the filling yarn. There are a variety of looms in use today and the refinements of their operating mechanisms are dictated by the complexities of the cloth patterns produced. Jacquard looms, for example, are used for figure work such as damasks. They require complicated control mechanisms, such as a very large "dobby," which manipulates the movements of the warp yarn so that dots and curves can be made in the pattern. (See Figure 17.10.)

In one study the main cause of fire in looms was found to be electrical faults. (Electrical problems caused more than 50 percent of all weave room fires — see Table 17.2.) Most of these faults are from vibrations at the loom causing breakdowns at connections in the wiring. Although the number of loom fires is high, the cost for each was found to be lower than the cost of fires in other areas of the plants surveyed (Table 17.3).

Plants using the concept of large roll cloth takeup on a lower floor, rather than at the looms themselves, may add to the chance of fire spreading throughout the weave room or from loom to loom because of the creation of large unprotected floor openings. Good cleanliness is important in weave rooms to control the fly and dust that are accumulated in the vicinity of the looms and their electrical equipment. A good procedure is to "blow down" or clean the looms at the time the warps are changed.

A special protection feature that has been found useful in weave rooms (particularly for Jacquard and plush weaving operations) is ¾-in. (19 mm) hose with spray nozzles, preferably equipped with 6-ft (1.89 m) applicators.

Cloth Inspection and Cleaning

An inspection and cleaning process follows the weaving operation. In this area the cloth is checked for defects, loose ends are cut or singed off, and the fabric made ready for shipment to a finishing plant. The hazards in cloth inspection areas are limited mostly to the use of torches for removing loose ends on the rolls of cloth. Some cloth rooms or inspection areas, however, have heat setting and finishing equipment that involves gas-fired ovens and ranges. The frequency of fires here is much greater, particularly when the stock "hangs up" and ignites in the dryers.

TABLE 17.2. *Causes of Weave Room Fires in Textile Mills, 1975 to 1983**

	Number Of Fires	Percent Of Total
Wiring and switches (electrical)	1,224	30.22
Stop Motion (electrical)	871	21.49
Motors (electrical)	326	8.04
Subtotal (electrical)	2,422	59.75
Friction (mechanical)	1,027	25.33
Overhead blowers	69	1.70
Unifil	159	3.92
Unknown	377	9.30
	4,054	100.00

*Based on forty-five plants in one company.

TABLE 17.3. *Comparison of the Cost of Weave Room Fires to Other Fires in Textile Plants, 1975 to 1983**

Total Fires	7,319
Weave room fires	4,054
Weave room fires, percent of total	55.39
Total cost, all fires	$2,836,693
Cost of weave room fires	$ 520,469
Weave room cost, percent of total	18.35
Average cost, all fires	$388
Average cost, weave room fires	$128
Average cost, other fires	$869

*Based on forty-five plants in one company.

Finishing

In the finishing plant, which may or may not be a part of the mill producing the fabric, the cloth may be bleached, dyed, printed, sanforized, have backings or coatings applied, or put through a variety of other processes to provide the desired effect in the finished cloth. Most of the finishing work involves the use of dip tanks, spreaders, rollers, and other heat and drying applications. (See Chapter 26 for information on dipping operations and Chapter 27 for discussions on dryer installations.) Shearing or slotting the surfaces of some fabrics is another operation often found as part of the finishing process.

From the finishing plant the fabrics go directly to cutting and sewing operations, or they may be stored for future shipment.

Knitting and Carpeting

There are two variations on the basic textile process worthy of special mention. These are knitting and carpet forming. Both processes start with yarn finished in a spinning mill and involve the steps and hazards previously described. In the case of knit goods, the fabric is made on a knitting machine instead of a loom. The inspection and finishing processes are similar to those for woven products.

The standard process for carpet forming is the tufting machine where the basic yarns are interlaced by tufting needles. The finishing process for carpeting, however, involves backing the carpet material with jute or a similar material and some form of latex, followed by a heat process for curing the latex and setting the fabric width. The heat process may be direct fired by some form of gas or other fuel or by steam drums. The malfunction of gas-fired equipment controls, buildup of lint, and electrical arcing are the most frequent causes of fire in carpet ovens. The drying ovens are frequently 200 to 300 ft (61 to 91.4 m) in length. A loss in such a large installation represents a considerable expense due to the amount of carpet involved (Figure 17.11).

Air Conditioning and Waste Recovery Systems

Two other parts of the processing system in a textile plant that deserve attention are the air conditioning system and the waste recovery system. Both of these are more subject to aiding the spread of fire than to being a source of ignition in themselves. When a fire enters one of these systems it can cause considerable down time and heavy damage to filter units.

Air conditioning systems: Good humidity control that provides increased moisture in the air can have a damping effect on the spread of fire over the surface of loose cotton and can also reduce the risk of ignition from mechanical sparks. Air systems have been added to older mills that either draw the air directly through wall openings into an adjoining air conditioning room or through floor openings connected by ducts to the conditioning room. Filters cleanse the air of dust and lint before it is humidified and returned to the process area. Lint tends to concentrate in the ducts if, through poor design, they have rough edges, poorly located turning vanes and dead air spots, adding to the hazard of dust fires. Ducts must be well-supported and dust-tight so that fire in the ducts cannot expose combustibles in the vicinity.

In some older mills without air conditioning systems, air moving ducts from machines in the preparatory area terminate in a dust room, usually located in the lower part of the mill. Good dust rooms are of noncombustible construction throughout and are protected by automatic sprinklers.

In new mills, picking, carding, spinning, and weaving operations are often an in-line process, and lint-laden air is drawn into the ducts

Chapter 17

FIGURE 17.11. *A finishing oven for applying backings, such as jute and latex, to rug material and for heat setting the fabric. The ovens are frequently 200 to 300 ft (61 to 91.4 m) in length. (West Point Pepperell)*

through grated floor openings under the machinery. The ducts are connected to a series of concrete tunnels that eventually discharge the air through a filter unit into an air conditioning room. While underfloor air return systems can promote cleanliness around machinery, surfaces of the underfloor tunnels frequently, because of low air velocities, become covered with lint deposits that are difficult to clean out. Then, too, if fire should originate in waste around the base of the machines, it would probably spread through the tunnel to endanger other machines, the air conditioning unit, and even the building itself. Underfloor return air systems represent one of the greatest potentials for spreading fire throughout a mill; consequently, automatic sprinklers in the tunnels are a good line of defense.

Air conditioning systems do, however, have an additional benefit. Insofar as they are responsible for most air movement in the areas they serve, they can also be used to effectively remove the smoke from a fire.

Pneumatic waste recovery systems: Two basic types of pneumatic waste recovery and cleaning systems are used with spinning and drawing frames: the unit system and the central air system. The unit system features suction tubes, called "flutes," which are connected (through a header) to an individual waste collector unit at the machine. A screen in the collector retains the waste material which is removed from the collector by hand. In the central air system, which is frequently a part of the air conditioning system, waste is collected from a number of machines through underfloor ductwork which leads to a centrally located fan unit and filter which separates the waste and either discharges the air to the outdoors or directs it to the air conditioner where it is cleaned, cooled, and humidified before being returned to the spinning room.

Waste removal units used at some high speed carding machines have separate air filtration units. One unit serves several cards. Fire protection for such an air filter unit is shown in Figure 17.12.

Electrical Installations

Faults in electrical systems and components are the single largest known cause of fires in textile mill process areas. The faults most commonly occur in electric motors, switch boxes, and on electric-powered traveling hoists and cleaners. A good maintenance program based on periodic inspections of all electrical installations will go a long way toward controlling the hazard.

In textile mills where dusty and lint-laden atmospheres are prevalent, the particular locations where such hazards exist are defined basically as Class III hazardous locations by NFPA 70, *National Electrical Code*. A Class III location is one that is hazardous because of the presence of easily ignitible fibers or flyings; however, they are not likely to be in such numbers as to produce ignitible mixtures in air. Class III locations are further described as either Division 1 or Division 2 locations. Division 1 locations are where easily ignitible fibers, such as cotton, cotton linters, and waste, are handled, manufactured, or used. This includes parts of most textile mills. A Division 2 location is an area where easily ignitible fibers are stored or handled, such as a cotton warehouse, but where no manufacturing is carried on. The conditions above are becoming less prevalent with the emergence of more sophisticated machinery and air cleaning systems.

Article 503 of the *NEC* contains general and specific requirements for electrical installations in Class III locations and should be consulted for detailed information on the appropriate types of acceptable installation practices and equipment.

Of particular concern in textile mills is the arrangement of wire, switch, and motor installations at machinery where vibrations and lint and dust accumulations can create hazardous conditions. Table 17.2 is a good indicator of areas where special attention should be placed in relation to electrical systems.

Dust-tight or vapor-tight lighting fixtures are appropriate for installation in manufacturing areas where large concentrations of lint and dust will be found, such as dust rooms, ventilating tunnels, baled cotton storage, and Jacquard loom rooms. Ordinary fluorescent lights of the open-type (but not high voltage cold cathode-type) are acceptable for overhead lighting in areas where the dust and lint tend to settle around the machinery at floor level. Temporary lighting by drop cords is not a good

Chapter 17

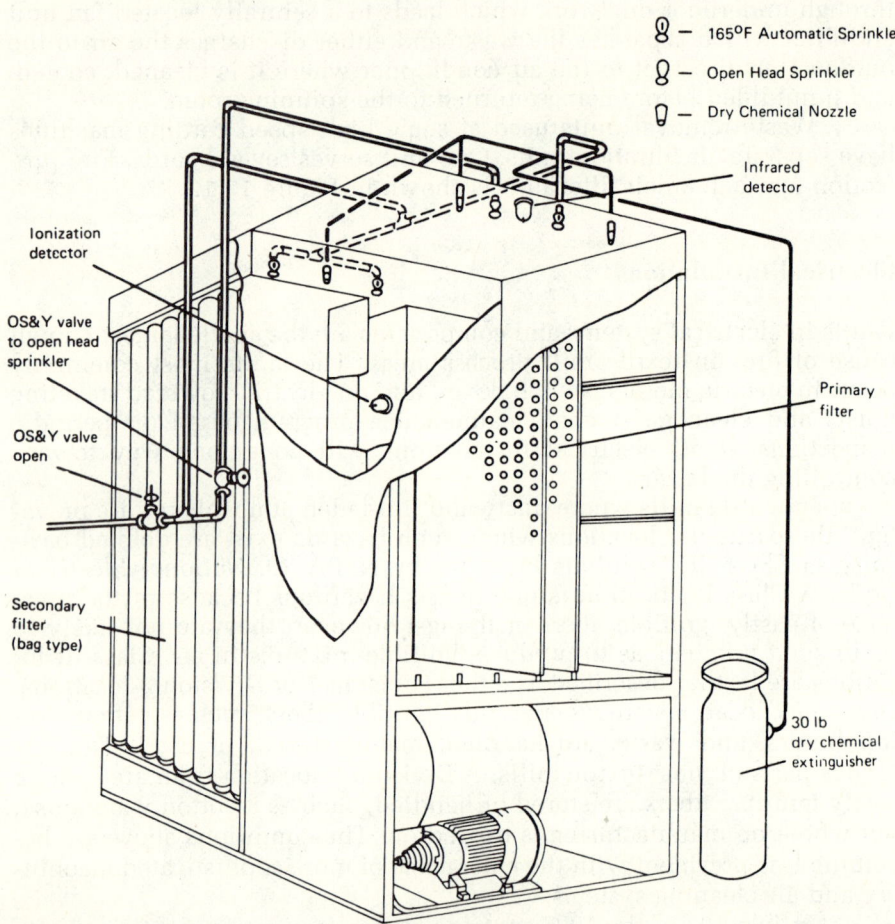

FIGURE 17.12. Arrangement of fire protection for a card filter unit. In this installation there are two basic extinguishing systems — sprinklers and dry chemical. The dry chemical system can be activated by the ionization detector or the infrared detector. Sprinkler protection is provided by automatic sprinklers backed up by open sprinklers manually controlled by an OS & Y valve. The filter unit is shown with doors removed and two primary filters in place. 165°F equals 73.9°C; 30 lbs equals 13.6 kg. (Factory Mutual System)

practice; the cords are often hung improperly and the danger of breakage can be high.

Traveling trolley and cleaning systems present the problem of keeping contact surfaces clean and lint-free (see Figure 17.13). Ground fault detectors are advisable in each system circuit so that a visual and audible alarm can be given and maintained as long as power is supplied to the system and the fault remains.

THE SAFEGUARDS

Construction

Ideal construction for a textile manufacturing plant is a one-story building of brick or concrete construction with sections of 20,000 sq ft (1858 m^2) or less separated by 1-hour rated fire resistant partitions. Larger areas are practical when enclosed and automated stock feed systems are used in conjunction with appropriate automatic detection and extinguishing systems. In any event, floors should be of fire-resistive construction and provided with sufficient floor drains and scuppers to handle any sprinkler runoff and hose discharge that may occur during a fire. Underfloor passages and tunnels have not proven satisfactory as they are prone to accumulating large amounts of lint and are relatively inaccessible for cleaning and fire fighting.

Warehouses: Buildings used for storage of baled fibers (cotton, jute, hemp, flax, wool, or synthetic fibers) are ideally divided into sections of 10,000 sq ft (929 m^2) by walls having a fire resistance rating of at least 1½ hr. Areas as large as 30,000 sq ft (2787 m^2) can be tolerated provided the buildings are of a superior type of construction (heavy timber or noncombustible) and are equipped with vents to carry off smoke to help in manual fire fighting. (NFPA 204, *Smoke and Heat Venting*, contains recommendations on types of vents and their sizing and location.) Properly pitched floors with drains and scuppers are necessary to draw off sprinkler and hose stream water during a fire. A well-designed warehouse also takes into account the additional loads imposed by the added weight of water which can be absorbed by the baled fibers. Unless adequate clearance is maintained between walls, columns, and bales, the bales also can swell and cause appreciable damage when they are wet.

If a portion of the warehousing facility is used for other activities, such as grading, or bale tie repair, the area should be separated from the remaining part of the structure by a 1-hour fire rated partition.

Fire Control Systems

Automatic sprinklers: All manufacturing, finishing, and storage areas in textile mills need automatic sprinkler protection installed on an ordinary hazard pipe schedule. (NFPA 13, *Sprinkler Systems*, contains guidance for installation of sprinkler systems in buildings. Definitions in that standard include what constitutes an "ordinary" hazard occupancy and an "ordinary" hazard pipe schedule.)

The temperature rating of sprinklers in the systems will vary according to the requirements of each location. For most manufacturing areas, for example, sprinklers rated at 165°F (74°C) would be required; higher rated sprinklers may be desirable in opener-picker and warehousing areas (see Chapter 48 for information on fire protection for storage occupancies). All sprinkler systems require local audible alarms and should

Chapter 17

FIGURE 17.13. *An overhead traveling cleaner unit in the carding area of a mill. Electrical faults in traveling cleaners are a common cause of fires. Cleaner unit circuits should be equipped with ground-fault detectors. (West Point Pepperell)*

be connected to a central station that is under surveillance on a regular basis.

Special protection: Automatic detection and suppression systems are a necessity in the opening through carding processes because of the speed of the moving stock and the interconnection of the processes, particularly in the totally enclosed automated systems with their large reserve bins and direct connection to the cards (automatic processes and special protection for them have been discussed previously). Additional automatic detection and suppression equipment, however, does not eliminate the need for sprinkler protection, not only in open areas where the automated equipment is located, but within the enclosures around the machinery which can effectively block discharge from the room sprinklers.

Portable extinguishers and small hose: Dry chemical and water-type portable extinguishers are well suited for textile mills. Sodium bicarbonator potassium bicarbonate-based dry chemical (ordinary type) are effective for combatting surface fires on cotton; however, the dry chemical discharge should be supplemented by water spray to extinguish smoldering embers or in case fire gets beneath the surface. For most conditions a garden hose nozzle with a spray discharge pattern on a ¾-in. (19 mm) rubber hose is effective on surface fires both in bale storage and cotton in process; however, larger 1½-in. (38 mm) hose is more suitable for storage areas.

Water pump tank and stored pressure water extinguishers are effective on fires involving small quantities of baled fibers. Spray nozzles would help to prevent scattering of loose cotton by the discharge. Carbon dioxide foam and multipurpose dry chemical extinguishers have limited application in textile mills but are not recommended for general use.

Personnel training: A good training program that familiarizes employees with the characteristics of burning baled fibers, fire fighting, and salvage is an important part of the suppression program. Surface fires in baled fibers are usually not difficult to control. The speed that flame can travel over the surfaces of the cotton requires rapid and broad coverage of extinguishing agent rather than concentrated extinguishing power; thus, employees must know how to use extinguishers and small hose skillfully. (See Chapter 3 for a detailed discussion of "Plant Emergency Training and Organization.")

Life Safety Features

Large numbers of similar machines in single open areas are common to textile plants and can create a maze of passages, alleys, and dead ends; thus, it is of utmost importance to keep access to exits well-marked, lighted, and clear of obstacles. For these reasons, and because of the presence in most plants of synthetic materials that may generate toxic gases in fire, it is prudent to equip the fire brigade with several units of self-contained breathing apparatus.

BIBLIOGRAPHY

REFERENCES

American Textile Manufacturers Institute, "Textile Hi-Lights."
Factory Mutual Engineering Corporation, "Textile Mills," *Loss Prevention Data* 7-1, January 1980, Factory Mutual System, Norwood, MA.
F. M. Eng. Corp., "Baled Fibers," *Loss Prevention Data* 8-7, March 1974.

NFPA CODES, STANDARDS, AND RECOMMENDED PRACTICES

Reference to the following NFPA Codes, Standards, and Recommended Practices will provide further information on the safeguards for textile manufacturing operations discussed in this chapter. (See the latest *NFPA Codes and Standards Catalog* for availability of current editions of the following documents.)

NFPA 10, *Portable Fire Extinguishers.*
NFPA 13, *Installation of Sprinkler Systems.*
NFPA 13A, *Care and Maintenance of Sprinkler Systems.*
NFPA 17, *Dry Chemical Extinguishing Systems.*
NFPA 20, *Centrifugal Fire Pumps.*
NFPA 26, *Supervision of Water Supply Valves.*
NFPA 27, *Private Fire Brigades.*
NFPA 30, *Flammable and Combustible Liquids Code.*
NFPA 49, *Hazardous Chemicals Data.*
NFPA 68, *Explosion Venting Guide.*
NFPA 70, *National Electrical Code.*
NFPA 70B, *Electrical Equipment Maintenance.*
NFPA 77, *Recommended Practice on Static Electricity.*
NFPA 80, *Fire Doors and Windows.*
NFPA 86A, *Ovens and Furnaces.*
NFPA 90A, *Air Conditioning and Ventilating Systems.*
NFPA 101, *Life Safety Code.*
NFPA 204M, *Smoke and Heat Venting.*
NFPA 214, *Water-Cooling Towers.*
NFPA 231C, *Rack Storage of Materials.*
NFPA 231E, *Storage of Baled Cotton.*
NFPA 325M, *Fire Hazard Properties of Flammable Liquids, Gases, and Volatile Solids.*
NFPA 505, *Powered Industrial Trucks.*
NFPA 701, *Flame-Resistant Textiles and Films.*
NFPA 702, *Flammability of Wearing Apparel.*

ADDITIONAL READING

Corbman, Bernard P., Textiles: *Fiber to Fabric*, 5th ed., McGraw-Hill, Book Co., NY, 1975.
"Designing to Limit Loss in High-Rise Rack Warehouses," Kemper Group Report, vol. 10, no. 3 (September 1981), pp. 2-9.
Harrington, J. L. and R. B. Hopkinson, *Rack Storage Protection*, Worcester Polytechnical Institute, Worcester, MA, 1977.
King, P. W. and J. Magid, *Industrial Hazard and Safety Handbook*, Butterworth Publishers, Inc., Woburn, MA, 1979.
McKinnon, G. P., ed., *Fire Protection Handbook*, Fifteenth Edition, National Fire Protection Association, Quincy, MA, 1981.
Olsen, Richard P., *Textile Industry*, Lexington Books, Lexington, MA, 1978.

Pajgit, O., ed., *Processing of Polyester Fibres*, Textile Science and Technology Series, Elsevier Science Publishing Co., Inc., New York, NY, 1980.

Szaloki, Z., *Textile Processing*: vol. 1: Opening, Cleaning and Picking, Textile Books, Service, Inc., Broadway, NJ, 1976.

Talbot, G., "Static Electricity," *Fire*, 70(871), 1978, pp. 397-398.

Thomas, C. L., "Fire Protection in Automated Textile Machinery," Paper No. 6/1-3, Textile Industry Technical Conference, 1976.

18

Clay Products Plants

Emil C. Hrbacek

Abundant in nature, easily obtainable, and readily shaped, clay has been and still is a common material for many uses. The making of useful and ornamental clay articles goes back to the early history of man. The first primitive clay products were merely sun-dried, and became soft when wet. Man soon learned, however, that when fired to a red heat, clay increased greatly in strength and was no longer affected by water. Hard-burned tablets inscribed with information about people living 6,000 years ago have been discovered in excavations at the site of Ur on the Chaldees, the City of Abraham.

Basically, there are four different types of clay products: heavy clay, whitewares, refractories, and special products.

Heavy clay products: These include common brick, face brick, paving brick, hollow tile, conduits, roofing tile, drain tile, flower pots, sewer pipe, wall coping, stoneware, and terra cotta.

Whitewares: These may be divided into two general classifications: (1)

Emil C. Hrbacek is Assistant Vice President, Loss Control Department, Alexander & Alexander Inc., St. Louis, Missouri. This study, originally conducted by Mr. Hrbacek and F William B. Larkin, has been revised for this *Handbook* by Mr. Hrbacek.

Vitreous products are those where the ware has been fired until practically all of the pores of the body are filled with a glassy bond and, as a result, the ware is impervious to water; vitreous products include porcelain and china tableware, sanitary ware, floor tile, and porcelain electrical insulators. (2) Semivitreous products are those of incomplete vitrification which are finished with a glaze to make them impervious to water; they include porcelain and chinaware, wall tile, and earthenware.

Refractories: These include refractory brick, special refractory shapes, and insulating brick.

Special products: These include refractory porcelain (usually vitrified), chemical porcelain (vitrified), and technical products (vitrification depending on use).

Today, clay is used for a great many products with varying physical and chemical characteristics.

RAW MATERIALS

Clay is a fine-grained mixture of mineral fragments formed by the disintegrating action of air, water, frost, vegetation, and chemicals on igneous rock, chiefly feldspar and feldspathic rock. Residual or primary clays overlay the parent rock, while sedimentary or secondary clays are those which have been carried from their place of origin to other locations by wind, water, or glaciers.

While clay does not have one definite chemical composition, all clays contain hydrated silicate of alumina, called the "clay substance." Clays also contain some oxides of calcium, iron, magnesium, potassium, sodium, and titanium which tend to act as fluxes and which affect the natural color of the clay products. Clays may contain sand and alumina in excess of that required for the clay substance. Some clays contain carbonaceous matter.

It is the physical nature of the hydrated alumina-silicates that gives clay its plasticity or workability. When moistened with water, most clays become plastic; i.e., they are readily formed into, and retain, a desired shape. Plasticity varies with different types, some being only slightly plastic (lean), and others very plastic (fat). The latter will take up more moisture than the former and is, therefore, more subject to shrinking and cracking.

Other important properties are particle fineness and distribution, power of suspension, dry strength, shrinkage, fusibility, fired strength, hardness, porosity, and color.

The most important materials used in the making of clay products are as follows:

Kaolinite: Hydrated alumina-silicate consisting of alumina (Al_2O_3), silica (SiO_2), and water of crystallization ($2H_2O$).

Kaolin: White-firing clay, usually from primary deposits, made up chiefly of minerals of the kaolinite type.

China clay: A commercial variety of kaolin. A relatively pure grade may be found where sand, flint, and mica have been removed by the natural washing effect of water. It is also prepared by artificial water washing and settling processes.

Ball clay: A sedimentary white-burning clay with extreme fineness of particle, and usually contains some organic matter which discolors it. Exceptionally plastic, it has high dry-shrinkage characteristics and is relatively strong in the dry unfired state. (This material was formerly dug from deposits in cubical form which rounded easily when handled.)

Buff- and red-firing clays: Materials together with shales used for making heavy clay products. The red-firing clays contain iron oxide and other impurities. Buff-firing clays require higher firing temperatures than the red-firing clays.

Shales: Fine-grained rock or clay strata hardened by intense pressure without chemical alteration.

Flint: Free silica obtained from sand, quartzite, and sandstone. Finely ground flint is added to the clay mix for making pottery bodies, where it acts as a nonplastic material in the dry unfired state and as a refractory in the firing stage.

Fire clays: Materials used to make refractories, and are not so classified unless they will withstand temperatures in excess of 2,900°F (1593°C). Typical chemical analysis shows the following: alumina, between 25 and 45 percent; silica, between 40 and 60 percent; fluxes, usually less than 4 percent. High-alumina clays contain more than 70 percent alumina. Kaolin, ball clay, and flint may also be used to make refractories.

Grog: Material that may be added to the clay mix to reduce porosity and shrinkage. It is a granular nonplastic material which may be either flint or fire clay that has been fired in a kiln and then reduced to granular form. Broken, cracked, or warped fire brick or other refractories may also be used for making grog.

Fluxes: Any materials that will lower the softening or liquefying temperature of another material. Most minerals in clay, other than alumina, act as fluxes in the firing.

THE PRODUCTION PROCESSES

The various steps in the processing of clay are as follows:

Winning: The removal of the clay from its original deposit, which may be from open pits or from underground mines.

Weathering: The exposure of the clay to outside air in shallow yard pits which improves it through natural slaking or the disintegrating action of sun, rain, and frost. Unwanted soluble salts are thereby leached out.

Crushing and grinding: Here the particle size of the fragmented minerals is reduced. For heavy clay products, workability, strength, shrinkage, density, hardness, texture, and color depend upon how finely the clays are ground. Strength, thermal conductivity, abrasion, resistance, slagging, and spalling are properties influenced by grain size. Very fine-grained particles are used for making chinaware and porcelains. Shales and rock-like raw materials are passed through various types of crushers, then ground to further reduce particle size. Dry-pan grinders, consisting of heavy wheel-type rollers mounted in vertical position and resting on a rotating horizontal pan, are extensively used. Wet-pan grinders of similar design are employed when the operation combines grinding, mixing, and tempering. Fine grinding is effected in such equipment as ball mills, tube mills, and Raymond mills.

Screening: After crushing and grinding, the various sized particles pass through any one of several types of screens used for separating and grading.

Tempering: The conditioning of the clay with water to the desired degree of plasticity or fluidity. Wet-pan grinders and pug mills are the type of equipment commonly used for tempering. The usual pug mill consists of blades or knives secured at right angles to a rotating shaft in a trough. The clay, to which water has been added, is cut and mixed by the action of the blades. A slight angle in the setting of the blades pushes the clay to one end of the mill.

Blunging and filtering: Finely ground clay is agitated by beater arms in an excess of water in an open tank to form "intimate" mixtures of clay and water for making fine chinaware and porcelains. A filter press removes the excess of water, after which the press cake may be reblunged. In some cases, mixing is done in an excess of water in a ball mill.

Product Forming

Clay is formed to the desired shape by methods best suited to the end product. Methods generally used include the following:

Slip casting: A method used for forming some pottery figurines and other complicated shapes which cannot be machine formed. Slip, a liquid clay mix having water content between 12 and 50 percent, is poured into split molds. The clay adheres to and takes the shape of the interior of the mold. After a predetermined drying period, the excess slip is poured off and the formed shape is removed from the mold. Slip casting equipment for toilet bowls is shown in Figure 18.1.

Soft mud: A method widely used because of the plasticity or workability

CLAY PRODUCTS PLANTS

FIGURE 18.1. Slip casting.

of the clay. Three general techniques are employed. They are hand molding, jiggering, and machine molding.

1. Hand molding may employ the potter's wheel, a method that has come down through the ages. A mass of prepared clay is placed on a revolving table and is hand shaped to the desired form.

   Another hand molding technique is used for making some ornamental terra cotta and certain special shapes, including fire brick and refractory. A mass of prepared clay called a "walk" is hand shaped to the general form of the mold and then thrown with some force into the mold. The mold is then dropped or bumped on the molding table to settle the clay, and the excess is removed by a cutting wire. An oil applied to the sides of the mold prevents sticking. Sand is sprinkled on the clay, and the mold then inverted on a pallet and the shaped form transferred to the drying process.
2. Jiggering combines the use of the potter's wheel and a mold. As the clay-filled mold is rotated on a table, a fixed tool cuts and shapes the top of the clay. At the same time, the clay takes the shape of the inside bottom of the mold.
3. Machine molding is used for making regular shaped products such as bricks. The clay mix is fed into a pug mill which mixes, kneads, and compresses the clay and forces it into molds from which the bricks are removed by automatic equipment. Oil may be used to prevent adhesion of the clay to the molds.

Stiff mud: A method of making bricks similar to the machine-molding, soft-mud technique except that the clay is extruded through a die in the form of a rectangular bar sized to the width and depth of the brick. Wire cutters slice the extruded bar to proper length. In some cases the clay is "de-aired" by high vacuum before it is passed through the die. Water content is from 15 to 30 percent. This method is also used for sewer pipe, electrical insulating rods, flower pots, crucibles, and special shapes, each in a machine specially designed to produce the desired form from stiff mud.

Dry-press: Clay mix with a water content of 4 to 12 percent is pressed into molds by the action of a mechanical toggle press, screw press, or hydraulic press. The clay is usually de-aired by high vacuum applied by hose connected to the compressing pod.

Drying

After the clay has been shaped, water is removed by evaporation. Shrinkage occurs at this stage, the amount depending on the type of clay. Too rapid drying can cause warping or cracking. Some products are merely air-dried, while others may be "fried" by the application of heat.

"Water smoking": This process, also called ordinary drying, which is done at temperatures up to about 400°F (204°C), drives off the free water used to make the clay workable. The formed shape can again be softened by water and can, therefore, be damaged by hose streams.

"Chemical water smoking": This process, also called dehydration, which occurs at temperatures from about 300°F to 1,800°F (149 to 982°C), removes the water in chemical combination with the minerals in the clay. The product after completion of this stage can no longer be softened by water, but is subject to damage by sudden chilling.

Hot floors: Handmade refractory shapes, and shapes that are repressed after partial drying are usually dried on floors heated by low-pressure steam. The best arrangement is floors of concrete or steel; however, some drying of this type is done in multistory buildings with slatted wood floor decks supported on wood joists. This method may also be used for drying glass pots, glass tank blocks, sewer pipe, and similar products.

Compartment or chamber driers: Totally enclosed steamheated compartments used where temperature and humidity control is essential. One use is for drying stoneware for the chemical industry.

Continuous driers: Tunnel-type driers are commonly used in the industry. The long, narrow tunnel is equipped with heating means, fans, and humidity and temperature controls; in some cases the waste heat from downdraft kilns is used. The products are carried on conveyors, cars, or belts, while dry hot air is forced through the tunnel by fans. Air travel may be in the same or in counter direction to the flow of material, the latter being generally the case.

Mangles: In this type of drier the product is placed on shelves suspended from an endless chain and is carried through a drier enclosure.

Hot dry air passed through the product may be in the direction of the travel of the material, or counter to it, the latter being the method generally used.

Rotary driers: This type, which is used for drying raw materials, consists of a nearly horizontal rotating cylinder slightly higher at the entrance end. Projecting blades are attached to the interior of the cylinder to help in the mixing effected by the rotating cylinder. Air flow may be counter or in the direction of the travel of the material, or both.

Humidity compartment driers: These are essentially laboratory equipment with more elaborate design than production driers. This type is equipped with temperature and humidity controls.

Kiln Firing

Following the drying operation, formed clay products are kiln fired to achieve the desired or acceptable degree of vitrification. The fuel may be wood, coal, oil, natural gas, or producer gas.

Oxidation: Occurs at temperatures from about 600°F to 1,750°F (316 to 954°C) when the combustible substances in the clay burn to produce porosity, and minerals such as carbonates, sulfides, sulfates, etc., are changed chemically.

Vitrification: Occurs at temperatures above 1,600°F (871°C), when the fusible minerals in the clay melt and flow into the pores of the clay, bonding the unfused parts together. The product thereby becomes denser and less porous, and some shrinkage occurs. The formation of the liquid is progressive, increasing as temperature increases. If the melted materials become too fluid, deformation of the product can result. Complete vitrification is reached when enough fluid is formed to practically fill the pores to produce a rock-like product. The degree of vitrification is controlled to achieve the desired product.

Periodic kilns: Used for batch operations wherein the dried shapes are piled or set in the firing space, the temperature raised to produce the required degree of vitrification, the kiln allowed to cool, and the product removed. These may be of the updraft or downdraft types.

Updraft kilns: Used extensively for whiteware and pottery. The usual type of kiln is cylindrical in shape and surmounted by a chimney. A number of fireboxes are recessed at the outside perimeter at or below the kiln floor; a part of the hot gases from each is led through underfloor tunnels to the center of the kiln, then passes upward through holes in the kiln floor into the firing space and through the dried shapes. Bag walls, semicircular or rectangular enclosures extending a short distance above the kiln floor and located directly above the fireboxes serve as flues to permit part of the hot gases to pass upward against the inside of the kiln wall. To minimize the rapid escape of the hot gases up the chimney, a

Chapter 18

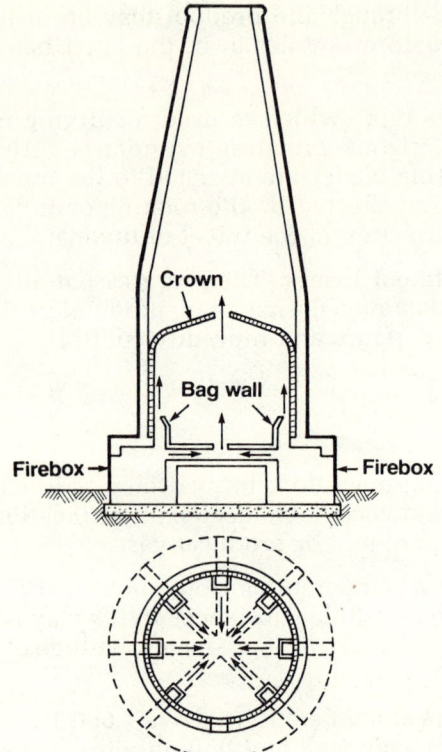

FIGURE 18.2. Updraft kiln.

refractory dome with small holes is located above the firing space; dampers may also be used to control the draft. Accurate temperature control is difficult with this type of kiln, and for this reason some updraft kilns are being replaced by tunnel kilns. A representative design of an updraft kiln is shown in Figure 18.2.

Downdraft kilns: Widely used for firing heavy clay products, including refractories and fire brick. The usual shape is round, with a dome top; however, some are rectangular. The fireboxes and bag walls are located essentially the same as in the updraft kiln. However, the latter are arranged to serve as flues to permit the hot gases from the fireboxes to pass upward against the inside of the wall of the kiln to the dome top where they are deflected to the center of the dome. An exterior chimney provides sufficient draft to pull the hot gases collected under the dome downward through the products and into underfloor tunnels and then up through the chimney. Vents in the dome can be opened to assist in the removal of steam and to allow the escape of heat during the cooling stage. In some cases the waste heat from the kilns is bypassed to tunnels and

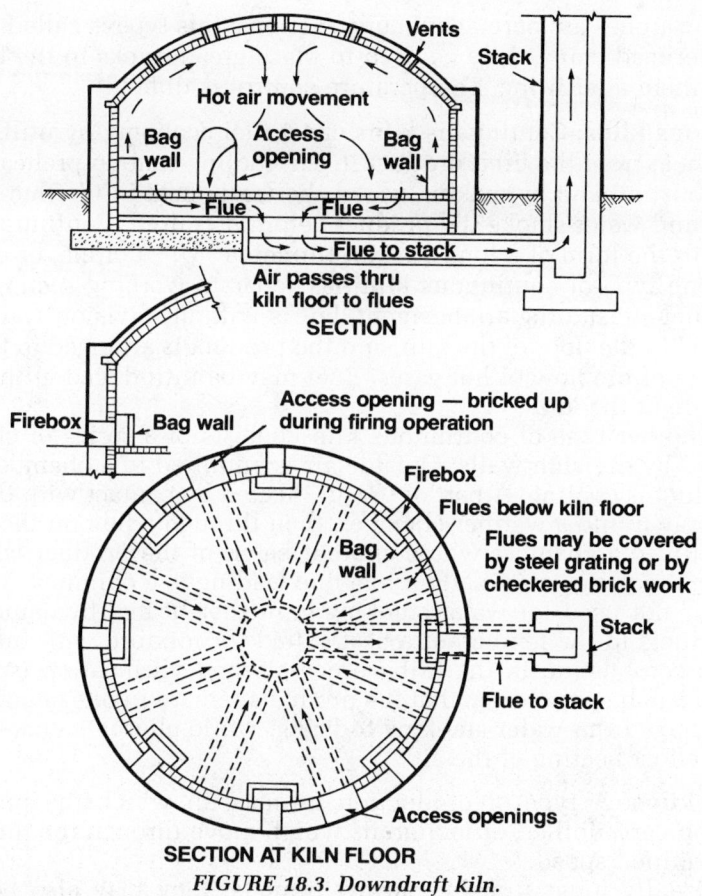

FIGURE 18.3. Downdraft kiln.

used for drying freshly shaped products. A representative downdraft kiln is shown in Figure 18.3.

Scove kilns: A periodic updraft type used for making common brick, usually from low-grade clays. The simplest type, built on the ground and using green (unfired) brick, is rebuilt after each firing. The kiln is constructed to permit the passage of hot gases through the stacked green brick being fired. Fires in green brick arches extend across the width of the kiln. Side walls may be of soft brick from previous firings. Side walls and the top of the stacked green brick are daubed with wet clay, called scoving, to make the interior of the kiln gastight. Products of combustion are drawn horizontally and upward to vertical chimneys. Shed-type roofs above the kiln provide protection against rain. Kiln sections, each containing a firing arch, are erected progressively, starting at one end of the shed; the first or original kiln section is rebuilt about the time the end section is placed in use. In some cases, the lower parts of the side walls and firing arches are permanent constructions; this type is called a clamp kiln. Overhead cranes may be used to place green bricks in the kiln and remove them after firing. Temperature control is difficult.

Continuous kilns: Continuous kilns effect fuel economy by utilizing the radiant heat from the fired product in the cooling state to preheat the air for combustion and by making use of the waste heat in the flue gases to preheat and water-smoke the product before it is fired. Continuous kilns may be in the form of a long straight tunnel, a "U," a circle, or an oval.

One type of continuous kiln has separate working sections, each with a fuel-consuming arrangement, but is without division walls. Draft is parallel to the floor of the kiln, and the product is arranged to take best advantage of the flow of hot gases. Fuel may be introduced at the top or at the side of the kiln.

Another type of continuous kiln consists of a series of chambers separated by division walls. Outside air is admitted to a chamber where the product is cooling. A part of the air, heated by contact with the product, is drawn into a warmer chamber, then through a flue on the outside of the kiln to a chamber where another batch of the product is stacked for water smoking, and finally into a flue leading to a chimney. That part of the air not used for water smoking is further heated by contact with hot products to the temperature best suited for mixture with fuel gas to produce combustion in the firing process when vitrification is desired. This is a batch operation in that the products remain in one chamber during the stage from water smoking to firing, while alternate chambers are used to effect heating of the air.

Tunnel kilns: A type of production furnace, in which the product is placed on cars, dollies, or monorails which move through the tunnel at a predetermined speed.

Usually long, straight, narrow tunnels, they may also be in the form of a circle, oval, square, or other shape, depending on the area available for construction. The kilns are made of refractory brick reinforced with steel straps. A hot zone is located at about the center of the tunnel, where temperatures are maintained sufficient for desired vitrification. The gases of combustion are mechanically drawn toward the entrance end of the tunnel and serve to heat progressively the incoming clay products. At the discharge or cooling end, air is blown into the tunnel to further the cooling of the products and to furnish a supply of heated air for combustion. Tunnel kilns have fireboxes in an enclosure attached to the exterior side of the kiln or in an enlargement of the cross-section of the tunnel. There are various types of firing, including those of zone-fired kiln, multiburner recuperative kiln, and regeneration cross-fired kiln. In some cases, the heat that is exhausted from the firing portion of the kiln is passed into the prefiring or heating zone, thereby achieving greater economy. A tunnel kiln is shown in Figures 18.4 and 18.5.

Muffle kilns: A type of tunnel kiln in which the gases of combustion do not come into direct contact with the product being fired. The fuel, usu-

FIGURE 18.4. Tunnel kiln.

ally gas, is burned in a space enclosed by thin refractory material having no openings to the interior of the tunnel. Heat is exchanged by circulating air through tubes surrounding the space in which the fuel is being burned; additional heat is transferred to the inside of the tunnel by radiation.

Glazing

Certain products are glazed to produce a glass-like coating that makes the porous clay body impervious to moisture. On whiteware, a glaze is applied for appearance and decoration. The glazes may be transparent, opaque, or colored. There are three general classes:

Salt glazes: Used as moisture protection for some heavy clay products such as sewer pipe and wall coping. Common salt (NaCl), introduced into the firebox in the later stage of firing, is vaporized by the heat. In the presence of moisture, the vapor reacts with the water (H_2O) to form sodium hydroxide (NaOH) and hydrogen chloride (HCl). The sodium hydroxide in contact with silicates and alumina forms a surface coating of sodium aluminum silicate. The resultant glaze is transparent.

Raw glazes: Composed of materials similar to those used for making commercial glass (sand, soda ash, and lime), together with smaller amounts

Chapter 18

FIGURE 18.5. Tunnel kiln entrance.

of other materials. Preparation of the glaze mix is by wet grinding and, therefore, water-soluble materials cannot be used. Fluxes may be feldspar, lead oxide, and lime. Firing follows the application of the glaze to the clay body.

Fritted glazes: Prepared by sintering or melting the mixed materials to form a glass. Ground then to a find powder, the glaze is known as frit. Application of the glaze precedes firing. Glaze materials include silica, boric oxide, and alumina. Fluxes include lime, magnesia, lead oxide, potash, and soda. Frit may also be used as an ingredient in fritted porcelain.

Before glazing, some wares are engobed to hide the texture of the clay body and, in some cases, its color. The engobe is a coating having properties intermediate between the clay body and the glaze.

Raw and fritted glazes are generally applied in the form of a water suspension; the product is dipped or sprayed. In some cases, application is by hand painting; and in others, by dusting the clay body with dry glaze.

Before the glaze is applied, the ware is "biscuit" fired in order to make it strong enough for handling. After application of the glaze, it is "ghost" fired to produce the glass-like surface coating on the ware. Sometimes, however, the glaze is applied to the unfired clay body.

Glazes are made so that the temperature of their fusion is below the fusion or melting temperature of the clay body.

THE FIRE HAZARDS

Many common fire hazards present in clay products plants constitute a considerable proportion of the known fire causes. The major hazards are: high temperatures; storage and use of solid and liquid fuels; overheating due to insufficient clearance between sources of heat and combustibles; indifferent maintenance of heated process equipment. Other more specialized hazards are: use of small particle combustibles in some clay mixes; use of straw and other readily ignitible combustibles for packing finished products; welding and careless smoking at locations where readily ignited combustibles are thoroughly dry as a result of prolonged exposure to heat from process equipment.

Preparation and Forming

The preparation of clay mixes involves the usual hazard of electrical and mechanical faults. Small particle carbonaceous materials such as sawdust, ground nut shells or other substances, added to the clay mix and burned out in the firing stage, are used to increase the porosity of some products. Suitable bulk storage, good handling methods, dust control, and good housekeeping are needed safeguards (See Chapter 50 for a complete discussion of Industrial Housekeeping Practices). Damp sawdust may heat spontaneously and should be safely disposed of.

Naphthalene has been used for forming voids in some kinds of brick. It is a combustible volatile solid and its storage and handling should be properly safeguarded. The sublimed naphthalene may be recovered and reused, in which case recovery equipment should be of safe design with adequate ventilation.

Oils, usually of the kerosene class, are used as a coating to prevent the sticking of clay to molds. Oil-soaked benches and floors contribute to a more rapid spread of fire.

Drying

The drying operation (water smoking) removes water added to clay mixes to produce the desired plasticity for forming products. Important hazards in water smoking are: electrical and mechanical faults; relatively high temperatures; inadequate clearances between heating and heated equipment and fixed and movable combustibles; indifferent maintenance of drying equipment; hazard of fuels; fuel burners.

Formed products may be dried by fan-circulated air heated by steam pipes or other heat sources. Waste heat from kilns may be used. Drying may be the first stage in a tunnel type of kiln. Where equipment for drying is other than noncombustible material, the prolonged exposure to heat is conducive to thorough drying of combustibles, presenting the hazard of easy ignition and rapid fire spread.

Hot floors for drying may be of wood construction in otherwise noncombustible structures, with steam pipes or other sources of heat in underfloor spaces. Such spaces should be kept free of combustible refuse and adequate clearances should be provided and maintained between wood and heat sources. Multi-story buildings with slatted wood floors at all levels may be used for drying some products. Heat may be from steam pipes or other heat sources located below floors. Prolonged exposure to heat, thoroughly dry structural materials, and unobstructed vertical draft can result in the rapid involvement of the entire building by fire.

Chemical Water Smoking

After excess water has been removed by drying (water smoking), the water chemically combined in the clay substance is removed by heat (chemical water smoking). This operation usually takes place in a kiln prior to firing for vitrification. Chemical-water-smoked products are no longer affected by water but can be damaged, when hot, by sudden chilling.

Kilns

Kilns provide means for producing the desired degree of vitrification and also provide heat for glazing operations. Kilns may be detached in yards, under shelters in yards, or inside plant buildings. Important hazards are: high temperatures; overheating; inadequate clearance to fixed combustibles; fuel storage; transfer of liquid or gaseous fuels to burners; degree of safety of burner controls. Cracks or holes in chimneys and waste heat ducts extending through combustible roofs and in poorly maintained kiln walls and tops can result in the ignition of nearby combustible structural materials. Movable combustibles should be kept well away from burners and kilns. Kilns may be damaged by sudden chilling by application of hose streams.

Fuels

Fuels may be coal, fuel oils, or fuel gases. Liquefied petroleum gas may be a standby fuel for use when gas supply from a public utility fails or where gas falls below required pressure. Producer-gas generating equipment is usually located close to the point of use to take advantage of its sensible heat, but should be located outside of plant buildings. The weight of fuel gases is a factor in relative hazard. Natural gas is lighter than air. Producer gas is slightly lighter than air and its high carbon monoxide content is toxic. Liquefied petroleum gases are heavier than air, can accumulate in pits and low places, and can remain for some time.

The storage, handling, and use of fuels should conform with nationally recognized standards and good practices. Important phases of good practices would include: suitable storage methods should be used to minimize spontaneous heating in large deep coal piles; aboveground tanks for liquid fuels should be at least the distance from buildings required by standards, and ground slope should not be toward buildings; gravity feed from large fuel oil tanks is not desirable; antisiphon devices in suction feed to oil burners should be provided where necessary; suitable safeguards should be provided at means for preheating heavy fuel oils; natural gas and liquefied petroleum gases should be odorized by equipment on premises where gases are not odorized at source of supply; gas-air mixers should not be located inside plant buildings unless the structure is large, has high ceilings, and is well ventilated.

Pipe lines for the transfer of liquid and gaseous fuels should be securely connected, leak-free, adequately supported, and protected against mechanical injury. Suitable and readily accessible shutoff valves should be provided and handwheels or operating levers should be in place and ready for instant use.

Combustion controls at liquid and gas burners are desirable as a safeguard against overheating, flooding of fuel oil at burners, and possible explosion of accumulated unburned fuel gases.

Smoking

Locations where smoking may be safely permitted should be determined. Where smoking constitutes a fire hazard, "No Smoking" signs should be posted and the no smoking rule should be rigidly enforced.

Welding

Where electric or acetylene welding must be done inside plant buildings, the hazards should be investigated prior to the start of welding operations. It should be remembered that in certain areas combustible structural materials may be thoroughly dry as the result of prolonged exposure to heat from process equipment. A watcher, provided with a suitable portable extinguisher, should be on duty during welding operations and the watcher should remain on duty until certain a spark or a bit of hot metal has not started a slow smoldering fire which may later burst into active flame. In some plants a permit must be obtained from the plant superintendent or plant safety director before welding is started — a good safety procedure. (See Chapter 24 for a complete discussion of cutting and welding.)

Chapter 18

THE SAFEGUARDS

The location of clay products plants is influenced by the available supply of clay suitable for products made; and plants are usually located relatively close to places where clays are obtained. Clay products plants may be within the corporate limits of a municipality where public protection is fair to good, or may be at locations remote from fire department stations, or in localities where public water supplies for fire fighting are limited, or may be beyond public water distribution mains.

All clay products plants should have adequate fire extinguishing equipment for all Class A, B, C fires. Regardless of the size of the plant, employees should be instructed and trained in the proper use of the fire fighting equipment provided and should be instructed in what to do in case of fire. Fixed ladders for fighting roof fires have been found useful. Providing easy access to buildings for fire fighting is an important part of fire control. Good maintenance of fire-fighting equipment is essential. (See Chapter 3, "Plant Emergency Training and Organization.")

Plant values and economics will influence the desirability and the good safety practices of private protection such as yard hydrants and fire hose, automatic sprinklers, open sprinklers as protection against fire in properties not under the control of plant management, watch service, and automatic fire alarms.

BIBLIOGRAPHY

NFPA Codes, Standards, and Recommended Practices

Reference to the following NFPA Codes, Standards, and Recommended Practices will provide further information on the safeguards for clay products plants discussed in this chapter. (See the latest *NFPA Codes and Standards Catalog* for availability of current editions of the following documents.)

NFPA 10, *Portable Fire Extinguishers.*
NFPA 13, *Installation of Sprinkler Systems.*
NFPA 14, *Standpipe and Hose Systems.*
NFPA 27, *Private Fire Brigades.*
NFPA 30, *Flammable and Combustible Liquids Code.*
NFPA 31, *Installation of Oil Burning Equipment.*
NFPA 54, *National Fuel Gas Code.*
NFPA 58, *Storage and Handling of Liquefied Petroleum Gas.*
NFPA 70, *National Electric Code.*

Additional Reading

"Hazard and Hazard Prevention in Solvent Evaporating Ovens," *Fire Prevention*, (127), 1978, pp. 22-25.

McKinnon, G. P., ed., *Fire Protection Handbook*, Fifteenth Edition, National Fire Protection Association, Quincy, MA, 1981.

Trinks, W., *Industrial Furnaces*, vols. 1 and 2, John Wiley and Sons, NY, 1967.

19

Plastics Products

Alfred J. Hogan, P.E.

The multi-billion dollar plastics industry is the fourth largest industry in the United States. Plastics are a basic material of use — on a par with metals, glass, wood, and paper. They have become increasingly important to advanced concepts in architecture, aerospace, communications, transportation — even medicine and the arts. The plastics industry originated in the United States in 1868 when John Wesley Hyatt developed — with celluloid — the first American plastic.

Currently the United States leads the world in the production of plastics, supplying more than half the total output. At least 6,000 companies in the United States make plastics; that is, they produce basic material, process or fabricate plastics into products or parts, or finish their goods. Companies that use plastics or these materials and services number well into the tens of thousands.

Plastic is any one of a large and varied number of materials consisting wholly or in part of combinations of carbon with oxygen, hydrogen, nitrogen, and other organic and inorganic elements. While it is solid in the finished state, at some stage in its manufacture it is liquid, and thus

---

Alfred J. Hogan is a Registered Professional Fire Protection Engineer and Consultant based in Cypress Gardens, Florida.

capable of being formed into various shapes, most through the application, either singly or together, of heat and pressure. Whatever their properties or form, however, most plastics fall into one of two groups: thermoplastics or thermosets.

Thermoplastic resins can be repeatedly softened and hardened by heating and cooling, without a chemical change taking place.

Thermoset resins, once heat treated, cannot be softened by further heating without causing a chemical change that would forever alter the material.

There are approximately 30 major classes of plastics or polymer groupings, with numerous variations of individual plastic products. This diversity has made plastics applicable to an extremely broad range of end-uses and products. Yet, paradoxically, this diversity has also made it difficult to grasp the concept of a single family of materials encompassing such a far reaching span of physical and chemical properties and fire hazard characteristics.

The plastics industry as a whole has three broad areas of processing: manufacturing, conversion, fabricating. Although each is basically distinct, they may be conducted in the same or at different locations.

Manufacturing, or synthesizing, of the basic plastic or feedstock sometimes includes compounding with colorants or other additives. The hazards of the synthesis plant are basically those of a chemical plant using some of the processes described in Chapter 31, "Chemical Processing Equipment."

Conversion of the plastics materials into useful articles by molding, extrusion, or casting usually involves heating the plastic so it will flow into a shape that is retained when the plastic is cooled. Although chemical reactions are not often a significant part of these operations, the plastic industry refers to them as "processing" or "converting."

Fabricating encompasses the largely mechanical operations of bending, machining, cementing, decorating, and polishing plastics.

The three areas accurately categorize the fire hazards that may be encountered in the plastics industry. It is important to recognize, though, that the synthesis, or manufacturing, of a given plastic may be much more hazardous than its molding or extrusion. It's also important to recognize that some plants nominally doing only converting and or fabricating may also be conducting chemical operations with flammable or reactive materials. One example is the overlapping of molding and synthesis operations in a plant that impregnates liquid polyester resins into reinforcing glass fibers for molding boat hulls or containers.

This chapter discusses the processing, or converting, phase of the plastics which includes recognizing, of course, that in some instances hazards associated with all three phases of the industry are carried on in one location.

TABLE 19.1. Examples of Plastics Materials

Thermoplastic		Thermoset	
ABS (acrylonitrile-butadiene-styrene)	Polyethylene	Epoxy	Silicones
a combination of three different monomers—acrylonitrile, butadiene, and styrene—suitable for tough consumer products; has good electrical insulating qualities; used to make automobile front grills; door handles, toys, and electric cart bodies.	tough; excellent chemical resistance and electrical insulating quality; near zero moisture absorption; used for packaging, molded housewares, and toys.	excellent chemical resistance, electrical properties, and dimensional stability; will bond metals, glass, hard rubber, etc.; used for aircraft components, tanks, and electrical and electronic assemblies.	long-term heat resistant, water repellent, electrical insulator, and mineral acid resistant; used in the electrical and electronics industry; insulation in meters and generators and induction heating apparatus.

Chapter 19

The Basic Hazards

Most plastics are combustible organic compounds that can burn under certain conditions. But aside from the inherent combustibility of plastic formulations influenced by the basic polymers used, the nature of plastic additives, the form the final product takes, and the conversion of feedstock plastics into finished articles involves the hazards associated with combustible dusts, flammable solvents, electrical faults, hydraulic fluids, and the storage and handling of large quantities of combustible raw materials and finished products. (See the chapters that specifically discuss these hazards.)

Terminology

Before discussing the converting of plastic feedstocks into finished products and the associated hazards, certain process terms and definitions are provided.

Additive: Any material mixed with a resin to modify its processing or end-use properties. The resulting mix usually is called a "plastic" to distinguish it from the "resin" or principal ingredient. Additives may be dyes, pigments, powdered fillers for stiffening, plasticizers for flexibility, fibers to reinforce, antioxidants, lubricants to aid flow into or release from a mold, or fire retardants. Generally they will be either liquid or solid and any color.

Binders: The resins in a plastic mixture that hold together all of the other ingredients.

Blowing agent: A material which releases gas upon heating so that the plastic in which it is mixed will expand into foam. Polystyrene is an example of such a foam.

Blown tubing: A thin film made by extruding a tube and simultaneously inflating it with air while hot; distention may be 20 times the diameter of the extruded tube.

Casting: Flowing liquid material into place with little or no pressure as contrasted with forcing material into place by molding.

Extrusion: The process of passing softened plastic under pressure through a die to make an essentially continuous profile; the equipment is called an extruder.

Fabrication: The making of articles by machining, cementing, heat-sealing or thermoforming of preformed sheets, rods, or tubes. The term is used in contrast to "processing."

Fillers: Materials that modify the strength and working properties of a plastic. They may be used to increase heat resistance and alter the dielectric strength. A wide variety of products are used including wood flour, cotton, sisal, glass, and clay.

Film: A general term for plastic not more than 0.01 in. (0.25 mm) thick, regardless of the process used to make it. An example of this is Saran Wrap™."Foil" is used today only to describe metal.

Finishing: This term has three distinct meanings: (1) removal of burrs or flash by filing, sanding, or tumbling, (2) buffing or waxing to polish surfaces, and (3) application of decorative or marking treatment, as by painting or metal plating.

Flash: Unwanted projections from molded articles resulting from flow of plastic into space between matching parts of a mold. (The term has no fire connotation.)

Foam: Plastic with many small gas bubbles. It is called cellular plastic. Rigid foams have rapidly reached large volume production as thermal insulation boards for construction, cups for hot and cold drinks, trays for prepackaged meats, and shock-resistant packaging; flexible foams are used for furniture padding, insulation of outer garments, and soft drape upholstery.

Foaming agent: The same as blowing agent.

Forming: The process in which the shape of plastic pieces, such as sheet, rod, or tube, are changed into a desired configuration.

Laminate: A composition of several layers of plastic firmly held together by partly melting and fusing the layers together, by an adhesive, or by impregnation.

Lubricant: Material added to improve feeding of powder or granules into molding or extrusion machines, to improve flow of molten plastic through machines and into molds, or to prevent adhesion of plastic to molds; the last of these uses is called "mold-release." Typical lubricants are zinc stearate, carnauba wax, and silicone oil.

Molding: The process of forcing plastic into a cavity to achieve a desired shape. The term is used in contrast to one form of casting that is taken to mean filling a mold with little or no force.

Monomers: The small starting molecules, usually gaseous or liquid, used to produce the polymer resins.

Plasticizers: Organic materials added to plastics to make the finished product more flexible or to facilitate compounding. Some plasticizers increase the combustibility of the plastic while others serve as flame retardants.

Plastics: Materials that contain as an essential ingredient an organic substance of large molecular weight, are solid in their finished state, and at some stage in their manufacture of processing into finished articles can be shaped by flow.

Polymerization: The process by which molecules of a monomer are made to respectively combine with other monomers. The result is a much longer chain-like molecule.

Processing: The converting of polymers into useful articles by molding or extrusion from granules, depositing film from solvent, or laminating resin and reinforcement. The term is used in contrast to fabricating. Most often the molding operation uses heat, but it is largely or entirely a physical rather than a chemical process.

Reinforced plastic: A plastic with a filler which significantly increases flexural, impact, or tensile strength. Additives are usually glass, asbestos, cotton, or nylon fibers. Reinforced plastics may be thermoplastic granules for injection molding, or use large areas of reinforcement as in layup molding or pulp molding.

Resin: A solid, semisolid, or pseudosolid organic material which has indefinite and often high molecular weight.

RAW MATERIALS

Many plastic feedstocks are derivatives of petroleum or gas recovered during the refining process; e.g., ethylene monomer (one of the most important feedstocks) is derived in gaseous form, from petroleum refinery gas, liquefied petroleum gases, or liquid hydrocarbons.

The monomer is subjected to a chemical reaction, known as polymerization, that causes the small molecules to link together into increasingly longer molecules. Chemically, the polymerization reaction has turned the monomer into a polymer (the reason why names of so many plastics materials begin with the prefix "poly"). The polymer, or plastic resin, must next be prepared for use by the processor, who will turn it into a finished product. In some instances, it is possible to use the plastic resin or feedstock, as it comes out of the polymerization reaction. More often, however, it is transformed into a form that can be more easily handled by the processor and more easily handled in processing equipment. The most common solid forms for the plastic resin are pellets, granules, flake, or powder. Feedstocks are also available as semisolids, e.g., pastes, or as liquids for casting.

THE PRODUCTION PROCESS

The ways in which plastics can be processed into useful end-products are as varied as the plastics themselves.

Though the processes differ, there are common elements. In the majority of cases, thermoplastic compounds in the forms of pellets, gran-

PLASTICS PRODUCTS

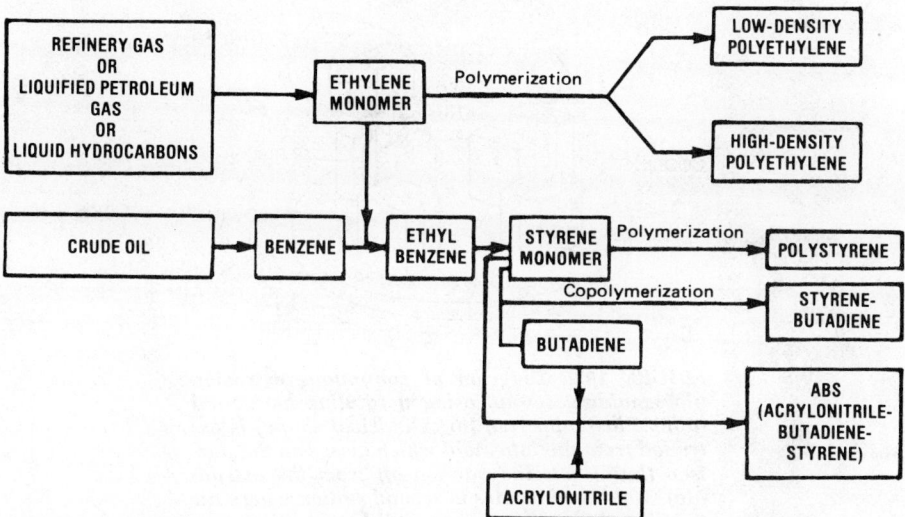

Figure 19.1. Flow chart shows ethylene's role in the manufacture of polyethylene, polystyrene, and styrene copolymers. (Society of the Plastics Industry)

ules, flake, and powder must be melted by heat so they can flow. Pressure is often involved in forcing the molten plastic into a mold cavity or through a die, and cooling must be provided to allow the molten plastic to harden. With thermosets, heat and pressure also are most often used. In their case, however, heat (rather than cooling) serves to cure (set) the thermosetting plastic in the mold under pressure. When thermoplastic or thermoset resins are in liquid form, heat and/or pressure need not necessarily be used, although in many casting techniques intended for high speed production, they do play a role.

The following descriptions of processes cover the basics of the major manufacturing systems. It should be recognized, however, that there are variations in virtually every process.

Blow molding: This process is generally used only with thermoplastics. It is applicable to the production of hollow plastics products such as bottles, gas tanks, and carboys. Basically, blow molding involves the melting of the thermoplastics resin, then forming it into a tube-like shape (known as a *parison*), sealing the ends of the tube, and injecting air (e.g., through a needle inserted in the tube) so that the tube, in a softened state, is inflated inside the mold and forced against the walls of the mold (see Figure 19.2).

Calendering: This process can be used to convert thermoplastics into film and sheeting, and to apply a plastic coating to textiles or other supporting materials. In calendering film and sheeting, the plastic compound is passed between a series of three or four large heated revolving rollers which squeeze the material between them into a sheet or film.

399

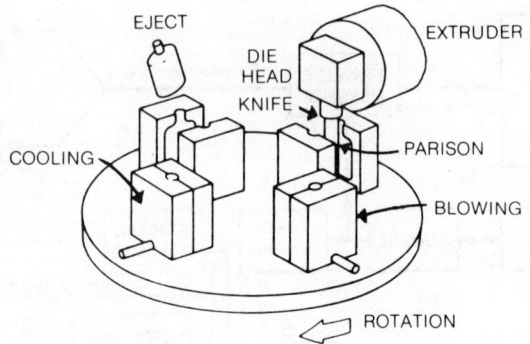

FIGURE 19.2. Diagram of continuous extrusion blow-molding set-up, using a rotating horizontal table. Plastic parison (in cylindrical shape) is extruded from die into mold which closes on the parison (knife cuts the parison off from the extrudate). Mold then rotates to second station where air is injected into the parison (still hot and therefore formable) to blow it out to the shape of the inside mold cavity. At third station, the blown part is allowed to cool and set. Mold finally rotates to last station where finished part (in this case, a bottle) is ejected from mold. (Society of the Plastics Industry)

Casting: This process may be used both with thermoplastics and thermosets to make products, shapes, rods, and tubes, by pouring a liquid monomer-polymer solution into an open mold or a closed mold where it finishes polymerizing into a solid. Pressure need not be used with the casting process, unlike the molding process. Another difference is that the starting material is usually in liquid form rather than solid form.

Coating: May be accomplished by using either thermoplastic or thermosetting materials. The coating may be applied to metal, wood, paper, glass, fabric, leather, or ceramics (see Figure 19.3).

Compounding: Mixing additives into previously formed resin on masticating rolls or calenders, as for rubber, or in kneading mixers or screw extruders of varied design, is called compounding. Rolls and kneaders are heated by high pressure steam or heat transfer fluids.

Screw extruders have largely displaced rolls and kneaders for compounding, because they provide better control, continuous output, and less exposure of hot plastic to air.

Compression molding: This is the most common method of forming thermosetting materials. Compression molding is simply the squeezing of material into a desired shape by applying heat and pressure to the material in a mold (see Figure 19.4).

Extrusion: This method is employed to form thermoplastic materials into

PLASTICS PRODUCTS

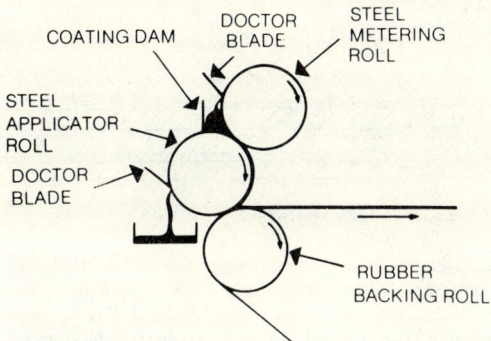

FIGURE 19.3. *A typical coating set-up, known as a 3-roll nip-fed reverse roll coater. Plastic feeds from dam through nip between steel metering and applicator rolls, rotating in the same directions. At bottom of the applicator roll, plastic is laid on top of the substrate (e.g., fabric, paper, etc.) as it comes in contact with the substrate at nip between applicator roll and backing roll (which carries the substrate up from the bottom of set-up). Doctor blade is used to scrape off excess plastic from applicator roll. (Society of the Plastics Industry)*

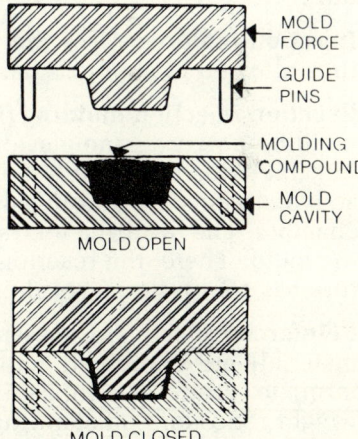

FIGURE 19.4. *Basics of a simple two-piece compression mold. Plastic molding material is loaded into lower half (cavity) of the heated mold (shown at top). Top half of the mold (mold force) is then lowered and the two halves are brought together under pressure (shown below). The softened molding material is thus formed into the shape of the cavity and allowed to harden with further heating. Mold is then opened and part is removed. (Society of Plastics Industry)*

continuous sheeting, film, tubes, rods, profile shapes, or filaments and to coat wire, cable, and cord (see Figure 19.5).

Foam plastics molding: This is the manufacturing process where foams can be used in casting, calendering, coating, rotational molding, flow molding, and even injection molding and extrusion (see Figure 19.6). The

401

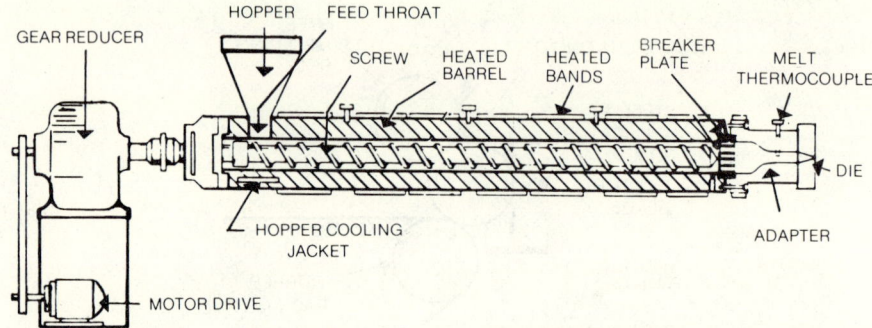

FIGURE 19.5. *In a basic single-screw extruder, plastics pellets (or powders) are fed through the hopper, through the feed throat, and into a screw that rotates in a heated barrel. The rotation of the screw (which is powered by the drive motor) conveys the plastic forward for melting and delivery through the breaker plate (reduces the rotary motion of the melt), through the adaptor, and into the die which dictates the shape and size of the final extrudate. (Society of the Plastics Industry)*

manufacturing of foamed products is discussed in more detail later in this chapter under "Flammable Solvents."

High pressure laminating: A process that uses thermosetting plastics to hold together reinforcing materials such as cloth, paper, wood, or glass fibers. Heat and high pressure are used to produce the laminated product.

Injection molding: Uses thermoplastic material that is softened by heat, then allowed to cool and harden (see Figure 19.7).

Reaction injection molding (RIM): This is a technique used primarily for molding polyurethane elastomers of foams into end-products with solid integral skins and cellular cores. Basically, two or more pressurized reactive streams are impinged together under high pressure in a mixing chamber. The resulting mixture is then injected, under low-pressure, into the mold. There, the reaction begins and continues until the liquid mixture has set up into a solid or cellular finished product.

Reinforced plastics processing: In this process resins (acting as binder material) are combined with reinforcing materials (usually in fibrous form) to produce composite products having exceptional strength-to-weight ratios and outstanding physical properties. The resins may be either thermosets or thermoplastics (see Figure 19.8).

Rotational molding: Like blow molding, is used to make hollow one-piece parts. Finely powdered plastic or molding granules are placed in a heated cavity that is rotated about two axes to distribute the plastic.

Thermoforming: Thermoplastic sheeting is heated to its softening temperature and the hot and flexible material is forced against the contours of a mold by mechanical means, e.g., tools, plugs, solid molds, etc., or by pneumatic means, e.g., differentials in air pressure created by pulling a vacuum between sheet and mold or by using the pressures of compressed air to force the sheet against the mold. This is referred to as pressure forming, vacuum forming, and plug-assist forming (see Figure 19.9).

Transfer molding: Most generally used for thermosetting plastics. It is similar to compression molding in that the plastic is cured into an infusible state in a mold under heat and pressure.

THE FIRE HAZARDS

Plants converting feedstocks into finished products are subject to a variety of hazards that can result in explosions and fire. The broad area of hazard involves the presence of combustible dusts, flammable and combustible liquids, high heat elements, hydraulic and heat transfer fluids, static electricity, and failure to observe good storage and housekeeping practices.

Dusts

Although when in solid massive form many plastics can be difficult to ignite and will not continue to burn on removal of exterior fuel, nearly all will burn rapidly in the form of dust, and if dispersed in air can be explosively ignited by a spark, flame, or metal surface above 700°F (371°C). Dust explosions should be considered possible when operations use pulverized plastic, convey larger granules through pneumatic conveying systems, or produce dust by machining or sanding in finishing work. Wood flour or finely ground dyestuffs also require safeguarding when being added to plastics.

Plastic pellets for injection or extrusion molding are commonly known as molding powder. In reality they are not pulverized material but cubes or cylindrical pellets. They are usually screened to remove finer particles to permit more uniform feeding to machines, and they are free from hazard of dust explosion. However, dust can be generated by abrasion of these particles when conveyed in a long pneumatic system.

Trimmings from injection molding are cut to small size for reuse alone or by blending with fresh (virgin) molding pellets. This cutting is called "regrinding," although it is a shearing and impacting action deliberately intended to avoid making dust and fine particles. Regrinding usually generates some fine powder, and dust hazards should be considered when much regrinding is done.

Compounding of resins with such additives as dyes, pigments, fillers, mold-release or flow-improving lubricants, plasticizers for flexibility, ultraviolet or heat stabilizers, or modifying resins is also a source of dust hazards. For rapid mixing, most of these are charged to mixing equipment as fine powders, and dust explosions are possible with any of

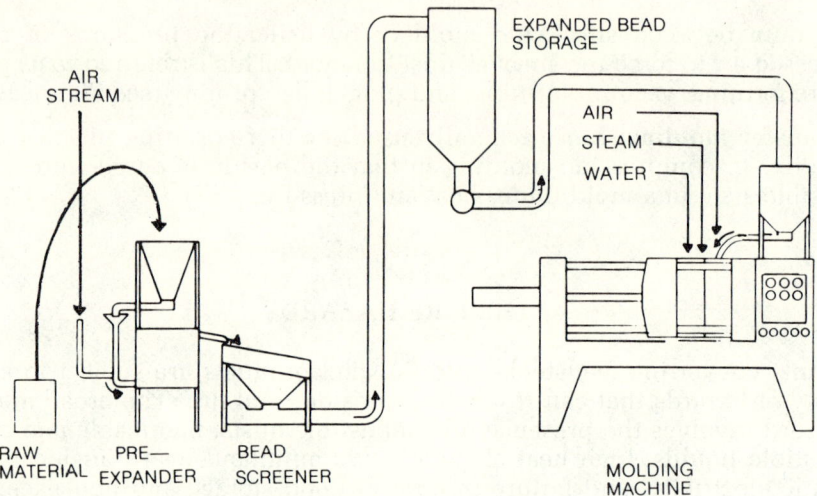

FIGURE 19.6. *Among the many variations in molding foamed plastics is this set-up for steam chest molding expandable styrene beads into products like foam cups, novelties, building products, etc. In this operation, the expandable beads, containing a blowing agent, are pre-expanded with steam, then screened to remove large clumps. The expanded beads are next blown into a storage hopper and allowed to dry and stabilize. From here, they feed into the final mold where steam is again used to complete expansion of the beads so that they fill the mold and fuse together. Water is used for cooling, prior to opening the mold and removing the finished foamed styrene part. (Society of the Plastics Industry)*

these ingredients. Some compounding, especially for reclaiming of once-processed materials, is done at molding machines in plastics plants of all types.

Generally, the basic chemical structure of the resin governs the explosibility of its dust. Incorporation of wood flour, cotton flock, or other combustible fillers usually increases the explosibility of dust. Incorporation of low percentages of fire retardants has but little effect on explosibility of dusts.

It is important that the escape and dispersion of dust into the atmosphere of a converting plant be kept to a minimum. Equally important is that provisions are made to reduce the possibility of ignition, relieve explosion pressure, and confine and control fire. Guidance in controlling dust explosion hazards in plastics plants is found in NFPA 654, *Prevention of Dust Explosions in the Plastics Industry*. Other applicable standards are NFPA 63, *Fundamental Principles for the Prevention of Dust Explosions;* NFPA 68, *Guide for Explosion Venting;* NFPA 69, *Explosion Prevention Systems;* and NFPA 650, *Pneumatic Conveying Systems for Handling Combustible Materials*. NFPA 650 is useful as a guide for evaluating pneumatic conveying systems. See also Chapter 45, "Air Moving Equipment," for information on hazards and design considerations associated with pneumatic conveying systems.

PLASTICS PRODUCTS

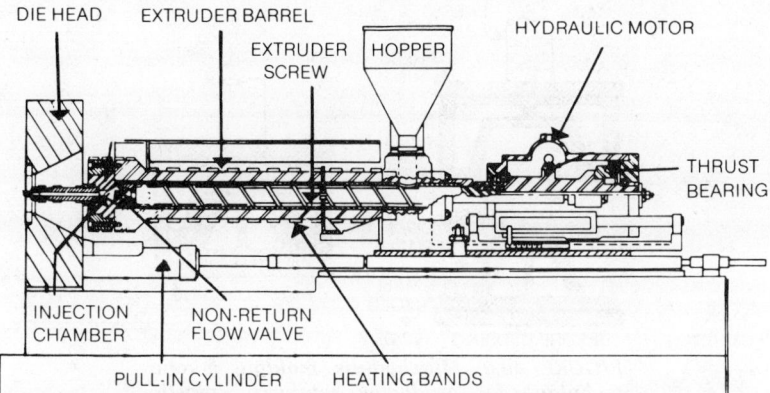

FIGURE 19.7. *Diagram of reciprocating screw injection molding machine. Plastics pellets feed through the hopper into the screw (much like the screw in an extruder) where they are compacted, melted, and pumped by the rotation of the screw past the non-return flow valve (allows material to flow right to left, but not from left to right) to the front of the screw where it is allowed to accumulate. At the proper time, the rotation of the screw is stopped and the amount of molten plastic in front of the screw is injected into the mold, using the screw as a plunger activated by the hydraulic injection cylinders. In the mold, the molten plastic flows throughout the cavity, completely filling it. The plastic is then allowed to cool and harden, the mold is opened, and the finished part removed. The back end of the machine shown above contains the motors and drives needed to power the machine. (Society of the Plastics Industry)*

Flammable Solvents

Flammable organic solvents are found in nearly every plastics plant. They may be used in very small quantities to apply adhesives, lacquers, or paints to molded or fabricated items; in large amounts to coat plastic on cloth, paper, leather, or metal, or on metal belts from which a dried film will be stripped. In these uses, the choice of solvents and hazards is similar to most lacquering operations. There may be increased hazard when solvents are applied to plastics, particularly when printing or coating on fast-moving films, because plastics usually have high electrical resistivity; they generate and retain static charges more readily than paper or cotton fabric.

Small plants are increasing their use of solvents for the preparation of both rigid and flexible foam plastic. The plastic is moistened with solvent and heated above the boiling point of the solvent in a closed mold or extruder. Upon release of pressure the boiling solvent expands the resin and produces a bubbled structure. Another type of foaming process heats a resin containing a chemical additive which evolves gas, usually nitrogen, carbon dioxide, or steam. Either the solvent or the chemical agent is called a blowing agent. The most common rigid foam is made from beads of polystyrene moistened with about 10 percent pentane or a similar hydrocarbon. The pentane does not soften and expand the resin,

Chapter 19

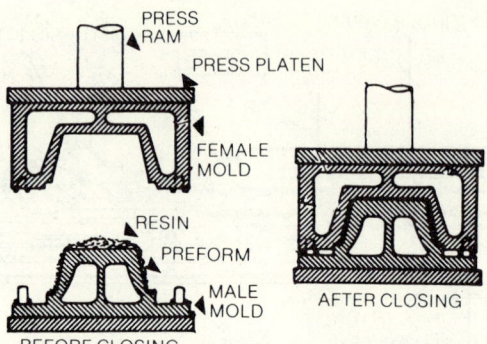

FIGURE 19.8. *Matched-die molding is one technique for producing reinforced plastic parts. Basically, it is a compression molding process in which resin and glass fibers are shaped into the finished product under heat and pressure between the two halves of a mold (male and female halves). The glass fiber reinforcements are laid over the male mold in the form of a "preform," a combination of glass and resin preformed before molding to the basic shape of the part to be molded. Additional liquid resin mix is added before mold is closed. (Society of the Plastics Industry)*

at room temperature, but does so at the low pressure steam temperature used in molding to shape, or in extruding as sheet. It is imperative that sources of electric spark or flame be avoided. Resin is usually shipped from the manufacturer with the hydrocarbon blowing agent already mixed with it. Containers will have free vapor of pentane, and should be opened outdoors or with good exhaust ventilation. Expanded articles should be aged under forced hot air to remove nearly all flammable blowing agent before shipping.

Flexible urethane foams for upholstery and garments, or rigid foams for pour-in-place construction or refrigeration insulation, are blown by steam and carbon dioxide generation which are by-products of the formation of the urethane resin from basic ingredients. Urethane foams of lower density may be made with a low-boiling hydrocarbon in the resin mixture to increase the expansion.

For elimination of the fire hazards from solvents in the foaming process, consideration should be given — for both styrene and urethane foams — to replacing hydrocarbon blowing agents with fluorinated hydrocarbons, similar to the nonflammable refrigerants.

Foams of polyvinyl chloride are used in garments and upholstery. These are usually called "expanded vinyl," because the extent of foaming is limited to provide an essentially continuous outer surface. These are usually made with the chemical blowing agent azodicarbonamide, also called azobisformamide. Above 300°F (149°C) it provides nitrogen

PLASTICS PRODUCTS

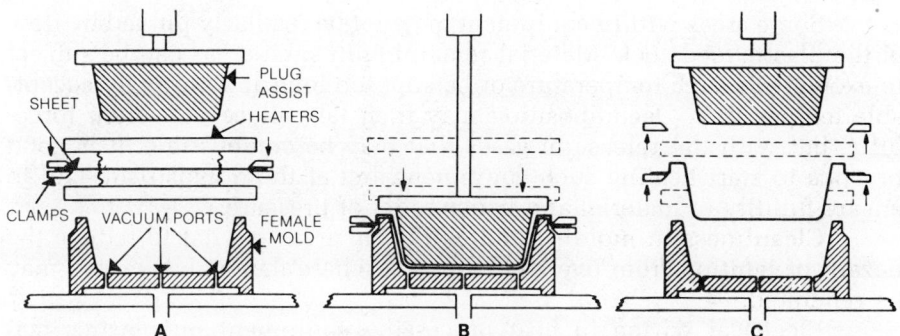

FIGURE 19.9. *This variation on the thermoforming of plastic sheet is known as plug-assist vacuum forming. In operation, the plastic sheet is clamped in place and heaters move in to heat the sheet top and bottom to soften it (A). Heaters are then withdrawn, and the frame holding the sheet is lowered down to contact the mold. At this point, the plug-assist is lowered into the softened sheet, stretching it down to the bottom of the mold cavity (B). After the plug-assist has reached its closed position, a vacuum is drawn through the ports to pull the stretched sheet completely into the cavity and finish the forming. Next, the plug-assist is withdrawn, the formed sheet is cooled, and the clamps are opened to remove the formed part from the frame (C). (Society of the Plastics Industry)*

gas at a controlled rate, is essentially nonhazardous in the proportions used, and does not give off flammable vapors. Drums of the reagent should be cooled with water when exposed to fire, but such reagents do not have nearly the rapid burning or explosive hazards of peroxygen materials.

Improper handling of flammable liquids has caused serious fires in plastics plants. Failure to recognize the importance of static spark prevention, explosion proof electrical equipment, and vapor removal systems have been the most frequent causes of flammable liquid fires. (See Chapter 37, "Handling and Storage of Flammable and Combustible Liquids.") NFPA Standards 30, 33, 35, 70, 77, and 91 contain further information on safeguards to observe in the handling and use of flammable liquids as they apply or may apply to the plastics industry.

Heating elements: Molding and extrusion operations, for both shaping and as compounding, have hazards associated with local overheating of electrical components. Operating temperatures normally range from 300° to 650°F (149° to 343°C), depending on which plastic is being processed. The upper temperature range is beyond that practical for heat transfer fluids, so electric resistance heating is almost universally employed. Heater bands are required to fuse the resin in the feed section at the upstream end of the extruder barrel. It is not uncommon for controllers to stick, permitting resistance heaters to exceed the temperature set for the thermocouple controller. In most cases, the character of extrudate will markedly change well before heater bands get hot enough to be a source of ignition.

Some areas within equipment may not be regularly purged by flow of the plastic feedstock. Material remaining in such areas can be subject to exclusively high temperature or be kept too long at a normally acceptable temperature. Decomposition may then take place, not often forcefully, but with the release of gases that may be combustible. It is good practice to start heating such equipment first at the downstream end to ensure fluidity of material and hence relief of pressure.

Cleanliness in molding and extruding areas is vital to reduce the hazard of ignition from overheated bands where flammable vapors may be generated.

Electrical wiring in heat processing equipment on plastics machines should be installed in accordance with applicable provisions of NFPA 70, *National Electrical Code*.

Static Electricity

Many operations in plastics plants generate static electricity. Because plastics are such good electrical insulators, static electricity on them can rapidly build up to spark discharge — a hazardous condition if dust or flammable vapors are present. Operations which can generate static are stripping of films from production or printing equipment, or rapid passage of films across rolls or guides. Belts for power transmission also are a significant source of static discharge. Because of their low water-absorption and high resistivity, plastics cannot have their static charge dissipated by high ambient humidity as practiced in cotton, wool, and paper mills. Attention should be given to grounding of equipment and ensuring that tinsel conductors firmly contact moving films or filaments. Care should also be given to separating vapor and dust hazards from machines where static electricity ignition sources could develop. NFPA 77, *Recommended Practice on Static Electricity*, is a good source for the safeguards to use in protecting against static electricity hazards.

Hydraulic Pressure Systems

Hydraulic systems are used to clamp molds and to provide pressure to rams or screws which force molten plastic into molds by compression, transfer, or injection molding. The molten plastic may be at pressures up to 20,000 psi (138 000 kPa), but the hydraulic systems are normally less than 2,000 psi (13 800 kPa). Petroleum fluids have been used in plastic operations where heating elements are generally below 600°F (315°C) (the same fluids have a poor record in die casting metals parts because molten metal is at a much higher temperature).

Fire resistant water-glycol or water-oil emulsions or "synthetic" fluids are available for hydraulic systems. Substituting fire resistant fluids for petroleum fluids should follow the recommendations of the manufacturers of hydraulic system components. (See Chapter 43, "Fluid Power Systems," for further information on hydraulic pressure systems.)

Storage Arrangements

The fire hazards of plastics in storage, whether as feedstocks for the conversion process or as finished articles, are determined by their chemical composition, physical form, and storage arrangement. The physical form may be foam, solid sheet, pellets, flake, random packed small objects, bags, or cartons. The storage of plastics generally should not exceed a maximum height of approximately 20 ft (6.1 m). The hazard of a particular plastic in any form of storage arrangement is the same whether it is encapsulated or nonencapsulated. If fire occurs, large quantities of smoke are usually generated, making manual fire fighting difficult and venting desirable in building construction consideration.

Plastics such as fluorocarbons, unplasticized polyvinyl chloride, and phenolics can be protected the same way as any Class III commodity (wood, paper, and natural fiber cloth), regardless of their physical form or storage arrangement. Pellets and small objects can be protected the same as Class IV commodities (commodity classifications are based on NFPA 231, *Indoor General Storage*).

Thermoplastics, such as polyurethane, polyethylene, and plasticized polyviny chloride; and thermosets, such as polyesters, present a fire hazard exceeded only by thermoplastics such as polystyrene and acrylonitrile-butadiene-styrene (ABS). These plastic materials will melt and break down (depolymerize) into their monomers, acting and burning like flammable liquids. High sprinkler discharge densities over relatively large areas are necessary to protect these types of plastics. In the form of foamed material, these plastics present the most severe fire hazard.

One-story buildings without basements are preferable for storage of plastics materials because of greater efficiency for fire fighting, ventilation, and salvage operations. Plastics should be stored, handled, and piled according to their fire characteristics. (See Chapter 48, "Industrial Storage Practices," for a discussion of the principles of good storage practices, involving the various classes of commodities, the various storage arrangements in use today, and for recommendations for fire protection for goods in storage.)

Housekeeping Practices

Housekeeping is basic to good firesafety. Good housekeeping practices reduce the danger of fire simply because they are time-proven methods of controlling the presence of unwanted fuels, obstructions, and sources of ignition. Approved containers should be provided and properly maintained for the disposal of refuse and rubbish. Spills of flammable liquids or combustible materials should be promptly cleaned up and properly disposed. Removal of combustible dust and lint accumulations from walls, ceilings, and exposed structural members is necessary. Smoking control, in addition to sensible regulations, also requires receptacles for spent smoking materials. Areas containing flammable liquids, vapors, or combustible plastics materials, processing, manufacturing, or storage must be clearly identified to prohibit smoking or the use of open flame devices.

THE SAFEGUARDS

Good fire protection starts with the firesafe design of the plant or warehouse, or inspection and modification of the existing facilities. Sprinkler-protected, noncombustible construction is appropriate for buildings occupied for storage, processing, and manufacturing of combustibles such as those involved in the plastics industry. Automatic sprinklers, standpipe and hose systems, and water-type portable extinguishers should be supplemented by fire extinguishers and special automatic systems suitable for flammable liquid fires and electrical fires, where these hazards exist.

Consideration should be given to the provision of roof vents, particularly in large one-story warehouses of manufacturing plants.

Building Construction

Long, narrow buildings provide greater ease in protection and fire fighting than large, square buildings. One-story buildings without basements are preferable to multistory buildings which may be subject to the spread of fire from lower to upper floors.

Large properties are best subdivided into fire areas to limit the spread of fire. Storage and manufacturing areas particularly need to be separated from each other by walls with sufficient fire rating to protect each area and occupancy from the other in case of fire. Preferably, fire walls should be without openings, but if openings are necessary, protection can be provided by self-closing or automatic fire doors suitable for openings in fire walls. Generally, a single fire area should not exceed 50,000 sq ft (4645 m^2).

Fire Control Systems

An alarm system that alerts building occupants, notifies fire suppression departments, and activates automatic suppression equipment is a necessary protection feature. Rapid extinguishment may be achieved by providing plastics processing equipment, conveyors, and manufacturing machinery subject to ignition or explosion with automatic fire-, smoke-, or explosion-detecting devices to initiate an alarm and to activate automatic suppressing systems (water spray, foam, dry chemical, carbon dioxide, or halogenated extinguishing agents).

Sprinklers are the most important single system for automatic control of fires in plastics plants. Among the advantages of automatic sprinklers is the fact that they operate directly over a fire and that smoke, toxic gases, and reduced visibility, often associated with fires in plastics, do not affect their operation. Automatic sprinklers, standpipes, and fire hose connections depend upon an adequate water supply delivered with the necessary pressure to control fires.

BIBLIOGRAPHY

REFERENCES

1983 Facts and Figures of the Plastics Industry, Society of the Plastics Industry, New York, NY, September 1983.

The Story of the Plastics Industry, Thirteenth Revised Edition, Society of the Plastics Industry, New York, NY, May 1971.

McKinnon, G. P., ed., *Fire Protection Handbook*, Fifteenth Edition, National Fire Protection Association, Quincy, MA, 1981.

NFPA CODES, STANDARDS, AND RECOMMENDED PRACTICES

Reference to the following NFPA Codes, Standards, and Recommended Practices will provide further information on the safeguards for plastic products discussed in this chapter. (See the latest *NFPA Codes and Standards Catalog* for availability of current editions of the following documents.)

NFPA 10, *Portable Fire Extinguishers.*

NFPA 11, *Foam Extinguishing Systems and Combined Agent Systems.*

NFPA 11A, *Medium and High Expansion Foam Systems.*

NFPA 12, *Carbon Dioxide Extinguishing Systems.*

NFPA 12A, *Halon 1301 Systems.*

NFPA 13, *Installation of Sprinkler Systems.*

NFPA 14, *Standpipe and Hose Systems.*

NFPA 15, *Water Spray Fixed Systems.*

NFPA 16, *Deluge Foam-Water Sprinkler and Spray Systems.*

NFPA 17, *Dry Chemical Extinguishing Systems.*

NFPA 30, *Flammable and Combustible Liquids Code.*

NFPA 33, *Spray Application Using Flammable and Combustible Materials.*

NFPA 35, *Manufacture of Organic Coatings*.
NFPA 40E, *Storage of Pyroxylin Plastics*.
NFPA 69, *Explosion Prevention Systems*.
NFPA 70, *National Electric Code*.
NFPA 77, *Recommended Practice on Static Electricity*.
NFPA 91, *Blower, and Exhaust Systems, Dust, Stock, and Vapor Removal or Conveying*.
NFPA 101, *Life Safety Code*.
NFPA 231, *Indoor General Storage*.
NFPA 650, *Pneumatic Conveying Systems for Handling Combustible Materials*.
NFPA 654, *Prevention of Dust Explosions in the Plastics Industry*.

ADDITIONAL READING

Beck, Ronald D., *Plastic Product Design*, 2nd ed., Van Nostrand Reinhold Co., New York, NY, 1980.
Bulletin No. 3, *Fire Hazards of Hydraulic Fluids Used in Processing Plastics*, Society of the Plastics Industry, New York, NY.
Bulletin No. 6, *Fire and Explosion Due to Electrostatic Charges in the Plastics Industry*, Society of the Plastics Industry, New York, NY.
Bulletin No. 9, *Planning Good Fire Protection for Plastics Plants*, Society of the Plastics Industry, New York, NY.
Bulletin No. 10, *Portable Fire Extinguishers for Use in the Plastics Industry*, Society of the Plastics Industry, New York, NY.
Bulletin No. 12, *Fire Safe Electrical Installations for Hazardous Locations in the Plastics Industry*, Society of the Plastics Industry, New York, NY.
Bulletin No. 13, *Flammable and Combustible Waste Disposal in the Plastics Industry*, Society of the Plastics Industry, New York, NY.
Bulletin No. 14, *A Plastics Plant Fire Squad*, Society of the Plastics Industry, New York, NY.
Bulletin No. 16, *A Basic Fundamental of Fire Protection for Small Plastics Plants*, Society of the Plastics Industry, New York, NY.
Bulletin No. 21, *Fire Safety in the Storage of Materials in Plastics Plants*, Society of the Plastics Industry, New York, NY.
Crawford, R. J., *Plastics Engineering*, Pergamon Press, Inc., Elmsford, New York, NY, 1981.
Dubois, J. Harry, ed., *Plastics*, 6th ed., Van Nostrand Reinhold Co., New York, NY, 1981.
Frados, Joel, *Plastics Engineering Handbook*, 4th ed., Van Nostrand Reinhold Co., New York, NY, 1976.
Gibson, A. E., *Processing of Polymer Composite Materials*, Pergamon Press, New York, NY.
Pajgit, O., ed., *Processing of Polyester Fibres*, Textile Science and Technology Series, Elsevier Science Publishing Co., Inc., New York, NY, 1980.

Rosato, D. V. and John R. Lawrence, *Plastics Industry Safety Handbook*, Society of the Plastics Industry, New York, NY, 1973.

Schwartz, Seymour and Sidney Goodman, *Plastics Materials and Processes*, Van Nostrand Reinhold Co., New York, NY, 1982.

Motor Vehicle Assembly

Richard Yapp

Motor vehicles, cars, trucks, and buses have been a major force in America's economic development since Henry Ford introduced the assembly line and revolutionized the industry 73 years ago. Worldwide, there are more than 427 million motor vehicles in operation; about 37 percent of these are registered in the United States, representing one passenger vehicle for every two Americans. Today, motor vehicles account for 84 percent of all personal travel, and for more than 57 percent of the transportation of manufactured commodities between cities.

Nationwide, nearly 12.6 percent of personal income is expended for the ownership and use of cars. One out of every six workers in the United States derives his income from motor vehicles and related industries. American motor vehicle manufacturers operate 309 plants located in 46 states. More than 50 domestic names and some 300 models of cars are manufactured at these plants.

Materials from the mines, fields, and forests of every continent make up the automotive industry's annual shopping list:

13 million short tons (11.8 million metric tons) of steel

Richard Yapp is the Fire Protection Engineer for the Body and Assembly Operations, Ford Motor Company, Dearborn, Michigan.

575 million pounds (261 000 metric tons) of copper and copper alloys
9 million pounds (408 metric tons) of cotton
One million metric tons (1.1 million short tons) of synthetic rubber
470 thousand metric tons (518 thousand short tons) of natural rubber
800 thousand short tons (726 thousand metric tons) of lead
16 million gallons (60 567 m^3) of paint

Many of these materials are initially processed by such basic manufacturing operations as foundry, stamping, forging, and machining. More than 400 thousand employees work in different branches of these industries to supply parts and subassemblies to motor vehicle assembly plants.

The overall process for producing motor vehicles consists of two major phases. During the first phase, parts such as engines, transmissions, instrument clusters, wheels, wiring harnesses, and a multitude of other components are manufactured, assembled, inspected, tested, packaged, and shipped from manufacturing plants to motor vehicle assembly plants. During the second phase, these components are received at assembly plants where they are warehoused and then distributed to assembly lines that subsequently converge to produce the completed motor vehicle. This chapter concerns itself with the second phase — the assembly of the motor vehicle, including the fire hazards involved.

Assembly techniques and the sequence of operations will, of course, vary somewhat between manufacturers or even between car lines of the same manufacturer. The description presented in this chapter is based on a composite of passenger car assembly operations. Trucks and buses generally are assembled in a similar manner.

RAW MATERIALS

There are very few industries that cannot list automotive manufacturing as a customer. Many materials (such as steel, glass, and paint) are readily recognized as part of a motor vehicle. The larger automotive manufacturing plants are self-sufficient "cities within a city" that generate their own heat and power, operate their own internal transportation systems including railroad sidings, operate fire departments, hospitals, and restaurants that feed many thousands of workers. However, this chapter is concerned with the vehicle assembly operations, or second phase, of the manufacture of motor cars and trucks. The assembly plant may consist of one or possibly two buildings in a large manufacturing complex, or it may stand alone many miles from the plants that manufacture the stampings, engines, transmissions, frames, electrical, and electronic components, etc., which the vehicle manufacturer produces in-house.

Hazards

Only those raw materials and subassemblies of special interest as fire hazards which are received and stored in the assembly plant will be discussed in this chapter. Two specific categories of materials present most of the fire hazards associated with automotive assembly operations. These two categories are plastics and flammable liquids. (See Chapters 9, 19, and 37, for a specific discussion of "Rubber Products Manufacturing," "Plastic Products Manufacturing, and "Handling and Storage of Flammable and Combustible Liquids," respectively.)

Plastics: Many plastic parts are large and bulky, occupying considerable space in the warehouse and creating a severe fire hazard. The large parts are represented by the instrument panel or the grill opening panel (the front of the vehicle which contains the grill, headlights, and turn signal lights). Sufficient pieces of each of these parts to support eight hours of production may occupy 160 sq ft (14.9 m^2) of floor space and be piled 16 ft (4.9 m) high. The normal float for a two-shift plant may be ten to fifteen times this amount. (See Chapter 48, "Industrial Storage Practices.")

Large plastic parts are shipped to the assembly plant in captive racks or wire baskets which are returned to the supplier plant for reuse. To protect the parts from shipping damage, it is commom practice to place separators or spacers between the parts during shipping. Some separators may be made of corrugated cardboard; the more common dunnage consists of formed blocks of polystyrene beadboard or molded sheet plastic shaped to allow the parts to nest.

Fork lifts stack these racks or baskets upon each other in piles that often reach heights of 15 to 20 ft (4.6 to 6 m). This arrangement limits the fire protection options available. Permanent racks with in-rack sprinklers are too costly, as are fully enclosed containers. Hydraulically designed ceiling sprinkler systems have been determined to be the most economical method of protection.

Plastic dunnage is also used to bulk-ship smaller automotive parts such as disc brake calipers, voltage regulators, and radios. The quantity of plastics in each unit load is sufficient to require greater-than-normal fire protection, even though the parts themselves present no serious fire hazard.

Rubber tires and polyurethane-foamed seat pads are especially difficult to protect because water discharging from sprinklers and hose streams cannot extinguish a fire in these commodities. (Protection for these components will be discussed later in this chapter.) A large assembly plant may use as many as 6,000 tires each working day, or about one standard boxcar load each production shift.

Flammable liquids: Several gallons (one gallon equals 3.785 liters) of flammable and combustible liquids are consumed by each motor vehicle produced. Consider, for example, the tremendous amounts of gasoline, lubricants, engine coolant, hydraulic fluids, paints, thinners, adhesives, sealers, and cleaning solvents used by an assembly plant that produces

FIGURE 20.1. *A modern paint-mixing facility. Paint is pumped from these mixing tanks to the spray booths in another part of the assembly plant.*

one thousand vehicles each day. These materials are received in tank cars, tank trucks, "Tote" tanks, barrels, and pails. Bulk liquids are generally stored in buried tanks, while the portable containers are kept in a detached or cutoff storage room.

Despite the volumes of hazardous materials involved, use and handling can be done in a firesafe manner. Qualified fire protection personnel working in cooperation with materials handling engineers and production management can do much to minimize the quantities of materials that must be kept out in the open plant areas. Small quantities may be kept in safety cans, approved storage cabinets, inside storage rooms, and within the confines of paint spray booths.

High-volume materials are generally pumped from the remote storage area to their point of use because this is more economical. Motor fuel, engine coolant, brake fluid, and hydraulic fluids are in this category. The more popular paint colors are used in sufficient volume to justify piping the paint from the mix room to the spray booth and return. The circulating pipeline is necessary to keep the paint pigment in suspension. However, if the quantity of material needed to fill the pipeline exceeds the weekly usage, then it is more appropriate to transport the paint to the spray booth in portable paint pots. The paint may then be pumped into a local pipe circulating system. One auto manufacturer has constructed inside storage rooms adjacent to the spray booth to house these pots.

Major subassemblies: As previously stated, major subassemblies are not dealt with in detail in this chapter. However, it is necessary to briefly mention a few of the more significant components. The larger subassemblies comprise the so-called power train: the engine, the transmission, and the drive axle. A relatively new component now used is the transaxle. This is a single component which includes the transmission and the differential which is a part of the conventional rear axle. This hybrid is installed in the front-wheel drive passenger cars now being introduced in even the larger sizes.

These three subassembly components share some common materials in their manufacture. Housings are made of cast iron or aluminum; parts that are subject to high torque are made from steel forgings; and wear surfaces are highly machined. Gears are first heat-treated and then machined by conventional methods.

These parts are fully assembled and then functionally tested before being shipped to the assembly plant. Engines are set in a test cradle, fuel and coolant lines are connected, and electrical power is supplied for starting and for the ignition system. The fuel may be natural gas or conventional liquid fuel. Transmissions are usually manufactured in another plant apart from the engines. After assembly, they are driven, for testing purposes, by an electric motor, and hydraulic fluid is circulated through a stationary cooling system. Lubricants and hydraulic fluid used during the testing may be drained out or remain in the part when it is shipped.

Smaller parts and subassemblies: Many smaller parts and subassemblies are manufactured by the automobile company or by the thousands of suppliers. Many of these parts are packaged in nonreturnable dunnage consisting of wooden pallets, wooden crates, or corrugated cardboard. Generally, these parts are stored in multitiered racks near their point of use. A few plants have installed automatic or manually operated stackers which may reach 40 or 50 feet (12.2 or 15.2 m) in height.

Materials handling: The forklift truck is now used extensively in the assembly plant and has permitted greater utilization of "air rights." This has radically changed the design of assembly plant buildings. Instead of the multistoried, reinforced-concrete structures so popular in the first half of this century, one-story buildings with steel-truss roof construction and bay dimensions of 50, 60, and even 100 ft (15.2, 18.3, and 30.5 m) between columns evolved. This development has resulted in larger and higher storage piles and overtaxed sprinkler systems.

Materials handling experts are divided in their views of the most efficient methods for handling the thousands of tons (1000 short tons equals 907.2 metric tons) of parts which must be received, stored, and distributed throughout the plant each production day. (See Chapter 46, "Materials Handling Systems.") They all agree that reduced handling means reduced cost and reduced damage. The ideal system would be to move the material only once from the railcar or truck dock directly to the assembly line. While this is possible with the smaller parts, it is not practical to store all of the bulky parts on the line. Operators do not have the time to walk away from their work stations to carry bulky parts more than a few feet. Thus, it is customary to set up reserve stock storage areas away from the assembly point and to "line feed" the materials periodically during the production workshifts. This generates considerable industrial truck traffic which, in turn, brings about a need for wide traffic aisles.

The reserve stock storage area concept results in large concentrations of highly combustible plastics, rubber, and soft trim materials. As more and more plastics are introduced as substitutes for metal parts, the fire loading escalates to a dangerous level. There is considerable resistance to the idea of cutoff or detached warehouses for these materials because of the increased handling costs and the inflexible nature that masonry fire walls have on expansion and plant rearrangement.

A new concept perfected by the Japanese may greatly change the American automotive industry. The Japanese call it "kambam" which is translated "just-in-time" production. The concept involves very close scheduling of parts as they move from one operation to the next. This concept, when applied to its fullest, extends back through the supply pipe line to the stock of raw materials stored at each supplier's facility.

Rather than building large warehouses for the storage of huge quantities of raw materials, subassemblies and finished parts, departments within a manufacturing facility schedule just enough ahead of production to supply the next operation "in time" to meet their needs. Thus, in-process inventories are greatly reduced.

While there is still a reluctance to change, American automotive managements see it as a way to reduce inventory costs and defective materials. Seminars and meetings have been held with parts suppliers to enlighten them and to encourage change. Suppliers are being encouraged to move their facilities close to the assembly plant to reduce the parts backed up in the pipe line and to assure "just-in-time" deliveries.

The effect on fire loss prevention is obvious. No longer will large masses of combustibles be stacked to the trusses. Fires in high piled stock may no longer be a threat in the automotive assembly plant.

Yard storage is generally limited to metal parts that are not affected by the elements; these include frames and chassis parts which are exposed on the finished vehicle. Some drum stock may be stored outdoors if it will not be damaged by extremes of temperature. The largest portion of the yard storage will be devoted to finished vehicles awaiting shipment, and to returnable durable dunnage and empty containers.

THE PRODUCTION PROCESS

The process of assembling a motor vehicle is generally divided into two major operations, body construction and final assembly. The dividing point between these two operations may vary from one manufacturer to another, but the sequence of individual functions is generally similar.

MOTOR VEHICLE ASSEMBLY

FIGURE 20.2. This automatic welding fixture opens at the finish of the weld cycle to release the assembled body. (Ford Motor Co.)

Body Construction

A stamping press is used to form sheet metal into the shapes and contours necessary to meet the style specifications of the vehicle's body designers. The stampings are then subassembled into the floor pan, fire wall, side panels, roof, and rear deck. The sheet metal subassemblies are placed in a framing fixture where the parts are joined together by welding.

These welding operations are semiautomatic, with workers needed only to load the subassemblies and to start the automatic welding cycle. The parts are clamped in place either by air pressure or hydraulic pressure, after which hydraulically operated resistance spot welders progressively fuse the various components into a structurally rigid metal auto body. The body, if not already mounted on a conveyor truck, is placed on metal skids which travel with the body until it reaches the final assembly line. Final construction operations consist of additional spot welding, either by handguns or by robots, gas welding of critical joints, solder application over seams, and grinding out metal defects. At this stage, the doors are bolted into place.

Next, the body is cleaned to remove metal-draw compound, filings, and oil by passing the body through an automatic washer. This is followed by an acid wash which prepares the metal to receive a prime coat of paint. After the body has been dried in a tunnel oven, various sealers and deadeners are applied to seal out potential water leaks and to prevent the metal panels from "oil canning." Excess sealer is wiped

Chapter 20

FIGURE 20.3. Robots perform spot welding on automatic body side panels. (Ford Motor Co.)

from the joints with a solvent and the body is ready for its first coat of paint.

Paint is applied in a number of ways. First, the body is primed by either spraying or dipping, using a solvent-base paint. A recent development which is replacing these conventional painting methods is the electrocoating process. The body is dipped into a bath of water-dispersed paint. Paint particles are kept suspended in the water by strong agitation, and an electrical charge is applied much like the electroplating process. The paint particles are attracted to the work piece and irreversibly deposited upon it.

It is the nature of the paint used in the electrocoating process to become water insoluble after it is deposited, and for the deposit to cease to build up once the prescribed thickness is obtained. Thus, a uniform coating is attained regardless of the shape of the work piece. The voltage setting controls film thickness.

Additional feed stock is added to the bath periodically to maintain the correct paint concentration of 8 to 12 percent. Small quantities of alcohols or amines are also added to control the pH and solubility of the bath. After one to two minutes of immersion time, the body (or other part) is rinsed and then baked at about 350°F (176.7°C), in a continuous, direct-fired oven.

Where the electrocoat process is used to prime the body, the previously mentioned sealer and deadener applications are delayed until this point in the painting cycle.

After the primed body has been inspected and painting defects repaired, one or more finished coats of color are applied. This material may be lacquer or baking enamel. The body (or other part) is conveyed through a tunnel-type spray booth where one or more coats of finish material are applied by a team of spray painters.

Many paint operations utilize the electrostatic spray method. The paint is electrically charged as it leaves the nozzle of the gun and is at-

FIGURE 20.4. *Equipment used to control the electrocoat bath consists of a deionized water process system, filters, and the heat exchangers shown here in front of the dip tank enclosure. (Ford Motor Co.)*

tracted to the grounded work piece. The advantages are: (1) paint wrap-around on metal edges, and (2) reduced overspray.

The different paint colors are supplied to the hand spray gun through plastic hoses which terminate in a quick disconnecting fitting. The operator selects the appropriate hose and couples it to the spray gun.

Some hand-spray stations have been automated by installing both horizontal and vertical reciprocating arms which carry the spray gun. A mechanical color changer may also be installed to select the proper paint color in accordance with a prearranged remote controlled coding system.

Ventilation within these large spray booths must be closely regulated. Too rapid an air movement will blow the paint away from the work piece, and too little air movement will result in overspray contamination between colors on adjacent work and complaints from operators. Thus, ample, well-controlled ventilation is necessary for this type of operation.

It is a common practice to provide a body storage bank equal to one or two hours worth of production requirement before the bodies progress to the next stage, the trimming of the interior and exterior. The body bank allows for temporary interruptions in body construction and painting. Some manufacturers initiate the assembly schedule at this point, selecting the sequence in which the various body colors ultimately arrive at the final assembly line. The metal body with a baked-on paint finish presents no increased fire hazard while in storage.

The trim operation involves adding window glass, external ornamentation, and the complete outfitting of the body interior with the ex-

Chapter 20

FIGURE 20.5. *These automatic paint spray guns operate with a 90,000-v electrostatic charge. The guns retract to follow the contour of the car body.*

ception of seats and carpets. These will be installed near the end of the final assembly line. For the most part, the trim will consist of plastics, some cloth, and small quantities of paper products. Other components installed during trim include window regulators, wiring, instrument cluster, radio, and sound-absorbent materials. The interior of the body contains considerable combustible materials; however, incidents of fire are minimal as there are few ignition sources.

The body is now ready to be "decked" on the chassis as it rides along the final assembly line. Compact model passenger cars generally do not have a frame. Instead, a front-end assembly is welded on to the body during construction and then painted along with the body. The integrated unit is provided with attachment points for the rear axle, front wheels, and engine, each installed during final assembly.

With some car lines, the body operations are performed at a separate plant remote from the assembly plant. The finished bodies (sometimes without interior trim) may be loaded on trucks for delivery to the assembly plant. It is customary to provide another storage bank for the finished bodies in order to help minimize interruptions to the final assembly schedule.

Scheduling: While the body is in the trim department, other major components are started through subassembly operations. The lead time is calculated so that these subassemblies arrive at the final assembly line at the precise moment that they are needed. Each subassembly operation is advised of the sequence in which to initiate the differing components to affect a proper matching of colors, styles, and accessories which have been specified for each unit produced. This information is transmitted over an in-house communications network which may use teletypewriters in each department. Once the scheduling system is initiated, components will match up properly as long as the "tickets" are kept in sequence. Some vehicles are built to customer's orders, while others are built to meet sales projections. Heavy trucks and "fleets" are, of course, built to order.

Obviously, any major interruption in one department, such as a fire, will quickly shut down the entire assembly plant. Rarely does an assembly plant possess parallel or duplicate facilities under the same roof. If lost production is to be made up, it must be done on overtime at one or more assembly plants.

Final assembly: When a vehicle is constructed on a frame, the appropriate frame is selected from a stockpile and hoisted onto a conveyor. Springs, axles, exhaust systems, and other underbody components are first installed. This assembly may be spray painted before it is started down the final assembly line.

Meanwhile, engines and transmissions are mated on the engine dress-up line. Other accessory items such as alternator, A/C compressor, power-steering pump, fan, and drive belts are installed. The assembled power plant arrives at the final line in time to be decked on the chassis. The power train is completed by connecting the driving axle and power plant together through the drive shaft.

The wheels may be the next component added. The appropriate tires are selected from storage while the wheels are being painted. Tires are mounted to the rims, and then inflated and balanced on a continuous conveyor line which is equipped with automatic and semiautomatic machinery. An overhead conveyor delivers the assembled wheels to both sides of the final assembly line where operators bolt them onto the new "unit," as the vehicle is called.

Up to this stage in the final assembly operation, the unit has been riding on its frame atop a waist-high, two-strand conveyor. If the new unit has no frame, it possibly was carried along by a clam-shell which embraces the body sides and is hung from an overhead "power and free" conveyor. The clam-shell rolls along on one or more sets of rollers which are engaged in both sides of the overhead I-beam rail. A chain, which is under power, rides along on a separate but adjacent rail to tow along the free-rolling load. By engaging and disengaging the "dogs" which link the power and free conveyors, the movement of the load may be alternately stopped and started.

Now that wheels have been installed on the new unit, it may be set down onto a two-strand, floor-level conveyor where it will ride down the final assembly line until it is driven off under its own power. Hood, front fenders, grill, headlights, battery, and other components are added by workers, each of whom has specific tasks assigned. The amount of work assigned to each employee is determined by the complexity of each task and the line speed. For example, at 45 units/hr and a line length of 960 ft (292.6 m), the cycle time for the unit to travel the length of the conveyor at 16 ft (4.9 m) per min would be 60 min. If the units were spaced 23 ft (7 m) apart on the line and workers were not to follow the unit for more than one space before returning to pick up the following unit, then each worker would be allocated 1½ min to perform his task walking or riding with the unit and returning back to his starting point to meet the next unit and repeat the cycle. Since some tasks may take more than 1½ min to complete, the task would be divided up among two or more workers.

To increase the line speed, it would be necessary to add more workers and reduce the total operation of each worker. It might also be necessary to lengthen the final line to provide more work space and stock storage along the line.

The final assembly is complete when the unit has been filled with radiator coolant, gasoline, brake fluid, A/C refrigerant, and when seats and carpets have been installed. At this time, the new car or truck can be said to come to "life." A worker enters the newly assembled vehicle and turns on the ignition switch. Most units usually start. However, in the event that further minor adjustments are necessary, it is occasionally necessary to push or pull the vehicle out of the way and into a repair stall.

New vehicles undergo inspection, repair, and then final acceptance by quality control inspectors. Most new units are turned over to an automobile haul-away company or railroad for delivery to the dealer. The entire operation as described in the preceding paragraphs takes about thirteen hours and involves nearly 1,000 direct-labor employees.

HAZARDS

Product Fires

Statistics are available (but of questionable accuracy) on fires in motor vehicles on the highway; however, no statistics are available on the number of fires occurring in finished vehicles before such vehicles reach the dealer's showroom. Although such fires are experienced, production management has shown a preference not to discuss or record them.

Probably the most common cause of fires in finished vehicles is electrical. Wiring is subassembled into wire harnesses with color coding or unique connectors installed to assure correct assembly. Mistakes in subassembly or final assembly may result in crossed circuits and the resultant overheating of the wiring system caused by excessive current flow. Also, metal clips are sometimes used to fasten the wiring harnesses to the metal parts of the vehicle. If the clip cuts through the wire insulation, the circuit becomes grounded and may result in a short circuit and overheating. Most of the resulting fires originate in the area of the instrument panel.

Another source of fires is from carburetors backfiring when engines are started with the air cleaner removed. Mechanics often prime the carburetor of a difficult-to-start engine with a small quantity of gasoline. The gasoline often becomes ignited when the starter is engaged. On occasion, fuel-line connections are not made tight enough and raw fuel will leak or spray onto the engine.

Some fires have occurred on fully assembled vehicles when mechanics used welding torches to make final repairs. Combustible upholstery, interior trim, and motor fuel are all easily ignited by the heat of a gas torch or electric arc. Damage is usually confined to the one vehicle involved in these fires, and the costs to either repair the damage or replace the destroyed vehicle are absorbed in the operating costs of the plant.

Motor Vehicle Safety Standard (MVSS) No. 302[1], issued by the U.S. Department of Transportation, applies to the flammability of materials used in motor vehicle interiors. This standard limits the horizontal burn rate when a sample is tested in the specified laboratory apparatus. The required reduced burn rate is intended to permit passengers to safely evacuate the burning vehicle. When these same interior trim materials are stacked for in-plant storage prior to use, the effect of the fire-retardant treatment is not perceivable. This has been demonstrated by full-scale fire tests. It is important to acknowledge that a horizontal burn test is not indicative of the vertical fire spread experienced in a three-dimensional fire.

General Operations Hazards

Common fire and explosion hazards related to industrial situations in general are discussed throughout this book. There are, however, many fire and explosion hazards that are unique to automotive assembly operations. These hazards, which evolve from the nature of automotive assembly operation, are identified and discussed in the following paragraphs.

Welding: Beginning with a vehicle's body construction, small sections of the auto body are welded together in small hand-clamped fixtures which present no unique fire problems. However, the larger fixtures utilize hydraulic clamps and hydraulic pressure to perform the resistance welding. Oil leaks are common, and other debris may collect in the lower portions of these complex fixtures. The ever-present arc is a ready source of ignition. Although flame-resistant hydraulic fluids are available, tool designers tend to resist their use because of higher costs and because they are not familiar with these materials. A fire in a body welding fixture is costly not only in material damage, but also in the extended down time

Chapter 20

FIGURE 20.6. *This complex machine welds the several sections of the automobile floor pan into one rigid structure. A sump below the machine collects any oil or water leaking from the many hose fittings. (Ford Motor Co.)*

required to effect repairs — unless, of course, the manufacturer can afford to have a spare fixture sitting idle.

Many small parts are brought into the welding area in their original shipping containers. Cardboard cartons, wire-bound crates, and even burlap bags are used. Welding sparks landing in these materials start numerous small fires. The obvious solution would be to require suppliers to ship parts in noncombustible containers. However, the greater initial investment in returnable containers and shipping costs would seem to override the negligible damage caused by these nuisance fires.

Generally, the sealers that are applied to the body seams either during or after welding have a high solids content and present no fire hazard in themselves. However, solvents are used to clean up spilled sealer from the welding fixture or floor. Chlorinated solvents cannot be used because of the health hazard and the tendency of the electric arc to break down any residual solvent and release hydrogen chloride gas. High flash solvents will not cut the sealer compound. Thorough cleanup of all solvent residue before welding resumes is essential to firesafety.

Some gas welding may be employed, but the extent is relatively minor and no unique problems exist. Gas torches are used to apply solder compound to cover unsightly joints. Again, the cost advantage of minimizing this operation tends to discourage the use of solder joints. Recently, there has been some interest in the use of plastic solder substitutes. It is general practice to perform these torch operations under an exhaust hood. Fires occur whenever cleanup of the exhaust system is overlooked.

Painting: The preparation of the metal surface to accept the prime paint consists of a process which goes under such proprietary names as "Bonderite" or "Parker." The metal is conveyed through, or dipped into, several stages to clean, rinse, and then acid treat the surface. The chemicals are water solutions of iron or zinc phosphate and an oxidizer. Following a final rinse, the body or other sheet metal parts are dried in an oven, cooled, and conveyed to either a sealer "deck" or the prime paint operation. The hazards of the sealer operation are the same as previously described, except that at this stage solvents are used to clean excess sealer from the parts.

Most of the paints used in automotive finishes are flammable liquids, lacquers, enamels, and solvent-base primers. As much as half of the fluid is a volatile solvent which flashes off in the spray booth or bake oven. Because this solvent constitutes an environmental pollutant, there is considerable interest in alternate finishing materials.

Electrocoat prime is a water-reduced resin feed stock to which pigment and stabilizers are added to produce an immersion bath through which work to be coated is conveyed. The flash point of the resin feed stock, as received, will range from 50°F (10°C) to more than 200°F (93.3°C). The resin feed stock may be shipped in steel barrels, portable tanks, or bulk tank vehicles. The method of storing will depend upon both flash point and shipping method. Once reduced with water, there is no flash point. Dry residue which may accumulate on the floor of the bake oven is combustible, as are the same drippings from other paints.

Currently, two-component urethane enamels are being applied in limited quantities. Because of the toxicity of the isocyanates used in this material, extreme care must be taken in order to prevent the release of volatiles into the atmosphere. A closed handling system minimizes the potential fire hazard. Another approach to the air pollution problem is to capture the volatile organic solvents emitted from the spray booth and bake oven. Large incinerators that burn off the effluent are in use at a few plants. At some locations, the waste heat is utilized to preheat the air forced into the spray booth.

Another recent development is the use of carbon absorption to capture the paint solvents emitted from the spray booth. After filtering out solid paint particles, the exhaust gases are passed through the carbon bed which removes much of the volatile material. Periodically, steam is forced through the bed to reactivate the carbon by driving off the volatiles. This process will not work on the hot oven exhaust because the heated volatiles are not absorbed. The captured volatile is useless as a reclaimed solvent and might be burned as fuel in a boiler.

Trimming: After the automobile bodies have been painted they are "trimmed." Since so much of the trim materials used in today's vehicles are composed of plastics, the stockpiling of such parts as instrument panels, heater and air conditioning units, tail light assemblies, and head liners has become the major fire protection concern of the industry. Large-scale fire tests have demonstrated that high-density sprinkler systems can control a fire in plastics, and one automotive manufacturer now specifies a hydraulic density of 0.60 gpm per sq ft [24.4 (L/min)/m^2] over plastic parts storage areas. The pile heights and pile array must also be controlled in order to limit the size and intensity of any fire that might occur.

Fueling: Gasoline is the most common fuel, with diesel oil and gaseous fuels representing a minimal percentage of the vehicles registered in the United States.[2] A large assembly plant may consume up to 20,000 gal (75 708 L) of fuel each production day. The fuel is metered into each vehicle as it approaches the end of the final assembly line. Occasionally, leakage of fuel occurs because a worker failed to complete a fuel-line connection, or because damage has occurred to the fuel line or tank. A drainage trench under the fueling area will carry away any spills. At some plants, a water flushing system with a fuel neutralizer added as required has eliminated the spill hazard. Other plants have installed carbon dioxide flooding systems.

Maintenance and Housekeeping

There is usually very little overall preventive maintenance performed during motor vehicle assembly operations, although much time is spent on breakdown maintenance. Thus, maintenance efforts are usually more concerned with key production equipment which often gets daily attention to assure that the "line" keeps moving. When management reduces indirect labor costs, personnel from departments such as maintenance (with large head counts) are often the first to be dismissed.

Areas that are susceptible to fire if neglected include exhaust systems, electrical equipment, and static grounding. Paint overspray collects inside the ductwork and fan shroud and, when it accumulates on the fan wheel, unbalanced vibrations result and quickly destroy the fan. When the fan wheel rubs against the housing, a fire soon develops; friction heat (not sparks) causes these fires.

Fires also occur in the test rolls where the new vehicle is checked out after assembly. Rubber from the new tires is ground off on the rolls and accumulates in the pit below. This fresh rubber dust is subject to spontaneous ignition.

Electrical equipment throughout the plant is constantly damaged by materials handling trucks. Light panels, conduit, fire alarm boxes, and important motor control centers must be guarded with heavy posts and highway guardrails. Notwithstanding such protection, damage is often heavy and usually only essential repairs are completed. It is unknown how many fires are caused by faulty and damaged electrical facilities.

Electronic equipment is now being installed at an accelerated pace in the automotive industry. In recent years there has been a marked improvement in the associated fire hazards. The time lost in restoring production fire losses is greatly reduced where small programmable controllers have replaced the large cabinets with electrical-mechanical relays and miles of wiring. The safety impact of these and other hi-tech systems used in the industry will be more recognizable in the near future.

Because large volumes of flammable liquids are used, the chance of fires from static electricity is great. The attention given to correct bonding and grounding varies from plant to plant, and since grounding is not essential to production operations, it generally does not get the attention it deserves.

Considerable quantities of combustible dunnage in the form of cardboard cartons, pallets, and wire-bound crates must be removed each day from the plant. The ban on open burning has eliminated one previously common fire cause. Waste disposal must receive continuous attention or the plant can become choked in its own debris.

The disposal of waste oils, solvents, and paint sludge has become a costly matter now that such materials can no longer be dumped on the ground or into a body of water. In-plant collection methods vary, and some accumulations awaiting disposal may present a serious fire hazard.

The Human Element

The human element contributes greatly to the degree of life safety at motor vehicle assembly plants. During times of high production, a large proportion of the many thousands of workers employed in assembly plants will be new to the work force, inexperienced, and unlikely to have a deep sense of loyalty to their employers. Because assembly-line tasks are easily learned, there is very little time devoted to training or instruction in fire prevention. Many supervisors find it necessary to function in other capacities (such as stock expeditors) and are thus limited as to the degree of supervision and training they can give their workers. Utility assemblers and relief workers have usually worked the assembly line long enough to have learned the operations and the inherent hazards. While such workers have no supervisory authority, many of them influence the newer workers and exercise an informal leadership. Union committee members are the formal link between the workers and management.

As is typical of industries where hazardous materials and flammable liquids are only incidental to the production operation, there is little awareness among assembly-line workers of the potential harm that can result from carelessness. All too often training is a "hit or miss" endeavor performed by the supervisor, plant safety engineer, or the plant fire "chief." Workers often look upon rules as a type of imposed harassment, and will deliberately break the rules as a form of rebellion against the company.

The effectiveness of local plant management is often measured in numbers of units produced and cost control effectiveness. This, in turn, encourages cost cutting in such "nonproductive" activities as maintenance, housekeeping, and fire prevention. Since many plant managers are as uninformed as their workers of the potential hazards present in the assembly plant, little attention is paid to fire prevention until an incident occurs to present a lesson. If a plant is lucky enough to not experience a fire loss of any magnitude, neither the workers nor the management will concern themselves with loss prevention. Thus, it is the responsibility of corporate or central management to communicate to the plant management and workers an awareness of the ever-present fire and explosion hazards and the techniques for preventing them.

Fire protection duties, including inspection of fixed fire protection systems, general conditions of the plant, and servicing of fire equipment, are generally assigned to a fire "chief" or "marshal" who is a member of the plant protection department which, in turn, is a part of the personnel department. While such persons are usually dedicated, they often lack the technical training necessary to cope with today's more sophisticated manufacturing operations. Often, these personnel are involved neither in the development of new manufacturing methods nor with the introduction of new materials; nor is their advice always solicited when changes are planned by the engineering department. There is a great need for closer coordination and cooperation with fire protection personnel.

The "human element" problems of an assembly plant are similar to those of most businesses:

Fire brigades: Production-line workers are not considered good candidates for the brigade because response to an alarm would stop the assembly line. Often, maintenance personnel are organized to assist the plant guards in fire fighting. Formal training ranges from excellent to nonexistent. (See Chapter 3, "Plant Emergency Training and Organization.")

Impairments: At some plants, maintenance tradesmen are not permitted to close sprinkler control valves or to shut down fire protection equipment without approval of the plant protection department. This centralizes the control of impairments and acts as a check against unapproved changes in the plant's protective systems. Unfortunately, this is not the general practice.

Smoking: Again, the practice varies greatly depending primarily upon the attitude of the labor relations personnel and their willingness to stand behind the enforcement of rules.

Nonproduction burning and welding: Like the control of impairments, some plants allow the maintenance welder "carte blanche" freedom to weld or use a cutting torch. Sometimes, even in plants with good controls, risks are taken when a conveyor jams and production is at a standstill. The control of outside contractors also varies from plant to plant, even within the same company.

LIFE SAFETY

Concern has been expressed in some quarters that the trend to larger industrial buildings and the introduction of new raw materials such as plastics will increase the life safety hazard at motor vehicle assembly plants. To date, this prediction has not materialized in the motor vehicle industry. To the credit of the industry, most assembly plants are fully sprinklered. Currently, however, the increased use of highly combustible plastics (which are replacing metal parts) may overburden the ordinary hazard sprinkler system. Only recently has this problem become apparent. The greatest risk to life is the very toxic nature of the products of combustion evolved from burning plastics. Production workers can avoid exposure by promptly evacuating a building; however, fire fighters may be seriously exposed unless proper protective gear is used.

Today, motor vehicle assembly buildings with over two million sq ft (185 806 m^2) of production and storage space are common. Modern materials handling methods make the use of multistoried buildings inefficient, and the trend is toward building most of the factory floor space at grade level. Powered materials handling equipment and the bulkiness of the motor vehicle as a product necessitate wide traffic aisles in assembly buildings, thus facilitating easy employee evacuation. While the travel distances spelled out in most codes may be exceeded, experience indicates that able-bodied workers have no problem exiting a building in a hurry.

Fifty years ago assembly-line workers stood shoulder-to-shoulder, and most assembly-line tasks were performed by human hands. Today, large, expensive machines perform most of the routine operations, and often the worker is needed only to oversee the repetitive motions of these machines. This change has resulted in a greater investment in capital goods for the manufacturer, and has considerably reduced the exposure of employees to situations in which they might be injured. It has also materially reduced the population density within the plant.

Modern plants have extensive communications facilities which augment the original fire alarm system. For example, telephones throughout plants are more common than they were 25 years ago. Also, many plant supervisory personnel and materials handling personnel now carry radios.

A more recent development that affects life safety and was brought on by concern for energy conservation and energy costs is the energy management system. In addition to the functions of peak shaving and remote start-up, data on the functioning of heating and ventilating systems, ovens and furnaces, and power distribution networks are fed to a central control which can alert management of impending trouble. At

Chapter 20

FIGURE 20.7. Interior and exterior of Building No. 10, the Packard Motor Car Co., Detroit. Note the sprinkler system and wiring method. (Albert Kahn Associates)

FIGURE 20.8. Chalmers Motor Car Co. (now owned by Chrysler Corp.). This is typical of early automotive factory construction designed by Albert Kahn in 1907. (Chrysler Corp.)

FIGURE 20.9. Interior view of Ford Motor Co. Highland Park (MI) plant showing roofed-over court between two 6-story buildings. Note balconies on which hoisted parts were deposited. Sprinklers are located at the roof line within exposed steel truss. (Albert Kahn Associates)

MOTOR VEHICLE ASSEMBLY

FIGURE 20.10. The Willow Run (MI) Bomber Plant operated by Ford Motor Co. during World War II was later an automotive assembly plant for Kaiser Fraser. It is now owned by the General Motors Corp. (Albert Kahn Associates)

FIGURE 20.11. Press shop designed by Albert Kahn for Chrysler Corp. Photo taken in 1938. (Albert Kahn Associates)

FIGURE 20.12. Modern assembly plant design is examplified by the St. Thomas (Ont.) Plant of the Ford Motor Co. built in 1968. The tower is used to store partially assembled passenger cars. The plant is fully sprinklered and has a strong water supply system. (Albert Kahn Associates)

some plants, the fire alarm system has been integrated with the energy management system.

THE SAFEGUARDS

The conventional mill construction so popular in the last century proved unsuitable for the automotive industry. In 1905, Detroit architect Albert Kahn built a reinforced concrete building for the Packard Motor Car Company, the first factory building of reinforced concrete construction in Detroit. Two years later, the Chalmers Motor Car Company built four parallel reinforced concrete buildings. These buildings were linked together to form open and closed courts. Eventually, these courts were roofed over and the inside window sash removed. This arrangement formed a series of high bay sections with intervening multilevel mezzanines.

At the Ford Highland Park Plant the same arrangement was repeated when new buildings were constructed in 1918. Materials were unloaded from rail sidings in the high bay sections and hoisted by overhead traveling cranes to cantilevered balconies on the floors above. Albert Kahn used the same scheme at the Ford Rouge Plant during World War I, and again in 1943 when he designed the huge Willow Run Bomber Plant in Ypsilanti, Michigan. This plant is now owned by General Motors.

During the 1920s, Kahn, realizing the inefficiencies of the multistoried factory building, began to construct automotive assembly plants entirely on one level and as much as ½-mile (805 m) long. These buildings had exposed steel frames and walls of unbroken expanses of glass set in steel sashes. Typical of this construction is the Chrysler Corporation Lynch Road Plant built in 1928. The work of Albert Kahn and his "Associates" can be found throughout the automotive industry, and many of his original-design buildings are part of the expanded and "modernized" assembly buildings still in use today.*

In 1953, a major fire at the General Motors Hydromatic Plant in Livonia, Michigan, created much interest in the combustibility of built-up metal deck roofs and the use of smoke vents. Only a small portion of this plant had automatic sprinkler protection. The unprotected combustible roofing was initially ignited by burning rustproofing compound, and the fire was further intensified when an overhead gas main collapsed with the sagging roof. The ensuing gas jet was hot enough to weld a collapsed steel truss to the cast-iron frame of a Gleason quench press. Many lessons were learned from this fire loss, not the least of which was the demonstrated ability of a well-organized company to quickly recover from the loss of a production facility.[†]

---

*Albert Kahn Associates also has built hospitals, schools, and office buildings such as the General Motors Building in Detroit and the Ford NAAO Building in Dearborn, Michigan.
†The author personally witnessed the fire and participated in the investigation which followed.

The evolved design of an automotive assembly plant tends toward a one-story structure with the roof supported by heavy steel trusses and columns spaced out 40 to 50 ft (12.2 to 15.2 m) apart. Mezzanines or a partial second floor are used for paint operations and body storage since the bodies are easily transported by overhead monorail conveyors. Factors such as limited building space may dictate that buildings rise higher. Stacker units to heights of 40 to 50 ft (12.2 to 15.2 m) have been used in a few plants for the storage of parts and semifinished vehicles.

The layout of the assembly operations are often changed to accommodate changes in process or to produce a new model vehicle. For this reason, the factory space must remain flexible; thus, fire walls are not popular with plant layout engineers. Because conveyors often extend over wide areas of the plant, separating walls would be pierced by many openings, rendering them ineffective.

Generally, assembly buildings are fully sprinklered. Full-time fire protection personnel test the systems periodically and make regular inspections of valves. The water is most often supplied by fire pumps arranged for automatic operation. Most serious fire losses experienced today are the result of highly combustible storage which is piled too high for ordinary sprinkler systems to control.

Fire protection deficiency at an assembly plant is not unlike that at most manufacturing plants. Basic protection hardware is available to meet the hazards that exist, and most plants have this hardware installed. The human element is the reason for the deficiencies most often found at the plant level, and lack of awareness of the hazards of fire is the weakness most often found at the management level.

BIBLIOGRAPHY

REFERENCES CITED

1. U.S. Department of Transportation, *Motor Vehicle Safety Standard No. 302*, U.S. Code of Federal Regulations, Section 571, Title 49, Government Printing Office, Washington, DC.
2. *1983 Facts & Figures*, Motor Vehicle Manufacturers Association of the United States, Inc., 300 New Center Bldg., Detroit, MI 48202.

REFERENCES

Motor Vehicle Manufacturers Association of the United States, Inc., 300 New Center Bldg., Detroit, MI 48202.
"Auto Makers Squeezed by Shift to New Materials," *Production*, January 1978, vol. 81, no. 1, pp. 60-67.

"The Legacy of Albert Kahn," The Detroit Institute of Arts, 1970.

Note: Most of the data was from the author's personal files and proprietary information of his employer.

NFPA CODES, STANDARDS, AND RECOMMENDED PRACTICES

Reference to the following NFPA Codes, Standards, and Recommended Practices will provide further information on the safeguards for motor vehicle assembly discussed in this chapter. (See the latest *NFPA Codes and Standards Catalog* for availability of current editions of the following documents.)

NFPA 13, *Installation of Sprinkler Systems.*
NFPA 27, *Private Fire Brigades.*
NFPA 30, *Flammable and Combustible Liquids Code.*
NFPA 34, *Dipping and Coating Processes Using Flammable or Combustible Liquids.*
NFPA 51B, *Fire Prevention in Use of Cutting and Welding Processes.*
NFPA 70, *National Electrical Code.*
NFPA 90A, *Air Conditioning and Ventilating Systems.*
NFPA 101, *Life Safety Code.*
NFPA 231, *General Storage, Indoor.*
NFPA 231D, *Storage of Rubber Tires.*

ADDITIONAL READING

Billinge, L., "Metal Working." Part I: Steel Making and Hot Forging, *Fire Surveyor*, vol. 10, no. 3 (June 1981), pp. 19-33.

Buehrer, D., "Check List for the Prevention of Fires Arising from Welding and Allied Processes," *Welding in the World*, vol. 14, no. 5-6 (1976), pp. 122-125.

"Fire Hazards of Electrostatic Paint Spraying," *Fire Prevention*, (124), 1978, pp. 19-20.

King, P. W. and J. Magid, *Industrial Hazard and Safety Handbook*, Butterworths Publishers, Inc., Woburn, MA, 1979.

McKinnon, G. P., ed., *Fire Protection Handbook*, Fifteenth Edition, National Fire Protection Association, Quincy, MA, 1981.

"Prevent Cutting/Welding Fires," *The Minnesota Fire Chief*, vol. 16, no. 5 (May/June 1980), pp. 16, 48, 54.

"Spray Booth Fire at Chrysler Plant Confined by Agressive Volunteers," *Fire Engineering*, vol. 131, no. 1 (January 1981), pp. 40, 42.

"How to Handle Flammable Liquids Safely," Justrite Manufacturing Co., Des Moines, Iowa, 1978.

"Fire Hazards of Paint Spraying," *Fire Prevention*, September 1979, pp. 22-25.

"Risk Assessment by Simulation — Firesim," Hansen-Tangen, E. and T. Baunan, *FIRE SAFETY JOURNAL*, vol. 5, nos. 3/4 (1983), pp. 205-212.

"Planning a Fire Inspection That Works for You," Bush, K.E., *American Health Care Association Journal*, vol. 9, no. 5, (September 1983) pp. 12-15.

21

Shipyards

John W. Van Brunt

Each shipyard presents a unique fire protection and prevention situation. The factors that contribute to dissimilarity among sites include structure, exposure, topography, operation, scheduling, personnel, and weather. In addition, local government regulations, including environmental and ecological attitudes, can substantially affect fire and other peril loss protection and prevention. The complexities of the many trade skills involved in the construction and repair of vessels also affect fire protection planning. However, it is essential to note that welding and steel erection techniques are broadly identified as open flame processes, and any loss prevention and protection program must first begin with this fact.

Shipbuilding has a considerable impact on the economy. The products range from small private hulls to huge passenger and cargo vessels more than 1000 ft (305 m) long and over 400,000 tons (1 132 674 m^3) displacement. Further, cargo hulls are fabricated for many specific uses: dry cargo vessels for bulk product transportation, a variety of tankers to deliver the world's energy products and other liquids, vessels to move

---

John W. Van Brunt (Retired) is the former Chief of the Shipyard Inspection Division for United States Salvage Association, Inc., New York City.

Chapter 21

FIGURE 21.1 Million of dollars' worth of equipment and materials are required to build a single ship. Here huge cranes lower the superstructure of a ship to the deck. (Quincy Shipbuilding Div., General Dynamics).

cargos in sealed specialty trailer and barge-type loading containers, etc. It should also be noted that each vessel constructed is generally a unique design for a designated purpose, and, in each construction program, planned fire loss prevention and protection must be designed to adequately complement the specific hull. Therefore, it can be readily understood that complex fire loss prevention analysis is required on an initial and continuing basis.

Product fabrication in the shipbuilding industry is extensive. Considering the broad area of its operations (wherever navigable waters are available and the variety of vessels produced, the total product value can reach hundreds of millions of dollars annually. In addition, like the automobile industry, repairs and renovation are continuing activities, and the ship repair segment contributes much to the total income of the shipyard industry. What is notable here is that many facilities engage in both new vessel construction and repair. While these operations are basically similar with regard to material and operational equipment usage each has specific fire perils that must be recognized. Building a vessel from keel to delivery can take months or even years to complete. Repair work is usually shorter in duration, and the various ship services remain operational. It is, therefore, obvious that fire protection and prevention must be planned differently in each instance. In the case of a new construction, a degree of permanency with regard to installing fire protection equipment and prevention procedures is possible, but for a vessel in repair permanent planning is generally not feasible. In addition, concern for other and, at times, incompatible shipboard operational procedures may continue during a repair lay-up.

The number of employees differs in each facility and can range from fewer than 100 to as many as 30,000. Also, many support services and components are fabricated elsewhere and delivered to the installation site. The impact on the steel industry, boiler and engine manufacturers, electronic component producers, paint suppliers, and many other vendors of goods essential to shipyard production is considerable. The impact is also substantial on the transportation industry, considering the volume of materials moved by every means of transportation to support a shipyard complex.

Historically, the shipyard industry has always been vulnerable to catastrophic fire occurrence. Initially, wooden vessels with hot pitch watertight sealant were obvious open flame hazards. When rivets became essential to steel construction, they had to be prepared by open forge activity, another extensive fire hazard. Then improved technology yielded full weld methods, but the hazards of open flame and heavy electrical energy production methods persist.

The very high frequency and severity of the fire loss record has been a serious concern to those who operate shipyards. It should be noted, however, that in recent years the severity and frequency of fires has been significantly reduced, and this is directly related to the joint efforts of management, labor leadership, and government (OSHA), as well as loss control analysis and engineering by management staffs, outside consultants, and insurance carriers. While fire remains a constant threat to property and life, the cooperative vigilance of those concerned has shown significant positive results.

Numerous fire hazards threaten shipyards. Many hazards result from combining noncompatible processes near each other. For example, a woodworking process under the same roof with metal-working activity requiring open flame or electricity in welding is dangerous. If these are not separated by fire stops or distance, the wood dust could accumulate on the welding electrodes and produce short-circuit electric arc or, by the concentration of flammable dust particles in the atmosphere, ignite with the lighting of the welding torch. The necessity of mixing, storing, and moving large quantities of paint and other finishing materials presents another formidable fire hazard. Modern solvents and paint formulas are, in many cases, highly unstable, toxic, subject to heavier-than-atmosphere properties, flowable, and highly volatile. (See Chapter 37 for a discussion of "Handling and Storage of Flammable and Combustible Liquids.") Insulating materials and adhesives are generally highly flammable and must be seriously considered in fire loss protection and prevention procedures. By controlling the handling of flammable components and establishing good loss prevention and protection procedures, occurrences can be minimized. Again, the key fire hazard is the open flame, and positive methods must be employed in order to prevent necessary, incompatible processes from mixing.

A good firesafety combat program in a shipyard will always need to be revised. That which is highly satisfactory for one scheduled construction program may be unsatisfactory for a subsequent construction schedule. Each program schedule change requires a study to determine the necessary maximum loss prevention and protection procedures. In the following text of this chapter, general and specific guidance is provided to accomplish this.

PHYSICAL PROPERTIES OF A SHIPYARD

There is no uniform standard construction arrangement for a shipyard complex. Differences in acreage, topography, operational procedures, production methods, geographic location, and numerous other considerations establish the physical structure of a shipyard. The facility may be located on a large area inland of a wide body of water or on a narrow plot along a limited width, flowing river. The management may commit to building or repairing small specialty hulls or large tonnage, oceangoing vessels. In some instances only new hull construction is produced, while in others only vessel repair is contracted for. In still others, new construction and repair is accepted, some commingled and some maintained at separate sections of the facility.

The operational purpose of a shipyard is to fabricate, repair, and deliver a vessel. This requires that methods be developed to construct the vessel, launch and outfit it or, in the case of repair work, safely secure the hull until it is returned to its owner. Considering the size and tonnage of the product, there are certain essential structural units peculiar to this industry.

In construction, the hull must be made watertight and buoyant. This is generally accomplished at positions designated as *ways, construction basins* or, many times, in the case of small vessels, a *marine railway*. Each has a specific conformity.

Ways

The vessel is constructed on a declining platform that is properly supported as the shape is erected. When it is ready for launch, the vessel is balanced on greased skids or placed in launching cradles and, by various trigger releases, moved into the launching basin or stream. It should be noted that *ways* of this nature can be of two different types: *end launching* or *side launching*. In end launching the vessel slides down an incline end to end, usually stern first. In side launching the vessel is committed to the water by a slide or drop. Generally, shipyards situated on

SHIPYARDS

FIGURE 21.2. *LNG tanker being guided out of the construction basin. (Quincy Shipbuilding Div., General Dynamics)*

narrow streams use side launch techniques while those having access to wide bodies of water use end launch techniques.

Construction Basin

This is a fixed position that is excavated, bulkheaded, and equipped with a coffer dam, or watertight gate, to dam out the waterway. The vessel is constructed in the basin on proper support blocks, and, when it is ready to float, the water is pumped into the basin until an equal depth with the waterway is established. The barrier gate is then removed, and the vessel floats out.

Marine Railway

This method, used only for small hulls, utilizes a support platform set on multiple wheel trucks on set tracks on a decline from dry land inboard to an outboard depth sufficient to enable the vessel to float. Usually the vessel is completed on level land, placed in a support cradle, skidded aboard the marine railway platform, and committed to the waterway. The marine railway is usually controlled by cables connected to power winches for a fully controlled launch.

Piers and Bulkheads

The necessity for a place to secure the vessel to land access is obvious. Depending on the individual site, securing and access structures consist of piers extended into the waterway or marginal bulkheads parallel to the waterway. These structures may be of driven piling and decked over open water or land-filled bulkheaded structures with solid decking. Included are the necessary facilities to properly secure a vessel to the structure. Fire protection and prevention procedures will vary based on type of construction, materials, and usage.

Other Buildings

The needs of the particular operation dictate the buildings' use and their distribution over the facility. Further, the objective use of a building for the purpose for which it was originally designed is a concern. If a facility has existed for many years, the possibility of converting a structure to a noncompatible occupancy is quite possible. There are instances where structures built and equipped for warehousing have been converted to trade fabrication occupancy without consideration of fire protection and prevention facilities and services. The clearances between buildings, combustibility of the structure, internal protection, such as sprinklers or the lack of them, and proximity of structures housing incompatible processes are essential considerations. Placing the paint mixing shop near an open flame production line could prove disastrous. Oxygen and pressure gas cylinders directly adjacent to processes that could damage them could also cause a major catastrophe.

RAW MATERIALS

The principal raw materials of a shipyard are the structural steel, aluminum, or wood necessary to construct a hull. Handling these bulk materials requires large open areas and adequate cranes, conveyors, and other equipment to efficiently move the material. In addition to bulk material, a variety of components from small nuts and bolts to complex fabricated electronic equipment can be found in warehouses. Turbines, boilers, motors, and other essential equipment are generally received ready for installation or packaged for assembly. Careful storage and handling procedures, which vary with the size of a given shipyard, and an emphasis on firesafety are essential in view of the unusually high value per unit and vulnerable nature of the assembled product.

SHIPYARDS

FIGURE 21.3. Open storage of steel plates. Conveyors with electromagnets are used to move steel into fabrication areas. (Quincy Shipbuilding Div., General Dynamics

FIGURE 21.4. Module sections being moved from fabrication area into open storage. (Quincy Shipbuilding Div., General Dynamics)

The largest of the raw materials is the hull structural material. With the exception of wood, the fact that these materials are of low fire support in nature and usually subject to open air storage of a limited exposure nature is prominent. Many shipyards fabricate module sections in the open and store them until required in the erection of the vessel. These sections are usually of limited fire support in nature, also. As far as fire services are concerned, the need to maintain open access and thereby avoid congestion is important.

Boilers, turbines, and other machinery components are generally prepared in buildings equipped with utility services and movement devices permitting orderly assembly. Occasionally this work is accomplished in the open or under a shed arrangement. The fire peril, the flammable nature of temporary weather coverings, and the susceptibility of the particular unit should be considered.

General stores in warehouses should be considered with respect to compatibility. In a general storage situation, including "Red Label," (flammable) liquids in the same area as nonsusceptible general materials should be avoided. (See Chapter 37, "Handling and Storage of Flammable and Combustible Liquids.") These flammable or unstable products should be separately stored in structures specifically constructed in accordance with the standards for safe containment. Likewise, items having toxic as well as extreme fire support properties should be separated. For example, polyurethane insulation materials, when exposed to flame, emit toxic fumes. Obviously, storing these materials near lumber and plywood is not a safe practice.

While paints and other finishing materials in sealed containers may not be overly susceptible to fire, the opposite is true for solvents. Good handling procedures dictate that all finishes be kept separate from general stores and stored in containment structures specifically built for the purpose of such storage. Of essential importance for raw materials storage is the need for adequate ventilation and containment of these flammable or unstable materials to avoid the activation of their inherent hazardous qualities and the resultant fire or explosion.

Efficient movement of materials in any shipyard is essential to successful operations. Proper handling and storage dictates that bulk supplies be stored in well-arranged noncombustible bins and vertical racks in order to maintain clear access. (See Chapter 48, "Industrial Storage Practices.") For bulk items, properly identified separate storage bays should be provided to allow clear access for handling equipment. The use of wooden pallets keeps products off the floor and provides ease of handling for motorized equipment. The use of motor-powered lift and movement vehicles (fork lifts) introduces an additional fire hazard. Where gasoline or pressurized cylinder gas (propane) is concerned, all filling should be done outside of the warehouse area and reserve cylinders should be stored away from the warehouse facility. If the handling vehicle is electrically powered, any material components stored in recharging bays should not be exposed.

Since materials handling and distribution is so varied, a singular fire protection and prevention program is not feasible. In general, planning should follow accepted standards for safe practices, including the necessary portable fire extinguishers and good maintenance on all handling equipment. Properly classified and maintained cranes and movement vehicles should be used. A deviation could eventually result in a fire-causing accident.

THE PRODUCTION PROCESS

Basically, the production processes of vessel construction are not similar to those of vessel repair. There are many common areas relative to fire protection and prevention services; however, it is necessary to describe each procedure in general terms to qualify the specific differences affecting firesafety.

Vessel Construction

Since this procedure involves the complete fabrication of a vessel, architectural and engineering plans and specifications are essential. The safe storage of these plans in vaults or fire-resistive storage cabinets is necessary. Duplicates of all pertinent data should be safely deposited away from the premises.

When all plans and specifications are agreed upon, and materials, components, and required equipment secured, subsection working drawings and specifications are distributed to each functional division, and the fabrication of the vessel will begin. Operational methods vary with each facility and craft and, as a consequence, fire protection and prevention services should be tailored to the situation. The following description, in general terms, explains the construction sequence with a reminder that variations most certainly will occur, depending on the facility and the vessel.

The erection site is prepared on a ways, in a building basin, or on a solidly stabilized construction land site, and the keel is laid. In some instances, particularly for small hulls, a form jig is fabricated, and the hull is constructed by laying on the steel plate welding seams. Then the hull is rolled over using crane rigging or hydraulic ring frames. In practically all large hull construction, however, the procedure begins with a keel, the primary longitudinal structure. The steel plates are shaped in accordance with the specifications either by computerized automatic cutting and shaping equipment or plywood patterns. Wooden templates — in most facilities certain plate shaping operations require template patterns — are produced by qualified personnel who prepare them in a *mold loft*. The mold loft is a large, level, wood-floored area specifically used for this purpose. Because of space requirements, it is normally located on an upper enclosed level of a large building in the shipyard complex.

The shaped plates are delivered to the erection site or to initial fabrication facilities designated as *platens*. In modern shipyard construction operations, modular fabrication, rather than singular plate preparation, is usually accomplished at the platens. It is in this stage that cross bracings (stiffeners) are welded to prepared plates, multiples of prepared sections are joined, and shaping of the subassembly occurs. Notable here is the obvious need for open flame and electric arc welding functions utilizing gas, air, oxygen, and electric power of sufficient voltage to operate the welding generators. (See Chapter 24, "Welding and Cutting.") A completed section is moved directly to the vessel in the sequence specified by the construction plant, or the section is temporarily stored before moving.

Various methods are used to transport sections from initial fabrication to erection, depending on the size of the section and the shipyard components. Some large operations skid or lift the sections aboard large flatbed transporters and then move them into position for crane lifts. Other operations exclusively use crane handling. The choice is dictated by the type of moving equipment, proximity to erection position, topography, and available space.

It should be noted that the stage during which plate finishing (cleaning and painting) is accomplished also varies with each facility. In several large complexes the steel is automatically shot or sandblasted, cleaned, and painted by an assembly line method prior to shaping or preassembly. In some shipyards sections are transported to specially constructed cleaning (blasting) and painting enclosures following module fabrication, and in others the work is accomplished in an open unexposed area reserved for the purpose.

While the vessel is taking shape at the erection site, it will be subject to the planned installation of components other than those associated with the hull at various stages of the fabrication. Since the vessel will be committed to flotation, all underwater components must be installed while the bottom is exposed in a dry state. This means that the propellers, rudders, shaft, sealant bearings, bow thrusters, etc., must be secured in this stage. Further, before the decks and some of the bulkheading is installed, the engine room and most of the operational machinery units must be installed. Many shipyards, following the modular erection concept, install service facilities, electrical conduits, service piping, and similar units prior to committing the modular section to the whole vessel. This permits immediate connection of all sectional services to the abutting module.

As the erection program proceeds, trade shops throughout the facility prepare their specific contributions to the whole vessel. The electric shop prepares all electrical units, including the subassembly of all service elements, control panels, and communications equipment, and directs each segment to the vessel as required in the master schedule. The pipe and metals shops fabricate piping, ventilation ducts, and other interior components.

Required machine work is committed to the machine shop, and

propulsion components (boilers, turbines, engines, etc.) are also separately prepared. These components are arranged to reach the vessel in the designed sequence. Each trade functions independently under the supervision of a project director. The support personnel, transport and riggers for handling material, and warehouse personnel for handling components are also committed to logistic compliance in order to avoid delay.

At a stage designated in the schedule, the vessel is either launched from the ways or floated out of the graving basin, depending, of course, on the particular facility. As discussed earlier, in end launching the vessel slides on greased skids or in a series of cradles on greased skids into the water end to end, usually stern first. A declivity, inboard to outboard, allows the vessel to launch by gravity. A trigger mechanism releases the vessel from its restraints and permits it to slide. Restraining lines or tugboats are used to retrieve the launched vessel.

Side launching, while similar, does have a distinction of its own. The vessel in position on a side launch ways rests in a series of cradles on greased skids and is restrained by cables attached to trigger releasing units. On release the vessel enters the water sideways and floats from under the cradles which are recovered. The vessel is then secured in the same manner indicated for end launching. In some side launching operations, the vessel is directly skidded into the water without cradles, or the ways tip by counterbalance and drop the vessel into the water.

Marine railways, as indicated previously, are generally used to commit small craft to flotation. The hull is either skidded aboard the railway platform or lifted by sling or crane and set on blocks on the platform. The platform is released to the water by controlled cable connected to power winches.

The incomplete vessel, now afloat, is secured to an outfitting position at a pier or bulkhead, and construction is completed. In that the balance of the construction is basically enclosed, with all trades involved working in confined areas, the fire hazard increases to critical proportions. The need for each trade to know what the other is doing at all times is very important. Incompatible trade action (e.g., spray painting during open flame welding) must be avoided without exception. All tools and equipment brought aboard should have no defects.

As the vessel nears completion, various motors, engines, generators, etc., are activated, climaxed by the initial activation of the propulsion components. This is a critical time, and all other activity is generally suspended during the initial start-up. Dock trials follow. After all structuring is complete, the vessel is committed to sea trials and then delivery.

Vessel Repair

In most instances a vessel is secured to a wet dock (pier or bulkhead). Fuel may have to be off-loaded and the tanks cleaned (gas freed), depending, of course, on the work to be accomplished. Usually, the contracted

FIGURE 21.5. *Vessel repairs being made in floating dry docks.*

work requires much removal from inside the hull; open flame disciplines become pronounced. Fire security surveillance is essential.

In the event a vessel requires underwater hull or component repair, dry docking is necessary. When dry docked, all fire services should be supplemented in order to maintain constancy.

THE FIRE HAZARDS

Earlier, reference was made to open flame processes in this industry. The fire hazard is always considered the principal peril exposure, and determining this exposure in all segments of the operation is essential. Further, the program development for fire loss prevention and protection is paramount. To establish this program, a thorough understanding of the fire hazards is imperative. An attempt to specify all of the contributing conditions, i.e., structure, procedure, processing, work attitudes, etc., would be almost impossible; however, in order to facilitate the recognition of fire hazards in the varied exposure conditions, they are grouped in the following manner: ways, piers and bulkheads, and buildings.

Ways

The materials used in the construction of shipyard ways determine the fire hazards and the procedures needed to safeguard against them, e.g., a totally wooden structure is much more vulnerable than a fully fire-resistive one. The type of activity accomplished on ways, erection of the basic hull with its interior components, is singular. Recognizing the fire hazards and designing methods and procedures to negate them is imperative.

Where the ways are totally combustible (wood), exposure to open flame and electric arc welding procedures is a hazard. Methods of providing gas, air, oxygen, and electric power require installation and monitoring according to accepted safety standards. Extremely vigilant supervision of the welding processes is necessary in order to avoid igniting the structure by molten welding residue or open flame (see Chapter 24, "Welding and Cutting"). The use of nonflammable coverings over burnable structural members directly exposed to the welding processes is desirable. Under no circumstances should any material storage or occupied structure, temporary or permanent, be permitted below the vessel being constructed on the ways. Further, excellent housekeeping on and below the ways structure, and aboard the vessel as it is formed, is of primary importance. An accumulation of flammable trash and debris can lead to a catastrophic fire.

Portable fire extinguishers and standpipes and hose on an adequate water main supply are constantly necessary, but where flammable structured ways are concerned, they are vital. It is also desirable to include substructure automatic sprinklers and periodic vertical draft stops of noncombustible materials transverse of the structure from the underside of the deck to the ground level. Parallel access to each side is desirable, and congestion should be eliminated. Most of these specific comments with regard to ways are also applicable to wood-structured construction graving basins. The same vigilance is required due to the flammable nature of the basic structure. It should be noted, however, that graving basins do not have open exposed substructures.

Since welding and burning (hot work) are the principal hazards, attention to good standard practices is required. Gas, oxygen, and air are provided under considerable pressure, and in the event of injury or accident, uncontrolled flow could result in a serious fire, hard to extinguish. Cutoff of flow procedure is extremely important. Further, if the flow is delivered through a fixed piping system to manifold takeoffs, each product connecting coupling should be incompatible with the coupling of any other in order to avoid an accidental connection error. If these products are provided in cylinders, singularly or collectively grouped, they should be restrained so that they remain upright.

Where welding generators are required, supply cables to the generators should be protected and without defect. All generators, whether individual or in group concentration, should be secured to avoid accidental movement and should have sufficient air flow around them to avoid overheating. All connections for cable takeoff leads should be of a safety type and without current leakage.

It is essential to hot work hazard control that generators, gas manifolds, and cylinders be located only on the open deck or overside of a vessel. Lead hoses and cables can be taken from these supply points to the welders torch or welding electrode under deck or in other enclosed areas. The disciplines specified in applicable standards should be rigidly enforced. In enclosed areas a competent fire watch should be provided. Welding torches should be removed to open areas on deck or overside during meal breaks and disconnected at shift changes. Since there necessarily will be a considerable number of hose and cable leads in use, they should be treed up above the deck in order to avoid injury.

Incompatible trades functioning in the same area should be avoided. Under no circumstances should welding or any open flame work be done while painting, solvent cleaning, insulating, woodworking, or any other activity that could support or contribute to fire is in progress. When enclosed painting, or any other operation that produces flammable or unstable vapor, is finished, the area should be inspected and tested for safety vapor levels by a marine chemist prior to permitting the resumption of open flame or electric welding processes.

In areas where painting and installing insulation using low flash point adhesives is done, all electric lighting, temporary and permanent, should be explosion-proof and subject to vigilant monitoring by competent electricians. With regard to all electrical installations, temporary and permanent, no one in any other trade should be allowed to alter, adjust, or extend any service.

In shipyards where heat is required during cold weather months, it should be provided by steam, hot water electricity, or another approved safety method. Wood kindling fires and other makeshift arrangements should be prohibited aboard the vessel and anywhere else in the facility.

Large concentrations of paints, solvents, adhesives, lumber, and other materials of varying flammable nature contribute to excessive fire hazard. Limiting the flow of such materials aboard or closely adjacent to immediate use will reduce the hazard factor. It is important to reduce congestion and access blocking by avoiding stockpiling and storage in work areas.

In many instances the need for temporary housing aboard exists. Housing should be of noncombustible fabrication, have safe utility installation, and be limited in number. If an exceptional fire hazard exists because of temporary housing aboard, this occupancy should be prohibited until it can be minimized by qualified protection and prevention procedures.

Housekeeping hazards are notable in any construction operation. A regular and frequent procedure for the removal of accumulated waste and debris is desirable. Crating and flammable packing materials brought aboard should be removed from the vessel and ways area immediately. Waste collection containers fabricated of noncombustible materials only are suitable. Smoking should be prohibited where fire hazards exist.

Maintenance and constant monitoring of all equipment is essential. All flame-producing or electrical function equipment should be subject to extreme vigilance. Gas, air, oxygen, and electric supply lines and leads should be examined and tested regularly for any leakage. Damaged units should be subject to immediate removal.

The flow of materials to the building ways presents unique re-

FIGURE 21.6. Because of the complexity and size of shipbuilding operations, care must be taken that proper housekeeping and maintenance procedures are observed. With scaffolding surrounding the hull, and materials and equipment cluttering the deck, quick access to the ship for fire fighting purposes is difficult. (Quincy Shipbuilding Div., General Dynamics)

quirements in order to schedule the orderly fabrication of the vessel. Limiting material aboard to current need is essential and requires supply planning and varied delivery methods. Cranes for handling should be properly powered and rigged with capacity as documented. Vehicles, including transporters, fork lifts, etc., should be properly maintained and efficiently operated by qualified personnel. If a motor vehicle of any kind is required aboard, it should be safety equipped with spark arrestors, toxic fume converters, or other suppressions in order to avoid excessive fire hazard.

When a vessel is prepared for launch, all flame-producing trade work should cease, all flammable, volatile, and other material not required for the launch should be removed, the skids should be properly greased, and the triggers installed. During the launch movement, first aid fire protection equipment should be immediately available, and personnel should be prepared to use it in the event of an occurrence of fire. Immediately following the launch, skid grease should be neutralized by removal.

When vessel construction is accomplished in a graving basin, the fire hazards and procedures to overcome them are substantially the same as those for a vessel on ways. It should be understood, however, that the graving basin is an inground, rather than aboveground (as with ways), structure with singular considerations. A pumping system is needed to water and dewater the basin. These pumps require constant monitoring and maintenance to avoid accidental water exposure or operational failure. Access from the parapet or apron surrounding the basin to the floor of the basin is essential.

The nature of the sunken structure limits access and requires constant housekeeping under and around the hull in construction. If space permits needed temporary structures, they should be noncombustible with utility services installed according to accepted safety standards. Gas manifolds and welding generator concentrations should be kept in clear space. Service lines to manifolds and generators, temporary or permanent, should be secured to avoid injury. Necessary first aid fire protection equipment, including onboard charged water lines to temporary pipe hydrants equipped with hoses, should be installed. High-level cranes, their housing and machinery, should be of noncombustible construction and conform to all established standards.

Piers and Bulkheads

The fire hazards associated with piers and bulkheads are similar in nature to those associated with ways and basins with some notable exceptions. If the pier or bulkhead is open on piles over water, vigilance with regard to the substructure is essential. If it is total wood construction (piles, stringers, support girders, and deck), substructure sprinklers and vertical draft stops are desirable. If it is fully noncombustible construction (reinforced concrete steel piles, support girders, and decking), the need for substructure sprinklers and draft stops is minimized. Vigilance with regard to flammable floating debris collecting under any open pier structures is a continuing concern. The permanent installation of utility services (air, oxygen, gas, electricity, steam, and water) along the structure, properly situated to avoid injury, is paramount. All service connections (manifolds, standpipes, and electrical) at various fixed positions should be accessible and identified. A color code system for all product takeoffs is desirable, and all connecting couplings should be different, thereby making a connection error impossible. The entire length of the structure should provide for fire access lanes, and open deck storage should be kept to a minimum. Master shutoff valves and switches at pier heads are desirable for all utility services. There should be hatches in the deck, properly identified and kept clear, for emergency access to the substructure. Provisions should be made for pipe nozzle openings at intervals in the deck. Where there are fixed rail installations for moving cranes, sleeve wells under the tracks for welding cable and hose lines are essential. Limiting the amount of paints, solvents, and other flammables along the pier and aboard the vessel secured to it is also essential.

Aboard the vessel it is absolutely essential that NFPA, OSHA, and specific local safety codes are complied with. No manifolds or generators should be permitted below the main deck. All lead lines off decks should be secured in trees. Fire watches, where required, and flammable vapor vigilance and certification before open flame or other hot work commences are absolutely necessary. Incompatible trade functions in the same area should be completely avoided. Housekeeping discipline, using nonflammable waste receiving containers, is very important. Security vigilance should be extended to onboard observation when access to the vessel is established.

As a vessel nears completion and operational services are activated, standards regarding fuel loading should be integrated into the firesafety program. Engine and other machinery testing should be vigilantly monitored, and all firesafety standards should be implemented. Proper exhaust ventilation should be maintained where required.

Further, it is desirable to provide for vessel access from the outboard side with extension ladders from the deck or access to a level whereby personnel could board from an attending vessel alongside. A plan to undock the vessel in emergency and control it by support vessels is desirable in each instance.

Buildings

There are many areas where components essential to the building of a vessel may be concentrated in various structures on the premises. Fire hazards in the structures are varied and directly related to the trade function involved.

Shipyard buildings should be of substantial masonry or other rated fire-resistive or noncombustible construction with standard enclosed floor openings, if any, and without concealed spaces. As much clear space as possible should be provided between buildings, and noncompatible occupancies should not be exposed to each other, inside or outside of the structure. Single occupancy by any trade is desired. High hazard functions (woodworking, paint mixing, etc.) should be confined in a single occupancy and, if possible, isolated from other production functions. The various trades and technical operations (pipe fabrication, electrical, riggers, etc.) should conform to the good practices outlined in the applicable standards. Buildings housing any trade should be structurally compatible with the purpose, and all services should be provided in accordance with standards. In instances where open flame work or electric welding is done, utility services should be properly installed and subject to continuous maintenance and service monitoring. Concentrations of flammables and volatiles should be isolated, and buildings should be provided with vaporproof electric circuitry and fume ventilation equipment. Warehouses, particularly those housing highly valued machinery, turbines, electronic control panels, and other hull-outfitting materials, should be used only for storage purposes and the components

Chapter 21

FIGURE 21.7. *A good example of noncombustible construction and large open areas in a fabrication shop. (Quincy Shipbuilding Div., General Dynamics)*

retained well-distributed therein. Flammable shipping containers should be removed, if practical, to avoid fire fuel. Good warehousing practices, such as spacing, maintaining accessibility, and avoiding congestion, are necessary. Housekeeping, as in all areas, must be vigorously pursued and rigidly enforced. (Chapters 48 and 50 discuss Storage Practices and Housekeeping.) Any flammable liquids required for lubrication or other functions should be limited and kept in proper safety containers. All materials-transporting equipment is to be fueled, operated, and maintained in accordance with the specific requirement for each unit. Specifically, motor-powered units, such as gasoline, propane, or electrically fueled forklifts, should be filled or recharged according to accepted safety standards. Fuel filling and propane tank charging and recharging should be done in the open, away from any operational structure. Recharging positions for electrically fueled vehicles should be located away from other materials or functions in a facility building.

Where automated steel plate conditioning and finishing operations are established, proper arrangements for ventilation should be provided. All electrical equipment should be grounded and installed according to vaporproof standards, and the facility should be a single occupancy. Open access to all areas of the shipyard building complex is essential, and congested blockage should be avoided.

FIGURE 21.8. *Automated welding techniques have reduced fire hazards in metal-fabricating shops and have made it possible to do most welding in protected areas. (Quincy Shipbuilding Div., General Dynamics)*

General

Staging should consist of noncombustible uprights such as fabricated pipe sections and fire retardant wood plank horizontals. If fire retardant wood is used, periodical testing to determine its retained fire retardant quality is suggested. Nonfire resistant staging should be kept to a minimum and removed immediately after its use. Temporary or portable buildings under ways, in graving basins, on decks, or closely adjacent and exposed to a vessel in construction should be of noncombustible construction with utility services installed according to accepted safety requirements. Where floating dry docks are involved, the fire hazard remedies indicated herein apply wherever appropriate. Should the dry dock be of wood construction, good firesafety is essential. In any situation where the possibility of electric spark or arc exists, proper grounds are required, including dry docks.

Vessel Repair Operations

Operations of this type follow the fire hazard controls associated with a particular trade function. In some areas, however, more pronounced situations are presented. Freeing fuel tanks and fuel and vapor lines before hot work is necessary for many repair operations. The services of a marine chemist to certify a vessel as gas-free and the maintenance of vapor testing are essential. Cutting into any vessel compartment with open flame equipment requires prior inspection of the other side of the bulk head in order to avoid burning into flammable materials. Maintaining the vessel's fire protection system in full operation or providing a supplemental temporary facility should be arranged before repairs commence. The same housekeeping, access, and congestion requirements are involved in repair operations.

THE SAFEGUARDS

The fire protection requirements of a shipyard will vary from very limited (for small facilities) to highly complex (for large ones). The extent and complexity of a property as well as the degree of publicly provided fire protection will affect private fire protection needs.

The shipyard, including ways and fitting-out areas, should be protected by an adequate layout of fire hydrants and hose houses on adequate sized, looped mains on an ample water supply. The requirements for each yard should be based upon area, construction, topography, available public fire department protection, etc. NFPA standards should prevail. Hydrant hose houses are to be equipped with adequate lengths of 1½-in. (38.1 mm) hose, and reducer connections to 1½-in. (38.1 mm) hydrant couplings are to be provided.

Vessels

The ground area below vessels on the ways should be equipped with an ample supply of properly rated hand extinguishers, protected against freezing where necessary, to cope with incipient fires. An ample supply of approved first aid fire extinguishers equivalent for Class ABC fires should be distributed throughout the vessel. Portable CO_2 or dry chemical extinguishers should be provided for use in machinery spaces or areas where solvents or other hazardous materials of a similar nature are exposed.

When practicable, vessels at the ways should be equipped with valved 1½-in. (38.1 mm) fire hose connections 200 ft (61 m) apart at main deck level, each with 200 ft (61 m) of 1½-in. (38.1 mm) c.r.l. hose to be supplied by the shore lines overside. Hose at the fire stations should be properly racked or boxed, use for any other purpose prohibited, and this protection should be maintained throughout the entire outfitting period. Small hand hose lines are recommended for use in the engine room, during construction of "reefer" spaces, or during installation of insulating materials.

Buildings

There should be an adequate supply of approved fire extinguishers and other first aid fire equipment distributed on the basis of one unit per 2500 sq ft (232 m^2) of floor space or fraction thereof.

It is desirable that all buildings of combustible construction and any building with combustible contents be equipped with an approved automatic sprinkler system. Tall, large open area buildings may prove to be ineffectively protected by sprinklers. The installation of adequately designed water curtain deluge systems to provide fire divisions may be more effective.

Fire lanes, to be kept accessible at all times, should be maintained in all shipyard areas and structures.

Fire Brigades

There should be a private fire brigade under the supervision of qualified officers. This may be a full-time salaried fire department or a brigade composed of shipyard employees with a nucleus formed by the watch staff. (See Chapter 3, "Plant Emergency Training and Organization," for a complete discussion of fire brigades.) The size of the brigade, extent of equipment, etc., will be governed by the size of the yard and should be organized in accordance with NFPA 27, *Private Fire Brigades*. In addition, close liaison should be established with the public fire department which should be encouraged to periodically hold familiarization drills within the entire facility.

There should be a centrally located fire station unexposed by, or protected against exposure from, other plant units. Equipment therein will be governed by the size of the yard but should include a supply of extra hose, spare fire extinguishers, emergency lighting, gas masks, and other appurtenances. Also, foam-generating equipment and supplies should be maintained if conditions warrant.

Fire Alarms

A proprietary fire alarm system operating through a shipyard central station is desirable. Call boxes on the system should be strategically located throughout the shipyard (ways, fitting-out, and vessels). A general fire alarm signal of sufficient intensity to be heard in all parts of the yard should be provided.

Where telephone alarms are used, an open circuit maintained specifically for emergency purposes should be connected to guard headquarters or another central station attended at all times. In any case, there should be a direct connection with the public fire department.

These permanent fire alarm measures should be supplemented by an additional method of informing all shipworkers of a fire or other emergency. An additional method is important because the necessary alarms may not yet be installed, may be temporarily out of service, or may require an auxiliary system. A simple workable signal is to provide a means to "blink the lights" throughout the vessel. The blinking lights will inform the workers that an emergency exists. Even those unfamiliar with the meaning of the signal will be alerted.

Watch Service

A watch force with suitable headquarters, supervised by a chief of guards, should be maintained for the protection of shipyard property. The size of the staff will be determined by the extent of the facilities but should nevertheless be adequate for the coverage of access gates, yard areas, buildings, and vessels under construction. When unattended by workers, main buildings, ways, and fitting-out areas should be patrolled by guards carrying approved portable watch clocks and reporting regularly to fixed key stations. In lieu of watch clocks, an electronic recording method of the tour is acceptable. When practicable, vessels on the ways, when unattended by workers, should be covered by watch clock patrols, preferably on an hourly basis. Vessels in an advanced stage of completion on the ways and moored at fitting-out areas should have a guard and/or fire patrol on duty whenever unattended by workers.

When guards form the nucleus of the fire brigade, or when there is a full-time salaried fire department, personnel should be familiar with the location and use of all fire fighting equipment. Periodic familiarization drills should be required and records of drills maintained. (See Chapter 3, "Plant Emergency Training and Organization.")

Guards should be responsible for the distribution, inspection, maintenance, and inventory of first aid fire fighting equipment in the yard and on the vessels. They should also be responsible for the observance of rules by employees, aboard ship and elsewhere in the yard, including those pertaining to housekeeping, smoking, and hazardous practices. Infractions should be reported daily, in writing, to the chief, and records kept. Periodic fire patrols should be maintained during working shifts on all vessels under construction but should be full time on vessels in an advanced stage of completion, and a detailed inspection of each vessel should be made on each change of shift.

Hydrants, hose houses, and sprinkler control valves should be inspected not less than once weekly and the conditions logged. However, the condition of fire equipment on vessels under construction should be inspected on a daily basis.

With repeated review of fire loss prevention and protection procedures throughout the shipyard, fire occurrences can be generally minimized. Relaxing of vigilance will almost certainly result in frequent and sometimes severe fire damages.

BIBLIOGRAPHY

REFERENCES

"Recommended Fire Protection and Fire Prevention Practices for Major Shipyards Engaged in New Vessel Construction," revised internal publication, May 31, 1977, American Hull Insurance Syndicate, New York, NY.
Vlachos, C. A., *The Normal Loss Expectancy*, edited by John W. Van Brunt, Vlachos & Co., 1935, and unpublished revision 1975.
Bailey, Michael S., *LNG By Sea*.
"Safety Engineering Standards," 1973, Crum and Foster Insurance Companies.
"Federal Register," January 26, 1976.
"Maritime Employment," June 19, 1974.

NFPA CODES, STANDARDS, AND RECOMMENDED PRACTICES

Reference to the following NFPA Codes, Standards, and Recommended Practices will provide further information on the safeguards for shipyards discussed in this chapter. (See the latest *NFPA Codes and Standards Catalog* for availability of current editions of the following documents.)

NFPA 10, *Portable Fire Extinguishers*.
NFPA 13, *Installation of Sprinkler Systems*.
NFPA 14, *Standpipe and Hose Systems*.
NFPA 15, *Water Spray Fixed Systems*.
NFPA 24, *Installation of Private Fire Service Mains and Their Appurtenances*.
NFPA 27, *Private Fire Brigades*.
NFPA 51, *Oxygen-Fuel Gas Systems for Welding and Cutting*.
NFPA 51B, *Fire Prevention in Use of Cutting and Welding Processes*.
NFPA 69, *Explosion Prevention Systems*.
NFPA 70, *National Electrical Code*.
NFPA 72D, *Proprietary Protective Signaling Systems*.
NFPA 87, *Construction and Protection, Piers and Wharves*.
NFPA 101, *Life Safety Code*.
NFPA 306, *Control of Gas Hazards on Vessels to be Repaired*.
NFPA 312, *Fire Protection of Vessels During Construction, Repair, and Lay-up*.
NFPA 601, *Guard Service in Fire Loss*.
NFPA 601A, *Guard Operations in Fire Loss Prevention*.

ADDITIONAL READING

Bartels, A. L., "Safeguards to Prevent Ignition by Hot Surfaces," *Electrical Review London*, vol. 203, no. 4 (1978), pp. 29-31.

Berg, A., "Fire-Resistant Walls for Use in Shipbuilding," U.S. Patent No. 4,027,444.

Buehrer, P., "Check List for the Prevention of Fires Arising from Welding and Allied Processes," *Welding in the World*, vol. 14, no. 5-6 (1976), pp. 122-125.

Magison, E. C., *Electrical Instruments in Hazardous Locations*, 3rd ed., Instrument Society of America, Pittsburgh, PA, 1978.

McKinnon, G. P., ed., *Fire Protection Handbook*, Fifteenth Edition, National Fire Protection Association, Quincy, MA, 1981.

"Prevent Cutting/Welding Fires," *The Minnesota Fire Chief*, vol. 16, no. 5 (May/June 1980), pp. 16, 48, 54.

Safety and Health in Shipbuilding and Ship Repairing: ILO Code of Practice, 2nd ed., International Labour Office, Washington, DC, 1978.

Weiby, P. and K. R. Dickinson, "Monitoring Work Areas for Explosive and Toxic Hazards," *Chemical Engineering*, vol. 83, no. 22 (1976), pp. 139-145.

ized*# 22

Food Processing

Jane I. Lataille, P.E.

Many industrial processes are represented by the products found on grocery store shelves. The food industry is indeed diverse; it encompasses agriculture, dairy production, grain milling (see Chapter 6, "Grain Mill Products"), vegetable and animal oil processing (see Chapter 7, "Vegetable and Animal Oil Processing"), convenience food manufacture, canning, freezing, pickling, baking, candy making, meat processing and distribution warehousing.

In 1981, consumers spent over $300 billion on food items (excluding alcoholic beverages and tobacco) in the United States.[1] Of this, over $250 billion was U.S. industry shipments.[2] The food processing industry employed 1.67 million people in manufacturing and 7.17 million in food trade in that year.[3] Agriculture adds 2.5 million more workers to these figures.[4]

The outstanding loss potential common to all food processing or handling facilities is contamination. This is defined as the introduction of foreign elements into a substance which render it unfit for use. Loss situations which cause relatively minor damage in other industries can cause extremely high amounts of damage to food items. This is because

---

Jane I. Lataille is an Executive Engineer with Industrial Risk Insurers, Hartford, CT.

health authorities often condemn items exposed to smoke, heat, or water, even though they may not be visibly damaged.

Each state is responsible for its own laws and programs concerning the edibility of food. Presently, forty-two of the states follow the Food and Drug Administration's recommended model law for food. (Drugs are federally handled in all states). Typically, health authorities choose areas considered susceptible to contamination by considering exposure, temperature and packaging. Samples from each area are sent for laboratory analysis. If a sample is judged inedible, all items in the sample area must be destroyed. The items to be analyzed are raw materials, intermediate products, additives, finished goods or any combination of these. It is this high susceptibility to damage which makes it important to properly identify and protect any hazards present.

Losses in the food processing industry are commonly caused by electrical sparks, mechanical friction, overheating, and failures involving fuel-fired equipment. Hazards vary among different processing plants but are primarily those associated with combustible dusts, flammable and combustible liquids, conveyors, refrigeration systems, fuel-fired equipment and general storage.

RAW MATERIALS

Fruits, vegetables, grains, beef, poultry, fish and milk are common raw materials in the food processing industry. Other materials are powders, such as flour, salt, sugar, corn starch and cocoa; liquids, such as water, oil, alcohol, molasses, and vinegar; and miscellaneous items such as butter, chocolate, flavorings, colorings, chemicals and spices. In addition to products' ingredients there are cooking oils and fats; various gases as used in carbonation, fumigation and ripening; and fluids associated with heating and cooling systems.

Dependence on crops can make plants subject to unusually high business interruption due to limited harvest seasons. Examples are plants which process oranges into juice, or freeze peas. In the extreme case, if a plant receives fruit once a year and suffers a fire which destroys all the fruit, one year's production is lost. Seasonal availability is therefore an important factor in determining loss potential. This potential is less severe if there are multiple harvests, if alternate crop sources are available, or if the plant handles more than one type of food product. To lessen potential for a total loss, crops can also be stored in more than one area of the plant.

Flour, salt, sugar, and other powdered or granulated raw materials are often pneumatically conveyed (from railroad cars, for example) into storage silos, bins, or hoppers. These materials are highly susceptible to damage by water and other liquids and have to be stored in well-sealed containers. In multiple-story plants they are usually stored on upper floors or in penthouses.

Cooking oils, alcohol, vinegar, and liquids used at room temperature are stored in tanks of various sizes and shapes. Molasses and other liquids requiring heating are stored in steam-jacketed tanks. Most liquids storage is found on lower floors, sometimes even in basements. Wherever stored, adequate curbing and drainage must be provided to reduce not only any existing loss potentials, but also the chance of damage to other materials, equipment, storage and the building itself. Generally, storage in basements is not advisable. Among the reasons for this are the chance of flooding from surface water run-off, water main breaks, and spills from upper floors; reduced accessibility to the fire department and difficulty of venting heat and smoke.

Many food processing plants have very high refrigeration demands. In the past, ammonia coolant was commonly used in refrigeration systems. Ammonia is combustible, explosive, and toxic. Ammonia-containing equipment should be protected from mechanical injury (for example, with guard rails) and its physical condition should be checked for deterioration. Freon has become a more commonly used refrigerant. While not as volatile as ammonia, it is still important that freon refrigeration and control systems be safely designed.

Refrigeration systems should be arranged to serve other systems as back-up when possible. This makes routine shutdown and servicing more convenient and reduces chance of spoilage when there is a loss in this equipment. The possibility of renting portable refrigeration systems should be explored so that little time is lost if a shutdown of regular equipment is required.

PROCESSES

Agriculture

Even while growing in the fields, fruits, vegetables and grains can burn. They are also susceptible to damage or total loss from wind, rain, hail, freezing, drought, disease, and vermin. After harvesting some crops are cut, dried, husked, peeled, sprayed, colored, waxed or stored before transport. The crops are then shipped either to wholesale and retail food traders or to plants for further processing. It is here that the food industry's unusual exposure to loss of large amounts of stock begins.

Dairy Production

The dairy industry depends heavily upon refrigeration. Dairy products must be kept within specified temperature ranges or they may be declared unfit for use. The consequences of loss of refrigeration capability are therefore important considerations in planning and protecting refrigeration systems.

In pasturization, milk is heated to kill bacteria and other microorganisms. It must then be cooled within a specified amount of time and kept within acceptable temperature ranges. Milk is also homogenized, or mixed, to give it a uniform consistency. Other dairy products are cream, skim milk, butter, cream cheese, dips, ice cream, buttermilk, yogurt, and cheese. These products must also be kept cooled for acceptable quality.

Convenience Food Manufacture

A relatively new facet of the food industry is convenience food manufacture. Examples of convenience products are instant soup, powdered milk, instant coffee, egg substitute, meat substitute, powdered drinks, instant potatoes, and instant rice. The advantages of these products include economy, ease of preparation, and long shelf life.

The processes most often used in producing convenience foods are flash drying and flash freezing. Here, food products are converted into their desired forms by rapid exposure to extreme heat or cold. Some products also have various chemical additives for preservation, texture or flavor improvement, coloring and moisture control.

Preserving

Since most crops can only be grown during limited times, it is necessary to preserve them to achieve year-round availability. Meat, poultry, and fish are also preserved. Canning, freezing and pickling are common means of food preservation. In both canning and freezing, the food is first cooked. Steam-jacketed vessels are usually used for doing this. Salt and vinegar are the primary materials used for pickling.

Baking

Continuous-type fuel-fired ovens are representative of food baking process equipment. The fuel, number of burners, number of zones, temperature, and amount of ventilation will vary. Occasionally, electric ovens are used. Baked products include bread, pastry, pies, cakes, cookies, and crackers. In pie and pastry making, fruit must be prepared and cooked before baking. Some ovens are arranged to spray vegetable oil into pans before baking or onto the product during baking. Some ovens have provisions for coating with sugar, salt or frosting, or for adding fillings.

Candy Making

Most candy making involves cooking or heating. Examples are melting chocolate to pour into molds, and cooking flavored sugar mixtures to

FOOD PROCESSING

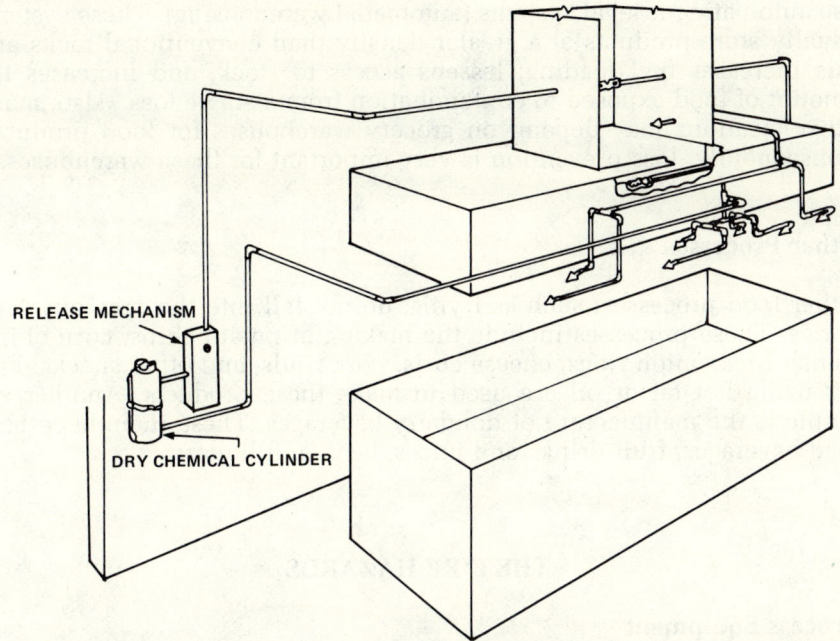

FIGURE 22.1. Layout of a wet chemical extinguishing system for a fat fryer. (©Reprinted with the permission of Wormald U.S., Inc.)

make hard candy. Process equipment is likely to include kettles, ovens, and cooling tunnels. Some candies go through more than one process, such as those in chocolate assortments. Many candies are covered with coatings, such as chocolate, icings, fruit mixtures, and glazes. The coatings are either melted and mixed or carried in a solvent.

Meat Processing

In processing, meat is cleaned and cut into the desired sections. Meat preserving consists mainly of drying, salting (soaking in brine), and smoking. The smoking is done in smoke chambers. Any other processing is usually done either by wholesalers and retailers or by plants using meat in their products, such as those making soups, stews, hash, canned meat, meat rolls, or even pet foods.

Warehousing

Some food facilities do not do any processing, but simply store large quantities of many different food products for distribution to grocery stores. As with non-food warehousing facilities, grocery warehouses can use automatic retrieval systems (automated warehousing). These systems usually store products at a greater density than conventional racks and this increases fuel loading, lessens access to stock, and increases the amount of food exposed to contamination from a single loss. Also, many other locations may depend on grocery warehouses for food products. Consequently, loss prevention is very important for these warehouses.

Other Processes

Other food processes, such as frying, do not fall into the previous categories. These processes include the making of potato chips, corn chips, french fries, onion rings, cheese curls, pork rinds, and other snacks. Fryers using hot fat or oil are used to make these products. Another example is the manufacture of nondairy beverages. These include carbonated beverages, fruit drinks and juices, beer, and liquors.

THE FIRE HAZARDS

Process Equipment

Refrigeration equipment presents hazards inherent to compressed gases, potential vapor release, moving mechanical parts, and electrical control systems. (See Chapter 47, "Electrical Installations for Industrial Locations," for a detailed discussion of the hazards associated with electrical control systems.) Ammonia refrigerant is volatile and toxic. Freon is less volatile but is still considered toxic. Both can contaminate food and otherwise damage equipment and stock. Vessels for cooking or melting are usually heated with electricity or steam (see Chapter 27, "Heat Processing Equipment"). Their fire hazards are usually limited to those of electrical systems and overheating. Boilers which produce the steam for these vessels have the fire and explosion hazards associated with fuel-fired equipment (see Chapter 42, "Boiler Furnaces").

Baking ovens also present the hazards of fuel firing. In addition, ovens have interior moving parts and fixed ductwork which become coated with oils and can be involved in a fire. Electrically heated ovens sometimes have concealed oil-filled transformers which are a concern. Another feature of ovens requiring attention is the presence of combustible insulation. The primary hazard introduced by cooling tunnels is that of combustible construction and insulation. Fryers and associated ductwork involve the heating of a combustible liquid. The degree of hazard increases as the temperature approaches the liquid's flash point.

A characteristic of food processing equipment which contributes to its loss potential is a high level of automation. Process equipment is often custom made and can take a long time to replace. It can also be so much more efficient than any other production method that business can-

FOOD PROCESSING

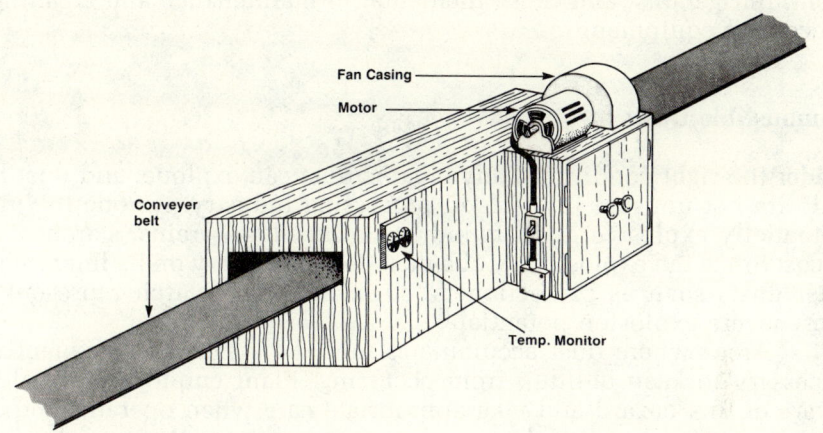

FIGURE 22.2. Cooling tunnels are sometimes made of wood.

not be made up when the equipment is out of service. For this reason, highly automated processes should be reviewed for duplication, alternate production methods and outside assistance. Where they exist, duplicate equipment lines should be separated so they will not be exposed to the same loss incident.

Gases and Liquids

Some gases used in the food industry are hazardous (see Chapter 38, "Handling and Storage of Industrial Gases"). Cocoa and other powdered products are treated with gaseous fumigants which, depending on their composition, could be flammable. Ethylene oxide is a flammable gas used to ripen fruit which has been shipped "green" to distribution centers. Whenever gases are used, their physical characteristics should be checked.

Combustible and flammable liquids are commonly found in food processing plants (see Chapter 37, "Handling and Storage of Flammable and Combustible Liquids"). Combustible liquids used include lubricating oils, cooking oils, certain alcoholic beverages, and solvent-based extracts. Alcohol, high-proof alcoholic beverages and solvents are the commonly used flammable liquids. In the candy industry, ether is used to dissolve the coating that makes jelly beans shiny. In many types of plants, alcohol dissolves dried, crushed materials or other liquids to make flavorings and colorings. One example of this is the production of vanilla extract. Waxes, which are liquids when heated, are also found in food plants.

Other gases and liquids in this industry are fuels for fired equipment, refrigerants, and those incidental to maintenance and cleaning of processing equipment.

Combustible Dusts

Under the right conditions, any organic dust can explode, and dust hazards are not unusual in the food processing industry. Among the many potentially explosive dusts found are sugar, flour, grains, starches, and cocoa. Since the explosibility of a dust depends partly on its fineness, the finer dusts such as confectioners' sugar and corn starch represent the most severe explosion potential.

Areas where dust accumulates must be cleaned as frequently as necessary to keep buildup from occurring. Plant employees should be aware of this hazard and take appropriate care when operating process equipment that handles dust-producing substances. An inspection program of some kind is generally desirable to monitor potentially hazardous areas.

Conveyors

Conveyors can be used anywhere in a process, from raw material transfer to processing, packaging, and storage (see Chapter 46, "Materials Handling Systems"). There are many hazards associated with conveyors. Their drives can jam, misalign, or overheat, and cause friction which can lead to fires. The conveyors themselves are a means of spreading fire from one area to another; so are the openings made in walls through which they pass. Many belt conveyors have combustible belts. Rubber or plastic belts produce large quantities of smoke when they burn, resulting in a high potential for contamination. Chain conveyors can carry plastic pans. Wide conveyors shield the areas below from ceiling sprinkler protection. Finally, conveyors are part of highly automated systems which cannot function without them.

Storage

Raw materials and food products are stored in various ways. Bins and tanks are often used for powders and liquids. Packaged products are commonly stored on pallets or in racks up to 80 ft (24.4 m) high or more, including automatic retrieval systems. Wood, cardboard and plastics are used to package food products, so the combustible loading in storage areas can therefore become very high.

Some food products, such as vegetable oil spray, are stored in aerosol cans (see Chapter 35, "Aerosol Filling and Storage"). Aerosol propellants under pressure will rocket cans during a fire and spread it beyond control. The storage of this type of product must be carefully planned.

FOOD PROCESSING

FIGURE 22.3. Erecting storage racks in a grocery warehouse. (Giant Food)

In addition to the various food products stored, food processing plants store the packaging materials used with their products. Typical packaging materials are wood or plastic pallets, cardboard carton flats, polyethylene shrinkwrap, wax-coated containers and styrofoam trays. The hazards of storing these materials are usually higher than the finished food products storage due to their high rate of heat release when burning.

THE SAFEGUARDS

Construction Considerations

Due to food products' high susceptibility to smoke contamination, noncombustible construction of buildings is preferred. Three-hour fire cut-offs should be provided between manufacturing and warehousing areas; to separate duplicate production lines; and to isolate fuel-fired equipment, refrigeration systems, and storage of more hazardous material such as flammable liquids, paper, cardboard, plastics, pallets and aerosols. Fire cut-offs need to be preserved where conveyors pass through them. To do this, conveyor lines are specially designed to break open at, or hinge away from, the point at which a fire door or shutter closes. When this is done, proper automatic operation of doors should be routinely

FIGURE 22.4. A combined dry-pipe and deluge system. (Giant Food)

checked for possible freezing or misaligning of conveyor parts required for complete door closure. The use of water spray systems is sometimes proposed as an alternative to cut-offs. This is less desirable because water spray by itself can't prevent the radiation of heat through an opening. Water spray systems are also dependent upon a water supply and are therefore subject to impairments.

Heating and cooling requirements in the food processing industry result in the widespread use of insulating materials in buildings and process equipment. Whenever possible, noncombustible insulation should be chosen. When combustible insulation is used, it should be one with low flame spread, fuel contributed, and smoke developed ratings as determined by tests in a recognized testing laboratory. Combustible insulation should also be protected by sprinklers and thermal barriers such as gypsum or sheetrock. This applies to the interior of freezers, coolers, ovens and other process equipment, as well as to buildings. Process equipment should also be noncombustible. Where wood cooling tunnels or other combustible equipment are used, sprinklers should be provided.

Sprinkler Systems

The various combustible materials commonly encountered in the food processing industry suggest the provision of appropriately designed

FOOD PROCESSING

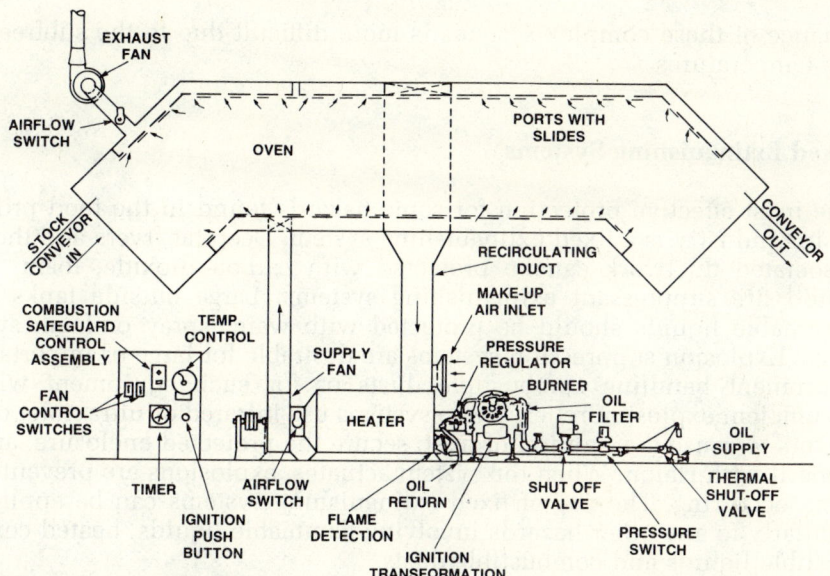

FIGURE 22.5. Conveyor-type oven with the controls labeled. (Industrial Risk Insurers)

sprinkler protection in all buildings. Sprinklers should also be provided for high storage racks, in process enclosures handling or built of combustible materials, under obstructions over four ft (1.22 m) wide (such as conveyor belts), and in coolers, freezers and heating enclosures. Hose connections supplied from sprinkler systems should be provided in storage areas and wherever large amounts of combustibles may be present. Water supplies should be sufficient to supply the largest demand area plus hose streams for the amount of time prescribed by the hazard in the protected area under consideration.

Despite low temperatures, it is possible for fires to occur within freezers. Providing sprinkler protection in them requires special design. The usual wet-pipe sprinkler system is replaced by a combination dry-pipe valve and deluge valve. This keeps the chance of water damage to a minimum by requiring two means of detection before the system operates. First, relatively low temperature rated — approximately 135°F (57°C) — heat actuated devices trip the deluge valve and allow water to flow to the dry-pipe valve. Next, the sprinkler heads open, tripping the dry valve and allowing water to flow to the heads. Alone, neither a break in the sprinkler piping nor a false operation of the heat actuated devices will allow water into the system. Hose connections are more complicated to install in freezers. They must be kept drained and must have a way of both releasing the air in the sprinkler piping and opening the deluge valve. These connections can also be installed on their own dry-pipe valve, but this valve will not trip when ceiling sprinklers open. Maintenance of these complex systems is more difficult due to the subfreezing temperatures.

Fixed Extinguishing Systems

The most effective protection for some hazards found in the food processing industry is a fixed extinguishing system. Deep fat fryers and their associated ductwork can be protected with carbon dioxide, foam, or liquid fire suppressant extinguishing systems. Large outside tanks of flammable liquids should be protected with water spray or foam systems. Explosion suppression systems are desirable for large or important equipment handling combustible dusts or for such equipment with insufficient explosion relief. These systems use infrared or ultraviolet detectors to sense the ignition of dust, secure the protected enclosure, and flood it with Halon. When the system actuates, explosions are prevented from occurring. The use of fixed extinguishing systems can be applied similarly to any other hazards involving flammable liquids, heated combustible liquids and combustible dusts.

Electrical Equipment

Electrical equipment in food processing plants should be suitable for the location in which it is installed. A review should be made of possible vapors or dusts which could be present in each area. Where liquids such as alcohol could be present, electrical equipment suitable for Class I, Group D locations is recommended. Where ether is used, the electrical classification is Class I, Group C. Ethylene oxide requires the use of Class I, Group B equipment. Class II, Group G locations are those where nonconductive dusts are likely to be present.
    All electrical equipment in a given classified area must be checked for its suitability. In addition to lights, switches and receptacles, this includes such commonly overlooked items as telephones and motors. Nonelectrical sources of ignition must be kept out of hazardous areas. These include direct heaters, fuel-fired equipment and devices with moving ferrous parts such as fans. Portable electrical equipment used on a temporary basis must also be suitable. This applies to testing equipment and to the vacuuming systems used for dust control.

Other Protection Features

Areas handling flammable liquids should be well ventilated to prevent significant vapor buildup. Enclosures which may contain dusts or vapors should be provided with explosion relief designed to minimize damage. Magnets are used wherever powdered materials enter a storage or distribution system to extract ferrous debris. Portions of systems handling powered materials from which dust can escape should be provided with permanent dust collection systems. For equipment handling either dusts or vapors, proper bonding and grounding practices must be observed.

High piled storage areas are a fire fighting challenge even with properly designed sprinkler protection. For these areas, there should be good outside hydrant coverage, as well as mechanically powered, automatically operated smoke and heat venting facilities. Venting should also be manually operable.

Interlocks are used on automated processes to run them safely. Fuel-fired equipment should be provided with standard combustion safeguards and air and fuel train controls. For various types of equipment, other safety interlocks are needed. Ovens and other heated enclosures should have high temperature cut-outs. Enclosures which could contain hazardous vapors need combustible gas detection systems. Boilers producing steam require excess pressure cut-outs and those producing hot water require high temperature cut-outs. Processes which must have a liquid flowing to run safely should monitor the liquid's flow and level. Conveyor belts are monitored for low speed or excessive friction to prevent their malfunction from causing a fire. Continuous feed ovens or dryers requiring ventilation are interlocked to shut down upon loss of ventilation. Potentially unsafe conditions in any process should be monitored and interlocked to avoid hazardous conditions.

BIBLIOGRAPHY

REFERENCES CITED

1. *Statistical Abstracts of the United States*, 1982-3, (US Dept. of Commerce, Bureau of the Census), p. 422.
2. *US Industrial Outlook*, 1983, (US Dept. of Commerce, Bureau of Industrial Economics), p. 37-2.
3. *Statistical Abstracts*, pp. 397 and 398.
4. Philip L. Martin, "Labor-Intensive Agriculture," *Scientific American*, vol. 249 No.4 (1983), pp. 54-59.

NFPA CODES, STANDARDS, AND RECOMMENDED PRACTICES

Reference to the following NFPA Codes, Standards, and Recommended practices will provide further information on the processes and protective systems discussed in this chapter. (See the latest *NFPA Codes and Standards Catalog* for availability of current editions of the following documents.

NFPA 10, *Portable Fire Extinguishers.*
NFPA 11, *Foam Extinguishing Systems.*
NFPA 11A, *Medium and High Expansion Foam Systems.*
NFPA 12, *Carbon Dioxide Extinguishing Systems.*
NFPA 12A, *Halon 1301 Fire Extinguishing Systems.*
NFPA 12B, *Halon 1211 Fire Extinguishing Systems.*
NFPA 13, *Installation of Sprinkler Systems.*
NFPA 14, *Standpipe and Hose Systems.*
NFPA 15, *Water Spray Fixed Systems for Fire Protection.*
NFPA 16A, *Installation of Closed-Head Foam-Water Sprinkler Systems.*
NFPA 30, *Flammable and Combustible Liquids Code.*
NFPA 34, *Dipping and Coating Processes Using Flammable or Combustible Liquids.*
NFPA 49, *Hazardous Chemicals Data.*
NFPA 61A, *Manufacturing and Handling of Starch.*
NFPA 61B, *Prevention of Fires and Explosions in Grain Elevators and Facilities Handling Bulk Raw and Agricultural Commodities.*
NFPA 61C, *Prevention of Fire and Dust Explosions in Feed Mills.*
NFPA 61D, *Prevention of Fire and Dust Explosions in the Milling of Agricultural Commodities for Human Consumption.*
NFPA 68, *Explosion Venting Guide.*
NFPA 69, *Explosion Prevention Systems.*
NFPA 70, *National Electrical Code.*
NFPA 77, *Recommended Practice on Static Electricity.*
NFPA 85A, *Prevention of Furnace Explosions in Fuel Oil- and Natural Gas-Fired Single Burner Boiler-Furnaces.*
NFPA 85B, *Prevention of Furnace Explosions in Natural Gas-Fired Multiple Burner Boiler-Furnaces.*
NFPA 85D, *Prevention of Furnace Explosions in Fuel Oil-Fired Multiple Burner Boiler-Furnaces.*
NFPA 85G, *Prevention of Furnace Implosions in Multiple Burner Boiler-Furnaces.*
NFPA 86A, *Ovens and Furnaces, Design, Location and Equipment.*
NFPA 96, *Vapor Removal from Cooking Equipment.*
NFPA 204M, *Guide for Smoke and Heat Venting.*
NFPA 231, *Indoor General Storage.*
NFPA 231C, *Rack Storage of Materials.*
NFPA 321, *Basic Classification of Flammable and Combustible Liquids.*
NFPA 325M, *Fire Hazard Properties of Flammable Liquids, Gases, Volatile Solids.*
NFPA 493, *Intrinsically Safe Apparatus in Division 1 Hazardous Locations.*
NFPA 496, *Purged and Pressurized Enclosures for Electrical Equipment.*
NFPA 497, *Classification of Class I Hazardous Locations for Electrical Installations in Chemical Plants.*

NFPA 497M, *Classification of Gases, Vapors and Dusts for Electrical Equipment in Hazardous (Classified) Locations.*

NFPA 650, *Pneumatic Conveying Systems for Handling Combustible Materials.*

ADDITIONAL READING

Food Processing and Packaging Equipment, Business Trend Analysts, Dix Hill, New York, NY, 19 ed.

Heldman, D. R. and R. P. Singh, *Food Process Engineering,* AVI Publishing Co., Inc., Westport, CT, 1981.

Linko, P., ed., *Food Processing Systems,* (vol. 1), Elsevier Science Publishing Co., Inc., New York, NY, 1980.

McKinnon, G. P., ed., *Fire Protection Handbook,* Fifteenth Edition, National Fire Protection Association, Quincy, MA 1981.

"Moving Fire: Fire Hazards of Belt Conveyors," *Record,* vol. 54, no. 6 (1977), pp. 18-21.

Schwartzberg, Henry G. and Daryl Lund, *Food Process Engineering,* American Institute of Chemical Engineers, New York, NY, 1982.

Talbot, G., "Static Electricity," *Fire,* 70 (871), 197, pp. 397-398.

23

Semiconductor Manufacturing

Kathleen Robinson

There is no historical parallel to the rapid development of the semiconductor industry, either chronological or technological. In little more than 30 years, the industry has virtually invented itself and become a billion-dollar business.

Unknown just a few decades ago, semiconductors are now found in a variety of goods, from small personal computers to large government information systems, from toys to satellites, watches to weapons systems, medical instruments to automatic bank tellers. In 1983 alone, the industry employed almost 200,000 people and shipments of semiconductor devices reached over $14 billion.

Along with its enormous growth potential, however, the semiconductor manufacturing industry has an enormous loss potential. This is true because its manufacturing processes and equipment are often easy to damage and expensive to replace.

The amount of acids and flammable liquids and gases found in the various operations may equal the quantity found in a small chemical plant. What is more, the cost of the manufacturing equipment may range from $30,000 to $3,000,000. Concentrated in the fabrication area, it gives

---

Kathleen Robinson is a Publications Editor with Industrial Risk Insurers, Hartford, CT.

the area an average value of over $1,000 per square foot. And a substantial amount of that equipment is made of or incorporates plastics, some of which burn almost as hot as gasoline.

Even the finished product is vulnerable to loss. Silicon wafers may contain as many as 1,000 chips, and each chip may cost up to several dollars. Any form of contamination, including fire, and smoke, will ruin them.

Contamination can also ruin the ultra clean environment in which semiconductors are made. Once contaminated, the area must be painstakingly cleaned before it can be used again.

In short, the dollar loss a controlled fire can cause in a well-protected facility may exceed $500,000. Facilities with little or no protection can suffer millions of dollars worth of damage.

THE PRODUCTION PROCESS

The semiconductor manufacturing industry's major product is the integrated circuit, an array of transistors and other components that are implanted in or deposited on a piece of semiconductor material and fastened to a frame that can be inserted into a printed circuit. The production of such a circuit involves several steps (see Figures 23.1 and 23.2).

Crystal Production

The first step in the process is making the semiconductor material in which the circuit information is to be implanted. This material must be electrically neutral so that it does not interfere with the electronic information that will be processed on it.

The most common semiconductor material is silicon, although other materials such as gallium arsenide and germanium may be used. Silicon is grown as crystals in water-cooled vacuum furnaces that are electrically heated to more than 2,552°F (1400°C).

After the crystals are grown, they are sliced into wafers, polished, cleaned, and given a thin surface of silicon dioxide. The wafers are then sent on to the fabrication area for further processing.

Mask Manufacturing

While the semiconductor material is being created, the masks used to make the circuits on it are also produced. A mask is a thin patterned sheet which shields portions of the semiconductor material as the information is imprinted on it. Because it contains only the pattern for a sin-

SEMICONDUCTOR MANUFACTURING

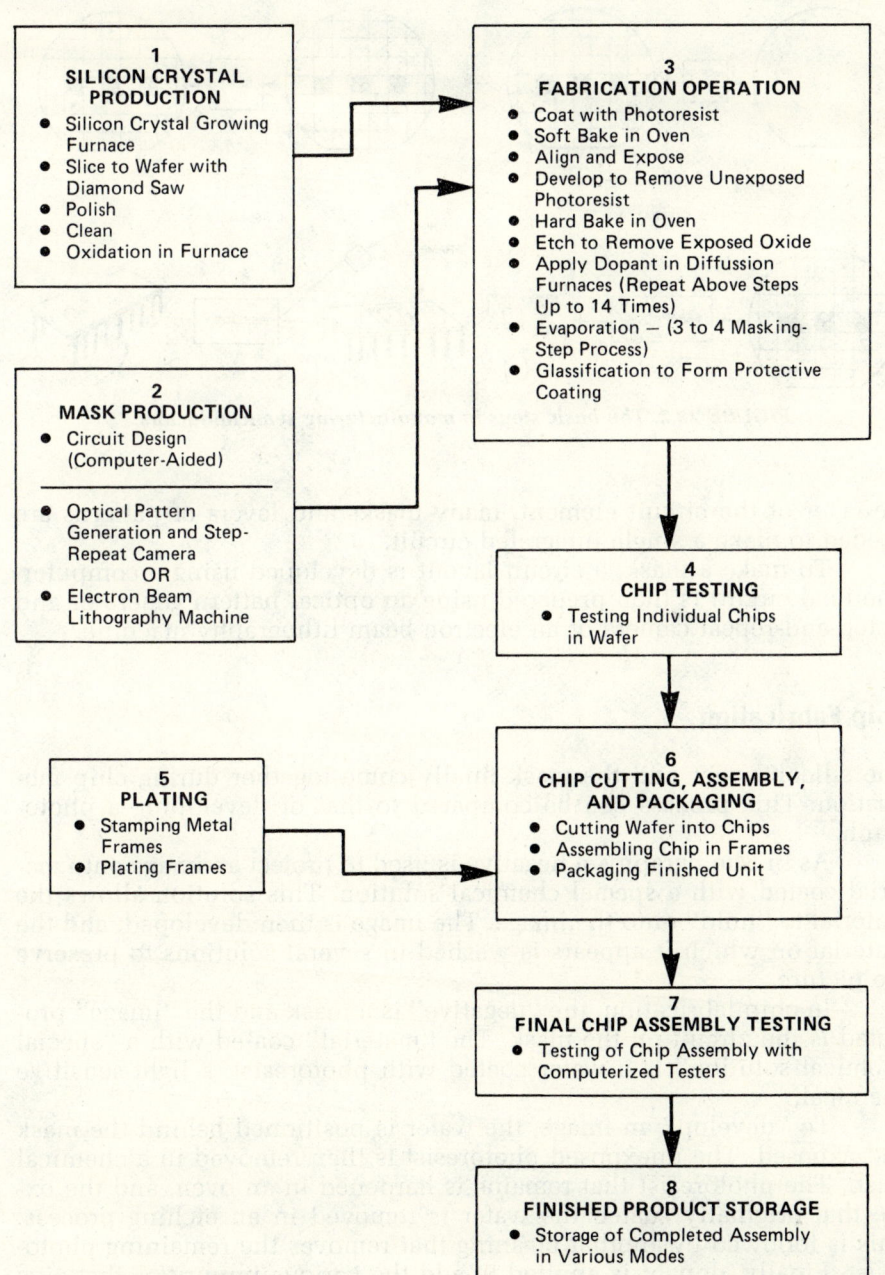

FIGURE 23.1. Flow diagram of the semiconductor manufacturing process.

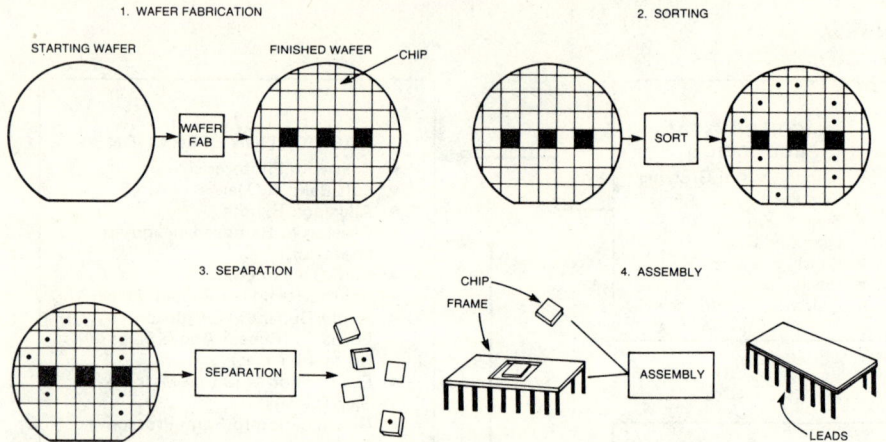

FIGURE 23.2. The basic steps in manufacturing semiconductors.

gle layer of the circuit element, many masks and layers of patterns are needed to make a single integrated circuit.

To make a mask, a circuit layout is developed using a computer. The final circuit is then prepared using an optical pattern generator and a step-and-repeat camera or an electron beam lithography machine.

Chip Fabrication

The silicon wafer and the mask finally come together during chip fabrication. This process can be compared to that of developing a photograph.

As in photography, a negative is used to project an image onto material coated with a special chemical solution. This solution allows the material to "hold" onto the image. The image is then developed, and the material on which it appears is washed in several solutions to preserve the picture.

In chip fabrication, the "negative" is a mask and the "image" projected is the circuit on the mask. The "material" coated with a "special chemical solution" is a wafer coated with photoresist, a light-sensitive chemical.

To "develop" an image, the wafer is positioned behind the mask and exposed. The unexposed photoresist is then removed in a chemical wash. The photoresist that remains is hardened in an oven, and the oxide that originally coated the wafer is removed in an etching process. This is followed by another cleaning that removes the remaining photoresist. Finally, dopant is applied to add the various impurities that give a circuit its desired characteristics.

This process is repeated up to 14 times with different masks until the chip is imprinted with the desired circuit. When the circuit is finished, a coating of metal such as aluminum is applied in another series of masking steps to provide electrical connections to outside circuits. Since a single wafer may contain as many as 1,000 chips, the surface of the wafer is scored to separate them from each other.

Chip Testing

Each chip in a wafer must be tested using computer-controlled test equipment. Defective chips are marked. The computer maintains statistics on the number and location of good chips per wafer and on the incidence of various defects.

Plating

While the chips are being tested, the frames into which they will go are plated using both standard and customized plating equipment. This is often done in a separate building.

Chip Cutting, Assembly, and Packaging

After the chips have been tested and marked, the wafers are cut apart. Defective chips are discarded, and the rest are fastened into plated frames. Fine wires connect the chips to the frame leads, and a plastic lid is fused onto the frame to protect both the chip and the wires.

Final Chip Assembly, Testing, and Storage

Various electrical tests are run on circuits in the completed chip assemblies using specially designed computerized equipment. Integrated circuits that must be highly reliable often undergo environmental functional testing. During this type of test, they are subjected to long-term temperature variations in specially designed ovens.

The finished product is placed in plastic or cardboard containers in cartons. These cartons are then stored in bulk, on pallets, in bins, or on racks.

THE FIRE HAZARDS

Most of the process hazards found in this industry are not unique to it. They are similar to hazards found in many other industries.

Chapter 23

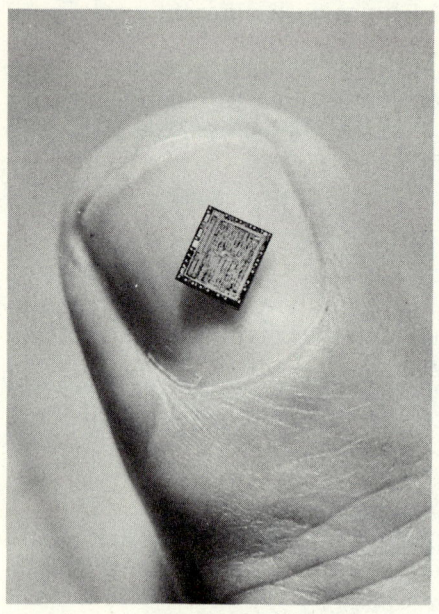

FIGURE 23.3. *A completed integrated circuit. (Honeywell)*

For instance, the problems associated with crystal production stem mostly from the furnace in which the crystals are grown. If cooling water comes in contact with the furnace contents, it could cause a steam explosion. A lack of cooling water could cause a burn-out. These hazards may be present wherever a water-cooled furnace is used. (See Chapter 27, "Heat Processing Equipment.")

Similarly, hazards common to all plating operations exist in semiconductor plating operations. Losses usually occur when electrical devices, such as immersion heaters, ignite combustible plastic or plastic-lined tanks, as well as the fume exhaust ducts and scrubbers.

Combustible plastics also contribute to the fire potential of the semiconductor industry. They are found not only in the tanks, fume exhaust ducts, and scrubbers of plating operations, but in the fabrication and chip testing areas, as well. In addition, plastic hoods and containers are used in both wafer cleaning and step-and-repeat camera operations.

In wafer cleaning, and step-and-repeat camera operations, plastics present even more of a problem than usual because they are used in conjunction with flammable solvents. Other hazardous materials, such as corrosive and flammable liquids and gases, are found in clean rooms and in the ovens in which wafers are baked at various times during chip fabrication. (See Chapter 37 for information on flammable and combustible liquids and Chapter 38 for industrial gases.)

Other hazards involve the expensive equipment used throughout the industry. Computers control much of the mask and fabrication operations and are used to test chips and the final chip assemblies (see

Chapter 40, "Computer Centers"). Hazards commonly associated with computer room operations include the danger of fire started by a short circuit. This type of fire can spread to the plastics used for wiring, insulation, and circuit boards, and can destroy the computer.

Valuable electronic equipment, such as the electron beam facility, is also used throughout the semiconductor manufacturing operation. This, too, may be easily damaged by an electrical fire and contaminated by products of combustion. It can take a long time to replace.

Chips and wafers are also vulnerable. These are especially open to contamination during both chip fabrication and mask manufacturing. Foreign material that gets onto a mask or wafer during fabrication can distort a circuit. To avoid contamination, parts of both processes are conducted in clean rooms, which are kept as nearly sterile as possible.

The storage of the finished chip assemblies presents the hazards commonly found when combustibles are grouped in bulk, in racks, and on pallets. Here, millions of dollars worth of product may be subject to a single loss.

While the hazards found in the semiconductor manufacturing industry are similar to those found in other industries, they cannot be dismissed lightly. These hazards can cause serious and costly losses, especially in an industry whose process equipment is so expensive.

THE SAFEGUARDS

Building Construction

Buildings that house production operations should be fire-resistive or of noncombustible construction. So should the interior partitions that separate the various processes. The flamespread rating for interior finishes, suspended ceilings, and piping insulation should not exceed 25 and their smoke-developed rating should not exceed 50 when tested in accordance with "*Methods of Test of Surface Burning Characteristics of Building Materials.*"[1,2] This test is referenced in NFPA 255, ASTM E-84, and UL-723.

The high values involved in this industry make the establishment of multiple production areas a good idea. These should be separated by fire walls with a minimum fire resistance rating of one hour. This will help keep fire, products of combustion, and water in one area from contaminating other areas. If the value of a fabrication area is very high, separate fabrication lines should be set up in different buildings or in areas separated by fire walls with a minimum three-hour rating. Ventilation, fume exhaust systems, and other utilities should be independent of those in adjacent buildings or areas.

Heating and air conditioning systems should be designed so that air stops recirculating if a chemical spill or fire occurs. This will keep contaminants from spreading throughout the entire building. To do this, separate air handling systems for clean rooms, arranged to convert automatically to full exhaust ventilation during a fire or spill, should be installed. The fabrication area should also be provided with some means of manually converting the system to the emergency mode.

Ducts for the heating and air conditioning systems should be made of noncombustible materials, such as steel and aluminum, or of Class O or Class I materials. Class O materials have flamespread and smoke-developed ratings of O. The flamespread rating of Class I filter materials does not exceed 25 and their smoke-developed rating is no more than 50. Any Class O or Class I material used should be tested in accordance with UL-181, *Standard for Factory-Made Air Duct Material and Air Duct Connectors*.[3]

Semiconductor manufacturing areas should use Class I air filters. Such filters do not contribute fuel to a fire when clean and give off only insignificant amounts of smoke when tested in accordance with UL-900, *Standard for High Efficiency Particulate Air Filter Units*.

Fire Protection Systems

Because automatic sprinklers are one of the most effective means of controlling and extinguishing fires, they should be installed in most buildings and areas, including interstitial spaces, combustible ductwork, and the space beneath laminar flow hoods. These systems must be able to handle fires that release quite a bit of heat very quickly.

The design of sprinkler protection used in process areas should be based on Ordinary Hazard, Group 3 occupancy. No system should cover an area larger than 52,000 sq ft (4830 m^2). Systems protecting bulk storage should be designed according to NFPA 231, *Standard for Indoor General Storage*, while those protecting rack storage should be based on information found in NFPA 231C, *Standard for Rack Storage of Materials*.

Protect testing areas, step-and-repeat camera enclosures, and production areas containing particularly valuable equipment with automatic total flooding Halon 1301 fire extinguishing systems, as well as sprinkler systems. Halon systems should also be installed under the raised floors of computer installations. If ordinary combustibles such as plastics or computer papers are not found in these areas, sprinklers can be eliminated.

Inside hose connections should be installed in all areas. They should be arranged to remain in service even if sprinkler protection in any one area is shut off.

Portable fire extinguishers are another protection requirement. They should be provided at 25-ft (7.6 m) intervals in fabrication areas and at 50-ft (15.2 m) intervals elsewhere in the building. Additional carbon dioxide or Halon extinguishers should be located in areas that contain small quantities of flammable liquids.

Dry chemical extinguishers should not be used in the facility. The

SEMICONDUCTOR MANUFACTURING

powder they discharge can contaminate electronic equipment, circuit masks, and wafers.

Finally, every semiconductor manufacturing facility needs a strong, reliable water supply. In fact, most facilities need two independent sources of water, each of which can meet anticipated sprinkler demand and provide another 500 gal (1892 L) per minute for hose streams. Several combinations of sources are acceptable. The one to use is the one best suited to the facility being protected.

Semiconductor facilities also need a water distribution system that can supply plant fire hydrants and sprinkler systems. Underground fire mains at least 8 in. (203 mm) in diameter should loop the facility, and control valves should divide the loop into sections. Fire hydrants should be installed at 250- to 300-ft (76 to 91 m) intervals.

Special Hazards

Flammable liquids and gases: Flammable liquids used in production areas should be kept in special enclosures whose openings are curbed and drained. (See Chapter 37 for a complete discussion of "Handling and Storage of Flammable and Combustible Liquids.") The walls of these enclosures should have a one-hour fire resistance rating. The enclosures should also be equipped with sprinklers, exhaust ventilation, and the proper electrical equipment.

Stainless steel safety cans should be used to dispense flammable liquids in fabrication areas. These are acceptable for handling electronic-grade liquids.

Flammable liquids may also be piped directly to the production area through piping that is equipped with easily accessible shut-off valves. Pressurized piping for liquids and gases should be equipped with excess flow control valves, too. These will shut off the flow if a pipe ruptures. Regardless of use, all piping should be color-coded or marked in some way to identify its contents.

Cylinders containing flammable, oxidizing, and corrosive gases should be stored outside the building until used and secured so they will not tip. Store oxidizing and flammable gases separately.

The number of process gas cylinders brought inside a building should be limited to the fewest practical. Once inside, they should be kept in ventilated, sprinklered gas cabinets. Electrical equipment associated with these cabinets should comply with the *National Electrical Code*®.

All bulk, gaseous hydrogen, and liquefied hydrogen systems should be designed according to the appropriate NFPA standards. Process areas in which hydrogen is used should be provided with combustible gas detectors that are interlocked to sound an alarm when the hydrogen concentration reaches 15 percent of the lower explosive limit

(LEL).* These detectors should also be arranged to shut off the gas when it reaches 25 percent of its LEL. A method of turning the gas off manually should be provided outside each fabrication area for use during emergencies.

Plating operations: Plating operations that use electrical immersion heaters present a special problem. These heaters are a common source of ignition. Should the plating solution in a heated tank accidentally evaporate to a low level, it will expose the heating element to the air. Since air is less effective than a liquid in conducting heat, the surface temperature of the element will begin to rise. Nothing will absorb the increased radiant heat, so it can ignite the wall of the tank.

To prevent immersion heater losses, heat exchangers that use steam or hot water as their heating medium should be installed in place of conventional electrical heaters. Tubing made of Teflon® or of corrosion-resistant metals are available to meet any application.

Fires caused by immersion heaters may also damage the fume exhaust systems that remove corrosive and toxic fumes produced by various processing operations. These systems are typically composed of plastic ducts. To prevent fire damage, the ducts should be flame-retardant, with a flamespread rating of no more than 25 and a smoke-developed rating no higher than 50 when tested in accordance with NFPA 255, *Methods of Test of Surface Burning Characteristics of Building Materials*. Automatic sprinklers should be installed inside ducts that are 8 in. (203 mm) in diameter or larger. Sprinklers should also be installed in fume scrubbers, in the plastic exhaust ductwork downstream of the scrubbers, and in each vertical connection to a work station.

There are several rules to follow when installing internal duct sprinklers:

1. Coat the sprinkler heads and pipe drops to protect them from corrosive fumes.
2. Take branch ducts to high-temperature equipment off the top of the header or above the duct's center line so that sprinkler discharge does not drain into the equipment. Sprinklers should not be installed at branch points where the discharge can spray directly into the branch duct, either.
3. Design sprinkler piping so that it can supply at least five sprinkler heads at a pressure of 30 psi (207 kPa). Internal sprinklers should be controlled by a separate valve, and a separate water flow detector should be installed to sound an alarm as soon as a sprinkler operates.
4. Provide drainage to keep the ducts from collapsing. Drains should be able to remove at least as much water as is discharged by five operating sprinklers.

Internal sprinklers may be unnecessary in metal ducts coated with

---

*The LEL of hydrogen is 4 percent.

PVC or epoxy resin if the coating is not too thick. Metal ducts should only be installed in areas which do not contain corrosive fumes, unless the metal has proven able to withstand such fumes.

Crystal furnaces: Furnaces in which crystals are grown should be equipped with a rupture disk and relief vent to prevent overpressurization. Cooling water flow alarms are also needed to warn of interruptions in furnace coolant and emergency cooling water connections. This will help protect the furnaces in the event of a primary cooling water failure.

Environmental ovens: Depending on the combustibility of the materials being tested, environmental ovens or chambers require special protection devices. These include products-of-combustion detectors and the Halon extinguishing systems they actuate, excess voltage and over-current protection, and an over-temperature cut-off that shuts off power to the oven if the temperature rises above an acceptable level.

Housekeeping

All production areas, especially mask manufacturing and fabrication sections, must be clean and free of any materials that are not actually used in the operations. Do not store cartons, paper, or packaging materials in these areas. Silicon wafers processed in fabrication areas should be stored in tightly covered containers. (See Chapter 50 for a complete discussion of "Industrial Housekeeping Practices.")

Surveillance

Surveillance of a facility during nonoperating hours will help prevent arson fires and provide early warning of emergencies. Watchclock stations should be installed throughout the facility, and recorded watchman service should be maintained in all unattended areas. Television surveillance can be used to supplement watchman service. A central station signaling system should be used to monitor sprinkler water flow, control valves, and other fire protection equipment. Gas detection alarms are best handled locally.

Management Support of Loss Prevention Programs

Many fires are the result of failures in the facility's fire protection philosophy rather than weaknesses in the facility's fire protection systems. If management does not organize and train its employees to participate in the plant's loss control program, they will not be able to perform properly during an emergency and a loss will occur.

Management's loss prevention and control programs should address such problems as training personnel to respond to fires and other emergencies (see Chapter 3) and to inspect and maintain loss control equipment. In addition, management should draw up procedures for restoring and repairing loss control equipment, and for controlling all phases of construction and renovation. Finally, management should make sure that duplicate master masks or data that can be used to produce master masks are kept on hand in an area that is remote from the mask manufacturing operation. This will help ensure continued production in the event of a loss.

BIBLIOGRAPHY

REFERENCES CITED

1. ASTM E-84, *Standard Test Method for Surface Burning Characteristics of Building Materials.*
2. UL-723, *Test for Surface Characteristics of Building Materials.*
3. UL-181, *Standard for Factory-Made Air Duct Material and Air Duct Connectors.*
4. UL-900, *Standard for High Efficiency Particulate Air Filter Units.*

NFPA, CODES, STANDARDS AND RECOMMENDED PRACTICES

Reference to the following Codes, Standards, and Recommended Practices will provide further information on the processes and protective systems discussed in this chapter. (See the latest *NFPA Codes and Standards Catalog* for availability of current editions of the following documents.)

NFPA 10, *Portable Fire Extinguishers.*
NFPA 12A, *Halon 1301 Fire Extinguishing Systems.*
NFPA 13, *Installation of Sprinkler Systems.*
NFPA 13A, *Care and Maintenance of Sprinkler Systems.*
NFPA 24, *Installation of Private Fire Service Mains and Their Appurtenances.*
NFPA 30, *Flammable and Combustible Liquids Code.*
NFPA 70, *National Electrical Code.*
NFPA 75, *Protection of Electronic Computer/Data Processing Equipment.*
NFPA 86A, *Ovens and Furnaces, Design, Location, and Equipment.*
NFPA 86B, *Industrial Furnaces - Design, Location, and Equipment.*
NFPA 90A, *Air Conditioning and Ventilating Systems.*
NFPA 91, *Blower and Exhaust Systems.*
NFPA 231, *Indoor General Storage.*
NFPA 231C, *Rack Storage of Materials.*
NFPA 255, *Method of Test of Surface Burning Characteristics of Building Materials.*

Additional Reading

Bartels, A. L., "Safeguards to Prevent Ignition of Hot Surfaces," *Electrical Review*, London, vol. 203, no. 4 (1978), pp. 29-31.

Davis, R. H. "Fire Protection Systems (for Computers)," *Data Management*, vol. 14, no. 11 (1976), pp. 11-14.

Deacon, F. L., "Designing Fire Protection to Limit Monetary Loss," *SFPE Technology Report No. 80-2*, Society of Fire Protection Engineers, Boston, MA, 1980.

Factory Mutual, "American Experience in the Fire Protection of Computers," *Fire*, vol. 71 (1979), pp. 498-499.

"Halon Prevents Major Central Processing Unit Fires," *Computer Decisions*, vol. 10, no. 8 (1978), pp. 56, 58.

Harrington, J. L. and R. B. Hopkinson, *Rack Storage Protection*, Worcester Polytechnical Institute, Worcester, MA, 1977.

King, P. W. and J. Magid, *Industrial Hazard and Safety Handbook*, Butterworths Publishing, Inc., Woburn, MA, 1979.

McKinnon, G. P., ed., *Fire Protection Handbook*, Fifteenth Edition, National Fire Protection Association, Quincy, MA, 1981.

Nailen, R. C., "Toxic, Flammable Chemicals, Gases, Breed Trouble in Electronic Plants," *Fire Engineering*, vol. 133, no. 10 (October 1980), pp. 54-57.

Perkins, C. and B. J. Berenblut, "Electronic Equipment: What Protection Is Required?," *Fire Surveyor*, vol. 9, no. 5 (October 1980), pp. 28-33.

Philbrick, S. E., "Selecting Cables for Fire-Risk Applications," *Electronics and Power*, vol. 26, no. 3 (March 1980), pp. 232-233.

Wolf, Helmut F., *Semiconductors*, John Wiley and Sons, Inc., New York, NY, 1971.

Yang, Edward J., *Fundamentals of Semiconductor Devices*, McGraw Hill, New York, NY, 1978.

Part Three

SPECIAL PROCESS FIRE HAZARDS

24

Welding and Cutting

L. G. Matthews

Welding, very broadly, encompasses any materials-joining process. As used in this chapter, *welding* means only fusion welding, i.e., melting together. Solid-state processes, such as forge welding and ultrasonic welding, are not included.

Similarly, in reference to cutting, melting is the significant feature. The correct term is *thermal cutting*, distinguishing it from mechanical severing processes such as sawing. (The term *burning* is frequently heard, but this is not preferred terminology, as it is usually used to describe oxy-fuel gas cutting.)

Both welding and thermal cutting require high-intensity energy sources — usually electricity or the combustion heat of a fuel gas. In this text it is not possible to cover *all* welding and cutting processes. They number in the hundreds; some are highly sophisticated and highly specialized in their use — for example, atomic-hydrogen welding and laser-beam welding. Obviously these less known and less practiced processes are not free of potential hazard, but the intent here is to present basic in-

---

L. G. Matthews, Director of the Safety Affairs Department (Ret.), Linde Division, Union Carbide Corporation, wrote this chapter. T. E. Willoughby, Associate Director, Linde Division of Union Carbide, reviewed the chapter for the second edition.

formation about the more common processes. These, by virtue of their characteristics and greater use, represent the preponderance of fire hazard situations.

As a source of ignition, welding and cutting account for about 6 percent of the fires reported in industrial properties,[1] as well as being the source of many fires in other properties. Means to prevent these fires are known and, for the most part, have been known for a long time. (The first edition of NFPA 51, *Installation and Operation of Oxygen-Fuel Gas Systems for Welding and Cutting*, was adopted in 1925; and in 1937, the now-disbanded International Acetylene Association published *Safe Practices for the Installation and Operation of Oxyacetylene Welding and Cutting Equipment*.) The necessary precautions are not difficult to understand. Through education, training, and on-the-job practice, the potential for welding and cutting fires can be reduced significantly. There are four things which must be kept in mind:

1. Two sides of the fire triangle are always present; namely, a source(s) of ignition and air to support combustion. The other controllable element is the combustible material.
2. The location where the work is being done is an important factor. Fires are seldom encountered in shops where the welding or cutting operation is a regular part of production, because the area is generally designed and protected in full recognition of the process hazards. However, welding and cutting torches are also commonly used for maintenance, repair, construction, and similar activities, which means they are used just about anywhere. Some years ago it was stated that about nine out of ten fires traceable to welding and cutting occurred when the work was done in a temporary location using portable equipment.[2]
3. The kind of process and equipment used affect the potential for fire. Cutting and certain arc-welding operations produce literally thousands of ignition sources in the form of sparks and hot slag. Arcs and oxy-fuel gas flames are inherent ignition sources, but they rarely, in themselves, cause fires except where they have overheated combustibles in the vicinity of the work.
4. The welder and the welder's supervisor are critical factors. Is the welder trained in the proper use of the equipment and mindful of the exposure? Who has knowledge of, and who has assessed, the risks involved in bringing a torch into the work area, and who has authorized its use there?

PROCESSES USING ELECTRICITY

There are several electrical welding and cutting processes pertinent to the scope of this text.

Arc Welding

This term applies to a number of processes that use an electric arc as the heat source for melting and joining metals. The arc is a useful tool because its heat can be concentrated and controlled quite effectively. Frequently, but not always, a filler metal must be used to obtain a good joint. The arc is struck between the metals to be welded and an electrode, which is maneuvered along the joint or which may remain stationary while the work is moved beneath it. The electrode may be consumable or nonconsumable. If the latter, a separate rod or wire may be used as the filler metal. Consumable electrodes supply their own filler metal by melting and, by decomposition of either a covering or a core, may shield the weld zone from unwanted atmospheric effects.

Shielded metal arc welding: This simple, well-known process, is widely used with ferrous-base metals. It produces coalescence* of metals by heating them with an arc between a covered metal electrode and the work. The hard electrode cover (powdered materials such as carbonates, oxides, and fluorides with binders) helps by producing slags and gases to exclude air and protect the weld. It may also assist in arc stabilization and in the addition of alloying elements to the weld. This process requires an alternating or direct current power supply, power cables, and an electrode holder. A primary electrical line capable of supplying up to about 10 kW (kilowatts) can be the power source. A gasoline engine-driven power source can be used where portability is required. Shielded metal arc welding can be done readily in remote, unusual, or confined locations. Consequently, this process is widely used in such industrial applications as construction, shipbuilding, and pipeline erection. Fairly simple, portable units find frequent use in maintenance and field construction work.

Gas metal arc welding: This process uses a continuous, solid wire filler metal, which functions as one terminal of the arc, and it employs a gas to shield the arc and the weld metal. The shielding gas depends on the base metal and process variations. It may be argon (inert), carbon dioxide (oxidizing or oxidizing and carburizing), or mixtures involving these gases with additions of helium or oxygen. Overall, this process can be used to join almost any metal in any configuration of joint.

Flux cored arc welding: Similar to gas metal arc welding in equipment used and types of applications, this process uses cored rather than solid electrodes. Minerals and alloys in the core assist in weld protection, and many such electrodes are intended for use with carbon dioxide-rich gases. Other kinds of cored electrodes are self-shielding and produce their own protective envelope without an auxiliary shielding gas.

Gas tungsten arc welding: This process is similar to gas metal arc weld-

---

*This is the preferred American Welding Society term used to describe "the growing together into one body of the materials being welded."[3]

Chapter 24

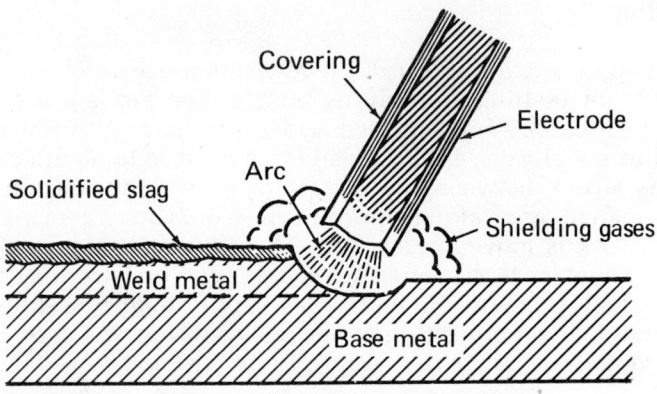

FIGURE 24.1. Shematic diagram of a typical shielded metal arc welding operation. (Lincoln Electric Company)

FIGURE 24.2. A portable AC electric generator and arc welder set. (Lincoln Electric Company)

ing and flux cored arc welding, but it uses a nonconsumable tungsten electrode for one pole of an inert gas-shielded arc. Filler metal may be used. Shielding gases are argon and helium. The tungsten electrodes are available in pure form and in *thoriated* and *zirconiated* varieties. Required equipment comprises a power supply, welding torch, source of inert gas, suitable pressure regulators and flowmeters, and connecting hoses. Arc welds of the highest quality are possible with this process.

Plasma arc welding: This process is characterized by the plasma state, wherein the temperature of argon gas is raised until the gas becomes at least partially ionized, enabling it to conduct an electric current. A plasma arc torch incorporates a tungsten electrode and an orifice through which a small flow of argon forms the arc plasma. Secondary gas flow through an outer nozzle cup arrangement that encircles the arc plasma orifice shields the arc and the weld. The shielding gas may be argon, helium, or mixtures of argon with hydrogen or helium. Filler metal may or

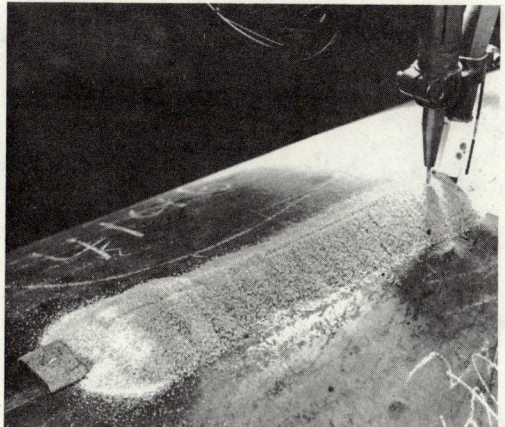

FIGURE 24.3. *Typical submerged arc welding.* (Linde Div., Union Carbide Corporation)

may not be used. Equipment required is not unlike that for gas tungsten arc welding with necessary differences in power supply and control, torch design, and gas supply and control.

Submerged arc welding: In this process the arc and molten metal are shielded by molten flux and a layer of unmelted flux granules. A continuously fed electrode is submerged in the flux, the arc is not visible, radiation and fumes are minimal, and fire hazard potential is likewise relatively minor compared to other processes. Most submerged arc welding operations are fully mechanized, although there is some semi-automatic equipment in use.

Resistance Welding

Welding heat for this process is created by resistance to the flow of electric current through the parts being joined. It is an in-line production joining technique, which uses fully mechanized and automatic or semi-automatic machines, as in the auto industry. Most commonly, resistance welding is used to join two overlapping metal sheets, which may have different thicknesses. Electrodes conduct electric current through the sheets, which are clamped or rigidly held to provide good contact and pressure for holding molten metal at the joint. The machines are rather sophisticated, and instantaneous power demand is very high.

FIGURE 24.4. Flash welding operation. (American Welding Society).

Flash Welding

Although frequently classed as resistance welding, flash welding is a separate process.[3] Heat is created by both resistance to current flow and by arcs at the interface. Force is applied after heating, resulting in the expulsion of metal, formation of a flash, and usually a significant shower of sparks. Applications usually involve butt welding rods or bars, end to end or edge to edge. One of the items to be welded is moved slowly toward the other, current flow occurs at contacting surface irregularities, and the flashing action begins. The action continues until a molten layer forms on both surfaces, the components are forced together rapidly, and molten metal and debris are squeezed out of the joint. The process is almost always automatic, and flash welding machines vary significantly in size, some being large enough to weld sheets in a steel mill.

Electroslag Welding

This process uses a slag, which is conductive while molten, to protect the weld and to melt base metal edges and filler metal. An arc is needed to start the process by melting the slag and preheating the work because the unfused, solid slag is nonconductive. Once the process is started, the arc is no longer necessary, and resistance to current flow through the molten slag produces the heat required to sustain the process.

WELDING AND CUTTING

FIGURE 24.5. Air carbon arc cutting. (American Welding Society)

Arc Cutting

This term applies to a group of processes that, like similar welding processes, melt the metals to be cut by heating them with an arc struck between an electrode and the base metal.

Air carbon arc cutting: This cutting process employs electrodes comprising a mixture of graphite and carbon, which are coated with a copper layer to improve current-carrying capacity. Metal melted by the arc is blown away by a jet of compressed air supplied by conventional compressors, usually at about 80 psi (550 kPa). Current is provided by standard electric welding power supply units. The process is applicable to most metals, both ferrous and nonferrous, and is frequently used in gouging, an operation involving the forming of a bevel or groove by metal removal.

Plasma arc cutting: This cutting process is closely allied to the plasma arc welding process and achieves cutting action with an extremely hot jet at high velocity, produced by forcing an arc and inert gas through a small orifice. Arc energy confined to a small area melts the metal, and the jet of highly heated, expanding gases forces the molten metal from the cutting area. High arc voltages, special power supplies, and water-cooled torches are required. Argon-hydrogen or nitrogen-hydrogen gas mixtures are commonly used. Additional energy may be provided when cutting cast iron or carbon steels by supplementing the gas stream with oxygen. The plasma arc process is usable with all metals — a significant feature — and may be performed manually or, more commonly, with mechanized equipment.

Other arc cutting processes do exist and are sometimes used — for example, shielded metal arc cutting, gas tungsten arc cutting, carbon arc cutting. Some are difficult to regulate and may not produce good cuts.

Chapter 24

FIGURE 24.6. A single plasma arc torch mounted on a large cutting machine cuts 5/16-in. (8mm) low carbon steel at speeds of 175 in./min (4.44 m/min). Linde Div., Union Carbide Corporation)

OXY-FUEL GAS PROCESSES

Oxy-fuel Gas Welding

This process uses the heat of combustion of a fuel gas and oxygen flame to melt the workpiece base metal and filler metal, if used. Oxygen, as distinguished from air, produces flame temperatures high enough to be useful, particularly with steels. Also, the flame (except for oxygen-hydrogen) is readily adjusted to provide a "neutral" environment, as opposed to *oxidizing* (excess oxygen) and to *reducing* or *carburizing* (excess fuel gas). This neutral flame condition is necessary to prevent contamination of the molten metal before it solidifies.

Historically, oxy-acetylene welding introduced a new era, with industrial usage beginning in about 1903.[4] Acetylene remains the pre-eminent welding fuel because of its unique properties. Recently, other fuel gases or mixtures, such as methyl acetylene-propadiene (stabilized), have found limited acceptance. Other hydrocarbons (propane, propylene, butane, natural gas-methane) are not generally suitable for application to ferrous metals. Hydrogen has a low flame temperature and heat content, and its flame, which is colorless, is difficult to adjust for oxygen-fuel gas ratio. Oxy-hydrogen welding does find application for welding lead — frequently but erroneously called lead burning. Table 24.1 shows maximum and neutral flame temperatures of some fuel gases.[3]

With the exception of acetylene, neutral flame temperatures are significantly lower than the maximums, so that it is not possible to melt and control the weld, except perhaps for thin sheet metal. At maximum temperatures, the flames are strongly oxidizing and not usable because

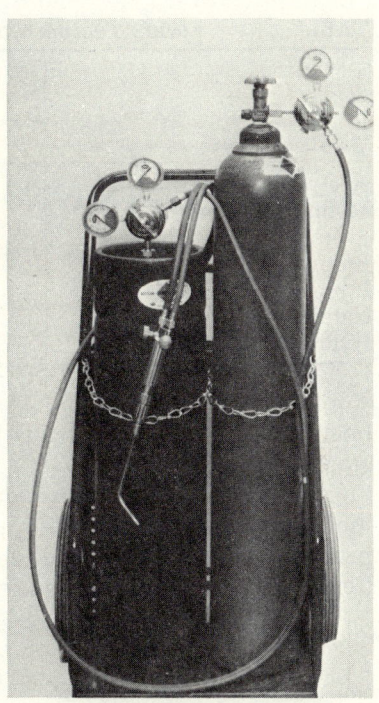

FIGURE 24.7 A portable welding outfit with oxygen and acetylene outfits chained to an easy-rolling cylinder truck (cutting attachments not shown). (Linde Div., Union Carbide Corporation)

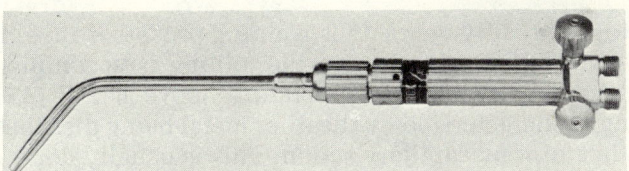

FIGURE 24.8. Welding torches of this design are by far the most widely used. They will handle any oxy-acetylene welding job, can be fitted with multi-frame heads for heating applications and accommodate cutting attachments that will cut steel 6 in. (152.4 mm) thick. (Linde Div., Union Carbide Corporation).

of deleterious metal oxide formation in the weld metal. Limited use of methyl acetylene-propadiene (stabilized) requires special procedures.

Combustion of acetylene in oxygen at a torch tip takes place in two steps. In the first step, acetylene + oxygen → carbon monoxide + hydrogen + heat. This takes place in a bright inner cone close to the torch. Additional oxygen supplied by the surrounding air completes the reaction (second step); namely, carbon monoxide + hydrogen + oxygen → carbon dioxide + water vapor + heat. This produces a brush-like outer flame. The heat generated in the second step serves largely for pre-

TABLE 24.1. *Flame Temperatures of Fuel Gases with Oxygen*

Gas	Maximum Flame Temp.		Neutral Flame Temp.	
	(°C)	(°F)	(°C)	(°F)
Acetylene	3,102	5,615	3,100	5,612
Methyl acetylene-propadiene (stabilized)	2,902	5,255	2,600	4,712
Propylene	2,857	5,174	2,500	4,532
Propane	2,777	5,030	2,450	4,442
Natural gas-methane	2,742	4,967	2,350	4,260
Hydrogen	2,871	5,200	2,390	4,334

heating. Heat for welding is effectively that from the first combustion step.

Oxy-fuel gas flames can be used to make butt welds, usually in a mechanized operation, without filler metal in the pressure gas welding process. Surfaces to be butted are heated and forced together, giving a forging action that can produce a sound joint.

Brazing and Braze Welding

Brazing: This is broadly defined[3] as welding processes in which the base metal is heated but not fused, and the joining is accomplished with a filler metal having a melting temperature *above* 842°F (450°C). Moreover, brazing is characterized by the filler metal being distributed through a closely fitted joint by capillary action. This is usually done with a manual or mechanized gas torch, usually on lap joints in relatively thin materials.

Braze welding: Braze welding is more significant in relation to fire prevention because of its frequent use in repair and maintenance activities. It differs from brazing in that capillary action in the joint is not a factor; the filler metal is laid down in a groove or fillet at the point of application. The edges of the joint are heated to a dull red color, flux is used, and filler metal is supplied by a bronze rod, hence the occasional erroneous use of the term bronze welding. The process is widely used on ferrous metals as well as on copper and nickel alloys. Sometimes it is the solution to the problem of joining dissimilar metals.

Hard soldering: This common but incorrect term is used to describe brazing with a silver-base filler metal. The operation is often performed with an oxy-fuel gas torch, as in assembling copper pipe joints with silver brazing alloy. Many times, however, the oxygen is not pure gas but is supplied as air, as in the air-acetylene torch. Thus, only one small gas supply cylinder may be needed, and portability is enhanced.*

Surfacing

The oxy-fuel gas flame (usually oxy-acetylene) is used, not in a true welding sense, but to deposit a layer of filler metal on a base metal to obtain certain surface properties or dimensions. Bronze may be used, but more frequently the process is one of *hard-facing*, where the deposited layer is a special alloy that will greatly prolong the life of parts subject to extreme wear or abrasion.

Oxy-fuel Gas Heating Operations

There are numerous industrial and commercial operations that regularly employ an oxy-fuel or air-fuel flame but do not involve a joining or severing operation. The flame and the nature and location of these operations nonetheless may represent an appreciable fire potential. Some such operations are:

- Forming — heating (ironwork, piping, etc.) to facilitate bending, shaping, or straightening.
- Annealing, flame-hardening, flame-softening — use of the flame in a controlled manner to obtain specific properties.
- Flame-priming — heating to remove scale and rust in preparing metal surfaces for painting.
- Flame-descaling — heating (generally a steel mill application) to remove scale from bars, billets, slabs, etc., to facilitate machining or inspection.
- Other applications — paint burning, glass finishing, leather edging, babbitting, antiquing of wood.

Oxy-fuel Gas Cutting

This term describes a process or group of processes named by the specific fuel gas used (for example, oxy-acetylene cutting, oxy-natural gas cutting) for severing metals by the reaction of high-purity oxygen with

---

*The air-fuel gas flame, particularly air-acetylene, also finds significant use in soldering, commonly called soft soldering, which by definition involves a filler metal that melts below 842°C). Plumbing-joint work, refrigeration piping, and heating and ventilation duct assembly are applications where the air-fuel gas flame is used. Equipment portability is an advantage, but it may also signify access to places where fire prevention is not easily accomplished and is easily overlooked because a single joining opoeration can be done in a matter of minutes.

Chapter 24

FIGURE 24.9. *A typical portable oxy-fuel gas cutting machine in operation. (Linde Div., Union Carbide Corporation)*

the metal at elevated temperatures. It can be applied to ferrous metals and titanium. Mild steel plates or sections as thick as 12 in. (305 mm) are easily cut. Alloy steels are more difficult to work with, and stainless steels are cut by introducing flux or metal powder (usually iron) into the oxygen stream.

The burning of iron in oxygen produces iron oxide, normally a solid, but the oxide melts at a temperature below the melting point of iron or steel and runs off as slag. Usual torch tip design has a ring of oxy-fuel preheat flames surrounding a jet of pure oxygen.

A variation of the process is oxy-fuel gas gouging, wherein a relatively low velocity oxygen jet permits gouging or grooving of a metal surface in a reasonably smooth, well-defined manner.

Of great importance to the steel industry is an oxy-fuel gas cutting operation similar to gouging and known variously as steel conditioning, scarfing, or "deseaming." The technique is applied to billets, blooms, and slabs. The term deseaming is usually applied to a hand operation involving removal of surface defects. In mechanized steel conditioning, the entire surface (one, two, or four sides simultaneously) is removed while the piece is still hot and with the machine set into the rolling mill line.

Oxy-fuel Gas Welding and Cutting Equipment

Basic elements are fuel-gas and oxygen supplies, pressure regulators, conduits (hoses or piping) to convey the gases, and a torch to mix and burn the gases in controlled fashion and to provide the jet used in cutting operations. In its simplest form, the equipment comprises a cylinder each of fuel gas and oxygen, a pressure regulator on each cylinder, hoses, and a torch.

FIGURE 24.10. Oxy-fuel gas gouging torch in operation. (Linde Div., Union Carbide Corporation)

At the other extreme, a steel mill, for example, might have a major installation for cutting and welding operations. Fuel gas might be supplied from large multi-cylinder manifolds (possibly truck-mounted and replaceable), from storage tanks for liquefied fuel, or from a public utility natural gas main. Oxygen may be supplied from the mill's own on-site oxygen-generating plant, from storage tanks for liquid oxygen or high pressure gas or both, or from multi-cylinder manifolds. A major installation may also have an extensive fixed piping distribution network and consuming devices running the gamut from simple hand torches to sophisticated steel-conditioning machines.

Auxiliary but necessary equipment includes standardized[5] fuel gas and oxygen hose connections, standardized hose,[6] protective equipment for service piping systems, and shut-off valves at points (station outlets) where gas is withdrawn from the piping system. Protective equipment for piping, which may be installed at one or more locations, should prevent the backflow of oxygen into the fuel gas supply system, the passage of flashback into the fuel gas supply system, and the development of pressures in excess of ratings of system components. In some systems, backflow prevention devices may also be required at station outlets.

Welding torches have inlet connections and valves for each gas at the rear of the handle. The gases, controlled by the inlet valves, are thoroughly mixed before issuing from the torch tip. Normally, a series of tips can be fitted to one handle so that flame size can be varied to accommodate a variety of work. Cutting torches are similar but provide passageways and separate valving for supplying and controlling the cutting oxygen jet at the center of the tip. Some welding torches are designed to accommodate a cutting attachment.

Mechanized cutting is common. Equipment varies from relatively simple, portable machines, used for straight-line work and perhaps cir-

Chapter 24

FIGURE 24.11. An oxy-fuel gas scarfing machine in operation at a large steel mill. (Linde Div., Union Carbide Corporation)

FIGURE 24.12. A large stationary oxy-fuel gas cutting machine in operation. (Linde Div., Union Carbide Corporation)

cles and some irregular shapes, to highly sophisticated multi-torch machines that can trace intricate shapes by photocell, or other electronic means, and accurately produce a number of parts of the same shape.

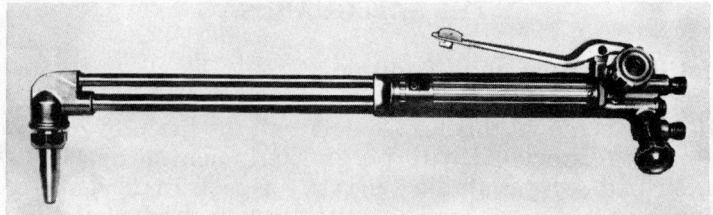

FIGURE 24.13. *A typical cutting torch. All control valvles are located at the rear of the body. (Linde Div., Uniion Carbide Corporation)*

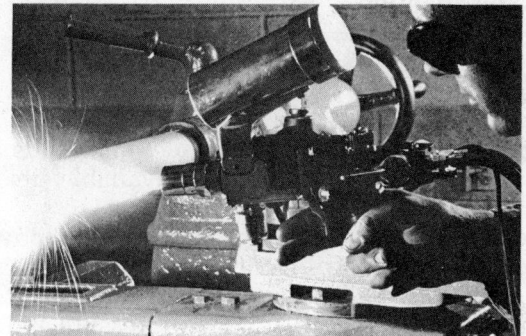

FIGURE 24.14. *Thermal spray gun. (American Welding Society)*

THERMAL SPRAYING

Closely allied to welding are several processes, known collectively as thermal spraying,* in which finely divided metallic or nonmetallic materials are deposited in molten or near-molten condition to form coatings. Special "guns" are used. In flame spraying, an oxy-fuel gas flame is used to melt the coating material, and an auxiliary compressed gas may be used to assist in atomizing and propelling coating material to the workpiece. Electric arc spraying employs an arc between two consumable electrodes of coating material, plus auxiliary compressed gas to atomize and propel the coating particles. Plasma spraying uses a plasma arc as the heat source and for propelling the coating material. In detonation flame spraying, the coating material is melted and propelled by the controlled explosion of fuel gas and oxygen.

---

*American Welding Society preferred terminology for "metal spraying" or "metallizing."

THE SAFEGUARDS

Equipment Preparation and Condition

It is beyond the scope of this text to detail all the fire prevention considerations that are associated with proper equipment design, installation, and maintenance — for example, the safety aspects in design of arc welding machines or the recognized safe practices in the installation of oxygen service piping systems. The bibliography at the end of the chapter will be useful in that regard. Nor is it sufficient to say merely that the equipment "shall be in satisfactory operating condition and in good repair."[1] Therefore, based on experience, some significant items have been selected for specific attention.

Oxy-fuel Gas Equipment

Where approved* equipment (torches, cylinder manifolds, pressure regulators, pipeline protective devices, etc.) is available, its use is recommended.

It must be recognized that oxygen is a far more powerful oxidizer than air; therefore, oxygen equipment must be kept clean (free of oil, grease, and other combustible contaminants). "Materials that burn in air will burn violently in pure oxygen at normal pressure and explosively in pressurized oxygen. Also, many materials that do not burn in air will do so in pure oxygen, particularly under pressure."[1] Consequently, it is widely recognized good practice to have equipment for oxygen service dedicated to oxygen service only.

Proper storage of gas cylinders is important. Details can be found in NFPA 51, *Oxygen-Fuel Gas Systems for Welding and Cutting*.

Cylinders should be moved and handled in accordance with recognized practices.[8,9] There are many such recognized practices, and their importance may vary with working locations and conditions. One such precept is almost always pertinent to safety and fire prevention; namely, cylinders should be supported or located in such a way that they cannot be knocked over.

Periodically, oxy-fuel gas cutting and welding equipment should be tested for leaks. Frequency of testing depends on the specific kind of equipment involved and how often it is used. Hose connections are known possible trouble spots. Also, experience has shown that far too many fires occur at fuel gas cylinder-to-regulator connections simply because someone failed to tighten the joint properly. The ignition of leaking gas at this point may in turn cause the release of cylinder safety devices, especially with acetylene cylinders, thereby releasing more gas and increasing the size of the fire.

---

*Official NFPA definition—"approved means acceptable to the authority having jurisdiction."

Only standard welding hose should be used.[6] They should be inspected frequently for burns, cuts, worn places, abrasions, and similar defects. Taped repairs are unacceptable. Replace the damaged hose, or if feasible, cut out the affected area and insert a proper splice.

When repairs to equipment such as torches and regulators are required (beyond simple operations such as repacking a torch valve), they should be done by trained, skilled mechanics. Improper repairs, especially to oxygen regulators, have been responsible for a number of fires and serious injuries.

Arc Welding Equipment

1. Where equipment meeting recognized American National Standards[10,11] is available, it should be chosen. Other criteria may be found in Paragraph 4.2 of Reference 8. Installation, including incoming power lines, and grounding of the machine frame or case, should comply with NFPA 70, *National Electrical Code*, with particular attention to its Article 630, "Electric Welders."
2. Proper storage and handling of cylinders of shielding gases[9] need to be observed.
3. At each work location, cylinders should be supported or located in such a way that they cannot be knocked over accidentally, and precautions must be taken that they are not grounded.
4. Cable sizes should be chosen that are adequate for current and duty cycles anticipated. Sustained overloading of inadequate cables may burn away insulation. Cables should be inspected frequently for wear and damage and, when necessary, properly repaired or replaced.

Precautions for the Work Area

Potential fire hazards may be present in the vicinity of all welding and cutting operations. The four key points covered at the opening of this chapter briefly restated are:

1. Two sides of the fire triangle — source of ignition and air to support combustion — are always present.
2. The majority of fires involve portable equipment used in temporary work locations.
3. The nature of the process and equipment used is significant.
4. The nature and thoroughness of fire protection supervision in the work area and the quality of the welder's training in fire prevention are vital concerns.

Dangerous sparks — globules of molten, burning metal or hot slag — are produced by both welding and cutting operations. Those from cutting, particularly oxy-fuel gas cutting, are generally more hazardous than those from welding because they are more numerous and travel greater

Chapter 24

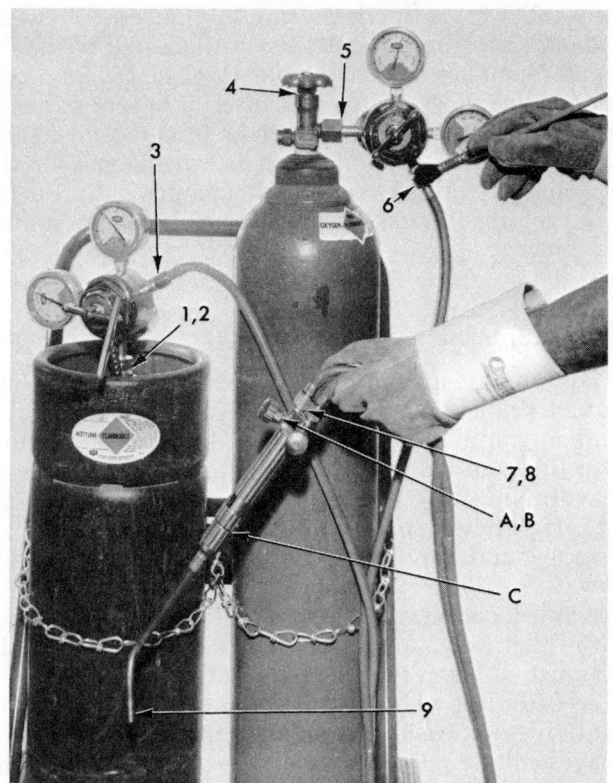

FIGURE 24.15. *Before operating an oxy-fuel gas outfit, the equipment should be checked for leaks. After pressurizing both hose lines (with torch valves tightly closed) test for leakage at the following points, using an approved leak-test solution or a thick solution of mild soap and water: (1,2) Acetylene cylinder connection and acetylene cylinder valve spindle, (3) Acetylene regulator-to-hose connection, (4) Oxygen valve spindle, (5) Oxygen cylinder connection, (6) Oxygen regulator-to-hose connection, (7,8) Hose connections at the torch, (9) Torch tip (for leakage past the torch valves). Later, after lighting the torch, check for leakage at the throttle valve stems (A,B) and at the welding head-to-torch handle connection (C). (Linde Div., Union Carbide Corporation)*

distances. In a sense, they are jet propelled by the oxygen or airstreams used in the cutting processes. Oxy-fuel gas flames and electric arcs are inherent and obvious ignition sources, as are hot workpieces or sections cut from the base workpiece. However, experience shows these to be less frequent ignition sources than sparks.

The key, therefore, to the physical environment aspect of fire prevention is isolation or protection of combustibles. Often combustibles are not obvious, but they may be exposed to sparks that fall through cracks or other openings in floors and partitions. If those sparks are of sufficient mass to retain heat for a time, they may ignite combustibles. The bare essentials of combustible control in the cutting or welding work area are the following:

1. Move all combustibles a safe distance away — at least 35 ft (10.7 m) horizontally — and be sure that there are no openings in walls or floors within 35 ft (10.7 m); or
2. Move the work to a safe location; or
3. If neither of the foregoing steps is possible, protect the exposed combustibles with suitable fire-resistant guards and provide a trained fire watcher with extinguishing equipment readily available.

These steps form only a partial solution to the problem of preventing cutting and welding fires. There are other important factors to consider. Are there any inconspicuous combustibles in an area proposed for cutting and welding operations? What conditions must be met before cutting and welding operations can take place? Who has the responsibility for authorizing the work to proceed? Are cutters, welders, and their supervisors properly trained in the use of their equipment and in emergency procedures should a fire occur? If an outside firm is engaged to do cutting and welding work, chances are that its employees will be unfamiliar with the premises and its contents. Have they been briefed on the conditions in the areas where they will work?

The difficulties of achieving an easily definable and totally effective fire prevention methodology for cutting and welding operations are apparent but not insurmountable. Procedures have been devised which, if fully implemented, can be effective in preventing welding- or cutting-initiated fires. The National Fire Protection Association's Committee on Cutting and Welding Practices developed NFPA 51B, *Standard for Fire Prevention in Use of Cutting and Welding Processes*, which was first adopted in 1962, and has been revised several times since. For those needing to know the full details, there is no substitute for the full text of the standard. However, based on the fundamental but necessary understanding that welders, their supervisors, and facility management share the responsibility for firesafety, the following streamlined version will serve quite well:

1. Management should establish areas designed and authorized for cutting and welding and/or designate a knowledgeable person to authorize welding or cutting in areas not specifically designed for such processes. This management designee should require trained fire watchers wherever there is potential for significant fire to develop. Fire watchers should also be required where appreciable quantities of shielded combustibles are less than 35 ft (10.7 m) away, where wall or floor openings within 35 ft (10.7 m) expose combustibles in adjacent areas, or where combustibles adjacent to opposite sides of partitions, ceiling, or roofs are likely to be ignited by heat from the work.

Management should select contractors with a view to their awareness of risks and the quality of their personnel. Management should also advise contractors of the presence of flammable materials or other hazardous conditions on the property work site.

2. The supervisor of welding and cutting operations in areas not designed for such processes (for example, the plant manager, plant maintenance foreman, contractor, or contractor's foreman) should be assigned the following responsibilities:

   Determine what combustible materials are present at the work site.

   If necessary, have the work moved, have the combustibles moved, or have the combustibles shielded.

   Secure authorization from management, preferably in the form of a written permit.

   See that the welder is aware of the authorization and conditions.

   See that fire watchers are available when required.

   Make a final check for fires one-half hour after completion of welding or cutting operations in cases where a fire watcher was not required.

3. Welders should have their supervisor's approval before starting to cut or weld, should handle their equipment safely, and should continue to work only so long as approval conditions are unchanged.

4. There are a number of precautions to be observed during welding and cutting operations:

   Welding and cutting must not be permitted in flammable (explosive) atmospheres; near large quantities of exposed, readily ignitable materials; in areas not authorized by management; or on metal partitions, walls, or roofs with combustible covering or with combustible sandwich-type panel construction.

   Floors should be free of combustibles such as wood shavings. If the floor is of combustible material, it should be kept wet or otherwise protected.

   If combustibles are closer to the welding or cutting process than 35 ft (10.7 m) and the work cannot be moved or the combustibles relocated at least 35 ft (10.7 m) away, they should be protected with flameproofed covers or metal or asbestos guards or curtains. This also applies to walls, partitions, ceilings, or roofs of combustible construction.

   Openings in walls, floors, or ducts within 35 ft (10.7 m) of the work should be covered.

   Cutting or welding on pipes or other metal in contact with combustible walls, partitions, ceilings, or roofs should not be done if close enough to cause ignition by heat conduction.

   Charged and operable fire extinguishers should be readily available.

   Trained fire watchers should be required. In the absence of fire watchers, an important minimum step would be to check the work area and adjacent areas carefully for at least one-half hour after completion of welding or cutting to detect possible smoldering fires.

SPECIAL SITUATIONS AND ADDITIONAL PRECAUTIONS

Containers That Have Held Combustibles

Every year inevitably brings a number of unfortunate and unnecessary accidents involving severe personal injuries or fatalities from explosions or fires arising from welding or cutting on containers that have held combustibles. Those frequently involved are drums that have been used for oil or chemicals, but smaller drums or cans, large tanks, and pipelines have also been the workpieces in such accidents. It is essential that flammable liquids, solids, or vapors be removed by some type of adequate cleaning procedure. Depending upon the application, it may be necessary or desirable to supplement the cleaning with inerting, water flooding, or periodic testing (for flammables) of the atmosphere within the container.

The cleaning procedure needs to be carefully considered, based on the flammable to be removed, and thoroughly carried out. It can be time-consuming and may well cost more than the used container is worth. With some materials, water washing or flushing may be sufficient; some may respond to steam cleaning; some may require alkaline cleaners (trisodium phosphate or caustic soda); some may call for more specialized methods. Heavy, viscous liquids and solids can be especially troublesome because they are difficult to remove completely and because residues left may be volatilized or decomposed into flammable products by the heat of the torch or arc. Details can be found in References 12 and 13.

Jacketed Containers and Hollow Parts

There are some similarities between the hazards of cutting or welding jacketed containers or other hollow workpieces and those of cutting or welding containers that have held combustible materials. Air confined inside an unvented hollow part will expand when heated, and pressure will increase. Since hot metal loses its strength rapidly, the container or hollow piece may burst with explosive force at the focus of the cutting or welding work. It is well to be suspicious of closed metal parts that seem unusually light. Drilling a vent hole before heating the part may be indicated.

Chapter 24

Hot Tapping

There are occasional situations where emergency repair or the complete impracticality of emptying and cleaning necessitates welding or cutting on a container while it holds flammable gas or liquid (for example, a natural gas transmission pipeline or utility distribution system). Schemes to safely accomplish such hot tapping have been developed.[14,15] Needless to say, any such work should be performed *only* by those specially trained and qualified, using recognized and authorized methods.

Public Exhibitions and Demonstrations

Special and enhanced fire safeguarding is in order when welding or cutting is done at trade shows and exhibitions where places of public assembly (auditoriums, hotel exhibit halls) and concentrations of people are involved. When welding or cutting is to be done in such occupancies, the fire department should have prior notification, operations should be under the control of a particularly competent person, gas cylinders should be charged to only one-half their maximum permissible content, storage sites for gas cylinders should have special restrictions, and fire extinguishing equipment should be appropriate to the situation.[1]

Personnel Protection

One aspect of welding and cutting fire protection that is sometimes overlooked or inadequately considered is the safety of the operator, helpers, or nearby workers. Clothing selection is important, and what is appropriate will vary with the nature and location of the work to be done and equipment to be used. Flame-resistant leather gloves are generally required, though other flame-resistant material may serve for light work. Woolen clothing is preferable to cotton. If used, cotton should be chemically treated to reduce its combustibility. Aprons of leather or some other durable, flame-resistant material may be needed, and cape sleeves or shoulder covers are sometimes worn during overhead work, with skull caps under helmets or with goggles. Leggings are also available for heavy work. High-top safety shoes are a general recommendation. Trousers should not be turned up or cuffed on the outside, front pockets on clothing should be eliminated, and sleeves and collars are usually kept buttoned to keep sparks from entering and lodging in such places. Outer clothing should be free of oil and grease. Welders should never carry matches in their pockets. They are easily set-off by stray sparks.

Oxygen must never be used to cool the welder, ventilate a confined space, or dust off clothing. As pointed out earlier, it accelerates combustion. Tests and experience have shown that oxygen-saturated

FIGURE 24.16. *Attention to clothing is an important factor in operator safety. Note the protective mask and gloves. (Linde Div., Union Carbide Corporation)*

clothing or clothing in an oxygen-enriched atmosphere will literally burn in a flash, with extremely serious and sometimes fatal results.

Several other operator-related precautions are concerned with health rather than fire but should not be overlooked. Proper eye protection from radiation is a necessity (see Reference 8). Some metals and coatings (for example, lead, cadmium, mercury, beryllium, zinc) will produce toxic fumes when heated. Depending upon the environment, careful attention to ventilation and/or provision of self-contained breathing apparatus may be required. Fumes and gases generated will also vary with the welding or cutting process. These considerations are doubly important where work is done in a confined space.

Manufacturers' Recommendations

Procedures given in manufacturers' instructions for setting up, connecting, lighting or starting, adjusting, and maintaining equipment have specific, safety-oriented purposes. They should be followed. In addition, many equipment suppliers have material safety data sheets, precautions, and safe practices publications that are valuable contributions to fire and personnel safety.

BIBLIOGRAPHY

REFERENCES CITED

1. NFPA 51B, *Fire Prevention in the Use of Cutting and Welding Processes.*
2. Matthews, L. G., "Cutting and Welding Fires — A Continuing Challenge," *Jour-*

nal of the American Society of Safety Engineers, vol. 1, no. 2, May 1956, pp. 23-26.
3. Weisman, C., ed., *Welding Handbook*, Seventh Edition, vol. 1, no. 1, American Welding Society, Miami, FL, 1976.
4. *The Oxy-Acetylene Handbook*, Union Carbide Corporation, Danbury, CT.
5. Pamphlet E-1, *Standard Connections for Regulators, Torches, and Fitted Hose for Welding and Cutting Equipment*, Compressed Gas Association, Arlington, VA.
6. *Specifications for Rubber Welding Hose*, Fourth Edition, Rubber Manufacturers Association and Compressed Gas Association, Arlington, VA.
7. NFPA 51, *Oxygen-Fuel Gas Systems for Welding, Cutting and Allied Processes*.
8. ANSI Z49.1, *Standard for Safety in Welding and Cutting*, American National Standards Institute, New York, NY.
9. Pamphlet P-1, *Safe Handling of Compressed Gases in Containers*, Compressed Gas Association, Arlington, VA.
10. NEMA EW-1, *Electric Arc Welding Apparatus*, National Electrical Manufacturers Association, Washington.
11. UL 551, *Safety Standard for Transformer Type Arc Welding Machines*, Underwriters Laboratories Inc., Chicago, IL.
12. A6.0, *Safe Practices for Welding and Cutting Containers That Have Held Combustibles*, American Welding Society, Miami, FL.
13. NFPA 327, *Cleaning or Safeguarding Small Tanks and Containers*.
14. ANSI B31.8, *Gas Transmission and Distribution Piping Systems*, Paragraph 841.27, American National Standards Institute, New York, NY.
15. SD-2201, *Welding or Hot Tapping on Equipment Containing Flammables*, American Petroleum Institute, Washington, DC.
16. Audio Visual AV-1, "Safe Handling and Storage of Compressed Gases," Compressed Gas Association, Arlington, VA.

NFPA Codes, Standards, and Recommended Practices

Reference to the following NFPA Codes, Standards, and Recommended Practices will provide further information on the safeguards for welding and cutting operations discussed in this chapter. (See the latest *NFPA Codes and Standards Catalog* for availability of current editions of the following documents.)

NFPA 10, *Portable Fire Extinguishers*.
NFPA 50, *Bulk Oxygen Systems at Consumer Sites*.
NFPA 51, *Oxygen-Fuel Gas Systems for Welding, Cutting and Allied Processes*.
NFPA 51A, *Acetylene Cylinder Charging Plants*.
NFPA 51B, *Fire Prevention in Use of Cutting and Welding Processes*.
NFPA 70, *National Electrical Code*.
NFPA 306, *Control of Gas Hazards on Vessels to Be Repaired*.
NFPA 327, *Cleaning or Safeguarding Small Tanks and Containers*.

Additional Reading

Accident Prevention Manual for Industrial Operations, Seventh Edition, National Safety Council, Chicago, IL, pp. 947-976.

A6.1, *Recommended Safe Practices for Gas Shielded Arc Welding*, American Welding Society, Miami, FL.

A13.1, *Scheme for Identification of Piping Systems*, American National Standards Institute, New York, NY.

Buehrer, P., "Check List for the Prevention of Fires Arising from Welding and Allied Processes," *Welding in the World*, vol. 14, no. 5-6 (1976), pp. 122-125.

Bulletin SB-4, *Handling Acetylene Cylinders in Fire Situations*, Compressed Gas Association, Arlington, VA.

CGA-V-1, *Compressed Gas Cylinder Valve Outlet and Inlet Connections*, Compressed Gas Association, Arlington, VA.

CGA-4, *Method of Marking Portable Compressed Gas Containers to Identify the Material Contained*, Compressed Gas Association, Arlington, VA.

Compressed Gas Association, *Handbook of Compressed Gases*, Reinhold Publishing Corporation, New York, NY.

C5.1, *Recommended Practices for Plasma Arc Cutting*, American Welding Society, Miami, FL.

C1.1, *Recommended Parctices for Resistance Welding*, American Welding Society, Miami, FL.

Jeffus, Larry, *Safety for Welders*, Delmar Publications, Albany, NY, 1980.

Kennedy, Gower A., *Welding Technology*, Second Edition, Bobbs-Merrill Co., Inc., New York, NY, 1982.

Manz, A. F., *The Welding Power Handbook*, Union Carbide Corporation, Tarrytown, New York, NY.

McKinnon, G. P., ed., *Fire Protection Handbook*, Fifteenth Edition, National Fire Protection Association, Quincy, MA, 1981.

Pamphlet E-2, *Hose Line Check Valve Standards for Welding and Cutting*, Compressed Gas Association, Arlington, VA.

Pamphlet E-3, *Pipeline Regulator Inlet Connection Standards*, Compressed Gas Association, Arlington, VA.

Pamphlet E-4, *Standard for Gas Regulators for Welding and Cutting*, Compressed Gas Association, Arlington, VA.

Pamphlet E-5, *Torch Standard for Welding and Cutting*, Compressed Gas Association, Arlington, VA.

Pamphlet G-1, *Acetylene*, Compressed Gas Association, Arlington, VA.

Pamphlet G-4, *Oxygen*, Compressed Gas Association, Arlington, VA.

Pamphlet G-4.4, *Industrial Practices for Gaseous Oxygen Transmission and Distribution Piping Systems*, Compressed Gas Association, Arlington, VA.

"Prevent Cutting/Welding Fires," *The Minnesota Fire Chief*, vol. 16, no. 5 (May/June 1980), pp. 16,48,54.

The Procedure Handbook of Arc Welding, Twelfth Edition, Lincoln Electric Company, Cleveland, OH.

Publication 2009, *Safe Practices in Gas and Electric Cutting and Welding* American Petroleum Institute, Washington, DC.

Publication 2013, *Cleaning Tank Vehicles Used for Transportation of Flammable Liquids*, American Petroleum Institute, Washington, DC.

Publication 2015, *Cleaning Petroleum Storage Tanks*, Second Edition, American Petroleum Institute, Washington, DC.

RWMA-16, *Resistance Welding Equipment*, Resistance Welder Manufacturers Association, Philadelphia, PA.

Resource Systems International, *Welding and Cutting Safety*, Reston Publishing Co., Inc., Reston, VA, 1982.

Safety in Welding and Cutting, Fourth Edition, American Welding Society, Miami, FL, 1973.

Stewart, John P., *Welder's Handbook*, Reston Publishing Co., Inc., Reston, VA, 1981.

Sullivan, James, *Welding Technology*, Reston Publishing Co., Inc., Reston, VA, 1982.

UL-123, *Standard for Oxy-Fuel Gas Torches*, Fifth Edition, Underwriters Laboratories Inc., Chicago, IL.

UL-252, *Standard for Gas Pressure Regulators*, Fourth Edition, Underwriters Laboratories Inc., Chicago, IL.

UL-407, *Standard for High-Pressure Gas Manifolds*, Third Edition, Underwriters Laboratories Inc., Chicago, IL.

Welding Handbook, (5 volumes), American Welding Society, Miami, FL.

Z117.1, *Safety Requirements for Working in Tanks and Other Confined Spaces*, American National Standards Institute, New York, NY.

25

Spray Finishing and Powder Coating

Don R. Scarbrough

Spray application of coatings is a process inherent to the manufacture of a significant percentage of all fabricated products. Regardless of the purpose for which the coating is applied — protective, decorative, lubrication, adhesive, structural, or a combination of these effects — flammable or combustible materials are commonly used in the process. In this chapter, the fire hazards attendant to these processes and the means, both preventive and protective, for controlling those hazards will be discussed as they relate to the process and to the equipment involved.

SPRAY APPLICATION PROCESS

Fluid Coatings

Within industry, the most commonly used atomizing device is the air spray gun. It is available in many variations to provide a wide range of

---

Don R. Scarbrough is Product Safety Staff Consultant for the Nordson Corporation of Amherst, Ohio.

fluid delivery rates, with special features to accommodate different fluid characteristics, and in hand-held or machine-mounted configurations. In simplest terms, the gun is composed of a valve to control the flow of fluid, a valve to control the flow of air, and an atomizing nozzle. Streams of fluid and compressed air are mixed within the nozzle to form a spray of fine droplets. A second air control valve is often provided to regulate the shape of the spray pattern.

Less widely encountered, yet commonly seen in high volume production processing, is the airless atomizer. As its name implies, this device generates a spray of fluid without the use of compressed air. It consists essentially of a valve to control the flow of the process fluid and a nozzle to shape the fluid stream in a manner that will result in the discharge of a high velocity, paper-thin fluid film. The film interacts with the quiet atmospheric air to produce a fine spray. The nozzle is usually made of tungsten carbide or some other extremely hard material to resist wear caused by high velocity fluid flow. Fluid pressures used with this type of atomizer range from about 300 to about 3,000 psi (2068 to 20 685 kPa). Airless atomizers are available as hand-held as well as machine-mounted spray guns. Fluid delivery capabilities of spray guns vary from a few ounces per minute to a few gallons per minute (a few millimeters to several liters), but the adjustment range of a given airless nozzle is limited. The nozzle must be replaced with one of different characteristics in order to produce a substantial change in spray pattern width or delivery rate. Compared to the air spray process, the airless spray gun offers substantially reduced overspray and energy requirements (compressed air), while sacrificing some degree of adjustability.

Air and airless atomizers are also used in electrostatic spray operations. For this process, the appropriate atomizer is built into a spray gun that is provided with a high voltage electrical input. Voltages applied to the gun range from about 35,000 V to somewhat over 120,000 V in currently available equipment. Electrical current drawn by the device is near zero, typically below 100 microamperes. Therefore, the term electrostatic is applied. There is less overspray with the electrostatic method than with the conventional air or airless spray methods because the electrically charged atomized particles are attracted to the grounded conductive or conductive-coated workpiece.

Each of the foregoing processes can be augmented by a process known as "hot spray." In this process, the fluid supply for the atomizer is preheated to some temperature between 100 and 200°F (37 and 93°C). In the lower portion of this temperature range, the primary benefit derived is stabilization of spray fluid viscosity. In the middle and upper portions, spray fluid viscosity is reduced as well as stabilized. Reduced viscosity allows atomization to take place at lower fluid pressure in the airless process or lower air pressure in the air spray process, thereby substantially reducing the amount of overspray in either process. With many types of coatings, considerable improvement in the finish is also realized. Although steam and hot water heat exchangers were used for this process years ago, virtually all industrial paint heaters currently being manufactured are the electric type.

A third type of atomizer in industrial use depends upon electrostatic forces for its operation and is commonly known as an electrostatic disc or electrostatic bell. In its most common configuration, a sharp-edged disc with a diameter in the range of 6 to 12 in. (150 to 305 mm) is mounted with its axis oriented vertically and is charged electrically to about 100 kV. The disc is spun about its axis while the coating fluid is poured slowly onto the surface of the disc. Centrifugal force spreads the fluid into a thin film and carries it to the sharp edge of the disc, where the film is disrupted by electrostatic forces and sprayed in a 360° circumferential pattern. Workpieces are carried through the process zone by a conveyor arranged in the form of a loop surrounding the disc and collect coating as they pass by it. An electric motor or air-driven turbine is used to rotate the disc. Alternative to the disc, a device operating on the same principle is constructed with a bell or cup having a diameter from 2 to 4 in. (51 to 102 mm). This device is most commonly seen in a machine-mounted application, but it can be used in a hand-held gun. While electrostatic atomizers typically function at low fluid delivery rates, they do function at high efficiency, depositing virtually all of the atomized fluid onto the workpiece.

Powder Coatings

A coating process that has gained broad acceptance during the 1970s is the application to workpieces of organic coatings in the form of dry powder. In this process, the powder is first suspended in air and then charged electrostatically from a dc power supply operating between 60 and 120 kV. The powder is then directed toward the intended workpiece upon which it is collected and held by electrostatic forces. Subsequently, the powder is formed into a continuous coating as it melts during passage of the workpiece through a process oven. The powder coating process differs substantially from fluid coating processes inasmuch as no organic solvents are used. Oversprayed coating materials that are discarded in fluid coating processes are recoverable in the powder coating process through use of simple dust collection devices and may be reintroduced to the process. This offers a substantial benefit in the form of essentially 100 percent material utilization.

Electrostatic application of powder is most commonly accomplished with spray gun techniques using both hand and machine-mounted guns. These spray guns are simple devices to which a mixture of powder and air is fed through a single tube. A separate cable connected to the gun provides the high voltage. Since no atomization takes place, the gun merely impresses a high voltage charge on the powder and shapes and directs the cloud of powder toward the workpiece.

Workpieces having dimensions less than 4 in. (102 mm) can be electrostatically powder coated with a device known as an electrostatic fluidized bed. In this process, powder is held in a container having an open top and a porous bottom through which an upward flow of air is constantly maintained. The volume of this airflow is adjusted to a point that causes the powder mass to levitate or "fluidize" within the container but not to be blown upward out of the container. An array of electrostatic charging electrodes near the surface of the fluidized powder imparts an electrical charge to the powder near the surface. Electrically grounded workpieces pass over the surface of the bed and collect a coating of powder. The coating is then cured in a baking oven. Workpieces may be transported through the coating and curing process by conveyor.

Although they are not widely used, there are techniques for applying dry powder coatings without the use of electrostatics. These processes typically involve preheating the workpiece to a temperature substantially above the melting point of the powder coating and then applying the powder to this heated surface, either by dipping the workpiece into a fluidized bed that has no electrostatic components or by spraying air-suspended powder directly onto the hot surface of the workpiece. The powder melts immediately upon contact and flows to form a film, which is subsequently cured in a bake oven. Hot workpiece coatings are usually quite thick and are applied for a protective or functional purpose.

SPRAY PROCESS EQUIPMENT AND COMPONENTS

Fluid Supply

Air spray guns draw their supply of coating fluid either from hose-fed systems or from small containers referred to as cups mounted directly upon the spray guns. Cups are commonly used for application of low volume, decorative finishes or for touch-up operations.

Air spray guns of essentially the same type may also be hose-fed in the configuration shown in Figure 25.1, an arrangement more commonly used for production operations that consume substantially greater quantities of fluid. In this case the fluid may be a pressurized container commonly referred to as a paint tank or pressure pot. Pressure pot capacities range from 2 to 60 gal (7.5 to 227 L).

In more permanent installations, coating fluid is held in an unpressurized tank from which it is drawn by a pump and supplied through a system of pipes to the hose at the paint spray area. Such piping systems are usually arranged to maintain constant paint circulation through a closed loop, thereby preventing sedimentation within the piping. Factories in which extensive painting operations are conducted often have a central "paint kitchen" in which all paints are prepared for spray in such tanks and then pumped to remote areas of the plant where spray processes take place. Fluids supplying airless spray guns are commonly drawn from an unpressurized tank into a pump which increases their pressure to that demanded by the process. Such pumps may be of piston

SPRAY FINISHING AND POWDER COATING

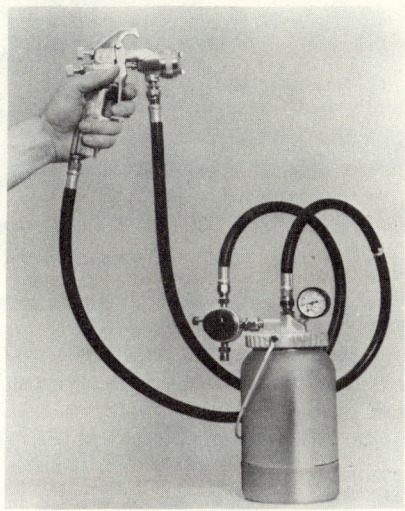

FIGURE 25.1. Air gun, hose-fed. (Graco)

or plunger type driven by an air cylinder or by an oil-powered hydraulic cylinder, or driven directly by an electric motor.

Before use, industrial coating materials are customarily modified by the addition of solvents to adjust viscosity for optimum results in the application process. A portion of the coating material is transferred from its original container to a second container in which it is blended with the solvent. This second container then is transported from the paint mix area to the paint spray booth where the fluid is transferred into the paint tank that supplies the spray guns. To diminish probability of a flammable liquid spill in this transportation process, a safety tank or safety container is used. Transportation of flammable liquids in open pails is poor practice.

Fluid Heaters

For hot spray processes, the temperature of the coating fluid is controlled by a paint heater. These electrically heated, thermostatically controlled devices have inputs ranging from 1.2 to 2.5 kW and are customarily used in tandem when higher capacities are required for specific applications. Designs capable of working pressures as high as 3,500 psi (24 132 kPa) are offered by several manufacturers. Heaters are located in a circulating paint system as near the spray operator as is practicable, usually mounted on the wall of a spray booth in a location where residues will not accumulate upon them (commonly on the outside wall).

Chapter 25

FIGURE 25.2. *Automatic air gun. (Binks)*

Electrical lines are contained in rigid conduit. When heaters are used with portable systems, electrical input is delivered through extra heavy-duty rubber insulated cable, and the heater is mounted on a cart adjacent to its associated pump.

Spray Guns and Devices

Among the various forms of industrial air spray guns, the two configurations most commonly seen are the pistol grip hand spray gun and the machine-mounted automatic air spray gun. Air is supplied to the hand spray gun through one hose, while fluid is either hose-fed or drawn from a gun-mounted container. The valves that control flow of air and fluid are both actuated by a manual trigger, which usually is configured to be operated by two fingers. Some heavy-duty guns, however, have a longer trigger, which will accommodate all four fingers of the hand.

Air spray guns intended for remote operation mounted either on a fixed gun stand or on a machine that will automatically move the gun typically appear very similar to the device shown in Figure 25.2. A third hose connected to the gun serves the air cylinder that actuates the gun in place of the trigger on the hand spray gun. A simpler form of automatic air spray gun has no actuating cylinder for triggering or valves to control air and fluid flow. This type of gun is intended to be controlled by external valves and may be used in applications where space is limited.

Hand and automatic spray guns for cold airless spray have a single hose fitting through which high pressure coating fluid is supplied to the gun. Those intended for the hot spray process have two hose connections to allow fluid to circulate through the gun and around a closed hydraulic circuit. Automatic spray guns have a third connection for the compressed air that controls the triggering cylinder.

Electrical grounding of hand airless spray guns to drain off static electricity generated during spraying is provided through a wire or conductive layer built into the fluid hose. A manual safety lock is usually provided on hand units to secure the trigger and prevent accidental ac-

SPRAY FINISHING AND POWDER COATING

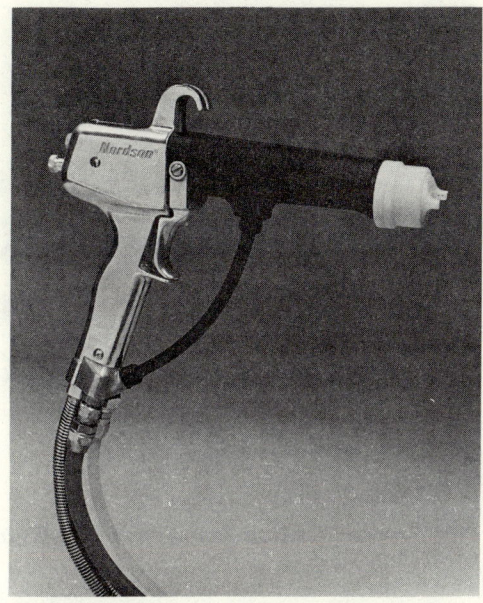

FIGURE 25.3. Hand air electrostatic spray gun. (Nordson)

tuation. Airless guns with adjustable spray pattern or fluid flow are very unusual. Those characteristics are controlled by nozzle selection.

Hand and automatic air electrostatic spray guns have electrically nonconducting extensions to insulate the electrical features of the guns from the grounded parts. (See Figure 25.3.) In addition to being connected to coating fluid and air supplies, these guns are also connected to high voltage power supplies. Voltage at the atomizer is in the 30 to 75 kV range for hand units and in the 30 to 120 kV range for automatic units. In hand spray guns of this type, the cable carrying the high voltage to the gun also supplies an electrical ground for the pistol grip and trigger. To prevent electrical sparking from accidental approach of the charged elements of the gun by a grounded object, the cable is usually terminated at the gun in an electrical resistance on the order of 75 to 250 M Ω (megaohms). Some automatic air electrostatic spray guns are not equipped with a high impedance termination, but rely on process control to maintain separation between gun and workpiece to reduce electrical sparking. Automatic units may have internal or external air and fluid valves; spray triggering and power supply operations are controlled remotely. Air and fluid flow in hand guns are controlled by the familiar trigger, while a separate switch that senses airflow when the trigger is pulled controls the electrostatic power pack.

Airless electrostatic guns require no atomizing air and are not provided with adjustments. In other respects, airless eletrostatic spray guns are similar to their air electrostatic counterparts.

FIGURE 25.4. Electrostatic hand gun and power pack. (Ransburg)

An electrostatic hand gun utilizing a spinning bell-shaped atomizer is shown in Figure 25.4, along with its connecting cable and power supply. A separate motor is built into the gun to spin the bell. Flow of fluid, introduced through a hose at the rear of the gun, and the power supply are controlled by the trigger.

Figure 25.5 shows an electrostatic disc surrounded by workpieces hanging from a conveyor loop. Immediately above the disc is the motor that drives it. Further up in the illustration is a device that moves the disc up and down, enabling it to coat workpieces over the entire height of the carriers. Separate electrical and fluid inputs are remotely controlled. Discs commonly are equipped with variable voltage power supplies that can be adjusted over a range of approximately 50 to 120 kV. Since considerable electrical energy is stored in the disc and its charging system, inadvertent electrical discharge can present both fire and electroshock hazards. Adequate safeguards must be maintained to prevent sparking.

SPRAY PROCESS ENCLOSURES

Vapors and airborne particulate material generated in the spray process must be controlled to prevent distribution into areas where their presence would constitute an unreasonable hazard. (See Chapter 45, "Air Moving Equipment.") Control may be accomplished through the use of a variety of power-ventilated enclosures or through simple physical separation of the spray process from other activities in the factory (open floor

SPRAY FINISHING AND POWDER COATING

FIGURE 25.5. *Electrostatic disc. (Ransburg)*

spraying). Power ventilation systems serving the various forms of enclosure consist of an exhaust fan, an overspray collector, and in some cases, a powered air make-up unit.

Spray Booths

A "spray booth" is defined as a power ventilated structure provided to enclose or accommodate a spraying operation, to confine and limit the escape of spray, vapor, and residue, and to safely conduct or direct them to an exhaust system (NFPA 70, *National Electrical Code*). Within this definition, a wide variety of configurations have been developed both in general purpose forms and in forms to meet special requirements of individual processes. While in most cases the booth structure does physically surround the spray operation, some configurations leave the spray operation unenclosed and accommodate its requirements by surrounding the operation with a controlled stream of air that is being drawn into the booth.

The most popular general purpose configuration for a spray booth is the "open face" or "open front" arrangement. It is a box-like structure that stands on the factory floor and has one open side to provide access for the operator. As dictated by process requirements, the width of the booth may range from 3 ft to more than 30 ft (0.9 to 9 m). Height and depth may also vary over a considerable range. If the coating process in-

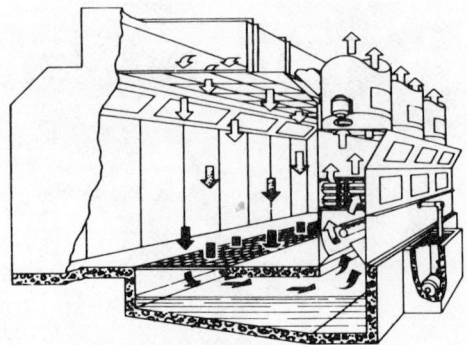

FIGURE 25.6. Tunnel booth. (Blinks)

volves use of a conveyor, the booth may have openings in the walls to allow entry and exit of workpieces. Interior illumination may be provided by approved explosion-proof fixtures inside the booth or by directing external lights through wired glass panels sealed into the walls and ceiling of the booth. The ventilation system associated with the booth may provide an airflow horizontal to the floor (cross draft) or vertical to the floor (down draft) as demanded by process requirements. Although uncommon, some booths have been constructed to provide vertical up-draft airflow.

Some spray booths are totally enclosed structures and may be used for coating workpieces that are carried in and out on carts or for such applications as automotive refinishing. Replacement airflow to support the exhaust system may be introduced through special ductwork or through a system of filters set in openings in the walls or ceiling of the booth.

For operations such as production finishing of automotive bodies, a configuration referred to as a tunnel spray booth is used (see Figure 25.6). This structure is virtually always arranged with vertical down-draft ventilation and a horizontal floor-mounted conveyor running the length of the tunnel. Make-up air is introduced through special ductwork and air diffusers in the ceiling.

Special Purpose Enclosures

Two categories of spray process enclosures that have been developed to a highly refined state as components of special machinery are the continuous coater and the decorating machine. While they fall within the definition of spray booth, they deserve separate discussion.

Continuous coaters are each individually engineered for a specific coating process. Within the coater enclosure is an array of airless spray guns, which may be mounted in fixed position or manipulated by internal machine components. These spray guns are arranged to apply coatings to work pieces that pass through the coater under high volume pro-

FIGURE 25.7. Continuous coaters. (Nordson)

duction conditions, typically at conveyor speeds between 100 and 900 fpm (30 to 274 m/min). The entry and exit portals through which workpieces pass are equipped with exhaust shrouds to capture any vapors or minor amounts of overspray that drift out from the main enclosure. Most of the overspray is collected in a sump at the bottom of the main enclosure and drawn off through a pumping system, which recycles the coating into the coating process. During recycling, solvent is added automatically to readjust the viscosity of the recovered coating material. The exhaust air stream is cleaned thoroughly by an integral venturi type scrubber and an associated settling tank/skimmer unit that removes collected solids from the water used by the scrubber. Coating fluid pumping, heating, filtration, and distribution valving functions to support operation of the coater are performed by equipment mounted on an adjacent control and support panel. Examples of a coater, a support panel, and a settling tank are seen in Figure 25.7.

An example of a decorating machine is shown in Figure 25.8. This particular machine is used in conjunction with masking devices to paint stripes or other patterns on workpieces similar to automobile side moldings. The mask is mounted in the opening beneath the row of cylinders at the front of the machine, and the workpiece is placed manually face down on the mask. When the operating cycle is started, air-driven pistons in the row of cylinders clamp the work to the machine. Automatically, spray guns inside the cabinet tilt in one direction and travel the length of the workpiece. At the end of the workpiece, the spray guns tilt in the opposite direction and travel back to the starting point. The air-driven clamps automatically release the work at the end of the operation.

An exhaust system built into the machine shown in Figure 25.8 draws fresh air in through the grill at the front, and internal filters remove overspray from the air-vapor stream before it is vented at the top.

FIGURE 25.8. Decorating machine. (Finish Engineering)

The machine shown here automatically cleans the mask. It is equipped with three masks — one in the service position, one in the washing position, and the third in a drying chamber where it is readied for reuse. Since the process is totally enclosed, the risk of escaping vapors or overspray and subsequent ignition is significantly diminished.

One other type of spray booth that should be mentioned here is adapted for spray application of cleaning solvents to remove grease and residue from machinery components. Such booths typically are of the open face type, having horizontal airflow and a drip-pan-like structure that serves as the floor of the booth. The pan catches the runoff of solvent, grease, and dirt from the workpieces and leads it to a reservoir for reclamation. Exhaust air is passed through a metal gauze demister prior to discharge to the exterior of the building or, in those cases where a solvent with a high flash point is used, back to the interior of the building. Solvents used in the booth may be applied either by a direct hose stream or by a coarse atomizing air spray gun specially adapted for washing parts. Generally, a small centrifugal or gear pump is used to reintroduce the reclaimed solvent to the spray process.

Spray Rooms

The size, shape, and weight of some workpieces may make the use of spray booths impractical for technical and/or economic reasons. When this is the case, an entire room dedicated to the spray process is a legiti-

SPRAY FINISHING AND POWDER COATING

mate alternative to the spray booth. Such rooms, called spray rooms, are classified as hazardous areas. Workpieces are brought into the spray room by cart, truck, forklift, or overhead traveling crane and are placed on the floor for spray coating.

Power ventilation removes the combustible vapors that are released during the process, but velocities are not high enough to capture particulate matter. Therefore, overspray is permitted to settle to the floor where it accumulates as a combustible residue until it is removed by a mechanical cleaning process (see Chapter 50). Vapor removal ventilation systems are most effective when the exhaust intake is located along one wall within 1 ft (0.3 m) of the floor and the make-up air is introduced along the opposite walls. As the air sweeps across the floor, it carries the vapors into the exhaust. When located in multiple occupancy buildings, spray rooms are usually required by code to be separated from the remainder of the building by fire-resistive construction.

Open Floor Spraying

A spray operation conducted without a spray booth and in an area not separated from general factory operations by a partition is called open floor spraying. Depending upon quantity of volatiles to be released in the operation and upon existing ventilation within the building, forced ventilation may or may not be provided to accommodate this type of spraying. Whether forced or not, adequate ventilation must be provided to prevent accumulation of flammable concentration of vapor. Particulate overspray materials settle to the floor where they accumulate as residues until removed by mechanical means. Ignition sources incidental to other factory operations are a greater concern whenever open floor spray techniques are used, and it is common to establish a buffer zone surrounding the spray area for about a 20-ft (6.1 m) radius to separate processes.

Overspray Collectors

Vapor and overspray removal ventilation systems typically include a fan to create an airflow, and a collection system which separates and collects particulate matter from the airstream and exhausts gaseous matter to the exterior of the building. (Chapter 45 discusses "Air Moving Equipment" in detail.) While such collectors typically are incorporated into the structure of a paint spray booth, they are at times used as components of less formal ventilation systems. These collectors function by directing the airflow in a manner that causes entrained particulates to impinge either onto a dry surface or into a shower of liquid where they will be removed from the air stream. Though many variations exist, overspray collectors can generally be placed into one of four categories.

Baffle maze: Perhaps the simplest of these systems is the baffle maze. It

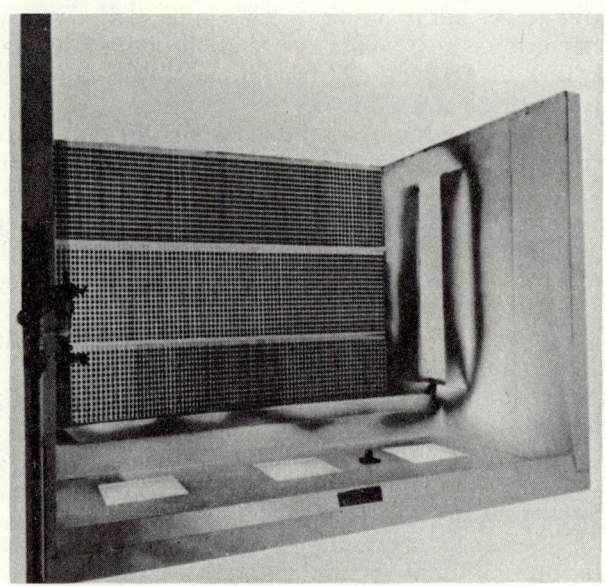

FIGURE 25.9. Andrea filter booth. (Binks)

consists of a series of flat panels arranged in a staggered pattern through which the airstream is directed. As the airstream undergoes several reversals of direction, a substantial portion of its particulate burden is removed by direct impaction and collection upon the surface of the dry baffle panels where it remains until removed by mechanical cleaning. Such systems are of limited efficiency and permit appreciable amounts of fine particulates to pass through, contributing to the accumulation of paint residues within the exhaust stack and on the exhaust fan. This type collector is seldom used in high volume coating operations. The maze shown in Figure 25.9 is composed of a multilayered accordion fold paper structure that can be discarded after becoming fouled.

Dry filter: Collectors using paper or fiberglass filter elements are called dry filters and are popular for low to intermediate volume spray operations. Considerable variety exists in the arrangement of the filter used in this type collector, but typically it is a replaceable element approximately 20 in. (0.51 m) square and perhaps 2 in. (51 mm) thick. Filter banks composed of these elements are constructed in sizes and shapes appropriate to accommodate the particular spray operation with which they are associated. Particulate residues are permitted to accumulate on the filter until a significant obstruction of the airflow through the filter is noted. At this time, the loaded filter elements are replaced by new elements. An alternate nonwoven filter configuration is one in which the filter medium is fed from a roll and advanced onto a take-up spool to remove residues.

SPRAY FINISHING AND POWDER COATING

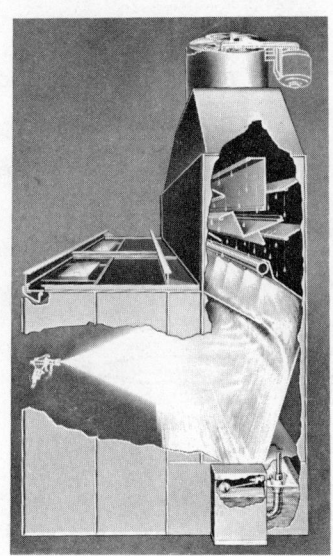

FIGURE 25.10. Water wash booth. (DeVilbiss)

Waterfall and cascade scrubbers: In factories where high volume spray coating operations are conducted for several hours a day, waterfall or cascade scrubbers are commonly used. In this type of collector, the exhaust airstream flows through a region where it is either scrubbed directly by sprays of water coming from nozzles or where the airstream follows a path through several stages of waterfall. Figure 25.10 is an example of one variation of this type of scrubber. Particulate matter that has been removed from the airstream by the scrubber is accumulated in a water tank from which it is removed either manually or by an automatic sludge removal device.

Chemical compounds must be added to the water used in such scrubbers to prevent paint residues from adhering to the walls or piping of the scrubber. A scrubber in which the concentration of chemical additive has not been adequately maintained probably will have clogged nozzles or accumulations of sludge on the cascade panels that will cause irregular distribution of the flowing water and ultimately create open channels through the scrubber. Air passing through these channels will not be adequately cleansed and will carry entrained particulate matter through the scrubber and into the exhaust system where it may be deposited on the lining of the exhaust stack, on exhaust fan blades, or on the roof of the building. Scrubbers that have been reasonably maintained will effectively remove particulates from the exhaust stream.

Venturi scrubbers: Perhaps the most efficient of the various forms of overspray collector is the venturi scrubber. These devices are constructed to direct the exhaust airflow through a channel in which there is a narrow throat (venturi) through which a high velocity spray of water is also

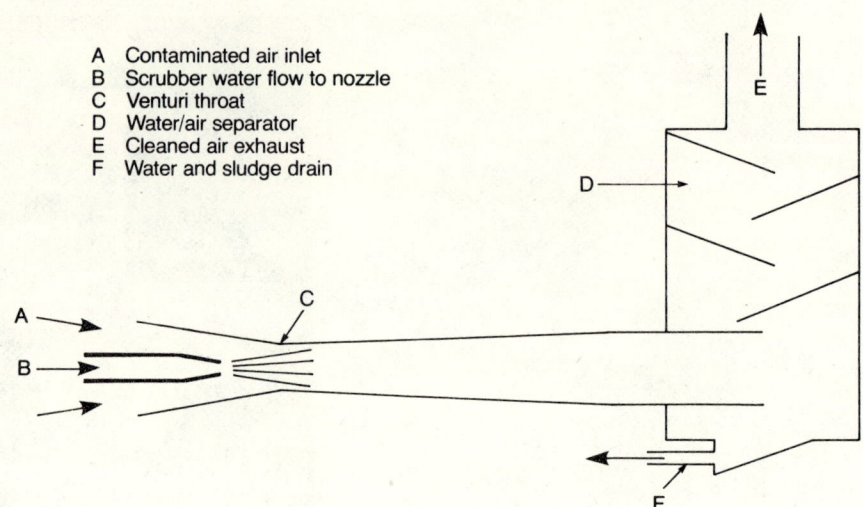

FIGURE 25.11. *Venturi scrubber. (Nordson)*

directed. Within the turbulent environment of this scrubber, virtually all particulate matter is extracted from the airstream and trapped in the water. The water is subsequently processed through a tank where residues are removed by settling and skimming. The same chemical compounds mentioned in the discussion of the waterfall-type scrubber must be added to the water used in this system to prevent plugging and blocking of water spray nozzles. Figure 25.11 is a schematic of a venturi scrubber.

POWDER COATING PROCESS EQUIPMENT AND COMPONENTS

Spray Process

The first element of the powder coating spray process is the device that mixes powder with air and supplies the mixture through a tube to the spray guns. This is commonly called a feeder and comprises a powder supply container and a pneumatic ejector. The powder may be contained in a fluidized bed or in a hopper having a bottom in the shape of an inverted cone. The ejector draws powder from the supply container, mixes it with air, and blows the mixture through a tube to the associated spray gun. Feeders may supply a single gun or several guns with a separate ejector for each gun.

A wide variety of chemical compositions are used to manufacture coating powders. The most common general classification is epoxy powder, but others include acrylic, polyester, vinyl, nylon, butyrate, polyolefin, and alkyd. These powders are shipped from the manufacturer to the user in containers of 25- to 100-lb (11.3 to 45.3 kg) capacity. Bulk containers of several hundred pounds (100 lb equals 45.3 kg) are used, but are less common. Unlike the coating materials used in organic solvent-based coatings, these powders present no unusual fire hazards in storage or handling. They may be classified as ordinary combustibles and, as such, may be stored without requirement for extra hazard protection. Usual containers for powder are fiberboard drums or corrugated boxes lined with plastic bags.

Many of the coating materials used in powder coating processes are sensitive to atmospheric moisture and are intolerant of high temperatures during storage. For these reasons, it is not at all unusual to find the powder coating process conducted within an air conditioned or a dehumidified room. Since it contains no flammable vapors, exhaust air from spray booths is customarily processed through a filter system and recycled into the room instead of being discharged to the exterior of the building.

Spray guns used for the electrostatic application of dry powders do not differ greatly in appearance from guns used for the electrostatic application of fluids (see Figure 25.12). Since no atomization is required, the functions of the spray gun are simply to control the shape of the spray pattern that is directed toward the workpiece and to impress a high voltage charge upon the powder cloud. The shape of the powder cloud can be adjusted by a sliding sleeve that is moved axially about the nozzle end or by introducing a swirling stream of air in the nozzle end of the gun. Powder and electrostatic power input connections may be made at the rear of the barrel as in Figure 25.13 or from below to the base of the grip and the forward portion of the barrel as in Figure 25.12. Connections to automatic guns typically are made at the rear.

Electrostatic power supplies associated with powder spray guns do not differ significantly in appearance from those used with fluid electrostatic coating operations. They are regulated power supplies that rectify stepped-up common ac line voltage inputs, either 115 or 230 V, to produce dc output ranging from 30 to 100 kV. Power supplies used with powder systems often are equipped with a variable output voltage control. Several automatic guns may be connected to a single power supply. For hand gun applications, however, a separate power supply is provided for each gun, and each supply has an interlock circuit that prevents the pack from being energized unless the trigger of its associated gun is actuated.

The interconnection between power supply and spray gun is usually made through a high voltage coaxial cable, which consists of a central conductor covered by heavy insulation inside an electrically grounded braided tube and an outer cover of plastic or rubber. Most manufacturers use metal for the inner conductor. However, one manufacturer uses a series of high value resistors as an inner conductor, and another, a high resistance liquid to suppress sparking in the event of cable damage. At the termination of the cable within the spray gun, the charging circuit is connected to the gun electrode through a 75- to 250-

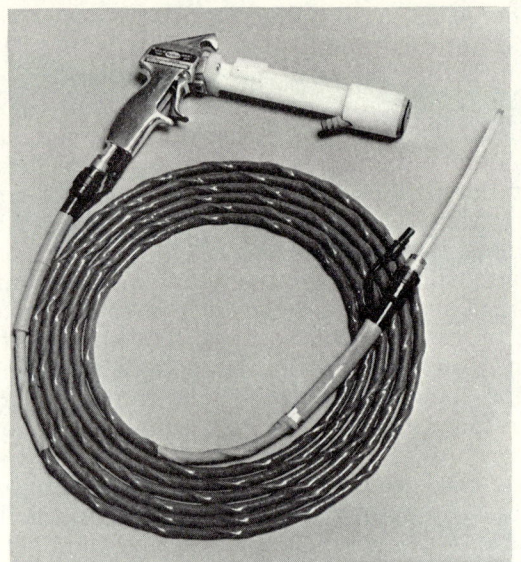

FIGURE 25.12 Hand powder gun. (Nordson)

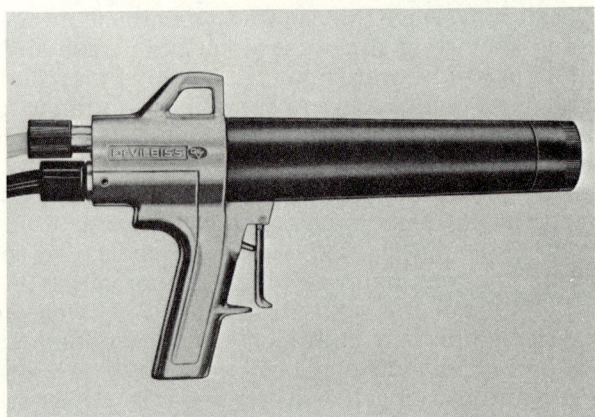

FIGURE 25.13. Hand powder gun. (DeVilbiss)

megaohm resistor. In operation, a faint blue glow called corona is developed about the end of the electrode, which projects through the front of the gun. Individual powder particles are charged as they pass through the corona. Modern spray guns may have an internal or external electrode. Some manufacturers incorporate the high voltage portion of the power supply in the body of the gun itself. Another uses no high voltage power supply at all but depends on friction to impress an electrostatic charge on the powder.

Powder spray operations may be conducted in conventional paint spray booths identical to those used for fluid spray operations, in a spray room, or in an open floor configuration. These approaches, however, are not commonly used because they include no provisions for recovery of overspray powder.

The most common facility in which electrostatic powder spray operations are conducted is a spray booth having a hopper type bottom through which exhaust air is drawn into a ductwork that leads to a powder collector. A cloth or fabric filter within the collector separates the powder from the airstream. The airstream leaves the collector through a fan and final filter, which prevents the escape of any powder that may have bypassed the collector filters. Powder separated within the collector falls to the bottom of the hopper from which it is extracted and subsequently reintroduced to the feeder for the spray guns. A basic outline of this arrangement is shown in Figure 25.14. With such an arrangement, virtually all the oversprayed coating material is recovered and recycled.

Less common as a recovery system is a spray booth constructed with a floor composed of a moving belt of filter fabric (see Figure 25.15). The main exhaust airstream for the booth is drawn downward through the fabric, through a duct, and through the fan. Subsequently, it passes through a final filter, which prevents the powder that has bypassed the belt filter from escaping into the room. As the filter belt progresses across the booth, it carries collected oversprayed powder to the intake nozzle of a high velocity suction system, which extracts the powder from the belt surface and carries it to a small cyclone or filter collector. Air exhausted from the cyclone along with the powder that bypasses the cyclone may be exhausted back into the spray booth where that powder is again collected upon the belt. Powder collected in the cyclone or filter collector is extracted for reintroduction to the gun feeder.

A recent development is the integration of the spray booth and powder recovery system into a single structure (Figure 25.16). In this equipment, cartridge-type filters and the exhaust fan are built into the base of the spray booth. The ductwork, which conventionally would separate the booth from the powder collector, is eliminated. Collected powder is pneumatically pumped from the hoppers immediately beneath the recovery filters directly back to the spray gun feeders.

Air exhausted from the main ventilation system of each of the systems just described is reintroduced to the room in which the spray booth is located rather than being exhausted to the outside of the building.

Electrostatic powder coating operations may also be conducted within a specially engineered enclosure commonly referred to as a pipe coater. The device consists of a steel enclosure, a ring of automatic electrostatic powder spray guns, and a conveyor for transporting the workpiece through the coater. The workpiece enters the coater at one end, passes through a cloud of coating powder, and leaves the coater at the opposite end. Before entering the coater, the workpiece is preheated to a temperature above the melting point of the powder, so that the powder will fuse to the workpiece upon contact.

Chapter 25

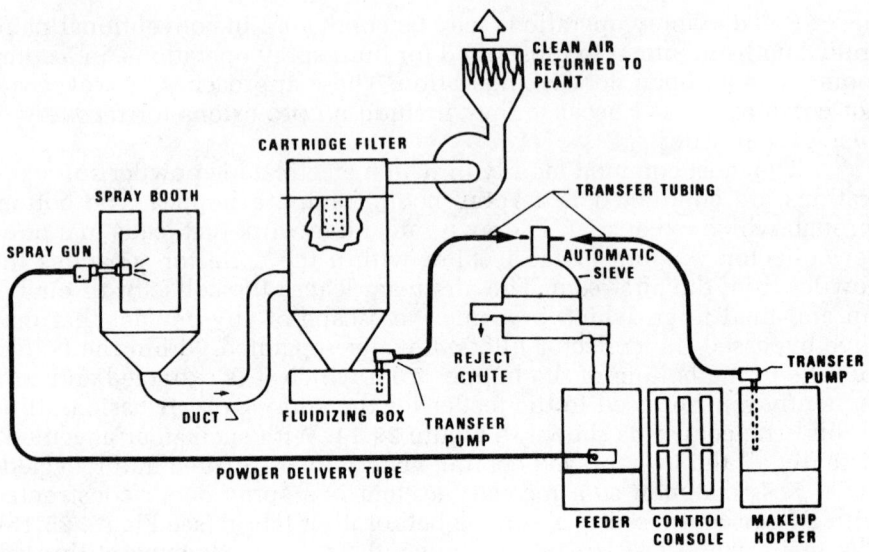

FIGURE 25.14. *Diagram of automatic recycle of a single-color powder coating system showing the booth, collector, fan, and filter. (Nordson)*

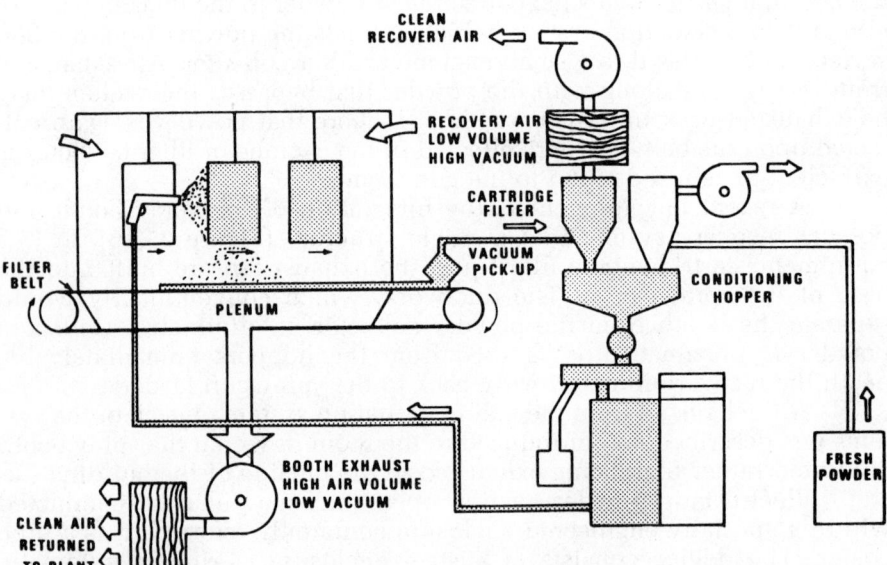

FIGURE 25.15. *Diagram of a recovery system showing the booth, belt, collector, and filter. (Nordson)*

The coater is provided with an exhaust system and powder recovery filter (baghouse) system similar to that used with the booth arrangement described in connection with Figure 25.14. Figure 25.17 shows one such device in use.

Fluidized Bed

Fluidized beds may be constructed in a wide variety of sizes ranging from something comparable to a coffee pot up to dimensions in excess of 8 ft (2.4 m) across and 15 ft (4.5 m) deep. They have vertical walls (typically steel) and a porous floor (textile or plastic). Air flows upward through the floor and through the powder contained within the enclosure. Air escaping the top of the bed does not carry with it any substantial quantity of powder. When the airstream is adjusted properly, the powder will behave as a liquid, flowing readily around, and contacting all surfaces of, any object that is dipped into the fluidized portion of the bed. The powder will immediately fuse to the surface of a workpiece that has been preheated to a temperature above the melting point of the powder. After the workpiece has been removed from the bed, it is placed in an oven to complete the curing of the coating.

An electrostatic fluidized bed has a series of high voltage electrodes mounted near the surface of the fluidized powder mass. Grounded workpieces, which may or may not be preheated, pass over the electrodes and are coated electrostatically by powder that is attracted to them. Subsequently, the coating is cured in a bake oven.

Both fluidized bed arrangements usually are provided with peripheral air exhaust systems that collect any powder grains escaping the bed and prevent their distribution into the general factory area. Since powders used with fluidized bed techniques are typically not as fine as those used with electrostatic spray techniques, they are not as likely to escape the immediate process area by accidental distribution as airborne dusts.

Cloud Chamber

One further electrostatic powder coating process is the cloud chamber. The apparatus for applying powder by this technique is generally a box-shaped enclosure with entry and exit portals through which workpieces are carried by a conveyor. The enclosure contains an apparatus that generates and maintains a billowing cloud of powder. The airborne particles are charged by high voltage electrodes and are attracted to the grounded workpiece as it passes through the cloud of powder. Powder is electrostatically collected on the workpiece and cured in an oven.

Chapter 25

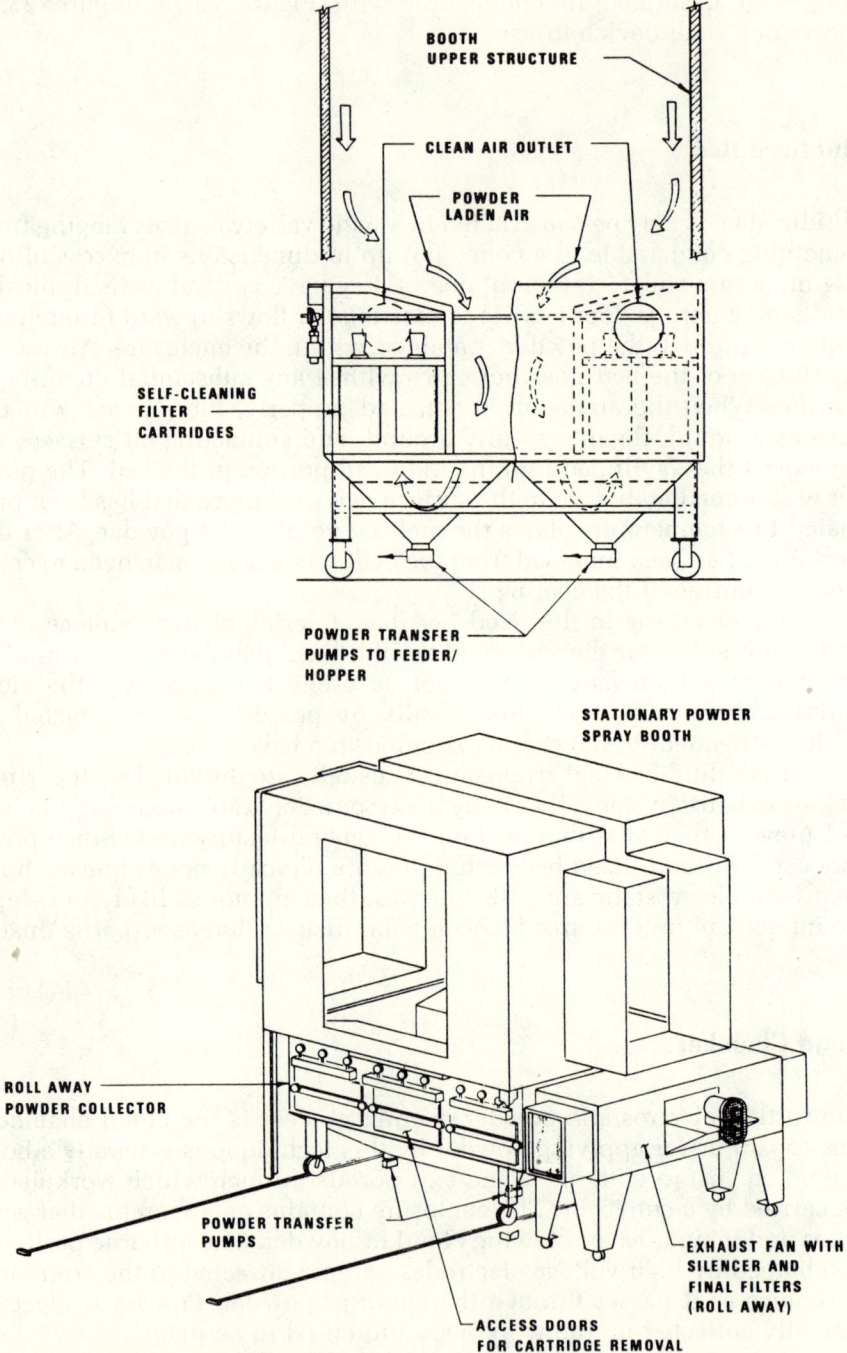

FIGURE 25.16. Integrated powder spray booth/recovery system. (Nordson)

SPRAY FINISHING AND POWDER COATING

FIGURE 25.17. Pipe coater. (Ransburg)

FLUID SPRAY PROCESS HAZARDS AND CONTROL

Materials and supplies used in organic spray finishing processes are usually flammable or combustible, are often toxic, and in some cases, may be highly reactive. For these reasons, spray finishing operations are considered hazardous, and suitable preventive and protective measures should be taken to minimize the hazard.

Fire Prevention

Key elements of fire prevention in fluid spray operations include hazard identification, storage and handling of flammable liquids, control of vapors and overspray, control of ignition sources, housekeeping, and training of personnel.

Hazard identification: The hazardous characteristics of materials should be identified and marked on containers as soon as they arrive at the plant. Rail cars and motor trucks carrying dangerous materials are required by the Department of Transportation to display placards indicating the nature of the hazard — for example, flammable liquid, poison, oxidizer. A more detailed method of identification is contained in NFPA 704, *Stand-*

ard *System for the Identification of the Fire Hazards of Materials.* The system provides for the identification of five degrees of health, flammability, and reactivity hazards.

Storage and handling: Bulk supplies of flammable liquids should be stored outdoors, away from buildings. Smaller quantities are brought indoors to a mixing room where they are prepared for use. The mixing room should be located adjacent to an outside wall where installation of pressure relief provisions is economical, and should be isolated from the rest of the building by fire-resistive construction. The room should have sufficient mechanical ventilation operating continuously to prevent the development of flammable vapor concentration in the explosive range. All flammable liquid containers of greater than 5 gal (18.9 L) capacity that are kept indoors should be equipped with a special plug that incorporates a pressure relief valve, a vacuum relief valve, and a flame arrester. (See Chapter 37 for a complete discussion of the "Handling and Storage of Flammable and Combustible Liquids.")

Prepared coating materials are transported from the mix room to the spray area in containers or through piping. Containers should be equipped with tightly clamped lids that will retain vapors and liquid in the event the container is upset. Piping should be identified as process pipes containing flammable materials, and a shutoff valve should be installed at every point where a hose is connected to the system.

To limit the scale of an accidental fire, quantities of flammable or combustible liquids kept in or near process areas should be limited to the amount needed for one shift's operations. The materials should be put in closed containers.

Flammable or combustible liquids should never be transferred from one container to another by the application of air pressure to the original shipping container. Drums and transfer tanks are not designed for this type of service. Pressurizing such containers may cause them to rupture and create a serious flammable liquid spill. Any pressure vessels used should be designed specifically for such use and be manufactured in conformance with the ASME *Code for Unfired Pressure Vessels.* Each such vessel should have a visible pressure gage and be provided with a relief valve to prevent its being pressurized above its design maximum working pressure.

Control of vapors and overspray: Limitation of vapors and overspray to the smallest practicable area is accomplished through the combination of process enclosures and power ventilation system use. Effective control prevents travel of the hazardous vapors away from the spray area to a location where there may be uncontrolled ignition sources. Ventilation systems should be checked frequently to assure they are operating and that specified flow rates are being maintained.

Ignition sources: All sources of ignition should be barred from, or controlled in, areas defined in Article 516 of NFPA 70, *National Electrical Code,* as Division 1 or Division 2. The careless use of smoking materials and improperly supervised welding or flame cutting operations pose significant threats to a fluid coating applications system. Enforcement of no smoking rules and establishment of a welding permit system are highly recommended for control of these problems.

To minimize chances of ignition, open flames, spark-producing equipment and any exposed surfaces exceeding ignition temperature of the material being sprayed should be prohibited in either Division 1 or Division 2 areas, and no equipment or process capable of producing sparks or hot particles may be located above or adjacent to those areas classified as hazardous unless partitions or other means are provided to prevent their entry into the classified area.

All electric wiring and equipment, unless specially designed and manufactured for use in hazardous areas, may be regarded as potential ignition sources. For this reason, only those devices and wiring types listed for this use may be installed or used in areas classified as hazardous. Electrical lighting fixtures installed above areas classified as hazardous should be totally enclosed or provided with a guard that will, in the event of a lamp or ballast failure, prevent hot particles from dropping into the hazardous area. Exhaust fans should be of nonsparking structure and conform to AMCA Class C requirements.

The most common source of flammable vapor ignition in electrostatic spray finishing operations is static electricity. Energized charging electrodes of electrostatic equipment send generous amounts of electrical charge into the air, from which it collects upon surrounding surfaces. Accumulation of this charge may raise a surface to a high voltage. Static electricity is also generated by many fluids as they flow through pipes, are poured from one container to another, or are sprayed from an airless nozzle. When the voltage on an object has developed sufficiently to break down the insulating medium between it and ground, it will arc to ground. If the energy of the arc is high enough and flammable vapors are present, ignition will occur. Electrostatic ignition can be prevented by electrically bonding together all the elements of a fluid transfer or spray system along with the workpiece and all other electrically conductive objects within 10 ft (3.04 m) of the charged elements of an electrostatic system and grounding the bond. Thus, static charges are bled to ground before they have a chance to accumulate. Additional information on bonding and grounding can be found in NFPA 77, *Static Electricity*. All electrostatic apparatus used with flammable or combustible spray materials must be labeled and listed.

Not to be overlooked in the grounding process is the spray operator. Human beings are also conductors of electricity. Gripping the grounded spray gun with the bare hand or wearing shoes having electrically conductive soles, provided the floor is grounded, are two methods for grounding the operator.

Electrostatic spray systems should be equipped with electrical interlocks to de-energize the electrostatic power supply any time the spray guns are not actually in use. All listed electrostatic spray guns include a high impedance charging circuit to limit the energy of a discharge arc to a value below that which is sufficient to cause ignition.

Housekeeping: Residues of spray materials are a solid form of fuel. A routine maintenance program should provide for the periodic removal of overspray residue from walls, floor, and ceiling of the spray booth, room, or area, as well as from conveyors and the interior of ventilation ducts (see Chapter 50). Residue buildup on conveyors may be particularly hazardous in that it may interfere with the bonding of workpieces to the conveyor by acting as insulation. Contaminated filters should be removed from the building as soon as they have been replaced, especially if they have been exposed to multicoating processes in which the combined coatings may be susceptible to chemical reaction or spontaneous heating. Alternatively, they should be immersed in water immediately upon removal and held immersed until disposed of.

Operator training: Operating personnel should undergo thorough initial training when they come on the job, and periodic additional training to maintain their appreciation of the hazards involved and to educate them in new processes. They should be aware of the hazards inherent to the materials particularly if the materials are chemically reactive. They should be familiar with the system of hazard identification that is employed. Operators should be drilled in proper operating, bonding and grounding, and maintenance procedures. They should know what to do if the equipment malfunctions or if fire occurs.

Fire Protection

Areas in which fluid spray processes are conducted and areas where coatings and solvents are mixed or stored should be separated from other plant operations by appropriate distance or partitions. Automatic sprinklers in spray booths, inside ventilation ducts, and at the ceiling of the building contribute considerably to the control of fire. In open spraying areas draft curtains extending down from the ceiling around spraying operations, and smoke and heat vents will slow the mushrooming of hot combustion gases along the ceiling and thus limit the number of sprinklers that will open. This will concentrate sprinkler discharge in the area where it is most needed.

Portable fire extinguishers and standpipe hoses are useful against fires involving small spray operations. For enclosed or semi-enclosed coating processes, such as in continuous coaters or automatic decorating machines, automatic fire extinguishing systems may be used to flood the interiors of the machines.

Structural considerations incorporated into the design of a spray booth generally follow the rationale that the booth and its exhaust ductwork should maintain structural integrity (i.e., not collapse) during the course of a fire of nominal proportions. If this structure remains intact, it is expected that the major portion of flame, heat, and fire gases produced will vent from the building through the exhaust ductwork, thereby diminishing the chances of fire spread in the building. In the hope of achieving these goals, spray booths and their exhaust ductwork are required to be constructed of steel, masonry, or other material of equivalent fire resistance. In addition to its own structure, support for the exhaust stack may be augmented by steel guy wires from the building roof structure. The probability that heat from a fire within a booth and stack will ignite nearby combustible materials can be reduced to an acceptable level by ensuring that reasonable clearances are maintained between any combustible structure or stored materials and the surfaces of the booth or stack (for recommended clearances, refer to NFPA 33, Chapter 5). Spray rooms are usually required to be separated from other occupancies within the same building by walls, floors, and ceiling structures having fire resistance ratings of at least 2 hours.

POWDER COATING PROCESS HAZARDS AND CONTROL

Powder coating operations using combustible organic powders are classified as hazardous processes because the powders can burn vigorously when suspended as airborne dusts and can explode when confined. The hazard level of powder coating processes, however, is substantially less than that associated with a flammable or combustible liquid coating operation of similar scale. There are no flammable vapors in the powder coating process, and residues resulting from overspray are not readily ignitable. The energy required to ignite a cloud of air-suspended coating powder is from 100 to 1,000 times higher than that required to ignite flammable vapors associated with fluid coating processes.

Fire Prevention

Fire prevention efforts associated with the powder coating process are similar to those used in the fluid coating process. Basically, preventive measures involve the storage and handling of powders, dust cloud control, and control of ignition sources.

Storage and handling: Coating powders kept in shipping containers are not customarily classified as hazardous materials and are commonly stored under conditions appropriate for ordinary combustible materials. They may be stored in the same room that houses the powder application equipment.

Reasonable care is required in the handling and movement of the powder containers because a fire hazard could be created if the containers are broken open and the powder distributed as an air-suspended cloud of dust. This potential hazard can be controlled through training programs to assure careful handling by material movement personnel and through prompt clean-up and removal of spilled materials.

Waste powder is commonly packaged into fiber drums or cardboard cartons lined with plastic bags and shipped to a landfill for proper disposal.

Coating operations: Virtually all spray powder coating processes are conducted in spray booths specially adapted for the collection of powders. These booths are generally more tightly enclosed and permit considerably less overspray to escape than booths used for fluid coating processes. Spray gun feeders and other powder handling equipment, such as sieves and blenders, are typically completely enclosed and provide minimum opportunity for accidental distribution of powder into the air. All connections in the hoses and piping associated with pneumatic transfer equipment should be routinely checked to confirm that they are secure and that ground connections are in place. Breakage of any such connections during operation could result in creation of a substantial hazard by blowing powder into the workspace air outside the process enclosure.

To prevent the escape of powder from the spray booth itself during operations, the booth must be large enough to accommodate the workpiece and must have adequate ventilation to provide effective capture velocity at all its openings. Minimum average velocities that are considered to be adequate through openings are in the vicinity of 60 fpm (183 m/min). Interlocks should be installed to permit operation of the powder application equipment only while the ventilation equipment is in operation and, similarly, to permit operation of the electrostatic power supply only during the time the powder is actually being applied.

Ignition sources: All sources of ignition should be barred from, or controlled in, areas defined in Article 516 of NFPA 70, *National Electrical Code*, as Division 1 or Division 2 for combustible dusts. The careless use of smoking materials and improperly supervised welding (see Chapter 24) or flame cutting operations pose significant threats to a powder coating applications system. Enforcement of no smoking rules and establishment of a welding permit system are highly recommended for control of these problems.

To minimize chances of ignition, open flames, spark-producing equipment and any exposed surfaces exceeding ignition temperature of the material being sprayed should be prohibited in either Division 1 or Division 2 areas. No equipment or process capable of producing sparks or hot particles may be located above or adjacent to those areas classified as hazardous unless partitions or other means are provided to prevent their entry into the classified area.

All electric wiring and equipment, unless specially designed and manufactured for use in hazardous areas, may be regarded as potential ignition sources. For this reason, only those devices and wiring types approved for this use may be installed or used in areas classified as hazardous. Electrical lighting fixtures installed above areas classified as hazardous should be totally enclosed or provided with a guard that will, in the event of a lamp or ballast failure, prevent hot particles from dropping into the hazardous area.

For those processes involving preheating of workpieces prior to coating, an automatic means should be provided to prevent those workpieces from being heated to dangerous temperatures, either by failure of primary temperature control or by overheating during a period when the conveyor is stopped.

Static electricity, as discussed in connection with fluid spray systems, is also a potential ignition source for airborne clouds of combustible powder. It can be eliminated as a potential ignition source by bonding and grounding all conductive objects, including the operator, in the vicinity of powder coating operations. This will bleed static charges to ground before they have had an opportunity to accumulate.

Static accumulations on electrically isolated conductive objects are the major cause of fires in electrostatic powder coating installations. Fire investigations repeatedly reveal that metal workpieces suspended from conveyor hangers that did not provide an effective electrical circuit between the workpiece and ground have passed through the coating process zone while discharging hot electrical sparks. This sparking ignited the powder cloud, and flame was sustained by the continuing flow of powder and air supplied through the spray guns. In an electrostatic fluidized bed, the entire volume of the bed can become involved in fire.

While some electrostatic application equipment is manufactured to be incapable of discharging a spark having sufficient energy to ignite a powder cloud, this feature is not universal. Some equipment will discharge incendiary sparks when approached too closely by a grounded object. Use of this type of equipment requires safeguards to prevent any grounded object from approaching the electrically charged high voltage elements closer than twice the sparking distance. Since such sparking may occur through an aerosol or vapor-laden atmosphere, the determination of "sparking distance" is a controversial subject and this value is not reliably established under many conditions. Discharge sparks that do occur during use of this type of equipment are often traceable to misracking of parts on a conveyor or swinging of the conveyor racks.

While not commonly identified as a source of ignition, mechanical sparking generated by ingestion of a small workpiece into the exhaust ductwork leading from the spray booth to the powder collector has been implicated in one collector explosion. To reduce the probability of this occurring, the use of grillwork at the intake of the exhaust duct is highly recommended, and magnetic devices can be added to catch tramp iron that may enter the process.

Fire Protection

Protective measures applied to powder coating installations are directed toward keeping the powder collector from becoming involved in a fire that has originated in the spray operation. The collector is equipped with pressure relief vents and ductwork to prevent its rupture in an explosion if it does become involved. The building that houses the coating operation is protected from heat and pressure effects that would be generated by an explosion.

In the powder application/recovery system shown in Figure 25.16, the confinement necessary for the generation of an explosion has been eliminated by design. Tests conducted under fire conditions indicate that ignition will simply produce a flame within the spray pattern of the application guns. This flame can be extinguished by interrupting the powder supplied to those guns.

To prevent propagation of a conventional powder spray booth fire through connecting ductwork to an associated separate powder collector, the ventilation system is usually arranged to keep powder concentrations in the exhaust ductwork below one-half the minimum explosible concentration (MEC). This airflow requirement is calculated against the maximum possible delivery of all spray guns in the booth and with the presumption that all powder from the guns is directed into the recovery system with none being collected on workpieces. Investigations of several spray booth fires have demonstrated the validity of this approach, as they have shown that flames within the booth are quenched when drawn into the ductwork where the fuel available is below the combustible concentration.

In those cases where fire has persisted in the spray booth for periods ranging beyond approximately 15 sec, however, the powder collector is threatened by induction through the ductwork of glowing embers that have formed in the spray fire. These glowing embers flying as solid particles through the duct are independent of powder concentration and are capable of igniting the powder collector.

Powder collectors connected to a booth by ductwork can be protected by an equipment arrangement involving a fast-acting flame detector within the spray booth, a system of interlocks to shut down application equipment, and a fast-acting damper to interrupt airflow in the duct between the spray booth and the powder collector. In operation, this apparatus will recognize any flame that has ignited in the spray booth, respond quickly (within ¼ sec) to the flame by shutting down all process equipment energy supplies (including electricity and compressed air), and by closing the ductwork damper. In shutting down the spray gun feeder immediately after recognition of a flame, the chances for ember formation are held to a bare minimum. Shutdown of all application equipment ceases all processes that may tend to keep powder suspended in air and, therefore, eligible for participation in the flame, and closure of the ductwork damper interrupts the airflow that would carry embers into the collector. This same equipment can be arranged to extend the response sequence by discharging an inerting gas into the dust collector if process circumstances seem to indicate that this is prudent.

For those application systems involving operational processes that require combustible concentrations of powder in the ductwork between the application area and the powder collector, a potential exists for almost instantaneous involvement of the collector whenever a fire occurs in the application area. Such systems should either be equipped with appropriately engineered suppression systems or be designed in a manner that will make the powder application and collection equipment fire and explosion resistant.

Airborne clouds of powder in a confined space burn at far higher rates than if unconfined. Such burning is often described as an explosion. Since the interior of the dust collector is a confined space, steps must be taken to accommodate the sudden pressurization and volumes of smoke and fire gases that evolve in the event of an ignition. Though some equipment has been designed to withstand such explosions, this approach is uncommon. A more usual technique is to install automatic pressure relief vents in the collector housing. These vents, upon ignition, open and allow the combustion products to escape, thereby limiting pressure development in the collector and minimizing collector damage. When the collector is located within a building, ductwork is usually required to direct the vented combustion products to the exterior of the building or to a safe location. In some cases, particularly if the building has a small volume, it is necessary to provide the building itself with pressure relief vents to prevent structural damage from building pressurization.

If powder escapes the process enclosure and is distributed throughout the immediate area, it will collect on horizontal surfaces and create a potential for a secondary explosion. The concussion of the primary explosion may shake loose powder that has collected on the building structure and nearby equipment and place it into air suspension where it could be ignited by flames or embers that persist from the initial combustion. If the quantity of powder thrown into suspension by the primary explosion were to be substantial, the secondary explosion could be considerably more serious than the primary.

Protection of a facility from secondary explosions is basically gained through maintenance of process equipment and procedures to prevent distribution of powder into the factory workspace and through scrupulous housekeeping to collect any amounts of powder that do escape the process enclosure.

Immediately after ignition of a spray gun powder cloud or a fluidized bed, the flame can usually be quickly extinguished by simply turning off the spray gun feeder or the air supply to the fluidized bed. No case has been reported in which flame has flashed back to the powder feed tube supplying the spray gun. It appears that the normal velocity of the feed systems exceeds typical flame front propagation velocities.

Automatic sprinkler systems are commonly required in factory areas where powder coating operations take place. Because sprinklers have rather long operating delays in relation to the duration of a powder fire, which is usually measured in seconds, it is common for powder fires to be extinguished by shutting down the equipment before sprinklers are able to function. In those cases where small fires have persisted in spray booth residues, portable fire extinguishers have been found to be adequate.

BIBLIOGRAPHY

REFERENCES CITED

McKinnon, G. P., ed., *Fire Protection Handbook*, Fifteenth Edition, National Fire Protection Association, Quincy, MA, 1981

Factory Mutual Engineering Corporation, "Spray Applications of Flammable and Combustible Materials," *Loss Prevention Data 7-27*, April 1977, Factory Mutual System, Norwood, MA.

NFPA CODES, STANDARDS, AND RECOMMENDED PRACTICES

Reference to the following NFPA Codes, Standards, and Recommended Practices will provide further information on the safeguards for spray finishing and powder coating operations discussed in this chapter. (See the latest *NFPA Codes and Standards Catalog* for availability of current editions of the following documents.)

NFPA 10, *Portable Fire Extinguishers.*

NFPA 13, *Installation of Sprinkler Systems.*

NFPA 30, *Flammable and Combustible Liquids Code.*

NFPA 33, *Spray Application Using Flammable and Combustible Materials.*

NFPA 35, *Manufacture of Organic Coatings.*

NFPA 68, *Explosion Venting Guide.*

NFPA 69, *Explosion Prevention Systems.*

NFPA 70, *National Electrical Code.*

NFPA 77, *Recommended Practice on Static Electricity.*

NFPA 704, *Identification of the Fire Hazards of Materials.*

ADDITIONAL READING

Bright, A. W., "Electrostatic Hazards in Liquids and Powders," *Journal of Electrostatics*, vol. 4, no. 2 (1977/78), pp. 131-147.

"Fire Hazards of Electrostatic Paint Spraying," *Fire Prevention*, (124), 1978, pp. 19-20.

Flash Point Index, 9th ed., National Fire Protection Association, Quincy, MA, 1978.

Gillies, M. T., ed., *Powder Coatings: Recent Developments*, Noyes Data Corp., Park Ridge, NJ, 1981.

Henry, Martin F., ed., *Flammable and Combustible Liquids Code Handbook*, National Fire Protection Association, Quincy, MA, 1981.

Meidl, James H., *Flammable Hazardous Materials*, 2nd ed., Macmillan Publishing Co., Inc., New York, NY, 1978.

Talbot, G., "Static Electricity," *Fire*, 70(871), 1978, pp. 397-398.

26

Dipping and Coating Processes

Nicholas L. Talbot and Paul H. Dobson

This chapter discusses the processes, equipment, hazards, and fire protection of dipping and coating operations that use flammable and combustible liquids. Such processes often involve liquids having tremendous heat energy and heat release capabilities, which can cause rapid property destruction when involved in a fire if not properly protected. The heat of combustion of flammable liquids is approximately two and one-half times that of an equivalent weight of wood.[1] In addition, heavy concentrations of smoke and toxic products of combustion often develop and make fire fighting extremely difficult as well as dangerous.

THE PROCESSES

Applications

Dipping and coating processes include, but are not limited to, finishing, impregnating, priming, cleaning, quenching, and other similar opera-

---

Nicholas L. Talbot is Resident Vice President, IRM Insurance, Dallas, Texas. Paul H. Dobson is Project Engineer with the Factory Mutual Research Corporation, Norwood, MA.

tions during which materials are immersed in, passed through, or coated by flammable or combustible liquids.

Processes vary widely and range from the cleaning of small parts in small quantities to an automated part-finishing process that uses a tank containing several thousand gallons (1000 gallons equals 3780 L) of a flammable or combustible liquid. As is the case with any potentially hazardous operation, a complete evaluation of the process and an understanding of the properties of the liquid(s) in use are essential.

Equipment

Dip tanks, roll coaters, curtain coaters, and flow coaters — the basic process equipment — present similar hazards. Related process equipment often used in conjunction with dipping and coating operations includes conveyors, pumps, piping systems, flammable or combustible liquid storage tanks, ovens, liquid heaters or heat exchangers, agitators, detearing equipment, and ventilation and exhaust systems, including energy recovery and pollution control devices. Ovens and dryers are discussed in more detail in Chapter 27. Support equipment must be selected and installed with full consideration of the inherent process hazards involved.

Dip tanks: Dip tanks are simply liquid containers of various sizes and shapes designed for the process involved. Tank sizes vary from a few gallons with small sq ft exposed surface area to several thousand gallons with a hundred or more sq ft of surface area. (one gallon equals 3.78 L; one sq ft equals 0.093 m^2). Figure 26.1 is an example of a typical dip tank process system. Dip tanks, their drainboards, and their covers should be constructed of noncombustible material such as heavy metal, reinforced concrete, or masonry. They should be designed for the process and liquid involved, with consideration given to static head of the liquid, corrosion, mechanical damage, and ease of maintenance and repair. Spill or exposure fires could weaken tank supports, possibly causing tank collapse. Therefore, supports for large tanks should be made of reinforced concrete or protected steel.

Flow coaters: In flow coaters (see Figure 26.2), the coating material is applied from nozzles or slots in an unatomized state onto the material being coated. The excess is collected in a trough or sump below the workpiece, returned to the reservoir, and recirculated to the nozzles by a pump. The nozzles may be fixed or oscillating, but the use of oscillating nozzles reduces the number required. The more nozzles used, the larger the reservoir and pump capacities required, and the greater the solvent loss will be. The workpiece usually enters and leaves through conveyor openings, and a drip tunnel is used in place of the drainboards that are used with dip tanks. The tunnel is enclosed on all sides, except for the conveyor opening, and is sloped toward the trough or sump. Airborne solvent concentration in the tunnel is sometimes used to keep the coating from drying before it gets to the oven.

DIPPING AND COATING PROCESSES

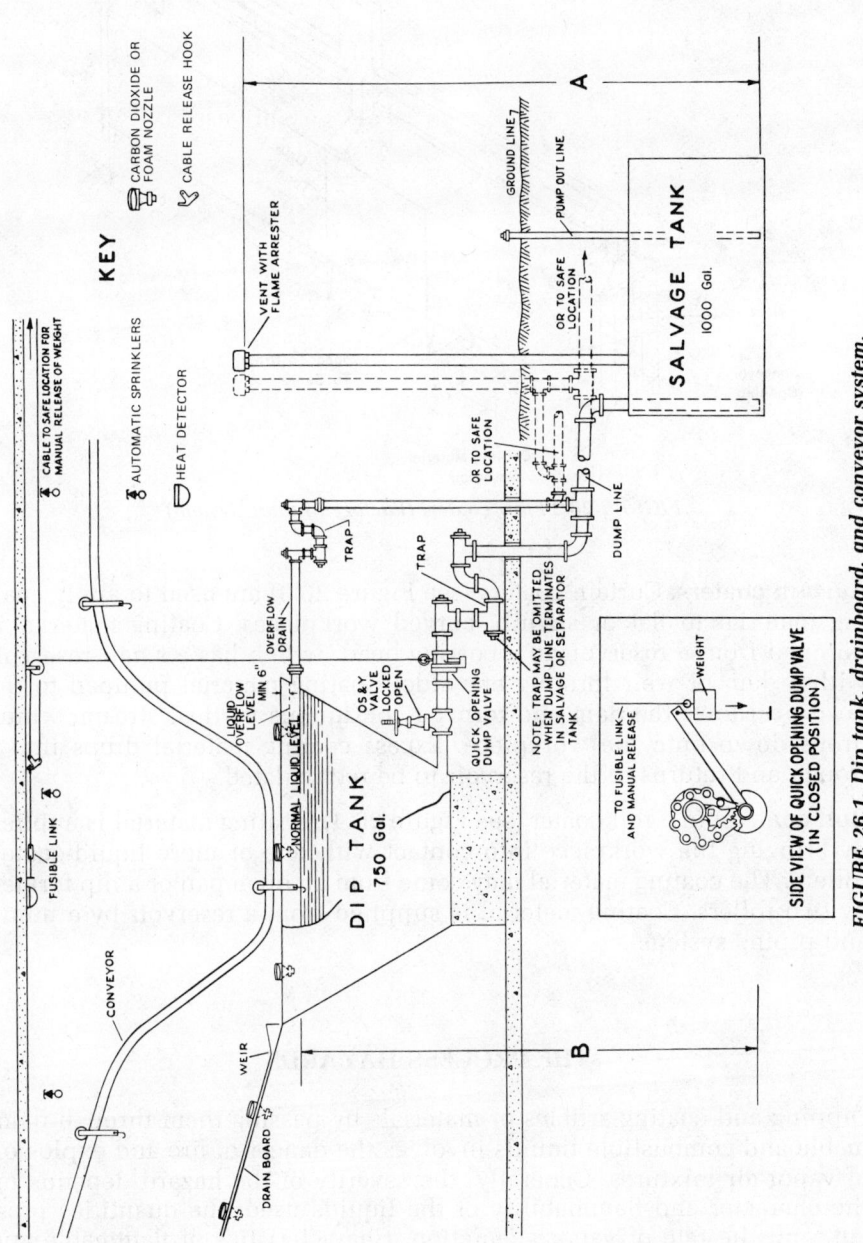

FIGURE 26.1. *Dip tank, drainboard, and conveyor system.*

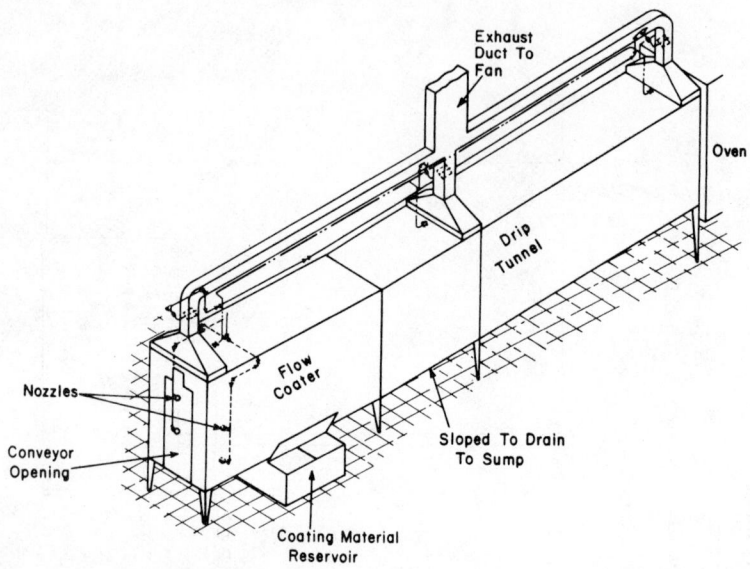

FIGURE 26.2 Flow coater. (Factory Mutual System)

Curtain coaters: Curtain coaters (see Figure 26.3) are used to apply coating material to flat or slightly curved workpieces. Coating material is pumped from a reservoir to a coating head, which has a small reservoir with a dam or weir forming one side. Coating material pumped to the head overflows the dam and forms a continuous vertical stream, which drops down onto the workpiece. Excess coating material drops into a trough and returns to the reservoir to be recirculated.

Roll coaters: In a roll coater (see Figure 26.4), coating material is applied by bringing the workpiece into contact with one or more liquid-coated rollers. The coating material may come from an open pan or a nip formed by two rollers. Coating material is supplied from a reservoir by a pump and piping system.

THE PROCESS HAZARDS

Dipping and coating articles or materials by passing them through flammable and combustible liquids involves the danger of fire and explosion of vapor-air mixtures. Generally, the severity of the hazard depends on the character and flammability of the liquids used, the quantities present, and the rate of vapor generation. Characteristics of flammable and combustible liquids are discussed in detail in Chapter 37, "Handling and Storage of Flammable and Combustible Liquids."

DIPPING AND COATING PROCESSES

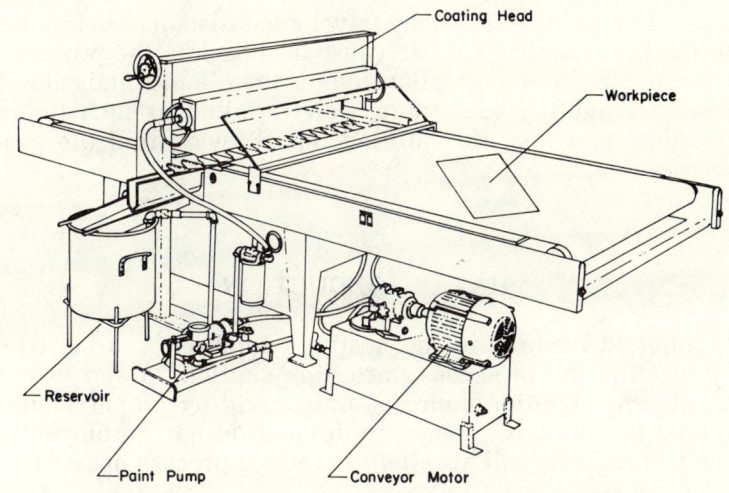

FIGURE 26.3. Curtain coater. (Factory Mutual System)

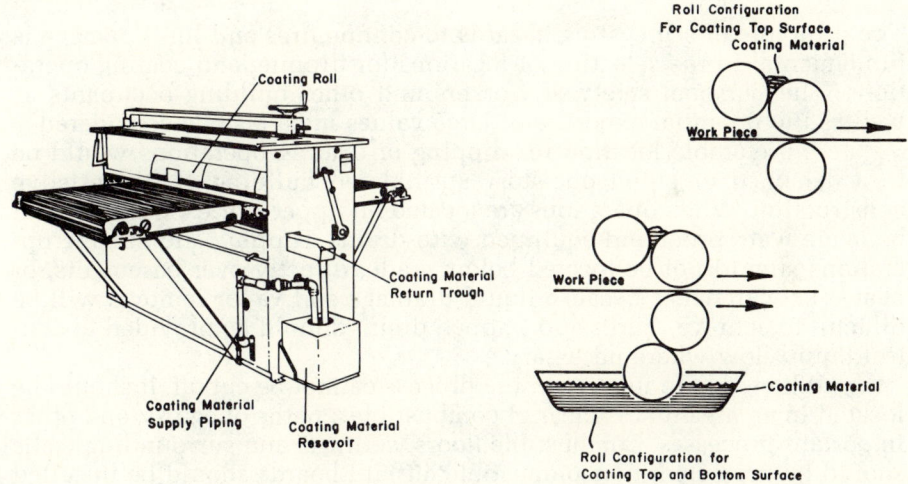

FIGURE 26.4. Roll coater. (Factory Mutual System)

Ease of ignition and the rate of flammable vapor generation from the liquid surface and from surfaces of freshly coated articles, floors, drain or drip boards, and other related equipment define the extent of the hazard. The intensity, persistence, and burning characteristics of the flammable vapor evolved and the probability of fire spread from radiated heat or from the flow of burning liquids because of a container rupture, boilover, or overflow complete the total hazard picture.

Flammable liquid vapors are usually heavier than air and, therefore, flow to low points. They may travel great distances before being exposed to ignition sources that can cause flashback to the process area.

Flammable liquids are often lighter than, and immiscible in, water. During fire fighting operations, water applied to such liquids may cause overflow and float the burning liquid away from the process to other areas.

HAZARD REDUCTION

Hazards inherent in dipping and coating operations can be reduced by properly locating the processes, installing ventilation and exhaust systems, eliminating ignition sources, and providing proper maintenance and periodic inspections. Special equipment design, employee training, and proper protection will also help to reduce process hazards.

Process Location

The principle of segregating hazards to confine fires and limit damage is fundamental in the selection of locations for dipping and coating operations. The personal safety of workers and other building occupants as well as the potential exposure of large values must also be considered.

A preferable location for dipping or coating operations would be in a detached or cutoff one-story sprinklered building of fire-resistive construction. When operations are located on upper stories, floors should be made waterproof and equipped with drains. Dipping and coating operations should not be located below grade, directly over basements, or near pits or trenches because liquid drainage and vapor removal will be difficult to achieve. Curbs and trapped drains should be provided to control liquid flow where necessary.

When, due to its nature, the process cannot be cut off, it should be located in an area that is clear of combustibles, paths of egress, and other important processes. Combustible floors, ceilings, and surrounding walls should be protected; noncombustible curtain boards should be installed around the perimeter of the process or protected construction area and extended downward as far as is practical.

As flammable and combustible liquid fires generally develop rapidly and release large amounts of heat, automatic roof vents are extremely desirable, especially in unsprinklered buildings where heat buildup would quickly involve the entire building. Roof vents allow heat and smoke to escape, improving the chances of fire control by automatic sprinklers or the fire department.

When processes are located in confined areas and highly flammable or unstable liquids are used, explosion relief venting may be provided to lessen potential losses. Venting can be in the form of lightweight walls and roofs or explosion-relieving wall panels, roof latches, and windows. Explosion relief is discussed in detail in NFPA 68, *Explosion Venting*.

Unfortunately, small dipping and coating operations are not generally considered in the same light as the larger operations, and, therefore, little consideration is given to their locations. However, it is often the small operation that produces a major fire loss. Careful evaluation of the process is necessary in developing location requirements for dipping and coating operations, and size is only one factor to be considered.

Figures 26.5, 26.6, and 26.7 show common location features of dipping and coating operations.

Ventilation

When processes involve liquids that produce vapors, ventilation is necessary to limit the vapor area to the smallest practical space possible. The vapor area created by dipping and coating processes is defined as "any area containing vapor at or above 25 percent of the lower flammable limit (LFL) in the vicinity of dipping and coating processes, their drainboards or associated drying, conveying, or other equipment during operation or shutdown periods."[2]

The extent of the vapor area of a process depends upon the properties of the liquid such as vapor pressure, flash point, boiling point, and evaporation rate, along with the characteristics of the process or wetted surface area exposed. Figure 26.8 clearly shows the difference in exposed wetted surface area between two dip tank operations using the same quantity of liquid. Note that, with all else being the same, the process shown in the upper part of the figure would generate a greater volume of vapor than that in the lower part; thus it would offer a greater fire potential.

Vapor areas for each process should be kept as small as is practical but should not extend more than 5 ft (1.5 m) from the vapor source. The extent of a vapor area can be determined with a combustible gas analyzer. A theoretical ventilation rate can be determined when the liquid usage rate is known.[3] Since vapor concentrations vary widely from one process to another, a single safety exhaust ventilation system cannot be designed for all operations. The prime objective is to limit the vapor area to the smallest space possible and to prevent dead air spaces or pockets where vapors can accumulate.

Because flammable liquid vapors are heavier than air, low-point peripheral ventilation systems are usually more desirable than overhead hood arrangements. Two types of design are shown in Figure 26.9. The least efficient system is the overhead hood that allows discharge of some vapors to surrounding work areas; such a system is usually acceptable for unoccupied locations.

When considering exhaust systems, the ratio of air to vapor is important. However, the effectiveness of exhaust air in picking up and diluting the vapor is often lost in design. Entrainment velocity with mix-

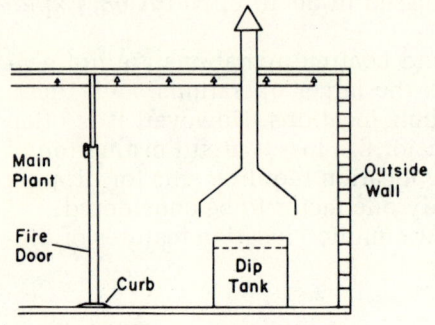

FIGURE 26.5a. Most satisfactory arrangement for location of processes. Tank located near an outside wall, cut off from main plant areas by fire-resistive construction. (Factory Mutualo System)

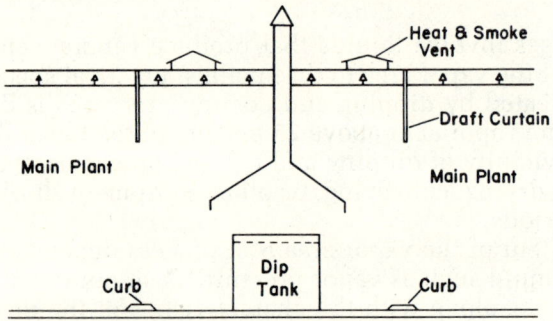

FIGURE 26.5b. Satisfactory arrangement. Tank located in main plant area cut off by draft curtains and curbing. Heat and smoke vents should limit opened sprinklers to those near the fire and reduce smoke concentrations. (Factory Mutual System)

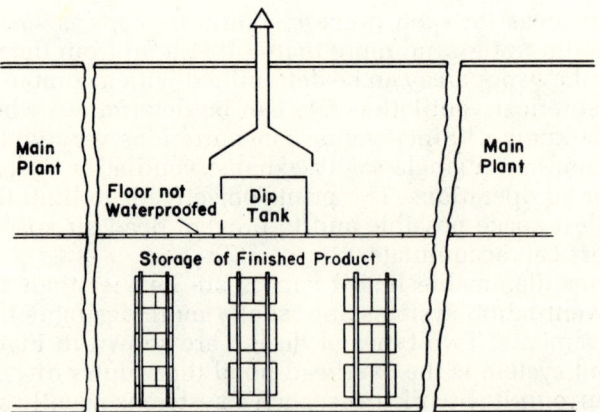

FIGURE 26.5c. Unsatisfactory arrangement. Tank not cut off from main plant area. Water from hose streams will leak down through floor and damage finished product. (Factory Mutual System)

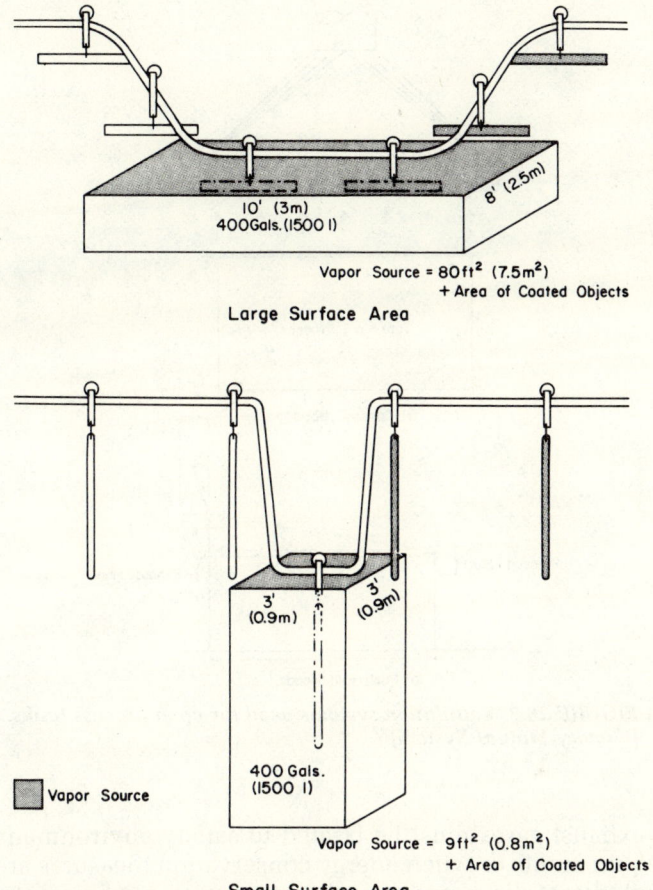

FIGURE 26.6. *The difference in vapor source area between two dip tanks with the same coating volume. (Factory Mutual System)*

ing action becomes the key to efficient use of exhaust air. Exhaust rates per sq ft (one sq ft equals 0.092 m) of wetted surface area usually range from 50 to 200 cubic feet per minute (1.4 to 5.7 m ^{3}/min). Effective entrainment of vapor often requires slot velocities of 1,000 to 2,000 cfm (28 to 56 m^3/min).[4]

In designing and evaluating ventilation systems for dipping and coating processes, there are several important considerations:

1. Automatic processes should be interlocked to shut down the operation in the event of an exhaust system failure.
2. The supply of makeup air should be sufficient to allow efficient operation of the exhaust fans and to minimize dead air spaces.
3. Ideally, each exhaust system should discharge directly outdoors.

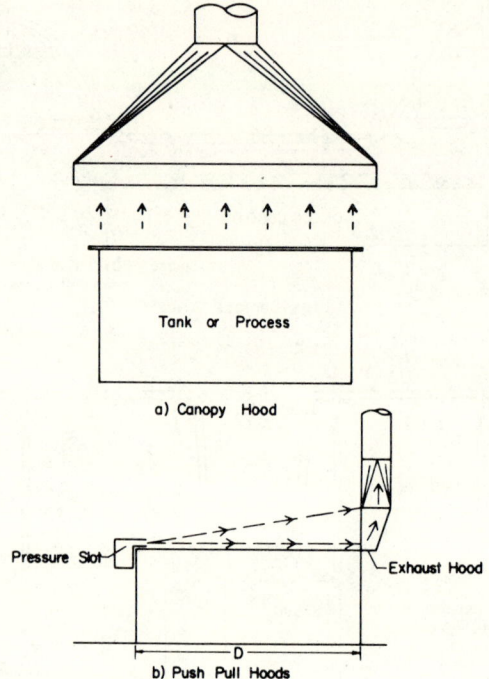

FIGURE 26.7. *Ventilation systems used for open process tanks. (Factory Mutual System)*

When exhaust gases must be treated to satisfy environmental protection requirements or when energy conservation measures are used, individual, direct discharge of each system may not be practical. Manifold systems increase the hazard; therefore, all such systems should be specifically approved and strict preventive maintenance measures taken. Reactive materials and coatings should not be used when ducts are connected to a manifold.

4. Exhaust air should not be used for makeup air in occupied spaces. It can be used in unoccupied areas, but only if it has been decontaminated to safe levels.
5. Exhaust ducts should not discharge near air intakes, nor should they discharge less than 6 ft (1.8 m) from a combustible wall or roof or 25 ft (7.6 m) from combustible construction or an unprotected opening in a noncombustible exterior wall.
6. Exhaust ducts should be equipped with ample access doors to facilitate cleaning.
7. Ducts should be constructed of fire-resistive material, such as steel or masonry, and should be substantially supported. They should be installed with adequate clearance from combustibles. Dampers should not restrict exhaust ventilation below minimum safe levels.

DIPPING AND COATING PROCESSES

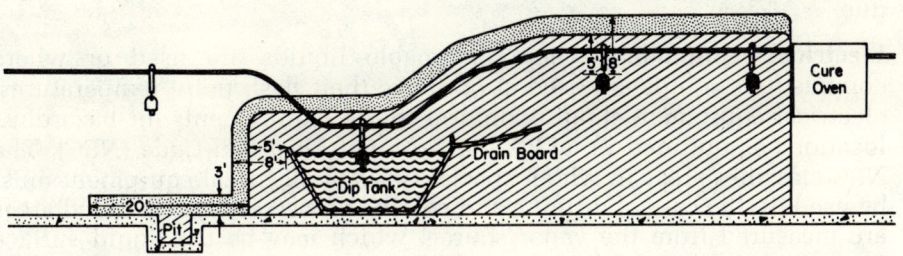

FIGURE 26.8. The Class I, Divisions 1 and 2, hazardous locations for a dipping operation. (Factory Mutual System)

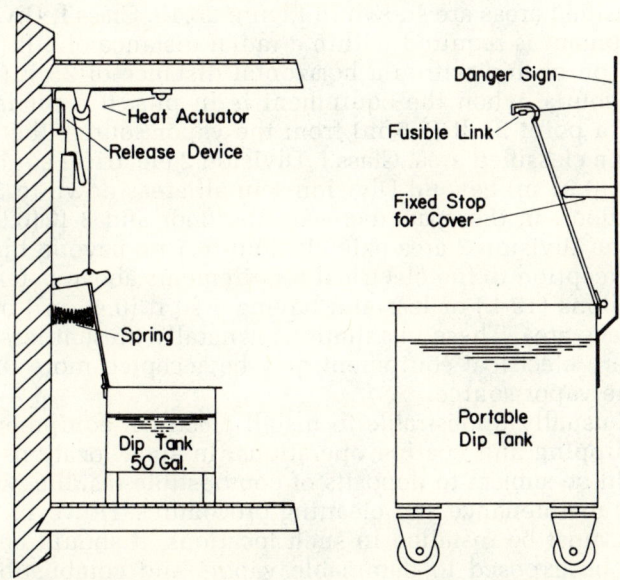

FIGURE 26.9. Automatic closing covers for small fixed and portable dip tanks. (Factory Mutual System)

Ignition Sources

The key to minimizing the hazard of fire or explosion from vapor-air mixtures in dipping and coating processes is maintaining concentrations well below the lower flammable limit (LFL) through properly designed safety exhaust ventilation systems. Without vapor (fuel) and air (oxygen), properly mixed, ignition cannot take place even though sources of ignition are available. As it is almost impossible to keep processes free of flammable vapor-air mixtures in the explosive range, sources of ignition must be eliminated or equipment must be designed for use in hazardous areas containing flammable vapors and/or combustible residue.

Electrical equipment: Where flammable liquids are used or where combustible liquids are used at or above their flash point temperatures, electrical equipment should conform to the requirements for hazardous locations as outlined in NFPA 70, *National Electrical Code (NEC)*. The NEC classifies areas in which special types of electrical equipment must be used. In the case of dipping and coating processes, the classified areas are measured from the vapor source, which may be the liquid surface (open tanks) or wetted surfaces (freshly coated workpieces, drain boards, floors). The space containing hazardous vapor concentrations normally extends laterally and upward a short distance from the vapor source and, since the vapors are usually heavier than air, downward to the floor.

Classified areas are shown in Figure 26.10. Class I, Division 1 electrical equipment is required within a radial distance of 5 ft (1.5 m) from the vapor source and within a horizontal distance of 25 ft (7.6 m) from the vapor source when the equipment is in pits. If a pit is not vapor-stopped at a point 25 ft (7.6 m) from the vapor source, the entire pit is considered a classified area. Class I, Division 2 electrical equipment is required 3 ft (0.91 m) beyond Division 1 in all areas down to 3 ft (0.91 m) above the floor. In the space between the floor and 3 ft (0.91 m) above the floor, the Division 2 area extends 20 ft (6.1 m) beyond the Division 1 area. An exception to the electrical requirements above is tanks containing five gallons (19 L) or less and having 5 sq ft (0.46 m^2) or less of exposed surface area. These installations generally present less of a hazard and ordinary electrical equipment may be accepted more than 8 ft (2.4 m) from the vapor source.

It is usually undesirable to install electrical equipment in the vicinity of dipping and coating operations. In such locations, the equipment would be subject to deposits of combustible residue, which would complicate maintenance and cleaning procedures. However, if electrical equipment must be installed in such locations, it should be a type that can safely be exposed to flammable vapors and combustible residues. Wiring in rigid conduit or in threaded boxes or fittings containing no taps, splices, or terminal connections is permitted in both residue and vapor areas.

Fixed lighting fixtures located outside of, but above, classified areas should be protected against physical damage, which may allow them to become sources of ignition of flammable vapors.

In areas where the electric service is supplied from overhead wires and lightning is a fairly common occurrence, protection against lightning-induced surge voltages should be provided. In the absence of lightning protection, high surge voltages could enter vapor areas and create high energy ignition sources. Suitable protection includes lightning arresters, interconnection of all grounds, and surge protection capacitors.

Under abnormal conditions, overheating and arcing can occur in electrical circuits and equipment. Therefore, it is important that electrical circuits and equipment servicing dipping and coating processes have the proper overcurrent protection, are grounded, and receive regular maintenance.

Dipping and coating processes are frequently changed when new coating materials are introduced, production demand is increased, and new equipment is installed. Quite often, unsafe electrical installations and poor maintenance habits creep into the production process. It is common to find extensions connected to additional fans and lighting that are not suitable for use in classified hazardous areas.

Other ignition sources: The presence of open flames, spark-producing processes or devices, and heated surfaces having temperatures high enough to ignite flammable vapors should be prohibited in flammable vapor areas.

Ovens located directly above or adjacent to dipping and coating operations have been the ignition source for many fires. This equipment should be located as far as practical from dipping and coating operations with noncombustible partitions provided when reasonable distances cannot be maintained. Ventilation exhaust systems of ovens and dryers located directly above, or adjacent to, dipping and coating processes should be interlocked with process ventilation systems, so that heating and process equipment cannot function unless all ventilation equipment is in operation. Indirect heating systems, which do not present ignition sources, are usually preferred in drying operations. More information on heating systems can be found in Chapter 27, "Heat Processing Equipment."

Dipping and coating processes often involve the use of conveyors, rollers, collectors, festoon dryers, and other devices that move the workpiece or liquid through the process. Flowing liquids and materials moving through a process can generate static electricity. A static charge can accumulate on process equipment and develop a difference in electrical potential sufficient to cause a spark containing enough energy to ignite flammable vapor-air concentrations. Static elimination by bonding and grounding process equipment or by humidification or ionization of the local atmosphere is essential in reducing the hazard of fire or explosion.

Of equal importance is the banning in hazardous classified areas of equipment and processes that are not essential to dipping and coating operations but that are potential ignition sources. Included in this category are cutting and welding operations, portable heaters, spark-producing (ferrous) tools, and smoking. Nonferrous tools are available as are nonferrous rotating parts for exhaust fans. If hazardous maintenance operations, such as cutting or welding, must be performed in the vicinity of dipping or coating processes, they should be performed in accordance with recognized good practices by competent personnel under strict supervision. The prohibition of smoking in process areas should be properly posted.

Special Design Considerations

Some hazards of dipping and coating operations can be reduced by special design features in the process equipment.

Many flammable and combustible coatings are lighter than, and immiscible with, water. In the event of fire, water from hose streams or sprinkler systems could cause an overflow of a large amount of liquid from the tank unless certain design precautions have been taken.

For example, the top of the tank should be at least 6 in. (152 mm) above the floor, and the liquid surface should be at least 6 in. (152 mm) below the top of the tank. Tanks having a capacity of more than 150 gal (568 L) or a liquid surface area of 10 sq ft (0.93 m^2) or more should be equipped with trapped overflow pipes that discharge to a safe location. Overflow pipes should be capable of handling the maximum delivery of the tank's fill pipes or of the automatic sprinkler discharge. In any case, overflow pipes should never be less than 3 in. (76.2 mm) in diameter.

Larger tanks — those having a capacity of 500 gal (1.9 m^3) or more — should be equipped with a bottom drain of sufficient size to empty the tank in 5 min. The flammable liquid should be drained to a vented salvage tank or other safe location. If gravity flow is not practical, pumps may be used to empty the tank. The drain may be operated manually or automatically. However, if a manual drain release is used, it should be accessible and at a safe location. If, due to increased hazard or by nature of the operation, bottom drains cannot be provided, the process should be cut off or fully diked, and both an automatic special extinguishing system and sprinklers should be installed.

Conveyor systems used in dipping and coating processes should be arranged to stop automatically in the event of a fire or failure of the exhaust ventilation system.

In processes where there is a possibility of flammable liquids being heated above their boiling points or to within 100°F (37.8°C) of their autoignition temperatures, suitable excess temperature limit controls are needed to prevent rapid vapor buildup and possible autoignition.

Controls should limit the surface temperatures of heated workpieces to at least 100°F (37.8°C) below the autoignition temperature of the coating material used. This limitation does not apply to quenching tanks, which are discussed in Chapter 28, "Oil Quenching."

Limit controls, such as liquid level devices, meters, and timers, should be used to prevent overfilling of tanks where automatic filling is designed into the process. If pumps are used, they should be interlocked to shut down in the event of fire.

Maintenance, Training, and Inspection

While location, ventilation, equipment design, and the elimination of ignition sources go a long way towards reducing the hazards in dipping and coating processes using flammable and combustible liquids, maintenance, training, and regular inspection of equipment are essential to safe operations. Unfortunately, due to production schedules, changes in processes, and turnover of personnel, these essentials may be neglected, resulting in unsafe operating conditions.

Areas in the vicinity of dipping and coating processes should be kept free of accumulations of combustible residues and unnecessary combustible materials. Spontaneous ignition, due to oxidation or exothermic reaction between various coating components, often occurs when excess residue accumulates in work areas, ducts, duct discharge points, or other adjacent areas. When excess residues accumulate in such locations, operations should be discontinued until conditions are corrected.

Combustible coverings (thin paper or plastic, for example) and strippable coatings are often used to facilitate cleanup of drippings and residues. The increased amount of combustibles introduced by the use of these materials is offset by the improved housekeeping and ease of cleaning that they provide. Suitable containers should be provided for the disposal of waste and rags used for cleaning.

The size and complexity of dipping and coating operations may vary widely but, in all operations, the personnel involved should be instructed in the potential safety and health hazards inherent in the particular process employed. The nature of the process and operational, maintenance, protection, and emergency procedures should be included. It is desirable to record such training and provide refresher instruction from time to time to ensure continuity and proper action in the event of an emergency.

Depending upon the size and nature of the process, periodic (usually at least monthly) inspection should be made of dipping and coating processes. Inspections should include noting the condition of equipment, covers, overflow pipe inlets and outlets, discharge, bottom drains and valves, electrical wiring and equipment, grounding and bonding connections, ventilation equipment, and extinguishing equipment. Inspections and protection equipment tests should be documented and conducted using a checklist. Defective equipment and unsafe conditions should be corrected promptly.

FIRE PROTECTION

In spite of the considerable care taken in the operation of dipping and coating processes that use flammable and combustible liquids, fires and explosions do occur, and fire protection and suppression equipment is necessary.

Fire loss experience has shown that damage in small processes is often as severe as in large operations. Less consideration is usually given to the location of small processes, and they may be in areas that expose high values or combustible construction. Some form of protection is usually required for all dipping and coating processes.

By far the simplest device for extinguishing a fire in a process tank is a self-closing cover, which cuts off the fire's air supply. Normally the cover is held open by a cable and fusible link. When exposed to the heat of a fire, the solder in the fusible link melts, releasing the cover and allowing it to drop tightly in place over the tank.

Automatic sprinklers are often provided at ceiling level and can provide adequate protection for small processes and where combustible coatings are used with limited liquid surface area exposure. Sprinklers are intended to help control the fire, prevent damage to the building, and allow effective manual fire fighting. Automatic sprinklers are often required as backup to special extinguishing systems.

Where processes involve considerable volumes, large liquid surface areas, or large wetted areas of low flash point materials, special extinguishing systems should be provided. In the case of dip tanks, special systems should be provided for tanks having capacities in excess of 150 gal (569 L) or surface areas greater than 10 sq ft (0.93 m^2). Suitable protection systems use foam, carbon dioxide, dry chemical, or a halogenated agent as the extinguishing medium. For liquids having flash points above 140°F (60°C), water spray systems may be used. All systems should operate automatically and conform to fire protection standards. The complexity and value of some processes and the absence of sprinklers as backup often make it desirable to provide redundancy in protection systems.

Dipping and coating process areas should be equipped with portable fire extinguishers suitable for use on flammable and combustible liquids fires.

Due to the flash fire conditions that usually prevail in flammable liquids fires, emergency personnel or fire brigades should be trained to act immediately to bring all protection elements to bear on the fire in hopes of achieving rapid control and limiting possible injury and damage.

BIBLIOGRAPHY

REFERENCES CITED

1. Factory Mutual Engineering Corporation, "Flammable Liquids — General Safeguards," *Loss Prevention Data* 7-35, January 1977, Factory Mutual System, Norwood, MA, p. 1.
2. NFPA 34, *Dip Tanks Containing Flammable or Combustible Liquids*, p. 4.

DIPPING AND COATING PROCESSES

3. NFPA 86A, *Ovens and Furnaces — Design, Location, and Equipment*, p. 74, (in revision).
4. *Industrial Ventilation — A Manual of Recommended Practice*, Fourteenth Edition, Committee on Industrial Ventilation, P.O. Box 16153, Lansing, MI 48901, pp. 5-57 to 5-60.

NFPA CODES, STANDARDS, AND RECOMMENDED PRACTICES

Reference to the following NFPA Codes, Standards, and Recommended Practices will provide further information on the safeguards for dipping and coating processes discussed in this chapter. (See the latest *NFPA Codes and Standards Catalog* for availability of current editions of the following documents.)

NFPA 10, *Portable Fire Extinguishers.*
NFPA 11, *Foam Extinguishing Systems and Combined Agent Systems.*
NFPA 12, *Carbon Dioxide Extinguishing Systems.*
NFPA 12A, *Halon 1301 Systems Fire Extinguishing.*
NFPA 12B, *Halon 1211 Systems Fire Extinguishing.*
NFPA 13, *Installation of Sprinkler Systems.*
NFPA 13A, *Care and Maintenance of Sprinkler Systems.*
NFPA 15, *Water Spray Fixed Systems.*
NFPA 34, *Dipping and Coating Processes Using Flammable or Combustible Liquids.*
NFPA 35, *Manufacture of Organic Coatings.*
NFPA 68, *Explosion Venting.*
NFPA 69, *Explosion Prevention Systems.*
NFPA 70, *National Electrical Code.*
NFPA 77, *Recommended Practice on Static Electricity.*
NFPA 91, *Blower and Exhaust Systems for Dust, Stock, and Vapor Removal or Conveying.*
NFPA 327, *Cleaning or Safeguarding Small Tanks and Containers.*
NFPA 497M, *Classification of Gases, Vapors and Dusts for Electrical Equipment in Hazardous (Classified) Locations.*

ADDITIONAL READING

Bartels, A. L., "Safeguards to Prevent Ignition by Hot Surfaces," *Electrical Review*, London, vol. 203, no. 4 (1978), pp. 29-31.
Bright, A. W., "Electrostatic Hazards in Liquids and Powders," *Journal of Electrostatics*, vol. 4, no. 2 (1977/78), pp. 131-147.
Flash Point Index, 9th ed., National Fire Protection Association, Quincy, MA, 1981.
Henry, Martin F., ed., *Flammable and Combustible Liquids Code Handbook*, National Fire Protection Association, Quincy, MA, 1981.

McKinnon, G. P., ed., *Fire Protection Handbook*, Fifteenth Edition, National Fire Protection Association, Quincy, MA, 1981.

Meidl, James H., *Flammable Hazardous Materials*, 2nd ed., Macmillan Publishing Col., Inc., NY, 1978.

Talbot, G., "Static Electricity," *Fire*, 70(871), 1978, pp. 397-398.

Weiby, P. and K. R. Dickinson, "Monitoring Work Areas for Explosive and Toxic Hazards," *Chemical Engineering*, vol. 83, no. 22(1976), pp. 139-145.

27

Heat Processing Equipment

J. M. Simmons

The use of heat processing equipment in industrial operations continues to grow and be considered an ordinary tool, useful for the manufacture of a wide variety of products. Many of the processes employ combustible gases, the misuse of which may result in destructive explosions. Because accidents involving heat processing equipment have been relatively rare, the latent dangers associated with them are not often fully recognized. Yet, accidents have occurred frequently enough to emphasize the fact that operations must be conducted with a careful regard for good safety procedures.

Heat processing equipment is used in the industrial environment to perform many functions. They include a variety of forges, furnaces, kettles, kilns, ovens, retorts, and others of many different configurations. They may be direct-fired, indirect-fired, or involve a heat transfer medium. Some typical uses for heat processing equipment are the drying of materials with various coatings (paints, lacquers, etc.) that frequently involves vaporizing large quantities of flammable solvents; the drying of

---

J. M. Simmons, P.E., is Senior Special Hazards Engineer at Factory Mutual Research Corporation, Norwood, MA. Mr. Simmons is Chairman of the NFPA Committee on Ovens and Furnaces.

fabrics that may involve accumulations of combustible lints; and heat treating materials in furnaces containing a special atmosphere, such as hydrogen gas, to improve the quality of metals and metal alloys. The sizes of these devices range from relatively small annealing furnaces with limited capacity to large single-pass carpet dryers several hundred feet long.

Because heat processing equipment is so varied in size, complexity, location, and usage, it has been difficult to draft a single set of rules applicable to every type of furnace or oven, even though the basic principles of operation and use may be similar. Microprocessors or programmable controllers are being incorporated on new units for improved production, quality, and efficiency. Careful design, operation and maintenance have proven to be key elements for continued production without major incidents. Few installations can ever be planned that are totally safe under all conditions of emergency. Even if they were completely safe, it is conceivable that production-anxious operating companies would not always allow a safety system to go through a complete cycle, if it were to affect the production rate.

Causes of practically all failures of heat processing equipment can be traced back to human failure either through deliberate circumventing of known safe practices or through ignorance of the ramifications of a procedure. The most significant specific failures have been found to be due to inadequate training of operators and maintenance technicians, faulty design of equipment, complacency on the part of users, and error in selection of process and combustion safeguards for the installation being protected.

FIRE AND EXPLOSION PROBLEMS

The fire problems associated with the installation and operation of heat processing equipment center around (1) the proximity and combustibility of the contents of the building or room in which they are located, (2) construction of the building, (3) setting, (4) ventilation, (5) location, (6) heat, gas, and smoke disposal, (7) maximum temperature required, and (8) handling of heated materials in connection with equipment. Fire in combustibles can be prevented by adequate separation from the source of heat or by insulation. Overheating can be prevented by temperature controls.

Explosion problems involve the presence of flammable vapor-air mixtures from gas or oil fuel that have not been completely burned in the heating system or from the materials being processed, such as freshly coated parts that give off flammable solvent vapors in the drying process. Explosions or fires are generally prevented by adequate ventilation and controls that keep the flammable vapor content below the lower flammable limit (LFL) of the vapor-air mixture.

Some special process ovens and furnaces contain hydrogen or

other flammable gases for such purposes as annealing copper and castings. Others evaporate flammable solvents in low oxygen special atmosphere ovens to permit effective recovery of the solvents and energy saving reduced ventilation rates. For safe operation of these special atmosphere devices, it is necessary to prevent the entrance of air into the processing enclosure under normal operating conditions. Both normal or unscheduled starts and stops of these devices require special safety procedures that rely heavily on the skill and training of operators and the adequacy of the controlling equipment.

CLASSIFICATION OF HEAT PROCESSING EQUIPMENT

For fire classification purposes, industrial heat processing equipment may be classified by the nature of the material and the environment in which it is being processed. NFPA standards governing the installation of various types of heat processing equipment classify ovens and furnaces into four categories. (The standards are NFPA 86A, *Ovens and Furnaces — Design, Location Equipment*; NFPA 86B, *Industrial Furnaces*; NFPA 86C, *Industrial Furnaces Using a Special Processing Atmosphere*; and NFPA 86D, *Industrial Furnaces Using Vacuum as an Atmosphere*. They are collectively referred to as "NFPA Ovens and Furnaces Standards.")

NFPA Classification System

The classification system for heat processing equipment as set forth in the NFPA standards are as follows.

Class A ovens or furnaces are heat utilization equipment operating at approximately atmospheric pressure wherein there is a potential explosion and/or fire hazard which may be occasioned by the presence of flammable volatiles or combustible material processed or heated in the oven. Such flammable volatiles and/or combustible material may, for instance, originate from paints, powder, or finishing processes, including dipped, coated, sprayed, impregnated materials or wood, paper and plastic pallets, spacers or packaging materials. Polymerization or similar molecular rearrangements and resin curing are processes which may produce flammable residues and/or volatiles. Potentially flammable materials, such as quench oil, water-borne finishes, cooling oil, etc., in sufficient quantities to present a hazard are ventilated according to Class A standards. Class A ovens may also utilize a special inert atmosphere to evaporate solvent within the oven.

Class B ovens or furnaces are heat utilization equipment operating at approximately atmospheric pressure wherein there are no flammable volatiles or combustible material being heated.

Class C furnaces are those in which there is a potential hazard due to a flammable or other special atmosphere being used for treatment of material in process. This type of furnace may use any type of heating system and includes any special atmosphere supply system. Also included in the Class C standard are furnaces with Integral Quench and Molten Salt Quench.

Class D furnaces are vacuum furnaces which operate at temperatures above ambient to over 5000°F (2760°C) and at pressures below atmospheric using any type of heating system. These furnaces may include the use of special processing atmospheres.

The difference between an oven and a furnace often is not clearly defined. A dictionary definition of an oven is "a compartment or receptacle for heating, baking, or drying by means of heat," while a furnace is "an enclosed chamber or structure in which heat is produced for heating a building, reducing ores and metals, baking pottery, etc." It has been a rule of thumb of industry to classify heating devices that do not "indicate color" — operate at temperatures of less than approximately 1,000°F (538°C) — as ovens. This rule does not always apply as evidenced by the fact that coke ovens operate at temperatures in excess of 2,200°F (1204°C), and some furnaces operate at temperatures below 1,000°F (538°C). Perhaps the only positive statement that can be made is that operations for which material is heated, either in ovens or in furnaces, extend through a wide range of temperatures and conditions.

Classification by Type of Handling System

Ovens and furnaces also may be further classified based on the method of handling a material in the equipment. The two principal types are: (1) the batch oven or furnace, sometimes referred to as "in and out," "intermittent," or "periodic" type, and (2) the continuous "conveyor" type oven or furnace.

Batch type: In the batch oven or furnace the temperature is practically constant throughout the interior. The material to be heated is placed in a predetermined position and remains there until the heating process is completed. It is then removed, generally through the opening by which it entered. Figures 27.1 and 27.2 are examples of this type of furnace and oven.

Continuous type: In the continuous furnace or oven the material being processed moves through the furnace while being heated. The straight-line furnace is probably the most common. Figure 27.3 is an example of this type. The material passes through a continuous oven or furnace on a conveyor or rolls. A web of a paper, cloth or sheet metal may be drawn through, suspended on a cushion of hot air jets. The oven or furnace may operate at a constant temperature throughout, or it may be divided into zones maintained at different temperatures.

A common variation of the continuous type is one with a moving hearth. The material to be heated is placed on the hearth and is removed

HEAT PROCESSING EQUIPMENT

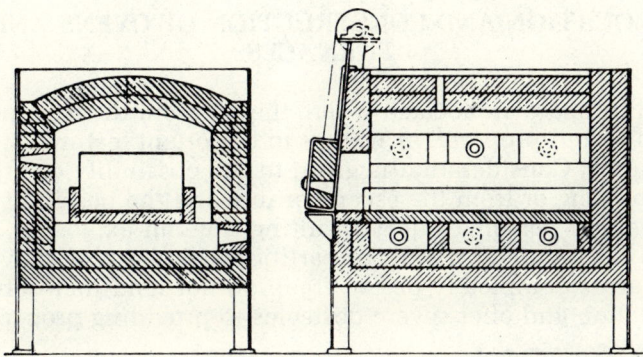

FIGURE 27.1. A large batch-type, under- and over-fired, semimuffle, heat-treating furnace.

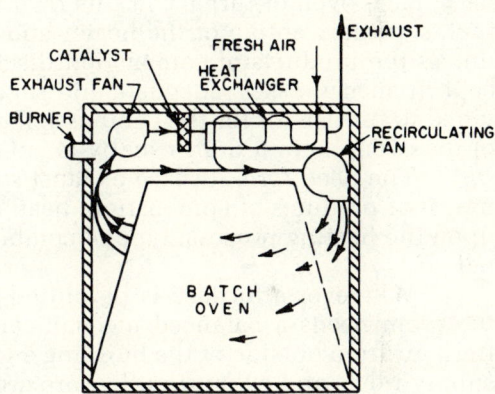

FIGURE 27.2. Batch oven with a catalytic heater for full air pollution control.

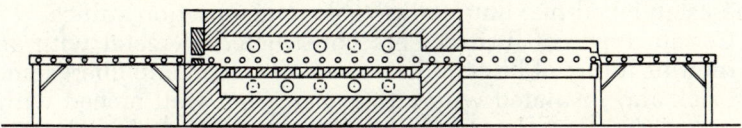

FIGURE 27.3. A continuous roller hearth furnace.

after the hearth has completed a revolution. This is known as rotating hearth or rotating-table furnace. Another method is to feed the material being processed through a furnace with a revolving hearth or tube by means of internal flights or screw thread.

575

LOCATION AND CONSTRUCTION OF OVENS AND FURNACES

Ovens and furnaces are located where they present the least possible exposure to life and property, as well as to important features of the building occupancy. Consideration is given to the possibility of fire resulting from overheating or from the escape of fuel and the possibility of building damage and personal injury resulting from an explosion. They may need to be surrounded by walls or partitions and be at or above grade, as basement areas completely below grade do not lend themselves to natural ventilation and offer severe obstacles to providing proper explosion release.

The construction of the oven or furnace and the building that houses it needs to be noncombustible. Explosion relief venting is provided where required. Combustibles in the vicinity of the oven are adequately separated or properly protected by insulation on the oven or furnace. Each oven or furnace has its own venting facilities. With gas or oil fuel, separate venting of the heater and the oven or furnace is provided unless the products of combustion discharge directly into the oven. Except in special cases, adequate and separate mechanical means are provided to furnish air for the combustion of the fuel and for the ventilation of the oven; natural draft usually is not adequate.

The need for a furnace exhaust system directly to outdoors for removal of products of combustion, heat, or toxic gases or vapors depends upon the heating process, type of combustion, and the hazard to personnel.

Wherever a furnace is permitted to exhaust directly into a room, the room needs a balanced mechanical ventilation system to intake in fresh air from outside of the building and to exhaust vapor-laden air. The supply inlets and exhaust outlets are arranged to provide a uniform flow of air throughout the area, without any dead air pockets. The room ventilation system is designed to adequately remove toxic contaminants at their maximum anticipated rate-of-release and keep concentrations below the established maximum allowable concentration values.

Certain types of furnaces are constructed of metal with an insulating interior lining of brick, other masonry or ceramic fiber; some types are of brick and insulated without covering, but well braced with metal supports, while others have an inner lining of metal, fire clay products or ceramic fiber. In some instances furnaces have air spaces or noncombustible fillers between the outer and inner walls. Most ovens have sandwich wall and roof panels constructed of sheet metal with noncombustible insulating fill.

Ovens and furnaces are well separated from valuable stock, important power equipment, machinery, and sprinkler risers, thereby securing a minimum interruption to production and protection in case of accidents to the oven or furnace. The installations are positioned so as to be readily accessible with adequate space above to permit installation of automatic sprinklers, the proper use of hose streams, the proper functioning of explosion vents, and performance of inspection and maintenance routines. Floors of ovens are insulated, and the space above and below ventilated, to keep temperatures at combustible ceilings and floors below 160°F (71°C).

HEATING SYSTEMS FOR HEAT PROCESSING EQUIPMENT

There are three basic methods of transferring heat to the materials being processed in ovens and furnaces: (1) by direct contact with the products of combustion or heated processing medium, (2) by convection and direct radiation from the hot gases, and (3) by reradiation from the hot walls of the furnace. In muffle furnaces (see Figure 27.1), the products of combustion are separated from the material being heated by a metal or refractory muffle and heat transfer is by radiation. In liquid-bath furnaces, i.e., salt baths used for tempering, hardening, etc., a metal pot containing a liquid heating or processing medium is heated and the heat transferred through the medium to the material being processed in the pot.

Oven Heaters

Oven heaters are of two general types, direct-fired and indirect-fired. With direct-fired heaters, the products of combustion enter the oven chamber and come in contact with the work in process. In indirect-fired oven heaters, the products of combustion do not enter the oven chamber; heating is by radiation from tubes or by convection of air passing over tubes and then into the oven. The indirect type is somewhat safer because dangerous fuel-air mixtures cannot readily fill the oven space. Nevertheless, the possibility of explosion from flammable vapors given off by the drying process still exists.

There are several arrangements of these two types of oven heaters; namely, direct-fired internal, direct-fired external, indirect-fired internal, and indirect-fired external heaters (see Figure 27.4).

Furnace Heaters

The arrangement of furnace heaters is usually the same as those for ovens — direct-fired internal, indirect-fired external, etc. Sometimes the terms used to express these arrangements are different. For direct-fired types, when products of combustion are under a hearth and then are carried up and into the heating chamber, the furnace is said to be under-fired. When the same thing occurs in a chamber at one side of the furnace and then passes over a bridge wall into the heating chamber, the furnace is referred to as a side-fired furnace. A furnace in which the products of combustion are produced in a space above the heating chamber and then pass through a perforated arch into the heating chamber is called an oven-fired furnace. If the same thing occurs at some distance above the heating chamber and hearth, and the products of combustion are deflected onto the hearth by an arched roof, the furnace is called a "reverberatory" furnace. A radiant tube heated furnace is an arrangement for indirect firing (see Figure 27.5).

Sources of Heat

The source of heat for an oven or furnace may be gas burners, oil burners, electric heaters, infrared lamps, electric induction heaters, or steam radiation systems.

Gas-fired: Gas fuel may be any type of gas that is in common industrial use. The important thing is that the burner, its adjustment, and the means of combustion are suitable for the gas that is to be burned. The burner may have a single nozzle, with burners located singly or in groups, or it may have multiple ports of the perforated pipe, ribbon, or slot type.

Oil-fired: Oven heating systems may be fired with fuel oil (nos. 2, 4, 5, and 6). Burners may be the mechanical pressure atomizing type or the type using an atomizing media such as steam or air.

Dual fuel-fired: Burner systems may be designed to fire either gas or oil; however, both fuels are seldom fired simultaneously. Oil burners are commonly ignited using gas pilots with an electric spark igniter. Some others use direct high energy electric spark igniters. For safety, the design and operating characteristics of oil burners are generally the same as for gas.

Electric heating systems: Electric heating systems for ovens and furnaces are of five types: resistance, infrared, induction, arc, and dielectric.

Resistance heat is produced by current flow through a resistive conductor. A resistance heater may be an "open" type with bare heating conductors or an "insulated sheath" type with heater conductors covered by a protecting sheath which may be filled with electrical insulating material (see Figure 27.6).

Infrared heat is transmitted by direct radiation generated by ceramic or metal alloy gas burners or from special incandescent electric lamps. The infrared radiation passes through air and transparent substances but not opaque objects, and therefore release their heat energy to these objects.

Induction heat is developed by eddy currents induced in the charge. Induction heaters have an electric coil surrounding the workspace, and heating is by electric currents induced in the work being processed.

Arc heat is developed by the passage of an electric current be-

HEAT PROCESSING EQUIPMENT

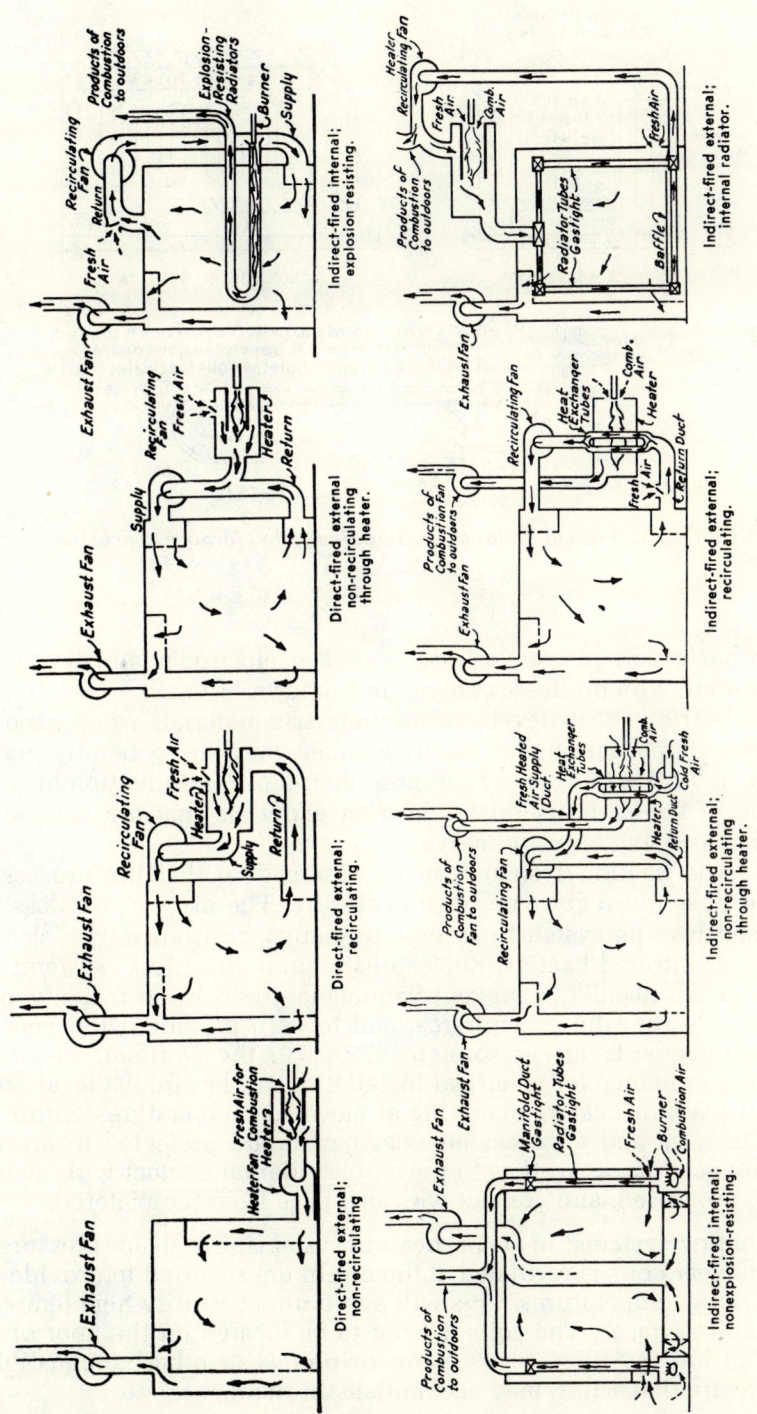

FIGURE 27.4. Types of oven-heating systems.

579

Chapter 27

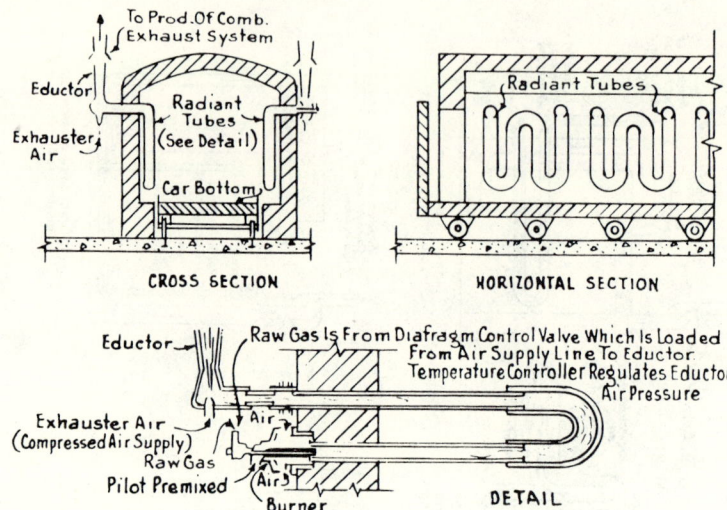

FIGURE 27.5. Typical indirect-fired radiant-tube furnace (carbottom type)

tween either a pair of electrodes or between electrodes and the work, causing an arc which releases energy in the form of heat.

Dielectric heat is developed in dielectric materials when exposed to an alternate electric field. The frequencies used are generally higher the order of 3 MHz (megahertz) or more than those in induction heating. This type of heater is useful for heating materials that are commonly considered as being nonconductive.

Electric heating systems can be arranged so that the processing operation does not require an oven enclosure. The use of "ovenless" or unenclosed heating systems can employ lamps, resistance-type electric elements, or infrared heaters. Enclosures around "ovenless" systems are provided when needed to prevent flammable, toxic, or corrosive vapors from escaping into the general area, and to help provide better ventilation and safeguards for personnel. NFPA 70, the *National Electrical Code*, gives guidance for electrical installations in hazardous locations.

All parts of heaters operating at elevated temperatures within an oven or furnace and all other energized parts are protected to prevent contact by persons, as well as to prevent accidental contact with materials being processed, and contact with drippage from the materials.

Steam heating systems: In steam heating systems, the steam pressure in heat exchanger coils is regulated at the minimum required to provide the proper drying temperatures. This will avoid unnecessarily high temperatures at coil surfaces. The coils are not to be located on the floor of the oven or in any position where paint drippings or other combustibles, such as recirculated lint, may accumulate on them.

HEAT PROCESSING EQUIPMENT

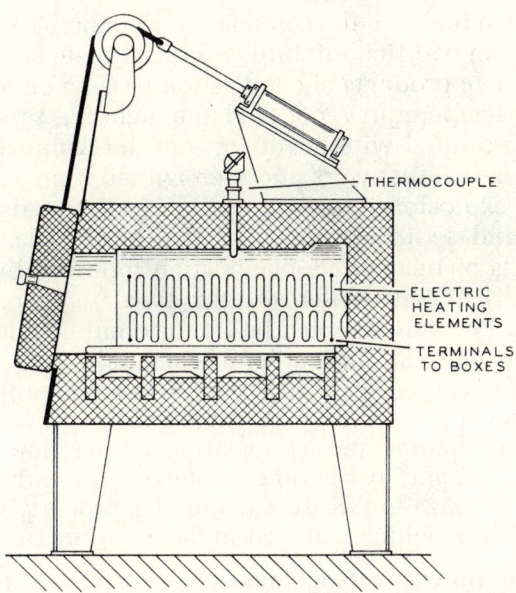

FIGURE 27.6. Electric furnace.

Hazards of Fuels

Gas and oil fuels present the hazard of unburned or incompletely burned fuel, which, when mixed with air, may be within the range of ignition and explosion. These hazards develop during lighting-off, firing, and shutting down the oven or furnace. It is necessary to treat each operation as a separate condition requiring certain specific operating procedures to avoid mishaps.

Lighting-off: Before torches, sparks, or other ignition sources are introduced, and until all burners are properly lighted, the operator must exercise every precaution, backed up by such automatic safety controls as are practical, to eliminate or to avoid producing dangerous unburned fuel accumulations.

Firing: In the firing phase, three conditions are given attention:

Safety requires continuous ignition and complete burning of the fuel before it passes beyond its normal combustion zone at the burner. To maintain this safe condition of a stable flame with complete combustion, the mixer and burner assembly must proportion the particular fuel and air properly throughout the combustion zone, to the correct concentrations, and the mixture velocity through the combustion zone must be neither too high, causing extinguishment by blowoff, nor too low, causing flame to flash back or "go out," at any firing rate within the turn-down range.

Total or partial failure of combustion air supply can result in an unstable flame that may lead to flame failure and introduction of unburned fuel into the combustion chamber. Too little combustion air results in an overrich mixture and incomplete combustion. The flammable incomplete products of combustion passing out of the combustion chamber at temperatures not high enough for prompt ignition may later become diluted with air down into the flammable range in the oven or recirculating duct work and then ignited from a number of sources, causing an explosion. Overrich combustion may also produce rapid smothering and extinguishment of the burner flame, and the flammable products of incomplete combustion followed by raw fuel may likewise become explosive when diluted by air later or in another part of the system. Precautions must be taken to cut off the fuel in event of failure of combustion air and to require manual reset.

Liquid fuels, such as fuel oil, must be broken up into fine globules by atomization for easy ignition and quick complete combustion. Improper oil temperature preventing proper flow, partial obstructions in burner tips, and loss of oil or atomizing medium pressures can cause improper atomization. Failure to atomize properly will usually result in unstable flame, which can lead to flame failure.

Shutting down: A dangerous accumulation of unburned fuel may occur in an oven and heating system following a shutdown if any manual fuel valves are left open or leaking and/or safety shutoff valves are not tight-closing. Subsequent ignition by hot refractory or an ignition source when starting up may produce an explosion.

OPERATION OF HEAT PROCESSING EQUIPMENT

Each of the four classes of heat processing equipment has operating requirements and conditions that must be observed and followed for efficient and safe operation. The following paragraphs discuss some of the more important considerations and hazards associated with each of the four classes of devices.

Class A Ovens and Furnaces

Adequate ventilation (in addition to that needed for the burner system) is of prime concern in the operation of Class A ovens and furnaces because of the presence of flammable vapor-air mixtures resulting from the materials being processed. Fires or explosions of vapor-air mixtures are, in general, prevented by good ventilation and controls that keep the flammable vapor content below the lower flammable limits.

Ventilation is required while an oven is in operation and flammable vapors are being given off. Control devices assure that the ventilating system is in operation and provide for preventilation of ovens to dissipate accumulated vapors, such as those resulting from fuel leakage or from a work charge left in an unventilated oven. Failure of the ventilating fan causes the shutdown of the heating system and of the conveyor, which carries material into a continuous oven. The following discussion of ventilation refers to that required for safety of operation, exclusive of combustion and recirculation requirements. It does not apply to ovens operating in conjunction with solvent recovery systems.

Proper ventilation means a sufficient supply of fresh air and proper exhaust to outdoors with a sufficiently vigorous and properly distributed air circulation to ensure that flammable vapor concentrations in all parts of the oven enclosure are safely below the lower flammable limit (LFL) of the vapor at all times. The basis for determining the quantity of fresh air required for safe ventilation is the amount of vapor produced by the flammable solvent used on the work in process.

In general, mechanical ventilation to the outdoors is required for all ovens in which flammable or toxic vapors are liberated, as well as for ovens heated by direct-fired gas or oil heaters. On all ovens, regardless of size, it is advisable to have a separate exhauster providing the required ventilation whenever appreciable amounts of flammable vapors are given off by the work.

Ovens in which no flammable or toxic vapors are released at any time do not require ventilation for safety, if heated by steam or electric energy, or by gas- or oil-fired indirect heating equipment.

Continuous conveyor-type ovens: The general rule for ventilating continuous conveyor-type ovens is to provide not less than 10,000 cfm (283 m^3) of fresh air at 70°F (21°C) for each gallon (3.78 L) of common solvent introduced into the oven. The basis for this general rule is that 1 gal (3.78 L) of common solvent produces a quantity of flammable vapor which will diffuse in air to form roughly 2,500 cu ft (71 m^3) of the leanest explosive mixture. Since a considerable portion of the ventilating air may pass through the oven without completely traversing the zone in which vapors are given off, and because of a possible lack of uniform distribution of ventilation air, and also to provide a margin of safety, four times this amount of air, or 10,000 cu ft (283 m^3) referred to 70°F (21°C), for each gallon of solvent evaporated is allowed. With certain solvents, where the volume of air rendered barely explosive exceeds 2,500 cu ft (71 m^3), the factor of safety decreases proportionately. Calculations should be made to avoid decreasing the factor of safety.

When a continuous-type oven is designed to operate with a particular solvent and where ventilating air may be accurately controlled, the required ventilation can be determined by calculation. See NFPA 86A for information on how to calculate the volume of vapor produced by 1 gal (3.78 L) of solvent and the volume of air required to provide sufficient dilution to prevent an ignitible mixture. As with the general rule for oven ventilation, the calculated rate of air change includes a factor of safety four times the volume of air required to prevent an ignitible mixture.

Batch process (box) ovens: The nature of the work being processed is the main factor in determining the ventilation rate in batch process ovens. Because of the wide variations in the materials, rate of evaporation and coating thickness, it is preferable that tests and calculations be made. However, tests and years of industrial experience have shown that approximately 380 cfm (10.76 m^3/min) referred to 70°F (21°C) of ventilation for each gallon (3.78 L) of flammable volatiles released from a batch of sheet metal or metal parts being baked after dip coating is a reasonably safe rate of air change. Tests of dipped sheet metal in batch process ovens also have shown that, in typical 1-hour bakes, practically all the solvent is evaporated in the first 20 min.

For other types of work, the figure of 380 cfm (10.76 m^3/min) referred to 70°F (21°C) is also used, unless the required ventilation rates can be calculated on the basis of reliable previous experience, or the maximum evaporation rate is determined by tests run under actual oven operating conditions. In the latter case, a margin of safety is introduced by requiring a rate of air change equaling four times the volume of air needed to produce an ignitible mixture. In any event, caution is needed in applying this estimating method to work of low mass which will heat quickly, such as paper, textiles, etc., or work coated with materials containing highly volatile solvents. Either condition may give too high a peak of evaporation rate for the estimating method.

Temperature correction: Temperature corrections must be made in using the above rules. The volume of a gas varies in direct proportion to its absolute temperature (0°F is equivalent to approximately 460 absolute on the Rankine scale).

For example, in order to supply 10,000 cu ft (283 m^3) of fresh air referred to 70°F (530 absolute) to an oven operating at 300°F (760 absolute), it is necessary to exhaust $760/530 \times 10,000$ (283 m^3) or 14,320 cu ft (405 m^3) of vapor-laden air.

In some cases, process requirements for ventilation are in excess of that needed to maintain safe conditions in an oven. When this is true, an approximate method of figuring ventilation may be adequate as a means of checking safety requirements. Except in these cases it is important that all factors, including solvent characteristics, type of oven, material being processed, oven temperatures, effect of temperature on LFL, etc., be carefully considered so that an adequate safety factor is assured.

Class B Industrial Furnaces

Class B furnaces are, in some ways, considered much the same as Class A ovens or furnaces. Because there are no flammable volatiles or residues present, there is nothing combustible in the construction or contents of the furnace. In many of these furnaces, the only hazard is from the gas or oil fuel. To attain higher temperatures with fuel-fired equipment, the quantity of fuel consumed will be relatively greater, and there may be a corresponding increase in the fuel hazard.

In many cases, little or no effective explosion relief venting can be provided, because of the heavy refractory, extensive insulation, and relatively strong construction required for the high operating temperatures. The emphasis must be placed on explosion prevention. This means preventilation before a source of ignition is introduced into the furnace. In some cases, completely automatic purging will be required; in others, it may be partly manual. In either case, safety controls are needed to interlock the ventilation, the fuel supply, combustion air, safety shutoff valve, and flame failure devices. An audible alarm may be provided to indicate the development of any unsafe condition.

Some furnaces have zones operating at different temperatures, which may have to be treated as separate units.

Modification of the usual requirements may be permitted with some of these ovens or furnaces, but if there is a possibility of damaging explosions in them from unburned fuel, the need for adequate safeguards should not be disregarded.

Class C Special Atmosphere Furnaces

Special atmosphere furnaces are used to improve the quality of metals and metal alloys by heating them in an atmosphere in which air has been replaced by other gases, some of which are combustible. In most cases, the special atmosphere gas is used to prevent oxidation of the metal during heating, but it may also be used to prevent the removal or addition of carbon. Some applications of special atmosphere gases include bright annealing of copper and steel, scale-free hardening and annealing of castings, brazing, and sintering. Examples of protective gases are hydrogen, charcoal gas, dissociated ammonia, and lithium vapor. Also commonly used are various hydrocarbon gases produced by partial or complete combustion reaction with air, occasionally in the presence of a catalyst.

Another method of producing similar atmospheres is to introduce a liquid hydrocarbon (e.g., methanol) into the hot furnace where it reacts and combines with the nitrogen atmosphere. Both the liquid and nitrogen are usually supplied from bulk storage.

Some heat treating furnaces contain an inert atmosphere (carbon dioxide, helium, argon, nitrogen) and so present no special fire hazard; however, these gases could present a health hazard. It is those which have a flammable atmsophere (hydrogen, dissociated ammonia, incompletely burned hydrocarbon gas, carbon monoxide, and methane) that present an explosion hazard and require special safeguards. Atmosphere furnaces are not limited to flammable gases but can also contain acid gases, such as chlorine and anhydrous hydrochloric. When the latter types of gases are used in a special atmosphere, extreme care must be used to keep air from entering the furnace.

Regardless of the type of special atmosphere, Class C furnaces have fuel hazards the same as for Class A and B ovens and furnaces explained earlier in this chapter. The subsequent discussion covers, in a general way, the protection needed for the special atmosphere generator and for the heat treating enclosure.

Hazardous conditions in Class C furnaces exist chiefly before the start of the process when the flammable atmosphere is replacing air in the furnace, when the process is finished and air is being admitted, and when, for some reason, the special atmosphere supply is interrupted and air permitted to enter. If in each of those instances an inert gas could be introduced into the furnace in sufficient quantity to prevent any combustible mixture of special atmosphere and air, there would be no danger of explosion. Automatic introduction of inert gas upon failure of the special atmosphere supply would be desirable; at the least, an audible alarm would notify the oven operator of the failure.

If the use of inert gas is not considered practical, alternate procedures are to burn out the air at the start of the process and the flammable atmosphere at the finish, or to evacuate the furnace by use of vacuum pumps or exhausters. With furnaces having an operating temperature above 1,400°F (760°C), burning may be done at the start by bringing the temperature up to 1,400°F (760°C) before the special atmosphere is introduced; at this temperature, when gradually introduced, the flammable gas will burn in the oven until the oxygen is used up, then the hazard is removed and operation may be started. At the end of the process, the heating should be continued so the temperature stays above 1,400°F (760°C); the flammable gas supply is then shut off and air gradually admitted; when burning stops, the flammable gas has been consumed. When the alarm indicates failure of the flammable gas supply or of the heating system, the oven operator immediately starts the admission of air to burn the flammable gas in the furnace as explained above; this is done before the furnace cools to below 1,400°F (760°C).

With furnaces operating at temperatues below 1,400°F (760°C), or startup, the furnace should be purged with inert gas, e.g., nitrogen, to remove the air before the flammable atmosphere is introduced. On shutdown, the reverse action should be taken, purging with inert gas to remove the flammable atmosphere before introducing the air.

Special Atmosphere Generators

Special atmosphere generators are a source for some atmosphere gases used in Class C furnaces. One type of generator (exothermic) produces the atmospheric gas by completely or partially burning fuel gas at a controlled ratio, usually at 60 percent to 100 percent aeration while another type (endothermic) produces the atmosphere at a controlled ratio of less than 50 percent. Atmospheres from an exothermic generator may be either inert or flammable depending on the generator's design and operating range, while those from endothermic generators are always flammable.

Another type of generator is an ammonia dissociator, which, by temperature reaction with a catalyst in an externally heated vessel, produces dissociated ammonia (25 percent nitrogen and 75 percent hydrogen) from ammonia.

It is important that special atmosphere generators be provided with safety controls that will assure the safe and proper functioning of the equipment. These would normally include interlocking of raw gas, air (if needed), burners for heating or processing, feed and discharge pressures, etc. Flame supervision is provided on the exothermic generator burner, to verify combustion of the gases for the special atmosphere going to the heat treat furnace. Safety shutoff valves are usually provided in feed and discharge piping, also devices to indicate the pressure or rate of flow of processed gas to the furnace, and also its analysis, if this might vary. The operator would thus be assisted in supplying the furnace with the desired special atmosphere.

The preferable location for the generator and its auxiliary equipment, such as surge tank, compressor, aftercooler, storage tank, etc., is in a separate, detached building of light, noncombustible construction.

Class D Vacuum Furnaces

Vacuum furnaces are used for heat treating metals; however, they are not limited to this industry. In a vacuum furnace a vacuum pump is used to displace oxygen, and in most cases, to reduce the water vapor content or dew point as well.

They are usually batch-type furnaces and are further classified as hot-wall and cold-wall furnaces with the latter in greater use at the present time. In the hot-wall furnace, the entire vacuum vessel is heated, usually not above 1,800°F (982°C) due to the reduction in the strength of materials at elevated temperatures. However, installation of a second vacuum vessel outside of the vacuum retort (within which a roughing vacuum is maintained during the heating cycle) permits construction of larger hot-wall furnaces with higher operating temperatures. Examples of cold-wall and hot-wall vacuum furnaces are shown in Figures 27.7 and 27.8.

Cold-wall furnaces contain a vacuum vessel that is water cooled, and usually the heating elements are located inside the vacuum vessel. The walls can be maintained at near ambient temperature during high temperature operations. As a result, large units operating at high temperatures — 4,000 to 5,000°F (2205 to 2760°C) — can be constructed. The two most common methods of heating cold-wall furnaces are resistance and induction, with the heating elements located within the vacuum vessel. The heating elements are usually water cooled, but there are cases where air has been used for element cooling.

Vacuum furnaces have the same hazards usually associated with electrically heated furnaces; however, the potential for more serious accidents, when they do occur, is probably greater. Some hazards other than fuel associated with vacuum furnaces are:
1. Explosions can be caused by water leaks in either heating elements

Chapter 27

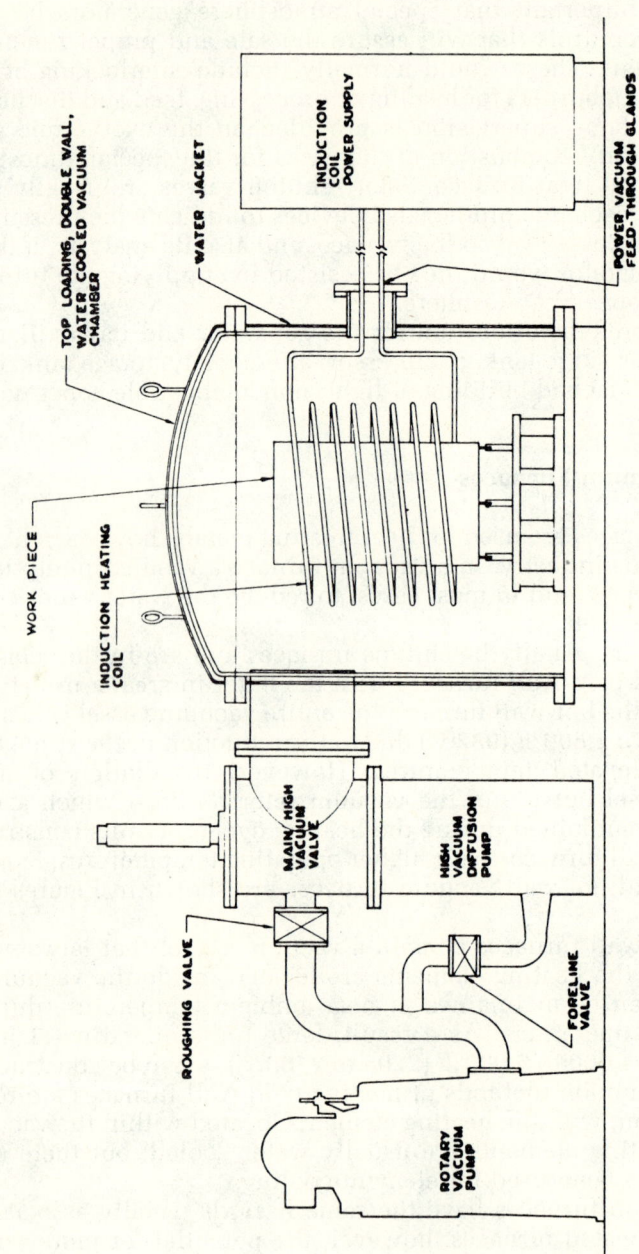

FIGURE 27.7. Example of cold-wall, induction-heated vacuum furnace.

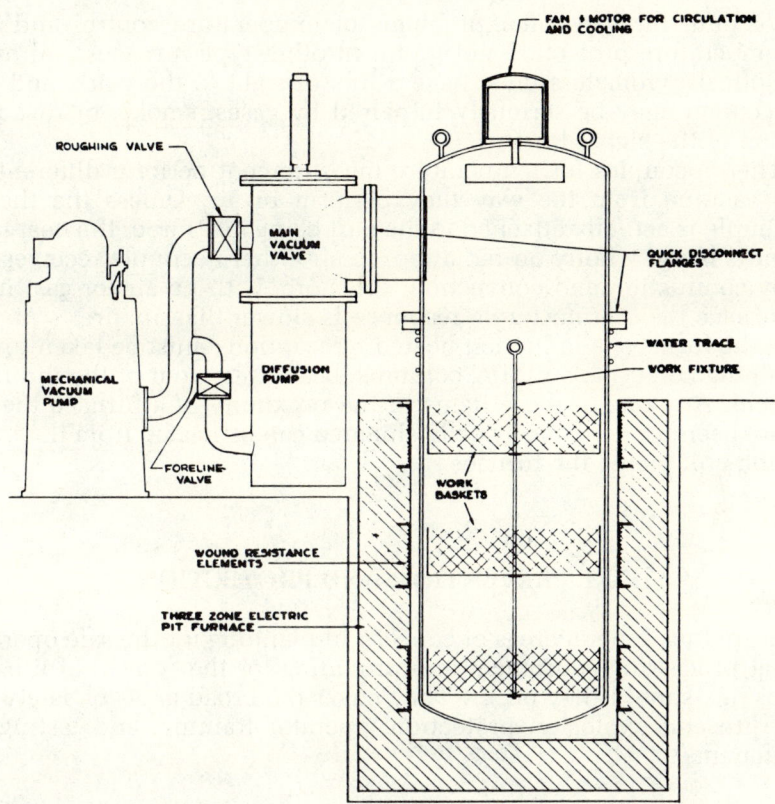

FIGURE 27.8. Example of hot-wall, single pump retort vacuum furnace.

or vessel jackets. Entrance of water into the furnace at temperatures at which these furnaces usually operate causes more than just a steam explosion.
2. The furnace wall can collapse due to failure of a relief valve on the water jacket.
3. Improper design of the vessel as related to strength of materials at high temperatures can cause collapse of the vacuum retort in a hot-wall furnace.
4. Vacuum pumps can pull fluids (water or oil) from hydraulic seal pots.
5. At the pressures at which this type furnace operates, even metals can vaporize, and condensed metallic vapors on electrical insulators can cause short-circuiting.
6. Improperly supported heat shields can sag at high temperatures and can cause short-circuiting by coming in contact with heating elements.
7. Sagging of heat shields will cause hot spots on furnace walls with the resulting weakening of the furnace wall.

Chapter 27

8. Vacuum furnaces offer problems of temperature control and over-temperature protection not found in other type furnaces and ovens. Optical pyrometers must have a line of sight to the work, and their accuracy may be seriously impaired by gases, smoke, or discoloration of the sight glass.
9. Thermocouples for temperature measurement perform differently in a vacuum from the way they perform in air. Unless the thermocouple is actually attached to the part being measured, the heat transfer is based wholly on radiation. In air a thermocouple receives heat by conduction and convection; therefore, with no air (or gas) in the furnace the thermocouple response is slower than in air.
10. If the furnace is induction heated, precautions must be taken to keep piping conduits, building columns, beams, etc., out of the induction field. Any one of these items in the proximity of a furnace that has not been properly electrically shielded can be heated from the induction coil inside the furnace.

FIRE PREVENTION AND PROTECTION

There are four general areas of concern in planning for the safe operation of heat processing equipment and providing for the control of fires and explosions should they occur. They cover the broad areas of safety controls, fire and explosion protection, operator training, and testing and maintenance.

Safety Controls for Ovens and Furnaces

It is essential that all ovens and furnaces processing flammable materials, or those involving flammable vapors, or heated with combustible fuels, be provided with adequate safety devices. Safety equipment must ensure sufficient preventilation and adequate ventilation during operation. This equipment must constantly supervise operating conditions so that safety will be assured and conditions that might result in fires or explosions will not be permitted to develop. Safety control means that an adequate number and suitable types of "fail-safe" devices, properly arranged, are provided to maintain safe conditions.

The safety control circuits, including appropriate interlocks and combustion safeguards, need to be arranged for reliability of operation and suitable sequencing. Use of solid state electronic controllers, microprocessors and programmable controllers needs careful evaluation to provide reliable protection and to minimize the possibility of human element error.

The type of safety control equipment employed with oven or furnace assemblies depends on the requirements of the particular operation. A list of the principal types of controls employed is given in

HEAT PROCESSING EQUIPMENT

TABLE 27.1. *Safety Control Equipment for Ovens and Furnaces*

Ventilation controls:
  Air flow switches
  Pressure switches
  Fan shaft rotation detectors
  Dampers
  Position limit switches
  Electrical interlocks
  Preventilation time-delay relays

Fuel safety controls:
  Safety shutoff valves
  Supervising cock (FM cock)
  Flame detection units (combustion safeguards)
  Flowmeters
  Firechecks
  Reliable ignition sources
  Pressure switches
  Program relays

Temperature controllers

Temperature limit controls

Continuous vapor concentration indicators and controls

Conveyor interlocks (with steam and electric-resistance heating equipment)

Electrical overload protection (with resistance and induction heating equipment)

Low oil-temperature limit controls (on oil-burner equipment using heavy residual fuel oil, such as No. 5 and No. 6, which require preheating)

Table 27.1. Most of these control units are tested by recognized laboratories and their listings indicate the appropriateness of the devices for varying situations (see Figure 27.9).

    The NFPA Ovens and Furnaces standards specify in considerable detail the safeguards that are needed for different ovens and furnaces and the types of devices that may be used. The applicable standards (86A, 86B, 86C or 86D) should be consulted, as the requirements for safety controls for one type of heat processing equipment do not necessarily pertain to another.

Fire and Explosion Protection

Automatic sprinklers and water spray systems are basic fire protection considerations for heat processing equipment containing or processing sufficient combustible materials to sustain a fire. However, the extent of protection required depends upon the construction and arrangement of

Chapter 27

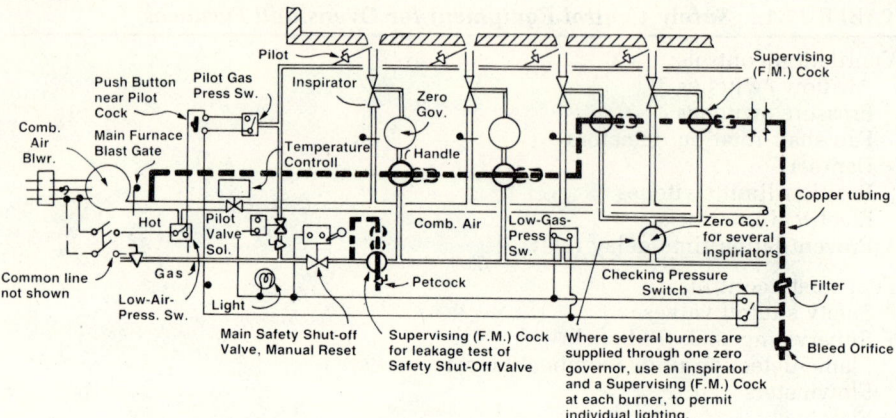

FIGURE 27.9. Supervising cock and gas safety control system.

the heat processing equipment as well as the materials handled. Fixed protection extends as far as necessary into the enclosure, if combustible material is processed, or if trucks or racks used are combustible, or subject to loading with excess finishing material, also if an appreciable amount of flammable drippings of finishing materials accumulates within the oven.

If desired, permanently installed, supplementary protection of an approved type such as carbon dioxide, foam, dry chemical, or halon may be provided. Such protection is not a substitute for automatic sprinklers.

Steam smothering systems are permissible only when oven temperatures exceed 225°F(107°C) and large supplies of steam are available at all times when the oven may be in operation. Complete standards paralleling those for other extinguishing agents have not been developed for the use of steam as an extinguishing agent, and until this has been done, the use of this form of protection is not as dependable or the results as certain as those qualities of water, carbon dioxide, dry chemical, halon, or foam.

Portable extinguishing equipment is needed near the oven, oven heater, and related equipment, including dip tanks or other finishing processes operated in conjunction with the oven. Small hose with combination nozzles is also provided so that all parts of the oven structure can be reached by small hose streams.

Ovens that may contain flammable gas or vapor mixtures with air are equipped with unobstructed relief vents for freely relieving internal explosion pressures. These vents are provided in the form of wall panels or gravity-retained roof panels designed to afford adequate insulation and possess the necessary structural strength, yet be as lightweight as possible. These explosion relief panels are proportioned in the ratio of their area in square feet to the explosion-containing volume of the oven, due allowance being made for openings or access doors or panels equipped with approved explosion-relieving hardware. The preferred ratio is 1:15, i.e., 1 sq ft (0.093 m^2) of relief panel area to every 15 cu ft (0.425 m^3) of oven volume. The lower the release pressure of the relieving panels, the lower the pressure build-up within the oven.

Operator Training

The most essential safety consideration is the selection of alert and competent operators. Their knowledge and training are vital to continued safe operation. It is important that new operators be thoroughly instructed and demonstrate an adequate understanding of the equipment and its operations. It is good practice to retrain regular operators at intervals to maintain a high level of proficiency and effectiveness.

Operating instructions are provided by the equipment manufacturer. These instructions include schematic piping and wiring diagrams, as well as:

1. Light-up procedures.
2. Shutdown procedures.
3. Emergency procedures.
4. Maintenance procedures.

Operator training includes:

1. Combustion of air-gas mixtures.
2. Explosion hazards.
3. Sources of ignition and ignition temperature.
4. Atmosphere gas analysis.
5. Handling of flammable atmosphere gases.
6. Handling of toxic atmosphere gases.
7. Functions of control and safety devices.
8. Purpose and basic principles of the gas atmosphere generators.

Operators need to have access to operating instructions at all times.

Testing and Maintenance for Ovens and Furnaces

It is important that the operating and safety control equipment of each oven is regularly checked and tested, preferably once a week; and at less frequent intervals, probably annually, a more comprehensive test and check are made by an expert. All deficiencies are to be promptly corrected. A regular cleaning program is observed to cover all portions of the oven and its attachments. Suitable access openings are provided for cleaning the oven enclosure and connecting ducts.

A program for inspecting and maintaining oven safety controls is given in the appendices of the NFPA Ovens and Furnaces standards.

BIBLIOGRAPHY

REFERENCE

McKinnon, G. P., ed., *Fire Protection Handbook*, Fifteenth Edition, National Fire Protection Association, Quincy, MA, 1981.

NFPA CODES, STANDARDS, AND RECOMMENDED PRACTICES

Reference to the following NFPA Codes, Standards, and Recommended Practices will provide further information on the safeguards for heat processing equipment discussed in this chapter. (See the latest *NFPA Codes and Standards Catalog* for availability of current editions of the following documents.)

NFPA 31, *Installation of Oil Burning Equipment.*

NFPA 54, *National Fuel Gas Code.*

NFPA 58, *Storage and Handling of Liquefied Petroleum Gases.*

NFPA 70, *National Electrical Code.*

NFPA 86A, *Ovens and Furnaces, Design, Location, and Equipment.*

NFPA 86B, *Industrial Furnaces, Design, Location and Equipment.*

NFPA 86C, *Industrial Furnaces Using Special Processing Atmospheres.*

NFPA 86D, *Industrial Furnaces Using Vacuum as an Atmosphere.*

ADDITIONAL READING

Factory Mutual Engineering Corporation, "Industrial Ovens and Dryers," *Loss Prevention Data* 6-9, October 1977, Factory Mutual System, Norwood, MA.

F. M. Eng. Corp., "Process Furnaces," *Loss Prevention Data* 6-10, November 1976.

Grace, C., "Fluid Choice Takes the Steam Out of Unsafe Process Heaters," *Process Engineering*, (5), 1977, pp. 85, 87-88.

"Hazard and Hazard Prevention in Solvent Evaporating Ovens," *Fire Protection*, (127), 1978, pp. 22-25.

King P. W. and J. Magid, *Industrial Hazard and Safety Handbook*, Butterworth Publishers Inc., Woburn, MA, 1979.

Reed, Robert D., *Furnace Operations*, 3rd ed., Gulf Publishing Co., Houston, TX, 1981.

Trinks, W., *Industrial Furnaces*, (vols. 1 and 2), John Wiley and Sons, New York, NY. (vol. 1: *Principals of Design and Operation*, 5th ed, 1961; vol. 2: *Fuels, Furnaces Types and Furnace Equipment: Their Selection and Influence Upon Furnace Operation*, 4th ed,. 1967.)

28

Oil Quenching

George M. Woods

Metallurgical requirements for heat treating metals vary widely. These requirements, individually and collectively, create the potential for loss. A critical step in some of the heat treating processes is the controlled cooling or *quenching* of the metal, which is achieved by immersing the work in a quench medium. This chapter is concerned with loss potential introduced by quenching operations using oil as the medium.

Heat treating requirements influence the loss potential contributed by a number of oil quenching conditions. Among these conditions are:

Special atmosphere requirements (a gas other than air blanketing the surface of the oil),
Quench medium temperature requirements,
Physical properties of the quench medium,
Quench medium volume limitations,
Size and configuration of process materials,

George M. Woods, HPR Staff Officer with the Kemper International Insurance Company, Long Grove, Illinois, wrote whis chapter. It was reviewed by Robert R. Brining, Mgr., Heat Treating & Welding Engineer, Caterpillar Tractor Co.

Chapter 28

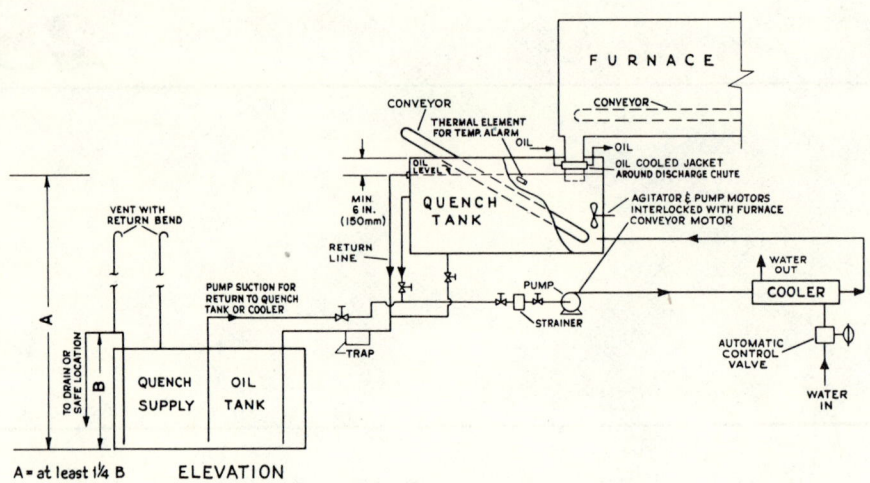

FIGURE 28.1. *A schematic of a typical continuous-type oil quench tank. (Factory Mutual System)*

Locations of furnaces and quench tanks, and
Mutual exposure between quenching and other processing or storage facilities.

Historically, quenching has been divided into batch operations and continuous operations. In practical application, batch operations can be performed either manually or automatically, while continuous operations are performed automatically. Combinations of the two are semiautomatic operations. The quenching process uses elevators, conveyors, hoists, and cranes, either individually or in combination, to immerse the work in, move it through, and remove it from the oil bath. Although all three steps are important to the process, the one that is the most critical to safety is the entrance of the work into the quench.

Some materials are maintained in a special atmosphere until they have been immersed in the oil quench; others are not. Each of these atmospheres produces its own degree of hazard. However, the general loss potentials with respect to the quenching oil are the same.

QUENCHING OILS

The selection of quench oil used depends upon the process involved, i.e., hardening, or tempering. However, from a fire protection and safety point of view, only the oil itself and the process equipment need be evaluated.

In most cases, mineral oils are used for quenching, but specific metallurgical requirements may dictate the use of mixtures of mineral, vegetable, and animal oils. In addition, wetting agents may be blended with certain oils or oil mixtures. In any case, the use of a quenching oil with a low viscosity is important. All materials being processed must be uniformly immersed in, and uniformly exposed to, the quench medium to prevent variations in the resultant physical properties.

Essential quench oil properties include the ability to remain stable over periods of extended usage and to retain fluidity. Quench oil consumption from vaporization and adherence to processed materials must be controlled. For unheated quenching, operating temperatures between 100°F (37.8°C) and 200°F (93.3°C) are considered normal. Standard quench oils for use in this temperature range usually have flash points somewhat above 300°F (149°C). For heated quenching, operating temperatures between 200°F (93.3°C) and 400°F (204.4°C) are common. Quench oils used at these temperatures generally have a flash point above 500°F (260°C).

Wherever possible, water or other nonflammable compounds should be used.

QUENCH TANKS

A quench tank not only should allow proper quenching under normal conditions, but also it should provide for minor variations in equipment control functions and operator error. Modern requirements for economy, efficiency, conservation (both materials and space), and competition have resulted in a reduction of the size of quench tanks to the point where there is almost no tolerance for variations in processing control.

It is vital to safety that quench oil remain in the tank. The design of tank freeboard, overflow drains, and liquid level control are all critical. A fire that is confined to the liquid surface within a tank is much more readily controlled and extinguished than a fire involving quench oil that has overflowed the tank. The ultimate loss in equipment and building facilities will be greatly reduced if fire is confined to the liquid surface within the tank.

Tank Freeboard

The distance from the quench surface to the top of the tank, with the tank loaded to capacity, is known as the *freeboard*. Freeboard design specifications should take into account the splashing effect to be expected when the maximum workload is immersed in the quench with maximum speed. In no case should the distance between the liquid level with the workload submerged and any openings in the tank wall be less than 6 in. (0.152 m).

TABLE 28.1 Quench Tank Overflow Drains, Minimum Pipe Sizes

Liquid surface area		Minimum pipe diameter, I.D.	
(m²)	(ft²)	(mm)	(in.)
0.9	10	76	3
7–14	75–150	102	4
14–21	150–225	127	5
21–30	225–325	152	6

Overflow Drains

Adequately sized, fully trapped overflow drains are important safety features for all oil quench tanks. They should direct the overflow to a safe location outside of buildings or into special tanks that will contain about twice the oil content of the quench tank. As a practical matter, small quench tanks frequently may be installed without overflow drains. However, all quench tanks with a liquid capacity of 150 gal (568 L) or a liquid surface area of 10 sq ft (0.93 m²) or larger, without exception, should be provided with overflow drains.

While overflow drains should be specifically designed for each tank, certain minimum sizes have been established and accepted. (See Table 28.1.) For large quench tanks, multiple overflow pipes are preferable to a single large pipe, provided the aggregate cross sectional area is equivalent to that of a single pipe.

Overflow connections may be subject to the accumulation of caked or dried material because of exposure to high temperature from stock in process or from open furnace doors. In these cases, drains should be provided with flared connections. In all cases, overflow connections should be located so that they are readily accessible for inspection and cleaning.

Emergency Drains

Under serious fire conditions, it may be necessary to empty a quench tank in order to reduce the amount of fuel available. This can be accomplished readily through adequately sized, fully trapped and valved bottom drains. These drains must be directed to safe locations, either outside the building or to properly sized and vented salvage tanks. In those cases where gravity drainage is not possible, special pumps may be provided for oil removal. Emergency drainage is strongly recommended for all oil quench tanks having a liquid capacity of 500 gal (1893 L) or larger. Gravity drainage or removal by pumping should be sized so that the quench oil can be removed within 5 min. Gravity drains should be sized according to the tank capacities and drain diameters given in Table 28.2.

Emergency drains must be used only under the guidance and con-

TABLE 28.2. *Gravity Drain Pipe Diameters, I.D.*

Tank capacity		Pipe diameter, I.D.	
(m^3)	(gal)	(mm)	(in.)
1.9–2.8	500–750	76	3
2.8–3.8	750–1,000	102	4
3.8–9.5	1,000–2,500	127	5
9.5–15.1	2,500–4,000	152	6
15.1	Over 4,000	203	8

trol of well-trained and experienced personnel. Improper usage can result in greater hazards under certain conditions. Whenever a flammable gas processing atmosphere is maintained above the quench oil, removal of the oil can create a negative pressure in the enclosure, which can draw in air. This can result in potential explosions and/or an increase in fire severity. Measures should be taken to insure that the oil cannot be inadvertently removed.

Tank Location

The safest location for quench tanks using combustible oils is at grade level. Boilover from tanks above grade can be expected to spread fire to floors below, thereby making fire control more difficult and significantly increasing the potential fire loss. Fires involving quench oils in below-grade locations will quickly force personnel from the area, making manual fire fighting ineffective and resulting in a potentially significant increase in fire loss.

Heat treating operations involving combustible quench oils should be housed in fire-resistive buildings and should be well separated from combustible materials, valuable stock, important power equipment, and important process equipment.

MATERIAL TRANSFER

Rapid and complete immersion of the work in process is essential to proper heat treating and safety. The method of immersion must result in minimal splashing and no overflow of the quench oil. It is also essential that the possibilities for partial immersion be eliminated or minimized. Partial immersion of the work is the most common cause of quench oil fires.

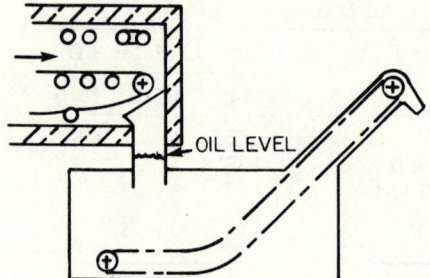

FIGURE 28.2. Bottom chute-type quench.

Chutes

Many furnaces are designed so that the work in process drops off the end of a conveyor, through a chute, and into the quench oil. The chute design and construction must allow the work to fall freely under all normal furnace feed conditions. This will involve proper chute sizing to accommodate the work, proper pitch to ensure continued stock motion through the chute, and smooth surfaces and seams to prevent the work from becoming hung up in the chute. The design of the system for feeding stock to the furnace must prevent an excessive amount of work from being dropped into the chute, causing it to become blocked.

Elevators

When stock in process is handled in baskets, elevators are commonly used to immerse the work in the quench oil. Of all the methods of immersion in use, baskets and elevators produce more partial immersions than any other method. The following are three critical concerns that require design consideration:

1. The elevating mechanism must be adequately and substantially supported by structural members. An elevator that drops because of the failure of supporting members will frequently fall unevenly and wedge itself into a partially submerged position.
2. Adequate guides must be provided for all elevators to ensure uniform and stable movement within the confined area of the quench tank. Without proper guides, it is possible for an elevator to become wedged. This can result in partial immergence.
3. Proper positioning of the workload on the elevator is essential. Suitable guides and stops must be provided to prevent shifting of the workload that can cause elevator jamming and partial immersion.

OIL QUENCHING

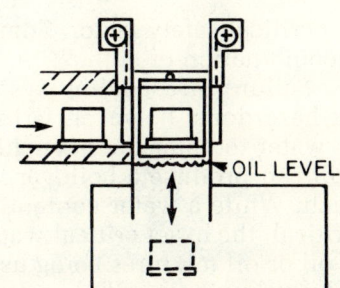

FIGURE 28.3. Dunk-type elevator quench.

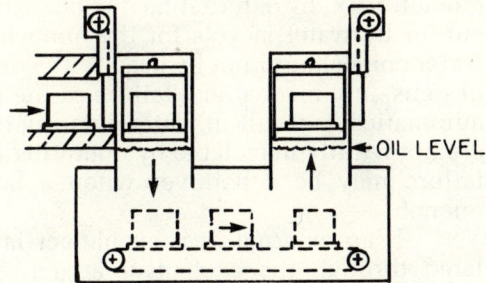

FIGURE 28.4. Dunk-type elevator quench with under-oil transfer.

Hoists and Cranes

A hoist or crane is usually required for moving large, specialized workloads into, and out of, the quench tank. Proper positioning of the hoist or crane over the tank is essential. Positioning can be accomplished by the use of stops and/or limit switches. Since hoists and cranes may permit some lateral motion, guides may be required to properly position the work as it enters the quench tank.

Unless the process allows oil to drain off of the work at the end of the quench period, an excessive amount of oil will be lost. Usually the work will still be warm at this time, and some oil may vaporize. Any ignition source can produce a fire involving the oil retained at this point in the process. Oil vapors will condense on cool surfaces above the drain area and thus contribute significantly to the potential fire loss. Noncombustible hoods and exhaust systems should be used where vapor condensation is determined to be a potential fire hazard.

OIL TEMPERATURE CONTROL

The control of quench oil temperature within specified design limits is absolutely essential to the heat treating process. Temperature control is also a critical safety factor. Some installations require cooling or heating or a combination of both.

Failure of a cooling system can cause the oil to overheat. Even more hazardous, however, is the type of cooling system failure that allows water to enter the quenching oil. An excessive amount of water in the oil can produce a boilover when a hot workload is immersed in the quench. While a water content of 5 percent by volume is considered to be critical, the exact critical water volume varies somewhat with the specific oil or oil mixtures being used. If the water content reaches 0.35 percent by volume, the oil is no longer considered safe to use.

Quench oil can be cooled by water circulating through coils in the quench tank, by external heat exchangers using water as the heat absorbent, or by water jackets for the quench tank. Water jackets and internal water coils should not be used with combustible quenching oils. In these designs, any mechanical failure of the coil or the quench tank shell will automatically result in water entering the quench medium. Such failures are not readily detectable by operating personnel; the first indication of a failure may be a boilover when a hot workload is immersed in the quench.

If an external heat exchanger is used, quench oil must be circulated through the exchanger at a higher pressure than the exchange medium water. If a leak should develop between the oil and water sides of the exchanger, oil will enter the water and be wasted. If the water pressure is higher than oil pressure, such a leak would result in water entering the oil, creating a potential for boilover.

The continuous flow of cooling water is essential for temperature control; therefore, it is imperative that waterflow be supervised. Supervision can be accomplished by operators observing the discharge from an open drain on the water side of the heat exchanger. Waterflow indicators are needed where a completely closed system must be used.

An oil quench tank is built as an integral part of many special atmosphere furnaces. In these cases, the vestibule above the quench tank is water-cooled, usually by water jackets. Many failures of the interior jacket walls have released cooling water into the quench tank below. As a result of these failures, safety considerations have dictated the use of special materials for vestibule jackets or the use of external coils. With special atmosphere furnaces, the use of a combustible gas for the atmosphere will also result in moisture development when the exit door is opened and the atmosphere is burned out.

Experience has also shown that water has been introduced into quench tanks with replacement oil. Though rare, it has happened and should be prevented.

Because of the various ways in which water can enter quench oil and become a significant potential hazard, approved moisture detectors should be used. Quench oil should be tested periodically for moisture content by qualified individuals. If moisture content reaches 0.35 percent by volume, quenching operations should be shut down and the oil replaced.

Other types of failures can also lead to overheating of the oil. If the oil level is too low and a large workload is immersed in the quench, the oil can overheat. Therefore, low oil level detectors should be used to sound an alarm and shut down quenching operations before overheating occurs.

Agitation is critical to maintaining a safe quench oil temperature and uniformity of temperature throughout the bath. If the agitation mechanism fails, localized overheating will occur, which may cause a fire at the surface of the quenching oil. Agitation failure and subsequent overheating can cause excessive vaporization, which can raise the pressure inside an enclosed special atmosphere furnace. When the pressure becomes high enough, the gas and oil vapor may be forced out of the chamber around doors and through vents. Frequently, these escaping gases ignite, damaging the facilities. Agitation systems should be supervised automatically, and their failure should result in the safe shutdown of quenching operations.

Where the process requires that the quench be heated, fuel-fired, electrically heated, or steam-heated immersion units are used as the heat source. All three methods of heating will create excessive temperatures at the interface of the heating unit and the quenching oil. A quench heating system should be prevented from operating if the oil level is too low, the agitation equipment is not functioning, or the oil temperature is too high. Where combustible oils are used, a temperature controller should be interlocked in the system to prevent quenching and to shut off the oil heating system when oil temperature exceeds a specified maximum limit.

All safety controls should provide visual and/or audible alarms. Where multiple safety interlocks are provided, a fault annunciator is useful in pinpointing a critical malfunction.

CENTRAL OIL SYSTEM

Many heat treating plants that have multiple quenching facilities utilize a quenching oil that is common to several operations. In these cases, efficiency and economy of operation may dictate the use of an oil supply system that is common to several quench tanks. When properly designed, central oil systems can contribute to the maintenance of reasonably dependable and problem-free oil quenching operations.

A proper design includes filtering to remove particulate contamination, water removal for safety of operations and prevention of boilover, and cooling to deliver the quenching medium to process tanks at an acceptable temperature. Water removal and dependable cooling of the oil are critical to the safety of the operation.

Water separation can be accomplished by settling and the use of centrifuges. However, these systems are not dependable enough to eliminate the need for moisture detection devices at the quench tanks or the requirement that oil in quench tanks be tested periodically.

In these installations the quench oil is cooled by air- or water-cooled heat exchangers. When water-cooled heat exchangers are used, the oil pressure through the exchangers must exceed the pressure on the water side to prevent water from entering the oil if a leak develops in the heat exchange surface.

Since central systems involve a constant removal of oil from the quench tanks and a constant replacement of the oil volume removed, it is essential for the sake of safety that oil flow be supervised. Any failure in the pumping system should result in an audible and/or visual alarm and safe shutdown of the quenching operation supplied by the central system.

THE SAFEGUARDS

All automatically created shutdowns of quenching operations should result in the workload being completely immersed in, or removed from, the quench. Partial immersion must always be considered hazardous.

Whenever movement of the workload into, or out of, the quench has been stopped by a malfunction of the equipment or by a jamming of the workload, qualified operating personnel must be permitted to override the safety interlocks to the extent that manual attempts can be made to achieve the complete immersion or removal of the workload. These manual efforts must be deemed critical and should be performed or directed only by qualified people. If exit doors must be opened, a fire condition should be anticipated. Extinguishing facilities adequate to protect the operator and prevent property damage should be available.

Regardless of the combustible quenching oil being used, the temperature of the oil must never be permitted to rise to within 50°F (10°C) of its flash point.

All safety controls and their interlocking functions should be tested on a regular schedule. The time interval between tests should be adequate for each installation but should not exceed six months.

To avoid foaming and boilover, replenishment oil should not be added while the quench temperature is 212°F (100°C) or higher.

Operating personnel must be well trained in both normal and emergency operating procedures. In addition, operating personnel should be trained in the effective handling of quench oil fires and in the use of fire extinguishing facilities provided.

Hydraulic control systems add to the fire hazard in the vicinity of high temperature equipment. Fire resistant hydraulic fluids should be used. When combustible hydraulic oils must be used, proper maintenance of the hydraulic equipment is vital to safety.

FIRE PROTECTION

Fire protection for combustible oil quenching operations requires a detailed evaluation of the fire potentials that are inherent to each specific area or department. There are three distinct areas of concern that must be considered.

Area protection is essential, regardless of the type of quench oil tanks involved. One of the most effective forms of area protection is the automatic sprinkler system. Experience has shown that ceiling-mounted sprinkler systems will limit building and equipment damage from quench oil fires whether they are confined to the quench tank surface or spread over a large area by a quench oil boilover. Sprinkler systems should be installed in accordance with the provisions of NFPA 13, *Sprinkler Systems*.

Specific protection for open quench tank surfaces and oil drainage areas is also important. Most oil fires can be extinguished by fixed carbon dioxide or dry chemical systems. In some operations various foam systems can also be effective. However, the suitability of foam will be determined by the quenching oil used and the temperatures involved. These systems are designed to operate automatically, well in advance of any potential sprinkler system discharge. As a result, cleanup is reduced, and remaining equipment can be quickly returned to operation. These special extinguishing systems should also be installed in accordance with standard requirements in NFPA 11, *Foam Extinguishing Systems*; NFPA 12, *Carbon Dioxide Extinguishing Systems*; and NFPA 17, *Dry Chemical Extinguishing Systems*.

Those situations that may require operating personnel to manually release jammed workloads and close furnace doors will dictate the provision of fire control and/or extinguishing facilities that are designed and installed for the protection of the operator who must perform in close proximity to the fire. In these instances, carbon dioxide or fixed water spray systems are the most effective. Water spray systems are discussed in NFPA 15, *Water Spray Fixed Systems*.

One of the most important fire protection features in every heat treatment shop is the manual fire extinguishing and control equipment and plant personnel properly trained in its use.

In addition to an adequate supply of portable, hand-carried fire extinguishers, the larger, wheeled extinguishers should also be available. These units will provide trained personnel with a capability of fire control not possible with smaller units, and, what is more important, they will provide a greater capability for life safety. (See Chapter 3, "Plant Emergency Training and Organization.")

Appropriately spaced hose connections with water fog or water spray nozzles should be considered an essential part of all heat treating area fire protection. These can provide prolonged periods of fire control and life safety beyond the limited supplies of portable extinguishers.

Hose systems should conform to NFPA 14, *Standpipe and Hose Systems*.

The selection and installation of fixed and portable fire extinguishing and control equipment should be accomplished in consultation with trained fire protection specialists.

BIBLIOGRAPHY

NFPA CODES, STANDARDS, AND RECOMMENDED PRACTICES

Reference to the following NFPA Codes, Standards, and Recommended Practices will provide further information on the safeguards for oil quenching operations discussed in this chapter. (See the latest *NFPA Codes and Standards Catalog* for availability of current editions of the following documents.)

NFPA 10, *Portable Fire Extinguishers.*
NFPA 11, *Foam Extinguishing Systems and Combined Agent Systems.*
NFPA 12, *Carbon Dioxide Extinguishing Systems.*
NFPA 13, *Installation of Sprinkler Systems.*
NFPA 14, *Standpipe and Hose Systems.*
NFPA 15, *Water Spray Fixed Systems.*
NFPA 17, *Dry Chemical Extinguishing Systems.*
NFPA 30, *Flammable and Combustible Liquids Code.*
NFPA 70, *National Electrical Code.*
NFPA 86C, *Industrial Furnaces Using Special Processing Atmospheres.* (See Chapter 10, Integral Quench Furnaces.)
NFPA 101, *Life Safety Code.*

ADDITIONAL READING

Flash Point Index of Trade Name Liquids, Ninth Edition, National Fire Protection Association, Boston, MA, 1978.
Henry, Martin F., *Flammable and Combustible Liquids Code Handbook*, National Fire Protection Association, Quincy, MA, 1981.
Kimura, H. and R. Maddin, *Quench Hardening in Metals*, Elsevier Science Publishing Co., Inc., New York, NY, 1971.
McKinnon, G. P., ed., *Fire Protection Handbook*, Fifteenth Edition, National Fire Protection Association, Quincy, MA, 1981.
Trinks, W., *Industrial Furnaces*, (vols. 1 and 2), John Wiley and Sons, New York, NY, (Vol 1: Principals of Design and Operation, 5th ed., 1961; vol. 2: Fuels, Furnace Types, and Furnace Equipment; Their Selection and Influence Upon Furnace Operation, 4th ed., 1967).

29

Molten Salt Baths

Abraham Willan

More than eighty years ago a method of heat treating metals by immersing them in molten salts was introduced in England. Because of the relatively inexpensive cost of the installation of such equipment and the rapid, precise method of heat transfer, the popularity of the method has grown considerably, even though it was more than thirty years later when the first installation appeared in the United States. In 1970, however, there were more than 5,000 molten salt bath installations, including automatic installations, in the U.S. One such bath contained more than 300,000 lbs (136 078 kg) of salt and was approximately 40 ft long by 20 ft deep by 8 ft wide (12.2 by 6.1 by 2.4 meters).

For purposes of definition, we generally assume that a molten salt bath is any heated container that holds a melt or fusion of one or more relatively stable chemical salts as a fluid medium into which metal, ceramic (i.e., glass), or polymer (i.e., Teflon) work is immersed for various types of treating. We must be careful to distinguish a molten salt bath from a heated chemical solution, which it is not. Very often, because of

---

Abraham Willan, retired from Pratt Whitney Aircraft Group, a Division of United Technologies, wrote this chapter. The chapter was revised by Q. D. Mehrkam, Senior Vice President, Ajax Electric Company, Huntingdon Valley, PA.

its temperature of operation and its function, the heated container is called a molten salt bath furnace. As such, salt baths are discussed in Chapter 11 of NFPA 86C, *Industrial Furnaces Using Special Processing Atmospheres.*

As work is immersed fully into a salt bath, the salt serves as a protective atmosphere as well as a heat transfer medium, since no air can come in contact with the heated metal. Heat is transferred to the work by direct contact with a heated medium of a very high heat capacity, so that transfer occurs more rapidly than by any other type of furnace heating. Moreover, because of the constant convection currents in the heated liquid medium, uniform temperature distribution is more easily attained in a molten salt bath furnace than in a normal heat treating furnace.

TYPES OF SALT BATHS

The various types of salt baths available today can be classified by heating method, by application, or by the particular salt mixture utilized. Usually, however, the salt used is directly related to the application.

Electrically Heated Baths

In the majority of cases, electrically heated salt baths are heated by electrodes, though in a few instances, resistance-type ribbon elements are used around a metal pot or container. The electrodes may be immersed in the bath through the top surface of the molten salt or submerged by being introduced through the sides of the furnace below the surface of the salt. For examples of each of these, see Figures 29.2 and 29.3. The electrodes generate heat in the salt melt by utilizing the salt's resistance to current flow. The molten salts are good high-resistance conductors. Electrical potential is applied to the molten salt by electrodes connected to the secondary winding of a special transformer. These electrodes are usually located so that they cause good salt circulation by electromagnetic forces when they are energized. Because of this circulation, temperature uniformity unattainable in ordinary furnaces is assured in molten salt furnaces. A metal vessel may be employed inside a bricked ceramic chamber to hold the molten salt. Often the vessel is dispensed with, and the brick or ceramic chamber is used alone, the theory being that the salt will form its own casing as it nears the cool exterior wall.

Fuel-Fired Salt Baths

There are two types of fuel-fired salt baths. In one instance the burners are fired directly outside a metal pot or retort containing the salt, similar to an electrical resistant-heated salt bath. In most instances the burners

MOLTEN SALT BATHS

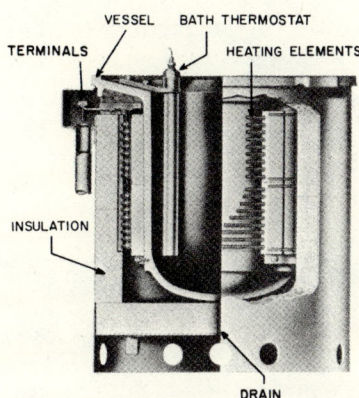

FIGURE 29.1. *Resistance-heated salt bath with pot. (Sunbeam Equipment Co.)*

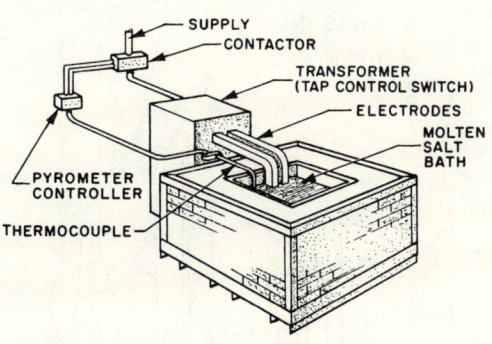

FIGURE 29.2. *Typical electrode immersion-type electrical heating arrangement.*

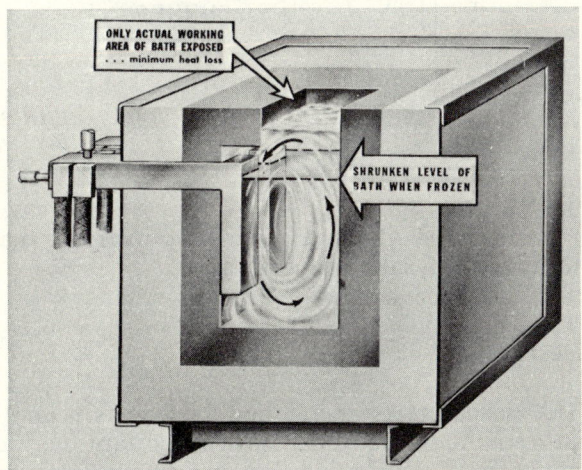

FIGURE 29.3. *Submerged electrode-type salt bath. (Ajax Electric Co.)*

are contained in metal tubes, known as radiant tubes, and the tubes are immersed in the molten salt. Generally these tubes are located in a plenum chamber through which the salt circulates due to the convection caused by the heat differential of the salt. (See Figure 29.4.)

APPLICATIONS

The heat treatment for which a salt bath is utilized determines the type of salt or mixture used, as well as the temperature of the operation. To describe in detail every type of application for salt bath furnaces is beyond the scope of this chapter. However, an attempt will be made to

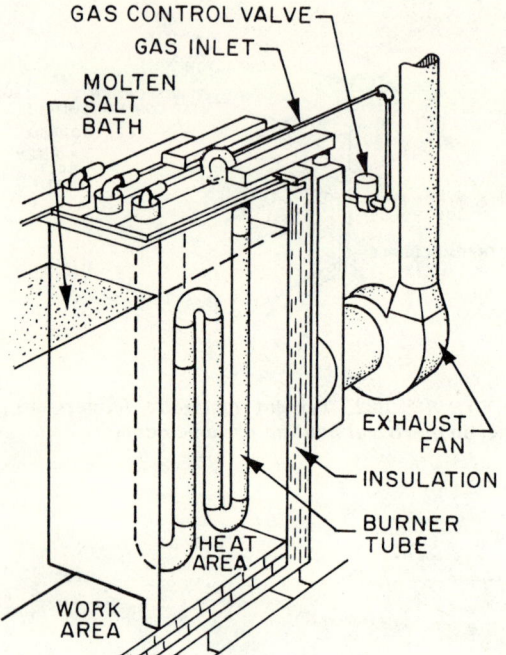

FIGURE 29.4. *Gas-fired radiant tube salt bath.*

touch briefly upon the major applications, their temperatures of operation, and the salts utilized.

Descaling

The descaling of metal parts previously heat treated in other furnace equipment is one of the most important uses for molten salt baths. Generally there are three different descaling processes: the reducing sodium hydride process; the oxidizing sodium hydroxide process; and the electrolytic process. In the first instance, the bath is a fused melt of sodium hydroxide and sodium hydride and operates at temperatures between 650°F and 780°F (343.3 and 415.6°C), depending on the material to be descaled. In the oxidizing process a fused melt of sodium hydroxide, sodium carbonate, and sodium chloride is used at temperatures between 700°F and 950°F (371.1 and 510°C). In both the reducing and oxidizing methods, the metal parts are water-quenched after immersion into the salt bath. The electrolytic process is principally utilized for a continuous operation, such as the descaling of strip stock. Here a fused melt of sodium hydroxide, sodium chloride, and sodium carbonate is used in two baths at a normal temperature of about 900°F (482°C). The strip, having been processed in an annealing furnace, is transported first through a bath where the direct current passes from cathodic grids to the strip and

FIGURE 29.5. *Liquid carburizing and isothermal heat-treating line. (Ajax Electric Co.)*

then through a bath where the current passes from the strip to anodic grids. Included in the descaling group of baths is the type that uses the process of stripping fixtures that hold parts during paint spraying operations, or paint dipping operations. Oxidizing-type salt baths are utilized for this process.

Liquid Carburizing

This is an operation in which salt baths are employed to case harden steel by diffusing carbon and small amounts of nitrogen into various steel surfaces. The salts used are fused melts of sodium cyanide, sodium chloride, and barium chloride, depending on the depth of case required. The operating temperature is between 1,450°F and 1,750°F (788 and 954°C). (See Figure 29.5.)

Cyaniding and Nitriding

These two operations are also performed with either sodium or potassium cyanides as the chief ingredient of a fused melt. Usually the operating temperatures are lower for these applications — in the case of cyaniding, 1450°F to 1500°F (788 to 816°C); nitriding, 950°F to 1,050°F (510 to 566°C).

Chapter 29

Neutral Hardening

The hardening of ferrous alloys without harmful surface effects, such as decarburization, pitting, or scaling, is performed in a molten fused bath where the principal components are sodium, potassium, and barium chlorides. Operating temperatures may be anywhere in the range of 1,400°F to 2,350°F (760 to 1288°C). The process may be followed by either an oil or a salt bath quench.

Hardening of High Speed Tool Steels

This process is usually performed in a series of molten salt bath furnaces at various temperatures. Thus, for example, a preheat bath at approximately 1,400°F (760°C) may be followed by a high-heat bath at 2,350°F (1288°C), which in turn is followed by a quench bath at 600°F (315°C). All of these baths use fused melts of barium chlorides, sodium chlorides, and potassium chlorides. (See Figure 29.6.)

Isothermal Quenching

This process involves the rapid cooling of metal parts to a selected temperature by quenching into a molten salt bath. The three major types of isothermal quench are austempering, martempering, and cyclic annealing. Each produces different hardness properties in metal alloys. Bath temperatures vary from 400°F to 1,550°F (204 to 843°C). The salts used are nitrate/nitrite salts for the lower temperatures and neutral chloride types for the higher temperatures.

Annealing

Annealing of low, medium, and high carbon steels in the form of rod and wire is achieved in molten salt bath furnaces. For medium carbon steels, the operation is usually carried out in a melt of carbonate and chloride salts at a temperature of 1,250°F to 1,300°F (677 to 704°C). For low carbon steels and coils of aluminum wire the melt is usually a nitrate type of salt operating at approximately 1,000°F (538°C). Chloride types of salts are used to anneal stainless steel products as well as nickel-chrome alloys at temperatures of 1,550°F to 2,150°F (843 to 1177°C).

Brazing

Brazing of ferrous and nonferrous alloys with silver, brass, and copper is another popular application for molten salt baths. Aluminum dip brazing is the most prevalent method of brazing aluminum assemblies. Braz-

MOLTEN SALT BATHS

FIGURE 29.6. Salt bath for hardening of High Speed steel. (Ajax Electric Co.)

ing operating temperatures range from 1080°F to 1120°F (582 to 604°C). For example, in aluminum dip brazing, the salt mixture is essentially a mixture of chlorides with smaller percentages of sodium and/or aluminum fluorides, which act as fluxes. In copper brazing, barium chloride mixtures are employed.

A special, but limited, application of salt baths is in a quenching operation that is an integral part of an atmosphere-type heat treatment furnace. This application usually involves alloy-hardening operations in which molten salt is preferable to oil as the quench medium. At temperatures between 350°F (177°C) and 750°F (399°C), such baths use nitrate and nitrite salts.

We see, then, that there are numerous applications in the metal-working industry for various salts and salt mixtures at many different temperatures. Table 29.1 lists the most common salts and their melting points. Table 29.2 lists the most common salt mixtures and their melting points. Other mixtures, which may be only slight variations of those given in the table, are sold as proprietary salts. Boiling points of any proprietary salts used on the premises should be obtained.

Chapter 29

THE HAZARDS

The hazards common to salt bath furnaces may be divided into the following three types:

1. The danger of fire by contact of molten salt with combustibles;
2. Explosion of the salt mixture due to physical or chemical reaction; and
3. The danger to the personnel operating the bath.

The possibility of fire is always present in areas adjacent to molten salt bath operations. Since salts are used at temperatures between 300°F (149°C) and 2,400°F (1316°C), accidental contact of these hot salts with any combustible material will cause a fire. The sudden ejection of salt, popping, spattering, or accidentally spilling will create a fire hazard if the salt comes in contact with anything combustible — a wooden floor, for example.

Ejection has often been the result of bath explosion, though it may be the result of other actions. Since molten salts have high heating potential and relatively little surface tension as well as low viscosities, any minor physical disturbance to the molten bath can result in salt spattering or being ejected from the bath container. Common physical disturbances include work or other objects falling into the bath, or malfunctioning of the mechanical agitator.

Overheating of internally heated furnaces will also cause a salt level rise and overflow. (Overheating does not normally cause violent reactions; however, if the temperature exceeds slat decomposition temperature, decomposed gases will be evolved.)

With externally heated equipment, failure such as a cracked pot leaking salt into the superheated combustion chamber may result in production of fumes, explosion, and burning of surrounding combustibles. Good preventive maintenance practices dictate replacement of equipment before failure occurs.

Salt eruptions can also be caused by chemical reactions. Introduction of liquids (water, oil, etc.) below the surface of the salt will almost instantly vaporize to a gaseous form with considerable increase in volume, causing an eruption of the bath.

Violent reactions will occur in nitrates when they are slightly overheated — 100°F (37.8°C) — above operating temperature. Nitrates will react with a mild steel container and go into rapid combustion at temperatures as low as 1,100°F (593°C). Nitrate salts alone may explode when overheated. Overheating liberates oxygen with the evolution of additional heat, which is an exothermic reaction. We see, then, that a chain reaction occurs. Accumulations of sludge or sediment in the bottoms of pots can cause nitrate salts to overheat when heat is applied externally through the bottom. Overheating can be caused by a malfunction of temperature controls in the heating system, by a hung-up workload, or by operator error.

Nitrate salts are commonly used for the removal of organic con-

TABLE 29.1. Melting Points of Common Chemical Salts

Salt	Melting point (°F)	Melting point (°C)
Barium chloride	1,764	963
Barium fluoride	2,336	1,280
Boric oxide (Anhydride)	1,071	578
Calcium chloride	1,422	773
Calcium fluoride	2,480	1,360
Calcium oxide	4,662	2,575
Lithium chloride	1,135	613
Lithium nitrate	491	255
Magnesium fluoride	2,545	1,397
Magnesium oxide	5,072	2,802
Potassium carbonate	1,636	892
Potassium chloride	1,429	777
Potassium cyanide	1,174	635
Potassium fluoride	1,616	880
Potassium hydroxide	716	380
Potassium nitrate	631	333
Potassium nitrite	567	297
Sodium carbonate	1,564	852
Sodium chloride	1,479	805
Sodium cyanide	1,047	564
Sodium fluoride	1,796	980
Sodium hydroxide	605	319
Sodium metaborate	1,771	967
Sodium nitrate	586	308
Sodium nitrite	520	271
Sodium tetraborate	1,366	742
Strontium chloride	1,603	873

taminations from tolls, fixtures, and work parts. Carbonaceous materials such as oil, paint, polymer, soot, tar, grease, etc. are safely removed under controlled conditions by oxidation of contaminants to gaseous by-products. Introduction of excessive amounts of organic materials into the bath can cause a system overload resulting in the bath rising, and burning of carbonaceous materials on the surface. Operating personnel must be aware of the type and amount of contaminated products that can be safely processed.

Accidental mixing of cyanide salts with nitrate salts must be avoided because of the severe danger of explosion.

The introduction of aluminum or magnesium parts into baths in which the molten salt temperatures approach or exceed the melting point temperatures of these metals can cause fire and explosion. The introduction of parts or fixtures that have been immersed in, and retain pockets of, degreasing fluids or cleaning solutions can cause an explosion by the formation of steam as was described for water.

TABLE 29.2 Melting Points of Common Salt Mixtures*

Mixture and proportion	Melting points (°F)	(°C)
Lithium nitrate 23.3, sodium nitrate 16.3, potassium nitrate 60.4	250	121
Potassium hydroxide 80, potassium nitrate 15, potassium carbonate 5	280	138
Potassium nitrate 53, sodium nitrate 7, sodium nitrite 40	285	141
Potassium nitrate 56, nitrite 44	295	146
Potassium nitrate 51.3, sodium nitrate 48.7	426	219
Sodium nitrate 50, sodium nitrite 50	430	221
Sodium hydroxide 90, sodium nitrate 8, sodium carbonate 2	560	294
Lithium chloride 45, potassium chloride 55	666	353
Barium chloride 31, calcium chloride 48, sodium chloride 21	806	430
Calcium chloride 66.5, potassium chloride 5.2, sodium chloride 28.3	939	504
Calcium chloride 67, sodium chloride 33	941	505
Potassium chloride 35, sodium chloride 35, lithium chloride 25, sodium fluoride 5	960	516
Potassium chloride 40, sodium chloride 35, lithium chloride 20, sodium fluoride 5	990	533
Barium chloride 48.1, potassium chloride 30.7, sodium chloride 21.2	1,026	555
Sodium chloride 27, strontium chloride 73	1,049	565
Potassium chloride 50, sodium carbonate 50	1,086	585
Barium chloride 35.7, calcium chloride 50.7, strontium chloride 13.6	1,110	599
Barium chloride 50.3, calcium chloride 49.7	1,112	600
Potassium chloride 61, potassium fluoride 39	1,121	605
Sodium carbonate 56.3, sodium chloride 43.7	1,177	637
Calcium chloride 81, potassium chloride 19	1,184	640
Barium chloride 70.3, sodium chloride 29.7	1,209	654
Potassium chloride 56, sodium chloride 44	1,220	660
Sodium chloride 72.6, sodium fluoride 27.4	1,247	675
Barium fluoride 70, calcium fluoride 15, magnesium fluoride 15	1,454	790
Barium chloride 83, barium fluoride 17	1,551	845
Calcium fluoride 48, magnesium fluoride 52	1,738	949

*Lowest constant melting points given; proportions are percentages by weight.

The buildup of sludge in any salt bath may cause hot spots in an external heated bath. Small parts that are carelessly dropped into the bath and are not removed contribute to the amount of sludge. Hot spots cause rapid pot attack. Pot failure may result in an explosion or the ejection of hot salts. Most cases of bath container deterioration have been attributed to lack of sludge removal.

The storage of salts can result in serious problems. Chemicals such as nitrates and cyanides must be stored in separate areas. Many salts are hydroscopic and, when stored in damp areas, will absorb moisture. When these salts are subsequently heated, moisture will be released below the surface of the bath and an explosion will ensue.

Intensive external reheating and remelting of a solidified salt bath may result in expansion sufficiently rapid to bulge or rupture the container. Such reheating can generate enough pressure to fracture the surface crust of the salt and scatter hot salts over a wide area.

When exposed to air having even a slight moisture content, the fumes from many salt baths become corrosive because chlorides, fluorides, and other salts will form acids by hydrolysis. These fumes are highly corrosive to adjacent building structures, wiring, and equipment and are, of course, detrimental to the bath operator.

The greatest hazard to personnel operating salt baths is being burned by the molten salt. Burns may be the result of salt ejection from the bath or of careless handling of parts, fixtures, or salts. Salts also may react chemically with the operator's skin. Ordinary garments may ignite upon contact with hot salts. Garments may also become impregnated with salts and become corrosive and flammable.

THE SAFEGUARDS

While there appear to be many hazards connected with the operation of molten salt bath furnaces, fire, explosion, and injury can be prevented by the constant observance of ordinary precautions.

Fires resulting from ejection, spills, or leaks from salt baths are usually of a fast-spreading nature and difficult to control. Fire fighters, who would respond in the event of fire, should be invited to inspect the premises and become thoroughly acquainted with the location and operation of the molten salt bath furnaces. Fire fighters also should be aware of the nature of the hazards involved with the chemical salts being utilized. In any multiple bath installation, the contents of each bath and the nature of the molten salt used should be carefully identified.

Clean dry sand may be used for diking purposes to confine and prevent the spread of the escaped melt. Where large quantities may be required, sand should be stored for easy access in protected bins with proper shovels for application. Carbon dioxide or dry powder-type extinguishers may be used to extinguish burning carbonaceous material in the immediate vicinity of the salt bath.

The location of a salt bath installation should be selected carefully. A liberal area should be allocated, and the area of operation should be spaced off to prevent congestion and interference with other operations. A preferred location is in a noncombustible building. (Noncombustible construction is described in NFPA 220, *Standard Types of Building Construction.*) The bath should be installed on a noncombustible pad in a cement-lined pit, or curbed area of sufficient capacity to contain the contents in the event of leakage during operation. The area around the salt bath should be kept free of combustible materials. All baths must be located so that the bath will not be exposed to liquid leakage from overhead piping, wall openings, air intakes, windows, or roofs or to seepage from the floor. Baffles, covers, and/or guards should be installed to prevent splashover from one tank to another.

All salts should be shipped and stored in tightly covered containers designed to prevent the absorption of liquids or moisture. These containers should be prominently marked with the identification of the salt or salt mixture they contain to prevent the accidental mixing of incompatible salts. Nitrate salts should be stored in a fireproof, damp-free room separated a reasonable distance from heat, liquids, and reactive chemicals. This room should be secured so that only authorized personnel are permitted entry. Nitrate and/or nitrite salts should never be stored in the vicinity of cyanide salts. Excess nitrate salts must not be stored adjacent to the salt bath furnace. Only the actual amount required to recharge the bath should be removed from the storage area, and this amount should be melted immediately. Paper bags or fabric sacks must not be used to transport salts from the storage area to the furnace. Metal drums or metal containers are suitable for this purpose, although fiber or cardboard drums can be used with care.

The safety precautions that apply to the heating systems of salt baths are identical to those recommended for the heating systems of other industrial furnaces. (See Chapter 27 and NFPA 86A, 86B, 86C for information on ovens and furnaces.) Some additional safety measures are necessary because of the hazardous nature of salt baths. Flame should never impinge on the wall of a salt pot or container when a gas- or oil-fired heating system is used. Flame should be tangential. The products of combustion should be vented and prevented from entering the salt mixture. Radiant tubes should be constructed of materials resistant to the corrosive action of the salts used. Materials used for electrodes in electrically heated salt baths should also be constructed of noncorrosive materials. Stray current leakage should be avoided to prevent electrolytic corrosion, which could lead to perforation of the wall of the salt container.

The buildup of sludge in any type salt bath should be avoided by constant maintenance. Sludge is caused by the deterioration of salt, the freezing of salt to a cold bottom, the introduction of foreign material into the bath, and the dropping of small parts into the bath. Several methods of removing sludge and foreign objects are available. The bath may be outfitted with a large screen, which can be removed at regular intervals and cleaned. Some equipment has been designed with a special plenum chamber equipped with agitators and a built-in removable sludge pan similar to that shown in Figure 29.7. The sludge is drawn into the pan by the agitators. For large installations, special sludge buckets on beams, similar to the clam buckets used for excavation purposes, have been employed to remove sludge. Another innovation is the automatic dump-

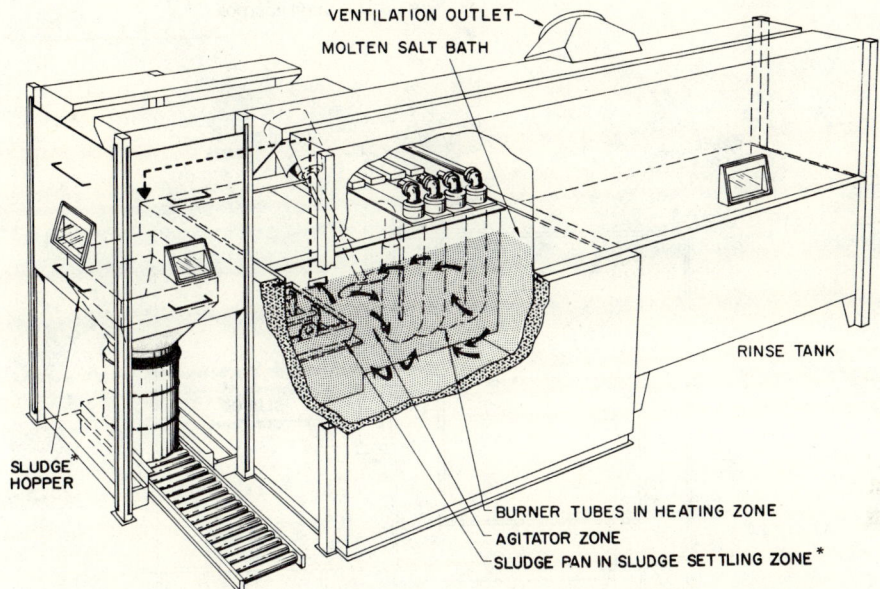

FIGURE 29.7. *Descaling salt bath showing sludge pan construction. (Kolene Co.)*

ing of fixter baskets (see Figure 29.8). This device replaces the need for manual desludging, thus avoiding direct exposure to heat and chemicals. At any rate, where it is practical, a regular sludge removal program should be instituted.

The careful selection of temperature-control equipment is very important. Automatic temperature controls are normally provided. A control thermocouple must be correctly located in the bath to accurately monitor hot spots. An overtemperature control should be provided, arranged to automatically shut off the heat source and actuate visual and audible alarms in the event of a malfunction of the normal operating controls and/or if the temperature of the bath exceeds the operating range. This control may be supplemented by distinctive visual and audible alarms that warn the operator of any excessive temperature rise, thus calling the operator's attention to an abnormal condition before the overtemperature device shuts down the furnace. All sensing couples must be protected from the corrosive effects of the particular salt used. For electrically heated salt baths, the use of a step-switch power transformer provides better control during idle periods when the bath is unattended. The heating load may be easily reduced by changing the setting to one of the lower voltage taps, thus circumventing any danger from faulty operation of thermostatic controls or relay circuits. When equipped with silicon controlled rectifiers, power input can be adjusted to a safe maximum level.

For the removal and control of any toxic, corrosive, and heated fumes, all salt bath furnaces should be equipped with hoods made of ma-

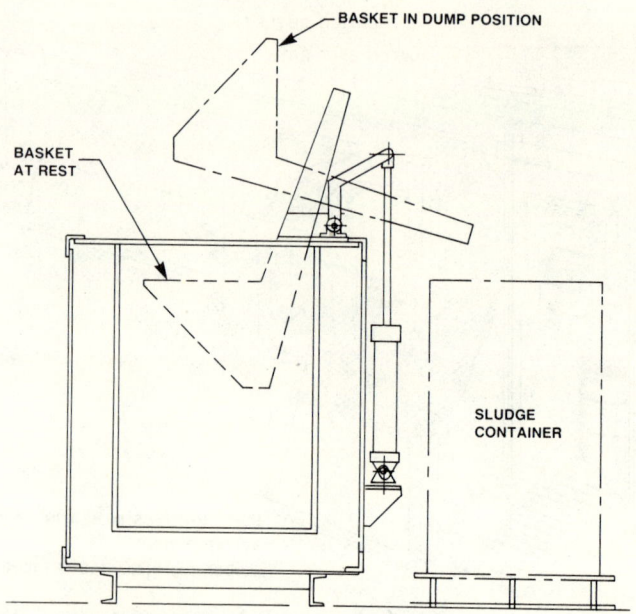

FIGURE 29.8. *Automatic basket dumper. (Ajax Electric Co.)*

terial that will not be affected by the corrosive nature of such fumes. In many cases, fume scrubbers or baghouse collectors should be installed to collect the fumes so that they will not corrode the external structure of the building or contaminate the atmosphere. Figure 29.9 shows a hooded salt bath.

Whenever the salt bath is not in use, its cover should be kept closed. If the bath is left idle over weekends, an attendant should inspect it for safety every few hours. If an electrically heated salt bath is left unattended for any length of time, the transformer step-switch should be set down to an idling position.

All fixtures used for dipping parts in baths, such as hooks, ladles, or baskets, should be of a solid, closed design without corners that could retain water. Closed piping or other hollow metallic articles in which air or water may be trapped should not be used. Any fixture used in one salt bath should be thoroughly cleaned and dried before it is immersed in another bath.

To preclude the possibility of the carry-over of water by wet work, all material drawn from quench tanks should be preheated to drive off moisture. Preheating may be accomplished for open small loads merely

MOLTEN SALT BATHS

FIGURE 29.9. Typical hooded salt bath construction. (Kolene Co.)

by holding the work above the heated salt tank for short times before immersion.

Wherever the furnace electrodes of an electrically heated salt bath are water-cooled, water flow or temperature devices should be provided to detect failure of the cooling water system. These devices should be interlocked to shut off power upon the detection of cooling system failure. Usually either a sensing instrument in the drain side of the system or a waterflow switch is used for this purpose.

In those cases where the salt bath is used as an internal quench tank within a furnace, and where a combustible atmosphere is used above the salt bath surface, measures should be provided to control the interface between the atmosphere and the salt surface to prevent carbon precipitation onto the salt surface. In addition, adequate salt circulation should be provided to prevent localized hot spots, whenever the salt is exposed to higher furnace temperatures.

A cold salt bath that has frozen should be remelted or liquefied by applying initial or starting heat as near to the top of the salt container as possible. All solidified salts expand when heated. The application of heat to the external surfaces will cause melting near the bottom and leave the top solidified. Build-up of excessive pressure may cause failure of the container and/or the subsequent ejection of hot salts. No attempt should

623

be made to break the surface crust, since the breaking of the crust will be accompanied by the immediate ejection of hot salt. Where purposeful solidification of salt is to take place in an electrically heated bath, this hazard can be avoided by inserting a round solid bar in the molten salt adjacent to the immersed electrodes. After the salt has frozen, the bar is removed. Melting can be started with a gas torch and completed by energizing the electrodes. The cavity created by the removal of the bar will serve as a vent duct for the escape of the gases formed. Other special devices for solidified salt baths are available. Gas burners that heat the top layers and gradually heat downward may be used with care. For bottom submerged electrodes, a coil set in between electrodes is recommended for melting. Furnaces designed with top submerged electrodes can be started from the top by preheating with a torch. In severe cases, the only possible way to empty a completely solidified cold salt bath may be to break up the salt with a jackhammer and begin anew. However, it should be ascertained that the bath is solidified completely prior to the start of any hammering. Pumps are available to remove salt while it is in the molten state. Considerable care should be exercised in removing any molten salt. It should be pumped directly into dry steel drums where it can freeze over immediately.

In the event that the operating temperature of a nitrate or nitrite salt bath exceeds the maximum limit, the following emergency action should be taken:

Shut off the heat supply to the salt bath furnace.
Allow only qualified maintenance people into the immediate area to determine the magnitude of the problem.
Alert standby services as required by the safety and fire departments.

After completing the preceding steps, if there appears to be an imminent failure resulting in an explosion or fire, notify the fire department and summon the plant fire brigade to stand by. Open all doors and windows to minimize the effects of an explosion.

Suitable equipment for handling fixtures and work should be selected carefully with due consideration to the corrosive nature and temperature of the bath. Thus, for example, aluminum or low melting point alloys should not be used in the construction of hoists and craneways. Only level loads that have been stabilized so that no tilting can occur should be handled. This is particularly important in the handling of salt sludge pans.

All operators and workers in the vicinity of molten salt baths should be provided with corrosion-resistant and heat-resistant shoes, gloves, aprons, hard hats, and face shields. Solutions for cleansing eyes and supplies for treating minor burns should be available at all times. Breathing apparatus should also be provided for emergency use against oxides of nitrogen or corrosive fumes such as chlorides and fluorides.

The best prevention against fire and explosion is the selection of only fully trained, competent operators who are constantly alert. Operators and their supervisors should be aware of the operating temperature of the bath, the salt and its chemical and physical properties, and the special hazards involved. They should be trained in all emergency procedures. Full instructions for the operation of the salt bath should be posted near the bath for immediate availability. Operators should be retrained periodically to keep safe operating procedures foremost in their minds. Operating instructions suitable for posting are available from manufacturers of molten salt bath furnaces; salt properties are available from salt suppliers.

Constant vigilance with safety in mind and good housekeeping by intelligent well-trained personnel can keep a salt bath operation as safe as any other metallurgical process that is conducted in an industrial furnace. The hazards are numerous and real, but constant awareness can help prevent an explosion or fire. Salt baths are safe if they are designed properly, operated reliably, and supervised attentively.

BIBLIOGRAPHY

NFPA Codes, Standards, and Recommended Practices

Reference to the following NFPA Codes, Standards, and Recommended Practices will provide further information on the safeguards for salt baths discussed in this chapter. (See the latest *NFPA Codes and Standards Catalog* for availability of current editions of the following documents.)

NFPA 70, *National Electrical Code.*
NFPA 86A, *Ovens and Furnaces, Design, Location, and Equipment.*
NFPA 86B, *Industrial Furnaces — Design, Location and Equipment.*
NFPA 86C, *Industrial Furnaces Using Special Processing Atmospheres.*
NFPA 101, *Life Safety Code.*
NFPA 220, *Standard Types of Building Construction.*
NFPA 491M, *Manual of Hazardous Chemical Reactions.*
NFPA 327, *Cleaning or Safeguarding Small Tanks and Containers.*

Additional Reading

Bartels, A. L., "Safeguards to Prevent Ignition by Hot Surfaces," *Electrical Review London*, vol. 203, no. 4(1978), pp. 29-31.

Janz, George J., *Molten Salts Handbook*, Academic Press, Inc., New York, NY, 1967.

Lovering, David G., ed., *Molten Salt Technology*, Plenum Publishing Corp., New York, NY, 1982.

McKinnon, G. P., ed., *Fire Protection Handbook*, Fifteenth Edition, National Fire Protection Association, Quincy, MA, 1981.

Mehrkam, Q. D., "An Introduction to Salt Bath Heat Treating," Reprint no. 182, July 1972, Ajax Electric Company, Huntingdon Valley, PA.

Molten Salt Electrolysis in Metal Production, IMM/North American Publications, Brookfield, VT, 1977.

Trinks, W., *Industrial Furnaces*, (vols. 1 and 2), John Wiley and Sons, New York. (vol. 1: Principals of Design and Operation, 5th ed., 1961; vol. 2: Fuels, Furnace Types, and Furnace Equipment: Their Selection and Influence Upon Furnace Operation, 4th ed., 1967.)

Ubbelohde, A. R., *Molten State of Matter: Melting and Crystal Structure*, John Wiley and Sons, New York, NY, 1979.

30

Machine Tool Processes

Raynal W. Andrews, Jr.

Machining consists of the cutting and shaping of metals using single point, multiple point, or bladed tools driven by power-actuated machine tools, which are usually provided with powered feeds. There are a limited number of manual operations that may be involved, such as threading, filing, and scraping; but the focus of this chapter is the power-driven machine tool.

Machine tools are used in all segments of industry for the shaping, dimensioning, or surface finishing of metal components to be used as machine elements, fasteners, or finished products. They are also used for the maintenance of plant equipment. The operations include turning, planing, shaping, slotting, milling, sawing, boring, trepanning, drilling, grinding, abrasive cutting, filing, threading, reaming, broaching, deburring, lapping, chamfering, spinning, and other metal-working functions. Occasionally materials such as plastic and wood are machined, but these are covered separately in Chapter 13, "Furniture Manufacturing," and Chapter 19, "Plastic Products."

Upon cursory observation of a machining operation it would ap-

---

Raynal W. Andrews, Jr., P.E., is an Engineer and Consultant based in Pittsburgh, Pennsylvania.

pear that there are few, if any, fire hazards about which to be concerned. This is deceiving for there are aspects of machining which require careful and constant attention. The fire record is full of experiences which point to the need for considered preplanning and preventive action. The principal hazards involve the possible combustion of the cuttings (chips) and fine particles (fines) which are produced as the workpiece is shaped and cut. Also of concern are the coolant/lubricants used for the lubrication of cutting tools. The principal sources of ignition include the heat generated by the work being done or friction from the cutting tool and spontaneous oxidation of materials used. It is important to remember that nearly all metals will burn in air under certain conditions, depending on size, shape, and quantity.[1]

RAW MATERIALS

The raw materials arriving at a machining plant may be in the form of cleaned castings, forgings, rough-cut plate, rolled or drawn bar or merchant mill stock, fabrications or weldments, tubes, and sometimes rolled or extruded shapes. In general, it can be safely assumed that these items have significant mass and so, in that form, can be considered to be noncombustible unless exposed to a massive fire. This also applies to the light metals which are known to be more readily combustible only when they have a large surface-to-mass ratio.

Other than cutting tools, which are a specialized commodity, the major operating supply is the coolant/lubricant. Different coolant/lubricants are employed for different metals, and they vary from alkaline aqueous solutions to various light petroleum distillates or compounded oils.

THE PRODUCTION PROCESS

Because machine shops may have a variety of machine tools on hand, each designed for a different function, it is difficult to talk of a production process, as such, in any but general terms. The materials being machined may be in their final form and require only grinding for a smooth surface, or they may be the raw metal or mill stock which must first be shaped at the beginning of a long manufacturing process. It is conceivable that no two pieces being machined in a shop will necessarily end up in the same finished product. On the other hand, all the work required to turn mill stock into a finished product may be performed under one roof.

MACHINE TOOL PROCESSES

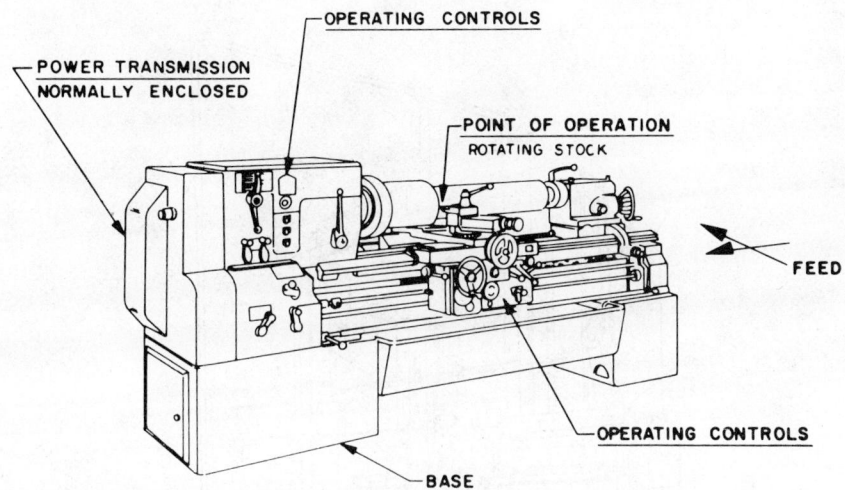

FIGURE 30.1. *Lathe, a single point cutting tool.*

The Machine Tool

The modern industrial machine tool may be a single-motor machine, such as a drill press, which performs a simple, repetitive operation or it may be a very large, multimotored automatic machine with a highly complex computerized numerical control system. Although these machines may be completely automatic, they are constantly attended, when operating, by a highly skilled operator.

The stock is brought in as needed by hand truck, forklift, or conveyor. The workpiece is then secured to the machine. This may involve attaching it to a chuck which grips the piece while it rotates. This is the case with the lathe shown in Figure 30.1. The single point cutting tool is fed longitudinally to the axis of the workpiece as it rotates. In other instances the work may be fixed to a table like the one shown in Figure 30.2 where the metal to be worked remains stationary on the table while the table rotates the work into the tools; or the table may feed the stock into the cutting tool in the manner shown in Figure 30.3.

The Cutting Tool

The geometry of the tool is established for the particular metal being machined depending on its physical character and machinability. Rake (the angle of the cutting edge) and side and back clearance (relief) of the tool are of extreme importance in minimizing frictional and metal-working heat. Side clearance must be such that the chips cut from the workpiece fall aside. Smoothly finished and smoothly rounded gullets between teeth are equally important to ensure the freeing of each chip.

Chapter 30

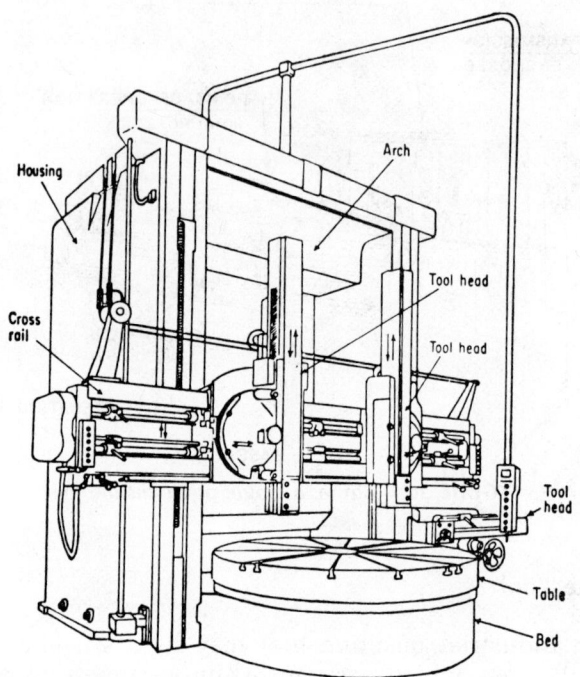

FIGURE 30.2. Heavy-duty vertical boring mill with slotted boring table for holding workpiece; table rotates into position before each cut.

Pointed or bladed cutting tools may be made of hardened high carbon steel, High Speed steel, hard ceramics, or steel tipped with hard metal carbides of tungsten, cobalt, tantalum, or other hard metallic elements. Those made of properly hardened High Speed steel, ceramic, or metal carbides retain sharp cutting edges longer than tools made of hardened high carbon steel. Grinding abrasives are usually alundum (fused aluminum oxide) or silicon carbide grains bonded with resin or a ceramic and pressed into the shape of wheels.

The importance of maintaining sharp edges on cutting tools cannot be overemphasized. Dull tools not only increase the requirement for power and influence the quality and tolerance of the work, but are often the ignition source for machine fires. This is especially true with the more combustible metals and with low flash point coolants. Dull tools also tend to produce fine particles which are more readily combustible or explosible than the larger chips.

A sharp cutting tool is particularly important when machining those metals which exhibit a tendency to weld or gall to the cutting edge. When welding or galling occurs, there is a tendency for the tool edge to

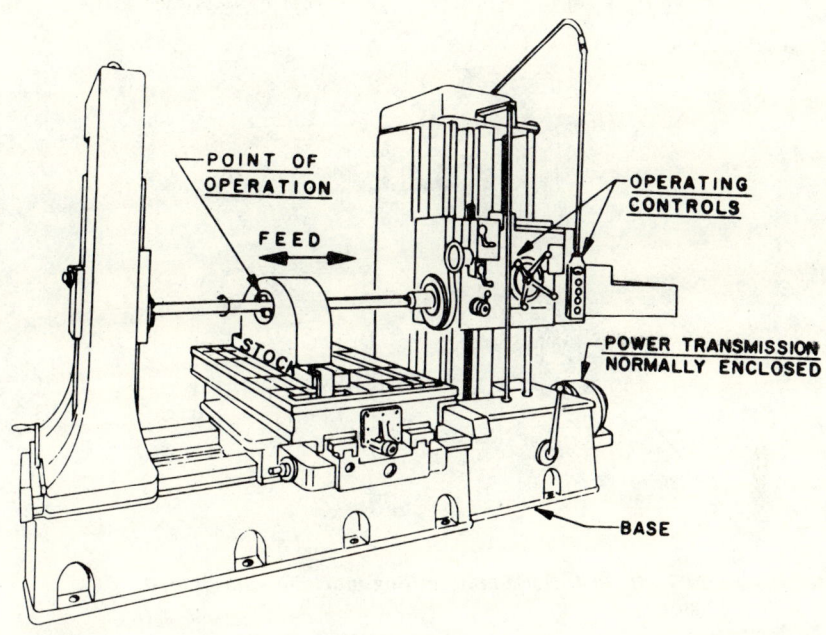

FIGURE 30.3. Horizontal boring machine using multiple point cutting tool.

fragment or break off upon a subsequent cut or revolution. It is often helpful to hone the cutting edges and the relief angles after grinding to remove as many surface imperfections as possible. The breaking tendency is especially noticeable when carbide tools are employed. The nitriding of bladed cutters such as milling cutters, taps, drills, reamers, and broaches appears to lessen this tendency to gall or weld.

Speed of cutting with an edged tool is customarily stated in linear fpm (feet per minute) or m/min. Each metal has an optimum cutting speed, dependent upon its metallurgical condition when presented to the cutting tool. This cutting speed is usually expressed as a range because light cuts may be made at faster speeds than heavy cuts. Faster cutting speeds are usually possible with harder cutting tools, such as High Speed steel, ceramic, or carbide-faced tools, because the cutting edge retains its sharpness for a longer period of time. For example, the cutting range for a single point cutting tool used for gray cast iron would be 70 to 110 fpm (21 to 33 m/min) if the tool used was made of High Speed steel; it would be 310 to 540 fpm (94.5 to 165 m/min) if a carbide tool were used. For cutting aluminum the range for High Speed steel would be 600 to 800 fpm (183 to 244 m/min); it would be 1,100 to 1,400 fpm (335 to 427 m/min) for a carbide tool.

Feeds are usually expressed in in. per revolution or per min, or in mm per revolution or per min. Thus the feed for the milling machine

Chapter 30

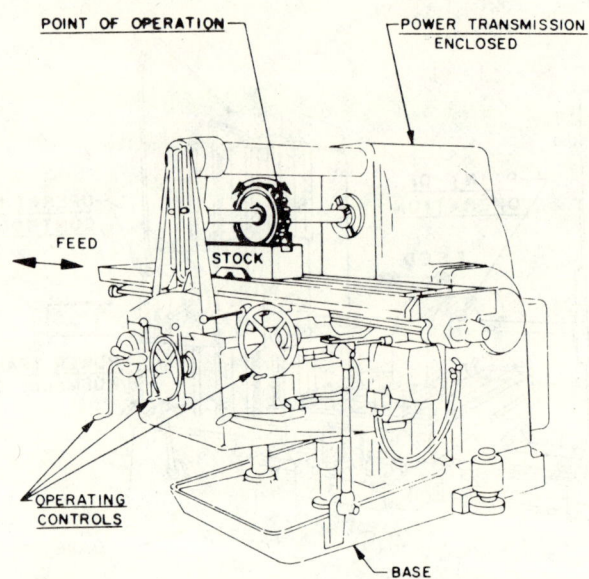

FIGURE 30.4. Horizontal milling machine with bladed cutter.

shown in Figure 30.4 may be expressed as the surface distance covered by a single revolution of its rotating cutter or the distance traveled, in this case, horizontally per min of cutting time. Feed and depth of cut must be selected based upon the objective of the machining operation. Heavy, deep cuts are employed where the purpose is to remove substantial quantities of metal to approach a desired finish dimension. Such cuts inevitably result in a coarse and somewhat torn surface, often exhibiting shallow cracks. Finishing operations usually use light cuts to produce a clean, unbroken, smooth, dimensionally accurate final finish. The surface grinder shown in Figure 30.5 is used to create a smooth finish.

In grinding operations, especially cylindrical or surface grinding, the limits of optimum rotating speed and speed of wheel or table travel vary depending on the metal being machined and the quality desired in the resultant finish. Grinding for an external or internal fit or for more precise dimensioning is entirely different from the grinding of a fine-grained, bright-finish rolling mill roll.

Alternate Means of Machining

There are also electrical processes used for the shaping and finishing of metals. These require no cutting edge as such.

Electrical discharge machining (EDM): EDM is sometimes employed for the machining of conformed cavities using a preshaped electrode. The

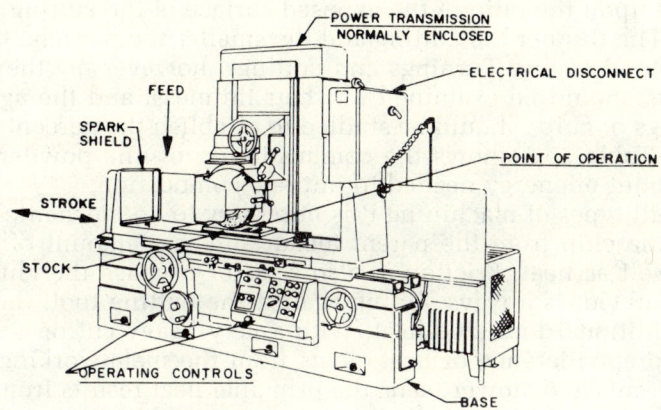

FIGURE 30.5. Surface grinder; rotating grinding wheel is usually enclosed as shown.

electrode is spaced 0.0005 to 0.0200 in. (0.127 to 0.508 mm) from the workpiece and is bathed in a continuous flow of dielectric fluid. A direct current of 0.5 to 400 amp at 40 to 400 V is pulsed from capacitors at 180 to 260,000 hertz. Each pulse partially ionizes the dielectric fluid and causes a submerged spark which melts and dislodges a metal particle. The dielectric fluid must be filtered to remove the particles before it is reused.

Electrochemical machining (ECM): ECM is a process in limited specialized use. It operates by anodic dissolution in an electric cell in which the workpiece is the anode and the tool is the cathode. The gap between the two surfaces is 0.001 to 0.030 in. (0.254 to 0.762 mm). An electrolyte, usually an aqueous solution of metallic salts, is circulated in a constant flow. A direct current of 50 to 2,000 amp at 4 to 30 V is imposed. At 10,000 amp the rate of metal removal is approximately 1 cu in./min (0.028 m^3/min). Current densities of 100 to 2,000 amp per sq in. (15.5 to 310 amp/Em^2) are employed.

Machine Cuttings

All machining operations result in the creation of chips which will vary widely in character, size, and configuration depending on the operation being performed, the feed and speed of cutting, and the intrinsic characteristics of the metal being cut. Frequently, cuttings and chips appear to be quite large and heavy, and for this reason it is often erroneously assumed that they do not constitute a fire hazard or an explosion hazard. However, a significant amount of fines may be obscured by the more obvious heavy chips.

Combustiblility depends upon the combustible character of the metal and upon the ratio of the exposed surface of the cutting or chip to its mass. The thinner the cutting and the smaller the particle, the greater the likelihood of fire. Turnings and cuttings, however big they may appear to be, should be examined for their thinness, and the aggregate of the cuttings or chips should be studied to establish the percentage of fine particles. Table 30.1 shows the combustibility of fine powders and the small amount of energy needed to inititate combustion.

In all types of machining it is necessary to do mechanical work to separate the chip from the parent metal, and that amount of work evidences itself as heat. Friction is also a factor because the chip, as it is being separated, is always sliding against the cutting tool, thereby producing additional localized heat. With a very heavy cut or a very rapid feed the preponderance of heat stems from the metalworking, whereas with light cuts and slower feeds the principal heat results from tool friction. The major portion of this heat ends up in the chips and cuttings. Often chips from the heavy machining of steel appear to have a blue or purple color showing that they must have experienced short-time heating at temperatures exceeding 500°F (260°C) followed by coolant or air quenching.

Some chips and cuttings may be subject to spontaneous ignition. As shown in Table 30.1 (Note 1), zirconium and uranium powders tend to ignite spontaneously. Spontaneous ignition of such apparently noncombustible materials as iron and steel borings and turnings has been recorded in scrapyards and in the holds of ships carrying scrap to foreign countries. Some of these fires have occurred where the heat of oxidation could not be dissipated to the surrounding atmosphere with sufficient rapidity to prevent the development of hot spots of sufficient temperature to initiate combustion.

Accumulations of chips, cuttings, or fine particles should never be permitted at the machine. They should be removed and placed in a noncombustible container for removal to the storage area, which preferably is located in a fire resistant building or in a shed located a safe distance from the shop building. Some machines are provided with a pan for catching chips and cuttings. These should be readily accessible to the operator for easy removal and placed so that they can be quickly withdrawn from under the machine in case of fire.

Pneumatic Handling and Collection of Chips

In machining operations where substantial quantities of chips are produced per unit of time, such as in high-speed milling, ingot scalping, etc., a pneumatic transport system is often used to deliver the metal scrap to a collection device, usually a cyclone collector. Collectors such as the centrifugal cyclone collector intercept and remove the heavier particles. The collection efficiency of such a collector is 75 to 85 percent at best, and if the exit velocity of the transport air is reasonably high, a large number of the finer particles will be carried through. If the amount of par-

MACHINE TOOL PROCESSES

TABLE 30.1 *Ignition and Explosibility of Metal Powders*

	Ignition Temp. (°C) Cloud Layer	Ignition Temp. (°F) Cloud Layer	Min Explosive Concentration		Min Ignition Energy Dust Cloud (millijoules)	Max Pressure		Max Rate of Pressure Rise		Explosion[2] Index
			(oz/cu ft)	(g/m^3)		(psig)	(kPa)	(psi/sec)	(kPa/sec)	
Aluminum	650—760	1202—1400	0.045	45	50	73	503	20,000	137900	>10
Magnesium	620—490	1148—914	0.040	40	40	90	621	9,000	62055	>10
Zirconium	20—190	68—374	0.045	45	15[1]	55	379	6,500	44818	>10
Titanium	330—510	626—950	0.045	45	25	70	483	5,500	37923	>10
Uranium	20—100[1]	68—212	0.060	60	45[1]	53	365	3,400	23443	>10
Iron	440—270	824—518	0.200	200	72—305	45	310	600	4137	0.1
Zinc	680—460	1256—860	0.500	500	960	48	331	1,800	12411	<0.1
Bronze	370—190	698—374	1.000	1000	—	44	303	1,300	3964	—
Copper	700	1292	—	—	—	—	—	—	—	<<0.1

NOTE: Data taken from Bureau of Mines RI 6516, Table 1, p. 4, of Reference 2. Data on Iron (Item 131) taken from Tables A1 (p. 12), A2 (p. 16), and A3 (p. 22).
[1]In this test less than 1 g of powder was used. Larger quantities ignited spontaneously.
[2]Index of Explosibility (RI 6516, p. 3).
None 0; Weak <0.1; Moderate 0.1–1.0; Strong 1.0–10; Severe >10.

ticulate effluent exceeds the limits set by the Environmental Protection Agency (EPA), OSHA regulations, or the local authorities, it may be necessary to install a second cyclone collector in series.

In the operation of the transport system it is probable that a substantial quantity of fine particles will be produced. A chip crusher, if used, will add to the quantity of fines already present. Also, the chips, as they travel through the system, will impact other chips and rub against duct surfaces and bends. This tends to further degrade the chips and create additional fine particles; thus, the fine dust content in the transport duct and beyond it will be significantly greater than that measured at either the point of origin or at the chip crusher.

Because of the flow of a significant amount of fine particles in the system, the system should be designed with explosion hazards in mind, especially when dealing with the more combustible metals. It should be designed so that at no time, even during peaks, will the fine dust concentration approach the lower explosion limit (LEL) of the metal dust. (Explosion limit values can be found in Reference 2.) The ducts should be designed and constructed so that, if an explosion does occur, the duct will not separate or open under the explosion pressure in such a manner as to expose plant personnel to the explosion flame. All ducts and collectors should be firmly grounded electrically, and the resistance to ground should be checked at appropriate intervals. A considerable number of major and minor explosions have been recorded in systems of this character. (See Chapter 45 for more information on air moving equipment.)

Fabric- or panel-type impingement collectors are not safe for the collection of fine metal dusts because they inherently provide the optimum conditions for rapid and complete combustion. Any combustible filter medium used would catch the particles of combustible or explosible metals and hold them in an envelope of oxygen (air containing enough oxygen to support rapid and complete combustion) so that all that would be necessary to initiate a fire or explosion would be a very small amount of energy, perhaps a small static discharge.

Wet collectors for fine metal dusts should be used only after the most careful consideration. Many dusts react with water to produce highly explosive hydrogen. The wetted sludge is also highly explosive unless it is kept submerged under a freeboard of water. (See Figure 30.6.)

There is now a trend in machine technology development toward extremely high high-speed milling of light metals. At least one experimental machine has been built and tested which achieves cutting speeds of up to 18,000 fpm (5486 m) and cutting rates of 1 cu in. (16.4 cm^3) of aluminum per min per hp (one hp = 746 W) employed. With a multibladed cutter it is apparent that the size of the individual chips must be very small. These developments must be accompanied by adequate provision for fire and explosion protection during the transport, collection, and disposition of the chips.

MACHINE TOOL PROCESSES

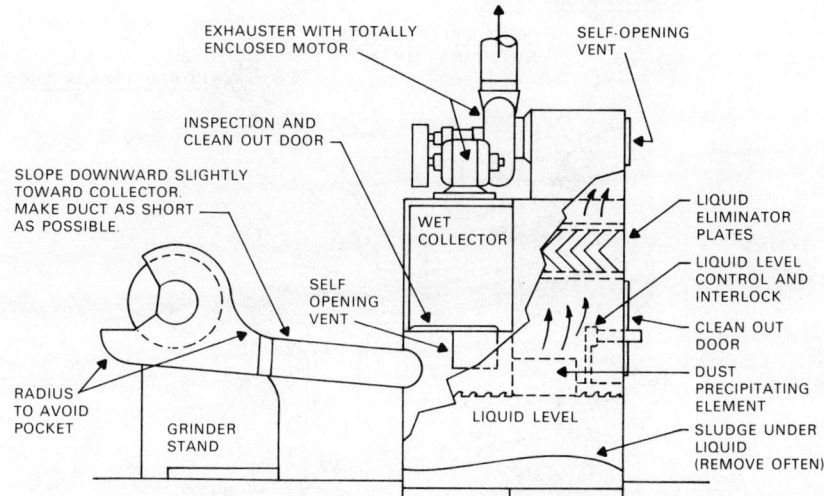

FIGURE 30.6. A shcematic diagram of a water precipitation-type collector for use in collecting dry combustible metal dust without creating explosive dust clouds or dangerous deposits in ducts or collection chambers. The diagram is intended only to show some of the features which should be incorporated in the design of a collector. It may be used for all combustible metal dusts.

Cutting Fluids

Cutting fluids are used in the machining of many metals and alloys to cool the cutting tool and to reduce the friction created as the chips slide across the tool. There are two types of fluid generally used, the choice being dictated by the metal being machined.

The simpler of the two is a water-based solution, usually alkaline, which functions principally as a coolant. Water-based cutting fluids should never be used when machining metals which are reactive with water because explosive hydrogen will be evolved and sensible heat will also occur.

More commonly used are oil-based coolant/lubricants. They range from low viscosity, relatively low flash point fluids to compounded oils which sometimes contain animal fats, fish oils, and/or complex synthetic-organic extra-pressure additives. Essentially all the oil-based and compounded cutting fluids are flammable to some degree. Compounded oils which are acidic or which contain organic additives that may oxidize and then become acidic with use should not be used for the machining of magnesium and certain other metals because the acid can react with the metal to release hydrogen. Uncompounded mineral oils should be used.

As can be seen in Figure 30.7, the coolant/lubricant is added at the point of operation where the tool and metal meet. Drippage of combustible cutting fluid should be absorbed or otherwise collected and then re-

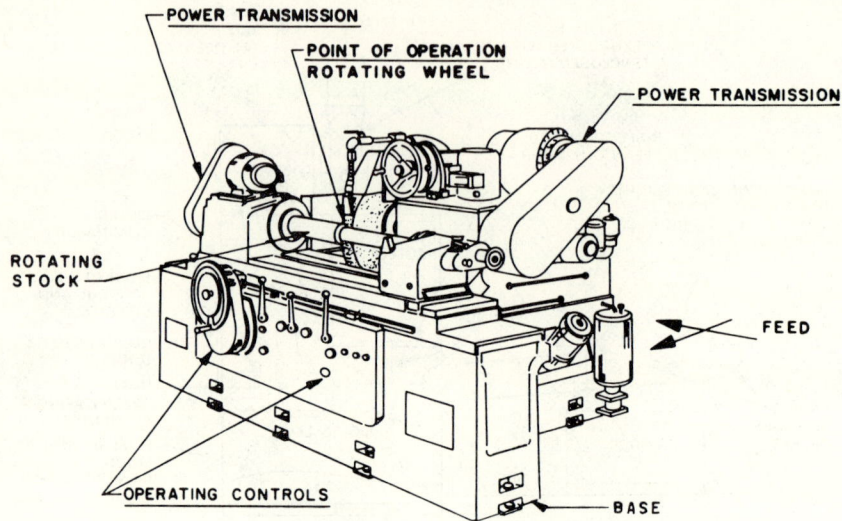

FIGURE 30.7. *Cylindrical grinder.*

moved from the workroom. Provision should also be made to collect and remove the residual cutting fluid carried by the scrap metal. Some times centrifuging is employed. The waste cutting fluid should not be allowed to accumulate but should be collected and disposed of in a safe manner. Draining it off into a sewer is not legally permissible, nor is it environmentally acceptable. (For more on the handling of flammable liquids, see Chapter 37.)

Hydraulic Machine Tool Actuators

Many machine tools employ oil-hydraulic devices for various control actuators. Oil pressures used in the machine tool industry range from 400 to 600 psi (2758 to 4137 kPa). That range was selected to minimize leakage from glands and joints which are sometimes located in inaccessible places. Flexible hoses are sometimes used, but they are subject to deterioration with misuse.

Many machine tool builders and operators use a petroleum-base hydraulic oil. The petroleum oils have a flash point usually between 400 and 450°F (204 and 232°C). Several organic substitutes are available for use. These materials have flash points not greatly in excess of that of the petroleum oils, but they do exhibit a tendency to self-extinguish as soon as the source of ignition is removed. The most effective of these was the family of askarels (polychlorinated biphenyls, more recently designated as PCB). These are being withdrawn from the market at the insistence of the Environmental Protection Agency (EPA). Also used are certain silicone oils and certain phosphate esters which do not have quite as good a

MACHINE TOOL PROCESSES

fire resistance but which closely approximate the askarels as fire resistant fluids.

Experience has shown that if there is a failure in a pressurized oil system which results in the spraying of a jet of oil which then contacts a hot surface and ignites, the neighboring personnel run for safety without shutting down the pump. The fire then continues by being constantly fed with a fresh supply of pressurized oil. Numerous complete fire losses can be attributed to this factor. (See Chapter 43, "Fluid Power Systems," for further information on the hazards and safeguards associated with hydraulic pressure systems.)

Cleaning of Machined Components

It is sometimes necessary to clean or degrease components after machining. This is often done in open tanks or pans using a moderately high flash point naphtha such as Stoddard Solvent. This constitutes a potential fire hazard because the solvent can be easily ignited by a flying spark from a neighboring operation.

Vapor degreasing using inhibited trichlorethylene or tetrachlorethylene (perchlorethylene) is sometimes employed. This is usually done by inserting the cold machined component into a degreaser tank below the vapor level. The solvent is boiled, and the vapor level is maintained through the use of condenser slots near the top of the tank. The cold machined part causes the vapor to condense on its surface. The condensate dissolves the oils and greases and runs off into the liquid portion of the unit.

It has been found, in a fire of record, that trichlorethylene will react with finely divided aluminum (regardless of the inhibitor) to produce aluminum chloride. The aluminum chloride then catalyzes the breakdown of the trichlorethylene into fine, soft, burning carbon with the evolution of great volumes of hydrochloric acid vapor. The latter corrodes all metallic surfaces with which it comes in contact. The vapor is obviously toxic.

MACHINING CHARACTERISTICS OF COMMERCIAL METALS

Because there are essentially no pure metals used in ordinary commerce and because the commercial metals are alloys to some degree, it is virtually impossible to characterize the machining qualities of any metal except in the most general terms. The several alloys are often amenable to various types of heat treatment or work-hardening so they have widely varying characteristics. The data included in References 3 and 4 will provide more specific information.

Ferrous Metals

Cast iron is relatively easy to machine except when in the chilled "white iron" stage. The chips are usually moderate in size and generally well broken. This is the result of the intercrystalline graphite which is always well dispersed throughout the mass.

Low or medium carbon steel is not too difficult to machine. High carbon steel and certain high alloys are somewhat more difficult. Heat-treated steel is also more difficult to machine. Turnings are often long, thin, curled ribbons of great length, whereas the chips from boring, milling, drilling, broaching, etc., are usually broken and often quite large. Very frequently the chips exhibit a blue or purple color, especially when heavy cuts are made, showing that, when created, they achieved a temperature well in excess of 500°F (260°C).

Copper and Copper-Based Alloys

From the viewpoint of machinability copper and copper-based alloys can be classified as follows:

1. Free cutting alloys. Free cutting alloys are relatively easy to machine. They produce short, brittle chips.
2. Moderate machinability. This group is represented by the nonleaded brasses with 60 to 85 percent copper and the nickel-silvers. The cuttings are long open coils or sometimes a closely wound helix. Usually the cuttings are relatively brittle.
3. Difficult machinability. This group includes nonleaded copper, low zinc brasses, nickel silver, phosphor bronze, aluminum bronze, cupronickel, and berylium copper. The cuttings are long, continuous chips, often tightly curled. The cuttings will bend but seldom break.

Water-soluble oil coolants are used for the free cutting alloys. For more difficult cutting, light mineral oil, sometimes mixed with lard oil, is used. Sulfurized, chlorinated, or sulfo-chlorinated coolants are used for the very difficult cutting, especially nickel-bearing alloys. When the latter compounds are used, subsequent degreasing and rinsing are required.

Aluminum and Magnesium

These metals are quite readily machinable, although the softer alloys tend to be somewhat gummy. Because they always have a surface film of abrasive oxide, they tend to dull cutting tools more rapidly. High silicon alloys are more abrasive, so they cause more rapid tool wear and create greater tool friction. Aluminum and magnesium alloys can be machined at significantly higher speeds than steel.

The fine particles produced with the chips constitute a significant

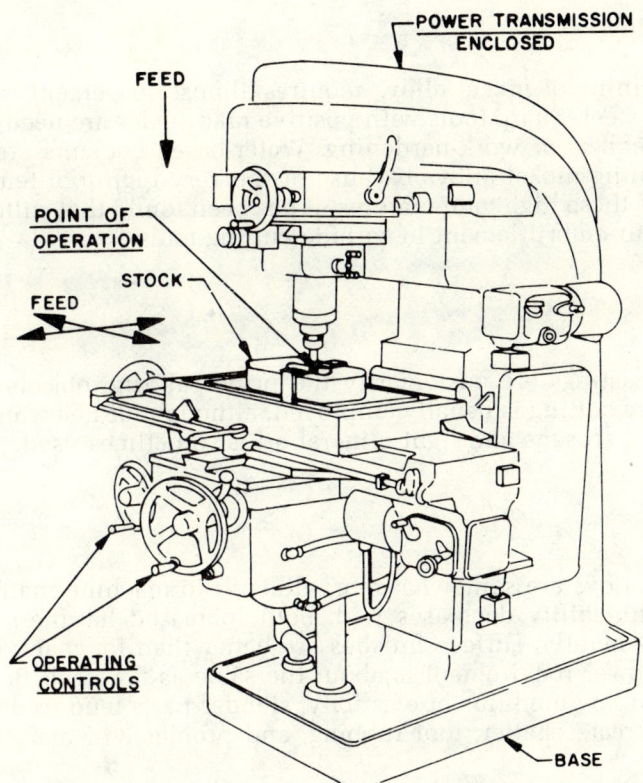
FIGURE 30.8. Vertical milling machine.

fire and explosion hazard, so the appropriate safeguards must be provided. A large and powerful milling-type aluminum-plate sawing machine, in operation in a plant in Davenport, Iowa, for example, was found to produce, in addition to a relatively large number of chips, a significant quantity of minute particles. These practically invisible fines became airborne and ultimately deposited on horizontal ledges and on the elements of the roof-supporting steel. If they had been allowed to accumulate, these fines could have constituted a fire and explosion hazard; thus, it was necessary to shut down the operation and conduct extensive cleaning at regular intervals.

Magnesium chips, and particularly the very fine particles, constitute a significant fire potential. Magnesium fines have been known to be ignited in the bed of a machine. When magnesium is machined, the machine operator should not only maintain scrupulous housekeeping practices, but should also have an appropriate fire extinguishing agent immediately available. This may be a pail of dry sand, dry soapstone, or a suitable dry proprietary extinguishant which is suitable for Class D fires.

Nickel Alloys

The machining of nickel alloys requires almost 40 percent more power than mild steel. Sharp tools with positive rake angles are needed to minimize the effects of work-hardening. Water-based coolants are preferred for machining nickel alloys because of the very high tool temperatures. Because of these high temperatures, it has been found that sulfonated oils cause sulfur embrittlement in carbide cutting tools.

Zinc

Zinc die castings are customarily the principal zinc objects to be machined. Dry cutting is usually employed, although for heavy cuts a 50-50 mixture of kerosene and light mineral oil is sometimes used.

Titanium

Titanium alloys are somewhat more difficult to machine than steel, and their machinability decreases with both increased hardness and alloy content. Generally, surface finishes are better than for many other metals. The power requirement is about the same as for steel. Because titanium has a low modulus of elasticity, slender parts tend to deflect more and thus create chatter, tool rubbing, and problems of maintaining tolerances.

Titanium chips tend to gall and weld to the cutting edge of the tool. It is characteristic of titanium to produce a high shear angle between the workpiece and the chip, resulting in a thin chip flowing at high velocity over the tool face. This, combined with the small contact area of the chip with the cutting edge, contributes to high tool tip temperatures. To minimize the galling and welding tendencies, the rigidity of the entire system (including the minimizing of backlash in the feed) is of extreme importance. Relief angles larger than those for steel should be used.

Highly chlorinated cutting fluids are very effective for the machining or grinding of titanium. There is some concern that chlorine promotes stress corrosion in titanium. When chlorinated cutting fluids are used, provision must be made for subsequent degreasing and rinsing.

Grinding of titanium with silicon carbide wheels may be done at surface speeds of 5,000 fpm (1524 m/min), although to minimize surface stresses and distortion it is more customary to use a speed of 3,000 fpm (914 m/min). Highly chlorinated oil is usually used. When aluminum oxide grinding wheels are used, the optimum surface speed is considered to be 1,500 to 2,000 fpm (457 to 610 m/min). The coolant may be a 10 percent solution of sodium nitrite-amine or other equivalent proprietary material.

There is a distinct fire hazard when grinding titanium or its alloys. A copious flow of cooling fluid should be used, considerably more than would otherwise be considered normal, to quench the sparking as much as possible. The cutting fluid should be continuously filtered to remove the fine titanium particles. The filter should be replaced more often than would be the case when grinding steel. The external surfaces of the grinder should be cleaned frequently to prevent the accumulation of fine combustible particles.

Dry soapstone, dry fine sand, or dry proprietary extinguishants suitable for Class D fires should be immediately available to quench any fires. Extreme care must be exercised in applying these materials so as to avoid the creation of a dust cloud of metallic particles which would be explosive.

Zirconium

Like titanium, zirconium has a tendency to gall and weld to the cutting tool. It has a high rate of work-hardening, and its abrasiveness contributes to rapid tool wear. To compensate for these characteristics, it is customary to use hard cutting tool materials and to grind the tool to a 15-degree positive rake or greater, as well as to employ a clearance angle of 6 to 10 degrees. Tools should be honed after grinding to provide a very smooth surface. Each successive cut should be deep enough to remove the work-hardened surface created by the preceding cut. It was found that at 220 surface fpm (67 m/min) cutting speed, the depth of the work-hardened surface was less than 0.003 in. (0.076 mm).

Zirconium is very pyrophoric. Tools should be flooded at the point of contact with a very generous flow of a suitable coolant/lubricant. Dull tools produce fine particles. It is desirable to retain a freeboard of coolant in the machine bed so that fine particles are submerged until they can be removed for disposal. Dry, fine dust will spontaneously ignite and can be explosive. Constant attention to immaculate housekeeping is imperative.

Uranium

Uranium is both pyrophoric and poisonous, so extreme care and immaculate housekeeping must be employed at all times. Like aluminum and magnesium, the fine particles can react rapidly with water to release explosive hydrogen. The machine chips if not adequately quenched with coolant will self-ignite and burn in air from the heat of machining.

Machining of extruded rod is similar to the machining of austenitic stainless steel in which the metalwork hardens under the tools. It is necessary to provide adequate ventilation to remove the oxide fumes. Individual exhaust pipes are usually required at each machine.

The grinding of uranium presents a great hazard of fire and explosion. Fine uranium powder self-ignites, and chips must fall into a freeboard of coolant in the bed of the machine to minimize this hazard. Extinguishants suitable for Class D fires should always be available in the immediate vicinity of any uranium machining operation.

Relative Power Requirement for Machining

Using the power required to machine nickel alloys as 100 percent, the following list shows the relative power required to machine certain commercial alloys and metals:

Nickel alloys	100%
Mild steel	63%
Titanium	63%
Cast grey iron	35%
Brass	23%
Aluminum	18%
Magnesium	10%

THE FIRE HAZARDS

The principal hazards to be encountered in a machining operation are:

1. Chip fires at the machine, where ignition is caused by the heat of metalworking, friction of the chip against the tool, or both.
2. Spontaneous oxidation of cuttings.
3. Combustion of oxidizable coolant/lubricants.
4. Fine particles which are either combustible or explosible.
5. Reaction of certain metals with water or other agents which often results in the evolution of hydrogen and heat.
6. Combustion of pressurized hydraulic fluids used for the actuation of machine tools and/or their accessories.
7. Combustion of oil vapors deposited upon building structure.
8. Combustion of oil-saturated floors.

The principal sources of ignition are:

1. Smoking materials.
2. Heat of cutting.
3. Spontaneous oxidation.
4. Hot particles from (a) grinding, (b) dressing of grinding wheels, and (c) welding and cutting.
5. Hot surfaces such as furnaces, torches, etc.
6. Electrical sparking or arcing.
7. Impact ignition of certain pyrophoric surface compounds which sometimes form during the earlier stages of fabrication. (An example

of this is magnesium nitride which sometimes appears on the surface of castings; it can explode under the impulse of a very minor impact.)

THE SAFEGUARDS

Building Construction and Protection

Buildings where machining operations are performed sometimes have oil-soaked floors, the source of the oil being the drippings of the coolant/lubricants which are employed. In addition, the high temperatures experienced at the tip of the cutting tool tend to vaporize some of this oily material. The vapors condense and precipitate out on walls, roof deck, roof truss members, and on any exposed horizontal surface.

Machining buildings should preferably be of fire-resistant construction with a fire-resistant roof deck. If the building is large, fire barriers in the roof support structure should be installed, where possible, to limit the spread of fire.

The need for sprinkler protection should be carefully considered. It will depend largely upon the combustible character of the building structure, especially floors and roof deck. Small fires at machines can be best controlled with portable extinguishers of the correct classification and, if proper housekeeping is practiced, the fire hazard from cuttings and turnings is minimal. Consideration must also be given to the dangers of applying water to certain burning reactive metals such as magnesium. Any advantage of sprinklers must be carefully weighed against the cost of repairing water damage to precision machinery and its sensitive electronic controls as well as the cost of loss of production while such repairs are being made.

Buildings where machining operations are performed should be regularly inspected for the accumulation of oily deposits or the accumulation of fine combustible metal particles. Cleaning should be undertaken whenever the need becomes apparent.

Housekeeping

Consistent with the basic rule of fire prevention, good housekeeping must be maintained at all times.

Cuttings and chips should not be allowed to accumulate at the machines or on the shop floor. They should be collected in noncombustible containers and should be removed to a safe storage location at regular intervals.

Spillage and drippage of cutting fluid should not be allowed to accumulate. It should be soaked up and the oily residue should be removed to a safe location for subsequent disposal.

Ledges, roof truss members, and other areas where metal dust can

Chapter 30

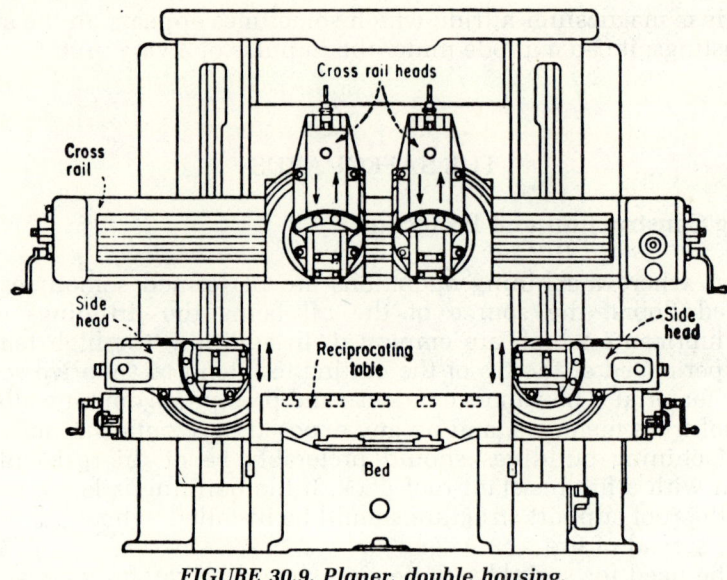

FIGURE 30.9. Planer, double housing.

accumulate or where oily vapors can condense and accumulate should be inspected at regular intervals and should be cleaned as necessary. Extreme care must be exercised to avoid the creation of even small dust clouds.

Oily waste and rags should be stored in metal containers and should be removed at least once each day to a safe disposal site. (See Chapter 50 for more information on industrial housekeeping.)

Electrical Controls

All electrical control equipment for control and, particularly, those controls for numerical control or computer-directed control of machines should preferably be housed in vapor-tight enclosures. The equipment itself should be inspected at regular intervals and should be cleaned when necessary.

Fire Protection

The more combustible metals should never be machined unless a suitable fire extinguishant is immediately available to the machine operator.

It is important to remember that extinguishants for Class A, B, and C fires are generally ineffective for burning metal or metal dust. Their use on burning metal can lead to a great intensification of the fire, the probable spread of the fire, with the strong likelihood of an explosion. Certain of these materials when used on burning metals can produce toxic products. Only extinguishants rated for Class D fires should be used on burning metal. When they are used, extreme care must be exercised to ensure that the emergent force of the extinguishant or the action of application to the fire is not sufficient to create an explosive metal dust cloud.

For their own life safety, it is most important that all workers thoroughly understand the hazards involved and that they be properly instructed so that they will fully understand the possible consequences of unconsidered or improper action under stress. In many instances it is highly desirable that the local fire fighters who would normally respond to an alarm be similarly instructed.

BIBLIOGRAPHY

REFERENCES CITED

1. McKinnon, G. P., ed., *Fire Protection Handbook*, Fourteenth Edition, National Fire Protection Asscociation, Quincy, MA, 1976, p. 3-119.
2. Bureau of Mines, *Explosibility of Metal Dusts*, Bulletin RI-6515, U.S. Department of the Interior, Washington, DC.
3. *Machining Data Handbook*, Metcut Research Associates, Inc., Cincinnati, OH, 1966.
4. "Machining," *Metals Handbook*, vol. 3, American Society for Metals, Metals Park, OH.

ADDITIONAL REFERENCES

Alico, J., *Introduction to Magnesium and Its Alloys*, Ziff-Davis, New York, NY, 1945.

Jacobsen, M., Cooper, A, and Nagy, J., *Explosibility of Metal Powders*, RI-6516, Bureau of Mines, U.S. Department of the Interior, 1964.

Maranchik, *Machining Data for Titanium Alloys*, AFMDC 65-1, Air Force Machinability Data Center, Cincinnati, OH, 1965.

Van Horn, K. R., ed., *Aluminum*, vol. 3, American Society for Metals, Metals Park, OH, 1967.

NFPA CODES, STANDARDS, AND RECOMMENDED PRACTICES

Reference to the following NFPA Codes, Standards, and Recommended Practices will provide further information on the safeguards for metalworking operations discussed in this chapter. (See the latest *NFPA Codes and Standards Catalog* for availability of current editions of the following documents.)

NFPA 10, *Portable Fire Extinguishers.*

NFPA 13, *Installation of Sprinkler Systems.*

NFPA 30, *Flammable and Combustible Liquids Code.*

NFPA 34, *Dipping and Coating Processes Using Flammable or Combustible Liquids.*

NFPA 48, *Magnesium Storage, Handling and Processing.*

NFPA 65, *Processing and Finishing of Aluminum .*

NFPA 70, *National Electrical Code.*

NFPA 70B, *Electrical Equipment Maintenance.*

NFPA 72E, *Automatic Fire Detectors.*

NFPA 75, *Protection of Electronic Computer/Data Processing Equipment.*

NFPA 79, *Electrical Standard for Metalworking Machine Tools and Plastics Machinery.*

NFPA 91, *Blower and Exhaust Systems for Dust, Stock, and Vapor Removal.*

NFPA 325M, *Fire Hazard Properties of Flammable Liquids, Gases, and Volatile Solids.*

NFPA 481, *Titanium Production, Processing, Handling and Storage.*

NFPA 482, *Zirconium Production, Processing, and Handling and Storage.*

NFPA 505, *Powered Industrial Trucks.*

NFPA 651, *Manufacture of Aluminum and Magnesium Powder.*

NFPA 801, *Facilities Handling Radioactive Materials.*

ADDITIONAL READING

Bartels, A. L., "Safeguards to Prevent Ignition by Hot Surfaces," *Electrical Review London*, vol. 203, no. 4 (1978), pp. 29-31.

Bellows, Guy, *Machining: A Process Checklist*, 3rd ed., Metcut Research Associates, Cincinnati, OH, 1982.

Feirer, John L., *Machine Tool Metalworking*, 2nd ed., McGraw Hill, New York, NY, 1973.

Follette, Daniel, ed., by Roy Williams and E. J. Weller, *Machining Fundamentals*, Society of Manufacturing Engineers, Dearborn, MI, 1980.

Heineman, Stephen S. and George W. Genevro, *Machine Tools: Processes and Applications*, Harper and Row Publishers, New York, NY, 1979.

Heritage, P. ed., *Machining for Toolmaking and Experimental Work*, 2nd ed., (3 vols.), International Ideas, Inc., Philadelphia, PA, 1977.

Holden, A. M., *Physical Metallurgy of Uranium*, Addison-Wesley, Reading, MA, 1958.

Kent's Mechanical Engineers Handbook, John Wiley and Sons, New York, NY.

Kibbe, Richard R. and John E. Neely, *Machine Tool Practices*, 2nd ed., John Wiley and Sons, New York, NY, 1982.

King, P. W. and J. Magid, *Industrial Hazard and Safety Handbook*, Butterworths Publishers, Inc., Woburn, MA, 1979.

Lustman, B., and Kurze, F., *The Metallurgy of Zirconium*, McGraw-Hill, New York, NY, 1955.

Marks, L. S., *Mechanical Engineers Handbook*, McGraw-Hill, New York, NY.

The Technical Staff of the Machinability Data Center, *Machining Data Handbook*, 3rd ed., (2 vols.), Metcut Research Associates, Cincinnati, OH, 1980.

Walker, John R., *Machining Fundamentals*, Goodheart-Willcox Co., Inc., South Holland, IL, 1981.

Warner, J. C., Chipman, J., and Stedding, F., *Metallurgy of Uranium and its Alloys*, NNES-IV-12A, National Technical Information Service, Springfield, VA.

White, Warren T., et al., *Machine Tools and Machining Practices* (2 vols.), John Wiley and Sons, New York, NY, 1977.

Williams, Roy, ed., *Machining Hard Materials*, Society of Manufacturing Engineers, Dearburn, MI, 1982.

31

Chemical Processes

Russell L. Miller

This chapter discusses chemical processes, process equipment, and their fire hazards. Traditionally, chemical processing equipment is classified according to the various unit operations being performed. A few of the unit operations that can be considered are: reaction, distillation, crystalization, and drying. In addition, there are variations of each unit operation. In distillation, for example, there are batch, continuous, steam, pressure, and vacuum units and an almost infinite variety of makes and models of stills and distillation columns. Thus, there are many kinds of chemical processing equipment to be found in industry. Consequently, the fire hazards of chemical processes is much too broad a subject to be examined in a single chapter; therefore, some basic principles that can be applied generally to individual analyses of the various processes and equipment will be discussed.

Russell L. Miller, safety consultant, Safety Loss Prevention Services, Inc., Kirkwood, Missouri, wrote this chapter. W. J. Bradford, P.E., loss prevention consultant, revised the chapter for the second edition.

Chapter 31

FIGURE 31.1. Chemical process reactors. (The Pfaudler Co., a Sohio Company)

OPERATIONS AND EQUIPMENT

Chemical processes alter the properties of raw materials through one or more stages of chemical conversion. They may involve liquids, gases, and solids — separately and in combination. The heart of a chemical process is the reactor, where chemical conversion takes place. There are also a number of support operations, which are largely physical or mechanical in nature. These include steps to prepare the raw materials for reaction as well as to separate and recover products and by-products from the process.

Reactors

Raw materials are brought together and chemical conversion takes place under controlled conditions in the reactor . The chemical change that takes place is accompanied by an energy change in the system. If energy is supplied to the system and absorbed by the reaction, the process is *endothermic*. If energy is released by the reaction, the process is *exothermic*. Even for exothermic reactions, it is usually necessary to supply some heat to provide energy of activation or to raise the reactant temperatures enough to develop a reaction rate that meets the needs of the system. Once a suitable reaction rate has been established in an exothermic reaction, heat removal must be provided to create an energy equilibrium (i.e., a controlled temperature). If heat removal is inadequate, the reaction temperature will continue to rise, increasing the reaction rate, increasing the energy release rate, and further increasing exponentially the temperature rise in a runaway reaction.

Continuous reactors: Continuous reactors continuously feed reactants into the system while simultaneously removing the reaction mass with its products and by-products. They may involve single-pass systems or continuous recirculation of primary reactants, solvents or carriers, and continuous controlled addition of other reactants, catalysts, and/or additives.

As a general rule, continuous systems can be designed to contain lower inventory (less fuel) than a batch system of comparable capacity. This is an advantage in limiting fire hazards. Some continuously operated systems, however, make use of large vessels and large inventories of hazardous fuels.

In continuous systems, one or more reactants are continuously added to the process. This provides a means of controlling energy generation by controlling the rate at which a reactant is fed into the process. Feed rates or reactant cutoff should be reliably linked by automated control systems to reactor system energy levels that are usually measured by temperature and/or pressure. Energy release rate is usually more easily and reliably controlled in a continuous reactor than in a batch reactor.

Continuous systems require pump or gas pressure to feed raw materials into the reactor, remove the product from the system, and possibly recirculate the reaction mass. Thus, there is some potential for spills to occur, which is not present in tightly closed batch reactor vessels. Pumps, packing glands, control valves, piping systems, and expansion joints offer some risk of leakage and spillage. Even in low-inventory systems, flow rates may be high, providing the potential for subtantial spills unless feed streams can be cut off rapidly. Sufficient remotely controlled block valves are generally required.

The accidental introduction of an excessive amount of oxidant into a continuous reaction system is possible, especially if one of the reactants is an oxidant itself, such as air or oxygen. An example of this type of system is a continuous liquid or vapor phase oxidation reaction. An excessive amount of oxidant in the process amounts to loss of control of the reaction rate. Accidental introduction of an unwanted oxidant into a process can result from operator error, instrument failures, or leakage in a system operating at reduced pressure.

Batch reactors: In batch reactors, raw materials are deposited in a vessel, the vessel is sealed, and the reaction is initiated by the application of heat. The reaction mass is discharged from the vessel upon completion of the reaction, and the process is repeated with a new batch of raw materials.

If the reaction is exothermic, the entire heat of reaction of the batch is available for release at whatever rate is provided by reaction control. Loss of heat control, usually by loss of heat removal capacity, leads to a runaway reaction. Common methods of heat removal from a batch reactor include the use of a cooling jacket, cooling coils, and circulation of the reactor contents through an external heat exchanger, or reflex of boiling reactants or solvent heat carrier from a condenser.

An external heat exchanger provides much heat removal capacity. The presence of an external heat exchanger usually indicates high rates of heat generation and the need for high rates of heat removal. Under these conditions, any loss of heat removing capacity will quickly create hazardous conditions in the reactor. An external heat exchange system may also indicate the need to maintain a low reaction temperature and a relatively small temperature differential between the heat exchange medium and the reactor mass, thus requiring more heat transfer surface than would be available in jackets or coils. This too might indicate a sensitive system that could quickly reach excessive temperatures or unstable conditions upon loss of cooling capacity. Most jacket- or coil-cooled reactors are less closely designed and may allow a wider range of control variability in the system.

Semibatch reactors: A variant of the batch reactor is the semicontinuous reactor in which reactants are added, as needed, to the reaction mass in the batch reactor to carry on the reaction. Products are allowed to accumulate until the vessel is full; then the process is stopped and the vessel is emptied. This system limits the amount of at least one reactant initially in the system, and limits the total potential reaction energy in inventory. Reaction energy generation can be controlled by reducing or stopping the addition of the reactant being continuously fed into the semicontinuous reactor. These increased safety factors may be somewhat offset by the added feed rate control requirement imposed by the system. Heat removing capacity is often closely designed to match the planned feed rate, leaving little margin for error and no protection for delayed reaction and accumulation of the reactant being added. Accidents have happened in such systems when operators, intending to be extra safe, reduced reactor temperature, not realizing that the total reaction rate was cut in half with each 10°C (50°F) drop in temperature. If the reactant is added to the process faster than it can be used, unreacted material will accumulate, increasing the danger of a subsequent runaway reaction.

Reactor pressure relief: Reactor pressure relief is especially important in batch reactors. In continuous systems, particularly in small tubular reactors, the maximum energy potential may be approximately limited to the feed stream energy content, with only a relatively small potential increase from the energy of reaction of the small amounts of material that might be present in a blocked reactor. In some continuous reactors, pressure relief must be provided to vent explosions that can result if feed rate control allows an explosive atmosphere to develop. An example of this is the explosion venting needed in a continuous reactor that partially oxidizes naphthalene to produce phthalic anhydride.

In a batch reactor, a large amount of potential energy is present in the inventory of reactants. Pressure relief venting capacity must be provided for relief of pressure developed if the total heat is released in a short time by a runaway reaction. Most chemical reactions approximately double in rate with each 10°C (50°F) rise in temperature.

Since some heat input and heat increase are usually needed to start a reaction and develop a self-sustained reaction rate, there must be a balance between a temperature that is too low to provide a measurable reaction and one that is too high for the heat removal system to handle. If a heat removal system is overpowered by the heat generation rate, the additional heat retained increases the batch temperature, which further increases the heat generation rate. The resulting rapid temperature rise generates pressure from increasing vapor pressure of the vessel contents. The vessel will rupture when the vent system can no longer remove expanding hot vapor and/or liquid as fast as increasing energy release increases temperature, vapor generation, and pressure. Reactor vent sizing requires knowledge of the kinetics of the reaction, the flow characteristics of the fluids to be vented, and the physical characteristics of the vent system itself. Precise design is difficult, although some mathematical models have been proposed. For practical field use or rule of thumb preliminary design concepts, a vent sizing chart gives some indication of vent requirements when the chart is used within its prescribed limitations.

In reactor systems, the greatest hazard added to the potential fire and explosion risk inherent in the presence of flammable materials is probably that of runaway reactions. Emergency venting must be provided as the ultimate backup to limit damage and loss from runaway reactions. Loss of agitation, power, or control might negate other safeguards provided in the system.

Fluid Flow

Fluid flow is achieved through the use of pumps, blowers, the force of gravity, or atmospheric pressure (the moving force in a jet suction system). Energy added to the fluid appears as pressure used to overcome the pressure drop from fluid friction in transfer pipes or ducts. Fire hazards in the fluid flow process are primarily related to the potential for spills to occur.

If fluid flow is interrupted by a closed valve or a plugged line and centrifugal pumps continue adding energy to the churning fluid, temperature and, therefore, pressure will rise. Pressure may rise high enough to cause a rupture in the system. Positive head movers working against a stopped line will increase pressure to the bursting point unless they are equipped with a pressure relief device or unless the burst points of the pump and fluid lines are greater than the maximum pressure that can be developed by the driver.

Raw Material Handling

A chemical process flow sheet generally begins with the receipt of raw materials. In design, the receiving, unloading, handling, and storing of materials are sometimes considered part of the service operations as distinguished from chemical process operations. However, from a fire control standpoint, the receipt of raw materials must be considered the first step in the process.

Processes using liquid raw materials often involve substantial quantities of fuel; therefore, spill control is essential to fire hazard control. Top unloading of rail cars and trucks by suction pumps has less potential for spills from broken or leaky lines than does bottom unloading or pressure transfers. Facilities to shut off lines in the event of a break, such as self-closing or remotely controlled valves, limit the amount of fuel spilled and thus limit the fire hazard.

Drainage lines and containment pits or dikes to control the flow of spilled flammable liquids are also elements of good raw material handling design. Adequate separation of fuel storage from ignition sources and other facilities improves loss control design. Some low-flash point materials may necessitate inerting the vapor space of the storage or receiving tanks to preclude the entry of oxygen. (See Chapter 37, "Handling and Storage of Flammable and Combustible Liquids.")

Static or electrical energy must be controlled by proper grounding. Flow rate control to limit fluid velocity in pipelines should be provided where necessary. If the stored chemicals are reactive, an inhibitor may be needed. Periodic checking and renewing of inhibitors according to the manufacturer's instructions may be needed to control energy release from polymerization or other reactions in storage.

Some dry raw materials may be handled in bulk and transferred from hopper cars to silos or storage bins by a pneumatic unloading system. If the material is combustible, i.e., an organic powder, accumulations of dust from spills or leaks or flammable dust clouds inside the storage vessels may create a fire hazard. In some cases, inerting may not be feasible. If not, proper siting of the storage vessels with adequate venting for controlled explosion pressure relief is a common approach to fire hazard control in dry powder handling systems. Vessels must be designed to withstand the pressure developed by the vented explosion.

Heat Transfer

Heat transfer equipment is essential for energy transfer into, and out of, chemical process equipment. The industrial fire hazard involving heaters and coolers is related to the quantity of fuel being handled, the quantity and rate of energy transfer being controlled, and the temperature at which the transfer is being made.

Heat exchangers having high rates of flammable fluid throughput, whether the flammable fluid is the process stream or the heat exchange medium, have a potential for larger spills of flammables in the event of a leak than do units having lower throughput rates. High pressure on either side of the heat exchanger increases the leakage potential, and high temperature relative to flash point and/or boiling point of the material handled increases the risk of fire in the event of a leak.

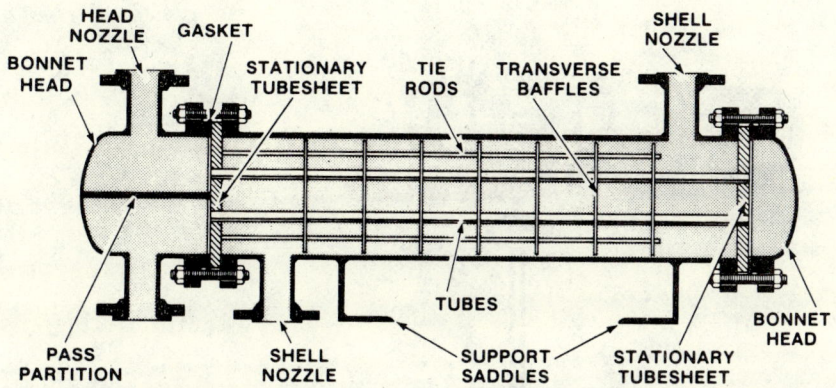
FIGURE 31.2. *The Pfaudler fixed tube sheet alloy heat exchanger. (The Pfaudler Co., A Sohio Company)*

The most common heat transfer devices seen are shell and tube units, jackets, coils, and air fin-tube units. Heat transfer from direct-fired equipment has as an added fire hazard the fuel for firing the unit and the ready ignition in the event of leakage.

Mixing

Mixing is another basic unit operation. Mixers that are installed as part of a process vessel, such as a reactor agitator, add little to the fire hazard that is inherent in the reactor and its contents. Packing glands on mixer shafts can be a source of friction, heat, and possible ignition, or a source for leakage of reactants or introduction of air into a system operating at reduced pressure. Mixer or agitator shafts and packing glands are often sources of slight leakage. If the process material leaked is flammable and easily ignited, a packing gland fire can sometimes result. This is particularly damaging if the small fire can cause additional vessel or joint failure and greater leakage of fuel to feed the fire.

Process equipment specifically built for mixing includes such items as dry blenders, including ribbon blenders or conical blenders, kneaders, rolls, mullers, and extruders. The mixing operation adds energy to the material handled, producing heat, but the amounts of heat and temperature rise are not normally significant. Some materials being mixed are reactive, and internal energy generation is a hazard if uncontrolled reaction is initiated.

Fire hazards arise out of the nature of the materials beings blended. Dry powders are difficult to contain in the system, and leakage of combustible dust may cause a dangerous accumulation of dry powder fuel in the operating area. Rigid control of dust and meticulous housekeeping are essential to controlling the fire hazard of exposed fuels. Contamination of material being handled or transferred from external sources, such

Chapter 31

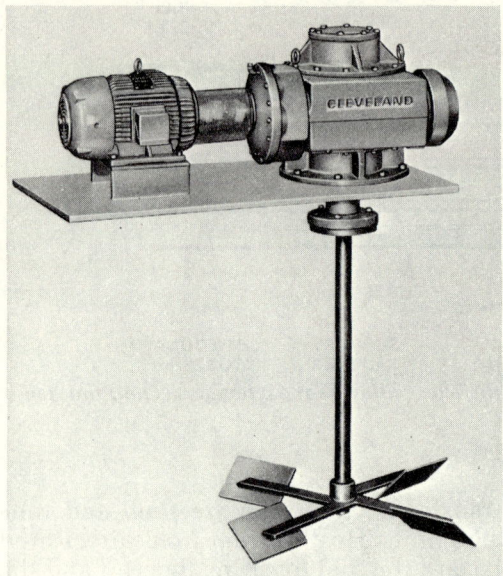

FIGURE 31.3. Paddle mixer and drive unit. (Cleveland Mixer Corp.)

as lubricants from bearings, moisture and dirt, can cause major problems. This is most important where oxidizing chemicals are present.

In most dry blending processes, it is difficult to exclude air from the equipment, especially during charging and discharging operations; therefore, dust explosions are an ever-present risk. Ignition can be caused by static, but is probably more often caused by heat generated in metal parts by bearing failure or metal-to-metal friction. Some mixing operations involve adding flammable fluids to the dry powders, and flammable vapors in the system enhance the ease of ignition.

Crushing and Grinding

Fire hazards in crushing and grinding equipment are related to the flammability of the materials and the fineness of the particles of fuel present. Additional information about grinding operations, equipment, hazards, and fire protection will be found in Chapter 34, which deals exclusively with that subject.

Mechanical Separation

Mechanical separation often follows crushing and grinding operations to separate materials of different sizes or remove solids from gas streams. Dry separators include cyclones, bag filters, electrostatic precipitators, screens, and filters. Wet separators, such as wet cyclones or scrubbers,

CHEMICAL PROCESSES

FIGURE 31.4. Conical dryer-blender. (The Pfaudler Co., a Sohio Company)

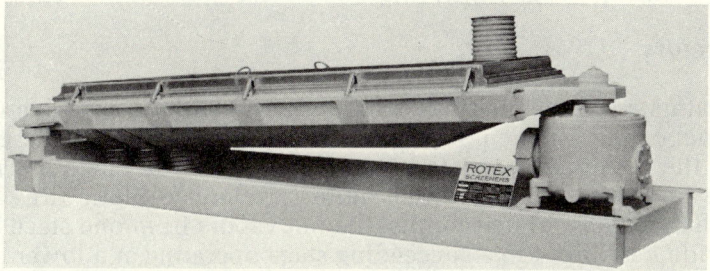

FIGURE 31.5. Shaker screen for size separation. (Rotex Inc.)

centrifuges, or settlers, may be used for mechanical separation of solids from a fluid stream.

Fire hazards are related to the combustibility of the materials handled, the potential for intermixing of fuel and oxidant inside or outside of the equipment, and the probability of ignition of any resulting flammamble mixtures. In most dry separation systems, dust control and containment are the principal problems. Bag filters, cyclones, collection hoppers, and duct construction may be of light gage metal, and an adequate explosion venting area to keep residual explosion pressures below equipment bursting strengths may be difficult to provide. Improper discharge from explosion vents of dry dust separation and handling equipment into enclosed spaces can result in a secondary explosion and very large losses.

Filtration

Filtration is another process used to separate solids from liquids. There are plate and frame, pan, drum (vacuum), inline cartridge, and leaf filters, to name a few types available. Combustible solid phase materials seldom present a major fire problem, as filter cakes are handled wet. If the liquid phase is combustible, a fire hazard exists due to the presence of fuel, possible spillage, and the residual fuel left in the wet filter cake. If the flash point is low, the liquid phase can be a serious hazard.

Distillation

Distillation is a major chemical processing step for separation, concentration, and purification of volatile liquid materials. Stills may be batch, continuous, pressure or vacuum, or steam operations. The major fire hazard is related to the flammability characteristics of the liquid and vapor being handled and the potential for fire if either is spilled into the atmosphere or, if air is introduced into flammable vapor, in the still.

Evaporators

Evaporators are somewhat like stills operated with no reflex return. Liquid is vaporized and removed from the system to change the composition of the residual materials left in the evaporator. There are multiple effect evaporators, which can be quite large and contain large inventories of flammables. In these units, the hot vapors from one stage are used to provide heat input to a succeeding stage operating at a lower temperature. Temperature differentials between stages may be obtained by concentration differences, pressure differences, or both.

Wiped film evaporators are continuous units. They are usually small and retain a small inventory of material. High flow rates of material to and from the units may be involved.

Crystallizers

Crystallizers cool a solution, removing heat of crystallization as solid crystals form in the system. The process is often preceded by a filtration step, or the separation of solid crystals from the liquid phase. Fire hazards are related directly to the flammability of the materials involved. Often, crystallization takes place from water solutions, and few fire hazards are involved. Even if the crystals formed consist of combustible or-

CHEMICAL PROCESSES

FIGURE 31.6. Swenson process evaporator. (Whiting Corp.)

FIGURE 31.7. Rototherm-E for evaporation, stripping, and heat exchange operations. (Artisan Industries, Inc.)

ganic material, little fire hazard exists until the crystals are separated, dried, and handled in other processing equipment. In a few cases, the liquid phase might be a flammable or combustible liquid, and the usual fire hazards associated with inventories of flammable liquids would exist (see Chapter 37).

FIGURE 31.8. Swenson salt crystallizers. (Whiting Corp.)

Absorption

Absorption is the chemical process by which some soluble components of a gas stream are dissolved in a liquid scrubbing agent. Absorption takes place by transfer of gas molecules through a gas phase film, a liquid phase film, and into solution in the liquid phase. Absorbers are normally packed on tray-type towers in which the gas stream is passed up through the falling liquid stream (countercurrent flow). Occasionally, gas and liquid may flow in the same direction in a concurrent flow device.

A similar but reverse process is stripping a dissolved gas from a liquid by passing the solution through a tower countercurrent to a gas stream having a lower partial pressure of the material being stripped than the equilibrium partial pressure over the contaminated liquid. Fire hazards of absorbers and strippers are directly related to the inventories and characteristics of the materials being handled. Energy changes are usually not large enough to be significant in the hazard control of the system, and the potential energy release from unwanted combustion inside or outside of the equipment is about the only industrial fire concern of such equipment.

Adsorption

Adsorption is the process by which gaseous components are attracted to and collected on the surface of a solid. Finely divided charcoal is an example of a common adsorbent. When a given volume of adsorbent has reached its capacity for collecting gas molecules, it will collect no more until the existing gas is driven off the surface of the solid — usually by heat. Hazards of adsorption are dependent upon the flammability of the

CHEMICAL PROCESSES

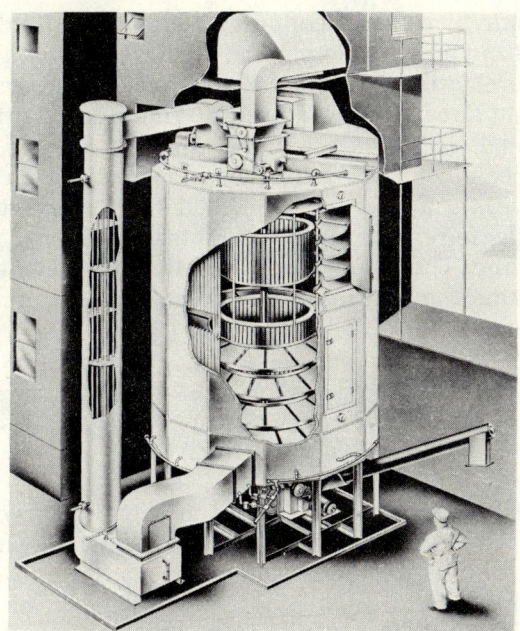

FIGURE 31.9. Wyssmont Turbo Dryer. (Wyssmont Company, Inc.)

gas being adsorbed and on whether or not adsorbent bed materials are subject to spontaneous heating.

Drying

Drying is often a final step before packaging a finished product. It involves many kinds of operations which usually remove a relatively small amount of water from a solid material. Equipment for these operations includes spray, fluid bed, vacuum, tray, belt, drum, and tunnel dryers. Heat is absorbed in the evaporation of water or other fluids from the solids and must be supplied by the input material stream or by heat transfer facilities in the dryer.

If the liquid phase being removed during the drying operation is flammable, a fire hazard can exist in the vapor space if the gas concentration enters the flammable range. Ordinarily, the liquid phase is water and the gas phase is largely air or inert gas and water vapor.

If the solid phase being dried is combustible, it poses a fire hazard that increases as dryness and/or temperature increase. If the solids are thermally unstable and temperature sensitive, as some organic plastics are, the dried powder can constitute a major fire hazard if allowed to spill and accumulate or if it becomes overheated from heat transfer facilities in the dryer. Spray dryers can contain an explosive cloud of dry combustible dust, creating a fire or explosion hazard in some systems. Knowledge of the characteristics of all the materials in the system and in the input and output streams should point to the industrial fire hazards that may be associated with the drying operation.

Some specialized operations involve passing an organic liquid, such as benzene, through a tower packed with a drying agent, such as calcium chloride, to remove small amounts of dissolved water. This, too, is a form of drying. The characteristics of the materials used in the process will determine the nature of the fire hazard and dictate proper protection measures.

PROCESS HAZARDS

Several potential hazards may be found in chemical processing facilities and in the operations performed. These include explosion and fire hazards, health hazards, and environmental hazards. The discussion here deals with explosion and fire.

Explosions

An explosion may be broadly described as a sudden uncontrolled release of energy caused by a violent expansion of gases. The process involves a chemical or physical transformation of the system into mechanical work, accompanied by a change in its potential energy.

In chemical explosions, heat causes gases to expand rapidly and, therefore, causes pressure to increase. The heat may be the product of deflagration of gases, dusts, or mists, or of chemical reaction, or of detonation of a solid, liquid, or gas. Detonation depends on a shock wave to establish and maintain the reaction. The explosion may result in the rupture of the vessel or container housing the exploding materials. High temperature, high pressure, impurities in materials, loss of cooling system capacity, poor initial mixing of reacting agents, or delay in the start of a reaction are typical causes of chemical explosions.

Physical (pressure release) explosions may result from overpressure or a defect in a container. Overpressure can be created by the generation of steam from a superheated liquid or the rapid expansion of a gas or vapor. A flaw or defect in a container may cause it to rupture at some pressure below its design burst pressure. Pressure explosions need not be limited to closed containers. An example of an unconfined physical explosion is the reaction that occurs when water is spilled into an open container of concentrated sulfuric acid. Water applied directly to this acid results in the evolution of heat and causes splattering.

Fires

A major cause of fires in chemical processes involves fuel spills from the equipment in the presence of an ignition source. Pumps account for a large number of fuel releases and should be so located that fire around them will cause minimum exposure to other equipment. Pipe line failure at valves or flanges due to vibration, erosion, or corrosion, while rare, can release very large amounts of fuel. This hazard must be controlled by installation of isolating block valves to reduce the extent of release. Such valves need to be accessible under fire conditions, or operable remotely. Small pipe for gauge connection or sampling are subject to mechanical abuse and failure. Small pipe should be reinforced by gussets to overcome this problem.

Fires can also be caused by the accidental or uncontrolled introduction of an oxidant into a process vessel containing a fuel. This can result from an operating error, instrumentation failure or from the accidental leakage of air into equipment operating at reduced pressure. Sometimes a dangerous mixture of fuel and oxidant occurs in a process involving controlled oxidation when control is lost and the proportion of oxidant to fuel is allowed to enter the flammable or explosive range. Except in cases where combustion is a desired part of the process, combustible mixtures of fuel and oxidant as well as ignition sources should be avoided.

Energy control is another important factor in chemical processes. Loss of control of the flow of energy into or out of a system or loss of control of the energy generation rate within the system can lead to overheating, overpressure, equipment failure, and spillage. Uncontrolled heat input to the process vessel can overheat the contents and lead to hazardous conditions. Loss of heat removal or cooling can quickly cause overpressure in the system if heat input is not cut off. Loss of control over exothermic chemical reactions can lead to a fire if the uncontrolled energy release causes overpressure and rupture of the vessel and release of the hot fuel into the atmosphere. Controlled oxidation is a special type of exothermic reaction in which loss of control of the fuel concentration, air concentration, or the reaction rate can lead directly to fire and explosion.

Control of exothermic reaction may be lost as a result of too much catalyst, loss of inhibitor, too much heat input, loss of heat output, loss of agitation, or accumulation of too much reactant.

Large amounts of energy are released, often in an uncontrolled manner, if thermally unstable or shock- or impact-sensitive materials are subjected to conditions causing them to decompose or degrade into compounds or elements having less energy content. Many chemical products or intermediates are known to undergo self-sustaining exothermic decomposition if heated to a certain temperature, but are considered to be safe if held under that temperature. In many cases serious losses have occurred after holding materials at "safe temperatures" for extended periods of time. The causes were either minor contamination which increased the rate of decomposition at normal temperatures or the subsequent discovery that slow exothermic decomposition occurred at "safe" temperatures which, with sufficient time, led to runaway reaction.

Hazard Analysis

All chemical processes are given some kind of hazard analysis before plants are built and whenever significant changes are made. Some analysis is quite informal while others are highly structured with complete documentation and attempts at quantification of risk inherent in the operation.

The first and most important consideration is the identification of potential hazards. The potential hazard of the raw materials, intermediates and final products must be completely understood. The effect of temperature and pressure outside of normal ranges, as well as contamination, must be studied. The possibility of creating a hazardous by-product and gradually increasing its concentration in the closed system must be considered. In other words, there must be complete hazard identification for all aspects of the process.

With this information the severity of the process hazards can be determined. Then the means to control the hazards can be identified.

Numerical quantification of the risk requires estimation of failure rates of equipment and people. A judgment must be made of the degree of maintenance of all important equipment and instruments, as well as the degree of training and judgment of operators. All this may be helpful but the most important aspect of hazard analysis is complete comprehensive hazard recognition.

FIRE PROTECTION

Fire Risk Control

Fire risk control demands that chemical processing plants and equipment be well designed and well maintained to minimize the potential for the occurrence of leaks and spills or loss of energy control. Redundant instrumentation, emergency controls, and fail-safe design are some of the tools of fire risk control. Venting and pressure relief systems are essential for the preservation of process equipment. Damage-limiting construction should be used where processes are housed in a structure. Equipment and piping should be well supported to prevent vibrations from causing leaks.

The principal hazard in chemical processes arises from the exposure of a fuel to an oxidant and an ignition source. Ideally, flammable or combustible process materials should be under control at all times. As a practical matter, however, one must recognize that leaks and spills will occur, and steps must be taken to minimize their effects. Equally important is the necessity for preventing an unwanted oxidant from entering process equipment that operates at low pressure.

Many modern chemical processing plants use space to minimize potential losses from fire and explosion. Individual processes and small groups of related processes are conducted outdoors on separate plots of ground. The amount of space between installations is determined by the nature of the processes and the materials involved. Access from all sides is available to each installation to facilitate fire fighting. Piping systems should be looped and valved so that defective sections or processes can be isolated from the rest of the plant. Drainage ditches are useful for conducting spilled flammable liquids away from the process equipment. Dikes are often used to confine spilled flammables, but they may contribute to the loss by keeping the burning material near the source of the spill, which may create further damage unless automatic fire control equipment is also used. Structural steel should be fireproofed to prevent early collapse of elevated units, because bare steel exposed to fire will fail in a few minutes unless protected.

All sources of ignition should be excluded from the process area. No open flames should be permitted near the process equipment. Pipes carrying heating or heat transfer media may need to be insulated to prevent them from becoming sources of ignition. Electrical components and equipment used near chemical processes should be appropriate for the hazard present.

Suppression Equipment

In those portions of the plant where only ordinary hazards exist — in offices, for example — automatic sprinkler systems could be used to provide the necessary protection. Portable fire extinguishers may be adequate in areas where only small amounts of hazardous materials are handled. Special extinguishing systems using carbon dioxide, dry chemical, foam, or halogenated agents may be installed to protect small open tanks of flammable liquids.

For protecting outdoor processes and exposures, automatic sprinkler or water spray systems designed to match the hazard may be used. In addition to fixed systems, a chemical processing plant may use a system of yard hydrants and monitor nozzles to protect exposures. The water supply should be sufficient to meet the demands of automatic extinguishing equipment and hose streams for a given anticipated fire duration.

Where a serious explosion hazard exists, it may be advisable to provide barriers that will protect extinguishing system valves and components from flying debris.

The nature of the materials used in a particular plant or process may make it necessary to have other special equipment available such as protective clothing, self-contained breathing apparatus, and acid resistant hose.

In some situations where the probability of an explosion occurring is high, automatic explosion detection and suppression equipment may be used to advantage. Such equipment is particularly suited to protecting vessels of limited volume that contain certain flammable vapors, gases, and dusts in flammable concentrations.

BIBLIOGRAPHY

REFERENCES CITED

Badger, W. L., and McCabe, W. L., *Elements of Chemical Engineering*, McGraw-Hill, New York, NY, 1931.

Bartknecht, W., "Explosion Pressure Relief," *Chemical Engineering Progress*, vol. 73, no. 9, September 1977, p. 97.

Boyle, W. J., Jr., "Sizing Relief Area for Polymerization Reactors," *Loss Prevention*, vol. 1, American Institute of Chemical Engineers, New York, NY, 1967, pp. 78-84.

Donat, C., "Pressure Relief as Used in Explosion Protection," *Loss Prevention*, vol. 11, American Institute of Chemical Engineers, NY, 1977, pp. 87-92.

Doyle, W. H., "Instrument Connected Losses in the CPI," *Instrumentation Technology*, October 1972, pp. 38-42.

Fire and Explosion Index Hazard Classification Guide, Fifth Edition, Dow Chemical Company, Midland, MI, 1980.

Freeman, Raymond A., "Problems with Risk Analysis in the Chemical Industry," *Plant/Operations Progress*, vol 2, no. 3, American Institute of Chemical Engineers, New York, NY, 1983, pp. 185-190.

Harmon, G. W., and Martin, A. W., "Sizing Rupture Discs for Vessels Containing Monomers," *Loss Prevention*, vol. 4, American Institute of Chemical Engineers, New York, NY, 1970, pp. 95-102.

Huff, J. E., "Computer Simulation of Polymerizer Pressure Relief," *Loss Prevention*, vol. 7, American Institute of Chemical Engineers, New York, NY, 1973, pp. 45-57.

Huff, J. E., "Emergency Venting Requirements," *Plant/Operations Progress*, vol. 1, no. 4, American Institute of Chemical Engineers, New York, NY, 1982, pp. 211-229.

Kletz, T. A., "Practical Applications of Hazard Analysis," *Chemical Engineering Progress*, vol. 74, no. 10, October 1978, p. 52.

Kline, P. E., et al., "Guidelines for Process Scale-Up," *Chemical Engineering Progress*, vol. 70, no. 10, October 1974, pp. 67-70.

McKinnon, G. P., ed., *Fire Protection Handbook*, Fifteenth Edition, National Fire Protection Association, Inc., Quincy, MA, 1981.

Prugh, R. W., "Application of Fault Tree Analysis." *Loss Prevention*, vol. 14, American Institute of Chemical Engineers, New York, NY, 1981, pp. 1-9.

Stull, Darrell R., *Fundamentals of Fire and Explosion*, American Institute of Chemical Engineers Monograph Series, no. 10, vol. 73, 1977.

NFPA CODES, STANDARDS, AND RECOMMENDED PRACTICES

Reference to the following NFPA Codes, Standards, and Recommended Practices will provide further information on the safeguards for chemical processing plants discussed in this chapter. (See the latest *NFPA Codes and Standards Catalog* for availability of current editions of the following documents.)

NFPA 10, *Portable Fire Extinguishers.*
NFPA 11, *Foam Extinguishing and Combined Agent Systems.*
NFPA 12, *Carbon Dioxide Extinguishing Systems.*
NFPA 12A, *Halon 1301 Fire Extinguishing Systems.*
NFPA 12B, *Halon 1211 Fire Extinguishing Systems.*
NFPA 13, *Installation of Sprinkler Systems.*
NFPA 15, *Water Spray Fixed Systems.*
NFPA 17, *Dry Chemical Extinguishing Systems.*
NFPA 24, *Installation of Private Fire Service Mains and Their Appurtenances.*
NFPA 30, *Flammable and Combustible Liquids Code.*
NFPA 49, *Hazardous Chemicals Data.*
NFPA 68, *Explosion Venting Guide.*
NFPA 69, *Explosion Prevention Systems.*
NFPA 70, *National Electrical Code.*
NFPA 325M, *Fire Hazard Properties of Flammable Liquids, Gases and Volatile Solids.*
NFPA 491M, *Manual of Hazardous Chemical Reactions.*
NFPA 495, *Manufacture, Transportation, Storage and Use of Explosive Materials.*

ADDITIONAL READING

Atallah, S., and Allan, D. S., "Safe Separation Distances from Liquid Fuel Fires," *Fire Technology*, vol. 7, no. 1, February 1971, pp. 47-55.
Bahme, Charles W., *Fire Officer's Guide to Dangerous Chemicals*, Second Edition, National Fire Protection Association, Inc., Boston, MA, 1978.
Bartknecht, W., *Explosions*, Springer-Verlag, Berlin, Heidelberg, New York, NY, 1981.
Bernhardt, Ernest C., ed., *Processing of Thermoplastic Materials*, Krieger Publishing Co., Melbourne, FL, 1974.
Bonyun, M. E., "Protecting Pressure Vessels with Rupture Discs," *Chemical and Metallurgical Engineering*, vol. 42, May 1945, pp. 260-263.
Chappell, W. G., "Calculating a Pressure-Time Diagram for an Explosion Vented

Space," *Loss Prevention*, vol. 11, American Institute of Chemical Engineers, New York, NY, 1977.

Chemical Engineering Magazine, *Safe and Efficient Plant Operation and Maintenance*, McGraw-Hill, New York, NY, 1980.

Cocks, R. E. and J. E. Rogerson, "Organizing a Process Safety Program," *Chemical Engineering*, vol. 85, no. 23(1978), pp. 138-146.

Coffee, R. D., "Dust Explosions: An Approach to Protection Design," *Fire Technology*, vol. 4, no. 2, May 1968, pp. 81-87.

Coffee, R. D., "Evaluation of Chemical Stability," *Fire Technology*, vol. 7, no. 1, February 1971, pp. 37-45.

Coffee, R. D., "Hazard Evaluation Testing," *Loss Prevention*, vol. 3, American Institute of Chemical Engineers, New York, NY, 1969, pp. 18-21.

Coffee, R. D., "Hazard Evaluation: The Basis for Chemical Plant Design," *Loss Prevention*, vol. 7, American Institute of Chemical Engineers, New York, NY, 1973, pp. 58-60.

Cousins, E. W., and Cotton, P. E., "The Protection of Closed Vessels Against Internal Explosions," Paper No. 51-PRI-2, American Society of Mechanical Engineers, New York, NY, 1951.

Creech, M. D., "Combustion Explosions in Pressure Vessels Protected with Rupture Discs," *Transactions*, vol. 63, no. 7, American Society of Mechanical Engineers, New York, NY.

Cubbage, P. A., and Marshall, M. R., "Explosion Relief Protection for Industrial Plants of Intermediate Strength," Institution of Chemical Engineers Symposium Series 39, April 1974.

Davenport, J. A., "A Survey of Vapor Cloud Incidents," *Chemical Engineering Progress*, vol. 73, no. 9, September, 1977.

―――, "Explosion Losses in Industry," *Fire Journal*, vol. 75, no. 1 (January 1981), pp. 52-56, 71-73.

Doyle, W. H., "Protection in Depth for Increased Chemical Hazards," *Fire Journal*, vol. 59, no. 5, September 1965, pp. 5-7.

"Explosion and Fire Hazards in the Storage and Handling of Organic Peroxides in Plastic Fabricating Plants," SPI-FPC 19, The Society of the Plastics Industry, New York, NY, June 1964.

Fawcett, H. H., and Wood, W. S., eds., *Safety and Accident Prevention in Chemical Operations*, 2nd ed., Wiley - Interscience, New York, NY, 1982.

Gibson, A. E., *Processing of Polymer Composite Materials*, Pergamon Press, New York, NY.

Grace, C., "Fluid Choice Takes the Steam Out of Unsafe Process Heaters," *Process Engineering*, (5), 1977, pp. 85, 87, 88.

Halpaap, W., "Special Appliance for the Chemical Industry," *Fire International*, 5(56), 1977, pp. 44-50.

Henry, Martin F., ed., *Flammable and Combustible Liquids Code Handbook*, National Fire Protection Association, Quincy, MA, 1981.

Howard, W. B., "Efficient Time Use to Achieve Safety of Processes," paper presented at EFCE 4th International Symposium on Loss Prevention, Harrogate, England, Sept. 1983.

Joschek, H. I., "Risk Assessment in the Chemical Industries," *Plant/Operations Progress*, vol. 2, no. 1, January, 1983, pp. 1-5.

King, R., "Plant Hazards," *Engineering* (London), vol. 216, no. 4 (1976), pp. 277-279.

Kirk, R.E., and Othmer, D. F., eds., *Encyclopedia of Chemical Technology*, Third Edition, 23 volumes, Interscience Encyclopedia, Inc., New York, 1978, 1983.

Lees, F. P. *Loss Prevention in the Process Industries*, 2 volumes, Butterworth & Co. Ltd., London, 1980.

Lewis, B., and Von Elbe, G., *Combustion, Flames, and Explosions of Gases*, Second Edition, Academic Press, New York, NY, 1961.

McElroy, Frank E., ed., *Accident Prevention Manual for Industrial Operations*, 7th ed., National Safety Council, Chicago, IL, 1980.

Nostrom, Gail P. II, "Fire/Explosion Losses in the CPI," *Chemical Engineering Progress*, vol. 78, no. 8, August, 1982, pp. 80-87.

Pajgit, O., ed., *Processing of Polyester Fibres*, Elsevier Science Publishing Co., Inc., New York, NY, 1980.

Perry, Robert H., and Chilton, C.H., eds., *Chemical Engineers' Handbook*, Fifth Edition, McGraw-Hill, New York, NY, 1974.

Pilborough, L., *Inspection of Chemical Plants*, Gulf Publishing Co., Houston, TX, 1977.

"Protection Against Ignitions Arising Out of Static, Lightning, and Stray Currents," RP-2003, American Petroleum Institute, Washington, DC, 1967.

Runes, E., "A CEP Technical Manual," *Loss Prevention*, vol. 6, American Institute of Chemical Engineers, New York, NY, 1972, pp. 63-67.

Sax, N. Irving, *Dangerous Properties of Industrial Materials*, Fifth Edition, Van Nostrand Reinhold Company, New York, NY, 1979.

Sestak, E. J., "Venting of Chemical Plant Equipment," *Engineering Bulletin N-53*, Factory Insurance Association, Hartford, CT, April 1965.

Sommer, E. C., "Preventing Electrostatic Ignitions," paper presented at a meeting of the American Petroleum Institute Central Committee on Safety and Fire Protection, Tulsa, OK, April 1967.

Windholz, Martha, ed., *The Merck Index*, Ninth Edition, Merck Company, Rahway, New Jersey, 1976.

Steere, N. V., ed., *Handbook of Laboratory Safety*, Second Edition, The Chemical Rubber Company, Cleveland, OH, 1971.

Stull, D. R., "Linking Thermodynamics and Kinetics to Predict Real Chemical Hazards," *Loss Prevention*, vol.7 , American Institute of Chemical Engineers, New York, NY, 1973, pp. 67-73.

Verralin, C. H., ed., *Fire Protection Manual for Hydrocarbon Processing Plants*, Second Edition, Gulf Publishing Company, Houston, TX, 1973.

Weiby, P. and K. R. Dickinson, "Monitoring Work Areas for Explosive and Toxic Hazards," Chemical Engineering, vol. 83, no. 22(1976), pp. 139-145.

Zabetakis, Michael D., "Flammability Characteristics of Combustible Gases and Vapors," Bulletin 627, U.S. Bureau of Mines, 1965.

32

Solvent Extraction

C. Louis Kingsbaker, Jr.

The recovery of fat from oilseeds by solvent extraction has grown into a large and highly technical industry. The Japanese and Germans were the pioneers in the solvent extraction field. The Japanese were processing soybeans in batch extractors at the beginning of this century. Following World War I, the Germans developed continuous solvent extraction technology into an important business located in the Hamburg area. The United States, however, did not utilize solvent extraction until 1936 using batch hydraulic presses and mechanical screw presses, which left about 5 percent oil in the cake compared to 0.5 percent in the more efficient solvent extraction process. By 1939, several extraction plants of German design had been built in the United States. At the outbreak of World War II, there were several partially constructed plants here that had to be abandoned by the Germans. It was then that the United States began to develop its own technology in the solvent extraction field. Now the United States is considered the world leader.

The soybean is the prime reason for the growth of the solvent extraction industry in the United States. From the 1920s, the soybean crop has grown to over 2.2 billion bushels (66 million short tons or approxi-

---

C. Louis Kingsbaker, Jr. is President, C.L. Kingsbaker, Inc., Atlanta, GA.

mately 60 million metric tons) in 1982. The processing of this crop is a huge undertaking. In 1950, a 200 short ton-per-day (181 metric tons) plant was considered large. Today, there are single plants processing 3,000 short tons (2722 metric tons) per day. A plant processing less than 1,500 short tons (1361 metric tons) per day is now considered uneconomical.

THE SOLVENT EXTRACTION PROCESS

Materials Processed

Many materials are processed by solvent extraction. Oilseeds are by far the largest in both number of varieties and tonnage. However, solvent extraction is also used to process animal scraps, remove and recover waxes from lignite, remove waxes and fats from tree barks such as Douglas fir, wash impurities from high density polyethylene, obtain flavors and fragrances from flowers, and separate drugs and medicines from plants. It also has many other unusual purposes.

The solvent extraction processing capacity is greater for soybeans than for all other oilseeds combined. Soybeans contain an oil content of approximately 20 percent by weight and are usually extracted directly, that is, without prior processing to remove oil before the extraction step. Other oilseeds containing more than 20 percent oil by weight are normally first screw-pressed to reduce the oil content to 10 or 20 percent by weight before solvent extraction. Facilities employing this step are called prepress solvent extraction plants. Some of the major oilseeds being processed by this procedure are cottonseed, rapeseed, flaxseed, corngerm, sunflower, safflower, peanuts, and copra. There are a few plants that process cottonseed directly without prepressing. While most plants process only one oilseed throughout the entire year, some plants process many different seeds on a scheduled basis; these are called switch plants.

End products for solvent extraction plants processing oilseeds are meal rich in protein used in animal feed, and vegetable oil, which, after refining and further processing, is sold as margarine, cooking oil, shortening, and salad oil. The oil also has industrial uses for the paint and lubrication industries.

An important by-product made in solvent extraction plants that process soybeans is a gum called lecithin used in the food, confectionery, and drug industries. Lecithin is made by adding water to filtered soybean oil from the extraction plant (to precipitate lecithin) which is separated in a centrifuge and then dried in a special evaporator. Six commercial grades of lecithin are sold.

Recently, many soybean solvent extraction plants began making special flakes that are used in the production of protein foods for human consumption. This is accomplished by special desolventizing equipment designed to control the nitrogen solubility index of the flakes. Some of these products are texturized vegetable protein, soybean concentrates, and soybean isolates.

Provision is made at the solvent extraction plant for storage of the raw material seed, solvent and the oil, and flake and lecithin products. These storage facilities must be designed for product purity and safety.

Solvents Used

The prime purpose of solvent extraction is to obtain the lowest residual oil content in the extracted material. The solvent used exclusively today in the extraction of oilseeds is hexane. In the earlier technology, three solvents were considered — hexane, heptane, and trichlorethylene. It is interesting to note that trichlorethylene removes the most extractables, followed by heptane, hexane, and petroleum ether in that order. There is a total difference of about 1 percent in extractables among these solvents. Hexane was selected through the process of elimination, even though it is highly flammable. Trichlorethylene, which is not flammable, was used earlier in this country. It was found to be toxic, and some animal deaths were attributed to incompletely desolventized meal that had been extracted by trichlorethylene. In addition, trichlorethylene is considerably more expensive, has a higher boiling point, and is more difficult to desolventize than hexane. Its density is about twice that of hexane, requiring stronger building structures to support the equipment. Since trichlorethylene is a chlorinated hydrocarbon, its use is now being restricted by the United States government due to possible health hazards to humans. Heptane never became widely used as a commercial solvent despite its relatively good solvency. Hexane has many advantages, including the fact that it is relatively easy to desolventize from the meal and oil products, making them nontoxic. It is essentially immiscible with water and it can be produced from petroleum at a relatively low cost. Its major disadvantage, the danger of fire and/or explosion, is overcome by careful design, construction, and operation of the solvent extraction plant. Despite all these precautions, some accidents still occur, usually due to carelessness and failure to follow good operating and maintenance procedures.

Other flammable solvents, such as mixtures of aromatic hydrocarbons, are used for extracting lignite, Douglas fir bark, and other special raw materials whose end products are not used by humans for internal consumption.

Description of the Process

The process for solvent extraction and solvent recovery described in this text is for soybeans. Unit operations for extraction of all oilseeds are similar, and normally the same equipment in the extraction area can handle the flakes or cakes from these materials. The preparation steps for dif-

Chapter 32

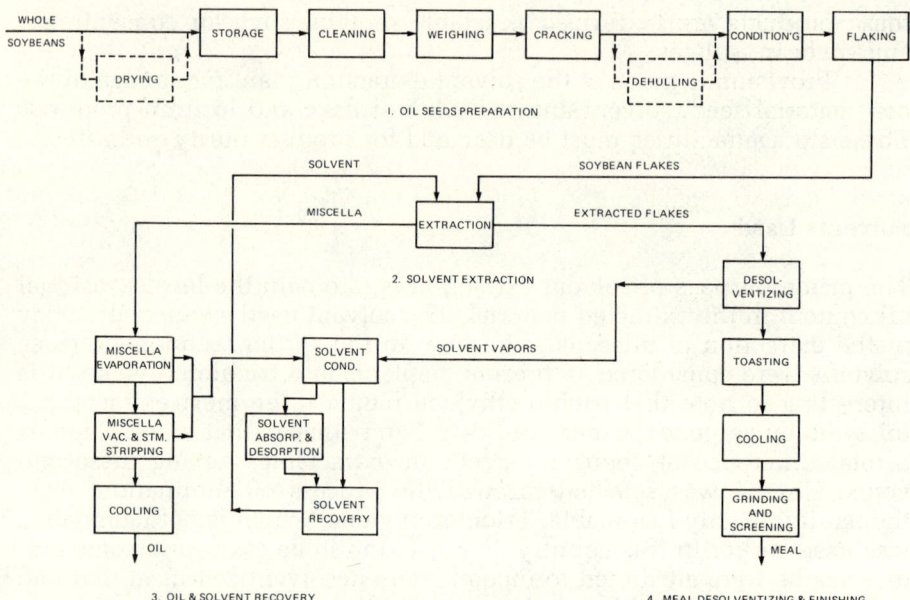

FIGURE 32.1. *Flow diagram of the soybean solvent extraction process.*

ferent seeds will vary, and continuous screw presses are used in the processing of oilseeds with an oil content exceeding 20 percent.

A block flow diagram of a typical soybean solvent extraction process, shown in Figure 32.1, outlines the basic unit operations.

Whole soybeans are brought into the processing plant by truck or railcar, checked for moisture content, and dried in a flue gas dryer to about 10 or 11 percent moisture content by weight for safe storage. The beans are cleaned to remove trash, dirt, and foreign material both before they are dried to eliminate fire hazards in the dryer, and also in the preparation step of the extraction process.

The solvent extraction process is a continuous operation and begins at a weigh scale. Normally, the processing rate is controlled by adjusting the cracking mill feed rolls and measuring the capacity at the weigh scale. The cracking mill breaks the bean into six to eight pieces with two pairs of grooved rolls and splits the hull covering the bean. (Two animal meal products are made in the extraction plant, one containing 44 percent protein and the other 49 percent protein. Forty-nine percent protein meal is the result of a dehulling step in which screens and aspiration are used to remove the bean cover from the bean pieces.)

Cracked beans enter a conditioner — a rotary steam tube unit — which heats the beans to between 140 and 165°F (60 and 73.8°C) and adjusts the moisture content to between 10 and 10.5 percent by weight, using controlled aspiration in the unit. The heating of the beans makes them thermoplastic, which facilitates the flaking operation. Flaking is accomplished by a smooth, single pair-type rolling mill. The normal flake thickness is 0.010 in. to 0.014 in. (0.254 to 0.356 mm). In this step, the oil cells are ruptured to aid extraction. Flake moisture of 10 to 10.5 percent is important in solvent extraction. Excess moisture retards extraction, and insufficient moisture causes flake breakage and the production of too many fine particles, which also results in poor extraction. It is essential to have proper flake production to obtain efficient extraction of the material as well as efficient operation of the entire extraction plant.

Hot flakes from the preparation building are continuously delivered to a sealed conveyor in the extraction building and into the extractor. Almost all commercial extractors in operation throughout the world are of the continuous percolation type. This type utilizes containers to hold the flakes. The liquid solvent used to remove the oil from the flakes enters above the flake bed and drains or percolates through the bed to a compartment below the flakes to be recycled by a pump to another stage of the extractor. Most percolation extractors have six or more extraction stages. The other type of extractor, called total immersion, mixes the solvent and flakes together during the entire extraction phase and separates liquid from solid by vacuum or centrifuge. The volume of oilseeds processed in the immersion extractor is very small and will not be discussed here.

There are many types of percolation extractors, some of which are the vertical basket, horizontal basket, rotary, perforated belt, and rectangular loop extractors. All utilize similar principles and are basically counter current, i.e., the flow of solvent in the extractor is opposite to the flow of flakes. Flake bed heights in these extractors range from 3 ft to 10 ft (0.91 to 3 m). Fresh, recovered hexane solvent, heated to about 140°F (60°C), contacts soybean flakes near the discharge end of the extractor to remove the remaining oil from the flakes. This solvent, which now contains some oil, is pumped from stage to stage in a direction countercurrent to the flow of flakes, becoming richer in oil with each successive stage. The oil extracted from the flakes and mixed with hexane normally has a concentration of 25 percent when it exits the extractor. The balance — 75 percent — is hexane. A mixture of oil and hexane is called miscella; thus, this liquid leaving the extractor is called 25 percent miscella. The amount of hexane solvent normally added to the extractor is one weight part of solvent to one weight part of flake entering the extractor. This is defined as solvent ratio and ranges from 0.85 to 1.1. The extractor normally operates at 140°F (60°C) and at either atmospheric pressure or a slight vacuum of ½-in. (12.7 mm) water gage. Variable speed drive is provided to control the extractor speed for changes of plant rate and also bed height.

After the flakes move beyond the solvent addition point, time is provided in extractors for hexane to drain from the flakes. Extracted and drained flakes are discharged from the extractor into a hopper and are conveyed to a single unit called a desolventizer-toaster (D-T) in which the solvent is stripped from the flakes with open or sparge steam. Toasting is accomplished by contact with a series of trays heated indirectly by

Chapter 32

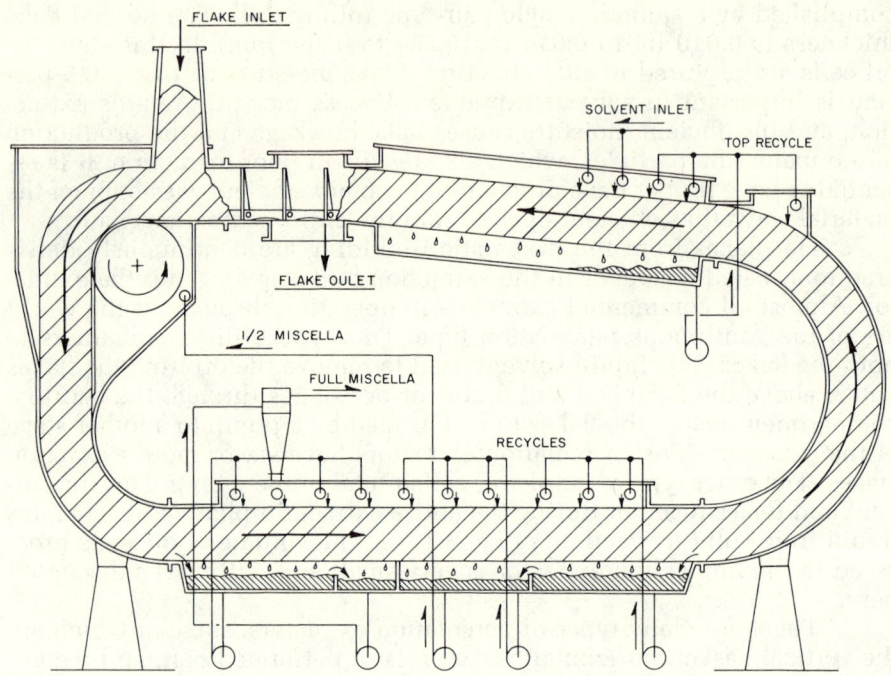

FIGURE 32.2. Rectangular loop extractor.

high pressure steam. The D-T is a vertical, cylindrical unit normally consisting of at least seven trays or decks. Flakes are kept in motion on the trays by rotating sweep arms, which are connected to a shaft located in the center of the unit. The flakes drop from tray to tray through openings in each tray. Solvent is removed by open steam in the top three trays of the unit, and the steam condenses in the flakes, raising the flake moisture content. Toasting or cooking in the bottom trays reduces the moisture content of the flakes, cooks the flakes to the desired color, and destroys enzymes in the soybean flakes that are injurious to animals who eat the meal made from the flakes. Usually about 15 percent excess sparge steam is used in the desolventizing step. The top of the D-T unit has a normal operating temperature of about 167°F (75°C). The flake outlet temperature from the bottom of this unit must be at least 221°F. Lower flake outlet temperatures indicate that not all solvent has been removed, which could present a potential safety hazard. The operating pressure at the top of the D-T unit should be lower than the pressure in the extractor. Therefore, there should be a vacuum of 1-in. (25.4 mm) water gage at the top of the D-T unit if the extractor operates at a vacuum of ½-in. (12.7 mm) water gage. This is to prevent steam vapor from flowing back to the extractor and also to stop the flow of solvent vapor downward through the D-T unit. Solvent and excess sparge steam vapors leaving the top of the D-T unit are scrubbed with liquid to remove fines and are condensed, and

SOLVENT EXTRACTION

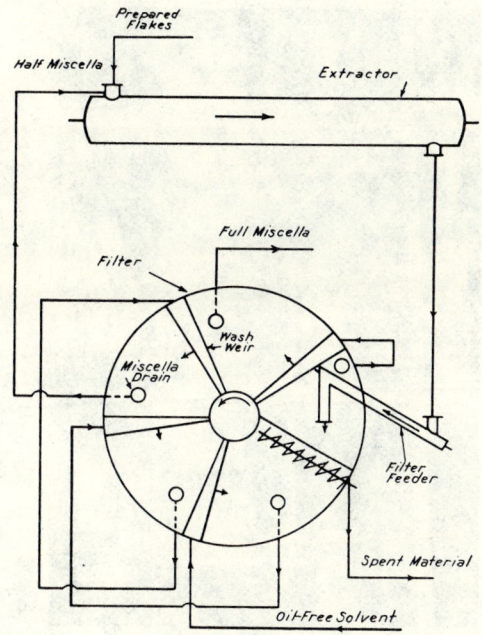

FIGURE 32.3. Filtration-type extractor.

the solvent is reused in the extraction step. There are several other types of desolventizers in use, such as the recycled vapor type, but these are not significant in number.

Extracted, desolventized, and toasted soybean flakes containing about 17 percent water by weight are conveyed to the meal dryer, which is a rotary steam tube unit, and the moisture content is reduced to 13 or 14 percent by the use of indirect steam and air aspiration. The meal must be cooled to between 115 and 120°F (46 and 49°C) and dried to 12 percent moisture or less for storage. This is done in another rotary unit by passing air through the flakes as they tumble inside the unit. Cooled and dried flakes are conveyed back to the meal finishing system, usually located in the preparation building, where they are sized for commercial sale by mechanical grinding and screening.

Miscella at a content of 25 percent oil by weight is concentrated to 92 percent in two long-tube, rising-film evaporators. The first stage evaporator operates at a vacuum of about 15 in. (381 mm) of mercury on the tube side and uses the vapors from the desolventizer-toaster to provide the heat. This method is used to conserve energy. Miscella from the first stage evaporator has an oil concentration of about 50 to 60 percent. The second stage evaporator operates at atmospheric pressure, using low pressure steam in the shell side. Final removal of solvent from the 92 percent miscella out of the second stage evaporator is done in a stripping column operating under vacuum ranging from 22 to 27 in. (559 to 686 mm) of mercury at a temperature of 210 to 240°F (99 to 116°C) and using open or sparge steam at the bottom of the column. The vacuum dries or

FIGURE 32.4. Typical desolventizer-toaster (with six trays) used for processing soybeans.

removes water from the oil, and the sparge steam strips out the final traces of hexane solvent. The stripping column is constructed of multiple pairs (fifteen to twenty) of "discs and doughnuts." Soybean oil leaving the stripping column is called crude oil and is pumped to storage either hot or cold. If lecithin product is desired, the oil is filtered, cooled, and about 3 percent water by weight is added to it. The mixture must be held in an agitated tank for about 15 minutes to allow the lecithin to separate. The mixture is centrifuged, and the lecithin is collected, dried, and sent to a special storage area. The oil is again dried under vacuum before being pumped to storage.

All the solvent removed in evaporation and stripping steps is condensed and returned to the solvent separator along with solvent from the flake-desolventizing step. The solvent separator normally is designed with a 15 min holding interval to allow water condensed in the process to separate from the solvent. The water flows from the separator to a waste water evaporator, where it is heated to at least 200°F (93°C) with open steam to remove any entrained solvent, before leaving the plant to a waste water trap and finally to a sewer. The separated solvent is returned to the extractor for reuse.

The extractor and all other tanks and condensers in the process are vented to a separate header and into a vent recovery system. This system, which uses an edible mineral oil to absorb vent gases before they exit the plant, normally consists of two columns filled with ceramic packing. One column absorbs hexane solvent vapors with a flow of cool min-

FIGURE 32.5. *Typical enclosed extraction building showing meal dryer to left and meal cooler to right.*

eral oil counter to the flow of vapor. The other column is used to strip the absorbed hexane solvent vapors from the mineral oil with open steam. A condensing system is provided to condense these vapors. The mineral oil is pumped and recirculated continuously from column to column for reuse. A blower is provided on top of the mineral oil absorber to remove the nonabsorbed water and air vapors that leave the column to the atmosphere through a flame arrestor.

Efficiency of the Process

Most designers of solvent extraction facilities make efficiency guarantees for the process. They are important not only for the profitability of the process, but also for the safety of plant operation. If these guarantees are exceeded, there is indication that there are potential safety problems, and steps must be taken to correct them. The guarantees given here are typical for the soybean processing industry and will not vary too much when other oilseeds are processed. In a process having a capacity that can be measured in U.S. tons per 24-hr day (one U.S. ton equals 0.9072 metric ton), the following conditions can be expected:

Solvent loss of 0.15 percent by weight of beans processed or 0.55 U.S. gal (2.08 liters) per U.S. ton of beans;
Residual oil content of flakes from extractor of 0.5 percent by

weight, corrected to 12 percent water by weight on a solvent-free basis;

Crude oil 0.15 percent of total moisture and volatiles by weight that will not flash at 300°F (149°C) in a closed-cup flash tester; and

Soybean flakes from a meal cooler will not exceed a 0.15 pH rise in urease enzyme activity, measured by the Caskey-Knapp Method.

HAZARDS OF SOLVENT EXTRACTION

The principal hazards in solvent extraction operations are the highly flammable solvent used in the process and combustible dusts associated with the storage and handling of oilseeds. Sources of ignition include arcing electrical equipment, static electricity, and open flame. Ignition can occur during normal operations, during maintenance operations, or when an operating emergency arises. Attempts to operate a facility beyond its design capacity can introduce hazards of abnormal operation.

Normal Operations

Solvent loss is a potential hazard at any time during normal operations. A hazard of a transient nature will be present for some finite interval during startup and shutdown procedures.

Solvent loss: In order to prevent fires, it is necessary to keep solvent liquid or vapor from leaving the extraction plant and controlled areas and reaching noncontrolled areas, such as the preparation building, office building, or boilers. Hexane solvent vapor is about three times heavier than air. Therefore, in case of a hexane liquid spill, the gas vaporized from the liquid will roll along the ground and, if not dispersed by wind, could possibly enter areas where a source of ignition might be present. In order to be ignitible, the hexane solvent vapor concentration in air must be within its flammable limits — 1.2 to 6.9 percent by volume. Therefore, specific hexane concentrations in air and a source of ignition are necessary for fire to occur.

Solvent loss in an extraction plant is a key indicator of possible hazards during normal operation. A solvent loss of 0.15 percent by weight of the material processed is considered within the bounds of efficient operation. If the loss exceeds 0.30 percent, reasons for the increase must be found and corrective action taken. Many processors use this rule in their operations and either reduce plant capacity or stop the process to determine the problem. Table 32.1 shows the five sources of solvent loss from the extraction process and the percentage of loss attributed to each.

TABLE 32.1. *Sources of Solvent Loss from the Solvent Extraction Process*

Source	Loss (% by weight)
Flakes from D-T unit	0.04
Leaks	0.04
Vent gas from vent system	0.04
Oil from oil stripper	0.02
Effluent from process waste water	0.01
Total	0.15

From a safety aspect, hexane losses in the vent gas from the vent system, oil from the oil stripper, and effluent from the process waste water are not critical. However, if solvent remains in the flakes after they have been removed from the D-T unit, a serious hazard exists, as the flakes are sent to the meal finishing system in the preparation building, which is not of explosion-proof design and where ignition could occur. A gas analyzer is not normally used because it must be located in a moist, dirty environment, which would soon render it inoperable. A simple check is to use a "pop" test on a sample of flakes from the D-T unit. Flakes are held in a quart-size can for 15 min, and the can is removed from the controlled area. As the lid is slowly opened, a match at least 8 in. (0.20 m) long is inserted into the can. If there is no flash, the solvent in the flakes is less than 0.04 percent. Modern plants install a low temperature alarm at the outlet of the D-T unit. If the temperature at this point drops below 194°F (90°C), an alarm sounds and the plant is shut down to correct the problem. Leaks are serious. They consist not only of liquid drips, which normally are a minor hazard, but also vapor leaks, which could be dangerous. Such a vapor leak could be caused by pressures in the vent system forcing the hexane vapors from the extractor back into the preparation building, which could cause a fire.

A rule of primary importance is to prevent flammable solvents from leaving the extraction building and entering the preparation building. If such a transfer occurs, steps must be taken immediately to rectify the situation.

Startup, shutdown, and purging: The startup or shutdown of an extraction plant introduces a transient hazard inherent in normal operations. Before startup, the equipment, especially the extractor, is full of air. As hexane solvent is added, the concentration of solvent vapor in the atmosphere passes through the flammable range of 1.2 to 6.9 percent by volume. The method originally used in the tall vertical basket extractors was to pump hot hexane liquid into the top of the extractor until the air fog inside the unit disappeared and the temperature at the top reached 104°F

FIGURE 32.6. *Typical outdoor-type extraction plant showing pagoda-type side and roof.*

(40°C). The extractor drive could then be started safely, since air had been displaced and the vapor content of hexane was above 6.9 percent. In one instance, a fire occurred in a small, rotary type extractor during startup. While the actual reasons were not revealed, it was reported that static electricity caused by the liquid falling from the top to the bottom of the extractor was the source of ignition. To eliminate any possibility of ignition caused by static electricity when using hot liquid to purge the extractor of air, there are two methods available. One is to design the plant so that hot vapors from the second stage evaporator can flow into the extractor to rapidly heat it to 104°F and (40°C), thus to replace the air. The advantage of this procedure is to shorten extractor heatup time from the usual 4 hrs, to about ½ hr. Stage pumps are not started until the extractor is hot. The second way is to replace the extractor air with carbon dioxide or nitrogen gas before adding hot liquid hexane to the extractor.

Most plants are provided with vapor-proof slide gates located upstream and downstream of the extractor, which are closed during the plant warmup stage of the startup to prevent flow of hexane solvent vapors back to the preparation building.

A normal or controlled shutdown is accomplished by emptying the extraction plant of all oilseed flakes and hexane liquid and purging the empty extractor or D-T unit with air until the unit is cooled. Carbon dioxide or nitrogen can also be used, but these gases must be replaced with air if entry into the unit is necessary. Quite often the extractor is purged with steam to vaporize residual hexane left in the unit. Steam purging is used if normal maintenance is required inside the extractor.

Occasionally, welding must be done in the extraction area. At such times, special precautions must be taken to prevent the possibility of a fire or explosion. Briefly, all equipment is emptied of solid and liquid material. The hexane from the plant is held in the solvent storage tanks, and the lines from the plant to these tanks are blanked off to isolate the tanks from the plant. All vessels in the plant are filled with water to replace hexane vapors. The extractor and D-T unit are steam purged, and extractor stage piping to the stage pumps is disconnected. Flake connections to the extractor and from the D-T unit are sealed off from the preparation building. A thorough check is made throughout the plant using a portable gas analyzer to locate possible flammable mixtures of solvent and air, especially in the place where welding is to be done. When the meter reading is negative, welding can be performed, but the surrounding atmosphere must be constantly monitored and rigid safety precautions maintained until the welding or burning is completed. Normally, it takes a minimum of 48 hrs to prepare the extraction plant for welding. (See Chapter 24, "Welding and Cutting.")

Inert gas, such as carbon dioxide or nitrogen, is a most satisfactory medium for purging an empty extractor of air before introducing a flammable solvent in normal startup procedures. Inert gas can also be used for purging during normal shutdown, but the extractor must then be air-purged to remove the inert gas. If traces of solvent vapor remain in the system, it is possible for the mixture to pass through the flammable limits as air is introduced. Steam purging is a surer method than inert gas purging for removing solvent vapors.

Abnormal Operations

The hazards become more serious and the risk of fire increases during periods of abnormal operations involving breakdowns, failures, or overloading of the system.

Emergency breakdowns: The most serious hazard in the operation of a solvent extraction plant is an emergency breakdown in the extractor when it is full of solvent-laden flakes and cannot be emptied. Quite often the problem can neither be determined nor corrected without someone having to enter the unit. Usually, the extractor is allowed to cool and drain for at least 3 hrs, and as much miscella as possible is pumped from it. During this period, a plan should be developed for identifying and correcting the problem. Any plan agreed upon should embody all necessary safety precautions.

Before an extractor containing solvent-laden flakes is entered for inspection and maintenance, an air purge is normally used. Steam purging is not practical, as it would take weeks to accomplish in deep beds of flakes saturated with hexane. Even after such a long time, one could not be sure that the flakes were free of solvent. Inert gas purging would not ensure that the working area would be free of a flammable mixture.

Anyone entering an extractor should wear an air mask, and personnel should be assigned to watch and help the worker. Support personnel should also guard against ignition sources, because it is possible to have solvent vapor and air mixtures in the flammable range.

If a D-T unit fails under load, the danger is that the flakes in the cooking trays may overheat and cause a fire. It is important to quickly turn off the heating steam in the cooking trays when such a failure occurs. Normally, sparge or open steam can be added to the D-T unit for several hours to desolventize the flakes. The doors on each tray are then opened one by one and the contents carefully raked out until the unit is empty and the problem corrected.

Failures: Emergency shutdowns can be caused by a failure of cooling water, steam, or electricity. If proper procedures are not followed, there is the possibility of a fire hazard. Solvent extraction plants are designed to provide audible or visual alarms in case of cooling water or steam failure.

Provision should exist for an emergency supply of cooling water in sufficient quantity to operate the condensers to assure safe shutdowns of the plant in the event of cooling water failure. This flow of water normally lasts at least 15 min. When water fails, the process must be stopped immediately, the flow of steam to the extraction plant shut off by closing the main steam valve, and the vaportight slide gates provided at the input and output of the extractor closed to prevent solvent vapors from flowing back to the preparation building.

A steam failure requires that the plant be stopped immediately, the main steam valve and the valves in all sparge steam lines closed, and the vaportight slide gates provided at the input and output of the extractor placed in the shutoff position.

In the case of an electrical failure, the equipment in the plant will be stopped by the lack of power, but it is necessary to shut off the main steam valve in the extraction plant and all sparge steam line valves and also to close the vaportight slide gates provided at the input and output of the extractor.

System overloads: Often a solvent extraction plant's capacity is increased by 25 percent or more above its design rating. This situation can develop into a hazard of abnormal operation if extraction facilities are not expanded. While the conveying capacity of the original plant can physically move the higher volume of flakes, normally there is not enough solvent-condensing capacity to recover the solvent. This results in higher vent pressures with the inherent danger that solvent vapors may be forced back into the preparation building or out of the D-T unit with the

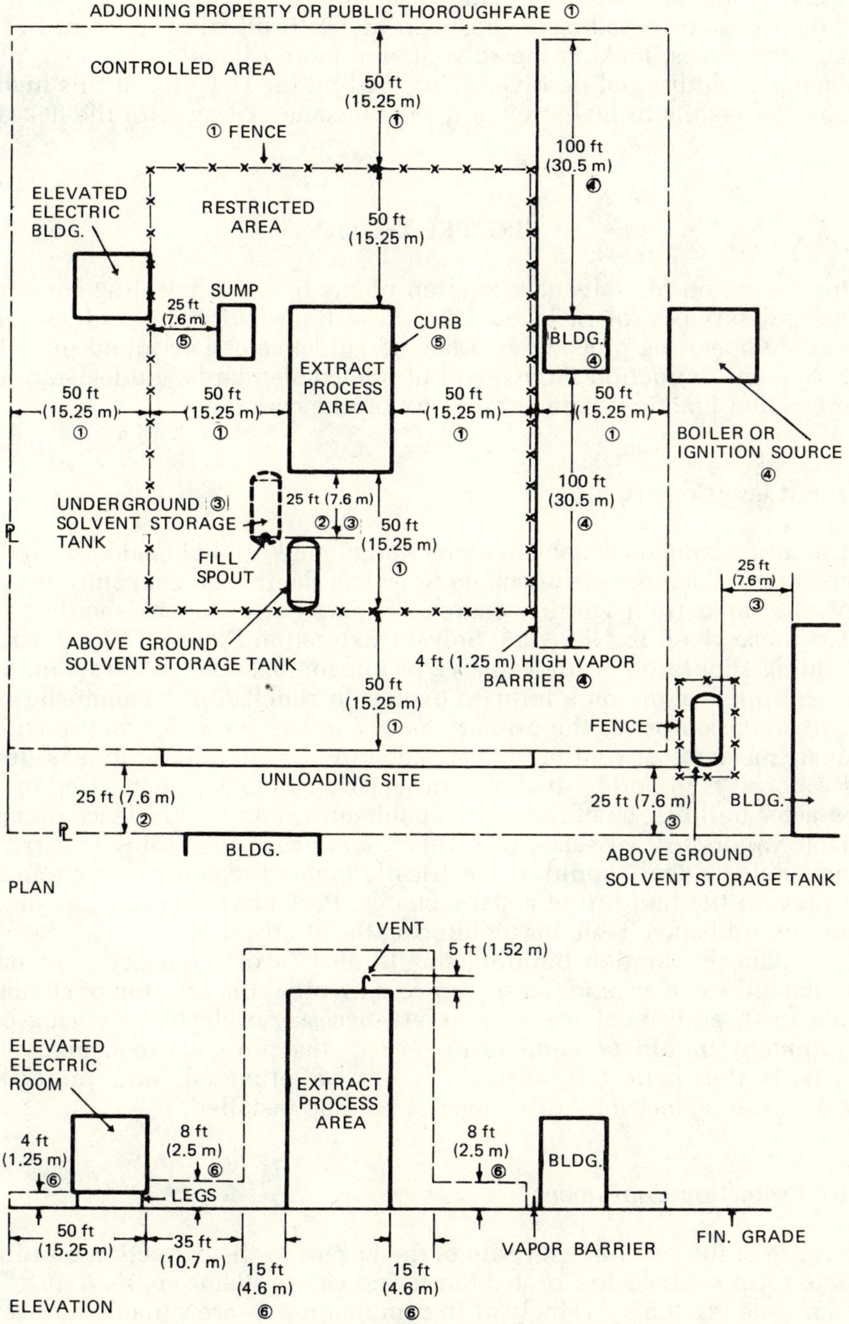

SOLVENT EXTRACTION

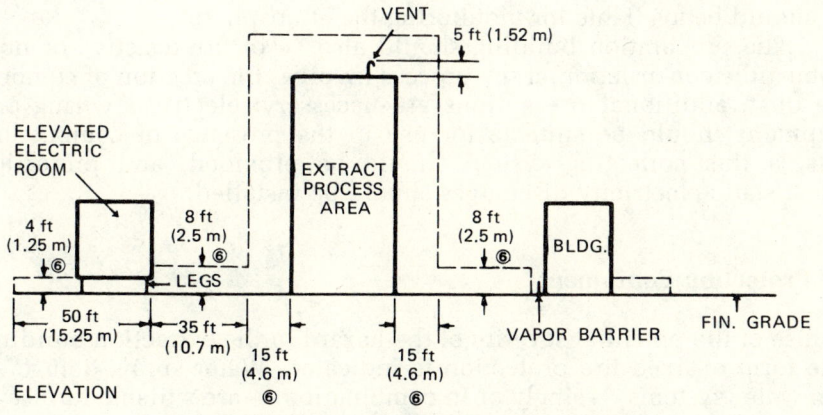

FIGURE 32.7. *A typical distance diagram*

687

flakes. At higher-than-rated plant capacities, the retention time available in the extractor is reduced, causing a higher residual oil content in the extracted flakes, making the solventizing more difficult. This fact, plus placing an additional desolventizing load on the D-T unit at this higher capacity, results in higher quantities of hexane exiting with the flakes.

FIRE PROTECTION

Fire protection for solvent extraction plants involves building construction and services, plant layout, fire protection equipment and systems, and safe operating procedures. Detailed guidance can be found in NFPA 36, *Solvent Extraction Plants*, and in related standards, guides, and recommended practices listed at the end of this chapter.

Layout and Construction

Structural elements of solvent extraction plants should be laid out so that separation distances are adequate to isolate flammable concentrations of solvent vapor from ignition sources. Separation distances should be at least those given in NFPA 36, *Solvent Extraction Plants*. The extraction building should be of fire-resistive or noncombustible construction and be equipped with some form of explosion relief. Since flammable solvent vapor comprises the primary hazard in the extraction building, adequate mechanical ventilation is needed to prevent a hazardous buildup of vapor. It is important that electrical wiring and equipment used in the extractor building be of the type suitable for use in the presence of flammable vapors. All vessels, pipes, tubes, and hoses containing or carrying flammable solvent should be electrically bonded together and grounded to prevent the buildup of a static charge. Portable flammable gas detectors should be available for monitoring the atmosphere.

The preparation building should also be of fire-resistive or noncombustible construction. If the process involves the creation of combustible dust, additional precautions are necessary: electrical wiring and equipment should be suitable for use in the presence of combustible dusts, a dust-collecting system should be provided, and protection against static electricity discharges should be installed.

Fire Protection Equipment

Because of the potential severity of the hazard in the extraction building, some form of fixed fire protection is indicated. Water spray, deluge, or foam water systems — singly or in combination — are suitable for use in the extractor building. Automatic sprinkler systems are appropriate for use in the preparation building.

The solvent unloading and storage area may be adequately protected by portable fire extinguishers of the proper types and sizes if the area is isolated from exposure hazards. Otherwise, additional protection such as a water spray system may be needed.

Portable fire extinguishers should be strategically located throughout the plant. The facility may also have standpipe and hose systems equipped with combination nozzles. A system of yard hydrants is an essential part of the protection for solvent extraction plants. Finally, the available water supply must be adequate to meet the needs of the fixed protection and yard hydrant systems.

Policies and Procedures

An enlightened management will recognize that a fire loss will be detrimental if not fatal to company operations. Accordingly, management will establish policies and procedures for normal operations, maintenance operations, and emergency situations that are designed to prevent or at least minimize fire losses.

It is imperative that solvent extraction equipment be operated in a manner prescribed by the manufacturer and that no attempt be made to bypass or defeat safety devices and systems. Normal operating procedures should be strictly regulated and well understood to minimize the possibility of employee error. Company policy should limit the output demanded of the extraction plant so that its design capacity is not overtaxed. Policy should also provide for shutdown in the event of excessive solvent loss and for regular maintenance and cleaning operations. A safety system audit program should be instituted for checking and testing of all plant safety equipment, safety instruments and safety alarms on a scheduled basis.

Most operators of solvent extraction plants schedule regular shutdowns of about one-half to one day's duration every month or two for maintenance and to clean condenser tubes that may be fouled by dirty cooling water or flake fines. The need for this cleaning becomes apparent when operating pressures begin to build up in the vent system. For example, if it is not possible to maintain a slight vacuum in the extractor and the D-T unit as described earlier in this chapter, the solvent-condensing capacity of the plant is reduced and the process should be shut down so that the condensers can be cleaned. It is vital that this be done periodically to prevent hexane vapors from being forced back from the extractor into the preparation building or from being forced out of the D-T unit with the flakes. If solvent loss increases from 0.15 to 0.30 percent or 1 gal per ton (3.78 liters/907.2 kg) of material processed, an abnormal hazard exists and action must be taken.

Fire prevention is a key factor in any fire protection program. Indicators of the importance attached to fire prevention in solvent extraction plants include the conspicuous posting of safety rules and emergency procedures throughout the plant, a safety education program for new employees, the appointment of a plant emergency organization, and a compulsory monthly safety program. Management's active participation in the program will demonstrate its genuine interest in maintaining a firesafe working environment.

If a fire should occur, plant operators need to know what they can do to minimize the loss. Appropriate action would be to activate fire extinguishing systems if they are not automatically controlled, activate the plant interlock system to shut down all solids-conveying equipment in the plant, turn off the main steam valve to the extraction plant, leave the area, and notify the fire department.

BIBLIOGRAPHY

NFPA CODES, STANDARDS, AND RECOMMENDED PRACTICES

Reference to the following NFPA Codes, Standards, and Recommended Practices will provide further information on the safeguards for solvent extraction operations discussed in this chapter.(See the latest *NFPA Codes and Standards Catalog* for availability of current editions of the following documents.)

NFPA 10, *Portable Fire Extinguishers.*
NFPA 13, *Installation of Sprinkler Systems.*
NFPA 14, *Standpipe and Hose Systems.*
NFPA 15, *Water Spray Fixed Systems.*
NFPA 16, *Deluge Foam-Water Sprinkler and Spray Systems.*
NFPA 24, *Installation of Private Fire Service Mains and Their Appurtenances.*
NFPA 36, *Solvent Extraction Plants.*
NFPA 61B, *Prevention of Fires and Explosions in Grain Elevators and Facilities Handling Bulk Raw Agricultural Commodities.*
NFPA 61C, *Prevention of Fire and Dust Explosions in Feed Mills.*
NFPA 68, *Explosion Venting Guide.*
NFPA 69, *Explosion Prevention Systems.*
NFPA 70, *National Electrical Code.*

ADDITIONAL READING

Collings, A. J., and S. G. Luxon, *Safe Use of Solvents*, Academic Press Inc., New York, NY.
De, Anil K., et. al., *Solvent Extraction of Metals*, Van Nostrand Reinhold, New York, NY, 1970.
"Hazard and Hazard Prevention in Solvent Evaporating Ovens," *Fire Prevention*, (127), 1978, pp. 22-25.

Jaffee, H. M., "Grain Elevator Protection, What's Being Done Today?", *Fire Journal*, vol. 74, no. 3 (May 1980), pp. 131-132.

Marcus, Y., ed., *Solvent Extraction Reviews*, vol. 1, Marcel Dekker, Inc., New York, NY, 1971.

McKinnon, G. P., ed., *Fire Protection Handbook*, Fifteenth Edition, National Fire Protection Association, Quincy, MA, 1981.

Ritcey, G. M. and A. W. Ashbrook, eds., *Solvent Extraction*, Pt. 1, (Process Metallurgy Series, vol. 1), Elsevier Science Publishing Co., Inc., New York.

———, Solvent Extraction, Pt. 2, (Process Metallurgy Series, vol. 1), Elsevier Science Publishing Co., Inc., New York, NY, 1979.

Sawyer, W., "Relationship of Flash Points of Solvents, Resin Solutions, and Paints," *Journal of Coatings Technology*, 49(627), 1977, pp. 52-55.

Tess, Roy W., ed., *Solvents Theory and Practice*, American Chemical Society, Washington, DC, 1973.

33

Lumber Kilns and Agricultural Dehydrators and Dryers

Reducing the moisture content of natural materials is a process common to the wood products and agricultural industries. Drying lumber to a predetermined moisture content is done in a variety of structures called kilns, while the moisture content of agricultural products is reduced in dehydrators and dryers, commonly referred to simply as dryers. The hazards of both kilns and dryers are basically the same; a large quantity of a combustible product is introduced into a structure and exposed to elevated temperatures that could, under certain conditions of malfunction, approximate the ignition temperatures of the materials themselves and lead to fires involving significant quantities of material. But because of the basic differences in the raw materials, the methods used in handling the materials, and the configuration and ways of heating the enclosures, kilns and dryers are treated separately in this chapter.

The material on lumber kilns was contributed by the Pulp & Paper—Wood Products Industry Committee, a standing committee of the NFPA's Industrial Fire Protection Section. The material on agricultural dehydrators and dryers was reviewed by Hal E. Bland, Manager of the Aeroglide Corp.

Chapter 33

LUMBER KILNS

Seasoning of wood by controlled heating and drying is an important step in the process of turning timber into a useful material for building construction, boat building, furniture making, and the manufacture of sporting goods. This seasoning is done in oven-like structures called dry kilns. Kilns make it possible to turn freshly cut, green wood into dry, accurately dimensioned lumber within a relatively short time. Lumber seasoned in the open air will take months to dry to a usable state. In a kiln, this is accomplished within weeks or days. Lumber kilns thus help to assure steady, adequate supplies of materials to meet the needs of industry.

Although called *dry kilns*, these wood dryers usually employ moisture in the form of steam jets or water spray. This moisture is added to maintain a uniform moisture content within the wood during the drying process. Because of its cell structure, wood tends to dry unevenly. Internal stresses cause warping, checking and splitting that result in waste and lumber unsuited for its intended use.

All wood, while growing, contains large amounts of water. The amount will vary with the species, and significant variations may be found not only within trees of the same species, but also within logs cut from the same tree. Moisture content is determined by weighing a sample of green wood and then oven-drying it. The dried sample is weighed and the difference is calculated as the moisture content. The sapwood of western red cedar, for instance, has a moisture content of 249 percent. This means that a piece of green wood weighs almost two and one-half times as much as it does after it has been dried. In contrast, the sapwood of hickory has a moisture content of 50 percent.

The basic function of lumber kilns is to remove the moisture from sawed lumber and thus condition it for its intended use. The length of time required to reduce the moisture content to an optimum of about two percent depends upon the species, its original moisture content, the dimensions of the pieces being seasoned, the type of kiln, and the volume of material.

Types of Kilns

A *dry kiln* is basically an oven with controlled heat and humidity. It may have more than one chamber and may operate as a batch dryer or as a progressive dryer. The wood being dried in a *batch*, or *compartment*, *kiln* remains stationary throughout the process. Temperature and humidity in all parts of the kiln are maintained as uniformly as possible at each stage and are varied as the wood dries.

Progressive kilns are equipped with one or more sets of rails for wheeled trucks or platforms. Lumber to be seasoned is stacked on the trucks and wheeled into the kiln while dried lumber is being removed from the other end. Such kilns are designed so that somewhat higher tem-

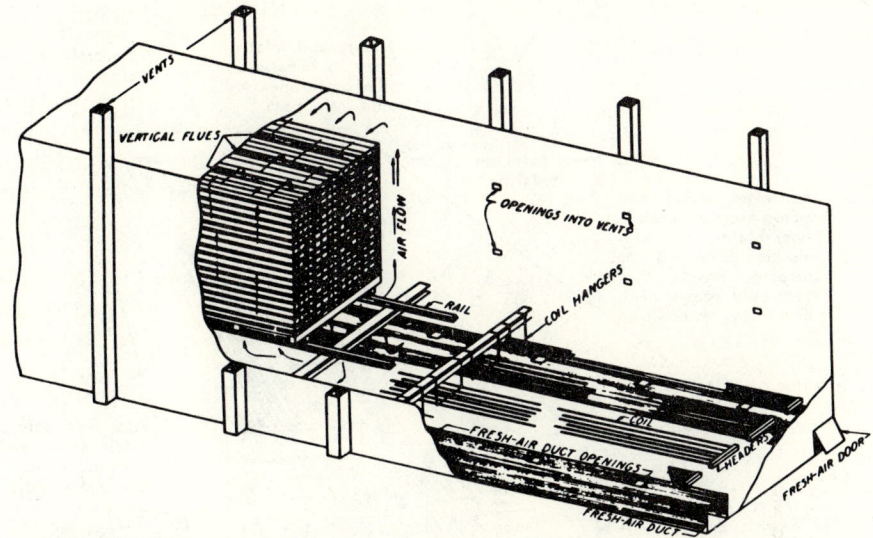

FIGURE 33.1. *A natural-circulation, steam-heated compartment kiln. Arrows indicate air movement during the early stages of the drying cycle.*

peratures are maintained at the dry, or discharge, end than at the loading, or green, end.

Progressive kilns may be up to 200 ft long (61 m) and 38 ft wide (11.6 m) and vary considerably in height. Kilns with fans located in pits below the track have a track-to-ceiling height of about 12 ft (3.7 m), while those with overhead fans have a height of about 17 ft (5.2 m). Compartment kilns may be as much as 28 ft (8.5 m) from floor to ceiling.

Kilns are also classified as natural circulation and forced circulation kilns (see Figures 33.1 and 33.2). Forced circulation kilns may be equipped with internal fans or external blowers.

In *natural circulation kilns*, heated air rises up through the stacked lumber by convection. Losing its heat, it travels down to the heating device located under the stack where it is reheated. During the first part of the drying cycle, steam coils along the sides near the floor of the kiln are heated. The air thus flows up along the sides, over the stacked lumber, and down through air passages built into the stack. When the wood's moisture content has been reduced to somewhere between 20 and 10 percent, the side heaters are shut off and the center ones turned on. Heated air now rises up through the center of the stack to equalize the drying. Vents in the walls or roof of the kiln exhaust the hot, moisture-laden air.

To speed up air circulation and drying speed, *forced circulation kilns* move air through the stacked lumber by blowers or fans. An *external blower* kiln is shown in Figure 33.2. Here the air is blown over the heating coils and into a central duct running the length of the kiln. Slots in this duct permit the heated air to move upward through a triangular flue built into the stack. The flue is closed at the top so the air is forced

Chapter 33

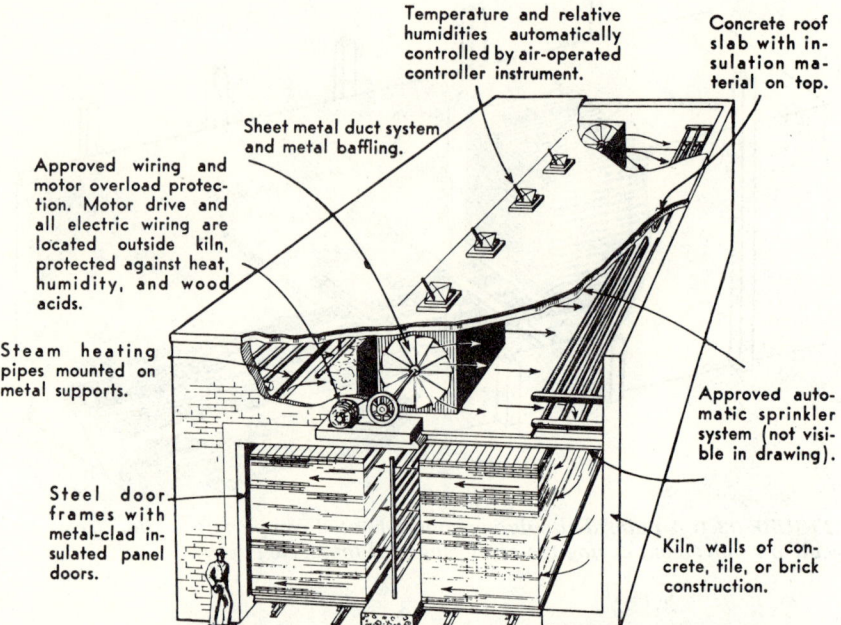

FIGURE 33.2. *A forced-circulation double-track compartment kiln. Note that automatic sprinklers are installed above and below the platform between the kiln and the overhead fan room.*

outward. It then moves down to return ducts located along the lower walls of the kiln. It is mixed with the entering air, reheated, and recirculated. To maintain the desired wet bulb temperature, steam is sprayed into the return air. Vents in the roof of the kiln permit hot moist air to escape.

The other type of forced circulation kiln utilizes *internal fans* to move heated air through the stacked lumber. The fans may be located above, below, or alongside the stacked lumber. Fan shafts may be directly connected to motors located inside or outside the kilns or they may be belt driven. Figure 33.3 shows an internal fan kiln with fans located beneath the floor. Gratings along the sides of the floor permit the heated air to rise at one side, flow through the stack and down the other side to be reheated and recirculated. The fans are reversible to change the airflow for optimum drying. Internal fan kilns normally require cloth or metal baffles to eliminate turbulence and keep the air flowing in the desired direction.

Much the same air-flow systems are used in both batch and progressive kilns. There are some minor differences, however. In progressive kilns, the air flows the length of the chamber and is discharged at the green end. Because the air has picked up moisture and is cooler by

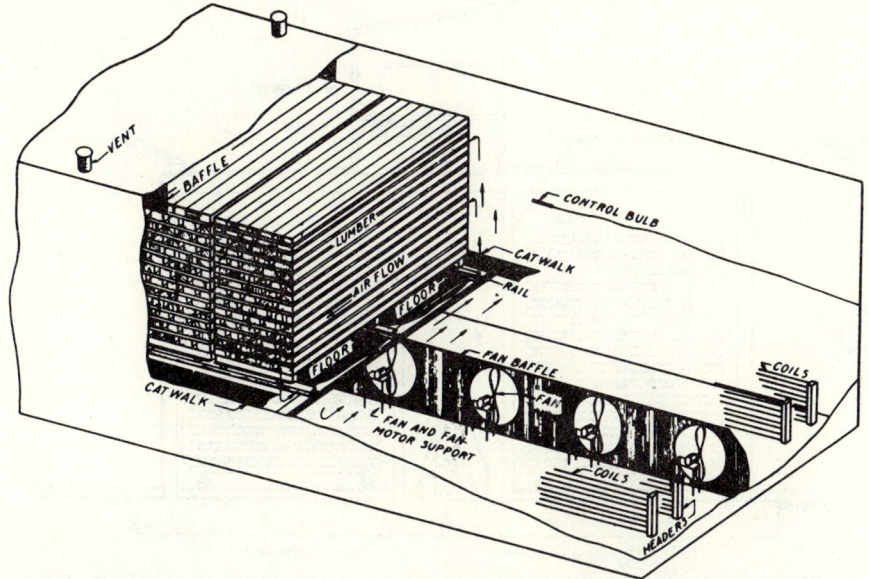

FIGURE 33.3. *A compartment kiln with internal fans and steam coils located under the grating at the floor level. Broken pieces of stickers and sawdust can fall through the grates and collect around the coils and fan motors to become a hazard, particularly if high pressure steam coils are used.*

the time it reaches the loading end of the kiln, the drying rate there is much slower than at the other end. Progressive kilns are generally built on a slight incline to facilitate the moving of the trucks.

Kiln Construction

Lumber kilns can be, and are, constructed with a wide variety of materials. Included in these are wood, brick, tile, concrete blocks, sheet metal, and cement-asbestos board. To reduce the fire hazard, kilns should be of fire-resistive or heavy timber construction.

Dry kilns, regardless of construction, are subject to rapid deterioration and require constant maintenance to keep them operating efficiently and economically. Unlike most other buildings, kilns are subjected to extreme variations in internal temperature and humidity. Such variations cause unusual expansion and contraction that reduce structural integrity and increase heat loss. Untended structural defects can also lead to premature failure of the structure.

The length and width of a kiln will depend upon its type, the desired capacity, and the method of stacking the lumber. If the lumber is end-stacked, a single track kiln will be from 12 to 16 ft (3.6 to 5 m) wide.

Chapter 33

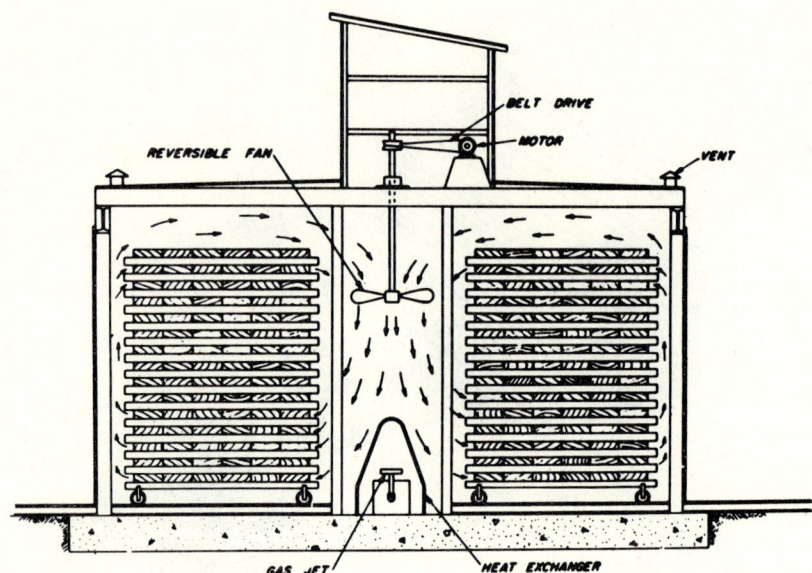

FIGURE 33.4. A double-track, internal-fan compartment kiln that is heated directly by a gas burner.

If the lumber is cross-piled, the width will increase to 18 to 22 ft (5.5 to 6.7 m). Lengths may vary from 20 to 225 ft (6 to 69 m).

As already indicated, the heat source for forced circulation kilns may be located above, below, or to the sides of the stacked lumber. In natural circulation kilns, the heat source has to be below the load in order to take advantage of the convection currents.

Heat Sources

Dry kilns require a constant source of heat to vaporize the water content of the wood. The heat is dissipated in raising the temperature of the wood and the water, in warming the fresh air introduced into the kiln, and in replacing the heat lost by radiation. The required heat is provided either directly or indirectly. Direct heating is accomplished by circulating hot gases, produced by burning gas, oil, sawdust, or other fuels, through the stacked lumber. It can also be done by heating large metal surfaces with an open gas or oil flame as in Figure 33.4. Air circulated by internal fans passes over the metal which acts as a heat exchanger. Then air flows over the lumber, carrying off moisture vapor.

Steam is a common source of heat for indirectly heated kilns, but hot gases and electrical resistance heaters are also used. Steam is circulated through pipes and hot gases through ducts. Air flowing over the pipes or ducts picks up the heat and carries it to the piled lumber. The temperature of steam varies with pressure. Thus, attaining a given kiln temperature with a low pressure system will require a larger radiating surface than it will with a high pressure system.

The Fire Hazards

Lumber kilns present high fire hazards because they treat readily ignitible wood with heat. This is especially true when direct heat systems or high pressure steam systems are used, and in those instances where the structure itself is of combustible materials. Although lumber kilns do not produce sawdust or other fine wood particles as part of the process, there always are enough small, dry pieces to kindle a fire if there is a sufficient ignition source. If such a fire begins, the oven is already charged with a heavy load of fuel and the resulting fire could be severe.

Direct-fired kilns present the greatest hazard because open ignition sources are close to the wood (see Figure 33.4). If the lumber is not carefully piled, it may shift enough to dislodge a portion of the load onto the heat source. Stickers, the small pieces of wood used to separate the lumber for air passage, are easily broken and dislodged and require special attention. Encased knots in the lumber may loosen and fall out during the drying process to provide a kindling supply.

Direct-fired kilns should be considered analagous to Class A ovens in that they should be equipped with all the combustion controls normally required for drying ovens where the heating fuel is introduced into the heating enclosure itself. Indirectly heated kilns utilizing steam coils as heat exchangers and having controlled humidity are usually considered to be of low hazard.

The Safeguards

To minimize the possibility of fire within the kiln, fallen stickers and lumber should be returned to the stack or completely removed. Sawdust and small pieces of wood should be collected frequently and removed. Oil or grease leaks from fan or blower bearings should be caught in drip pans and the pans should be emptied at regular intervals to prevent overflow.

The important requirements for firesafety are automatic sprinkler protection, sound construction, good housekeeping, automatic humidity control, good air circulation, and ventilation. Kilns should be situated at safe distances from storage yards, sheds, and mill buildings. Ideally, they should be of fire-resistive construction and equipped with a complete automatic sprinkler system connected to an adequate water supply. The sprinkler protection should extend to fan houses and control rooms. Hydrants or hose connections should be located on the exterior for manual fire fighting.

DEHYDRATORS AND DRYERS
(AGRICULTURAL PRODUCTS)

Dehydrators and dryers for agricultural products, commonly referred to as dryers, utilize heat to reduce the moisture content of the products treated. The hazards of dryers involve primarily (1) the possibility of igniting combustible materials near them, (2) the utilization of the fuel or electricity as a heat source, and (3) the ignition of stock being dried. The general recommendations for the design, installation, and utilization of these dehydrators and dryers are contained in *Construction, Installation, and Rating of Equipment for Drying Farm Crops*, a Standard of the American Society of Agricultural Engineers. NFPA 61B, *Grain Elevators and Bulk Grain Handling Facilities*, which deals with grain includes specific recommendations on dryers. Dryers for other products have generally the same design features as those discussed here or are designed as Class A ovens (see Chapter 27, "Heat Processing Equipment").

Types of Dehydrators and Dryers

Agricultural product dryers are of three general types, depending on the arrangement and operation of the drying chamber. These are continuous, batch, and bulk dryers. Continuous dryers include:

1. Drum dryers for milk, puree, and sludges.
2. Spray dryers for milk, eggs, and soup.
3. Flash dryers for chopped forage crops.
4. Gravity dryers (may also be batch type) for small grains, beans, and seeds (see Figure 33.5).
5. Tunnel dryers (may also be batch type and may be further classified according to flow of air and whether or not intermediate heating is used) for fruits, vegetables, grains, seeds, nuts, fibers, and forage crops (see Figure 33.6).
6. Rotary dryers for milk, puree, and sludges (see Figure 33.7).

Batch dryers (see Figure 33.8) may be either fixed or portable and include pan dryers for sugar, puree, sludges, and other products.

Bulk dryers (see Figure 33.9), as the name implies, involve drying the product in a bin, crib, or compartment in which it is to be stored. They are used to dry seeds, grains, nuts, tobacco, hay, and forage.

Methods of Heating

Dryers used for agricultural products may be direct-fired (where products of combustion contact the material being dried) or indirect-fired. The heaters may be oil-fired, gas-fired, solid-fuel-fired, electrical, or the dryers may be heated by waste heat or a heat transfer medium such as steam. The general requirement for the installation of burners and the storage of

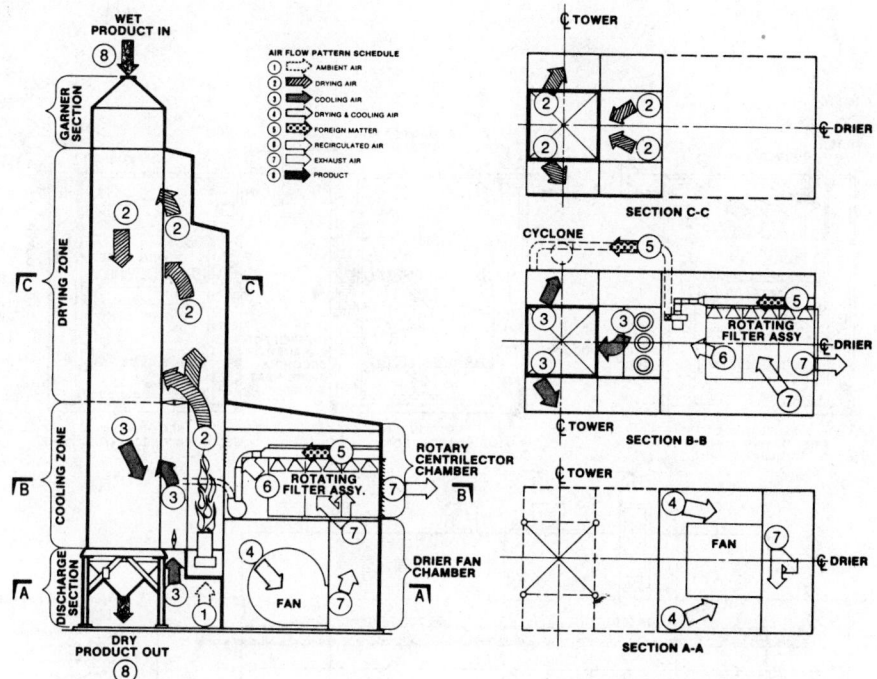

FIGURE 33.5. *A tower-type gravity dryer. (Aeroglide Corporation)*

fuel is the same as for other heat-producing devices. Detailed recommendations are given in NFPA 31, *Oil Burning Equipment;* NFPA 54, *National Fuel Gas Code;* NFPA 58, *Liquefied Petroleum Gases;* and NFPA 70, *National Electrical Code.*

If gas-fired infrared heaters or infrared lamps are used, the focal length should be such that unsafe temperatures are not reached on the surface of the product being dried. Electrical infrared lamps are undesirable for use in dryers because combustible dust can collect on the lamps.

Solid-fuel furnaces (other than those burning coke and anthracite coal) are not desirable where the products of combustion can enter the drying chamber. Indirect solid-fuel dryers need temperature-controlled heat-relief openings to the outside air.

Dryer controls: Some suggested controls for dryers, excluding those on the heating equipment, include the following:

1. A method for automatically shutting down the dryer in the event of fire or excessive temperature in the dryer.
2. In dryers where the product is fed automatically from the dryer to a storage building, a thermostat in the exhaust air which, in the event of excessive temperature, (a) shuts off heat to the dryer and stops the

Chapter 33

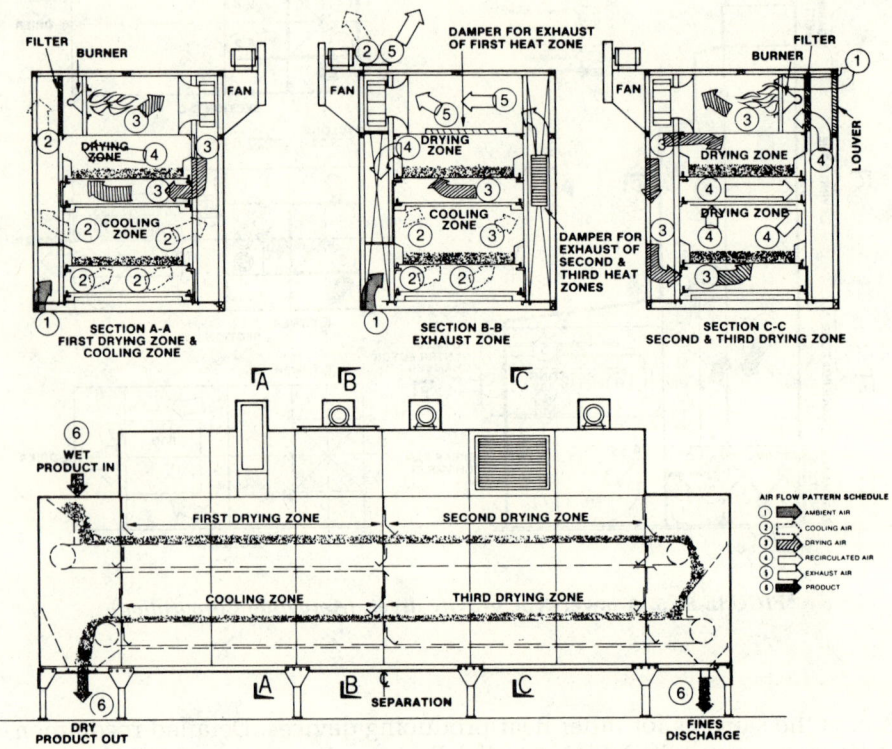

FIGURE 33.6. A tunnel dryer. (Aeroglide Corporation)

flow of air (except when the product being dried is in suspension), (b) stops the flow of the product through the dryer, and (c) sounds an audible alarm.

3. A thermostat in combustible dryers, or dryers with combustible trays, which, when the temperature of the combustible reaches 165°F (73.9°C), shuts off heat to the dryer but permits unheated air to pass through the dryer and sounds an audible alarm.
4. A device to shut off heat to the dryer in the event air movement through the dryer is stopped.
5. A high-limit thermostat located between the heat-producing device and the dryer.

Burner controls: In general, the burner controls for dryers are the same as for other devices that are automatically fired. It is important to install

LUMBER KILNS AND AGRICULTURAL DEHYDRATORS AND DRYERS

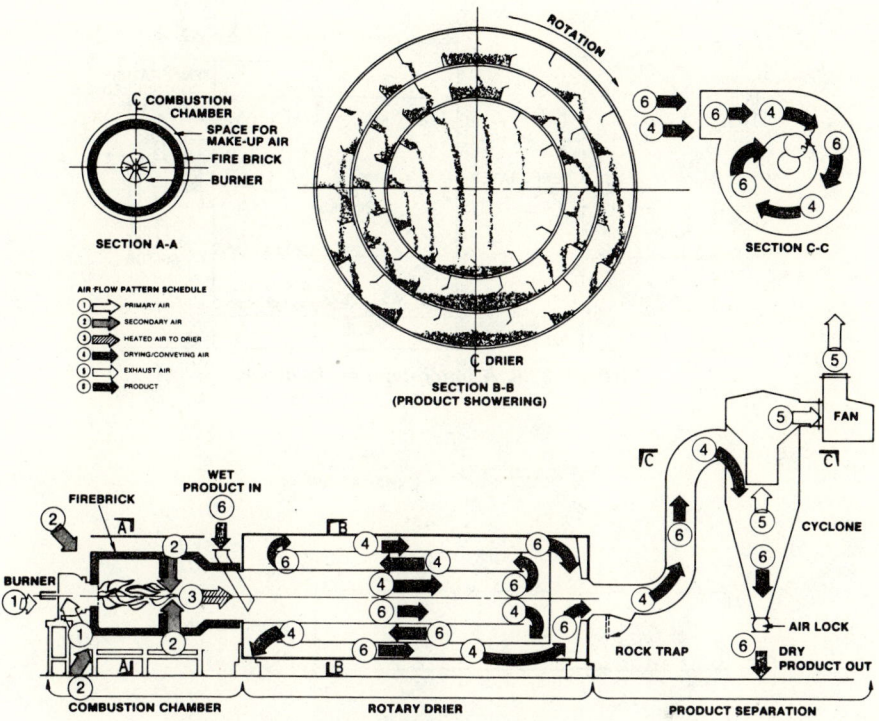

FIGURE 33.7. A rotary dryer. (Aeroglide Corporation)

a manual, quick-operating shutoff valve in the supply line of gas- and oil-fired burners, and for the controls to be arranged so that, following automatic shutdown due to an unsafe condition, manual restart will be necessary. Other control safeguards include flame-failure protection, automatic shutoff on pilot extinguishment, and pre-ventilation of the combustion chamber before start-up. All safety controls are arranged to "fail-safe."

Construction and Installation of Dryers

As dryers operate at elevated temperatures, it is standard practice to construct them of fire-resistive or noncombustible construction. If combustible materials must be used, they must not be located where they will be subjected to sustained temperatures in excess of about 165°F (73.9°C). Expansion joints are provided where necessary to prevent damage due to expansion and contraction.

Secondary air openings for direct-fired dryers are screened with ½-

Chapter 33

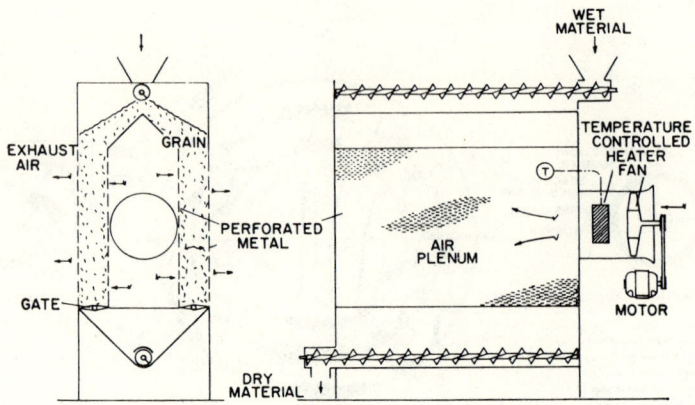

FIGURE 33.8. A batch-type grain dryer.

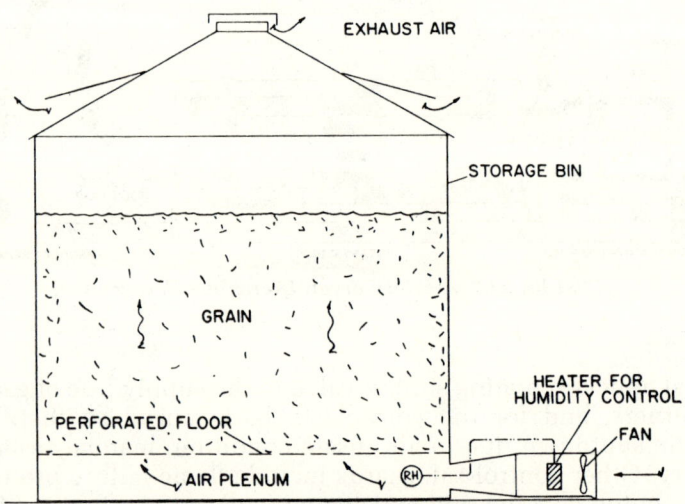

FIGURE 33.9. A bulk-type grain dryer.

in. (12.7 mm) mesh screen to prevent materials from entering the combustion chamber. Primary air openings require screens not larger than ¼-in. (6.3 mm) mesh.

An ample supply of easily opened access panels is necessary for inspection, cleaning, and fire fighting.

When stock is moved through the dryers in a manner that results in the generation of static electricity, all conductive parts of the dryers are electrically bonded and grounded.

Like any heat-producing device, a dryer must have adequate clearance from nearby combustibles to prevent overheating.

If there is a combustible dust hazard in the building where the dryer is located, the heating device and blowers are installed in a dust-free room or area separated from the remainder of the building.

Ducts to convey heated air to the dryer and exhaust air from the dryer to the outside are noncombustible.

Extinguishing Equipment

The most satisfactory method of protecting a dryer enclosure is by the installation of water-spray nozzles or automatic sprinklers within the enclosure wherever it is practicable. An exception to this is the direct-fired rotary dryer, which may be damaged by the internal application of water. A carbon dioxide system is satisfactory for protecting direct-fired rotary dryers.

For manual extinguishment of fires in and around most dryers, a standpipe hose is particularly useful. Water-type portable fire extinguishers may also be used. For fires in heating equipment, extinguishers rated for Class B (flammable liquid) and Class C (electrical) fires are applicable.

Cooling of Dehydrated Products

Any product being dehydrated or dried requires adequate cooling before being packaged or stored. The degree of cooling required to prevent subsequent ignition will depend upon the properties of each material (whether it will heat spontaneously and subsequently ignite) and the method of packaging or storing.

BIBLIOGRAPHY

REFERENCES

McKinnon, G. P., ed., *Fire Protection Handbook*, Fifteenth Edition, National Fire Protection Association, Quincy, MA, 1981.

Rasmussen, E. F., *Dry Kiln Operator's Handbook*, Agricultural Handbook No. 188, Forest Products Laboratory, Forest Service, U.S. Department of Agriculture, Washington, DC, March 1961.

Standard for Construction, Installation, and Rating of Equipment for Drying Farm Crops, The American Society of Agricultural Engineers, St. Joseph, MI, December 1962.

NFPA Codes, Standards, and Recommended Practices

Reference to the following NFPA Codes, Standards, and Recommended Practices will provide further information on the safeguards for lumber kilns and agricultural dryers discussed in this chapter. (See the latest *NFPA Codes and Standards Catalog* for availability of current editions of the following documents.)

NFPA 10, *Portable Fire Extinguishers.*
NFPA 13, *Installation of Sprinkler Systems.*
NFPA 15, *Water Spray Fixed Systems.*
NFPA 31, *Installation of Oil Burning Equipment.*
NFPA 54, *National Fuel Gas Code.*
NFPA 61B, *Prevention of Fires and Explosions in Grain Elevators and Facilities Handling Bulk Raw Agricultural Commodities.*
NFPA 70, *National Electrical Code.*
NFPA 86A, *Ovens and Furnaces — Design, Location and Equipment.*

Additional Reading

Dust Explosion and Fires: A Manual, National Particleboard Association, Silver Spring, MD, 1977.
Feirer, John L., *Wood: Materials and Processes*, Bennett Publishing Co., Peoria, IL, 1980.
Jaffee, H. M., "Grain Elevator Protection, What's Being Done Today?", Fire Journal, vol. 74, no. 3 (May 1980), pp. 131-132.

34

Grinding and Milling Operations

Delwyn D. Bluhm, Ph.D., P.E.

A ny industrial process that reduces a combustible material and some normally noncombustible materials to a finely divided state presents a potential for a serious fire or explosion. Grinding, or pulverizing, is the process by which materials are reduced to very small particles The hazards arise when these particles are suspended in air in critical concentrations.

MATERIALS GROUND

Agricultural products, such as wheat, corn, and other grains, present explosion hazards both during storage and during processing into flour or starch. Wood flour, finely divided sawdust, sulfur, coal, and some plastics present the same hazards as magnesium and aluminum. An explo-

---

Dr. Delwyn Bluhm, Senior Engineer, is Manager of Engineering for Ames Laboratory, and Energy & Minerals Resources Research Institute operated by Iowa State University, Ames, Iowa.

sion or fire can originate in the process equipment itself or in the ambient environment into which dust may escape and accumulate.

Actual occurrences, reinforced by laboratory tests, have demonstrated that finely divided particles of almost any readily oxidized material suspended in air in proper concentration will ignite and explode or burn. Table 34.1 classifies fifty-nine different materials into three groups according to the range of explosibility. The table includes nine different plastic dusts but does not cover the whole range of explosive plastic dusts. Table 34.2 gives the explosion characteristics of a variety of agricultural and industrial dusts. A more complete list appears in the *Fire Protection Handbook*.[3] Materials other than those mentioned may also be explosive under proper conditions of particle size and oxygen availability.

Table 34.1 cites low, moderate, and high maximum rates of pressure rise for selected explosive materials.

The maximum rate of pressure rise is calculated from the steepest part of the pressure versus time curve. As with the maximum explosion pressure, appropriate adjustments need to be made if the dispersing air effectively increases the ambient pressure. The average rate of pressure rise is also calculated from the pressure versus time curve. An approximate correlation often exists between the average and maximum rates of pressure rise, but the latter is preferred when assessing practical needs. The average rate of pressure rise can be affected by slow development of the explosion and is usually one-half to one-third the value for the maximum rate of pressure rise.

The United States Bureau of Mines has developed three additional measures of the relative explosion hazard: the Ignition Sensitivity, the Explosion Severity, and the Index of Explosibility. Each of these is a dimensionless value derived by comparing the measured explosibility parameters for a given dust, as described above, with those of Pittsburgh coal dust. The latter was chosen from tests conducted over many years in full scale experimental mines.

1. Ignition Sensitivity = $\dfrac{\text{(min ign temp} \times \text{min ign energy} \times \text{min explo conc) Pitt. coal dust}}{\text{(min ign temp} \times \text{min ign energy} \times \text{min explo conc) dust sample}}$

2. Explosion Severity = $\dfrac{\text{(max explo pressure} \times \text{max rate pressure rise) dust sample}}{\text{(max explo pressure} \times \text{max rate pressure rise) Pitt. coal dust}}$

3. Index of Explosibility = ignition sensitivity × explosion severity

Known values for these quantities are included in Table 34.2. In the absence of a sound theoretical basis for predicting explosion hazards of dusts, the Index of Explosibility provides a useful evaluation of relative explosibility. An empirical correlation between the U.S. Bureau of Mines indices and a descriptive categorization, the explosion hazard rating, is shown in Table 34.3.

It is important to distinguish the processes used to reduce the size

TABLE 34.1. *Classification of Explosive Material According to Rates of Pressure Rise*

Class A Materials (Low maximum rates of pressure rise†)	
Metal Dusts	**Miscellaneous Dusts**
Antimony	Anthracite
Cadmium	Carbon Black
Chromium	Coffee
Copper	Coke, low volatile
Iron (impure)	Graphite
Lead	Leather
Tungsten	Tea

Class B Materials (Moderate maximum rates of pressure rise‡)	
Metal Dusts, or Powders	Polyethylene
Iron (carbonyl, electrolytic or H$_2$, reduced)	Polystyrene
	Urea Resins
Manganese	Urea—Melamine
Tin	Vinyl Butryal
Zinc	**Miscellaneous Dusts**
Grains, Spices, etc.—Dusts	Bituminous Coal
Alfalfa	Cork
Cocoa	Calcium Lignosulfonic Acid
Grain Dust and Flour	Coumarone Indene
Mixed Grains	Dextrin
Rice	Lignin
Soy Bean	Lignite
Spices	Peat
Starch	Powdered Drugs
Yeast	Pyrethrum
Plastic Dusts	Shellac
Cellulose Acetate	Silicon
Methyl Methacrylate	Sulfur
Phenolformaldehyde	Tung
Phthalic Anhydride and its resins	Wood Flour

Class C Materials (High maximum rates of pressure rise§)	
Metal Dusts	*Sorbic Acid
Aluminum	*Titanium
*Stamped Aluminum	*Zirconium
Magnesium	Some Metal Hydrides
Magnesium—aluminum alloys	

*These are exceptionally fast.

†≤7300 psi/sec (50000 kPa/sec) measured via the Hartman apparatus.

‡7300 to 22000 psi/sec (50000 to 151000 kPa/sec) measured via the Hartman apparatus.

§>22000 psi/sec (151000 kPa/sec) measured via the Hartman apparatus.

of materials by grinding from those processes employing abrasive wheels, disks, or drums to surface or shape articles of wood, metal, or plastic. The

TABLE 34.2. *Explosion Characteristics of Various Dusts*

(Compiled from the following reports of the U.S. Department of Interior, Bureau of Metal Powders; RI 5971, Explosibility of Dusts Used in the Plastics Industry; RI Chemicals, Drugs, Dyes and Pesticides; and RI 7208, Explosibility of

Type of Dust	Explosibility Index	Ignition Sensitivity	Explosion Severity	Maximum Explosion Pressure	
				psig	kPa
Agricultural Dusts					
Alfalfa meal	0.1	0.1	1.2	66	455
Coca bean shell	13.7	3.6	3.8	77	531
Coffee, raw bean	<0.1	0.1	0.1	33	228
Cornstarch, commercial product	9.5	2.8	3.4	106	731
Cork dust	>10.0	3.6	3.3	96	662
Grain, dust, winter wheat, corn, oats	9.2	2.8	3.3	131	903
Peat, sphagnum, sun dried	2.0	2.0	1.0	104	717
Pyrethrum, ground flower leaves	0.4	0.6	0.6	95	655
Rice	0.3	0.5	0.5	47	324
Soy flour	0.7	0.6	1.1	94	648
Wheat flour	4.1	1.5	2.7	97	669
Yeast, torula	2.2	1.6	1.4	123	848
Carbonaceous Dusts					
Charcoal, hardwood mixture	1.3	1.4	0.9	83	572
Coke, petroleum	0.1†	0.1†	—	—	—
Graphite	0.1†	0.1†	—	—	—
Lignite, California	>10.0	5.0	3.8	94	648
Metals					
Cadmium, atomized (98% Cd)	—	—	—	7	48
Iron, carbonyl (99% Fe)	1.6	3.0	0.5	43	296
Lead, atomized (99% Pb)	—	—	—	—	—
Magnesium, milled, Grade B	>10.0	3.0	7.4	116	800
Manganese	0.1	0.1	0.7	53	365
Tantalum	0.1	0.1	0.7	55	379
Thermoplastic Resins and Molding Compounds					
Cellulose acetate	>10.0	8.0	1.6	85	586
Methyl methacrylate polymer	6.3	7.0	0.9	84	579
Polyethylene, low-pressure process	>10.0	22.4	2.3	80	552
Polystyrene molding compound	>10.0	6.0	2.0	77	531
Thermosetting Resins and Molding Compounds					
Phenolformaldehyde	>10.0	9.3	1.4	77	531
Urea formaldehyde molding compound, Grade II, fine	1.0	0.6	1.7	89	614
Special Resins and Molding Compounds					
Lignin, hydrolized-wood-type, fines	>10.0	5.6	2.7	102	703
Rubber, synthetic, hard, contains 33% sulfur	>10.0	7.0	1.5	93	641
Shellac	>10.0	25.2	1.4	73	503

*Numbers in this column indicate oxygen percentage while the letter prefix indicates the with carbon dioxide as the diluent gas.
†0.1 designates materials presenting primarily a fire hazard as ignition of the dust cloud is
‡Guncotton ignition source.
§No ignition.

Mines: RI 5753, The Explosibility of Agricultural Dusts; RI 6516, Explosibility of 6597, Explosibility of Carbonaceous Dusts; RI 7132, Dust Explosibility of Miscellaneous Dusts.)

Max Rate of Pressure Rise		Ignition Temperature				Min Cloud Ignition Energy joules	Min Explosion Conc		Limiting Oxygen Percentage* (Spark Ignition)
		Cloud		Layer					
psi/sec	kPa/sec	°C	°F	°C	°F		oz/cu ft	g/m³	
1,100	7585	530	986	—	—	0.320	0.105	105	—
3,300	22754	470	878	370	698	0.030	0.040	40	—
150	1034	650	1202	280	536	0.320	0.150	150	C17
7,500	51713	400	752	—	—	0.040	0.045	45	—
7,500	51713	460	860	210	410	0.035	0.035	35	—
7,000	48265	430	806	230	446	0.030	0.055	55	—
2,200	15169	460	860	240	464	0.050	0.045	45	—
1,500	10344	460	860	210	410	0.080	0.100	100	—
700	4827	510	950	450	842	0.100	0.085	85	—
800	5116	550	1022	340	644	0.100	0.060	60	C15
2,800	19306	440	824	440	824	0.060	0.050	50	—
3,500	24133	520	968	260	500	0.050	0.050	50	—
1,300	8964	530	986	180	356	0.020	0.140	140	—
200	1379	670	1238	—	—	‡	1.000	1005	—
—	—	§	—	580	1076	—	—	—	—
8,000	55160	450	842	200	392	0.030	0.030	30	—
100	690	570	1058	250	482	4.000	—	—	—
2,400	16548	320	608	310	590	0.020	0.105	105	C10
—	—	710	1310	270	518	—	—	—	—
15,000	103425	560	1040	430	806	0.040	0.030	30	—
4,900	33786	460	860	240	464	0.305	0.125	125	—
4,400	30338	630	1166	300	572	0.120	<0.200	<201	—
3,600	24822	420	788	—	—	0.015	0.040	40	C14
2,000	13790	480	896	—	—	0.020	0.030	30	C11
7,500	51713	450	842	—	—	0.010	0.020	20	—
5,000	34475	560	1040	—	—	0.040	0.015	15	C14
3,500	24133	580	1076	—	—	0.015	0.025	25	C17
3,600	24822	460	860	—	—	0.080	0.085	85	C17
5,000	34475	450	842	—	—	0.020	0.040	40	C17
3,100	21375	320	608	—	—	0.030	0.030	30	C15
3,600	24822	400	752	—	—	0.010	0.020	20	C14

diluent gas. For example, the entry "C17" means dilution to an oxygen content of 17 percent

not obtained by the spark or flame source but only by the intense heated surface source.

TABLE 34.3. *Correlation Between Descriptive Categories for Dust Explosions and U.S. Bureau of Mines Indices.**

Type of Explosion	Ignition Sensitivity	Explosion Severity	Index of Explosibility
Weak	<0.2	<0.5	<0.1
Moderate	0.2–1.0	0.5–1.0	0.1–1.0
Strong	1.0–5.0	1.0–2.0	1.0–10
Severe	>5.0	>2.0	>10

*(From Jacobson et al., U.S. Bureau of Mines RI5753, "The Explosibility of Agricultural Dusts")

latter processes could produce explosive dusts under certain conditions. However, those operations normally are conducted in environments where there is little likelihood that dust concentrations will reach the lower explosive limit (LEL). This chapter, therefore, is concerned only with the hazards of milling or grinding operations involving large quantities of potentially flammable or explosive materials.

CLASSIFICATION OF GRINDING EQUIPMENT

Equipment for reducing particle size of a given material may be classified into four groups according to the method of grinding used. A material may be reduced between two solid surfaces; it may be reduced by impact on one solid surface; it may be reduced by action of the surrounding medium; or it may be reduced by such nonmechanical means as thermal shock, explosive shattering, or electrohydraulic processes.

Grinding may be either a wet or a dry process. This chapter will limit its treatment to dry processes in which the hazards are greatest.

Most grinding is done by machines that pass the material between two solid surfaces. Such machines can be classified into five different types -tumbling mills, ring-roller mills, roller mills, hammer mills, and attrition mills.

Tumbling mills: Tumbling mills consist of a cylindrical or conical shell charged with balls of steel, flint, or porcelain, or with steel rods. As the shell revolves about its horizontal axis, the balls or rods tumble about, grinding the material to be reduced against the wall of the shell or between themselves (see Figure 34.1). The size of the balls or rods and the duration of operation will determine particle size. Some tumbling mills are compartmented by perforated partitions that allow material to pass from one compartment to another for finer grinding.

Ring-roller mills: A second type of grinding machine is known as a ring-roller mill. Mills of this type consist of a grinding ring or plate moving between rollers (see Figure 34.2). The ring may be either horizontal or vertical, and either it or the roller may rotate, grinding the product be-

GRINDING AND MILLING OPERATIONS

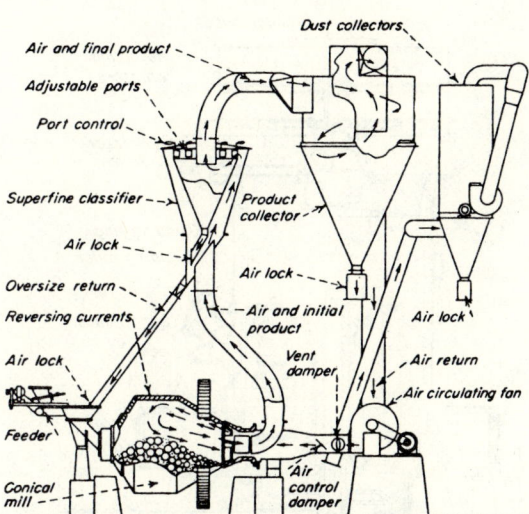

FIGURE 34.1. A Hardinge conical mill with reversed-current air classifier.

tween the two surfaces. A variation of the ring-roller mill is the bowl mill in which a bowl and ring revolve around stationary rollers to grind the product (see Figure 34.3). There is no metal-to-metal contact, and the space between the bowl and rollers can be set to produce the required particle size.

Roller mills: Roller mills differ from ring-roller mills in that the material to be ground passes between two or more rollers revolving in opposite directions at different speeds (see Figure 34.4). A scraper blade at the discharge end removes the finely ground material. Most dry materials are ground between rollers having corrugations which determine the final particle size. Corrugations may be either sharp or dull, and they may be used in various combinations to achieve the desired results.

Hammer mills: Hammer mills have hammers, or beaters, attached to a rotating shaft (see Figure 34.5). The hammers may be hinged or fixed to the shaft and of almost any shape. The fineness of the finished product is determined by the speed of the rotating shaft, the clearance between the hammers and grinding plates, the number and size of the hammers, the feed rate, and the size of the discharge openings. Two variations of the hammer mill are the disintegrator and the pin mill.

The disintegrator has a vertical rotating shaft with hammers that run at close tolerances to a cylindrical screen (see Figure 34.6). Material is fed in parallel to the rotating shaft, and the centrifugal action of the hammers discharges the ground material into a primary chute.

The pin mill is a high speed mill with two disks in which pins are set in alternating circular rows. Either one or both of the disks may rotate. If both rotate, they do so in opposite directions. The material to be ground is broken up between the pins (see Figure 34.7).

Chapter 34

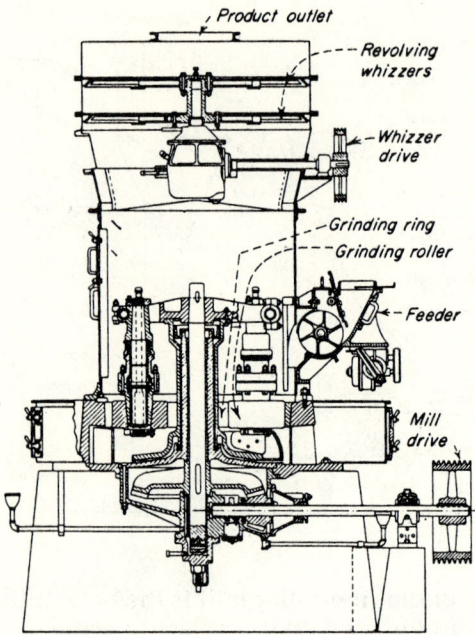

FIGURE 34.2. *A Raymond high-side mill with internal whizzer classifier.*

Attrition mills: Attrition mills use metal or abrasive grinding plates that rotate at high speed in either a horizontal or vertical plane. One disk may be stationary, or the two may rotate in opposite directions. The material enters at the axis and is discharged at the periphery of the grinding plates. One type of attrition mill, known as the Buhrstone mill, uses two circular stones instead of metal or abrasive disks between which the material is ground.

Jet mills: Another type of mill, the jet mill, differs from the others in that the material is not ground against a hard surface. Instead, a gaseous medium is introduced. The gas may convey the feed material at high velocity in opposing streams, or it may move the material around the periphery of the grinding and classifying chamber. The high turbulence causes the particles of feed material to collide and grind upon themselves.

Because of variations in their characteristics, different types of mills are used for different products. For grinding wheat and rye into high-grade flour, roller mills are most generally used. Dull corrugations are most often used, but sharp rolls are required for grinding hard wheats, corn, and feed grains. For flour with a controlled protein content, high-speed hammer or pin mills are used. Disk attrition mills, often of the Buhrstone type, are also used for grinding wheat and other grains for flour.

In industrial processes, grinding mills have a wide variety of ap-

GRINDING AND MILLING OPERATIONS

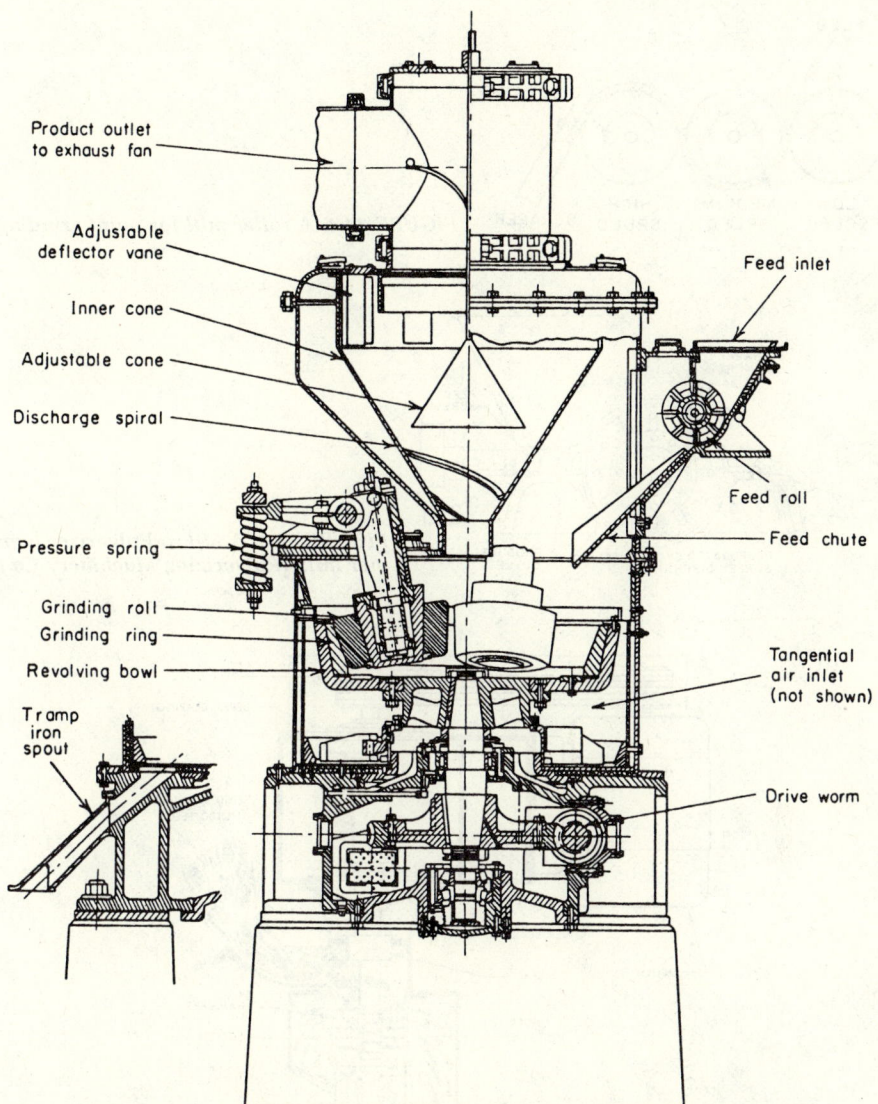

FIGURE 34.3. A bowl mill. (Raymond Div., Combustion Engineering Inc.)

plications, including the pulverizing of coal for firing furnaces, boilers, and rotary kilns. Graphite is ground in varying grades in different types of mills; the jet mill produces the extremely fine powdered graphite needed for pencil leads. For pulverizing dye stuffs and paint pigments, hammer mills or pebble (tumbling) mills are generally used. Hammer mills are often used to grind plastics, but because some resins, gums, and

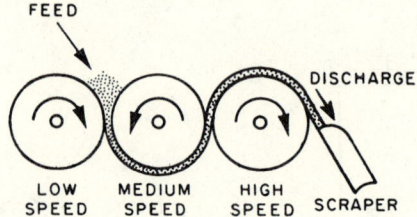

FIGURE 34.4. *A roller mill for paint grinding.*

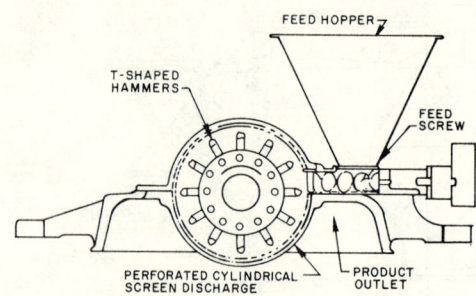

FIGURE 34.5. *A Mikro-Pulverizer hammer mill. (Pulverizing Machinery Co.)*

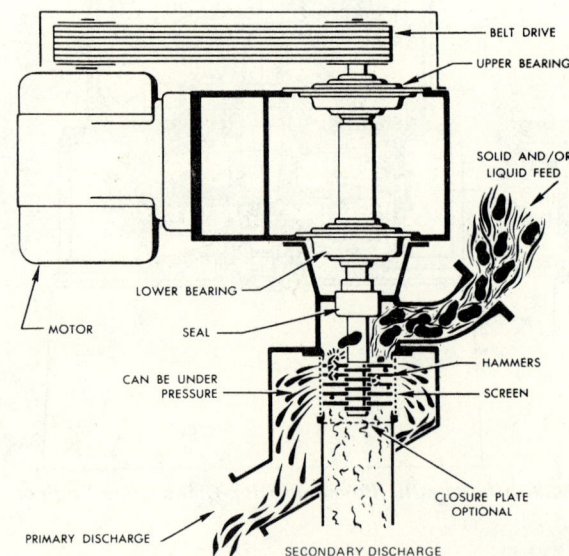

FIGURE 34.6. *Reitz disintegrator. (Reitz Mfg. Co.)*

rubbers tend to soften under the heat of friction, the mills must be water- or air-cooled.

Regardless of the material being processed, grinding at cryogenic temperatures can provide some distinct advantages. In any size or type

GRINDING AND MILLING OPERATIONS

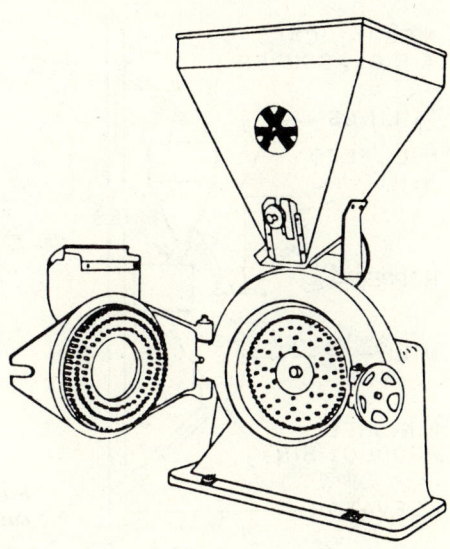

FIGURE 34.7. An Alpine-Kolloplex pin mill. (Alpine American Corp.)

of mill, a smaller particle size can be produced when grinding is done at low temperatures; thus, when fineness is a criterion, cryogenic grinding may be indicated as the method of choice. Cooling is achieved by passing liquid nitrogen in coils around or through the feed hopper or conveyor, so that the temperature of the product is reduced before it enters the grinding chamber. Cooling can also be done by spraying liquid nitrogen into the chamber. This second method may provide the additional advantage of inerting the atmosphere within the chamber, thus reducing or eliminating the fire or explosion hazard.

GRINDING HAZARDS

The hazards of grinding operations lie in the fact that the process produces very fine particles of readily oxidized materials that may be mixed with process air or environmental air in flammable or explosive concentrations. To separate the properly sized material from the coarser particles being fed into the mill, a system of classifiers is used. For the most part, classifiers are based on the force of gravity and the principles of air drag and particle inertia. A continuous flow of air passes through the mill, regulated so that particles of the desired fineness are carried through to a collecting bin or compartment, while coarser particles fall out and return to the mill for further grinding. This principle is illustrated in Figure 34.8.

There are several different types of air classifiers, some of which may be external to the mill as in Figure 34.8. Others may be internal as

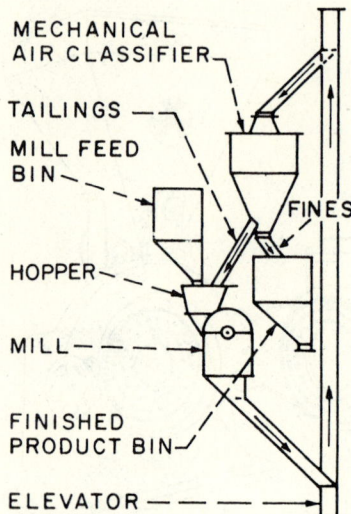

FIGURE 34.8. *A hammer mill in a closed circuit with air classifier.*

in the ring-roller mill illustrated in Figure 34.2. Here the whizzers are rotating blades whose centrifugal action throws the coarser particles outward, permitting them to drop back onto the grinding surfaces. The finished product is carried through an outlet by the airstream.

Theoretically, the process airstream should be fully and tightly enclosed, so that it does not carry the finished product or unwanted dust into the atmosphere surrounding the mill. In practice, this is rather difficult to accomplish, and dust builds up in the structure housing the mill. There are, then, two distinct hazards. One is that an explosion may occur within the milling system, and the other is that an explosion may occur in the structure. All that is required is a critical concentration of the material in air and an ignition source.

The lower limit of flammability will vary from one material to another. For some, such as phthalic anhydride, shellac, aluminum stearate, and phenothiazine, it may be as low as 0.015 oz/cu ft (15 g/m^3). For zinc, it may be as much as 0.5 oz/cu ft (500 g/m^3). Upper flammability limits for dusts, that is, the concentration above which an explosion will not occur, are poorly defined, not usually reproducible, and not yet determined for many dusts. Therefore, any concentration above the lower limit should be considered potentially explosive. The minimum electric spark energy required for ignition of a dust cloud can vary from as little as 10 mJ (millijoule) to as much as 1900 mJ.

Ignition sources, too, can vary widely. Since a bit of metal accidentally entering the mill may cause a spark as it strikes against one of the grinding surfaces, mills should be equipped with magnetic or mechanical separators to remove any tramp metal from the feed system. Sparking may also be caused by tools made of ferrous metals. Other pos-

GRINDING AND MILLING OPERATIONS

sible ignition sources are static electricity, hot surfaces, friction, open flames, welding arcs, and smoking.

CHARACTERISTICS OF DUST EXPLOSIONS

Dust explosions, in some respects, are similar to vapor and gas explosions, but they do differ in some important ways. Like a gas, dust must be mixed with air or another supporter of combustion, and a source of ignition is generally required to cause an explosion. Rarely have dust explosions resulted from spontaneous oxidation and heating. Reaction rates and rates of pressure rise are usually higher in vapor and gas explosions than in dust explosions. However, complete combustion of dust in a given volume of air will frequently develop energy and pressure greater than those developed by the combustion of a gas. Dust explosions, then, are sometimes more disastrous than gas explosions. This, in part, is due to their slower rate of development and longer duration. The slower rate of development results from the fact that the combustion of dust is a surface reaction, and the diffusion of oxygen toward the reacting surface is necessarily slower and less complete than it is in a flammable gas. (See Chapter 6, "Grain Mill Products," for more details on dust explosions.)

Requisite Conditions for Dust Explosions

In order for a dust explosion to occur, four conditions must be satisfied simultaneously:

1. A combustible solid in a finely divided state must be dispersed in an oxidizing medium — usually oxygen in air.
2. The concentration of the dust in air must be within the explosible range.
3. An external source of ignition of sufficient energy and duration to initiate the explosive chain reaction for that particular dust must be present.
4. The chemical reaction must occur in a confined volume.

The rapid chemical reaction, or flash fire, characteristic of explosions, will occur if only the first three conditions are satisfied. However, the rapid buildup of excessive pressures, inherent in the working definition of dust explosions, will result only when the reaction occurs in an enclosed space.

Characterization of Explosion Hazards of Dusts

Parameters employed to describe the relative explosion hazards of various dusts include:

1. lower and upper limit of dust concentrations within which an explosion is possible;
2. minimum ignition energy (minimum electric spark energy required for ignition of the dust cloud);
3. minimum ignition temperature as measured by a furnace apparatus;
4. maximum oxygen concentration permissible to prevent ignition;
5. maximum explosion pressure attained during the course of the explosion;
6. maximum rate of pressure rise (sometimes also the average rate of pressure rise).

Values for these parameters (see Table 34.2) are not fixed but depend on various factors — namely, particle size and shape, ambient temperature and pressure, moisture content of the dust, degree of turbulence in the suspension, and size of the ignition source. Experimental work has made possible some qualitative observations of the effects of these factors, but quantitative relations are not available.

Definitions of dust in terms of particle size vary but normally dust is defined as particles with diameters of 0.1 to 1000 microns (one micron equals 10^{-6} meters).

The minimum explosible concentration and the minimum ignition energy generally tend to decrease with a decrease in average particle size (i.e. the explosion hazard increases with a decrease in particle size). For average particle sizes less than 50 microns, the effect is much less pronounced. According to Palmer,[2] uniform dispersion of the dust in laboratory test equipment may become more difficult as particles become very small. Due to the greater cohesiveness of very fine particles, some may exist as agglomerates rather than as individual particles leading to an apparent reduction in explosibility. A different method of dispersion or a more vigorous ignition source may break up the agglomerations resulting in an increased explosion hazard. Tests performed in large scale coal mines are said to confirm this.

Little experimental evidence is available concerning the effect of particle shape on explosibility. Particles may resemble fibers, needles, or flakes, as well as spheres. Studies employing atmozied spherical particles and flaked aluminum particles indicate that a significant difference in explosibility can occur with changes in particle shape. Size and shape, which are generally measured before ignition, may be altered during preignition stages and subsequently affect propagation of the explosion. Changes may occur as a result of melting, vaporization, expansion to form hollow spheres, and fragmentation.

The effect of ambient temperature on explosibility would be particularly applicable to industrial situations such as dryers, but no information is available on measurements in plant scale units. Theoretically, it would be expected that the minimum ignition energy would decrease as the ambient temperature increased, other factors remaining constant. If the final temperature reached by the combustion products is limited by the molecular dissociation of product gases, then the maximum explosion pressure would be expected to decrease as the ambient temperature increases, and the rate of pressure rise would increase as the ambient temperature increased. Laboratory tests employing coal dust revealed that the minimum ignition energy decreased with increasing ambient temperature as expected, but only over a limited temperature range; thereafter, it increased. Palmer[2] also notes that the minimum ignition energy would depend on previous exposures to high temperatures.

If the pressure in the reaction vessel is atmospheric when the explosion is initiated, rises in pressure during the course of the explosion are not considered to be changes in ambient pressure. Laboratory tests employing methane/air mixtures resulted in proportional increases in maximum explosion pressures with increased initial pressure, but smaller increases were obtained for coal dust/air mixtures. The maximum explosible concentration decreased with increasing ambient pressure, while the minimum explosible concentration remained relatively unchanged. Explanations for these effects have not yet been explored.

An increase in moisture content of dusts tends to increase the minimum explosible concentration, the minimum ignition energy, and the minimum ignition temperature when measured in small scale apparatus. These effects are apparently due to the absorption of heat in vaporizing water and to the decrease in dispersibility of the dust. Nineteen percent water content (on a weight basis) was found to prevent ignition of starch dust.

Tests with coal dust/air suspensions have indicated that explosions tend to increase in severity (i.e., maximum explosion pressure and/or maximum rate of pressure rise increases) with increases in the size of the ignition source. The nature of the ignition source also affects the explosibility of dusts. Organic dusts tend to be ignited more readily by heated coils, but metal dusts react more readily to spark ignition.

Experiments indicate that some dusts produce stronger explosions than others. Metallic dusts, such as stamped aluminum powder, milled and stamped magnesium, and atomized aluminum, produced the most violent dust explosions. Phenolformaldehyde resin, cornstarch, soybean protein, wood flour, and coal dust, respectively, followed the metallic dusts.

The character and severity of any dust explosion will be affected by several factors. One of these is particle size. For any given material, the finer the particle size, the more violent the explosion will be. Less energy will be required to ignite the dust, and it will remain in suspension for a longer time, increasing the total force exerted. Figure 34.9 shows the relationship of particle size to the explosibility index.

Turbulence is another factor that contributes to the severity of a dust explosion. Turbulence speeds up the diffusion of oxygen to the reacting surfaces and promotes stronger explosions. The smaller the particle size and the greater the turbulence, the more the dust resembles a gas or vapor in its explosive characteristics. Relatively little investigation has been carried out with regard to the effects of turbulence on

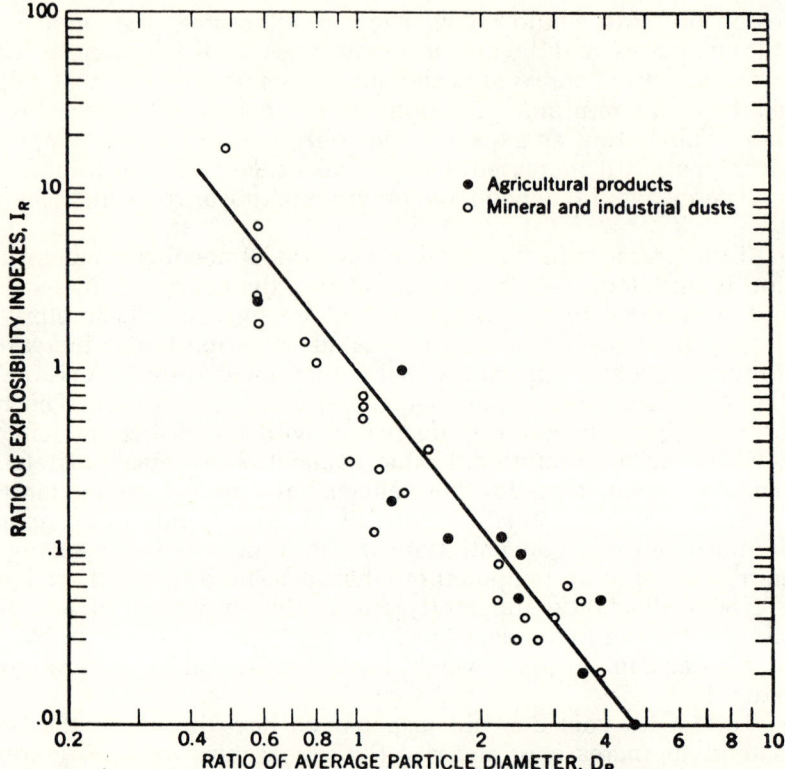

FIGURE 34.9. *Effect of particle diameter on relative explosibility index. Relative average particle diameter is the ratio of the mean particle size of the dust to the mean size of the through No. 200 sieve sample. Relative explosibility index is the ratio of the index computed for the dust to that computed for a through No. 200 sieve sample. (From Bureau of Mines Report of Investigations 5753)*

explosibility. Tests with coal dust/air suspensions revealed an increase in maximum explosion pressure and maximum rate of pressure rise with increased turbulence. Both parameters passed through maxima and then decreased, although more slowly in the case of the maximum explosion pressure. Sources of turbulence may include the presence of obstacles and rapid volume expansion due to flame.

PROTECTION AGAINST EXPLOSIONS

Combustion and explosion, or more properly deflagration relative to dusts, are practically identical processes. The difference lies in the speed with which the oxidation reaction takes place. NFPA 68, *Explosion Vent-*

ing, defines deflagration as "burning which takes place at a flame speed below the velocity of sound in the unburned medium." It defines explosion as "the bursting of a building or container as a result of development of internal pressure beyond the confinement capacity of the building or container." Inasmuch as this is usually the result of detonation, the term explosion, as generally used herein, refers to the entire process.

Because an explosion takes place almost instaneously, it cannot be brought under control by containment as can many fires. The effects, however, can be minimized by certain protective measures. These are prevention, venting, inerting, and suppression.

Prevention

The simplest means of preventing an explosion is to keep a critical concentration of dust from developing and to eliminate any potential sources of ignition.

In grinding operations, there are two types of locations where dust accumulation may become critical. One is within the milling or grinding equipment itself. The other is in the surrounding environment, that is, the structure in which the equipment is housed.

The design and construction of the structure may vary considerably, depending on the product being ground. It is impractical to construct a building that will withstand the pressure generated by a dust explosion. An ordinary 12-in. (0.3 m) thick brick wall can be destroyed by an internal pressure of less than 1 psi (6.9 kPa). Most dust explosions produce much higher pressures. Typical pressures range from 13 psi (90 kPa) for zinc to 89 psi (614 kPa) for stamped aluminum, with most being above 30 psi (207 kPa).

If a dust explosion hazard exists, the structure should be designed with panels, vents, windows, or other closures that will open at the lowest practical pressures to minimize structural damage.

Materials used should be noncombustible or fire-resistive. Any interior walls intended to serve as fire walls should be capable of providing at least 3 hours of fire resistance under standard methods of fire testing. Interior stairs, lifts, or elevators should be enclosed in shafts of noncombustible materials and have fire-resistive ratings of at least 1 hour. Such enclosures should be protected by automatically closing fire doors, and any openings in fire walls should be similarly protected.

The number of horizontal surfaces that might collect dust and are difficult to reach or are inaccessible for cleaning should be minimized. Wherever practical, such surfaces should be built up to an angle of at least 60 degrees from horizontal so that dust tends to slide off rather than accumulate. Good housekeeping is a necessary part of explosion prevention. It means keeping both equipment and structure clean so that dust cannot accumulate. Dust should be removed by vacuum systems. Brushing or sweeping will disperse the dust and increase turbulence, which increases the possibility of an explosion occurring and increases its potential severity.

Equipment should be made of metal and be as dust-tight as possible. It should be designed so that there is continuous suction at openings during grinding, dumping, transfer, and similar operations. The collected dust should be conveyed through tightly constructed ducts or chutes to well-designed dust collectors located in a safe place, preferably outside the structure.

Potential ignition sources should be eliminated or adequately shielded, for many dusts can be ignited by low energy sparks. Welding, cutting, or any other operation that uses an open flame or arc should not be permitted unless the work area is dust-free. Smoking should be prohibited. Torque-limiting or fluid drive couplings should be designed to dissipate heat readily. Moving equipment, elevators, belts, and conveyors should be grounded or of nonconductive material to eliminate the possibility of static sparks. All electrical wiring should conform to the requirements of NFPA 70, the *National Electrical Code*, for hazardous locations containing combustible dusts.

Adequate fire extinguishing equipment, both fixed and portable, should be provided. Hose nozzles should be the spray type. Solid streams can cause turbulence and dispersion of dust into the air, increasing the explosion hazard.

Venting

Venting does not prevent explosions, but it does serve to limit the maximum pressure resulting from a deflagration in order to limit damage to the enclosure. Vents are openings in the equipment or structure that allow heated explosion gases to escape more readily. The most effective vents would be free and unrestricted openings; however, they are not always practical.

Vents should be closed in such a way that they will open under the lowest practical pressure. Typical venting arrangements include rupture diaphragms, hinged or blow-out windows and panels, and weakly constructed walls and roofs.

The venting area required, calculated from empirical formulas, depends upon the expected pressure and its rate of rise, the type of closure, and the volume and strength of the enclosure. Determination of vent area for an enclosure is mostly empirical. Vents should be located where there will be minimum damage from the shattering or blowing out of the vent closure.

Inerting

One method of preventing a dust explosion is inerting — the replacement of oxygen in the grinding process with an inert gas such as nitrogen or carbon dioxide. Such equipment as grinders, pulverizers, mixers, driers, conveyors, dust collectors, and filling machines can be protected by this method.

The amount of oxygen that must be replaced to provide a safe concentration depends on the type of dust, particle size, concentration, turbulence, the diluent gas, and the intensity of the ignition source. To prevent ignition of carbonaceous dusts by a spark, for example, oxygen should be reduced to 8 percent by nitrogen or to 11 percent by carbon dioxide. However, if a stronger ignition source is present, the oxygen should be reduced to 3 and 4 percent, respectively.

Many factors will affect the use of inerting to prevent explosion. Among them are protection of personnel, the hazard to be protected, the required reduction in oxygen concentration, the availability and cost of the inert gas supply, and the necessary control equipment.

Suppression

Suppression is a technique of stopping an explosion before it develops destructive pressures. The factors involved are much the same as those used in the extinguishment of fire, namely: cooling, limiting the supply of oxygen, and inhibiting flame spread.

Despite the rapidity with which combustion proceeds in an explosion, there is a short period of time before which the destructive force is evolved. During this time, the initial pressure rise can be detected by suitable sensing devices, which automatically trigger the release of the suppressing agent, normally an inert gas or liquid that inhibits the combustion process.

Suppression systems can be used in confined spaces such as reactors, mixers, pulverizers, mills, driers, storage bins, ovens, bucket elevator transport systems, and pneumatic transport systems. The effective application of such systems requires careful consideration of many factors. Among them are the characteristics of the dust, the rate of pressure rise, ignition sources, and the characteristics of the suppressant. Though the principles of explosion suppression apply to all installations, each must be individually designed to cover the wide range of variables that will be encountered.

BIBLIOGRAPHY

REFERENCES CITED

1. Johnson, et. al., *Explosibility of Agricultural Dusts*, U.S. Bureau of Mines Report on Investigations 5753, 1961.
2. Palmer, K. N., *Dust Explosions and Fires*, Chapman and Hill, London, 1977.

3. McKinnon, G. P., ed., *Fire Protection Handbook*, Fifteenth Edition, National Fire Protection Association, Quincy, MA, 1981.

NFPA CODES, STANDARDS, AND RECOMMENDED PRACTICES

Reference to the following NFPA Codes, Standards, and Recommended Practices will provide further information on the safeguards for grinding and milling operations discussed in this chapter. (See the latest *NFPA Codes and Standards Catalog* for availability of current editions of the following documents.)

NFPA 61A, *Manufacturing and Handling Starch.*

NFPA 61B, *Prevention of Fires and Explosions in Grain Elevators and Facilities Handling Bulk Raw Agricultural Commodities.*

NFPA 61C, *Prevention of Fire and Dust Explosions in Feed Mills.*

NFPA 61D, *Prevention of Fire and Dust Explosions in the Milling of Agricultural Commodities for Human Consumption.*

NFPA 63, *Dust Explosions in Industrial Plants.*

NFPA 68, *Explosion Venting Guide.*

NFPA 69, *Explosion Prevention Systems.*

NFPA 70, *National Electrical Code.* •

NFPA 651, *Manufacture of Aluminum and Magnesium Powder.*

NFPA 654, *Prevention of Dust Explosions in the Chemical, Dye, Pharmaceutical, Plastics Industry.*

NFPA 655, *Prevention of Sulfur Fires and Explosions.*

NFPA 664, *Prevention of Fires and Explosions in Wood Processing and Woodworking Facilities.*

ADDITIONAL READING

Allen, J., ed., *Grinding*, vol. 1, International Ideas, Inc., Philadelphia, PA, 1968.

Barlow, D. W., et. al, eds., *Grinding*, vol. 2, International Ideas, Inc., Philadelphia, PA, 1972.

Bartnecht, W., *Explosions: Course, Prevention, Protection*, Springer-Verlag, New York, NY, 1981.

Davenport, J. A., "Explosion Losses in Industry," *Fire Journal*, vol. 75, no. 1 (January 1981), pp. 52-56, 71-73.

Explosion Hazards and Evaluation, Elsevier Science Publishing Co., Inc., New York, NY.

Factory Mutual Engineering Corporation, "Grain Storage and Milling," *Loss Prevention Data 7-75*, August 1976, Factory Mutual System, Norwood, MA.

F. M. Eng. Corp., "Combustible Dusts," *Loss Prevention Data 7-76*, August 1976.

Grobel, Edward, "Cryogenic Grinding Gives Process Flexibility," *Cryogenics and Industrial Gases*, vol. 9, no. 4, July/Aug. 1974, pp. 27-30.

Henderson, S. M. and R. L. Perry, eds., *Agricultural Process Engineering*, 3rd ed., AVI Publishing Co., Westport, CT, 1976.

Jaffe, H. M., "Grain Elevator Protection, What's Being Done Today?", *Fire Journal*, vol. 74, no. 3 (May 1980), pp. 131-132.

Jensen, Rolf, et al., *Fire Protection for the Design Professional*, Cahners Books, Boston, MA, 1975.

Milling: *Methods and Machines*, Society of Manufacturing Engineers, Dearborn, MI, 1982.

Milling One and Milling Two, State Mutual Book and Periodical Service, Ltd., New York, NY, 1983.

Perry, Robert H. and Chilton, Cecil H., *Chemical Engineers' Handbook*, Fifth Edition, McGraw-Hill, New York, NY, 1973.

Plaster, C., et. al, eds., *Milling*, vol. 1, 2nd ed., International Ideas, Inc., Philadelphia, PA, 1977.

Richey, C. B., et. al., *Agricultural Engineer's Handbook*, McGraw-Hill, New York, 1961.

Spencer, A. G., *Milling*, vol. 2, International Ideas, Inc., Philadelphia, PA, 1969.

Tuve, Richard L., *Principles of Fire Protection Chemistry*, National Fire Protection Association, Boston, MA, 1976.

Weiby, P. and K. R. Dickinson, "Monitoring Work Areas for Explosive and Toxic Hazards," *Chemical Engineering*, vol. 83, no. 22(1976), pp. 139-145.

35

Storage and Filling of Aerosol Products

Henry C. Scuoteguazza

Aerosol products play a large part in the nation's economy. It would be an unusual household or business that did not possess at least one aerosol product at any time as the variety is extremely wide. Products so packaged include food such as whipped toppings, release agents for cooking and industrial processes, deodorants, hair sprays, carburetor cleaners, paints, pesticides, window cleaners, deicers, and engine cleaners as well as many others.

Containers and Propellants

The typical aerosol container is a small, welded-joint, high-strength metal can with a capacity of up to one quart containing up to 16 oz. (473 mL) of liquid. The top and base of the container are domed to withstand the pressure which may be as high as 240 to 400 psi (1655 to 2758 kPa). A spring-loaded plunger nozzle cap is pressed to release the pressurized product or mixture of liquefied gas and product (see Figure 35.1).

Henry C. Scuoteguazza is Standards Coordinator for Factory Mutual Research Corp., Norwood, MA.

Chapter 35

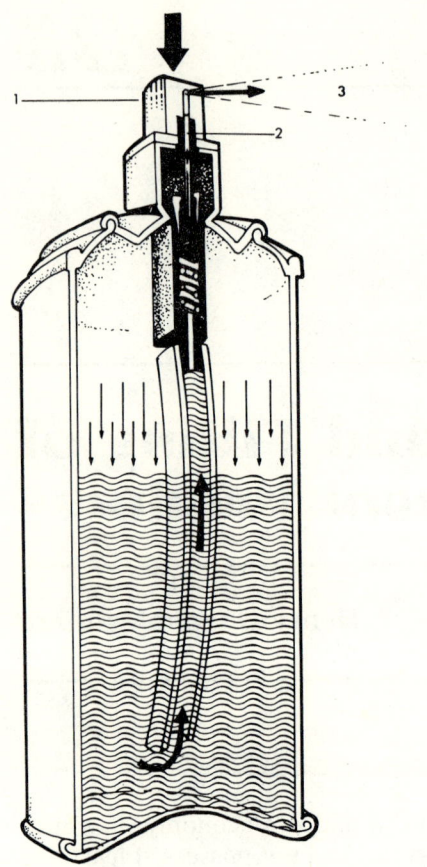

FIGURE 35.1. An aerosol can. When the plunger (1) is pressed, a hole in the valve (2) allows a pressurized mixture of product and propellant (3) to flow through the plunger's exit orifice.

Current propellants are chiefly hydro-carbons, such as propane and butane, in various mixtures depending upon the product and its uses. Other less common propellants are nitrogen, carbon dioxide and nitrous oxide (also methylene chloride). The hydrocarbons are highly flammable, and, although nitrous oxide is nonflammable, it is an oxidizer and can contribute to the fire hazards of aerosols.

Hydrocarbon propellants have little odor and are of low toxicity. Propane and butane have been classified by the FDA as "generally recognized as safe for food additives."

The quantity of propellant in an aerosol product ranges from 0.5 to 90 percent of the total weight of the contents. There is some feeling that the flammability of the base product is a major concern and that the propellant adds little to the overall hazard unless the propellant accounts for most of the can's contents.[1] Despite this, hydrocarbons are extremely flammable and require careful storage and handling.

Hydrocarbon propellants are predominantly propane, isobutane,

TABLE 35.1. Fire Hazard Properties of Hydrocarbon Propellants

	Flammable Limits*		Flash Point		Autoignition Temperature	
	Lower	Upper	°F	°C	°F	°C
Propane	2.1	9.5	−156	−104	940	504
Isobutane	1.8	8.4	−117	−83	890	477
n-Butane	1.8	8.4	−101	−74	860	460

*percent in air by volume

n-butane or a combination of any of these. Specifications for them are much more exacting than for commercial grade liquefied petroleum gases. Under normal conditions, they are chemically inert and do not react with most materials dispensed in aerosol packages. They are gaseous at ambient temperatures and pressures and they will condense to liquids at moderate pressures or low temperatures. They maintain constant pressure at a given temperature until all liquid is converted to gas.

The vapor pressure of these materials and the fact that they produce large quantities of gas when vaporized makes them suitable for propellant purposes. The vapor pressure at 70°F (21.1°C) is 108 ± 6 psig (745 ± 41 kPa) for propane; 31 ± 2 psig (214 ± 13.8 kPa) for isobutane; and 17 ± 2 psig (117 ± 13.8 kPa) for n-butane. They can be mixed to deliver any vapor pressure between the two extremes and thus are suitable for a wide variety of products. The liquid-vapor expansion ratios are about 270 to 1 for propane, 221 to 1 for isobutane and 228 to 1 for n-butane.

The basic fire hazard properties of these hydrocarbon propellants are listed in Table 35.1.

FIRE HAZARDS

Aerosol products have been directly involved, and perhaps responsible for, extensive, costly and sometimes fatal, fires. In one instance, an aerosol can which was thrown into an incinerator exploded, blew off the incinerator door and ignited rubbish. The damage was estimated at $35,000. In another instance, aerosol cans with flammable propellant were stocked close to a heater. Leaking propellant ignited and bursting containers made fire fighting efforts futile. The damage was estimated at $800,000.

At a can-filling plant an explosion involving isobutane killed four employees and injured 18 others. Property damage was estimated at $1,000,000.

Perhaps the most costly fire involving aerosol products was the fire that destroyed the 27-acre (10.9 ha) distribution center of the K Mart Corporation in Falls Township, PA. The fire "involving palletized storage of petroleum-liquid-based aerosol containers overwhelmed the sprinkler

Chapter 35

FIGURE 35.2. These aerosol cans were located in the area of fire origin of the K Mart fire.

systems. The facility was protected with hydraulically designed ceiling sprinkler systems."

"An extremely fast-developing fire spread through the aerosol storage on June 21, 1982; rocketing cans spread the fire throughout this and adjoining areas, overtaxing fire protection systems, resulting in early roof collapse and broken sprinkler piping."

"There was no loss of life. The loss estimate is in excess of $100 million."[2]

Classification of Aerosols

To simplify the consideration of the fire hazards involved, aerosols can be divided into two groups:

1. Flammable products with flammable or nonflammable propellants.
2. Nonflammable products with flammable or nonflammable propellants.

Flammable products can be further subdivided as water miscible or water nonmiscible.

Factory Mutual Engineering Corporation Groups aerosols into three levels according to the percentage by weight of the base product with flammable or nonflammable propellants.[1]

I

Maximum of 25 percent water miscible or nonmiscible flammable products (i.e., 75 to 100 percent nonflammable products).

II

25 to 100 percent water miscible flammable products, 25 to 55 percent nonmiscible flammable products (remaining 45 to 75 percent is nonflammable product).

III

Greater than 55 percent nonwater miscible flammable products.

Level II products containing methylene chloride should be treated as Level III.

Products containing more than 75 percent nonflammable liquids are classified as Level I aerosols. Level I aerosols include such products as shave cream, spray starch, oven cleaners, rug shampoos, air fresheners and some insecticides. Their storage hazard is about the same as ordinary combustible goods in cartons. When a can containing Level I aerosol fails, the nonflammable contents tend to quench the flammable contents.

Level III water miscible products include most personal care products such as deodorants (except oil-based antiperspirants), hair sprays, antiseptics and anesthetics and other products such as some furniture polishes and windshield deicers.

Level II aerosols include automotive products such as undercoaters, and engine and carburetor cleaners, furniture polishes, paints, lacquers, lubricants, and some insecticides.

Tests indicate that mixtures containing 55 percent petroleum liquid and 45 percent nonflammable liquid have burning characteristics significantly below those of 100 percent petroleum liquids. If the product has from 25 to 55 percent flammable liquid and the balance is nonflammable, it can be treated as a Level II aerosol. Addition of methylene chloride may increase the fire hazard.

Fire Behavior of Aerosols

The behavior of filled aerosol cans in fire tests confirms what has often been reported about actual fires. Level II aerosols produced intense fires and ruptured containers rocketed through the test area, occasionally trailing burning liquid. Sometimes flaming packaging material or plastic caps were propelled beyond the test array.

Level III aerosols burned more intensely and rocketed with trailing burning liquid more frequently than Level II aerosols. In addition, they produced dense black smoke that completely obscured visibility in 4 to 5 min. In a 30-ft (9.1 m) high test bay, fire in a single pallet load of Level III aerosol sent the ceiling temperature to 1900°F (1038°C), exceeding 1000°F (538°C) for one minute. The fire opened 36 high-temperature sprinkler heads discharging 0.30 gpm per sq ft (12.2 (L/min)/m^2).

In a rack storage test with both level II and Level III aerosols, there was aisle-jump, with ignition in Level III tests. In both rack and pallet storage, can ruptures dislodged surrounding containers and exposed others inside the load, increasing the fire severity.

As indicated by the fire incidents cited previously, there are three phases in the production and distribution of aerosol products. They are:

1. Transport, transfer, and storage of bulk propellants.
2. Filling or "gassing" the aerosol containers.
3. Storage of charged containers in warehouses.

Aerosol Filling Plants

Because the hydrocarbon aerosol propellants are highly flammable, container-filling plants require special design considerations and operating procedures. Gases present hazards both when they are confined and when they escape. Compressed gases and liquefied gases are affected somewhat differently when heated.[3] A compressed gas simply attempts to expand and the laws of Boyle and Charles predict its behavior. A liquefied gas has a more complicated behavior. The result of heating is the net effect of the simultaneous combination of three effects.

First, the gas phase is subject to the same effect as a compressed gas. Second, the liquid attempts to expand and compresses the vapor. Third, the vapor pressure of the liquid increases as the temperature increases. The combined result is an increase in pressure when the container is heated. Containers of liquefied gases can represent high levels of potential energy release. Container failure can release this energy rapidly and violently resulting in a BLEVE (Boiling Liquid Expanding Vapor Explosion).

THE SAFEGUARDS

In addition to their fire and explosion hazards, hydrocarbon propellants are colorless and odorless and can cause asphyxiation if they displace air. The Department of Transportation (DOT) has promulgated regulations governing all aspects of transportation in interstate commerce. NFPA 58, *Liquid Petroleum Gases, Storage and Handling*, sets the standards for storage and handling of liquefied petroleum gases.

The rules and regulations are both extensive and specific, and acceptable procedures for transferring gases from transport vehicles or containers require:

1. Transfer into outdoor containers or into specially designed structures.
2. Transport vehicles must be at least 10 ft (3 m) from the storage tank.
3. A concrete bulkhead or an equivalent structure, or a shear fitting, in-

stalled between the transport vehicle and the storage tank to retain the integrity of key valves and piping should a mishap occur.
4. The personnel involved in the transfer from a shipping to a storage container must be present throughout the unloading and they must be thoroughly familiar with the rules and regulations.
5. Storage containers should be located outside of any buildings and separated by distances determined by the container size.
6. Valves, piping, regulators, and hose or flexible connections should all meet the requirements of NFPA 58. Hoses used inside buildings should not exceed 6 ft (1.8 m) in length, but should not be so short that they are subject to short radius bends near connectors.
7. Pressure relief devices should be installed on all non-product containers and vented to the outside.

Because release of hydrocarbon propellants within the charging (gassing) room can create an explosive atmosphere, the room should be located and constructed to reflect a combustion explosion hazard. Often the room is a separate building apart from the main production building. The separation should be a minimum of 5 ft (1.5 m) with conveyor openings as small and few in number as possible. The rooms may be located within the main building, but if so, they should be constructed and vented in accordance with the provisions of NFPA 58.

If separate from the main building, the filling room should be at least 10 to 25 ft (3 to 7.6 m) from the bulk storage facilities (depending upon building construction), 25 ft (7.6 m) from a property line which can be built upon, and at least 25 ft (7.6 m) from ignition sources. The wall facing the main building should have a minimum pressure rating of 100 psf (488 kg/m^2). If local wind conditions require increasing the vent release pressure above 20 psf (97.6 kg/m^2), the resistant wall should be strengthened to five times that of the relieving wall. Venting area should be at least 1 sq ft per 30 cu ft (0.1 m^2/m^3) of enclosed volume. Explosion venting areas should be designed to vent at a minimum static pressure of 20 psf (97.6 kg/m^2).

Even in a well-maintained and efficiently operated charging room, enough flammable vapor is released to create a locally explosive atmosphere unless there is a ventilation system which maintains a concentration below the lower flammable limit (LFL). As noted in Table 35.1, for hydrocarbon propellants the LFL is about 2 percent by volume in air.

Because hydrocarbon propellant vapors are heavier than air, the ventilation system should be designed to flow uncontaminated intake air over the floor to pick up vapors and discharge them up and out an exhaust stack. It is wise to maintain a slight negative pressure within the charging room to prevent migration of the vapors to other areas.

Equipment failures or other accidents may release quantities of vapor greater than can be handled by the general ventilation. The system, therefore, should be designed to automatically provide a high ventilation level when the monitoring system indicates a vapor concentration of 20 percent of the LFL.

Local ventilation systems such as hoods over the vapor source can capture almost 100 percent of the released vapors during normal charging operations.

The charging room should be equipped with gas detection devices as part of a system that will shut off the propellant supply, purge the atmosphere, and isolate ignition sources. Such systems are generally electric or electronic. As such, they have the potential of providing an ignition source unless properly selected and installed. NFPA 70, *National Electrical Code*, and NFPA 493, *Instrinsically Safe Apparatus*, describe the selection and installation of the protective system.

Despite the extent and efficiency of fire and explosion preventative measures, there remains the possibility of fire or explosion. As added safety measures, then, fire detection and extinguishing equipment should be installed.

For a charging room, such equipment may be any of the usual systems, i.e, wet-pipe sprinklers, dry-pipe sprinklers, carbon dioxide, foam, dry chemicals, and explosion suppression. The most common extinguishing system is the open-head dry-pipe deluge system with rate-of-temperature-rise detection.[4] This system reacts quickly and allows a water application of at least 0.35 gpm per sq ft [14(L/min)/m^2]. It is effective no matter what the condition of the room and involves no physical danger to personnel.

While carbon dioxide and dry chemical systems are fast acting and effective extinguishers, they do have some disadvantages. Application must be delayed until personnel can be evacuated. Gas fires that are extinguished before the escape of gas has stopped are potential explosions. Foam is of value only for fires in flammable products. Such systems have a limited amount of extinguishing material and, if installed, require a sprinkler system in the event of reignition.

Explosion suppression systems have been effectively employed in charging rooms. Such a system provides protection by extinguishing flame before damage occurs. Ultraviolet sensors react in milliseconds, releasing a Halon agent or water which traps the spark or flame before explosive pressures can develop. Halon 1301 is about twice as effective as water in an explosion suppression system. However, a sprinkler system is desirable in the event of reignition of the flammable material after the Halon vapors have been purged from the area.

Disposal of defective cans is an essential part of aerosol product manufacture. After filling, containers are passed through a warm water bath to detect leaks. Leaking containers are removed and crushed in a separate facility. Safety precautions here should be on the same order as those in the filling room.

Product development and quality control laboratories associated with aerosol product plants also present hazards of fire and explosion. They should be built, equipped and operated in accordance with the various safety codes. Among these are the NFPA standards listed at the end of this chapter.

STORAGE AND FILLING OF AEROSOL PRODUCTS

TABLE 35.2. Arrangement and Protection of Palletized and Solid-Pile Storage of Aerosol Containers[2]

Level	Max Pile Height	Sprinkler Spacing	Temperature Rating & Sprinkler Size	Sprinkler Demand	Hose Stream Demand*	Max. Height Storage to Sprinkler Clearance	Duration Sprinklers & Hose Stream
II	5 ft (1.5 m)	100 ft² (9 m²) max	286°F (141°C) ½ in. (12.5 mm) or 17/32 in. (14 mm)	0.30 gpm/ft² over 2500 ft² (12 mm/min over 230 m²)	500 gpm (1.9 m³/min)	30 ft (9.1 m)	2 hr
II	18 ft (5.5 m)	80–100 ft² (7.4–9 m²)	160°F (71°C) 0.64 in. (16.3 mm) "large drop"	15 heads at 50 psi (344 kPa)	500 gpm (1.9 m³/min)	5 ft or less (1.5 m or less)	2 hr
III	5 ft (1.5 m)	100 ft² (9 m²) max	286°F (141°C) 17/32 in. (14 mm)	0.60 gpm/ft² over 2500 ft² (24 mm/min over 230 m²)	500 gpm (1.9 m³/min)	30 ft (9.1 m)	2 hr
III	10 ft (3 m)	80–100 ft² (7.4–9 m²)	160°F (71°C) 0.64 in. (16.3 mm) "large drop"	15 heads at 75 psi (517 kPa)	500 gpm (1.9 m³/min)	10 ft or less (3 m or less)	2 hr

*For cut-off rooms of less than 2000 ft² (185 m²), hose stream demand is 250 gpm (0.95 m³/min).

Chapter 35

TABLE 35.3. *Protection of Rack Storage of Pressurized Containers of Flammable Products[2]*

Level	Ceiling Sprinkler Arrangement	In-Rack Sprinkler Arrangement[1]	Ceiling Sprinkler Demand	In-Rack Sprinkler Demand	Hose Stream Demand[2]	Duration-Sprinklers and Hose Streams	Clearance Storage to Sprinklers
II	286°F (141°C) rated heads 100 ft² (9.3 m²) max. spacing ½ in. (12.5 mm) or 17/32 in. (14 mm) orifice	165°F (74°C) or less sprinklers 8 ft (2.4 m) apart max. One line at each tier except top. Locate in longitudinal flue spaces of double-row racks.	0.30 gpm/ft² (12 mm/min) over 2500 ft² (230 m²)	30 gpm (0.11 m³/min) discharge per head minimum. Base on operation of hydraulically most remote: (1) 8 sprinklers if one level. (2) 6 sprinklers each of 2 levels if only 2 levels. (3) 6 sprinklers on top 3 levels if 3 or more levels.	500 gpm (1.9 m³/min)	2 hr	15 ft (4.6 m). Need barrier with sprinklers beneath if clearance exceeds 15 ft (4.6 m)
III	286°F (141°C) rated heads 17/32 in. (14 mm) orifice 100 ft² (9.3 m²) max. spacing	165°F (74°C) or less sprinklers 8 ft (2.4 m) apart max. Install in longitudinal flue and on face. Stagger face sprinklers with sprinklers on opposite side of rack.	0.30 gpm/ft² (12 mm/min) over 2500 ft² (230 m²) 0.60 gpm/ft² (24 mm/min) over 1500 ft² (140 m²) to 2500 ft² (230 m²)	Same as Level II	500 gpm (1.9 m³/min)	2 hr	5 ft (1.5 m) or less 5 ft (1.5 m) to 15 ft (4.6 m). If greater than 15 ft (4.6 m), need barrier with sprinklers beneath.

[1]Provide approved rack storage sprinklers with built-in water shields. Locate longitudinal flue in-rack sprinklers at least 2 ft (0.6 m) from rack uprights. Provide at least 6 in. (150 mm) between sprinkler deflectors and top of storage in tier.
[2]For cut-off rooms of less than 2000 ft² (185 m²), hose stream demand is 250 gpm (0.95 m³/min).

Product Warehousing

Although open-head dry-pipe sprinkler systems seem to be preferred for fire protection in gassing rooms,[4] wet, or preaction, systems are recommended for aerosol storage.[1] Sprinkler spacing and size and temperature rating will depend upon the classification of the aerosol product and the storage method. Level II aerosols in palletized or solid pile storage should have 1 sprinkler per 100 sq ft (9.3 m^2) delivering 0.30 gpm per sq ft [12.2(L/min)/m^2] over 2500 sq ft (232 m^2) as a minimum. Other values for Level II and Level III are shown in Tables 35.2 and 35.3 as developed by Factory Mutual Engineering Corporation.

Level I aerosols should be protected as Class III commodities in both palletized and rack storage. Any aerosol products which cannot be identified should also be protected as Level III aerosols.

Whether in rack or palletized storage, Levell III aerosols should be in a cut-off area with walls or ceiling-high partitions having a 1-hr fire resistance rating. Wet-pipe sprinkler protection should be in accordance with the figures in Tables 35.2 and 35.3.

Rack storage of Level II aerosols should also be in a dedicated area with sprinkler protection as shown in the Tables. For small quantities and picking areas, Level II aerosols may be stored on the first tier with in-rack sprinklers provided specifically for that level and the one above. It is important to have in-rack sprinklers in every tier to prevent the fire from spreading throughout the rack.

BIBLIOGRAPHY

REFERENCES CITED

1. Factory Mutual Engineering Corp., "Storage of Aerosol Products," *Loss Prevention Data 7-29S*, Norwood, MA, May 1983.
2. Best, Richard, "$100 Million Fire in K Mart Distribution Center," *Fire Journal*, vol. 77, no. 2 (March 1983), pp. 36-42, 80.
3. McKinnon, G. P., ed, *Fire Protection Handbook*, Fifteenth Edition, National Fire Protection Association, Quincy, MA, 1981, pp. 4-36.
4. Chemical Specialties Manufacturers Association, *Hydrocarbon Propellants: Considerations for Effective Handling in the Aerosol Plant and Laboratory*, Washington, DC, 1979.

NFPA CODES, STANDARDS, AND RECOMMENDED PRACTICES

Reference to the following NFPA Codes, Standards, and Recommended Practices will provide further information on the safeguards for aerosol filling and storage facilities discussed in this chapter. (See the latest *NFPA Codes and Standards Catalog* for availability of current editions of the following documents.)

NFPA 11A, *Medium and High Expansion Foam Systems.*
NFPA 12, *Carbon Dioxide Extinguishing Systems.*
NFPA 12A, *Halon 1301 Fire Extinguishing Systems.*
NFPA 13, *Installation of Sprinkler Systems.*
NFPA 15, *Water Spray Fixed Systems.*
NFPA 17, *Dry Chemical Extinguishing Systems.*
NFPA 18, *Wetting Agents.*
NFPA 30, *Flammable & Combustible Liquids Code.*
NFPA 33, *Spray Application Using Flammable and Combustible Materials.*
NFPA 45, *Fire Protection for Laboratories Using Chemicals.*
NFPA 58, *Liquefied Petroleum Gases, Storage and Handling.*
NFPA 69, *Explosion Prevention Systems.*
NFPA 70, *National Electrical Code.*
NFPA 77, *Recommended Practice on Static Electricity.*
NFPA 91, *Blower and Exhaust Systems, Dust, Stock and Vapor Removal or Conveying.*
NFPA 231C, *Rack Storage of Materials.*
NFPA 493, *Intrinsically Safe Apparatus in Division 1 Hazardous Locations.*
NFPA 704, *Identification of the Fire Hazards of Materials.*

ADDITIONAL READING

Bright, A. W., "Electrostatic Hazards in Liquids and Powders," *Journal of Electrostatics,* vol. 4, no. 2(1977/78), pp. 131-147.

Chemical Engineering Magazine, *Safe and Efficient Plant Operation and Maintenance,* McGraw-Hill, New York, NY, 1980.

Cocks, R. E. and J. E. Rogerson, "Organizing a Process Safety Program," *Chemical Engineering,* vol. 85, no. 23 (1978), pp. 138-146.

"Concern Over Warehouse Fires Leads to Tests on Aerosols," *Fire,* January 1981, pp. 410-412.

Davenport, J. A., "Explosion Losses in Industry," *Fire Journal,* vol. 75, no. 1 (January 1981), pp. 52-56, 71-73.

Deacon, F. L., "Designing Fire Protection to Limit Monetary Loss," *SFPE Technology Report No. 80-2,* Society of Fire Protection Engineers, Boston, MA, 1980.

Dennis, Richard, ed., *Handbook on Aerosols,* ERDA Technical Information Center, 1976.

"Fire and Explosion Hazards with Aerosols," *Fire Prevention,* December 1980, pp. 22-24.

Henry, Martin F., *Flammable and Combustible Liquids Code Handbook,* National Fire Protection Association, 1981.

Hinds, Williams C., *Aerosol Technology: Properties, Behavior, and Measure-*

ment of Airborne Particles, John Wiley and Sons, Inc., New York, NY, 1982.

Johnsen, Montfort A., *Aerosol Handbook*, 2nd ed., Dorland Publishing Co., Mendham, NJ, 1982.

King, R., "Plant Hazards," *Engineering (London)*, vol. 216, no. 4 (1976), pp. 139-145.

Magison, E. C., *Electrical Instruments in Hazardous Locations*," 3rd ed., Instrument Society of America, Pittsburgh, PA, 1978.

Marple, V. A. and B. Y. H. Liu, eds., *Aerosols In the Mining and Industrial Work Environments*, vols. 1-3, Ann Arbor Science Publications, Woburn, MA, 1983.

Nailen, R. L., "Hazardous Area Electrical Equipment," *Fire Engineering*, vol. 131, no. 6 (1978), pp. 32, 34, 36.

Pilborough, L., *Inspection of Chemical Plants*, Gulf Publishing Co., Houston, TX, 1977.

Talbot, G., "Static Electricity," *Fire*, 70(871), 1978, pp. 397-398.

$10 Million Spent for Aerosol Plant Fire Safety," *Fire Engineering*, vol. 134, no. 12 (December 1981), p. 42.

Weiby, P. and K. R. Dickinson, "Monitoring Work Areas for Explosive and Toxic Hazards," *Chemical Engineering*, vol. 83, no. 22 (1976), pp. 139-145.

36

Radioactive Materials

Robert G. Purington

Since Antoine Henri Becquerel's discovery of radioactivity in 1896, the practical uses of radioactive materials have played an increasingly more important role in our daily lives. For example, we all experience background radiation in our natural environment — including cosmic rays and radiation from the naturally radioactive elements, both outside and inside our bodies. Many of us derive beneficial results from radiation machines used for medical examination and treatment purposes. The academic and scientific communities continue to explore the nature of this form of energy and its potential for benefiting humanity. The practical uses of radioactive materials have become an essential part of almost all aspects of modern industry.

Robert G. Purington, a registered professional fire protection engineer, is retired fire chief at the Lawrence Livermore Laboratory in Livermore, California. He has published a number of articles on fire department operations involving radioactive materials and is author of the books *Fire Fighting Hydraulics* and *Handling Radiation Emergencies*.

Chapter 36

USES OF RADIOACTIVE MATERIALS

The general uses of radioactive materials in industry range from the production of electricity by nuclear power plants to smoke detection. There is practically no technical industry that does not use, or benefit from, radioactivity. To illustrate the widespread use of radioactive materials, some major applications are: flow measurement; leak detection; friction and wear studies; measuring solubility of materials; evaluating the efficiency of industrial mixers; detection of minute quantities of a material; material tracking; separation, filtration, distillation, and electrolysis; research; density gages; height, level, and thickness gages; measuring surface moisture; radiography; radioactive catalysts; fiber modification; food preservations; growth inhibition; insect control; polymer modification; luminescence; seed mutation; static elimination; sterilization of pharmaceuticals and medical equipment, X-ray fluorescence; smoke detection; heat sources; power sources; propulsion; and medical diagnosis and treatment.

One of the most extensive uses of radioactive materials, and certainly one of the most controversial, is nuclear power generation. Stripped of its essentials, a nuclear power plant is a relatively simple device for creating a controlled chain reaction with uranium, using the heat that results to boil water and create steam, and then funneling the steam into a turbine to generate electricity. There is nothing simple about the components that make up a nuclear power plant, however. A typical plant is usually a high-pressure giant whose radioactive heart packs the energy equivalent of many hundreds of pounds (100 pounds equals approximately 45 kg) of TNT and combines thousands of tons (one ton equals 0.9072 metric ton) of reinforced concrete, a labyrinth of pipes, tanks, cooling towers and valves, and spiderweb upon spiderweb of electrical circuitry. These components add up to one of the most complex and delicately balanced mechanisms on earth. (Nuclear power plants are discussed in greater detail earlier in this book in Chapter 5, "Nuclear Energy Plants.")

In spite of increasing government regulations, protests from environmental groups, and the negative publicity that resulted from the Three Mile Island nuclear power plant accident in 1979, the nuclear power generation industry continues to expand. In mid-1983 there were 76 nuclear power plants operating in the United States, including one operated by the U.S. Department of Energy. Also in 1983, there were an additional 67 plants under construction or on order. When completed, total installed capacity will be more than 130,000 megawatts of electricity.[1]

Another indication of the status of the nuclear industry is the production of uranium ore which, in 1981 in the United States alone, reached a total production figure of over 13 million tons (11.8 million metric tons).[2] Most of the uranium ore produced in the United States comes from Colorado, New Mexico, Texas, Utah, Wyoming, South Dakota, and Washington.

Some of the products of processes involving radioactive materials are: chemicals and chemical by-products; instruments; pulp, paper, and paper products; primary metals and fabricated metal products; transportation equipment; pharmaceutical and cosmetic products; food products; glass products; rubber products; petroleum products; medicinal products; educational materials; plastics; and plastic products.

Characteristics of Radioactive Materials

What are the characteristics of radioactive materials that make them so useful? First, radioactive materials emit radiation (the emission and propagation of energy through matter or space by means of electromagnetic disturbances and streams of fast-moving particles) when they decay. With proper equipment, ionizing radiation (any radiation — alpha, beta, and gamma — that directly displaces electrons from the outer domains of atoms) can be detected and measured. This is because the radiation produces ionization in other materials. Furthermore, the presence of minute quantities of radioactive materials can be detected. For example, compare the use of standard chemical testing techniques to the use of radioactive tracing techniques. Sensitive chemical tests can detect one molecule in 10 million, whereas with the use of radioactive tracer techniques, one molecule in 10 *billion* can be detected. Thus, radioactive tracers can be 1000 times more sensitive than normal chemical techniques. Using this principle, tiny quantities of radioactive materials can be added to a process, including living entities, and the flow of the materials traced through the process. This is called "tracing techniques." Tracing includes the measurement of flow, friction, and wear; the gaging of thicknesses and densities; and the determination of the presence of hidden, abrasive, or corrosive materials.

An example of the benefits derived from using tracing techniques in agricultural research is their use in determining the best location for fertilizer in order to enhance the growth of plants. Fertilized areas marked as containing radioactive fertilizer or as containing fertilizer with some radioactive components are located in relation to the roots of plants. By measuring the rate at which the plant itself emits radiation, it is possible to determine the optimum placement of the fertilizer for the best uptake of the plant nutrients.

Another characteristic of radiation is that it can pass through opaque materials. However, the density and thickness of the material — as well as the type of radiation — determine the amount of radiation that will pass through the material. Alpha radiation has very low penetrating capability and therefore cannot be used for measurement through materials. Beta radiation has moderate penetrating power and can be detected by sensitive detectors. On the other hand, gamma radiation — which is similar to X-rays — can penetrate considerable thicknesses of even very dense materials. (See Figure 36.1.) This quality can be used to measure the thickness of materials, find voids in metal castings, and detect weak spots in welds. For example, the thickness of rolled sheet material can

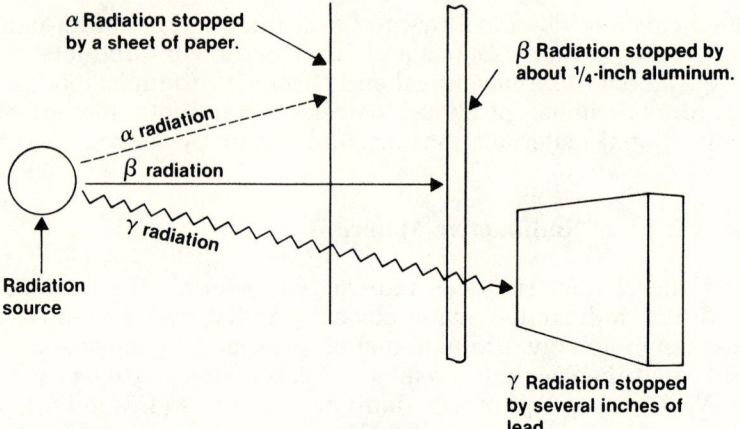

FIGURE 36.1. *Penetrating power of various kinds of radiation (¼ in. equals 6.35 mm).*

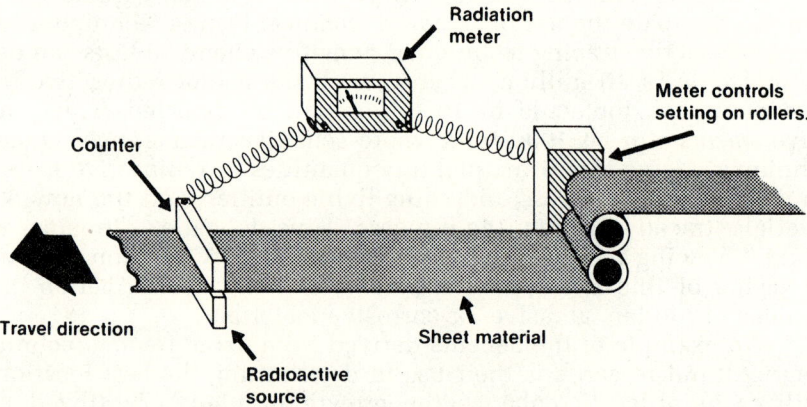

FIGURE 36.2. *Gaging thickness of sheet material using a radioactive source to control the thickness in producing sheet material.*

be measured by placing a radiation source on one side of the material and a radiation detector on the other side. The detector can then measure the amount of radiation passing through the sheet material. The greater the amount of radiation detected, the thinner the material. (Thicker material causes a decrease in the amount of radiation passing through while thin material permits more radiation to pass through.) This technique can be adapted to automatic control by sending an electrical signal to a control device that adjusts the thickness of the materials. (See Figure 36.2.)

Through the use of X-rays (or gamma radiation), radiation can be used to photograph items located inside containers or through solids. This is called radiography — the taking of pictures with radiation other than light. This involves a source of radiation, either an X-ray machine

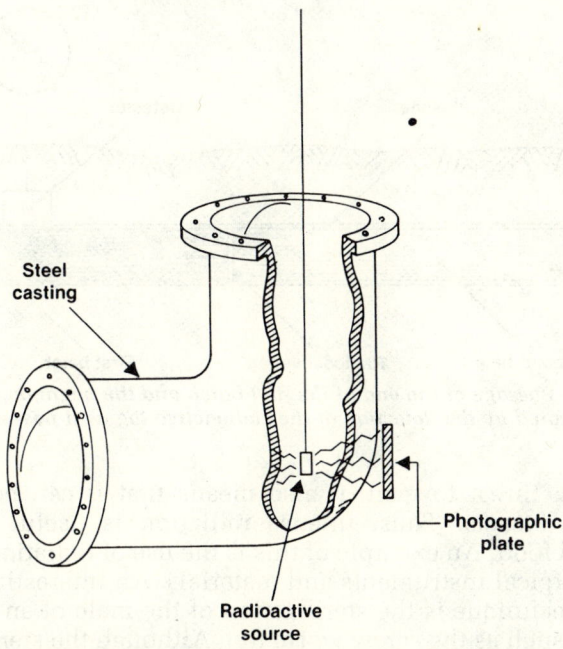

FIGURE 36.3. Radiographic testing of a steel casting. Flaws in the casting show up in the developed photographic plate.

or a radioactive material, being placed on one side of the object to be radiographed, and photographic film being placed on the other side. By properly adjusting the exposure time (and the strength of the radiation) an image appears on the developed film. The image indicates the mass (density and thickness) of the material through which the radiation passes. Since the amount of radiation passing through varies according to the density and thickness of the material, thicker parts (or more dense parts) are less exposed on the film. Common applications of this technique include medical and dental X-rays, and the radiography used to detect imperfections (such as voids) in metal castings. Also, the continuity of welds can be determined through the use of radiography. (See Figure 36.3.)

Still another example of the use of radiation detectors is their use in detecting the transportation of differing batches of oil through pipelines. Pipelines are used by different companies, and it is difficult to tell when the batch belonging to one company ends and another company's starts. To solve this problem, the leading edge of one company's batch of oil is tagged with a radioactive material; the progress of the batch of oil can then be measured by using radiation detectors. (See Figure 36.4.)

In the process of passing through materials, radiation can change the chemical composition of the materials. Radiation can damage and, if intense enough, can kill living organisms. While this means that radia-

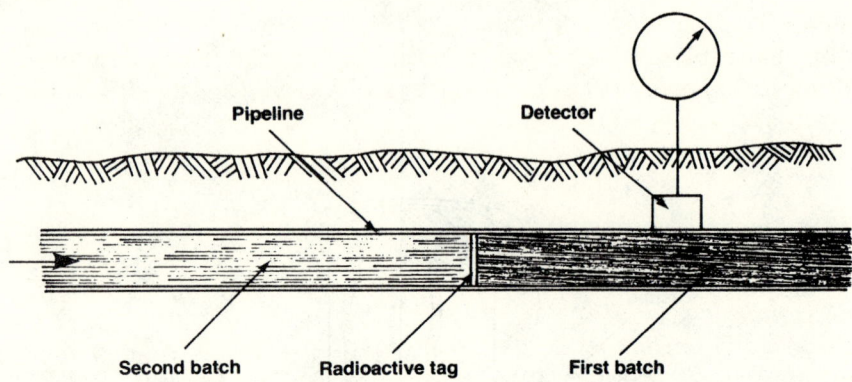

FIGURE 36.4. *Passage of the end of the first batch and the beginning of the second batch is indicated by the detection of the radioactive tag as it passes the detector.*

tion can be a threat to man, it also means that it can be used to kill bacteria and insects. Thus, intense radiation is useful in sterilizing materials and food. An example of this is the use of radiation for the sterilization of surgical instruments and materials. An interesting application of a similar technique is the sterilization of the male of an unwanted insect species (such as the screw worm fly). Although the sterile males continue to compete with the fertile males for mates, the resultant sterile eggs don't hatch. Thus, by releasing a large number of sterile males, the population of the unwanted species decreases.

Because radiation also affects chemical reactions, it can be useful in causing polymerization of materials produced by the plastics industry. Another illustration is the production of ethyl bromide, an important raw material used in the manufacture of drugs. Ethylene and hydrogen bromide gases are mixed in a large tank, the center of which contains a cobalt-60 source. Radiation from the cobalt-60 causes the two gases to form a liquid which can then be removed from the tank. This method is not only less expensive than former methods, but the product is purer.

As radioactive materials decay, the amount of the material is reduced and the amount that is left can be accurately calculated as well as measured. This fact is used in the "carbon dating" of substances.

Decaying radioactive materials, along with the absorption of the radiation by other materials, produces heat which is utilized in nuclear batteries. Nuclear batteries are especially useful for operating devices in remote locations such as in satellites, remote weather stations, in pacemakers, and for providing power for the operation of light beacons.

Another aspect is that some materials can fission, or split apart, releasing tremendous amounts of energy. This type of energy, in the form of heat, is the basis of nuclear power plants. Fission is also used for nuclear explosives.

Fusion, which is the combination of two light elements, also releases great amounts of energy. Currently, fusion is the basis of most modern nuclear weapons. However, fusion may someday be the source of inexhaustible usable power. Research is presently being conducted to extract fusion energy at a rate that can be recovered for the production of power.

HAZARDS

The primary hazard of radioactive materials stems from potential contamination. Contamination, the spread of radioactive materials to unwanted areas of the environment, can threaten living organisms and can result in costly cleanup. Consequently, a major objective of fire protection at industrial locations involved with the handling of radioactive materials is the containment of these materials. Another major fire protection effort is the protection of the individual worker from radiation. Radiation protection is a specialized subject that goes beyond the intended scope of this chapter. Details may be found in publications relating to radiation protection. However, for the purposes of this chapter, it should be noted that radiation protection techniques can impact on fire protection methods and emergency operation — a fact that will be discussed in greater detail later in the chapter.

Generally, radioactive materials are not a fire hazard because of their radioactivity, although a few are pyrophoric, e.g., plutonium and uranium. When in a finely divided state, these materials may ignite spontaneously. Even then, the quantity present is usually so small that other than being a potential source of ignition, they are not a significant fire problem. Where pyrophoric properties are present, special precautions must be taken.

The presence of nonradioactive combustibles together with the radioactive substances indicates a need for important fire protection consideration. This includes combustibles of all sorts. Some radioactive materials require shielding for personnel protection. While lead is often used for shielding, combustible shielding materials are also widely used. Packing materials and other combustibles, as found in many occupancies, present one of the greatest potentials for fire and the subsequent release of radioactive materials.

PRODUCTION

Generally, the production of radioactive materials and radiation can be placed in three categories: (1) uranium (fuel for nuclear reactors), (2) radioisotopes, and (3) radiation-producing machines.

Uranium

Uranium is a naturally occurring radioactive element with the atomic number 92 and, as found in natural ores, an average atomic weight of approximately 238. The two principal natural isotopes are uranium 235 (0.7 percent of natural uranium), which is fissionable and can be used as a nuclear fuel, and uranium 238 (99.3 percent of natural uranium), which is fertile. Natural uranium also includes a minute amount of uranium 234. Uranium is the basic raw material of nuclear energy. Uranium occurs naturally as uranite (uranium oxide), coffinite (uranium hydrous silicate), and carnotite (potassium uranium vanadate). Uranite is also known as pitchblende.

The ore is mined using regular mining techniques (with some special precautions taken to protect the workers from the radioactive daughter products and from inhaling the dust). The ore is then milled, which involves crushing, grinding, leaching, and precipitation. The result is uranium oxide (called yellow cake) which contains 98 percent uranium. The yellow cake is refined by dissolving it in nitric acid to produce uranyl nitrate, after which it is mixed with tributyl phosphate and kerosine, and then settled out and washed in pure water to form uranium trioxide (called orange oxide). The orange oxide is changed to uranium dioxide by hydrogenation, and then treated with hydrogen fluoride gas to form uranium tetrafluoride (called green salt). The uranium tetrafluoride is treated with fluorine gas to form uranium hexafluoride.

At this stage, the uranium has such a low concentration of the fissionable U-235 that it cannot be used as a nuclear fuel. Thus, the uranium hexafluoride is passed through a gaseous diffusion plant to concentrate the U-235 — usually from two to three percent — for normal nuclear reactor use. The result is called enriched uranium, and the residue is called depleted uranium. The enriched uranium is then converted back to uranium dioxide from which it can be fabricated into reactor fuel elements.

Radioisotopes

Radioisotopes (radioactive isotopes) are generally produced in a nuclear reactor by bombarding an appropriate target material with slow neutrons from the reactor. The material is transmuted by a nuclear reaction into the desired radioisotope. For example, bombarding a target of sodium-23 with slow neutrons results in the radioisotope sodium-24. The use of radioisotopes in industry can be denoted as falling into two general categories — tracing and radiography.

Some short half-life radioisotopes are useful as biomedical tracers. However, most medical facilities don't have ready access to a nuclear reactor and materials have to be shipped. If transportation time is excessive, most of the short half-life materials will have decayed by the time they arrive. To help solve this problem, a radionuclide generator (called an isotope cow) has been developed. It consists of a "parent" radioactive species of relatively long half-life which, when decaying, produces the desired short-lived "daughter" product. One example is molybdenum-99 (67-hr half-life), which decays to produce technetium-99 (6-hr half-life) which is useful in medical research. The "daughter" isotope is "milked" by pouring the appropriate liquid reagent, or substance, through a column of the molybdenum-99.[3] Radioisotopes can also be produced by bombarding the target with charged particles or neutrons in one of the several types of particle accelerators.

Radiation Machines

Another method for producing radiation is by use of "radiation machines" which produce or make use of subatomic particles or electromagnetic radiation, or both. Included are X-ray machines, linear accelerators, Van de Graaff generators, and circular accelerators. These machines produce radiation only when running. Otherwise, the fire protection problem is similar to that of any large electrical installation.

Transportation

The transportation of radioactive materials is regulated by the U.S. Department of Transportation (DOT). Transportation may be by any means of carrier. Containers for shipping radioactive materials must undergo rigorous tests before they are approved. Table 36.1 summarizes the tests for these packages. Containers fall into two general classifications: Type A and Type B. Type A containers are designed so that normal transportation conditions will not cause the loss of the radioactive materials or loss of shielding. (See Figure 36.5.) Type B containers must withstand both normal transportation conditions and accidents with only limited loss of shielding and no loss of radioactive materials. (See Figure 36.6.) In addition, there are special containers that require special permits from DOT. In most shipments, several containers are used, one inside the other, before they are placed in the package. (See Figure 36.7.) A special container called a "bird cage" is sometimes used for shipping fissile materials. (See Figure 36.8.) A fissile material is any material fissionable by neutrons of all energies including (and especially) thermal (slow) neutrons as well as fast neutrons (for example, uranium 235 and plutonium 239). Another special type container is used for shipping spent fuel elements from nuclear reactors. (See Figure 36.9.)

Labeling: Each package is labeled according to the quantity and type of material being shipped. The package must have two labels on opposite sides of the package. Vehicles carrying radioactive materials with "Radioactive Yellow III" labels (see Figure 36.10) must also have "radioactive" placards.

TABLE 36.1. *Tests for Packages for Shipment of Radioactive Materials*

Tests	Type A	Type B
Water spray that keeps the package wet for 30 min and after 1½ to 2½ hrs, a 4-ft (1.2 m) free drop onto an unyielding surface.	X	X
A 1-ft (0.3 m) free drop onto each corner. (Applies only to wood or fiberboard packages of 100 lbs (45 kg) gross or less.)	X	X
Impact of the hemispherical end of a 13-lb (5.9 kg) steel cylinder, 1½ in. (38.1 mm) diameter, dropped 40 in (1 m) onto the vulnerable side of the package.	X	X
A compressive load of either 5 times the weight of the package or 2 lb/in.^{2} (14 kPa) times the maximum horizontal cross-section, whichever is greater.		
Direct sunlight at ambient temperature of 130°F (54.4°C). Ambient temperature of −40°F (−40°C) in still air and shade.	X	X
Reduced pressure equal to 0.5 (51 kPa) atmospheres.	X	X
Vibration normally incident to transportation.	X	X
A free drop of 30 ft (9.1 m) onto a flat unyielding surface as to cause the most damage.		
A drop of 40 in. (1 m) onto the end of a rigidly mounted 6-in (152 mm) diameter steel shaft.		X
Whole surface thermal exposure at 1475°F (802°C) for 30 min and not cooled artificially until at least 3 hrs after the test.		X
Water immersion under 3 ft (0.9 m) of water for 8 hrs. (For fissile materials packaging only.)		X

There are three kinds of labels, as shown in Figure 36.10. Radioactive White I is for packages not exceeding 0.5 mrem per hour on the surface. Radioactive Yellow II is for packages not exceeding 10 mrem per hour on the surface and 0.5 mrem per hour at 3 ft (0.91 m) from the package.

Radioactive Yellow III is for packages exceeding these limits and for quantities of fissile material such that criticality safety must be controlled during shipment by special arrangements between the shipper and the carrier. The number of millirem per hour (mrem/hr) at 3 ft (0.91 m) is known as the *transport index*. This index determines the type of package that must be used for a particular shipment.

Emergencies: In the event of an accident during the transportation of radioactive materials, every effort should be made to get the shipping documents. They provide information to judge the degree of hazard in the event that the package has ruptured. If the package is intact, emergency action should proceed without regard for the radioactivity. If the package integrity cannot be assessed and if time permits, keep people out of the area until expert advice is available. Preplanning should always

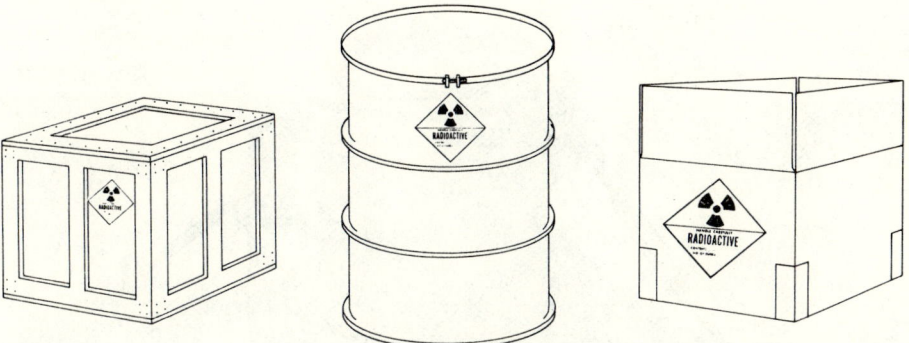

FIGURE 36.5. *Three different styles of Type A shipping packages for radioactive materials.*

be done if possible. Unfortunately, preplanning for transportation accidents is difficult. However, by preplanning for emergency action with the users and possessors of radioactive material, much can be learned about the nature and extent of possible problems in transportation facilities.

PROTECTION FROM RADIOACTIVE MATERIALS

As mentioned earlier in this chapter, although radioactive materials are not in themselves a significant fire problem, it is necessary to make provisions to safeguard people and the environment from the effects of radiation. Adequate fire protection is an important aspect of such protection. However, the methods used to protect personnel from radiation hazards can complicate both emergency operations and the fire protection measures. Hazards to personnel fall into three general categories: (1) external radiation, (2) contamination, and (3) criticality.

External Radiation

The threat of external radiation results mainly from gamma radiation and X-rays. While beta emitters may also be considered as an external radiation hazard, their effect is rather short range. Also, other particles (such as protons) from particle accelerators can cause radiation injuries. Fortunately, the latter are generally confined to research installations. The three ways to protect people from external radiation hazards are: (1) time, (2) distance, and (3) shielding.

Time: Since the effect of radiation exposure is accumulative, the amount of time a person remains in a radiation field has an important bearing. Therefore, the more limited the time spent in a radiation field, the more

Chapter 36

FIGURE 36.6. *A Type B package is shown strapped to a pallet for shipment.*

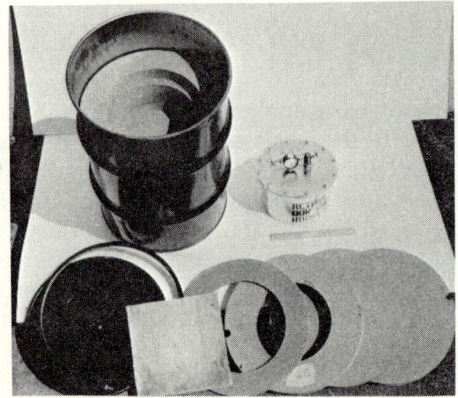

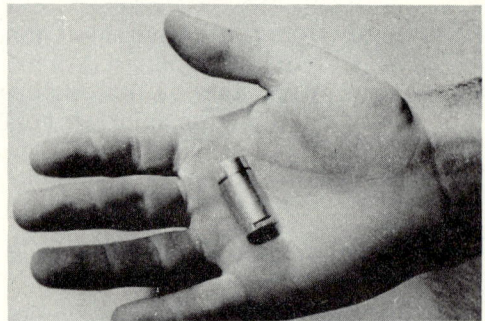

FIGURE 36.7a. *(upper left.) Various inner shipping containers for radioisotopes.*
FIGURE 36.7b. *(upper right.) The inner and outer units of a shipping container for radioactive materials.*
FIGURE 36.7c. *(left.) A sealed container for a small radiation source.*

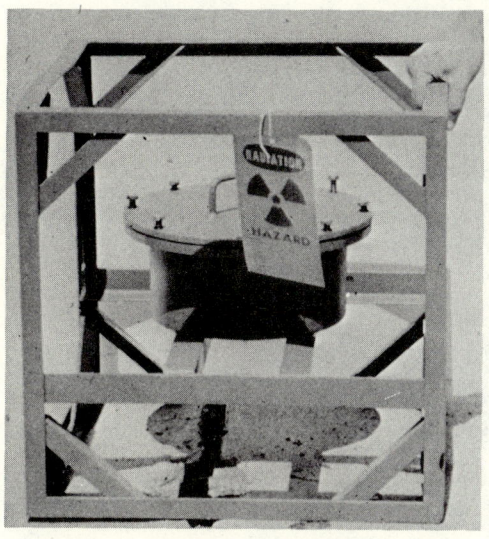

Figure 36.8. A "bird cage" container that prevents adjacent containers from coming too close to each other for criticality safety.

limited the exposure. Exposure time is usually regulated by administrative controls.

Distance: As with ordinary light, external radiation is attenuated by the air, and its effect decreases with distance. It naturally follows that the greater the distance from the radiation source, the less the exposure rate. Remote operations are often used to assure adequate distance. This means using remote handling devices such as long-handled tongs and manipulators.

Shielding: While gamma rays and X-rays pass through solid materials, they are also attenuated by such materials; i.e., the strength of the radiation decreases with the passage through materials. Generally, shielding depends on the density and thickness of the shielding materials. Thus, lead — a particularly dense material — is an effective shielding substance. Other materials are also used for shielding purposes, including water, mineral oil (in hot-cell windows), high-density concrete, and earth. Neutrons require special shielding since the absorption of neutrons does not follow the same rules as do gamma rays and X-rays. The type of shielding for neutrons depends on the velocity of the neutrons (their energy). Generally, neutron shielding ability is determined by the "cross section" of the material used; i.e., the probability of neutron capture. Substances such as water and paraffin, with high hydrogen contents, have large cross-sections and are useful for neutron shielding.

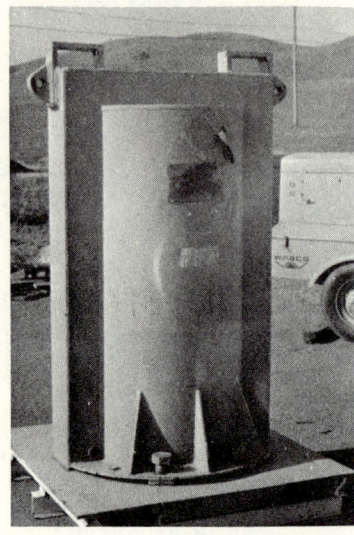

FIGURE 36.9. *A container for shipping spent fuel elements.*

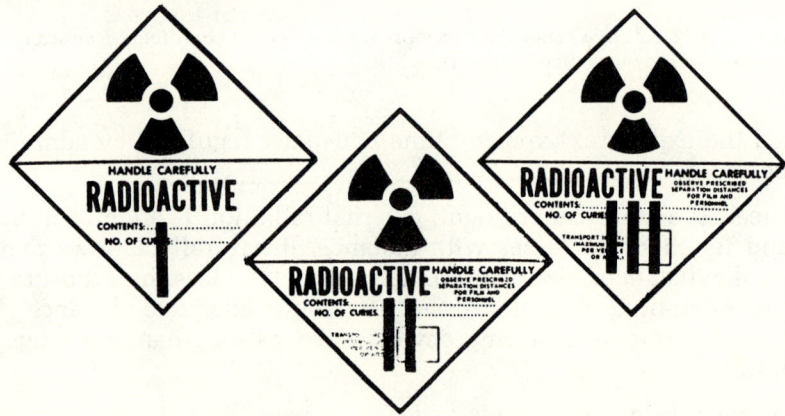

FIGURE 36.10. *Labels for radioactive shipment.*

To provide protection from high-energy gamma radiation by means of shielding and distance, hot cells and remote operating devices (manipulators) are often used. The hot cells (used for the most hazardous radioactive materials) are constructed of good shielding materials and have windows that are often several feet thick. Between the glass on each side of the windows is a core of water or mineral oil. Hot cells often have fixed extinguishing or detection systems. (See Figure 36.11.)

FIGURE 36.11. *A typical hot cell installation. Note the manipulators for handling radioactive material within the cells. Portable extinguishers adapted for protection of the interior of the cells can be seen near the tops of the viewing ports.*

Contamination Control

The end result of effective contamination control is the containment of the radioactive materials. A major objective of contamination control is to prevent radioactive materials from entering the human body. It must be noted that the unwanted spread of radioactive materials is considered highly undesirable, even though the materials are in a state that presents no personnel hazards. Thus, "spills" — including those resulting from fires -can result in costly cleanup operations (called decontamination).

The extent of contamination control depends on a number of factors, including: the radioactive substance involved; the chemical and physical characteristics of the material; the physical state of the materials (i.e., liquid, solid, or gas); the degree of subdivision of the substance (e.g., powders); and the packaging of the material.

Different isotopes emit different kinds of radiation and with different strengths. Each isotope has its own chemical characteristics which affect the contamination control efforts. For example, tritium gas is much less of a hazard when breathed than if tritiated water (containing tritium) enters the body. This is because the tritiated water is retained in the body for a longer period of time, thus resulting in greater damage to the body.

The physical state of the material is also important. For example, because radioactive gases can diffuse in the air if released, they are difficult to contain and must be kept in closed systems. On the other hand, solid masses of radioactive substances are generally easier to contain. An exception is plutonium which, even in a solid state, can result in the spread of contamination when left out in the open. This is because plutonium left out in the open reacts chemically with the air to form plutonium oxides. Plutonium oxides thus produced are very fine powders and are easily picked up and dispersed by air currents. Consequently, plutonium — even in solid form — must be handled with some degree of containment.

Fine powders can be easily spread if released. When airborne, they can cause respiratory problems and result in widespread contamination. Sealed sources, which are radioactive materials that are sealed and bonded to encapsulating materials or formed as ceramic materials, present little hazard of contamination. Sealed sources generally fare well in fires.

The efforts taken to contain radioactive materials are complicated, and there are many publications on this subject. For purposes of this chapter, the general methods described herein are: (1) containment, and (2) ventilation control.

Containment: Containment includes the handling of radioactive materials in special enclosures such as hoods and glove boxes. (See Figure 36.12.) Hoods are used for low-hazard radioactive materials, usually open in the front, and depend upon the flow of air into the hood to prevent the movement of radioactive materials outside the hood. Some hoods have fixed extinguishing systems with either manual or automatic actuation. Glove boxes are constructed of metal, wood, or plastic. From a fire standpoint, the gloves and glove ports are the weakest part of a glove box. Containment may also include special packaging (such as plastic bags).

Ventilation control: Ventilation control is one of the most powerful methods for keeping radioactive materials in their proper place. However, ventilation control sometimes complicates the fire protection problem. The general objective of ventilation control is to provide air flow in a manner that eliminates the unwanted spread of the radioactive material. Usually, this means that areas containing radioactive materials are at a negative pressure (less than the pressure of the surrounding area) with respect to outside areas. This pressure differential causes the air to flow towards the radioactive area, thus keeping the radioactive material inside. Consequently, even if the radioactive materials are released, they will tend to stay inside.

Air being moved from radioactive areas must be filtered. In large

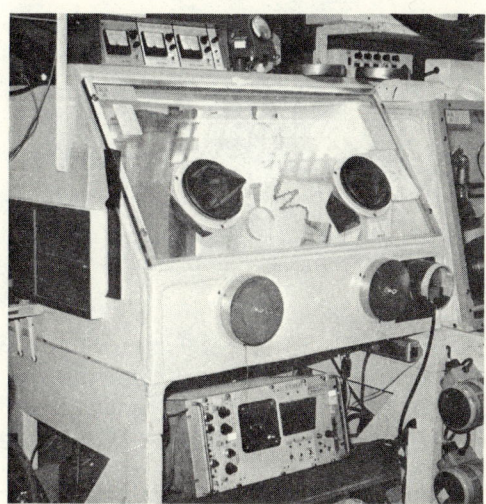

FIGURE 36.12. *A typical glove box installation. Note the covers over the unused glove ports.*

industrial installations, huge banks of high-efficiency filters are used. Smaller installations may use only a few filters. (See Figure 36.13.) Since filters and the materials they pick up are often combustible, some method of fire protection must be provided.

Criticality

The fissioning of materials can result in the release of neutrons and gamma radiation. If the rate of fissioning becomes self-generating (a chain reaction), it is called criticality. Accidental chain reactions are undesirable, and special efforts must be taken to prevent them. These efforts generally center around keeping the amount of fissile materials below the critical mass (geometry of the material affects the critical mass), and regulating the presence of moderators and reflectors. Moderators are substances that can slow neutrons down so that the neutrons can cause additional fissions. Reflectors are materials that reflect neutrons back to the fissile material; they can also affect the rate of fission. Water is a good moderator and a good reflector. As such, it is often eliminated from areas where fissile materials are kept. However, experience has shown that the danger that results from water used for fire fighting, sprinklers, and hose lines is much less than formerly believed.

FIRE PROTECTION

With a few exceptions, fire protection for facilities containing radioactive materials is usually similar to that provided for most industrial oc-

FIGURE 36.13. Several filters connected to glove boxes (glove boxes are out of the picture). Note the combustible flexible ventilation hose. This type of hose is not recommended.

cupancies. A prime objective of fire protection in facilities containing radioactive materials is to prevent the release of radioactive substances as a result of fire. The degree of protection required depends on the amount and nature of the material. For example, no special protection is needed for facilities using sealed sources, whereas the ultimate protection is necessary for facilities handling plutonium.

Water

In most cases, water — especially from automatic sprinklers — is the best way to control fire, including fires involving radioactive materials. Nevertheless, the use of sprinkler systems in radioactive areas is often debated, even though experience has shown that one of the best fire protection techniques for these facilities is automatic sprinklers. The arguments against sprinklers are usually unfounded, even when fissile materials are present. The materials are usually in some type of containment (such as hoods or glove boxes), and are therefore shielded from the water. Opponents of sprinkler systems cite the possibility of the spread of contamination by means of the water used to control the fire. Even if this is possible, the threat of airborne contamination resulting from uncontrolled fires is greater. Consequently, some contaminated water is a small price for quick and effective fire suppression by means of automatic sprinklers. If the threat of the spread of contaminated water is serious, sumps, drains, berms, and other means of water containment should be provided. However, care should be taken if fissile materials are present since the criticality hazard may be greater when such substances are dissolved or suspended in water. In such cases, the use of sumps and other collection methods should be carefully evaluated.

It is important to note that water from sprinklers seldom reacts violently with radioactive substances. Occasionally, sodium or other water-reactive substances are used in nuclear reactors or in research institutions. In these cases, a careful evaluation of the fire protection requirements and alternate fire suppression methods must be specified.

Generally, it is best to protect an entire facility with sprinklers. In those areas where water might produce serious hazards, the sprinklers can be replaced with plugs. If the hazard is subsequently reduced, the sprinklers can be easily reinstalled. In summary, automatic sprinklers are recommended — with very few exceptions — for facilities handling radioactive materials.

Detection

In some facilities, automatic fire detection should be provided. Detection may be a desirable adjunct to automatic suppression systems (but never a substitute) for early detection of incipient fires. Thus, fires may be controlled manually before the activation of the suppression systems.

Construction

Buildings containing radioactive materials should be fire-resistive or of noncombustible construction. Single-story facilities are usually recommended with no below-grade areas where contaminated water from fire suppression operations can collect. On the other hand, subgrade areas may be specified to provide collection basins for the contaminated water. The choice depends on the result of the hazard analysis.

If the interior finishes are designed for ease of decontamination, care must be taken to assure that materials with high flame spread characteristics are not used.

Ventilation Systems

Ventilation systems require special consideration. In extreme cases (such as facilities that handle plutonium) the ventilation system should continue to operate, even during fires. Consequently, ventilation systems in these facilities should be designed so they don't "pump" smoke to unaffected areas. Fortunately, these systems are usually the "once through" nonrecirculating type.

Ordinary fire dampers in ventilation ducts may be undesirable for protection of fire separation penetrations. In such cases, the entire ventilation system can be enclosed with fire-resistive materials and considered as part of the "fire envelope." Another possibility is "spray dampers" consisting of spray sprinklers located inside the ventilation ducts. The spray heads cool the hot fire gases flowing in the duct.

Fire suppression systems are often needed to protect the filter banks. While water spray is usually best, provisions must be made to keep the water away from the filters since the water may render the filters ineffective. Carbon filters are sometimes used for filtering radioactive gases. However, these filters require special protection since, once they start to burn, they are difficult to extinguish. Normal heat and smoke removal techniques can't be used in facilities handling radioactive ma-

FIGURE 36.14. *A typical glove box showing the gloves extending from the ports in the viewing window at left. Note the portable fire extinguisher adapted for discharging through a fixed piping arrangement into the box.*

terials because the smoke may be contaminated. Thus, smoke vents and windows are seldom recommended.

Compartmentation

Keeping the areas that handle radioactive substances as small as possible, yet consistent with efficient operation, is important. This may not always be possible in facilities processing radioactive materials. However, steps to reduce the spread of fire (and contamination) from one area to the next are essential. Large, open areas are undesirable from both the fire spread and the contamination spread point of view.

Housekeeping

Good housekeeping is an important aspect in the reduction of the fire potential in areas where radioactive materials are handled. The accumulation of combustibles is often a by-product of processes and packaging operations, especially in research facilities. Regular disposal of combustibles and special attention to the reduction of combustibles is essential. Also, the use of flammable liquids and gases should be tightly controlled. Processes requiring the use of flammable liquids should have the same type of fire protection normally used for flammable liquids.

Life Safety

Generally, life safety problems at industrial locations that handle radioactive materials are similar to those of other industrial locations. Adequate detection of radiation is required for personnel protection, and detection methods include the use of portable detectors, hand and foot detectors, and air monitors. Where fissile materials are present, criticality detectors are often used. In such instances, the use of automatic alarms with both local alarms and signals transmitted to the central alarm headquarters should be considered.

Adequate exits should be provided and, if the hazard potential is extreme, there should be more than one way out. The blocking of exits by equipment should be avoided, and congestion, often a problem in these facilities (particularly in research institutions), should be guarded against. Sufficient space should be provided around all equipment used for processing radioactive materials, and there should be ample aisle space.

Glove boxes, hoods, hot cells, and other containment facilities may require special fire protection systems or built-in detection systems. (See Figure 36.14.) A particular problem is the possible overpressuring of the glove box by the extinguishing agent. It is necessary that adequate lighting be provided, and in large buildings emergency lighting should be required.

BIBLIOGRAPHY

REFERENCES CITED

1. "Nuclear News," vol. 26, no. 10 (August 1983), pp. 97-99.
2. *Statistical Data of the Uranium Industry*, GJO-100(82), January 1982, USDOE, Grand Junction Area Office, CO.
3. Glasstone, Samuel, "The Uses of Isotopes and Radiation," (Chapter 17), *Source book on Atomic Energy*, Third Edition, D. Van Nostrand Co., Inc., Princeton, NJ, 1967, pp. 672-674.

NFPA Codes, Standards, and Recommended Practices

Reference to the following NFPA Codes, Standards, and Recommended Practices will provide further information on the safeguards for radioactive materials discussed in this chapter. (See the latest *NFPA Codes and Standards* for availability of current editions of the following documents.)

NFPA 13, *Installation of Sprinkler Systems.*
NFPA 90A, *Air Conditioning and Ventilating Systems.*
NFPA 101, *Life Safety Code.*
NFPA 801, *Facilities Handling Radioactive Materials.*
NFPA 802, *Nuclear Reactors.*
NFPA 803, *Nuclear Power Plants.*

Additional Reading

Asimov, I. and T. Dobzhansky, *Genetic Effects of Radiation,* (Understanding The Atom Series), U.S. Atomic Energy Commission, Division of Technical Information, Oak Ridge, TN, Sept. 1966.

Baker, P. S.; D. A. Fuccillo, M. W. Gerrard, and R. H. Laffety, *Radioisotopes in Industry,* (Understanding The Atom Series), U.S. Atomic Energy Commission, Division of Technical Information, Oak Ridge, TN, 1965.

Burchsted, C. A. and A. B. Fuller, *Design, Construction, and Testing of High-Efficiency Air Filtration Systems For Nuclear Application,* Oak Ridge National Lab., Tenn. Report No. ORNL-NSIC-65, Jan. 1970 (Availability: NTIS).

Fire Protection Guide on Hazardous Materials, National Fire Protection Association, Quincy, MA, 1978.

Firgerio, N. A., *Your Body and Radiation,* (Understanding The Atom Series), U.S. Atomic Energy Commission, Division of Technical Information, Oak Ridge, TN, 1966.

Glasstone, S., *Controlled Thermonuclear Reactions, an Introduction to Theory and Experiment,* Van Nostrand, Princeton, NJ, 1960.

————, *Source Book on Atomic Energy,* 3rd. ed., Van Nostrand, Princeton, NJ, 1967.

Guide and Checklist For Development and Evaluation of State and Local Government Radiological Emergency Response Plans in Support of Fixed Nuclear Facilities, U.S. Nuclear Regulatory Commission, Office of International and State Programs, Washington, DC, March 1977. Report No. NUREG-75-111. (Availability: NTIS).

King, R. W. and J. Magid, *Industrial Hazard & Safety Handbook,* Butterworths Publishers, Inc., Woburn, MA, 1979.

McKinnon, G. P., ed.,"Nuclear Reactors, Radiation Machines, and Facilities Handling Radioactive Materials," in *Fire Protection Handbook*, 15th Ed., National Fire Protection Association, Quincy, MA, 1981.

Operational Accidents and Radiation Exposure Experience Within the United States Atomic Energy Commission, 1943 - 1975, U.S. Atomic Energy Commission, Division of Operational Safety, Washington, DC, Report No. WASH-1192 (Rev.) (Availability: NTIS).

Osborne, T. S., *Atoms In Agriculture, Applications of Nuclear Science to Agriculture*, (Understanding The Atom Series), U.S. Atomic Energy Commission, Division of Technical Information, Oak Ridge, TN, 1963.

Particle Accelerator Safety Manual, Brobeck (William M.) and Associates, Berkely, Calif., Oct. 1968. For National Center for Radiological Health, Rockville, MD, Report No. MDRP-68-12.

Phelan, E. W., *Radioisotopes In Medicine*, (Understanding The Atom Series), U.S. Atomic Energy Commission, Division of Technical Information, Oak Ridge, TN, 1966.

Purington, R. G. and W. Patterson, *Handling Radiation Emergencies*, National Fire Protection Association, Quincy, MA, 1977.

Singleton, A. L., *Sources of Nuclear Fuel*, (Understanding the Atom Series), U.S. Atomic Energy Commission, Division of Technical Information, Oak Ridge, TN, 1968.

Urrows, G. M., *Food Preservation By Irradiation*, (Understanding the Atom Series), U.S. Atomic Energy Commission, Division of Technical Information, Oak Ridge, TN, 1968.

Yao, C., J. DeRis, S. N. Bajpai, and J. L. Buckely, *Evaluation of Protection from Explosion Overpressure In AEC Gloveboxes*, Factory Mutual Research Corp., Norwood, MA, Dec. 1969. Report Nos. C00-1393-1; and FMRC-16215.1 (Availability: NTIS)

Part Four

GENERAL OCCUPANCY FIRE HAZARDS

37

Flammable and Combustible Liquids Handling and Storage

Richard D. Stalker

Flammable and combustible liquids are found in virtually every industrial plant. The quantities of these liquids in any given plant can vary from a few ounces (one oz equals 30 m L) in aerosol cans to several thousand gallons (1000 gallons equals approximately 3.78 m^3) in bulk storage tanks. This chapter discusses the hazards associated with flammable and combustible liquids and the safeguards that should be observed in their storage and handling to avoid their exposure to ignition sources.

NFPA 30, *Flammable and Combustible Liquids Code*, is the standard that is widely used as the basis of legal requirements for the storage and handling of flammable and combustible liquids. A familiarity with NFPA 30 will give the reader the detailed information needed to intelligently apply the principles of good storage and handling practices discussed in the following pages.

Flammable liquids are defined as those liquids having a flash point below 100°F (37.7°C) and having a vapor pressure not exceeding 40 psi (276 kPa); combustible liquids are defined as those liquids having a flash point at, or above, 100°F (37.7°C). (Table 37.1 gives further information

---

Richard D. Stalker is Property Protection Manager for the Stauffer Chemical Company of Westport, Connecticut.

TABLE 37.1. *Classification of Flammable and Combustible Liquids*

Combustible Liquids
(flash points at or above 100°F (37.8°C)

Class II Combustible Liquid:	Flash point at or above 100°F (37.8°C) and below 140°F (60°C).
Class IIIA Combustible Liquid:	Flash point at or above 140°F (60°C) and below 200°F (93.4°C).
Class IIIB Combustible Liquid:	Flash point at or above 200°F (93.4°C).

Flammable Liquids
(flash points below 100°F [37.8°C] and vapor pressures not exceeding 40 psia at 100°F (37.8°C)

Class IA Flammable Liquid:	Flash point below 73°F (22.8°C) and a boiling point below 100°F (37.8°C).
Class IB Flammable Liquid:	Flash point below 73°F (22.8°C) and a boiling point at or above 100°F (37.8°C).
Class IC Flammable Liquid:	Flash point at or above 73°F (22.8°C) and below 100°F (37.8°C).

on the classification of flammable or combustible liquids.) Flammable liquids are volatile in nature and are, for the most part, continually giving off vapors that cannot be seen by the naked eye and that are heavier than air. Combustible liquids, when heated to temperatures above their flash points, take on many of the characteristics of flammable liquids and may become just as hazardous as the more volatile liquids. The one main difference between the two classes of liquids is the ability for the vapors from the liquids to flow from one area to another. Vapors from flammable liquids can easily travel along the surface, down stairs, elevator shafts, or air ducts to areas far removed from their source. Vapors from combustible liquids, on the other hand, will not travel away from their source unless the atmospheric temperature remains above the flash point of the liquid.

Because of their inherently hazardous characteristics and widespread use, flammable and combustible liquids are quite frequently contributing factors in major fire losses. A spark or other source of ignition which might otherwise be harmless will readily result in fire or explosion in the presence of ignitible mixtures of flammable vapors. Based on loss reports submitted to NFPA on major fires (property damages of $500,000 or more) at manufacturing plants from 1973 to 1982, a flammable or combustible liquid was identified as the primary fuel accidentally ignited in 25 percent of the fire and explosions. Factors influencing the involvement of flammable and combustible liquids in fires are:

1. Lack of adequate personnel training in safe operating procedures;
2. Hazardous operations not isolated from other operations;

3. Improper use of equipment and flammable liquids;
4. Poor property maintenance and supervision;
5. Lack of adequate fire control systems.

Since the specific characteristics of flammable and combustible liquids vary, as do their required handling and storage precautions, this chapter does not attempt to cover the subject in all of its many details (NFPA 325M, *Fire Hazard Properties of Flammable Liquids, Gases, and Volatile Solids*, lists the known properties of some 1,500 different materials). It is intended only to provide basic general information regarding the hazards and the precautions for the storage and handling of flammable and combustible liquids in general industry, as opposed to those associated with the manufacturing of the liquids.

For more detailed information on specific materials or operations, it is advisable to consult the appropriate National Fire Protection Association standards, product trade association guidelines, underwriters' manuals, and other specific handbooks (see Bibliography).

HAZARDOUS CHARACTERISTICS

In evaluating the hazards associated with the storage and handling of flammable and combustible liquids, one must remember that it is not the liquids themselves that burn or explode, but rather the flammable vapors resulting from the evaporation of liquids when exposed to temperatures above their flash points. Since, by definition, most flammable liquids are normally stored and handled at temperatures above their flash points, they continually give off vapors that are easily ignited when the vapor-air mixture is within the flammable (explosive) range.

Although the flash point of a liquid is commonly referred to as being the most important criterion of the relative hazard of flammable and combustible liquids, it is by no means the only criterion that should be used in evaluating the hazard. The ignition temperature, flammable (explosive) range, rate of evaporation, vapor density, viscosity, specific gravity, solubility in water, and boiling point all have a bearing on the hazards of a liquid. The flash point and other characteristics which determine the relative susceptibility of a liquid to ignition have practically no influence on its burning rate after the fire has been burning for a short time. On the other hand, characteristics such as rate of evaporation, viscosity, solubility in water, etc., are of prime importance in determining how a fire will behave after ignition and during fire fighting operations.

Ignitible mixtures occur when the concentrations of vapors in air are within a definite percentage range, which is commonly referred to as the flammable (explosive) range. The lower limit of the range is known as the lower flammable limit (LFL) and is the minimum concentration below which propagation of flame does not occur upon contact with a source of ignition. The upper limit of the range is known as the upper flammable limit (UFL) and is the maximum vapor-air concentration above which propagation of flame does not occur. For example, the flammable limits for 92 Octane gasoline are 1.4 percent (lower) and 7.6 percent (upper).

Fires and less violent explosions occur when the vapor-air mixtures are near the lower or upper limits of flammable range. When the mixtures are near the median range more intense explosions occur, if the mixtures are ignited within a confined space.

Flammable Liquid Fires

A flammable liquid fire has been defined as the combination of flammable liquid vapors and air with the evolution of heat and light without significant pressure development. The heat of combustion of a flammable liquid fire is about 20,000 Btu/lb (46 520 kJ/kg) or approximately 2½ times that of wood. Flammable liquids are easily ignited and difficult to confine; combustible liquids are more difficult to ignite and easier to confine and extinguish.

The burning rates of flammable and combustible liquids vary with environmental conditions, heat of combustion, latent heat of vaporization, and barometric pressure. For example, the lighter, more volatile liquids, such as gasoline and low flash point solvents, will burn rapidly. When confined to a tank or open container, they will burn at a rate of 8 to 10 in. (203 to 254 mm) of liquid per hr. Heavier, less volatile, combustible liquids, such as fuel oil, will burn at a much lower rate of 5 to 7 in. (127 to 178 mm) of liquid per hr. The normal heat release from a confined flammable or combustible liquid fire is about 10,000 Btu per min per sq ft [approximately 114 000 (kJ/min)/m^2] of burning surface.

Fire in unconfined liquids resulting from spills, leaks, or overflows on relatively level surfaces will also burn with a heat release rate of about 10,000 Btu per min per sq ft [approximately 11 400 (kJ/min)/m^2] of burning surface area. In an unconfined spill each gallon of liquid will cover approximately 20 sq ft (1.86 m^2) of level surface, for the lighter more volatile liquids. The vapors, however, may spread over a much greater area prior to ignition, resulting in a much larger area of involvement.

Spray fires result from leaks in systems under pressure, i.e., hydraulic oil lines on presses and similar production equipment, liquid transfer piping systems, spray finishing equipment, etc. The spray from such a leak is generally much easier to ignite, even at temperatures well below the flash point of the liquid. The liquid in a spray fire will burn nearly as fast as it is released, liberating as much as 120,000 Btu per gal (33 524 kJ/L) for the more volatile liquids. If ignition does not take place immediately after the leak occurs, an explosion can take place with the low flash point liquids in confined areas.

FLAMMABLE AND COMBUSTIBLE LIQUIDS HANDLING AND STORAGE

Explosions

There are three basic types of explosions to be concerned about when evaluating the hazards of storing and handling flammable or combustible liquids in industrial plants: combustion explosions, detonation explosions, and boiling liquid expanding vapor explosions (BLEVEs).

Combustion explosions: These involve the rapid combination of a flammable liquid vapor and air with the evolution of heat and light and an increase in pressure. Such explosions occur when a mixture of flammable vapor and air is within the explosive range and is ignited. Combustion is very rapid with a flame propagation rate of about 7 feet per second (2.13 m/s). In small-scale tests, heat release rates as high as 10,000,000 Btu per min per gal (2.8 million kJ/L) of liquid have been determined. If unvented, the pressures developed during an explosion may reach 6 or 7 times the initial absolute pressure.

Detonation explosions: These differ from combustion explosions primarily in the rate of heat release, which is many times greater in a detonation explosion. The shock wave created by a detonation travels through the explosive mixture at a velocity of 1 to 5 miles per sec (1.6 to 8 km), depending upon the chemical and physical properties of the mixture.

Boiling liquid expanding vapor explosions: BLEVEs, in the context of flammable liquids, can occur when a confined liquid is heated above its atmospheric boiling point by an external source of heat or exposure fire, and is suddenly released by the rupture of the container due to overpressurization by the expanding liquids. A portion of the superheated liquid immediately flashes to vapor and is ignited by the external heat source, releasing heat at a rate much lower than that of the combustion explosion, but generally for a longer period of time.

Explosion hazards primarily exist in enclosed areas, such as small rooms, process equipment, or storage tanks, whenever one of the following conditions exist:

1. The liquid has a flash point (closed cup) of less than 20°F (6.7°C)
2. The liquid has a flash point of less than 110°F (43.3°C) and is heated to more than 60°F (15.5°C) above its flash point
3. The liquid has a flash point of 300°F (149°C) or less and is capable of being heated above its boiling point.

Unheated liquids having flash points above 20°F (6.7°C) normally do not present explosion hazards, except in processes involving large surfaces of liquids, because their rates of evaporation are low enough so that normal rates of air exchange generally prevent the formation of mixtures within the explosive range.

STORAGE

The principal hazard associated with the storage of flammable and combustible liquids is the accidental discharge of the liquid into the surrounding environment. This accidental discharge is frequently the result of: (1) overpressure failure of the container when subjected to exposure fires, (2) rupture of the container due to mishandling, (3) container leakage due to being punctured by forklift trucks, etc., or (4) rupture of transfer piping. The release of these liquids during a fire adds to the intensity of the fire, hindering fire fighting operations and often resulting in the rupture of additional containers or piping.

Flammable and combustible liquids at industrial plants are normally stored in tanks, drums, or small containers in cartons on pallets. The hazards associated with each method of storage are outlined in the following paragraphs.

Tank Storage

For economic reasons, bulk quantities of flammable and combustible liquids are generally stored in tanks which may be installed underground, aboveground, or even inside buildings under special conditions.

When using properly designed, installed, and maintained storage tanks, the greatest hazards are probably those associated with the transferring of the liquid to or from storage, rather than the storage itself. The hazard of storage does not necessarily depend upon the quantity of liquids being stored, but rather is more dependent upon other factors, such as the characteristics of the liquid, the design of the tank, foundations and supports, size and location of vents, related piping and connections, as well as operating procedures.

Buried Tanks

Horizontal tanks buried outside buildings provide the safest method of storing flammable and combustible liquids and are the preferred method of storage when the liquids are delivered by tank vehicles. Tanks may also be buried underneath buildings if they are properly installed and the fill and vent connections are piped outside the building walls.

When installing buried tanks there are several precautions that need to be taken into consideration:

Location: Leaking contents from underground tanks has been known to travel several miles from the source, frequently finding its way into basements, sewers, underground water streams, etc. Therefore, the site selected for the location of buried tanks should be one that minimizes the probability of leakage from the tank and migration from the tank site.

Underground tanks must be located where they will not be subject to damaging loads imposed by building foundations, vehicular traffic, or vibrations from production equipment. If such locations are not available, the tanks must be adequately protected to prevent the loads from damaging the tanks and piping connections.

Ground-water conditions also have to be taken into consideration, and additional anchorage may be needed to prevent flotation of the tank where ground-water conditions are poor or the site is subject to flooding.

Corrosion protection: The normal life expectancy of properly installed, buried metal tanks is about 20 to 25 years. However, this may be shortened to 4 or 5 years if the soil is corrosive or if there are stray electrical currents in the vicinity. To prolong their life, buried metal tanks should be appropriately protected to prevent corrosion. (See NFPA 30, *Flammable and Combustible Liquids Code*.)

Inventory records: In order to detect leakage from buried tanks, it is essential that accurate inventory records be kept on each buried tank. Tanks suspected of leaking should be hydrostatically tested, using the same liquid being stored in the tank. Testing with air on top of stored product has been proven to be dangerous and should be avoided. NFPA 329, *Underground Leakage of Flammable and Combustible Liquids*, contains detailed information on testing for leaks in buried tanks.

Aboveground Tanks

The largest practical size for a buried tank is one having a capacity of 30,000 gal (113 m^3). Therefore, industries having a need for greater storage capacity generally use aboveground tanks rather than multiple buried tanks.

Aboveground tanks come in a variety of designs, but these can be divided into three general categories:

1. Atmospheric tanks for internal pressures of 0 to 0.5 pounds per square inch gage (0 to 3.4 kPa)
2. Low pressure storage tanks for internal pressures of 0.5 to 15 psig (3.4 to 103.4 kPa)
3. Pressure vessels for internal pressures above 15 psig (103.4 kPa).

Figure 37.1 shows some of the more common types of aboveground storage tanks. Cone roof tanks are generally used for the storage of liquids having flash points above the normal storage temperature and where there is no requirement for vapor conservation. Floating roof tanks, lifter roof tanks, vapor dome tanks, and cone roof tanks with internal floating roofs are used for the storage of lighter more volatile liquids and for vapor conservation purposes. Pressure tanks and pressure vessels are normally used for liquids having high vapor pressures and for vapor conservation purposes.

Tank construction: The preferred materials for use in the construction of aboveground tanks are steel and concrete because of their relatively high resistance to heat conditions that would be expected during exposure fires. Tanks built of materials with low melting points, such as aluminum, plastics, etc., will readily fail under exposure fire conditions and thereby intensify the situation.

Details for steel tank construction are outlined in the following standards:

1. *Welded Steel Tanks for Oil Storage*, Standard No. 650, 1978, American Petroleum Institute, 2101 L Street, N.W., Washington, DC 20037.
2. *Bolted Tanks for Storage of Production Liquids*, Specifications No. 12B, 1977, American Petroleum Institute, 2101 L Street, N.W., Washington, DC 20037.
3. *Field Welded Tanks for Storage of Production Liquids*, Specifications No. 12D, 1977, American Petroleum Institute, 2101 L Street, N.W., Washington, DC 20037.
4. *Shop Welded Tanks for Storage of Production Liquids*, Specifications No. 12F, 1977, American Petroleum Institute, 2101 L Street, N.W., Washington, DC 20037.
5. *Standard for Aboveground Tanks for Flammable and Combustible Liquids*, UL 142, Underwriters Laboratories Inc., 207 E. Ohio Street, Chicago, Illinois 60611.

Installation: The rupture of an aboveground tank or a leak in the tank shell or piping connected to the tank, at a point below the liquid level, could result in all the liquid above the leak being released into the surrounding environment. Therefore, if possible, aboveground tanks should be installed on ground that slopes away from important buildings, plant utilities, other flammable or combustible storage tanks, and other hazardous operations or processes. Where this cannot be achieved, the tanks can be isolated by means of dikes, drainage systems, etc. NFPA 30, *Flammable and Combustible Liquids Code*, contains detailed information on tank spacing, dikes, drainage systems and other pertinent information associated with the installation of aboveground tanks.

Aboveground tank fires: The history of tank fires tells us that the probability of a fire starting within a properly designed, constructed, and maintained aboveground storage tank is remote. Fires rarely occur in tanks containing the higher flash point combustible liquids. The majority of tank fires involve the lighter, more volatile Class I materials and are usually the result of releases from leaks, ruptures, or overflows. Ignition of the vapors, which may take place at some distance from the tank, results in the fire flashing back and exposing the tank. When exposed to a spill fire, the shell of a large vertical steel tank, above the liquid level, will generally soften and collapse into the tank without splitting the tank shell.

Internal explosions: Internal explosions occur when the vapor space of a storage tank is in the flammable range and an ignition source is present. The two most frequent ignition sources are static electricity and light-

FLAMMABLE AND COMBUSTIBLE LIQUIDS HANDLING AND STORAGE

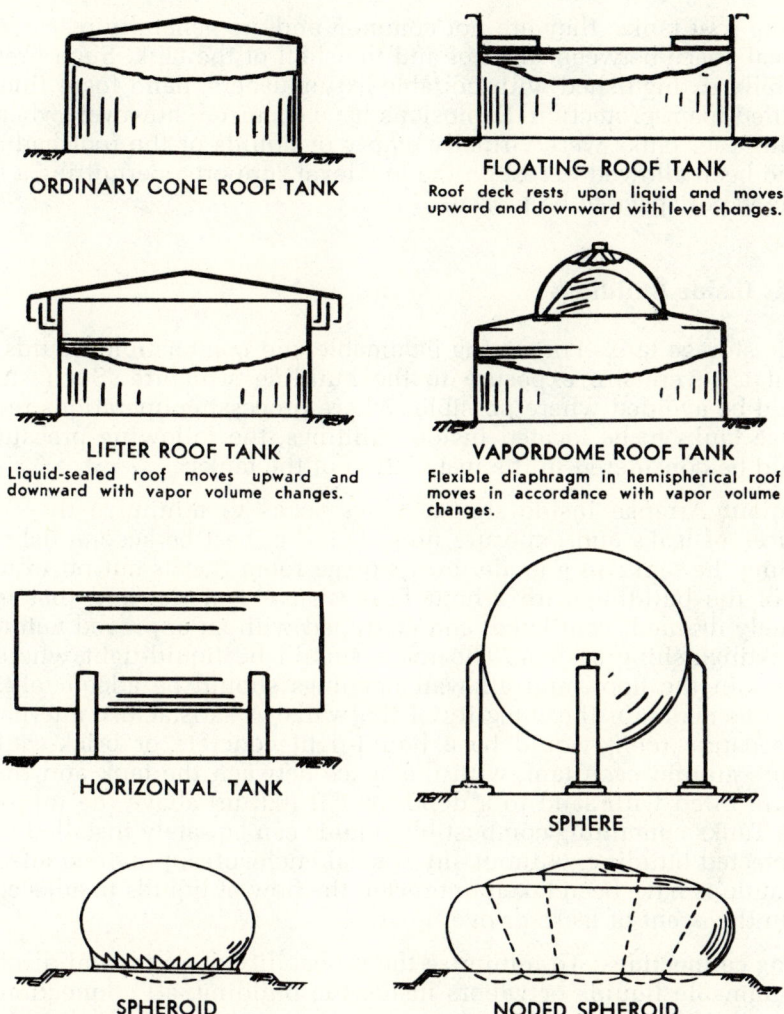

FIGURE 37.1. *Common types of tanks for storage of flammable and combustible liquids.*

ning. Liquids that are stored at temperatures near or above their flash points are most likely to have a vapor space that is in the flammable range and, therefore, are the most susceptible to ignition and explosions.

Liquids with low flash points will be most susceptible when the surrounding air temperature is reduced for extended periods, such as during the winter months in colder climates. On the other hand, the higher flash point liquids will be susceptible when heated to temperatures near or above their flash points.

Due to their design, floating roof tanks provide the greatest degree of firesafety for aboveground installations. Although fires do occur in floating roof tanks, they are not common and are generally restricted to the seal space between the roof and the shell of the tank. Such fires are generally extinguished with portable extinguishers, hand foam lines, or installed foam protection. Explosions have occurred, however, when the floating roof tanks were virtually empty of liquids or the roof had sunk or had been allowed to rest on the low level supports, permitting a flammable vapor space to be present.

Tanks Inside Buildings

Inside storage tanks containing flammable and combustible liquids represent a severe fire exposure to the building structure. Such storage should be avoided where possible. Where processing operations require storage tanks to be located inside buildings, the following precautions should be considered in the installation of the tanks:

Location: Arrange inside storage tanks so as to minimize the consequences of leaks and exposure fires. This can best be accomplished by locating the tanks in a grade level storage room that is cut off from the rest of the building with 2-hour fire-resistive material and that is adequately drained, ventilated, and equipped with an approved automatic fire extinguishing system. The room should be liquid-tight where the walls join the floor, and all wall openings should be adequately protected to maintain the integrity of the walls. A satisfactory alternate to the separate room would be a liquid-tight concrete or brick-wall enclosure around each tank with the space between the tank and the enclosure filled with sand to a depth of 1 ft (0.3 m) above the top of the tank. Tanks containing combustible liquids can be safely installed inside a protected building, without any special enclosures, provided adequate precautions have been taken to prevent the flow of liquids to adjacent areas in the event of leaks or overflows.

Piping connections: To minimize the probability of accidental discharge of flammable liquids or vapors inside the building, all connections for tank openings and piping systems must be liquid-tight and all vents and fill pipes for tanks containing Class I and Class II liquids must terminate on the outside of the building, at least 5 ft (1.5 m) from any building opening.

Ideally each tank should be equipped with some means of preventing overflow into the building during filling operations.

Portable Tanks

Portable tanks are defined as closed vessels having a liquid capacity between 60 and 660 U.S. gal (277 and 2500 L) and are not intended for fixed installation. Approved tanks are equipped with pressure relief devices that are designed to limit the internal pressure to 30 percent of the bursting pressure of the tank, and are, therefore, considered to be more desirable than 55-gal (208 L) drums for shipping and storage of flammable and combustible liquids. However, portable tanks are not suitable for the storage of Class IA point below 73°F (22.8°C) and boiling point below 100°F (37.8°C) materials since these materials would normally be vaporizing and venting through the relief device at room temperatures.

The use of portable tanks is widespread throughout industry for handling liquids in quantities greater than 55 gal (208 L). This is especially true in the automotive, chemical, food processing, and the paint industries where the portable tank is most widely used.

Location: As with larger fixed tank installations, good location of portable tanks containing flammable and combustible liquids takes into consideration the effects of leaking containers, as well as the effects of exposure fires. The *Flammable and Combustible Liquids Code* contains detailed information regarding storage arrangements to minimize the effects of a fire involving portable tanks.

Portable tank fires: The fire loss record for portable tanks has been excellent, and the probability of a fire starting in a portable tank is very remote. This is not to say that portable tank fires do not occur, but rather that those which do occur are generally controlled and extinguished with minimal damage. As with large fixed installations, most portable tank fires occur either during transfer operations or as the result of an exposure fire involving adjacent flammable or combustible materials. When subject to an exposure fire, the portable tank will release flammable vapors through the relief vent as the temperature of the liquid in the tank is increased by the heat from the exposure fire. It is this release of flammable vapors through the vent that creates the greatest fire hazard with portable tanks. Frequently, the discharge of the vapors from the relief vent is so directed that, when ignited, there is direct flame impingement on either the tank involved or adjacent tanks, resulting in premature failure of the tanks, intensifying the fire condition.

Internal explosions: Internal explosions rarely occur in portable tanks, and those that do occur are generally the result of the flammable vapors in the tank being ignited by a static spark during liquid transfer operations. Boiling liquid expanding vapor explosions (BLEVEs) have also occurred when portable tanks were subject to exposure fires and the relief vents were incapable of adequately relieving the internal pressure of the portable tanks.

Container Storage

For the purpose of this discussion a container is defined as any vessel of 60 U.S. gal (227 L), or less, used for transporting or storing flammable or combustible liquids. The types of containers that are most commonly found in industry are pressurized aerosol containers, metal shipping containers ranging in size from less than 1 gal to 55-gal (3.78 to 208 L) drums, and safety containers. In recent years the use of nonmetallic containers, such as fiberboard drums and corrugated-paper cartons lined with polyethylene, as well as high density polyethylene containers, has become increasingly more popular for certain types of liquids.

Unopened containers of flammable and combustible liquids normally present only a moderate fire hazard. It is when containers start to leak or become exposed to excess heat that the hazard becomes very severe. For example, when subjected to excess heat from an exposure fire the liquids in the containers expand, overpressurizing the containers, causing them to rupture and release their contents. This release of liquid adds to the intensity of the fire and may result in the rupture of other containers. Naturally, the larger the container, the greater the hazard. If exposed to fire, a metal drum of liquids can burst with explosive violence, spreading flaming liquids over a large area. Exploding drums have been known to be rocketed several hundred feet from the scene of the original ignition. On the other hand, nonpressurized containers of 5 gal (19 L) or less have a tendency not to explode with any violence; instead, they usually come apart at the side seam or the lid pops open at a relatively low internal pressure, minimizing the hazard of spreading liquids throughout the area. The behavior of small pressurized containers, such as aerosols, is dependent upon the design of the container and the flammability of the contents. Aerosol containers of Class II and Class III materials with relief devices will normally relieve with little or no additional hazard, whereas containers of Class I materials, with or without relief devices, will explode and be rocketed throughout the area. Additional full-scale testing is being conducted by the aerosol industry to further define the hazards associated with storage of aerosols.

The loss experience associated with the storage of flammable and combustible liquids in nonmetallic containers is limited, and there is insufficient test data available at this time to provide a clear understanding of what happens to these types of containers when subjected to exposure fires. Preliminary studies indicate that the high density polyethylene containers have a tendency to soften and burn above the liquid level of the contents at relatively low fire exposure temperatures, without exploding. Fiber drums and corrugated-paper cartons lined with polyethylene also have a tendency to rupture during the very early stages of an exposure fire, increasing the intensity of the fire.

The loss potential associated with container storage of flammable and combustible liquids is lessened if the containers are segregated from production areas, other important buildings, and ordinary combustible materials by either separation distances or construction. If the containers must be kept inside, the preferred method of storage is in cutoff rooms or in attached buildings, rather than in inside rooms, because these arrangements give easier access for the fire department and permit explosion venting where it is needed. Some storage arrangements that follow the precepts of protection by good separation distances or construction (given in order of their preference and coded to Figure 37.2) are:

FLAMMABLE AND COMBUSTIBLE LIQUIDS HANDLING AND STORAGE

1. Outside yard areas that are at least 50 ft (15 m) from any important structures, buildings, ordinary combustible storage, or property lines.
2. Detached buildings of lightweight construction located at least 50 ft (15 m) from any important structures, buildings, or property lines.
3. Detached buildings of lightweight construction equipped with automatic fire detection and extinguishing systems and located 10 to 50 ft (3 to 15 m) from any important buildings, structures, or property lines.
4. Attached one-story addition cut off from main building with standard 4-hour fire wall.
5. Attached one-story addition equipped with automatic fire detection and extinguishing systems and cut off from main building with 2-hour fire partitions.
6. Inside main building in a corner with the two interior walls and the ceiling having a fire resistance rating of at least 3 hrs and a pressure resistance rating of 100 pounds per square foot (488 kg/m^2), and the exterior walls designed of lightweight pressure-relieving construction.
7. Same as (6), but equipped with automatic fire detection and extinguishing systems and the interior walls and ceiling having a 2-hour fire resistance rating.
8. Same as (7) but located in the upper stories of multistory building and equipped with adequate drainage and liquid-tight floors.
9. Inside main building at or above grade level with one exterior wall of lightweight pressure-relieving construction and interior walls and ceiling having at least a 2-hour fire resistance rating. Area, also, equipped with automatic fire detection and extinguishing systems.
10. Totally within main building, at or above grade level, with automatic fire detection and extinguishing systems, and with enclosure walls and ceiling having fire resistance rating of at least 2 hours.
11. Unprotected areas within main buildings or any basement areas. These are the least desirable and should be avoided if at all possible, especially for Class I flammable liquids.

Storage of small amounts of flammable and combustible liquids in shipping containers can be stored within approved/listed safety cabinets in the general work area (see Figure 37.3).

The use of approved safety cans for the storage of flammable and combustible liquids, except Class IA liquids, generally represents no fire hazard, and these can be stored in the general work area with no restrictions. Since the safety cans are designed with pressure/vacuum relief and since Class IA liquids normally vaporize at ambient room temperatures, there is a possibility of creating a flammable vapor-air mixture within the room if Class IA liquids are stored in safety cans.

TRANSFERRING AND DISPENSING

Operations involving the transferring, dispensing, and handling of flammable or combustible liquids that are heated above their flash points are generally considered to be among the more hazardous operations associated with the use of flammable and combustible liquids in an industrial plant. Operations involving unheated combustible liquids generally do not present any unusual hazards, except when involved in high pressure piping systems.

The primary objectives of any transferring or dispensing operation should be to prevent the escape of liquids to the work area and to keep to a minimum the quantity of liquid that could escape in the event of an accident.

For the purpose of this discussion, "transfer" is defined as the movement of liquid from one container, tank, or tank vehicle to another; and "dispensing" is defined as that operation involved in the distribution of flammable and combustible liquids to their points of use in an industrial plant.

Transfer of Liquids

Flammable and combustible liquids are normally transferred by pumps, gravity flow, hydraulic displacement, or compressed gas displacement. Pumping systems are most commonly used for the transfer of large quantities, and pumping through a closed piping system is considered to be the safest method of transferring large quantities.

Pumping systems: Positive displacement pumps are preferred because they provide a reasonably tight shutoff and prevent siphoning of the liquid when not in use. To prevent excess pressure on the system, it is necessary to provide a relief valve on the discharge side of the positive displacement pump. The relief valve discharge should be piped back to the supply source or to the suction side of the pump when handling the lower flash point liquids.

Centrifugal pumps are acceptable, but cannot provide a tight shutoff if taking suction under head. This could result in siphoning liquid from the storage tank when the pumps are not in use.

Pump construction, packings, and trim should be suitable for the liquid being handled. Pumps should preferably take suction under a lift and be located in an area where a fire at the pump would not expose the storage tank or important process equipment and buildings.

Gravity systems: Gravity transfer is required for many industrial process operations, especially when handling very volatile liquids that may cause vapor lock in pumping systems. Gravity systems are not desirable when associated with a large supply source and should be used only when required by the operation. Gravity systems are constantly under pressure and are more difficult to arrange for prompt shutoff than are pumping sys-

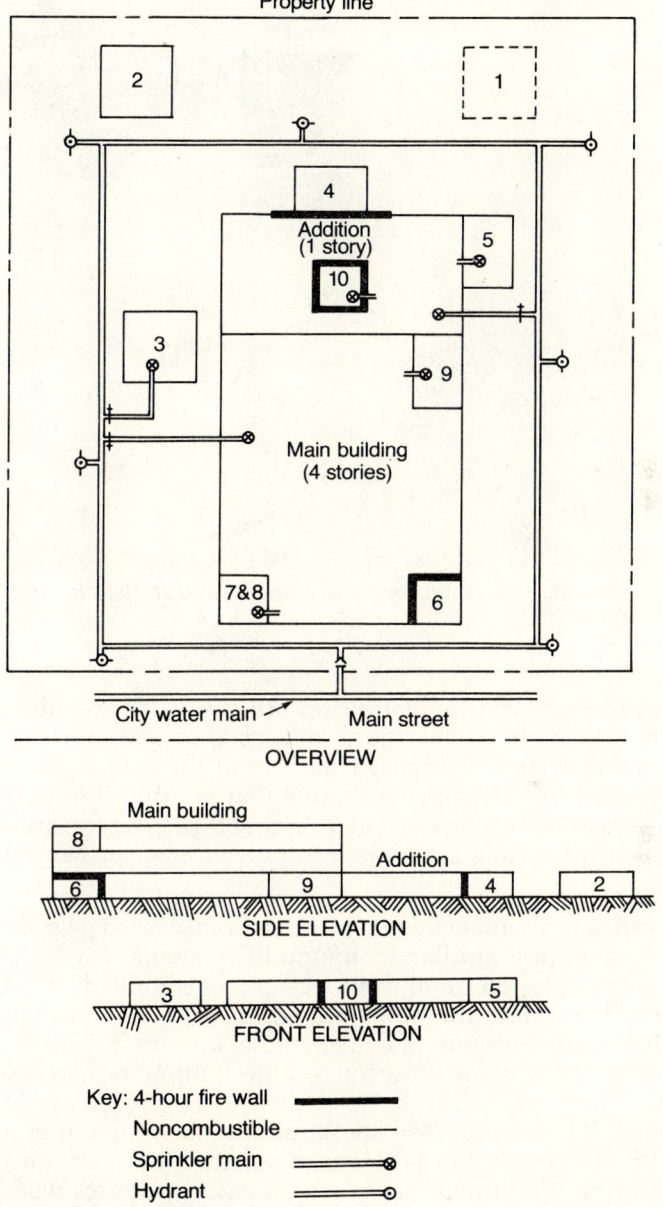

FIGURE 37.2. *Recommended storage locations numbered in order of preference.*

tems; therefore, gravity systems are more subject to major accidental spills.

FIGURE 37.3. Approved/listed storage cabinet. (Justrite Mfg. Co.)

Hydraulic systems: Hydraulic transfer utilizes water pressure to force the flammable or combustible liquid out of the container and into the transfer piping system. The disadvantages of these systems are: (1) They cannot be used for transferring liquids that are miscible with water, (2) the containers need to be designed as coded pressure vessels, and (3) a complex control system is required to prevent over-pressurizing the system.

Compressed gas displacement systems: Compressed gas displacement transfer systems are similar to hydraulic systems, except they utilize compressed gas rather than water as a transfer medium. Due to the compressible nature of the transfer medium and the fact that the system is under constant pressure, a considerable amount of liquid can be discharged from the system in event of pipe failures or careless valve operations.

Transfer by compressed air should not be done under any circumstances. In addition to the problem of constant pressure on the system, the probability of a violent vapor-air explosion increases when flammable and combustible liquids are under pressure in the presence of air.

Dispensing operations: Dispensing operations generally involve the transfer of liquids from fixed piping systems, drums, or 5-gal (19 L) cans into smaller end-use containers or equipment reservoirs and are generally carried out in areas where ignition sources may be present. Dispensing of flammable liquids at the point of use within the main plant area

FLAMMABLE AND COMBUSTIBLE LIQUIDS HANDLING AND STORAGE

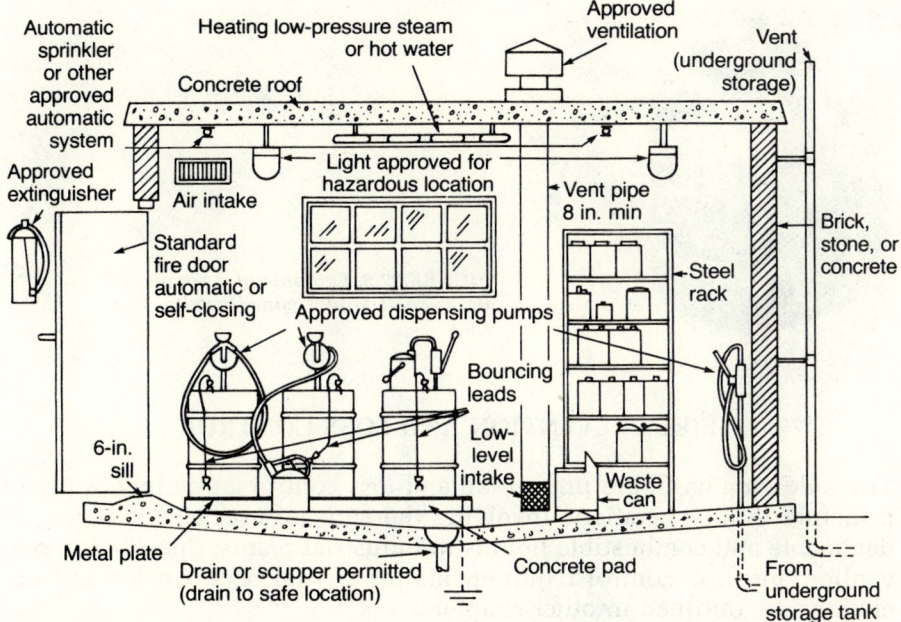

FIGURE 37.4. Sketch showing recommended dispensing area. One inch equals 25.4 mm. (National Safety Council)

normally results in the release of a certain amount of flammable vapors which may create a fire potential.

The preferred arrangement for dispensing flammable liquids in an industrial plant would be to set up an area, one that is adequately protected and ventilated, as a dispensing area and bring all end-use containers to this area to be filled instead of dispensing at the individual work stations (see Figure 37.4).

When dispensing from a fixed piping liquid transfer system, it is desirable to discharge directly into the container through a closed connection, using a dead-man valve that closes automatically when released by the operator.

The preferred method of dispensing liquids from drums or 5-gal (19 L) cans is by means of laboratory tested hand pumps drawing through a dip leg that extends from the top of the container. This minimizes the probability of spills and leakage (see Figure 37.5). Gravity dispensing from drums is acceptable if approved or listed self-closing faucets and drum vents are used (see Figure 37.6).

When there is a need to handle small quantities of flammable and combustible liquids outside the dispensing area, approved or listed safety cans should be used to minimize the fire risks. Open pails should never be used for the transferring or dispensing of flammable liquids.

FIGURE 37.5. Example of a nonmetallic Type II safety can. (Justrite Mfg. Co.)

FIRE PREVENTION AND LOSS CONTROL

The following basic fire prevention and loss control guidelines apply in principle to all operations involving the storage, use, and handling of flammable and combustible liquids in industrial plants. Specific fire prevention and loss control requirements for special occupancies and operations are outlined in other chapters.

Training of Personnel

As mentioned earlier in this chapter, the lack of adequate personnel training is one of the major factors influencing the involvement of flammable and combustible liquids in industrial fires. Therefore, the safety of any operation involving these materials depends largely upon proper operator action, which can only be accomplished by an effective employee training program. Such a training program should include, but not be limited to:

1. Thorough indoctrination of all employees, including supervisors, in the hazards associated with the storage, transfer, and use of flammable and combustible liquids.
2. Thorough indoctrination of all employees in the normal operating procedures, as well as procedures for handling emergency conditions.
3. Indoctrination of employees in the importance of maintaining excellent housekeeping in the work area at all times.
4. Indoctrination of employees in the importance of keeping flammable liquids and vapors confined to closed equipment and containers.
5. Indoctrination of employees in the importance of limiting the quantities of liquids in the work area to that amount needed for efficient operations.
6. Establishment of a series of routine check points to be observed by employees for prompt detection of abnormal conditions.

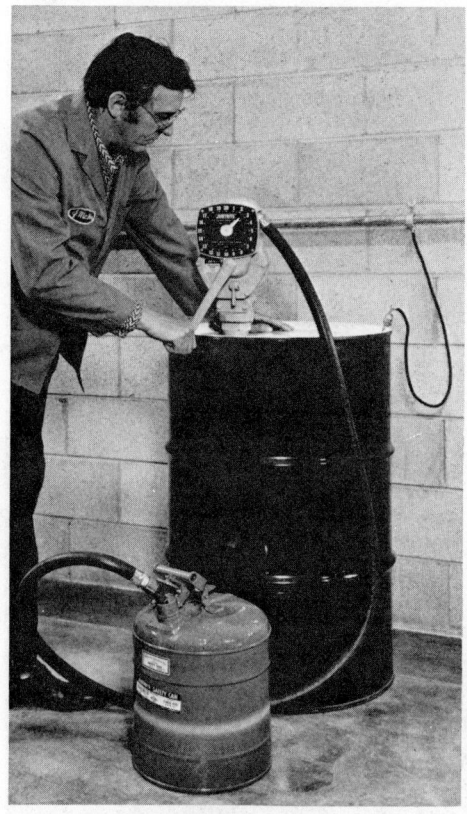

FIGURE 37.6. Dispensing from drum using approved/listed hand pump. (Justrite Mfg. Co.)

7. Training of employees in the need for constant attendance during all transfer operations.
8. Training of employees in the proper procedure for control and cleanup of leaks and spills.
9. Training of employees in the proper procedures for disposal of waste material.

Confinement of Liquids

The major objective of an effective flammable liquids loss prevention program is to confine the liquids and vapors within the equipment, vessels, piping, or containers, as well as to minimize the effects of a spill or leak by draining the escaping liquid to a safe location. To achieve this goal:

1. Use only vapor- and liquid-tight equipment with the minimum number of openings necessary for effective operations.

Chapter 37

FIGURE 37.7. Recommended arrangement for gravity dispensing from drums. (Justrite Mfg. Co.)

2. Use only equipment designed for handling flammable or combustible liquids.
3. Equipment subject to explosion hazards should preferably either be designed to withstand the maximum pressure expected to be developed during abnormal conditions, or be designed to relieve excess internal pressures through explosion vents that discharge directly outdoors to a safe location.
4. Equip open vessels and vessels that have loose fitting covers with overflow drains and emergency bottom dump drains that are piped to a safe location or salvage tank.
5. Handle small amounts of liquids only in approved or listed safety containers.

6. Equip dispensing areas, inside storage rooms, and process areas involving large quantities of flammable and combustible liquids with adequate drainage systems to prevent the flow of liquids into adjacent work areas.

Ventilation

Ventilation is a loss prevention measure essential to preventing flammable liquid fires and combustion explosions. The purpose of ventilation is to confine, dilute, and exhaust to a safe location the vapors that are released during normal operations. Abnormal vapor releases, as might occur with a break in a piping system, the rupture of a drum, or overfilling a tank, cannot be adequately safeguarded by ventilation.

In addition to creating a fire hazard, most flammable and combustible liquid vapors also create a health hazard. The ventilation required for health safety is generally more than adequate for fire and explosion safety.

The following guidelines are considered to be the minimum fire-safety ventilation requirements necessary to prevent hazardous accumulation of flammable vapors in the work area:

1. In confined areas utilizing flammable liquids or combustible liquids heated above their flash points, provide continuous mechanical ventilation at the rate of at least 1 cubic foot per minute per sq ft ($0.3 \text{ m}^3/\text{m}^2$) of floor area.

   The ventilation system should be designed to provide an air-sweep of the entire floor area with the inlets to the exhaust system located near floor level. The exhaust system should be ducted to a safe location outdoors, using the most direct path possible.

   Provisions will be needed for the introduction of make-up air which should be installed in such a manner so as not to short-circuit the ventilation at floor level.

   The ventilation system should be interlocked so that operations utilizing flammable and combustible liquids cannot be carried out unless there is adequate ventilation. This can best be accomplished by interlocking an airflow switch in the exhaust duct with the equipment power supplies.

2. In large unconfined production areas with individual work stations utilizing small amounts of flammable or combustible liquids heated above their flash points, such as might be found in automotive and electronic assembly plants, provide spot ventilation at the work stations. The ventilation should be provided at a rate adequate to maintain the vapor-air mixture at a safe concentration at the work station and within a 5-ft (1.5 m) radius of the work station. Exhaust inlets should be near floor level and should be arranged to provide an air-sweep of the entire affected floor area including any equipment pits where flammable vapors may collect.

3. No special firesafety ventilation is required for unheated combustible liquids.

Control of Ignition Sources

Another very important basic loss prevention guideline is the careful control or elimination of all potential sources of ignition in areas that might contain hazardous accumulations of flammable or combustible liquids. The following are some of the precautions that should be taken to minimize the probability of ignition:

1. Electrical equipment and wiring in the area should be suitable for the hazard. NFPA 30, *Flammable and Combustible Liquids Code*, and NFPA 70, *National Electrical Code*,* specify the requirements for electrical equipment in hazardous areas.
2. For operations involving the heating of liquids, use indirect heating methods and provide the necessary safety controls and interlocks to prevent overheating of the liquids.
3. Do not locate any equipment having open flames, hot surfaces, or radiant heat sources in areas where flammable or combustible liquids are used, handled, or stored.
4. Prohibit the use of friction and spark producing equipment in areas where flammable liquids are used.
5. Provide adequate grounding and bonding of all equipment handling or using flammable liquids to minimize the accumulation of hazardous static charges. NFPA 77, *Static Electricity*, provides the guidelines for grounding and bonding of equipment.
6. Establish an effective preventative maintenance program to ensure that all equipment and safety controls are working satisfactorily.
7. Prohibit smoking, open flame torches, and cutting and welding in hazardous areas.
8. Establish a program to ensure that all vessels, tanks, piping, and process equipment are properly drained and purged of flammable and combustible liquids prior to performing maintenance operations.

Protection

A wet-pipe automatic sprinkler system with an adequate and reliable water supply is the preferred basic fire control system for all areas where flammable and combustible liquids are stored, used, or handled. The sprinkler system design should be based on the hazard of the operation

---

*The *National Electrical Code* divides hazardous locations to three classes depending upon the kind of hazardous materials involved; i.e., I — flammable gases or vapors, II — combustible dusts, and III — ignitible fibers. Each class is further divided into two divisions according to the degree of severity of the hazard. Locations where flammable vapors are a hazard are either Class I, Division 1 or Class I, Division 2 (less severe hazard) locations.

being protected. The following are some basic guidelines to be used in providing automatic sprinkler protection:

1. The sprinkler system for small inside heated storage areas or dispensing rooms should be a standard wet-pipe system designed to provide a discharge density of 0.5 gpm per sq ft [20.3 (L/min)/m^2] of floor area using 165°F (74°C) sprinklers. For unheated storage and dispensing areas, the preferred system would be a water deluge system.
2. Large areas of containerized storage of flammable and combustible liquids may require special sprinkler installations, e.g., sprinklers within the racks holding the containers as well as at the ceiling level, depending on the method of storage. NFPA 30, *Flammable and Combustible Liquids Code*, gives guidance on designing protection for the liquids in warehouses. (See Chapter 48, "Industrial Storage Practices.")
3. Storage tanks, vessels, and process equipment containing in excess of 500 gal (1893 L) of flammable liquids or heated combustible liquids within buildings should be protected with deluge water spray systems that are designed to provide a discharge density of at least 0.25 gpm per sq ft [10.2 (L/min)/m^2] of vessel surface area.

All tank foundations and process equipment supports should be of fire-resistive or protected steel construction.

In small confined areas or inside special equipment or vessels it may be desirable to provide special extinguishing systems to supplement the automatic sprinkler systems and minimize production downtime. Such special extinguishing systems include fixed foam systems, total flooding low pressure carbon dioxide systems, total flooding Halon systems, and dry chemical systems. The selection of a specific type of special extinguishing system should be based on (1) the effectiveness of the agent in extinguishing a fire involving the material being protected, (2) the minimum downtime that can be tolerated for cleanup and restoration of production, and (3) the cost of the system. (See the Bibliography for NFPA standards covering installation of the extinguishing systems discussed above.)

Portable foam-making equipment and a supply of foam-making materials are desirable for the protection of aboveground storage tanks containing flammable liquids. Portable fire extinguishers are necessary in the event of small liquid fires or fires in other combustibles. An adequate number of portable extinguishers of the appropriate type and capacity should be located where they will be accessible under fire conditions. (Guidance on good practices to follow in providing extinguishers in locations where flammable and combustible liquids are stored and used can be found in NFPA 10, *Portable Fire Extinguishers*.)

Hydrants and small fire hose with adjustable stream nozzles should be provided in the areas where flammable and combustible liquids are used, stored, or handled. Hose streams are necessary for (1) cooling adjacent tanks and structures, (2) extinguishing fires in flammable and combustible liquids, (3) extinguishing fires in ordinary combustibles, and (4) washing down spills to remove hazardous material to a safe location. (Guidance on the installation of standpipe and small hose systems and outside private fire protection systems, including hydrants, can be found in NFPA 14, *Standpipe and Hose Systems*, and NFPA 24, *Installation of Private Fire Service Mains and Their Appurtenances*.)

An effective maintenance, test, and inspection program should be established for all fire protection and control equipment to ensure that it is in satisfactory operating condition at all times.

BIBLIOGRAPHY

NFPA CODES, STANDARDS, AND RECOMMENDED PRACTICES

Reference to the following NFPA Codes, Standards, and Recommended Practices will provide further information on the safeguards for the storage and handling of flammable and combustible liquids. (See the latest *NFPA Codes and Standards Catalog* for availability of current editions of the following documents.)

NFPA 10, *Portable Fire Extinguishers*.
NFPA 11, *Foam Extinguishing Systems and Combined Agent Systems*.
NFPA 12, *Carbon Dioxide Extinguishing Systems*.
NFPA 12A, *Halon 1301 Fire Extinguishing Agent Systems*.
NFPA 12B, *Halon 1211 Fire Extinguishing Systems*.
NFPA 13, *Installation of Sprinkler Systems*.
NFPA 14, *Standpipe and Hose Systems*.
NFPA 16, *Deluge Foam-Water Sprinkler and Spray Systems*.
NFPA 17, *Dry Chemical Extinguishing Systems*.
NFPA 24, *Installation of Private Fire Service Mains and Their Appurtenances*.
NFPA 30, *Flammable and Combustible Liquids Code*.
NFPA 31, *Installation of Oil Burning Equipment*.
NFPA 70, *National Electrical Code*.
NFPA 77, *Recommended Practice on Static Electricity*.
NFPA 325M, *Fire Hazard Properties of Flammable Liquids, Gases, and Volatile Solids*.
NFPA 327, *Cleaning or Safeguarding Small Tanks and Containers*.
NFPA 329, *Underground Leakage of Flammable and Combustible Liquids*.
NFPA 385, *Tank Vehicles for Flammable and Combustible Liquids*.
NFPA 395, *Storage of Flammable and Combustible Liquids on Farms and Isolated Construction Projects*.
NFPA 650, *Standard for Pneumatic Conveying Systems for Handling Combustible Materials*.

Additional Reading

Accident Prevention Manual for Industrial Operations, 8th ed., National Safety Council, Chicago, IL, 1980.

Bahme, Charles W., *Fire Officer's Guide on Hazardous Chemicals*, National Fire Protection Association, Quincy, MA, 1978.

Brookes, Vincent J., and Morris B. Jacobs, *Poisons: Properties, Chemical Identification, Origin and Use — Signs, Symptoms, and Emergency Treatment*, 3rd ed., Krieger Publishing Co., Melbourne, FL, 1975.

Carpenter, R. A., C. C. Bolze, and L. E. Findley, "A System for the Correlation and Physical Properties and Structural Characteristics of Chemical Compounds with their Commercial Uses," Midwest Research Institute, Kansas City, MO (reprint from *American Documentation*, vol. X, no. 2).

Chemical Engineering Magazine, *Safe and Efficient Plant Operation and Maintenance*, McGraw-Hill, New York, NY, 1980.

Cocks, R. E., and J. E. Rogerson, "Organizing a Process Safety Program," *Chemical Engineering*, vol. 85, no. 23 (1978), pp. 138-146.

Compilation of Labeling Laws and Regulations for Hazardous Substances, Chemical Specialties Manufacturers Association, New York, NY.

Coward, H. F., and G. W. Jones, "Limits of Flammability of Gases and Vapors," Bulletin No. 503, 1952, U.S. Department of the Interior, Bureau of Mines, Washington, DC.

Deacon, F. L., "Designing Fire Protection to Limit Monetary Loss," *SFPE Technology Report No. 80-2*, Society of Fire Protection Engineers, Boston, MA, 1980.

Factory Mutual Engineering Corporation, *Handbook of Industrial Loss Prevention*, 2nd ed., McGraw-Hill, New York, NY, 1967.

Fawcett, H. H. and W. S. Wood, *Safety and Accident Prevention in Chemical Operations*, John Wiley and Sons, New York, NY, 1982.

Fire Protection Guide on Hazardous Materials, National Fire Protection Association, Quincy, MA, 1978.

Flash Point Index of Trade Name Liquids, 9th ed., National Fire Protection Association, Quincy, MA, 1978.

Goring, G., "Sprinkler Protection of Storage Risks," *Fire Protection*, vol. 8, no. 2 (June 1981), pp. 20-25.

Halpaap, W., "Special Appliance for the Chemical Industry," *Fire International*, 5(56), 1977, pp. 44-50.

Handbook of Organic Industrial Solvents, 2nd ed., National Association of Mutual Casualty Companies, Chicago, IL, 1961.

Hawley, G. G., ed., *The Condensed Chemical Dictionary*, 10th ed., Van Nostrand Reinhold Co., New York, NY, 1981.

Henry, Martin F., ed., *Flammable and Combustible Liquids Code Handbook*, National Fire Protection Association, Quincy, MA, 1981.

How to Handle Flammable Liquids Safely, Justrite Manufacturing Co., Des Moines, IL, 1978.

Hygienic Guide Series, American Industrial Hygiene Association, Detroit, MI.

King, P. W. and J. Magid, *Industrial Hazard and Safety Handbook*, Butterworths Publishers, Inc., Woburn, MA, 1979.

King, R., "Plant Hazards," *Engineering (London)*, vol. 216, no. 4 (1976), pp. 277-279.

Kirk, R. E., and D. F. Othmer, eds., *Encyclopedia of Chemical Technology*, 3rd ed., 12 vols., Wiley-Interscience, New York, NY, 1978-1980.

McKinnon, G. P., ed., *Fire Protection Handbook*, Fifteenth Edition, National Fire Protection Association, Quincy, MA, 1981.

Meidl, James, *Flammable Hazardous Materials*, 2nd ed., Glenroe Publishing Co. Inc., Encino, CA, 1970.

Mellan, I., *Industrial Solvents Handbook*, Noyes Press, Park Ridge, NJ, 1977.

Nailen, R. C., "Toxic, Flammable Chemicals, Gases Breed Trouble in Electronic Plants," *Fire Engineering*, vol. 133, no. 10 (October 1980), pp. 54-57.

National Safety Council, "Fire Protection for Combustible Materials," *National Safety News*, vol. 119, no. 6 (June 1979), pp. 75-82.

Perry, J. H., and C. H. Chilton, eds., *Chemical Engineers' Handbook*, 5th ed., McGraw-Hill, New York, NY, 1974.

Pilborough, C., *Inspection of Chemical Plants*, Gulf Publishing Co., Houston, TX, 1977.

Safety and Fire Protection Committee, Manufacturing Chemists Association, Inc., *Guide For Safety in the Chemical Laboratory*, 2nd ed., Van Nostrand Reinhold Co., New York, NY, 1972.

Sax N. I., *Dangerous Properties of Industrial Materials*, 5th ed., Van Nostrand Reinhold Co., New York, NY, 1968.

Schieler, Leroy and Denis Pauze, *Hazardous Materials*, Delmar Publishers, Albany, NY, 1976.

Stevens, A. M., "Controlling a Violent Friend?," *Professional Safety*, vol. 25, no. 2 (Feb. 1980), pp. 17-20.

Threshold Limit Values, American Conference of Governmental Industrial Hygienists, Cincinnati, OH.

Tile 46, "Shipping," Parts 146 to 149; Title 49, "Transportation," Parts 171 to 178, *Code of Federal Regulations*, U.S. Government Printing Office, Washington, DC.

Turner, Charles F. and Joseph W. McCreery, *The Chemistry of Fire and Hazardous Materials*, Allyn and Bacon, Inc., Boston, MA, 1981.

Van Dolah, R. W., et al., "Flame Propagation Extinguishment and Environmental Effects on Combustion," *Fire Technology*, vol. 1, no. 2 (May 1965), pp. 138-145.

Vervalin, C. H., ed., *Fire Protection Manual for Hydrocarbon Processing Plants*, vols. 1 and 2, Gulf Publishing Co., Houston, TX, 1973.

Weast, R. C., ed., *Handbook of Chemistry and Physics*, 63rd ed., Chemical Rubber Co., Cleveland, OH, 1982.

Weiby, P. and K. R. Dickinson, "Monitorying Work Areas for Explosive and Toxic Hazards," *Chemical Engineering*, vol. 83, no. 22 (1976), pp. 139-145.

Weiss, J., *Hazardous Materials Chemical Data Book*, Noyes Press, Park Ridge, NJ, 1980.

Welker, J. R., O. A. Pipkin, and C. M. Sliepcevich, "The Effect of Wind on Flames," *Fire Technology*, vol. 1, no. 2 (May 1965), pp. 122-129.

Welker, J. R., and C. M. Sliepcevich, "Bending of Wind-Blown Flames from Liquid Pools," *Fire Technology*, vol. 2, no. 2 (May 1966), pp. 127-135.

Welker, J. R., and C. M. Sliepcevich, "Burning Rates and Heat Transfer from Wind-Blown Flames," *Fire Technology*, vol. 2, no. 3 (August 1966), pp. 211-218.

Zimmerman, O. T., and Irvin Lavine, *Handbook of Material Trade Names*, Industrial Research Service, Dover, NH, 1953-65 (plus supplements).

38

Handling and Storage of Industrial Gases

F. Carl Saacke

The characterization of gases as "industrial gases" is primarily a matter of convenience. It is a general term which can be applied to the variety of gases used in manufacturing processes — either for direct product fabrication or to create and maintain a suitable production environment. In all cases, specific physical or chemical properties are tailored to particular storage and usage requirements. These same gases may also be suitable for other uses, e.g., as sources of heat or power generation, for medical applications, etc.; therefore, the term "industrial gases" is not meant to be a definitive one.

More often than not, the gases used in an industrial setting are manufactured elsewhere, so they must be transported to, and stored at, the consuming facility. With the exception of some high-volume consumers who may be supplied by pipelines, industrial gases are transported, stored, and used in containers of a relatively small volume. In fact, in many instances the availability of industrial gases in rather small and usually portable containers is a major consideration in assessing their

---

The late F. Carl Saacke, C.S.P., who wrote this chapter was a Safety Consultant based in Rockville Centre, New York. George R. Spies, Manager Product Safety, The BOC Group, Montvale, New Jersey, revised this chapter for the second edition.

TABLE 38.1. Physical and Chemical Properties of Industrial Gases at 70°F (21.1°C) and/or 760 mm

Classification	Gas	Molecular Weight	Specific Gravity (air = 1)	Specific Volume cu ft/lb	Specific Volume m³/kg	Vol. Gas/Vol. Liq. cu ft	Vol. Gas/Vol. Liq. m³
Fuel	Acetylene	26.04	0.9	14.7	0.918	—	—
Oxidizer	Air	28.96	1.0	13.3	0.833	728	20.6
Fuel	Ammonia	17.03	0.6	22.6	1.297	859	24.3
Fuel	Butane	58.12	2.0	6.5	0.406	—	—
Inert	Carbon Dioxide	44.01	1.5	8.7	0.546	—	—
Fuel	Carbon Dioxide	28.01	1.0	13.8	0.862	706	20.0
Oxidizer	Chlorine	70.91	2.4	5.4	0.312	—	—
Fuel	Ethylene	28.05	1.0	13.8	0.862	481	13.6
Oxidizer	Fluorine	38.00	1.3	10.2	0.637	961	27.2
Fuel	Hydrogen	2.02	0.07	192.0	11.990	850	24.1
Fuel	Methane	16.04	0.6	27.1	1.693	636	18.0
Fuel	MPS (MAPP®)	—	1.5	8.8	0.550	—	—
Fuel	Natural Gas (Liq.)	—	0.6/0.7	—	—	—	—
Inert	Nitrogen	28.01	1.0	13.8	0.862	696	19.7
Oxidizer	Oxygen	32.00	1.1	12.1	0.754	861	24.4
Fuel	Propane	44.10	1.5	8.7	0.543	—	—
Fuel	Propylene	42.08	1.5	9.1	0.568	—	—

utility; but this also is one of the major hazards associated with industrial gases.

Because gases have the least amount of matter per unit of volume, it is necessary to "concentrate" gas volumes for practical transportation, storage, and use. This is done by either compressing the gas to a rather high pressure or by converting it to a liquid state. The term *compressed gas* in this text applies to any material or mixture having in the container an absolute pressure exceeding 40 psi (275 kPa) at 70°F (21.1°C).* This pressure of gas in any process or container is sufficient to create a significant hazard if released unintentionally. A *liquefied gas* is a gas which, under its vapor pressure, is partially liquid at the temperature involved. Gases classified as liquefied gases are usually those that are in liq-

---

*This definition is taken, in part, from the Hazardous Materials Regulations of the U.S. Department of Transportation (49 CFR 173.300).

TABLE 38.1. (continued)

Boiling Point		Auto-ignition Temp.		LFL (air)		UFL (air)		Toxicity
°F	°C	°F	°C	°F	°C	°F	°C	TLV-TWA[5]
−118	−83.3	571	299.4	2.5	−16.4	81.0	27.2	asphyxiant
−318	−194.4	—	—	—	—	—	—	life supporting
−28	−33.3	1,204	651.1	16.0	−8.9	25.0	−3.9	25 ppm
+31	−0.6	761	405.0	1.9	−16.7	8.5	−13.1	800 ppm
—	—	—	—	—	—	—	—	5,000 ppm
−314	−192.2	1,128	608.9	12.5	−10.8	74.0	23.3	50 ppm
−30	−34.4	—	—	—	—	—	—	1 ppm
−155	−103.8	842	450.0	3.1	−16.1	32.0	0	asphyxiant
−307	−188.3	—	—	—	—	—	—	1 ppm
−423	−252.8	932	500.0	4.0	−15.6	75.0	23.9	asphyxiant
−259	−161.7	999	537.2	5.3	−14.8	14.0	−10.0	asphyxiant
−36/−4	−38/−20	850	454.4	3.4	−15.9	10.8	−11.8	1,000 ppm
—	—	900/1,170	482/632	4.5	−15.3	15.0	−9.4	
−320	−195.6	—	—	—	—	—	—	asphyxiant
−297	−181.1	—	—	—	—	—	—	life supporting
−44	−42.2	871	466.1	2.2	−16.5	9.5	−12.5	asphyxiant
−53	−47.2	927	497.2	2.0	−16.7	11.1	−11.6	asphyxiant

uefied form at 70°F (21.1°C). Some gases will liquefy when compressed with the gas temperature being the same as normal atmospheric temperature. Others must be cooled to rather low temperatures for liquefaction and, if not, will remain in the gaseous state regardless of how much they are compressed. Liquefied gases can be stored indefinitely.

In practice, industrial gases are stored and used in all three forms: (1) in 100 percent gaseous form at normal temperatures, (2) in both the liquid and gaseous form at normal temperatures, and (3) in both the liquid and gaseous form at low temperatures. In transportation regulations and in general terminology, these forms are often referred to as "compressed gases." While technically correct, this term does not distinguish between the liquid and gaseous states, and hazard evaluation and control is very much contingent upon recognition of the states actually present.

The hazards of industrial gases relate to their chemical properties and physical properties. In their application, industrial gases manifest chemical properties that run the range of possible chemical reactivity. Obviously, so do the hazards. However, the ability to burn, to cause other materials to burn, and to prevent combustion or other chemical reaction represent the great majority of industrial gas applications; accordingly, this chapter will emphasize these chemical properties.

Despite the intrinsic variety and complexity of industrial gas hazards, experience has clearly shown that most accidents are caused by failure to recognize some rather simple safeguards. This chapter, therefore, will not detail the more unusual hazardous aspects which, while important, are influenced only to a limited extent by the users of industrial gases. Instead, it will emphasize the user activities that will have the greatest impact upon the safe use of industrial gases.

PROPERTIES OF COMMON INDUSTRIAL GASES

Flammable Gases

A flammable gas is a gas that can burn with the evolution of heat and a flame. Flammable compressed gas is any compressed gas of which: (1) a mixture of 13 percent or less (by volume) with air is flammable, or (2) the flammable range with air is under 12 percent.

A flammable gas may be either *endothermic* or *exothermic* in reaction. An endothermic compound is one that absorbs energy during its formation and which, when further strongly energized, can decompose explosively without reacting with an oxidizer. Such reactions, with or without an oxidizer, are exothermic reactions because of the evolution of heat from the decomposing compound. Under high ignition energies and/or low heat-conducting passages or containers, flame velocities after ignition can progress to detonation speeds — over one mile (1609 m) per second — and create violent explosions and damage. Expert attention to equipment design is a necessity in handling, transporting, and using such compounds. An exothermic compound is one that creates thermal energy during its formation and that consequently will absorb energy when taking part in a chemical reaction. Most flammable gases react with oxidizers to create heat. The pressures they develop under restraint are a function of the change in gas volume due to the reaction and the increase in gas temperature.

Acetylene: Acetylene is a colorless, lighter-than-air, highly flammable endothermic gas. Pure acetylene is odorless, but the acetylene in general use has a distinctive odor due to commercial impurities inherent in its generation from calcium carbide or derivation from other hydrocarbons. Its low flammablity limit (LFL) and its wide flammablity range make it extremely easy to ignite. When combined with oxygen, it has a high heat of combustion, which leads to its widespread use in welding and cutting. Acetylene has the highest flame temperature available from a flammable gas reaction. It is extremely chemically reactive, and is a basic source for many synthetic rubbers, vinyl compounds, plastics, and other chemicals. It is readily generated by adding calcium carbide to water. It is either piped to its point of use, or compressed into special cylinders that are filled with a highly porous (and usually monolithic) mass saturated with acetone in which the acetylene becomes increasingly soluble under pressure. Listed and approved welding and cutting equipment is sized and designed to allow acetylene to be used safely and without fear of explosion as long as the usage pressures are maintained at or under 15 psig (103.4 kPa) and the supplier's instructions followed. However, acetylene can decompose explosively without oxygen or air at all pressures under excessively high energy inputs and when used in large-diameter piping and other large spaces.

Acetylene, in the presence of moisture, can react with copper, silver, and mercury to form metallic acetylides that are extremely shock sensitive, explosive compounds. If detonated, they can initiate acetylene decomposition. Consequently, copper alloys (such as brass) which are widely used in welding and cutting equipment are usually restricted to a maximum copper content of 65 to 70 percent. Protection against acetylide formation or decomposition is achieved by using the lower copper content alloys in relatively dry atmospheres that inhibit the formation of acetylides.

Acetylene is also used to ripen fruit during transportation and storage. Carbide is dropped into water, proportioning the calcium carbide according to the atmospheric volume to be treated to keep the acetylene concentration well below its LFL.

A hazardous condition occurs when an acetylene cylinder is decomposing internally, venting pure hydrogen through the pressure relief openings and burning with an almost invisible flame. The worker who notes that the cylinder is getting hot and who tries to move it to a safer location may get severely burned unless he looks for open fusible metal plugs and takes care to determine that no hot gases are venting through such openings.

Ammonia: A colorless, alkaline, toxic, liquefied flammable gas that is lighter than air and has a pungent odor, ammonia is widely used as a refrigerant gas, cleanser, fertilizer, source of nitrogen for explosives, and as a chemical processing agent. Because of its high LFL and its narrow flammablity range, small leaks are not likely to form hazardous flammable mixtures in air. For this reason, the liquefied gas is packaged and shipped as a nonflammable gas. In the past, sulfur candles were used to locate leaks around refrigerating equipment; this practice was extremely dangerous because of the possibility of encountering a large leak. Hydrochloric acid fumes or moistened litmus or phenolphthalein paper are equally effective for locating leaks, and cannot cause ignition.

Butane (see liquefied petroleum gas): Butane is one of the heavier hydrocarbons of low vapor pressure which can develop negative (suction) pressures at the container valve outlet at temperatures as high as 31°F (0.5°C), approximately the freezing point of water. Frost on a container not in use should be taken as a warning.

Ethylene: Ethylene is a flammable, liquefiable gas, slightly lighter than air and with a characteristic sweet odor and taste. Ethylene has a critical temperature of 49°F (9.4°C), above which it cannot be liquefied. While it can be charged into high-pressure containers in the liquid phase, it is transported in such containers as a compressed nonliquefied gas. However, it can be stored and transported in bulk tankers if kept refrigerated at temperatures where its vapor pressure allows it to be contained in low-pressure vessels.

Ethylene is also frequently used to ripen fruit during or after shipment to market at concentrations well below the LFL. It has been known to be used in excessive quantities in this application to the point where truck and warehouse atmospheres have reached the flammable range and have ignited with explosive results.

Hydrogen: A highly flammable, extremely light elemental gas, hydrogen has a very wide flammability range, a fairly wide detonatable range, a low ignition temperature, and minimum ignition energy. It is odorless, colorless, nontoxic, and burns with a nonluminous flame (which is often invisible in daylight) unless colored by dust, particulate matter, or chemical vapors. Leaking or venting hydrogen occasionally self-ignites from static electricity. Hydrogen vents should release into a safe location where ignition is not objectionable. Nevertheless, nitrogen inerting is often provided in plant stacks where a stack ignition might be objectionable to neighbors. Hydrogen is stored and shipped as a high-pressure gas or a low-pressure liquid despite its very low boiling point. Liquid hydrogen must be vacuum insulated to minimize boiling. In the event of a loss of vacuum, oxygen-rich air, which can condense, can create a strongly combustible atmosphere in contact with combustible insulation. When a hydrogen-air mixture is ignited in the atmosphere, the amount of constraint required to develop explosion pressures decreases in proportion to the increase in the atmospheric volume of flammable mixture formed.

Liquefied petroleum gas (LP-Gas, LPG): Liquefied petroleum gas is the generic name for a number of low-pressure, liquefied, hydrocarbon gases or mixtures thereof derived from petroleum. The most common are butane and propane. They are flammable, noncorrosive, and practically nontoxic. Although readily liquefied at atmospheric temperatures, and stored and shipped as liquids, they are used in the vapor phase as a fuel with air or oxygen. Their odor is often artificially produced by the addition of an odorous sulfur compound.

In extremely cold climates, the vapor pressure of liquid petroleum gases can drop below atmospheric pressure. Before attempting to connect or use LPG at temperatures below 0°F (-17.7°C), users should be sure there is a positive pressure at the cylinder valve outlet. At normal temperatures, skin contact with the liquid phase can freeze tissue because of

FIGURE 38.1 Trailer tube transport of high pressure [2,400 - 2,640 psig (16 548 - 18 202 kPa) gas, in this case hydrogen. Pressure relief vents are positioned vertically to discharge upwards.

the chilling effect of the vaporizing liquid. While leak tests are ordinarily performed with soapy water, the presence of frost often provides a visual indication of the presence of something more than a bubble leak.

Methylacetylene-propadiene, stabilized (MPS or MAPP®): Methylacetylene-propadiene, stabilized, is an acetylene derivative stabilized against exothermic decomposition at higher temperatures, pressures, and impact energies. The gas is flammable, heavier than air, and has a disagreeable odor that serves as a warning of leakage. It has a low ignition temperature and, unlike acetylene, a relatively narrow flammablity range. Methylacetylene-propadiene, stabilized, is stored and shipped as a liquid in low-pressure containers similar to liquefied petroleum gases. As the liquid vaporizes, it maintains relatively uniform combustion properties. (Although MPS has a slight tendency to form mildly reactive methylacetylides with copper and high copper alloys under certain conditions, no incident involving methylacetylides has yet been recorded.)

Oxidizing Gases

An oxidizing gas is a gas or gaseous compound that can decompose to provide a gas such as oxygen, fluorine, chlorine, etc., that can react with a fuel to generate heat and light.

Chlorine: Chlorine is a greenish-yellow, heavy, nonflammable elemental gas with a distinctive, disagreeable odor. It is a severe irritant to skin, eyes, and respiratory passages, and may be fatal at high concentrations. Dry chlorine is relatively noncorrosive to metals (other than powdered metals or fine wire) at atmospheric temperature. At elevated temperatures, metals can burn in chlorine. Chlorine can cause spontaneous combustion in the presence of acetylene, ammonia, and some organic materials such as hydrocarbons, alcohols, and esters. It is shipped as compressed liquefied gas under its vapor pressure of 85.5 psig (590 kPa) at 70°F (21.1°C).

Fluorine: A highly toxic and corrosive gas oxidizer with a distinctive sharp odor, fluorine reacts violently with moisture and organic materials. It is explosive on contact with water vapor and powdered metals. Although corrosive to glass and most metals, it can form a metal fluoride on the surface to inhibit further corrosion. Fluorine is packaged in steel or nickel cylinders as a nonflammable, nonliquefied gas at a pressure not exceeding 400 psi (2758 kPa) at 70°F (21.1°C), without pressure relief safety devices. Backflow of moisture into a fluorine container can cause metal to melt, ignite, and fail with expulsion of molten metal. Fluorine can severely burn skin and eyes on contact. Fluorine leaks are detectable by ammonia vapors or moistened potassium iodide on filter paper.

Oxygen: Oxygen is a colorless, odorless, elemental gas present in the atmosphere at a concentration of 20.946 percent by volume. Oxygen is the gaseous component that supports animal life and makes combustion possible. It is usually shipped as a nonliquefied gas at 2,200 - 2,640 psig (15 169 - 18 202 kPa) at 70°F (21.1°C), or as a cryogenic liquid. A single spark (static sparks excluded) in oxygen-rich air (over 40 percent oxygen) or pure oxygen can cause a combustible fibrous material to burst into flame on contact. In the presence of oxygen-rich air or pure oxygen, combustibles such as clothing will burn vigorously and spread the flame rapidly so long as the products of combustion do not blanket the combustible from the oxidizing atmosphere. Almost all nonmetallic or metallic materials will burn in oxygen as long as the materials are not oxides or compounds of oxidizers, or are resistant to further oxidation. With a density only one-tenth heavier than air, oxygen diffuses rapidly in air at atmospheric pressure. When added to nitrogen under pressure, oxygen can stratify unless there is sufficient turbulence to create a uniform mixture.

Compression heating, sometimes called adiabatic heat of compression, is a frequent cause of ignition of nonmetallic materials used in pressure regulators in high-pressure oxygen service. The heating effect, except for its magnitude, is similar to the diesel engine ignition of fuel oil-air mixtures. The temperature-pressure curve shown in Figure 38.2 represents the ideal temperatures developed by the compression of oxygen from atmospheric pressure without loss of heat. Deviations from the ideal are caused by heat transfer to the side walls of the gas passage, a function of time, turbulence, and mixing. However, the main protection against ignition of the nonmetallic seat is the ability of the seat plug to absorb substantial heat before reaching ignition temperature. So long as the seat material retains its smooth surface and is not contaminated with low flash point material, and so long as the gas being compressed is of low volume (close-coupled regulator-to-valve), the combustible material will not be heated to its ignition temperature by the hot but rela-

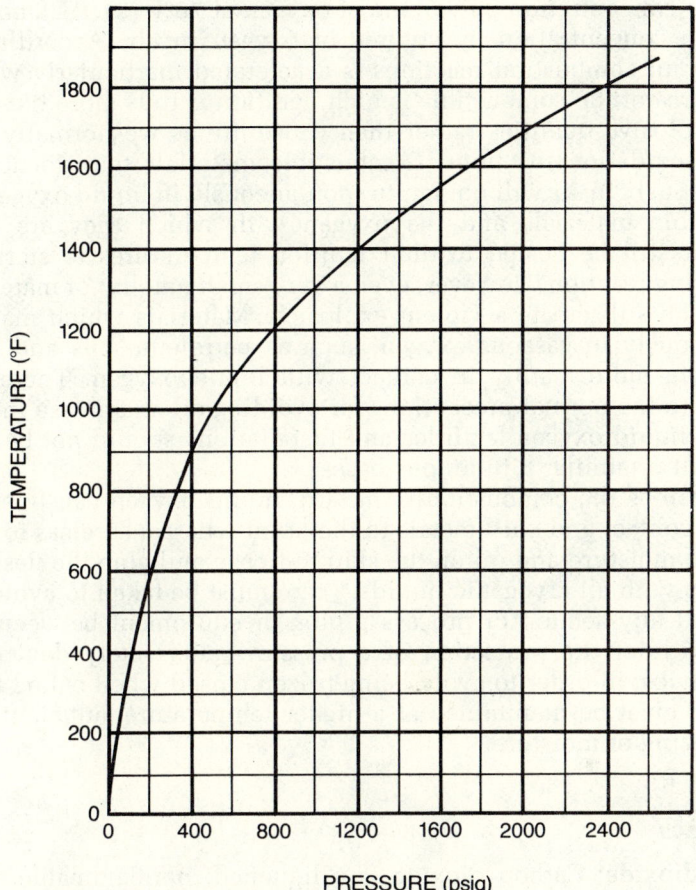

FIGURE 38.2. Theoretical temperatures resulting from the adiabatic compression of oxygen. 5/9 (°F-32) = °C; 1 psig = 6.895 kPa.

tively small slug of gas. Long, large, and straight gas passages and frayed, combustible, or contaminated seat plugs should be avoided.

Heat of combustion is the amount of heat generated by the combustion of a unit weight or volume of a material. Such properties are recorded in Btu per pound or cubic feet (kJ/kg or m^3). The magnitude of the heat of combustion determines the magnitude of the damage created by the combustion of the material. In the case of oxygen, in which almost all materials are combustible, a low heat of combustion will often govern the choice of material, metallic or nonmetallic, to be used in the design of oxygen process equipment.

Oxygen (liquid): Oxygen liquefies at -297°F (-183°C) and atmospheric pressure to form a pale, blue, very cold liquid. In this form it has 861 times the concentration by volume of oxygen at 70°F (21.1°C), and 4,109 times the concentration by volume of oxygen in air. Accordingly, its reactivity in combustion reactions is accelerated, particularly when the surface area of the combustible is high. Ignition is thus more likely to involve explosive qualities rather than cause fire as we normally categorize fuel-oxidation reactions. Combustible materials that do not ignite spontaneously in air will not ignite spontaneously in liquid oxygen. Such combustible materials and the oxygen with which they are in contact must still be heated to their ignition temperatures to start a self-propagating reaction. However, even a very small amount of material and oxygen can still create a violent explosion. Materials which may ignite spontaneously in gaseous oxygen, such as petroleum oils and greases, may ignite more readily in contact with liquid oxygen. Therefore, in addition to the normal precautions for avoiding oil and grease in oxygen systems, liquid oxygen facilities and installations should not be located on asphalt or similar bituminous bases.

Besides the combustibility hazard, liquid oxygen can freeze flesh solid on contact and chill metals to the point where a careless touch can cause the moisture and oil on the skin to freeze and glue the flesh to the metal. As with all cryogenic liquids,* care must be taken to avoid filling liquid-full any section of process piping or equipment between closed valves without the protection of a pressure-relief safety device. Such safety device, in order to avoid being frozen closed when called upon to function, must be maintained at a higher temperature sufficient to prevent freezing of moisture.

Inert Gases

Carbon dioxide: Carbon dioxide is a liquefied, nonflammable gas. Because its vapor pressure increases rapidly with temperature, it is furnished in high-pressure containers. Carbon dioxide is colorless, odorless, and heavier than air. It is normally inert and is classified as an asphyxiant. On contact with moisture, however, it forms a weak acid and hence can be irritating to eyes, nose and the respiratory system. Because of its physiological effects, the American Conference of Governmental Industrial Hygienists[1] has assigned a TLV-TWA (Threshold Limit Value-Time Weighted Average) to the gas. Carbon dioxide is widely used as a fire extinguishing agent, as a refrigerating agent, and as an inerting means. When the liquid vaporizes, its refrigerating effect will change some liquids to solids. Droplets of liquid carbon dioxide will form snow and carry a static charge. When inerting a chamber in the absence of a prior purge, the static can ignite the atmosphere it is intended to inert.

---

*A cryogenic liquid is a liquefied gas at a temperature of -150°F (-101°C) or less. Unless kept cold by refrigeration, cryogenic liquids cannot be stored indefinitely without venting.

HANDLING AND STORAGE OF INDUSTRIAL GASES

FIGURE 38.3. Tank used for transporting liquid oxygen.

FIGURE 38.4. Tractor-trailer unit used for transporting carbon dioxide.

In fixed fire extinguishing installations, operating personnel must be protected from the hazard of asphyxiation.

Nitrogen: A colorless, odorless, elemental, nontoxic, nonflammable gas classified as asphyxiant, nitrogen is useful as a purge or inerting gas to clear flammable gases from a vapor space or dilute them to a concentration at which they will not burn. Nitrogen is packaged, shipped, and stored as a nonliquefied compressed gas in high-pressure cylinders or trailer tubes, or as a cryogenic liquid in Dewar flasks, insulated cylinders, tank trucks, tank cars, and storage tanks. At 70°F (21.1°C), nitrogen is slightly lighter than air.

FIGURE 38.5. Dewar flasks and liquid cylinders used for liquid nitrogen.

GENERAL STORAGE AND USE CRITERIA

Shipping

Bulk industrial gases are produced as liquefied gases, but are packaged in accordance with their vapor pressure-temperature properties. Thus, they are packaged as nonliquefied or liquefied gases in uninsulated containers, or as liquefied or cryogenic gases in refrigerated or vacuum-insulated containers.

The most frequent shipping hazard occurs when an uninsulated, flammable, liquefied gas tank is involved in a fire caused by a motor vehicle or railroad accident, and the tank walls — exposed at low-liquid levels — become overheated and rupture. The resulting BLEVE* type of explosion is violent and, if it occurs near a residential or business district, can seriously injure people and damage property. Water spray, and plenty of it, is still the best protection against fires involving present low-pressure transport equipment in service. A less frequent, but no less hazardous, condition arises when a bulk container of an asphyxiating (non-life-supporting) or corrosive gas is ruptured. Such gases, particularly when shipped in large volumes and when chilled by the sudden drop in pressure, hug the ground and can be extremely hazardous to any nearby homes that are in the path of the ensuing vapor cloud.

Refrigerated or cryogenic liquids are primarily shipped to distributors in vacuum-insulated railroad tank cars or motor vehicle tanks. At the distributors, they are transferred to storage stations. The product is then

---

*BLEVE — Boiling Liquid Expanding Vapor Explosion

FIGURE 38.6. *Insulated storage sphere and tractor-trailer unit used for liquid hydrogen.*

shipped to industrial sites as a cryogenic liquid or as a compressed gas. Shipments of cryogenic liquids such as liquid hydrogen, liquid oxygen, and liquid nitrogen are usually made in liquid cylinders transported by motor vehicles or tank trucks. Cryogenic tanks, because of their thermal insulation, resist fire damage well; however, if 10 percent of the tank's insulation is destroyed in a fire, and the relief valve venting is designed for a fully insulated vessel only, an explosive rupture can theoretically occur.

Shipments of oxidizing, inert, and several of the flammable gases such as hydrogen, carbon monoxide, and methane can be made by compressing the cold liquid through heat exchangers into high-pressure gas cylinders or trailer tubes. The high-pressure gas is shipped in tube trailers, cylinder cradles, or in individual cylinders. One unique product is carbon dioxide. Since carbon dioxide can be solidified at -110°F (-79°C), it is often cooled and compressed into 60-lb (27.2 kg) dry ice cakes (cubes) that are wrapped and delivered as an atmospheric-pressure solid and later converted into a warm liquid by heating in a closed pressure vessel (after loading through a large-end closure). As it warms up, the dry ice vaporizes, developing enough pressure to convert the remaining solid into a liquid, thereby serving as a source of carbon dioxide gas for inert atmospheres.

Delivery

The transfer of bulk gas is accomplished by replacing storage-bank cylinders or by replacing trailers carrying manifolded cylinders or trailer tubes. Such containers and stations must be labeled or placarded with the printed name of the gas so that the responsible attendant will always connect the replacement supply to the correct station supply line. Color coding must not be relied upon as the primary method of identification. Further, when disconnecting a trailer-mounted manifold, a standardized procedure should be followed; for example, the use of a chain-supported sign that contains a large, readily visible message stating "Warning! Line Connected" might be used. Such a sign should be hung on the cab door of the truck or tractor to remind the driver to disconnect before driving away so that the line piping is not torn down and broken. The local attendant should police this procedure.

Storage

National and consensus standards for welding and cutting[2,3,4] require the separation in storage of fuel gas cylinders from oxygen cylinders by at least 20 ft (6.1 m) or a fire resistant wall of ½-hour minimum rating. This policy should be extended to the separation of all fully charged nonflammable and flammable compressed gas containers unless suitable automatic sprinkler protection is available. In addition, readily combustible materials should be separated from all compressed gas containers.[3] While automatic sprinkler protection of storage containers is not usually necessary at consumer sites, the storage site of flammable gases should be within reasonable reach of an adequate fire hose.

Storage sites for bulk oxygen and hydrogen are covered in NFPA 50, *Bulk Oxygen Systems at Consumer Sites*, NFPA 50A, *Gaseous Hydrogen Systems*, and NFPA 50B, *Liquefied Hydrogen Systems at Consumer Sites*; for liquefied petroleum gases, see NFPA 58, *Liquefied Petroleum Gases, Storage and Handling*, and NFPA 59A, *Liquefied Natural Gas, Storage and Handling*.

Piping Installation

Installation of industrial gas systems should be made in accordance with applicable federal, state, and local ordinances. There are NFPA standards for the design and installation of gas systems and for the storage and handling of many industrial gases. (See the appropriate NFPA standards listed in the bibliography of this chapter.)

Chapter 5 of NFPA 51, *Oxygen-Fuel Gas Systems for Welding and Cutting*, includes a section titled "Piping Protective Equipment" which is concerned with the judicious use of safety and control devices; this section could well serve as a model for most gas services. The safety devices are: (1) flashback arrestors (of particular use where the gas mixtures with air, such as acetylene or hydrogen, are capable of detonating reactions); (2) pressure-relief valves; and (3) back-flow check valves. The control devices are shutoff valves and pressure regulators. Essentially, all three safety devices should be installed in the main, branch, or outlet piping to ensure that all continuous paths from one outlet to any gas supply or other outlet are suitably protected (unless an exception is agreed upon in an industry consensus standard). It is important to note that, although check valves may be helpful as flashback preventers by preventing the mixing of flammable gases with an oxidizer, they are not designed for 100 percent reliability, cannot be maintained with 100 percent reliability, and cannot stop the propagation of detonating flashbacks in endothermic compounds like acetylene or ethylene oxide.

CGA Pamphlet G1.4, *Acetylene Transmission for Chemical Synthesis*,[5] requires all acetylene piping and equipment to be electrically bonded and grounded. This is done to ensure the prevention of lightning discharges from arcing inside piping across insulated flanges which, due to the nature of acetylene, may initiate a propagating explosive decomposition. Such a practice is equally applicable to other endothermic compounds such as methyl acetylene and ethylene oxide.

INDUSTRIAL GAS EQUIPMENT

Pipe Threads

Taper thread connections are gastight only when a suitable sealing material is used. Currently, Teflon® tape is in widespread use for connections on piping used for most gases at low and high pressures since it is quite resistant to oxygen in this service and relatively impervious to hydrogen and helium. However, Teflon tape has high lubricity and may loosen when nearby threaded parts become disengaged. Viscous sealing compounds can maintain gastightness even when the parts are not securely fastened, although this can be hazardous if the parts become disconnected inadvertently under pressure. Also, the application of viscous materials can be messy and contaminate the internal system undesirably.

Straight Threads

Straight threads are usually sealed with nonmetallic washers, although flexible-shaped metal washers can hold the highest pressures. At high pressures, the compression stress in a nonmetallic sealing material can cause cold flow and leakage — particularly when a drop in temperature due to the weather or high flow conditions shrinks the washer more than the metals compressing the washer. Thin washers reduce this risk. At high pressures, suitably designed "O" ring closures are usually gastight. Bleed holes adjacent to the sealing faces are often used to prevent disconnecting under pressure.

Flanges

Flanged connections require solid gaskets or viscous sealing compounds. Although viscous sealing compounds are still used by some manufacturers, consumers prefer to use gaskets since they can be sized to prevent extending into gas passageways where they can be picked up in the gas stream and can cause trouble.

Industrial Gas Valves

Industrial gas valves vary widely in type and should not be disassembled in any manner while in use. Tightening threaded parts under pressure can strip threads and lead to violent expulsion of valve parts and the uncontrolled release of fuel or oxidizing gases. Relief valves, on which the settings are pretested, are particularly dangerous to tamper with. Using wrenches on valves not fitted with handwheels can lead to excessive torque or friction which can ignite gases like acetylene and metallic or nonmetallic components of oxidizing gas valves. Diaphragm-type valves are less likely to leak and are widely used on small, low-pressure valves. In high-pressure service they have not yet proven practical because of the high diaphragm thrust against the valve stem.

Pressure Relief Valves

Wherever high-pressure compressed gases or liquefied gases are involved, there is always the hazard of excessive pressures. These pressures can develop from a malfunctioning pressure regulator, or from liquefied gases at high temperatures caused by sunlight, furnaces, or fires, or from cryogenic lines warming up between closed valves. For protection, individual pressure-relief valves or safety devices incorporating a pressure-relief function are needed. This pressure-relief protection is in addition to any pressure-regulator relief device. Pressure-relief devices can be actuated by pressure, heat, gas flow, or some combination of these conditions.

Pressure-actuated pressure-relief devices: These devices are either rupture discs, spring-controlled relief valves, or water seals. Relief valves will vent less gas under operating pressure peaks; however, under fire conditions they can permit the pressure vessel to explode when the liquid level in the vessel drops excessively before the fire is brought under control. The spring compression in relief valves should be as tamperproof as possible and yet provide a ready means of readjustment. Such devices are particularly useful with liquefied gases because excessive pressures from overcharging can be developed while the gases are still at low temperatures. Cryogenic liquid pressure-relief valves must not be permitted to freeze or to leak in the closed position. Hydraulic valves or water seals must have their liquid levels maintained and weather-proofed to prevent freezing and, with critical processes, require isolating valves for maintenance in order to avoid shutting down the process.

Temperature-actuated pressure-relief devices: Temperature-actuated pressure-relief devices are fusible metal plugs that soften or melt at predetermined temperatures, or bimetal (Belleville) springs that reverse their curvature with a "snap" at a predetermined temperature and pierce themselves against a cutter that causes a full-flow rupture of the disk. Some large copper disks bulge enough under excessive pressure to pierce themselves against a cutter without any spring reversal action. But these are classified as pressure-actuated and should be suitable under all conditions. All temperature-actuated devices are effective in fires and at high temperatures. However, temperature-actuated devices are ineffective against pressure increases that are caused by the overfilling of an enclosed volume with more liquid than can be accommodated under normal temperature conditions. Some fusible metal devices are located on the underside of cylinders and should be inspected periodically for damage from handling and for leakage.

Combination pressure-relief safety devices: These devices are assemblies with both pressure-actuated and temperature-actuated components. They are suitable for the protection of full and some partially full containers in fires, and have the added advantage of being less subject to disk failure; should the disc fail prematurely, the result is usually a small leak rather than the sudden release of the full gas volume. With nonliquefied gases, there is no problem of excessive pressure from volumetric overfilling. Thus, these devices are used with the medical gases where a loud noise could be a hazard to patients, or in large-volume supply systems that involve high-pressure compressed gases (such as hydrogen) where the sudden release of a substantial amount of the gas, if ignited, could cause an explosion.

Excess Flow-Actuated Check Valves

These valves are designed to close in the event of an equipment rupture which could cause the release of a large volume of industrial gas. However, they must remain open to permit a gas flow that is as large as that required by the system design. Consequently, excess flow-actuated check valves are designed to function in a relatively narrow pressure-drop region between some maximum design flow and the minimum flow on equipment rupture. They are particularly useful with bulk delivery tankers and tank cars, and are usually assembled internally at the base of the vessel to avoid damage from collision and to function in case the tanker or tank car is moved from the storage supply tank before disconnection. Where nonportable (bulk) gas supply sources are used, excess-flow check valves are desirable to prevent the venting of large volumes of fuel gas in the event of a pipe or fitting failure.

Check Valves

Where portable gas supply containers are replaced as they are emptied, the installation of check valves at the supply source (to avoid feedback into, and contamination of, the supply containers) is good practice and required by NFPA 51, *Oxygen-Fuel Gas Systems for Cutting and Welding*.[3]

Flash Arrestors

Flash arrestors are needed wherever fuel gas-oxidizer mixtures are being burned or reacted, and where there is even a remote possibility of the reaction rate building up to an explosion. At low pressures, a hydraulic valve or water seal sometimes incorporates in one assembly the combined properties of a flash arrestor, check valve, and pressure-relief valve.

Sealing Caps or Plugs

Dust and dirt, regardless of their nature, are to be avoided with industrial gas equipment. Foreign matter can interfere with the sealing of valves and cause flammable gas leaks or loss of gas supply. It can also interfere with the functioning of safety devices such as to cause inoperative pressure-relief valves, and can plug line filters to block gas flow. In oxygen lines, it can increase friction heat when in contact with combustible valve or regulator elements, or increase the chance of an internal fire if the contaminant itself is readily combustible (like lubricating oil or lint). Wherever possible, open, unused gas lines should be sealed (against atmospheric dirt and grime) with brightly colored plastic caps or plugs, until ready for use.

Filters

Strong, conical, self-cleaning filters are used in large pipe lines to trap scale and other debris without plugging or breaking the line. Smaller equipment uses glass wool, pressed wire, or sintered metal filters. Operating personnel should be aware that when the pressure drop across a filter becomes excessive, it is necessary to have the filters cleaned or replaced by a qualified maintenance person.

Oxygen Service

Oxygen valves and regulators cannot be made bubble-tight and readily adjustable without using some elastomeric parts, such as rubber or plastic seat plugs, washers, and diaphragms. These parts must have enough flexibility and plasticity to accommodate small discrete particles that would otherwise prevent proper gas control properties. However, care must be taken in selecting these materials because nonmetallics have relatively low heat conductivity, specific heat, and ignition temperatures. Caution must also be exercised in the choice of lubricants. Lubricants can migrate (even when applied only to "metal-to-metal" sliding surfaces). Only those having extremely low heats of combustion in oxygen should be used. These lubricants should be applied as sparingly as possible because even they can be ignited. Experience shows that the fluorocarbon oils and greases are good in this service.

Noninterchangeable Connections

Since the 1940s, the compressed gas industry has adopted non-interchangeable valve outlet or regulator inlet thread standards for the various groups of compressed gases. The most important objective is to avoid charging flammable or oxidizing gases into storage supplies of inert gases, or to avoid intermixing flammable and oxidizing gas supplies. (For details, see ANSI/CGA V-1,[6] *Compressed Gas Cylinder Outlet Connection Standard.*) But even with "noninterchangeable" connections, some users make mismatches, and, instead of investigating the resistance to tightening or recognizing the implication of excessive looseness — particularly those in a hurry — cross-thread and overtorque the connections to wrench tightness (but not gastightness). Some even use makeshift means in an attempt to connect excessively loose fits. Poor connections and improper adapters are the major sources of leaks of compressed, flammable, and oxidizing gases. Users should be encouraged to avoid adapters and use only proper fittings in good condition.

With the cryogenic liquids, the gas industry has also adopted noninterchangeable connection standards. Although no cryogenic connection can be foolproof, since a joint of sorts can be made with any cryogenic gas by forming a block of water-ice around the fittings to be joined, the noninterchangeable connections should be strictly observed. Where pipe threads are used, no standardization is considered practical in view of the universal use and availability of reducing fittings.

GENERAL OPERATING PRACTICES

Safe Handling of Containers

The safe handling of industrial gases is covered by industry pamphlets for the individual gases. The CGA Pamphlet P-1 titled *Safe Handling of Compressed Gases in Containers*[7] recommends safe practices — for users of gas products — that are applicable to most industrial gases. The major practices not otherwise covered in this chapter are summarized as follows:

1. Do not transfill or mix gases in supply containers. This avoids the possibility of mixing gases that can react with each other, creating mixtures that are explosive, developing highly corrosive compounds, overfilling the container, impairing the container valve and safety devices, interchanging life supporting and asphyxiating gases, and identifying the cylinder contents improperly. Where the gases are used in chemical processes, it is desirable to assure the prevention of unintentional transfilling or mixing by installing reliable check valves between the gas supply and its distribution piping.
2. If the container is contaminated, dented, gouged, or exposed to a fire, or if the valve or safety devices are damaged, empty the container in a safe area and, when returning the empty container, attach a tag advising the gas supplier of the defect.
3. Do not tamper with the supply container, with its valves, safety devices, or with the container markings. All maintenance or repairs should be performed by the supplier. Personnel handling the containers should be instructed to call the gas supplier for advice whenever a problem arises with respect to the container or its gas content.
4. Do not damage or contaminate container valves.
5. Do not expose containers to corrosive atmospheres or materials. Salt water mist and chemical plant atmospheres can be abnormally corrosive.
6. Keep portable containers capped when they are being moved, and fasten them securely to the cart or vehicle being used to move them.
7. Do not use containers as rollers or supports.
8. Do not allow containers to touch energized electrical conductors.
9. Avoid container temperatures over 130°F (54.4°C) and under -20°F (-28.9°C). At excessively low temperatures under shock conditions, containers can fail in a brittle manner.
10. Do not allow containers to suffer harmful dents, gouges, or abrasions, and do not permit the use of such containers; avoid violent impacts by or against containers.

Of major importance to cylinder-filling and shipping personnel in the safe handling of containers is a knowledge of the "filling densities" or "loading densities" specified for specific gases — in some cases, groupings of similar gases — in codes, standards, and regulations. Filling density is the percent ratio of weight of gas in a container to the weight of water that the container will hold at 60°F (15.5°C).[8] Maximum filling densities of liquefied gases are specified in the DOT regulations to assure a vapor space above the liquid in a fully charged container at 60°F (15.5°C) with sufficient allowance for expansion of the liquid at temperatures as high as 130°F (54.4°C). If filling density limitations are exceeded, the cylinder can become liquid-full at such low temperatures that, on warming, the pressure increase no longer is directly proportional to the increase in absolute temperature (Charles' Law), but rises

TABLE 38.2. *Maximum Permitted Filling Density (LPG Stored at Normal Temperatures in Uninsulated Containers)*

Specific Gravity at 60°F (15.6°C)	Aboveground Containers		Underground Containers All Capacities
	0 to 1200 US Gal (1000 Imp. gal, 4.5 m^3) Total Water Cap.	Over 1200 US Gal (1000 Imp. gal, 4.5 m^3) Total Water Cap.	
.496–.503	41%	44%	45%
.504–.510	42	45	46
.511–.519	43	46	47
.520–.527	44	47	48
.528–.536	45	48	49
.537–.544	46	49	50
.545–.552	47	50	51
.553–.560	48	51	52
.561–.568	49	52	53
.569–.576	50	53	54
.577–.584	51	54	55
.585–.592	52	55	56
.593–.600	53	56	57

with excessive rapidity because a liquid invariably expands faster with increasing temperature than does the container in which it is charged.

Since liquefied gases in containers are two-phase systems of different densities, and since the vapor pressure is a function of the liquid temperature and not of the weight of gas contained, the quantity of liquefied gas in a container cannot be determined by pressure, but must be determined by weight. This information, which is of particular importance to plant cylinder-filling and shipping personnel, also provides additional reasons why users should never refill compressed gas containers or mix gases in containers.

Table 38.2 illustrates how filling densities are expressed for liquefied gases (in this case, LPG) stored at normal temperatures in uninsulated containers. It should be noted that larger quantities of higher specific gravity materials are permitted (indicating that these liquids tend to expand less), that larger containers can be filled more than smaller containers (indicating that it takes longer for them to absorb heat from atmospheric temperatures or sunlight), and that underground containers can be filled even more (indicating that their ambient temperatures are relatively constant and well below summer atmospheric temperatures).

Specific Gravity of Vented Gases

Industrial Gases that may be deliberately vented from containers or that may vent from containers, such as by the action of safety relief devices or through inadvertent leaks, may vary in specific gravity. Specific gravity (air = 1) is the ratio of the density of the gas or vapor to the density of air at 70°F (21.1°C) and 1 atm (atmosphere).

Gases heavier than air will tend to fall to the floor and may accumulate in pits and low areas. Under still air conditions they may travel distances of several hundred feet (100 ft equals 30.4 m) to reach such hazards as sources of ignition if they are flammable.

Gases lighter than air tend to rise and may accumulate in higher areas such as near ceilings. This tendency is less pronounced than in heavier gases, however, because lighter gases also tend to diffuse more rapidly in air. The rate of diffusion is inversely proportional to specific gravity. Gases with specific gravities less than 0.9 usually dissipate rapidly in the ordinary atmosphere.

When gases are deliberately vented or may vent from industrial gas containers, the consequences of their behavior due to specific gravity must be considered before entering a potentially hazardous area.

Working in Confined Spaces

The hazard of leaks from flammable, inert, or oxidizing gas supplies poses an additional major problem when working around industrial gas equipment in confined spaces. Closing shutoff valves is not enough. Either all gas lines must be disconnected and physically blanked with a suitable positive closure, or a trio of block and bleed valves used. Should such positive disconnecting not be possible, continuous monitoring of the atmosphere is necessary so that all workers will be warned to leave if a significantly hazardous atmosphere develops. Where flammable, inert, or oxidizing gases are occasionally intentionally vented (or spilled as a liquid) near work areas (particularly in trenches or pits), vent-free time periods must be agreed upon for work periods, atmospheric monitoring must be continuous, and workers must evacuate immediately upon signal.

Acetylene — 15 psig (103.4 kPa) Maximum

Porous fillers and solvents used in acetylene cylinders are designed and tested to store and transport acetylene safely at pressures of 250 psi (1723 kPa) at 70°F (21.1°C) and higher. Acetylene combustion equipment in the United States is, by custom, designed to supply acetylene safely at pressures no higher than 15 psig (103.4 kPa). Accordingly, acetylene regulator delivery pressure gages are colored red at pressures above that limit to serve as a warning. The pressure rating of 15 psig (103.4 kPa) is not a magic limit above which it is unsafe to use acetylene and below which its usage is foolproof. It is, however, a value which, because it has been established by custom and experience, can be used as a reasonable guideline above which the safety factor is dangerously reduced. When a regulator gage reads 15 psig (103.4 kPa), it is being subjected to an internal pressure equal to the sum of the gage reading, plus the atmospheric pressure surrounding it; in most cases, one atmosphere. Working at pressures greater than one atmosphere with acetylene should be prohibited to avoid a flashback that can propagate past the gas mixer with a chance of burning up or rupturing the hose, regulator, generator, or cylinder. When high gas flows above the capacity of a cylinder already delivering 15 psig (103.4 kPa) of acetylene are required, cylinders should be manifolded rather than modified to furnish pressures above 15 psig (103.4 kPa). Acetylene cylinders should never be used in caissons or other pressurized chambers because the risk of explosion is too great.

"Cracking" Cylinder Valves

A long standing industry custom is the practice of quickly opening and closing a cylinder valve to check its operating quality and to blow out any loose dust in the outlet connection. This procedure is particularly recommended in the case of oxygen cylinders to help avoid the ignition of contaminants by the adiabatic heat of compression. This practice is not recommended for hydrogen or acetylene which may ignite spontaneously. It can also cause an acetylene cylinder to spit acetone. For obvious reasons, the valve should be pointed in a safe direction with nothing combustible and no one in front of the outlet. "Cracking" a cylinder valve will not clean an outlet connection reliably, although it will blow out large deposits of accumulated debris that might otherwise be time consuming to remove.

Identification

The gas content of piping and containers is identifiable primarily by the printed label. ANSI A13.1, *Scheme for the Identification of Piping Systems*,[9] requires the name of the gas contained to be placed at each valve and outlet fitting together with (in the absence of a uniformly painted pipe line) a color-banded label identifying the color-band code used in tracing the line for operating and maintenance purposes. ANSI/CGA C-4, *Method of Marking Portable Compressed Air Containers to Identify the Material Contained*,[10] requires portable compressed gas containers to be identified primarily by labeling the name of the gas. Containers that do not carry the name legibly printed on the gas supplier's label, stencil, or tag should not be used.

DOT regulations have recently improved the placarding requirements for vehicles transporting hazardous materials to help make a flammable or combustible lading more readily identifiable under all positions of the transport vehicle in a wreck or fire. All labeling and placarding of hazardous materials now requires that the United Nations (UN) identification number of the product contained or being transported be shown. Identification numbers for all DOT hazardous materials may be found in reference 8 (Table 172.101, CFR 49). Recommended response to hazardous materials emergencies is contained in reference 12 (DOT P 5800.2).

INSPECTION FOR LEAKS

Bubble Testing

Bubble testing is a method involving the application of a thin film of clear liquid of high surface tension to a metal surface suspected of leaking under gas pressure. The test solution is usually a weak, clear, soap solution. When the metal to be tested is dry and the pressurizing gas is dry, leaks with widely varying leak rates are identifiable. If the metal that is to be tested by this method has recently been hydrostatic-proof pressure tested, only the largest leaks will be identifiable. The water from the pressure test can plug small capillaries so that they will not leak gas even under high pressure until the metal surfaces (internal and external) are dried. Bubble testing is a simple, reliable method for detecting leaks. When properly done, it offers visual proof of a leak, or lack of one. It also determines the location and magnitude of leaks.

Pressure-Recession Test

The pressure-recession test is a method involving the measurement of changes in pressure with respect to time and temperature. Measurements are usually made over a period of at least two days to minimize the effect of temperature changes on the volume of gas being pressure tested. The pressure-recession test is a rather crude test, considering the variations in temperature over the system, the pressure-gage inaccuracies, and the need to estimate the effective system temperature. While it readily proves the presence of large leaks, this test involves excessive amounts of time to prove that pressure changes were due to a leak and not temperature changes. If performed on wet pipe, potential leaks will not be identifiable until after the pipe connections have either been dried out or been given time to dry out.

Tracer Gas Test

Where flammable gas pipelines are underground and a leak is suspected, nitrous oxide is used as a tracer gas in detecting the leak. Such testing involves the injection of small concentrations of nitrous oxide, after which the right of way is surveyed for the tracer gas which is readily detected at concentrations well below the LFL of all flammable gases.

Audio Test

Where compressed gas pipelines are aboveground and not easily reached, audio amplifiers have been used to test for leakage. The amplifier is tuned to the frequency range of most gas leaks. This test is not frequently used.

Atmospheric Analysis

When work is to be performed in confined spaces in the presence of gas distribution lines, off gas, or venting gas from industrial processes, or decomposing vegetable (organic) matter, atmospheric tests may be made and then evaluated to determine if the results are representative of the entire work area or only of the local zone tested. Wherever variations from normal atmospheric concentrations are found, unless all gas sources have been completely blocked or disconnected, it should be assumed that a leak exists, one which may change in magnitude, so continuous monitoring may be required.

Inspection for Corrosion (Potential Leaks)

Checking connections for gastightness may be the primary leak prevention procedure. It is also important to ensure that piping, fittings, containers, and equipment subject to corrosion remain gastight. The water-jacket test for the periodic retesting of compressed gas containers limits the allowable corrosion by limiting the increase in permanent and elastic expansion under retest pressure. With process piping as well as compressed gas containers, hammer testing will locate vibration-damping scale or other materials so that the equipment can be cleaned and the wall thickness measured by probing or ultrasonic testing. Ultrasonic testing equipment is in widespread use and displays direct readings of metal thickness computed from harmonic frequencies under which the metal sections will vibrate when in contact with a piezoelectric crystal. Mechanical inspection with micrometers or depth gages is sometimes possible. Visual inspection is often used, but principally to locate questionable areas. The depth of corrosion cannot be estimated visually with any accuracy although the inspector can distinguish between general, pit, and linear corrosion. Automated, X-ray inspection is useful with critical sections for estimating corrosion and locating flaws.

MAINTENANCE AND REPAIR

The need for qualified maintenance and repair persons in the field of industrial gases cannot be stressed too strongly. Such personnel must be trained in means of avoiding the hazards that can be created by:

(1) mismating of threaded connections, (2) excessively loose fits, (3) insufficient thread engagement, (4) unreliability of some plastic seals with straight-threaded connections, (5) leaking valves and connections and the flammability of fuel gases, (6) compatibility of nonmetallic and metallic materials with oxygen, (7) lack of cleanliness, (8) the difficulty of eliminating all ignition sources in the vicinity of industrial or commercial gases and gas mixtures, (9) mistaken reliance on color coding to identify a gas product, (10) improper repair parts, (11) excessive corrosion, and (12) low temperature "burns" and other defects. Such hazards may cause injury to maintenance and repair personnel as well as cause property damage. Maintenance personnel must see that the gas services involved conform to national, state, or local codes that are applicable. They must be equipped with, and know how to use, emergency fire protection equipment.

FIRE PREVENTION, PROTECTION, AND EMERGENCY HANDLING

Stopping Leaks

Leaks along valve stems can be stopped by tightening packing nuts. All other leaks such as those at connections and in distribution lines are corrected by shutting down, bleeding, disconnecting and cleaning connections, or replacing defective components before reassembling. Leaks on low-pressure components can sometimes be plugged by clamping a resilient patch over the leak until the equipment can be bled down in a safe area and the leaking member repaired or replaced. With the exception of leaking safety devices that require replacement, leaks through valves and regulators can be plugged.

Purging

Where flammable gases are present in areas where they can develop into a flammable or explosive hazard, the safest way to remove them is to purge the area with inert gases such as carbon dioxide or nitrogen, and then ventilate the area with fresh air or, depending upon future work activities, leave the inert gas atmosphere. Once the atmosphere has been returned to normal, and suitable ventilating or monitoring procedures established to assure maintenance of a normal breathing atmosphere (i.e., freedom from flammable gas contaminants, oxygen enrichment, or oxygen deficiency), workers can return to work.

Eliminate Ignition Sources

Besides keeping flammable gases from mixing with oxidizers, it is equally important to try to eliminate the third element of the fire triangle — the heat or ignition source. This is difficult to achieve.

Cutting, welding, and grinding operations provide a ready source of ignition in the form of sparks (see Chapters 24 and 34). Sparks, which may include droplets of molten or hot metal, are a frequent source of gas ignition because they scatter over a wide area in direct proportion to the grinding speeds or cutting oxygen velocities employed. Fuel gas leaks are a fire source particularly sensitive to ignition by sparks. Cutting slag (which includes about 30 percent molten iron), if present in sufficient quantities, can even cause a fusible plug safety device in a flammable or toxic gas cylinder to release the cylinder contents. Welding and cutting operations should be prohibited in areas where flammable gases or vapors are being produced or stored. Of course, welding, cutting, and grinding are not the only sources of metal sparks. Impact tools are potential sources. Spark "resistant" tools can be helpful, although impact blows with such tools can strike sparks from stone and many metals, and are friction sources as well.

Other ignition sources can be found in flames, electric arcs and hot surfaces from process equipment, electrical and mechanical controls, portable appliances, metal impact and wear, smoking materials, and static electricity. Both automatic as well as operating controls must be de-energized when facing a potential hazard from flammable or oxidizing gases. Smoking, of course, should be forbidden. The use of conductive materials, intercoupling, ionization of the air, and grounding can help to reduce static. Static, however, is the most difficult ignition source to eliminate since it is generated while people work, even with all power sources disconnected. Lightning protection and avoidance of overhead power lines also require attention.

Protection from External Fires

Fire protection for flammable, oxidizing, or inert industrial gases in storage areas or process equipment may utilize dry-pipe or deluge water sprinkler systems, or systems utilizing carbon dioxide or halogenated hydrocarbons such as bromotrifluoromethane (Halon 1301). These systems are usually automatically actuated by heat sensors. Protection can also be provided by subdividing large fuel supplies into smaller fuel gas storage units, by spacing the inert and oxidizing gases a suitable distance from the flammable gases, by using fire resistant barriers or, where flammable liquids are involved, by installing appropriate dikes and drains where needed to keep possible spills and overflows from the gas storage areas. Fires in portable, flammable gas supplies not in storage are best controlled or extinguished by hose lines or portable extinguishers. Before deciding to extinguish any burning gas flame, one must be sure that when the flame is extinguished: (1) either the gas supply valve can be immediately closed or the gas supply (if portable) removed safely to the out-of-doors, or (2) that the venting of unburned gas will not be allowed to create an explosion hazard that may be more dangerous than the burning gas. The Compressed Gas Association has a bulletin[11] which deals with the handling of acetylene cylinders in fire situations. This bulletin can also be used as a conservative guide for the disposing of leaking, portable flammable gas containers.

Protection from Internal Fires

In the presence of the combustion (within process, storage, transportation, or usage equipment) of a flammable gas mixture or the dissociation of an endothermic compound (such as acetylene or ethylene oxide), there are hazards from a possible rupture of equipment or the venting of hot flammable gases through open pressure-relief devices. Further, both combustion and dissociation reactions can produce flames that, under certain conditions, are difficult to see. In handling such equipment in emergencies, personnel should be familiar with the location of pressure-relief devices so that they don't inadvertently pass in front of a device that might be venting from an internal exothermic reaction. In the event that any industrial gas equipment feels excessively hot to the touch, it should be cooled with water and kept cool until water cooling is no longer needed.

Nonliquefied Gas Containers

Automatic water spray sprinklers are ideal protection for process equipment and storage supplies involving flammable and nonflammable gases. Nonliquefied gas containers (with their high ratios of cylinder weight to the weight of gas contained) and nonflammable gas containers are reasonably fire resistant and may not require fixed water protection. However, gas containers should not be expected to withstand fires without water protection unless the user is familiar with the action of the pressure-relief device and the venting arrangements under the pressure-temperature conditions involved.

Nonliquefied gases, for example, are packaged in high-pressure containers that are usually equipped with a bursting disk. Such a container when *full* and exposed to a general fire will be subjected to a pressure rise that will rupture the disk and release the entire gas content before the container can fail, regardless of the intensity or duration of the fire. But the same cylinder and safety device with only a 25 percent charge when exposed to a general fire may explode before the disk can rupture to release the internal pressure. In either case the venting of flammable gases can add significantly to the intensity of an indoor fire so that flammable gas storage areas and manifolded flammable gas cylinders are often protected indoors by a sprinkler system.

Low-pressure and high-pressure portable gas containers are used for packaging, transporting, storing, and supplying some of the more frequently used industrial gases.

Liquefied Gas Containers

Whether portable or storage vessels, liquefied gas containers are equipped with pressure-relief valves which should open and close periodically during a fire. When the liquid level drops significantly due to venting, the exposed unwetted walls can heat rapidly and, if the fire continues, the container might explode under the pressure retained by the relief valve. The possibility of rupture thus depends upon the duration of the fire (proximity of combustibles) and the ability of the valve to continue to hold a minimum pressure despite the fire. Uninsulated transport tankers, when involved with fuel and tire fires capable of maintaining a hot fire, will often rupture at low liquid levels. Smaller vessels and cylinders may not fail if the relief valve seal softens or melts and blows out to release all the fuel gas to the atmosphere. Spraying water on a relief valve is not advisable unless the entire tank can be kept wetted.

Cryogenic Liquid Containers

Cryogenic liquid containers are highly insulated against heat transfer from the ambient atmosphere to minimize vaporization of the cold liquefied gas. In a fire, because of the insulation, there is no cooling of the outer jacket by the liquid contents so the jacket's usefulness in maintaining the quality of the insulation can be ended by melting (in the case of relatively thin aluminum jackets in a sustained fire), or by oxygen scarfing (in the case of steel-jacketed liquid oxygen tanks where venting oxygen impinges on the hot steel shell). Design and maintenance both play an important role in the performance of transport and storage containers in fires where there is no water spray sprinkler protection.

Venting Bursting Disk

If a cylinder, tank, or storage vessel is not involved in a fire and a bursting disk functions to release a flammable, oxidizing, or asphyxiating gas, it may be equally important (depending upon possible volumes and locations) to determine if the disk rupture was due to overfilling, overheating, excessive storage time, or a need for insulation repair. If the disk is used on a nonliquefied gas that is not involved in a fire, the system has been either overfilled, overheated, or the device has been improperly assembled. If a relief valve (set at a lower pressure than the disk) has failed to function, it would be desirable before replacing the disk to determine if the valve is so close to the cryogenic liquid that it froze in the closed position.

Venting Relief Valve

With liquefied gas containers not involved in a fire, a venting relief valve is usually a sign that the container has been overfilled (allowable filling density exceeded) or overheated (usually by close proximity to a radiator or furnace), or that closed valves have trapped a section of piping that is filled with a liquefied gas. With the cryogenic liquids, the venting sign may well indicate that the vacuum insulation needs maintenance unless the standby time without gas consumption has exceeded the design rating.

BIBLIOGRAPHY

REFERENCES CITED

1. American Conference of Governmental Industrial Hygienists, Threshold Limit Values for Chemical Substances and Physical Agents in the Workroom Environment with Intended Changes for 1983-84, ACGIH, 6500 Glenway Ave., Bldg D-5, Cincinnati, OH, 45211.
2. OSHA Regulations, 49 CFR Part 1910.250/252, Subpart Q — Welding, Cutting, and Brazing, U.S. Government Printing Office, Washington, DC, 20402.
3. NFPA 51, *Oxygen-Fuel Gas Systems for Welding and Cutting*, 1983.
4. ANSI Z49.1, *Safety in Welding and Cutting*, American Welding Society, 550 NW LeJeune Rd., Miami, FL, 33126.
5. Compressed Gas Association, *Acetylene Transmission for Chemical Synthesis*, CGA G1.4, 1235 Jefferson Davis Hwy., Arlington, VA, 22202.
6. Compressed Gas Association, *Compressed Gas Cylinder Valve Outlet Connection Standard*, ANSI/CGA V-1.
7. Compressed Gas Association, *Safe Handling of Compressed Gases in Containers*, CGA P-1.
8. U.S. Department of Transportation, Hazardous Materials Regulations, Parts 100-199, U.S. Government Printing Office, Washington, DC, 20402.
9. ANSI A13.1, *Scheme for the Identification of Piping Systems*, American National Standards Institute, 1430 Broadway, New York, NY.
10. Compressed Gas Association, *Method of Marking Portable Compressed Gas Containers to Identify the Material Contained*, ANSI/CGA C-4.
11. Compressed Gas Association, *Handling Acetylene Cylinders in Fire Situations*, CGA SB-4.
12. U.S. Department of Transportation, Hazardous Materials — Emergency Response Guidebook, Labelmaster, 7525 N. Wolcott Ave., Chicago, IL, 60626.

NFPA Codes, Standards, and Recommended Practices

Reference to the following NFPA Codes, Standards, and Recommended Practices will provide further information on the safeguards for the handling and storage of industrial gases discussed in this chapter. (See the latest *NFPA Codes and Standards Catalog* for availability of current editions of the following documents.)

NFPA 50, *Bulk Oxygen Systems at Consumer Sites.*
NFPA 50A, *Gaseous Hydrogen Systems at Consumer Sites.*
NFPA 50B, *Liquefied Hydrogen Systems at Consumer Sites.*
NFPA 51, *Oxygen-Fuel Gas Systems for Welding and Cutting.*
NFPA 51A, *Acetylene Cylinder Charging Plants.*
NFPA 51B, *Fire Prevention in Use of Cutting and Welding Processes.*
NFPA 58, *Storage and Handling of Liquefied Petroleum Gases.*
NFPA 59, *Storage and Handling of Liquefied Petroleum Gases at Utility Gas Plants.*
NFPA 59A, *Liquefied Natural Gas, Production, Storage and Handling.*

Additional Reading

Bahme, Charles W., *Fire Officer's Guide on Hazardous Chemicals*, National Fire Protection Association, Quincy, MA, 1978.

Chemical Engineering Magazine, *Safe and Efficient Plant Operation and Maintenance*, McGraw-Hill, New York, NY, 1980.

Compressed Gas Association, *Handbook of Compressed Gases*, 1981 CGA, Arlington, VA, 22202.

Federal Emergency Management Agency, *Planning Guide and Checklist for Hazardous Materials Contingency Plans*, FEMA, Washington, DC, 1981.

Fire Protection Guide on Hazardous Materials, National Fire Protection Association, Quincy, MA, 1978.

Hawley, Gessner J., *Condensed Chemical Dictionary*, 10th ed., Van Nostrand, Reinhold Co., New York, NY, 1981.

Hazardous Materials Guide, J. J. Keller and Associates, Inc., Neenah, WI, 1984.

Isman, Warren E., and Gene P. Carlson, *Hazardous Materials*, Glencoe Publishing Co., Inc., Encino, CA, 1981.

Manufacturing Chemists Association, *Manual L-1, Guide to Precautionary Labeling of Hazardous Chemicals*, 1825 Connecticut Avenue, N.W., Washington, DC 20009.

The Matheson Company, Inc., *Matheson Gas Data Book*, East Rutherford, NJ.

McKinnon, G. P., ed., *Fire Protection Handbook*, Fifteenth Edition, National Fire Protection Association, Quincy, MA, 1981.

Meidl, James, *Explosive and Toxic Hazardous Materials*, Glencoe Publishing Co., Inc, Encino, CA, 1970.

Chapter 38

———, *Flammable Hazardous Materials*, 2nd ed., Glencoe Publishing Co., Inc., Encino, CA, 1978.

Meyer, Eugene, *Chemistry of Hazardous Materials*, Prentice Hall, Inc., Englewood Cliffs, NJ, 1977.

National Safety Council, *Accident Prevention Manual for Industrial Operations*, 8th ed., Chicago, IL, 1980.

Sax, N. Irving, *Dangerous Properties of Industrial Materials*, 5th ed., Van Nostrand, Reinhold Co., New York, 1979.

Schieler, Leroy and Denis Pauze, *Hazardous Materials*, Delmar Publishers, Albany, NY, 1976.

Senesky, J., "Safe Storage and Handling of Compressed Gases," *Plant Engineering*, vol. 33, no. 10 (Oct. 1979), pp. 143-148.

Turner, Charles F. and Joseph W. McCreery, *The Chemistry of Fire and Hazardous Materials*, Allyn and Bacon, Inc., Boston, MA, 1981.

U.S. Department of Transportation, *Emergency Response Guidebook*, Government Printing Office, Washington, DC, 1980.

Weiss, J., *Hazardous Materials Chemical Data Book*, Noyes Publishing Company, Park Ridge, NJ, 1980.

39

Liquefied Petroleum Gases at Industrial Plants

H. Emerson Thomas

Liquefied petroleum gases (LP-Gas, LPG) are used in industrial plants as (1) a basic fuel, (2) a standby to the use of natural gas, or (3) a propellant for aerosol products. As a basic fuel, it is utilized in five main ways: heat treating applications, building heating, fuel for lift trucks, with oxygen for cutting operations, and in research laboratories, cafeterias, etc.

There are two sources of liquefied petroleum gas — natural gas as it comes from the ground, and natural gas as a product of petroleum refining. Natural gas can be of two types: (1) "dry," which consists primarily of methane and ethane; and (2) "wet," which is generally composed of methane, ethane, propane, butane, pentane, hexane, heptane, etc. Propane and butane, both derivatives of "wet" natural gas, are easily liquefied on extraction from the natural gas. They can be stored and shipped in liquid form under moderate pressure, but they convert back to vapor at normal atmospheric temperatures.

The second source of LP-Gas is refining of petroleum. In addition to propane and butane, propylene and butylene are also constituents of

---

H. Emerson Thomas is a design, engineering, and construction consultant to the LP-Gas industry. He is president of Thomas Associates, Inc., Westfield, New Jersey.

the refinery process. These latter two gases are somewhat similar in characteristics, and for most uses are commingled with or used similarly to propane and butane from the natural gas source. Some physical properties of commercial propane and butane are given in Table 39.1.

Note that the initial boiling point of propane at sea level is −51°F (−46.1°C) and butane 15°F (−9.4°C). The heating value per standard cu ft (one cu ft equals 0.028 m^3) of propane is 2,516 Btu (2657 kJ) and of butane 3,280 Btu (3464 kJ). They both require more air per unit for combustion than does natural gas.

LP-Gases are odorless, and a warning agent, generally ethyl mercaptan, that imparts a "rotten egg" odor to the gas has to be added. The range of flammability is relatively narrow in its percentage of vapor in air-gas mixture—being 2.15 to 9.6 percent for propane and 1.55 to 8.6 percent for butane.

HISTORY OF LP-GAS IN INDUSTRY

LP-Gas was first extracted about 1914, and its commercial development was slow until about the mid-1920s when it was primarily used in residences, and to some extent in commercial establishments. Small amounts however, were used in industry, mainly for heat treating. Originally it was exclusively distributed in cylinders generally of 100 lbs (45.5 kg) product capacity. For industrial use a multiple number of cylinders, which were built to Interstate Commerce Commission specifications, were manifolded together. In some cases there would be as many as 50 to 100 cylinders in banks. This number of cylinders was necessary to provide sufficient heat transfer surface (the container shells) for the required vaporization rate to supply the needed quantity of gas without setting up undue refrigeration and loss of pressure due to too rapid boiling in individual cylinders.

Around 1929, tank truck delivery was inaugurated. This enabled industrial sites to install relatively large tanks so that a single tank could store the product rather than the customary banks of cylinders. At the time that LP-Gas was being introduced for industrial uses, the principal gas available to industry was manufactured gas which had disadvantages in industrial heating processes. Because LP-Gases are sulfur free, they were found to be useful in many direct applications where manufactured gas had to be used in an indirect application. An example of the growing usefulness of LP-Gas was in the bright annealing of copper pipe. After annealing with manufactured gas the product had to be pickled to get a bright finish; with LP-Gas the pickling process was eliminated.

The superior characteristics of the LP-Gases encouraged many industries to convert from manufactured gas to LP-Gas. This conversion gave them a superior gas heating product while permitting them to materially cut their costs. However, as natural gas became available to in-

TABLE 39.1. Approximate Properties of LP-Gases

	Commercial propane (NLPGA avg)	Commercial butane (NLPGA avg)
Vapor pressure in psig at:		
(a) 70°F	132 psig	17 psig
(b) 100°F	205 psig	37 psig
(c) 105°F	216 psig	41 psig
(d) 130°F	300 psig	69 psig
Vapor pressure in kPa gage at:		
(a) 20°C	930 kPa	103 kPa
(b) 40°C	1550 kPa	285 kPa
(c) 45°C	1720 kPa	345 kPa
(d) 55°C	2070 kPa	462 kPa
Specific gravity of liquid at 60°F (15.56°C)	0.509	0.582
Initial boiling point at 14.7 psia	−51°F	15°F
Initial boiling point at 101 kPa	−46°C	−9°C
Weight per gal of liquid at 60°F	4.24 lb	4.81 lb
Weight per cubic meter of liquid at 15.56°C	509 kg	582 kg
Cu ft of vapor per gal at 60°F	36.39	31.26
m^3 of vapor per liter at 15.56°C	0.271	0.235
Cu ft of vapor per lb at 60°F	8.58	6.51
m^3 of vapor per kg at 15.56°C	0.534	0.410
Specific gravity of vapor (air=1) at 60°F (15.5°C)	1.52	2.01
Ignition temperature in air	920–1120°F	900–1000°F
Ignition temperature in air	493–604°C	482–538°C
Limits of flammability in air (percent of vapor in air-gas mixture):		
(a) lower	2.15	1.55
(b) upper	9.60	8.60
Total heating values after vaporization:		
(a) Btu per cu ft	2,516	3,280
(b) Btu per lb	21,591	21,221
(c) Btu per gal	91,547	102,032
(d) kJ/m^3	93470	121280
(e) kJ/kg	50020	49140
(f) kJ/liter	25430	28100

dustrial centers, plants that had converted to LP-Gases were then converted to natural gas. In most cases, the LP-Gas system was retained as a standby system.

Standby LP-Gas systems were useful because of the possibility of interruption of regular natural gas service being brought in by transmission pipelines, often from great distances. As natural gas loads grew larger, especially for home heating, there was a peak demand in the winter during which pipelines could not carry the full volume demand. As a result, industrial plants would be temporarily shut off from the natural gas supply and would be forced to rely upon the LP-Gas Air/Standby plant. Mixing equipment in the standby plant was engineered to permit an exchange of LP-Gas for natural gas without adjusting ovens, furnaces, heaters, etc. Other industrial uses were as engine fuels, with oxygen for metal cutting operations, and as aerosol propellants. (Chapter 24 covers welding and cutting practices and the fuels used with them; aerosol propellants are covered in Chapter 35.)

LP-GAS STANDARDS

In the pioneering days of 1929 and 1930, the LP-Gas industry recognized the need for proper standards for the storage and handling of LP-Gases. The Compressed Gas Association developed a set of standards on good practices to follow in the handling and storage of LP-Gases that were first adopted by NFPA in 1932. That initial standard, along with several other LP-Gas related standards, eventually became NFPA 58, *Standard for the Storage and Handling of Liquefied Petroleum Gases*.

The basic approach to developing NFPA 58 is to provide safeguards in LP-Gas storage systems that will not allow uncontrolled discharge of the product. The standard's provisions are developed, so far as possible, on the concept that automatic safeguards are built into the system to compensate for human errors. Consequently, the requirements of NFPA 58 are quite precise and are based on long field experience with LP-Gas installations. They cover the design, construction, installation, and operation of most LP-Gas systems with some exceptions (the exceptions are the subject of, or included in, other NFPA standards, e.g., NFPA 51, *Oxygen-Fuel Gas Systems for Welding and Cutting*).

LP-GAS STORAGE FACILITIES

At industrial locations perhaps the most visible evidence of the presence of an LP-Gas system is the storage facility. As the applications for LP-Gas grew, larger and larger storage containers within industrial sites were required, and many of these were in built-up areas. A typical industrial plant installation today involves one to ten 30,000-gal (113.5 m^3) water

FIGURE 39.1. A propane-air standby plant with two 30,000-gal (113.5 m³) tanks. The vaporizers and blender are at left, and the truck unloading station and test flare are at right protected by crash posts. (Thomas Associates, Inc.)

capacity* tanks, a transport truck or tank car unloading station (or both), vaporizers, and, in the majority of cases, gas-air mixers including a permanent field flare (see Figure 39.1). The field flare is used for gas-air mixture testing and adjustment purposes as well as for training personnel. Some systems incorporate temporary flares which were installed on a makeshift basis and may present safety problems.

In some plants, where the only use of LP-Gas is for fueling industrial (forklift) trucks, the storage facility may be smaller, generally tanks from 1,000- to 18,000-gal (3.8 to 88.1 m³) water capacity; however, additional equipment is required in order to fill the small containers for the lift trucks. This storage arrangement involves plant personnel in filling the individual cylinders for the lift trucks; LP-Gas industry personnel are only involved in deliveries to the bulk storage container.

Selection of the storage site within a plant yard involves many considerations. For one thing, the bearing capacity of the soil must be checked to ensure that the ground where the tanks will stand can support the weight of the tanks and their full water capacity. At some later date the tanks may have to undergo hydrostatic testing (filling them full of water which is much heavier than an equal volume of LP-Gas).

Equally important, and the subject of regulation, is the location of the storage containers and associated equipment in relation to buildings on the site, the setbacks from lot lines, and storage containers for other flammable gases and liquids. Good separation distances have a twofold purpose: they offer protection to other structures from fires involving an LP-Gas storage facility, and they help to protect the facility itself from exposing fires. The requirements for minimum separation distances between individual LP-Gas storage containers and associated equipment,

---

*Water capacity is the amount of water, in either pounds (grams) or gallons (liters or cubic meters), at 60°F (15.6°C) required to fill a container liquid-full of water.

Chapter 39

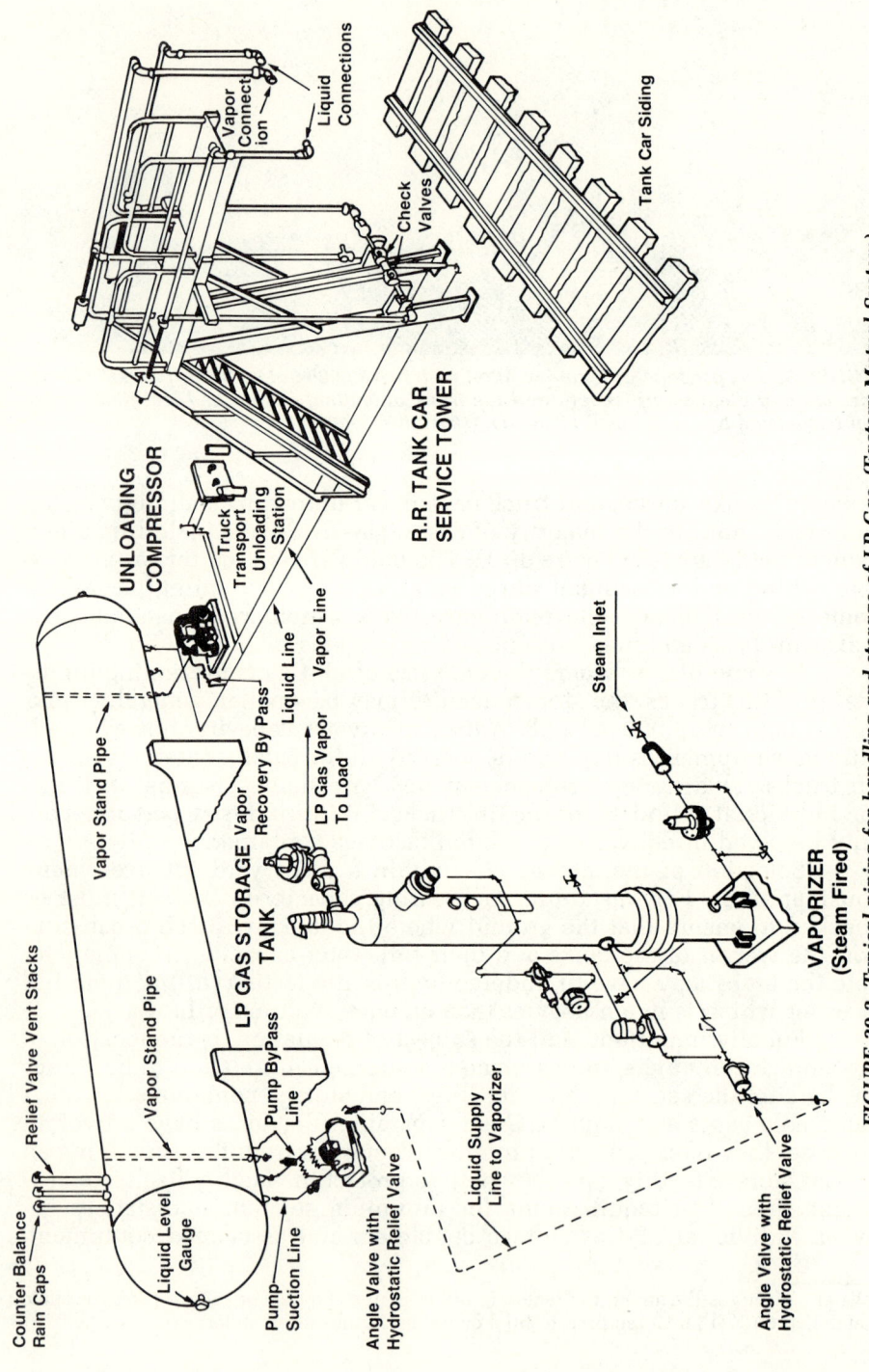

FIGURE 39.2. Typical piping for handling and storage of LP-Gas. (Factory Mutual System)

LIQUEFIED PETROLEUM GASES AT INDUSTRIAL PLANTS

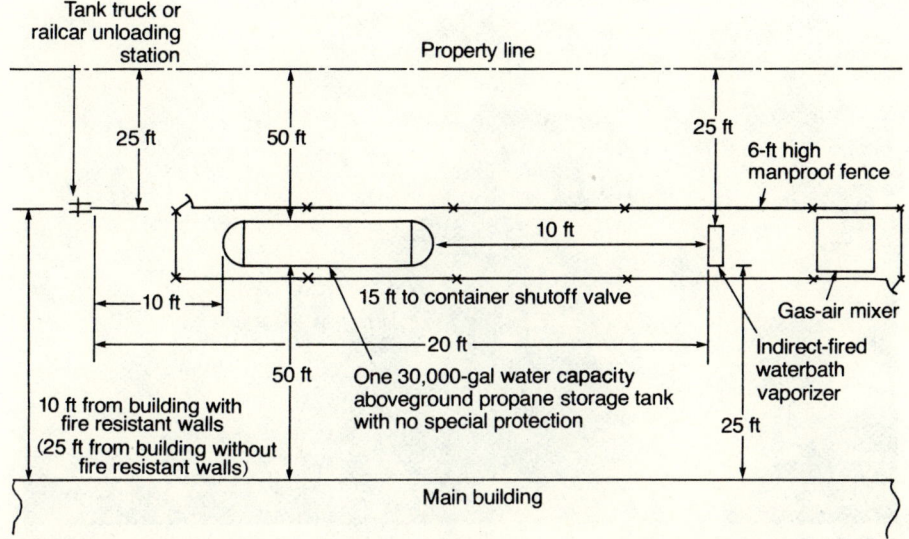

FIGURE 39.3. *The minimum distances (not shown to scale) for exposure protection of propane equipment at an industrial site. The distances shown are requirements given in NFPA 58, Storage and Handling of Liquefied Petroleum Gases. (Thomas Associates, Inc.)*

buildings, other containers, and set-backs from lot lines are quite specific and encompass a variety of special conditions. They are all contained in detail in NFPA 58. A rule of thumb which could be applied would be: the larger the facility's capacity, the greater the separation distance.

Large tanks usually rest on concrete or masonry foundations (saddles) formed to fit the contour of the container. In some instances, the tanks are supported by metal saddles placed on top of concrete or brick piers. When the tanks rest on concrete saddles, corrosion can be a particular problem at the points where the tanks and saddles meet. Before a tank is lowered on to its saddles during installation, a continuous tar impregnated felt pad is placed on the saddles to prevent the shell of the tank from actually coming into contact with the saddles themselves (see Figure 39.4). Corrosion is always a danger in outdoor installations; thus, periodic inspections are required. The containers should also be kept well painted as a preventative measure.

Listed or Approved Equipment

If storage containers and their appurtenances (safety relief devices, piping connections, valves, etc.) are not of a proper type or installed properly with particular attention paid to the joints to prevent leakage, the in-

FIGURE 39.4. *A close-up view of a tar impregnated felt pad that prevents the shell of an LP-Gas storage tank from coming in contact with the concrete saddle on which it rests. The pad protects the tank from the corrosive effects of the concrete. (Thomas Associates, Inc.)*

tegrity of the whole LP-Gas system can be compromised. "Proper type" means devices that have been tested and listed by a recognized testing laboratory, such as Underwriters Laboratories Inc., or approved by an authority having jurisdiction.

Excess Flow Valves

One of the early effective developments for preventing gas from escaping containers was the excess flow valve. An excess flow valve, however, does not operate except when the flow is in excess of its rated capacity; consequently, a downstream leakage that is less than the valve's capacity would continue to flow until the flow was shut off at the source. The majority of cases of uncontrolled discharge of LP-Gas from stationary containers have happened at loading or unloading points, particularly involving tank trucks. There are two major problems: (1) often a driver fails to disconnect the hose before driving away and breaks the hose connections or the piping, or (2) a loading or unloading hose bursts because it was improperly attached to the end fittings and came loose.

FIGURE 39.5. *A vapor-equalizing line at a truck unloading station with an emergency shutoff valve (located behind concrete bulkhead), remote cable release, and excess flow valve (in the pipe leading to the shutoff). The fusible release is at the top of the emergency shutoff valve handle. (Thomas Associates, Inc.)*

Emergency Shutoff Valves

An emergency shutoff valve installed at the loading or unloading point provides a positive shutoff in an emergency. When a hose or swivel-type piping 1½ in. (38 mm) or larger is used for liquid transfer, or a 1¼ in. (32 mm) or larger vapor hose or swivel-type piping is used for vapor transfer, an emergency shutoff valve is located in the transfer line where the hose or swivel-type piping is connected to the fixed piping part of the system that leads to the plant. The principal emergency features of the shutoff valve assembly are temperature-sensing elements (one of which must be part of the valve assembly) at the valve that will automatically shut the valve in a fire emergency. A temperature-sensing element is located not more than 5 ft (1.5 m) in an unobstructed direct line from the nearest end of the hose or swivel-type piping connection. In addition, the emergency valve is installed in the plant piping so that any break resulting from a pull occurs on the hose or swivel-type piping side of the connection; the valves and piping on the plant side of the connection remain intact. The concrete bulkhead or similar anchorage is one method of resisting a pulling movement which causes a break, if any, to be at the proper point. A weakness fitting or shear fitting installed in the piping is another means of protection. The emergency shutoff valve can be operated manually at the installed location but is also arranged so that it can be shut off from a remote location. A good installation of an emergency valve is shown in Figure 39.5.

An emergency safety valve installation is required by NFPA 58 on all single container systems of over 4,000 gal (15.1 m^3) water capacity or stationary multiple container systems with an aggregate water capacity of more than 4,000 gal (15.1 m^3) utilizing a common manifold liquid transfer line. If the flow is only in one direction, however, a backflow check valve may be used in place of the emergency shutoff. It will be found in the fixed piping downstream of the hose or swivel-type piping.

FIRE PROTECTION FOR LP-GAS STORAGE SYSTEMS

Good fire protection for a storage installation beyond that inherent in the safety devices installed in the storage system itself is important. Varying conditions result in different needs for different LP-Gas installations. NFPA 58 requires fire protection where an installation has storage containers with an aggregate capacity of more than 4,000 gal (15.1 m^2) subject to exposure from a single fire. The level of protection required is arrived at after a competent firesafety analysis involving evaluation of the container site, the exposure to or from other properties, water supply available, the probable effectiveness of plant fire brigades (see Chapter 3), and the time of response and probable effectiveness of the local fire department. For the latter reason, it is important that plant management coordinate plans with the local fire and police department for handling emergencies involving the plant's LP-Gas system. Familiarity on the part of the public protection agencies with the actual layout of the facility, as well as the proper techniques in handling LP-Gas fires, is extremely valuable in correctly handling an LP-Gas emergency.

The fire protection requirement analysis and preplanning for emergencies are vital because of the potential for fires and a Boiling Liquid Expanding Vapor Explosion (BLEVE). An important factor in fire fighting is to quickly apply water to the fire-exposed surfaces of a container in its "unwetted" area (the surface area of the tank above the liquid level within the tank.) Flame impingement on such "unwetted" surface areas quickly heats the steel and reduces its strength resulting in the internal pressure bulging the steel and creating a rupture.

Experience has shown that application of water by hose streams in adequate quantities (and in a safe manner) as soon as possible after flames contact the unwetted area is an effective way to prevent container failure from fire exposure. The majority of large containers exposed to sufficient fire to result in container failure have failed in from 10 to 30 min after the start of the fire when water was not applied. Water in the form of a spray can also be used to control unignited gas leakage. The NFPA training film "Handling LP-Gas Emergencies" is an excellent guide for training fire fighters in handling LP-Gas emergencies, including the correct methods to apply hose streams.

Aside from quick response and application of water by well-trained fire fighters, special types of fire protection can be considered for the tanks if the fire protection analysis reveals a need for them. Monitor nozzles installed so that the surfaces of the storage containers exposed to fire are wetted is one method. Automatic as well as manual operation of the nozzles is a requirement. Still another method is a fixed water spray system that can be operated both automatically and manually. The source for guidance in installing both monitor nozzles and water spray systems is NFPA 15, *Water Spray Fixed Systems*. A problem with a water spray fixed system is that the system may be knocked out of service in an emergency; an actuality that must be considered in making the fire protection analysis.

Two other special methods of protection are somewhat closely related. One is mounding earth or sand around the tanks so that they are covered for a minimum thickness of 1 ft (0.30 m). The other is burying the tanks in the ground. Again NFPA 58 is quite specific on requirements for installing tanks underground.

Insulating the tanks themselves against fire exposure is another alternative. Quite often a firesafety analysis may show that for storage tanks of 60,000 gal (227 m^3) water capacity or less, a good insulation system may be the most practical solution for special fire protection.

INDUSTRIAL LIFT TRUCKS

The use of LP-Gas fueled industrial lift trucks is common, and the fuel requires special and careful handling to ensure safety. Some plants receive their supply of filled LP-Gas cylinders from an outside supplier; other plants have their own storage and filling facilities. In either event a proper location for filling and storing the cylinders that gives protection to the cylinders from mechanical injury and protection from possible fire exposure, particularly for filled cylinders, must be provided. A good practice is to store the cylinders outdoors in a rack or cabinet; however, a total of 300 lbs (136 kg) of fuel may be stored inside in designated areas.

When an industrial plant fills its own cylinders, only well-trained individuals should be assigned the job, and it is best to have them trained by the supplier of the product. Weighing scales used in the filling operation are fitted with automatic shutoff devices so that the scale will automatically shut off the flow of gas before the cylinder is overfilled (see Figure 39.6). An overfilled cylinder can lead to the release of gas in the plant due to relief valve operation.

Some industrial lift trucks use a liquid feed to the engine while others use vapor feed. It is important that the cylinder be properly equipped for the actual service to be used. The cylinders are usually marked "liquid" or "vapor" by the outlet valve to indicate the type of use for that particular cylinder. When painting a cylinder, the marking that designates its type of use should not be obliterated.

The valves on the lift truck cylinders should be equipped with handles so they can be readily turned "on" and "off"; they should not require a wrench to turn them. While in storage or being transported, all lift truck cylinder valves should be equipped either with plugs or caps in order to keep dirt, dust, plant fumes, etc., out of the valve outlet.

Generally, industrial plants have chemical fumes, dust, contam-

Chapter 39

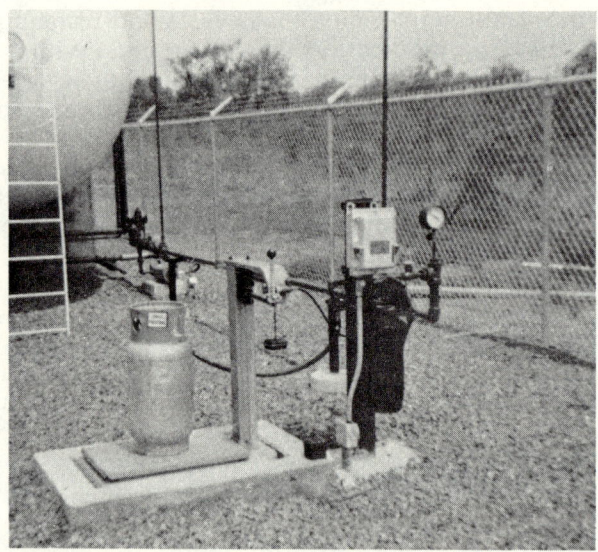

FIGURE 39.6. A cylinder-filling scale installation with an automatic shutoff to prevent overfilling of the container. (Thomas Associates, Inc.)

inated air, and other matter that can become imbedded in valves and other equipment. This is especially true for safety relief valves on lift truck LP-Gas containers. The result can be a dangerous elevation of the relieving pressure and the eventual rupture of the container. Therefore, special maintenance attention is required. A cap that will blow off on operation of the valve must always be in place over the safety relief valve.

It is important that cylinders be positioned properly, securely fastened, and good leak tight connections made when they are placed on lift trucks. Equally important is that the proper type of tank is used, either "liquid" or "vapor," as required for the particular lift truck.

Some lift trucks use ASME tanks rather than DOT cylinders. They are usually equipped with a fixed dip tube liquid level gauge connected to a bleed valve. The tube is fabricated to the proper length to be at the maximum fill level of the container. It is very important not to fill beyond this level. The operator filling the container must be alert and shut off the flow of liquid into the container the instant the bleed valve indicates a change from vapor to liquid. A slight delay can result in an overfilled container and thus pose a hazard. DOT cylinders, on the other hand, must be filled only by weight in accordance with DOT requirements. Design specifications for the various types of ASME and DOT containers will be found in NFPA 58.

A factor that is often overlooked with regard to the use of LP-Gas cylinders in industrial plants is the failure to requalify or retest the ICC or DOT cylinders within twelve years of the date of manufacture and every seven years thereafter. A cylinder twelve years past the original manufacture date or past the stamped retest date should not be filled until this requirement is met.

IN-PLANT FACILITIES

In addition to the fire protection needed for the storage facility, other portions of the LP-Gas system in an industrial plant deserve attention. The gas piping should be installed so that it is not exposed to mechanical damage. It must be checked periodically for leaks and to see that hangers, supports, etc., do not allow strains on the piping. Controls on ovens and furnaces should be of the proper type and listed by a recognized testing laboratory or approved by the authority having jurisdiction. A scheduled maintenance program for the controls and the units they serve is desirable. In addition, operators of the ovens and furnaces should be trained in the utilization and maintenance of the gas system.

Another factor is the need for proper ventilation. What constitutes proper ventilation depends upon the way LP-Gas is used in the area in question. When considering ventilation bear in mind that LP-Gases are heavier than air and thus tend to flow to lower areas followed by a complete dissipation throughout the entire structure over a period of time.

LP-Gas is increasingly being used as a propellant in aerosol cans. When aerosol filling operations are being carried on, special precautions in ventilation and handling must be designed into the equipment and buildings housing the operation. NFPA 58 gives guidance on construction and ventilation of structures and rooms housing LP-Gas distribution facilities. (See Chapter 35, "Aerosol Filling and Storage," for a complete discussion of the hazards associated with aerosol cans.

DELIVERY OF LP-GAS TO INDUSTRIAL PLANTS

Deliveries, other than cylinder-type containers, are usually made to an industrial plant by railroad tank car or by transport tractor trailer units. Industrial plant management should require that any transport unit bringing LP-Gas onto the plant site be built according to DOT specifications. Further, particular attention must be given to the maintenance of the unloading equipment on the vehicles and tank cars, particularly the unloading hose, to eliminate the possibility of hose rupture during the unloading. The unloading hoses should each be equipped with a backflow check valve. Management should also insist that drivers of tank vehicles entering the premises be trained in handling LP-Gas.

If the plant furnishes the unloading hose it should be maintained properly and replaced as needed. In any event, hose should be replaced

FIGURE 39.7. *An LP-Gas tank car unloading rack with a jumbo tank car containing aerosol propellant in position (but not coupled) for unloading. Note the monitor nozzle in the foreground for wetting down tank cars in the event they are exposed to fire. (Thomas Associates, Inc.)*

after five years of use. When not in use, hose should be protected from the direct rays of the sun.

Industrial plant supervisors and employees in the areas where LP-Gases are stored, mixed, piped, and utilized should know the characteristics of the product and be acquainted with its odor to be aware of possible leaks. They should also be knowledgeable of system shutoff valves in case of leakage or malfunction. If there is an in-plant fire brigade, the members should be trained to handle emergencies that might involve the LP-Gas system. In addition, the local fire department should become familiar with the LP-Gas system and valving in order to handle emergencies in an effective manner.

BIBLIOGRAPHY

NFPA CODES, STANDARDS, AND RECOMMENDED PRACTICES

Reference to the following NFPA Codes, Standards, and Recommended Practices will provide further information on the safeguards for LP-Gas Installations at industrial sites discussed in this chapter. (See the latest *NFPA Codes and Standards Catalog* for availability of current editions of the following documents.)

NFPA 15, *Water Spray Fixed Systems.*

NFPA 58, *Storage and Handling of Liquefied Petroleum Gases.*

NFPA 59, *Storage and Handling of Liquefied Petroleum Gases at Utility Gas Plants.*

NFPA 505, *Firesafety Standard for Powered Industrial Trucks Including Type Designations and Areas of Use.*

ADDITIONAL READING

Bray, G., "Applying Water to LPG Spill Fires," *Fire*, May, 1980, pp. 667-668.

———, "Water Spray Protection of LPG Hazards," *Fire Protection*, vol. 7, no. 4 (December 1980), pp. 3-10.

Clifford, E. A., "A Practical Guide to LP-Gas Utilization," *LP-Gas Magazine*, 5th edition, Duluth, MI, 1973.

Compressed Gas Association, *Handbook of Compressed Gases*, 2nd ed., Van Nostrand Reinhold Publishing Corp., New York, NY, 1981.

"HSE Recommendations Following Aerosol Fire," *Fire Prevention*, August 1981, pp. 23-24.

LP-Gas Safety Handbook, National LP-Gas Association (continuous updating).

Matthews, D., "Industrial Fire Hazards of LPG," in *LPG: The Hazards and Precautions*, NFPA Conference, March 5-6, 1980, pp. 41-44.

McKinnon, G. P., ed., *Fire Protection Handbook*, Fifteenth Edition, National Fire Protection Association, Quincy, MA, 1981, pp. 3-393-55.

National Fire Protection Association, *LPG: The Hazards and Precautions*, March 5-6, 1980 Conference, also in *Fire Protection*, April, 1980, pp. 26, 30.

Patterson, C. B., "Powered Industrial Trucks, Appraising Their In-Plant Fire Safety," *Fire Journal*, vol. 66, no. 5 (Sept. 1972), pp. 103-104.

Rasbash, D. J., "Review of Explosion and Fire Hazard of Liquefied Petroleum Gas," *Fire Safety Journal*, vol. 2, no. 4 (July 1980), pp. 223-236.

40

Computer Centers

Robert Riley

Today, computer systems are used in virtually every kind of organized activity. It is estimated that over 400,000 small-to-large-scale systems typically housed in "computer rooms" are in use in the United States.[1]

The very rapid growth in computer utilization has undergone many changes. What has been called data processing equipment is now increasingly referred to as information processing, information handling, or information technology equipment. A Stanford University study released by the U.S. Department of Commerce, Office of Telecommunications, found that 46 percent of the gross national product is integrally linked with information activity and all its ramifications, and nearly half the labor force holds some kind of an "information" job, earning 53 percent of labor income.[2]

More than 2.5 million people are employed nationwide as computer programmers, operators, systems analysts, and systems managers. The (data) information processing industry comprises some 10,000 companies offering mainframes, peripheral equipment, software, and various

---

Robert Riley is a Manager of Installation Planning for IBM Corporation, White Plains, New York.

contract services. The desk-top terminal, which puts a person directly in touch with a distant computer system, has become a common sight at airports, auto rental locations, banks, warehouses, and offices of every description.

By 1981, the information processing industry was expected to grow to a more than $1 trillion industry.

The primary reasons for the rapid proliferation of computer equipment are because of the constantly changing nature of the equipment itself and the function and efficiency of computers in industry and government.

During the past 20 years, the cost of performing a unit of computing work on some computers has fallen from $1.26 to $0.007 (seven-tenths of one cent) — and it continues to fall. The computer industry improves price/performance at a rate of 25 percent annually, making it ever more profitable for users to develop and implement new computer applications.

From the user's point of view, the computer has proved to be an excellent productivity tool and, therefore, a good investment. In the federal government, the Internal Revenue Service (IRS) and General Accounting Office (GAO) have reported 10 percent productivity increases due primarily to computer-based systems. Other industries credited with above average productivity gains in recent years include air transportation, gas and electric utilities, and communications — all heavy computer users.

In both large and small computer systems, operators and operations very quickly become dependent on the equipment. This dependence is evident when a minor problem forces the system to "go down" and operators must wait until their terminals can be used again — a delay that is usually no more than a matter of minutes. A serious fire — a catastrophe that might knock out the central computer for days — has become "unthinkable."

In fact, the "unthinkable" was faced some time ago in critical operations such as air traffic control where human lives are at stake, and suitable solutions were found through appropriate fire protective measures and related techniques. Similarly, government, industry, and commercial organizations of all types and sizes, who cannot afford to have their equipment go down for any extended period of time, have prepared for such situations.

Because both users and data processing equipment manufacturers have been conscious of this requirement from the very start, the industry has an excellent fire prevention record. In the 20 years since the computer began to be used heavily, there have been approximately six major fires, fires of such a magnitude that computer processing could not continue. Discussions of these fires throughout this chapter will help emphasize the protective steps needed to guard against them. These fires are well-known within fire prevention circles, and almost every one was closely involved with the "raw materials" of data processing.

COMPUTER CENTERS

FIGURE 40.1. Modern computer room with storage components. (IBM)

RAW MATERIALS

Data processing machines typically consist of electronic and electrical components housed within metal and/or plastic covers. Power supply and communication and interconnecting cables are usually under a raised floor.

The basic machines — central processors, consoles, tape or disk memory units, or high-speed printers — are not inherently combustible. Experience has shown that they can develop component malfunction which can lead to smoke and fire. Quick action using hand fire extinguishers readily controls such situations. The specific unit may have to be taken out of service, but computer processing continues.

The remaining question in such cases is whether the affected unit has to be repopulated with new or rebuilt internal components. One manufacturer routinely removes and repairs such units in leased or rental installations and makes the decision in a repair facility. If the machine has been purchased and is therefore owned by the user or a third party, the manufacturer will provide owners with expert opinions as to whether repopulation is needed.

The real problem, with regard to fire prevention, is that the raw materials of data processing include paper, cards, plastics of various types, microfilm, and other combustible records-related materials.

In the Pentagon fire of July 1959 — one of the first of the half-dozen major computer fires — the trouble started in the ceiling of the magnetic tape storage area. The tapes were stored in clear plastic containers which

FIGURE 40.2. *Modern computer room with terminals. (IBM)*

were combustible. The fire, and heat, then spread to the adjacent computer room and forced operations to stop.

At the Program Information Department of a major computer manufacturer, fire started in a storage area containing cards, paper, and cardboard boxes. The storage area was directly below the computer room. Very little flame got up into that room (some did pass through an air conditioning duct), but the heat was sufficient to "cook" and disable the equipment.

It becomes evident, therefore, that computer center fire protection and suppression is as much concerned with what is near the computer room, as with what goes on inside that room.

THE PRODUCTION PROCESS

Input information is entered into the computer system either for immediate use or to be stored in memory units for later use. "Memory" takes various forms. In some older systems data is mechanically encoded on punched cards and/or punched paper tape. With newer systems the data is electronically encoded on magnetic tape. Input data can also be transmitted over communication lines, in the form of electrical signals, directly into the computer for processing or directly into memory units.

The system processes this input data according to its programs of instruction and delivers it in the form of output. Delivery can be back over telephone lines to a remote terminal, where the information is displayed or printed, or to the memory unit of another remote computer.

Output is also delivered, within the computer room, onto magnetic tape and disk units, punched into cards, and/or printed on paper forms. The variety of output is almost limitless, including payment checks, portions of new insurance policies, legal documents, picking/packaging instructions in a warehouse, monthly bank account statements, students' report cards, etc.

THE FIRE HAZARDS

It is worth noting that some forms of output previously mentioned are printed on combustible paper. This is where data or information processing poses a possible fire hazard.

As the various machines perform their tasks, a malfunction could occur causing some components to overheat or start smoking, but these types of failures are readily detectable and controllable with proper detection systems and portable fire extinguishers. The major hazard, however, stems from the surrounding area and the fact that a computer installation processing data uses paper forms, paper records, magnetic tapes and cardboard boxes in which paper records are frequently stored. If recommended protective standards are not met and these materials are allowed to accumulate while waiting to be used or distributed, the hazard becomes greater.

In some cases, offices are actually part of the computer room, posing additional hazards. Also, magnetic tape storage areas and paper records storage areas frequently "look out" onto the computer room. They are so placed to facilitate the movement of tape reels, paper records, and related documents.

Some users, especially smaller organizations, install fire resistant vaults within the computer room to house interchangeable magnetic tape reels and similar high-value components. Typically, the vault is open while computer operations proceed. Should a fire break out, that vault will provide protection only if someone closes it before fleeing — and, even then, only those components still in the vault will be protected.

THE SAFEGUARDS

Site Selection

The logical conclusion, borne out by experience, is that fire protection must be approached from a "total" point of view. It begins with the selection of a site. Whether the new computer room — or information proc-

Chapter 40

FIGURE 40.3. Tape storage room. (IBM)

essing center — is to be in a building of its own or occupy space in an existing building, it should stand free of high hazards.

One disabling computer room fire began in a building devoted to the repair and maintenance of auto parts. A highly flammable solvent was ignited by a spark. This building was adjacent to five others, including the computer center. The fire spread to all the buildings.

Site selection should carefully evaluate activities in any adjacent buildings or on adjacent floors as to low, minor, and high hazards. The data processing center should preferably be located in a low hazard location or, at most, one with minor hazards.

Data processing installations have been targets for sabotage and arson, and site selection and construction should take this into account. Both should minimize opportunities for penetration of the computer center by an explosive or incendiary device, and access should be restricted to those persons necessary to equipment operations.

Planned Total Protection

Typically, the computer room will be planned to house the processors and their peripheral equipment plus whatever additional equipment and supplies must be located within the room. Adjacent to the computer room will be various storage and work areas, including supervisory and other offices. Fire protection should be planned not only for the main room but for the adjacent support areas as well.

Preplanning should take in the range of considerations, including, but not limited to, restricting the amount of paper stock housed in the computer room to the absolute minimum required for efficient operations. Combustible materials to be kept in the computer room should be housed in totally enclosed metal file cases or cabinets.

Similarly, computer room office furniture should be of metal construction or other fire-resistive materials.

A key point worth emphasizing is that the concentrated attention devoted to the main computer room area should also be applied to the adjacent support areas. That, in brief, is the key lesson of past experience.

Construction

The building should be of fire-resistive, noncombustible, or sprinklered construction. The materials used within the computer facility should have a flame-spread rating of 25 or less.

The wall separating the computer room from the support areas should have a 1-hour rating and should extend from slab to slab.

Among these support areas will typically be a paper forms and records storage area and a magnetic tape storage area. The walls separating these from other areas should have a 2-hour rating.

Wall constructions for other support areas — which may include offices, mechanical and electrical equipment rooms, and input or data entry operations — would depend on the materials kept in those areas and the degree of protection they would require.

Fire Detection Equipment

Fire detection equipment should be installed in all areas critical to human safety and computer operations, including air-conditioning and electrical equipment, uninterrupted power supply, transformers, and switchgear. It should be used, along with appropriate fire suppression equipment, in paper and tape storage areas, maintenance areas, data entry areas, and other areas where paper stock and other combustibles are used frequently in quantities.

Modern detection equipment is very sensitive to the products of combustion, sensing smoke and gases before flames develop. It constitutes an early warning system that can provide excellent protection in the main computer room and in all supporting areas.

At this point it is appropriate to discuss another of the few fires which have disabled computer systems. It involved an arsonist who set fire to the computer installation at which he worked because he had been utilizing it illegally on weekends and became fearful of being caught.

The computer room was protected by fire detection equipment designed to set off an alarm in the area. The wires went into a typical control cabinet. Since the cabinet had to be opened periodically for maintenance, the key was left alongside to make things easier. This also made

Chapter 40

FIGURE 40.4. Air conditioning unit. (IBM)

FIGURE 40.5. Electrical support equipment. (IBM)

it easy for the arsonist to open the cabinet and turn off the alarm; any chance that it might have been heard by someone walking past on the outside was eliminated.

The arsonist set the fire, which burned for some time before it was detected, disabling the entire operation. It was not, however, effective enough to wipe out the evidence of arson. Fire investigators readily spotted that it had been deliberately set and that the detection system had been turned off; and the subsequent investigation led to the arsonist.

Alarms that can easily be turned off cannot be trusted; and detection systems that only set off alarms within the area are not much good if fire should start when no one is around. The detection system, in addition to setting off a local alarm, should also be wired into security locations that are manned around the clock, such as a security guard or watchman location in the building, or the local fire department. The system should be designed so that any attempt to tamper with it will alert the authorities.

The detection system should indicate on one or more panels exactly where the incipient fire is located. Within the computer room, this should be by section. In other areas, it could also be by section or specific equipment, such as switchgear.

The capability of quickly spotting where the smoke or gas is coming from can significantly decrease the time required to bring an incipient fire under control and also minimize any damage that might be caused by suppression materials. It can also hold costs down.

Fire Suppression Equipment

Fire detection and suppression equipment should go hand-in-hand, and automatic suppression systems utilizing Halon 1301 are recommended. In the concentrations required to control fire in a data processing environment, Halon 1301 has no harmful effect on people and no damaging effect on electric equipment. It is not even necessary to turn the power off.

A factor to be considered is that although Halon 1301 has been proven safe and effective, experience has shown that when it is released as part of an automatic suppression system it makes a loud noise that can frighten people. Further, depending on the size of the area to be covered, the quantity of Halon 1301 released could be quite expensive. The early warning detection system setting it off might be sensing no more than smoke coming out of one machine, a situation that can easily be controlled with a hand extinguisher.

In view of these facts, some installations have adopted the following procedures: An automatic suppression system is installed using Halon 1301, but a time delay is built in so that when smoke or gas is detected by the equipment, an alarm that permits people to get out of the area before the Halon is released is sounded. The extent of the delay depends on the number and location of exits; typically, it is from 15 to 45 seconds. If people are unfortunately caught in the area, they know the gas is about to be released and fright is minimized.

Another procedure has a built-in delay of about 30 seconds after the alarm sounds. During this time, one responsible individual unlocks the control panel, switches the automatic to manual control, and holds the gas from being released. An additional 30 to 45 seconds are allotted while others check the problem. If the incipient fire is readily put out with hand extinguishers, the Halon 1301 is not released. With this procedure, people have time to leave the area, they are alerted, and gas is released only if it is necessary.

An adequate number of portable hand extinguishers should be on hand to control electrical fires, with as many people as possible trained in their use.

Foam and dry chemicals are not recommended for data processing equipment fires because of the extensive cleanup they require. In most cases, only small incipient fires that readily respond to portable CO_2 or Halon 1211 extinguishers, which leave no mess, have to be handled.

Automatic sprinkler systems may be feasible for magnetic tape and paper storage areas. Water does not badly damage tape containers, and damage to paper stock may not be critical. It depends on the nature and the quantity of these two combustibles, where and how they are stored, and the amount of money management wants to spend on their protection. If they are essential to continuous computer operations, they require maximum protection.

In this connection, it should be noted that stored paper forms should be carefully evaluated. They may include costly preprinted forms on which the organization's invoices or other critical documents are printed. Frequently, the commercial printers producing those forms require more than three weeks lead time. If data processing output depends on these forms, equipment protection alone is insufficient. There can be no operation without them.

NFPA 75, *Protection of Electronic Computer/Data Processing Equipment*, gives detailed guidance on requirements for protection of computer and data processing equipment where special housing and fire protection measures are essential because of the size or complexity of the installation.

Good Housekeeping

A final point worth emphasizing is good housekeeping. This includes minimizing combustibles kept in the main computer room and insisting upon clean and safe work areas. (See Chapter 50 for a complete discussion of "Industrial Housekeeping Practices.")

In a data processing center good housekeeping also includes a periodic inspection under the raised floor areas. The raised floors that accommodate cabling, controlled air cooling, and other services vital to the operation of the system must be kept clean and free of other materials.

The total data processing area, including the computer room and adjacent support areas, will probably be equipped with several different types of hand extinguishers. A sign should be placed alongside each portable extinguisher clearly indicating what it contains and for what it is to be used.

Successful fire prevention is as much a matter of insisting on good housekeeping principles as it is the application of methods such as automatic suppression systems. The excellent loss record achieved by data processing users and manufacturers proves that with planning data processing can be protected from major fires.

BIBLIOGRAPHY

REFERENCES CITED

1. EDP Industry Report, May 1978.
2. *Communications*, February 1977.
3. *Business Week*, January 9, 1978.

NFPA CODES, STANDARDS, AND RECOMMENDED PRACTICES

Reference to the following NFPA Codes, Standards, and Recommended Practices will provide further information on the safeguards for computer centers discussed in this chapter. (See the latest *NFPA Codes and Standards Catalog* for availability of current editions of the following documents.)

NFPA 10, *Portable Fire Extinguishers.*
NFPA 12, *Carbon Dioxide Extinguishing Systems.*
NFPA 12A, *Halon 1301 Fire Extinguishing Systems.*
NFPA 12B, *Halon 1211 Fire Extinguishing Systems.*
NFPA 13, *Installation of Sprinkler Systems.*
NFPA 70, *National Electrical Code.*
NFPA 75, *Protection of Electronic Computer/Data Processing Equipment.*
NFPA 90A, *Air Conditioning and Ventilating Systems.*
NFPA 101, *Life Safety Code.*
NFPA 231, *Indoor General Storage.*
NFPA 232, *Protection of Records.*
NFPA 232AM, *Archives and Records Centers.*

ADDITIONAL READING

"Computer Centers — Measures to Reduce Losses," *Brand Brandweer*, vol. 1, no. 3 (1977), pp. 53-56.
Davis, R. H., "Fire Protection Systems (for Computers)," *Data Management*, vol. 14, no. 11 (1976), pp. 11-14.
Factory Mutual Corp., "American Experience in the Fire Protection of Computers," *Fire*, vol. 7 (1979), pp. 498-499.
Facts About Protecting Electronic Equipment Against Fire, Ansul, Marinette, WI, 1984.

"Halon Prevents Major Central Processing Unit Fires," *Computer Decisions*, vol. 10, no. 8 (1978), pp. 56, 58.

McKinnon, G. P., ed., *Fire Protection Handbook*, Fifteenth Edition, National Fire Protection Association, Inc., Quincy, MA, 1981.

Osborn, Richard W., ed., *Tapping In to the NEC*, National Fire Protection Association, Quincy, MA, 1982.

Perkins, C., and B. J. Berenblut, "Electronic Equipment: What Protection is Required?", *Fire Surveyor*, vol. 9, no. 5 (October 1980), pp. 28-33.

Philbrick, S. E., "Selecting Cables for Fire-Risk Applications," *Electronics and Power,*" vol. 26, no. 3 (March 1980), pp. 232-233.

"Support for Sprinkler Systems in Controversial Settings," *Fire*, February 1981, pp. 455-456.

41

Laboratories

Norman V. Steere

Laboratories in industrial occupancies serve several different purposes, including analysis of raw materials, quality control of processes and products, and product or process improvement. Some industrial laboratories may also be devoted to basic and applied research or to experimental study in a particular branch of science.

Processes in industrial laboratories encompass many forms of energy and many industrial processes. The hazards in laboratories are usually minimal because small quantities of materials are involved, and there is emphasis on controlled conditions to achieve precise measurements. However, laboratories can have serious fire hazards from excessive quantities of flammable or reactive chemicals, uncontrolled ignition sources, and inadequate procedures or equipment for handling hazardous materials.

The basic goal of fire protection in laboratories is to safeguard lives and property from the dangers of fire and explosion. Life safety considerations extend beyond the laboratory occupants to visitors, the general public who may be exposed to the effects of a laboratory emergency, and

---

Norman V. Steere, P.E., is a Laboratory Safety and Design Consultant, Norman V. Steere & Associates, Inc., in Minneapolis, MN.

to fire fighting personnel who respond to the emergency. Property to be protected includes not only the facility itself, but also the apparatus, samples, and records of tests being conducted. Fire risks to apparatus, samples and test records are not often considered by laboratory management, nor is the loss of productive time if fire damages or destroys facilities and apparatus.

The principles of fire protection in laboratories include: limiting the availability of fuel and oxidizers, limiting the amounts of highly reactive materials, controlling ignition sources, detecting fires in their early stages and limiting their growth, evacuating occupants, and extinguishing fires by manual or automatic means.

THE HAZARDS

Providing effective and economical fire protection measures for a laboratory requires a careful assessment of potential hazards. Laboratories and experimental apparatus are so diverse and complex that it is impractical to describe every hazard or combination of hazards that may exist. One useful way to assess fire and damage potential is to evaluate the types of hazards that may not be causes of fires or fire emergencies but that should be considered in fire protection programs.[1]

Some hazards that are not likely to cause fire but may endanger occupants and personnel engaged in fire fighting or emergency control operations are radiation energy hazards, biological hazards, mechanical hazards, and many chemical hazards.

Mechanical Hazards

Mechanical energy generated or stored in pressure vessels or compressed gas cylinders concerns fire protection personnel because of the potential for an explosive release of energy. Possible failure of glass vessels under vacuum can result in violent implosions and high velocity fragments.

Mechanical energy can also be transferred traumatically by the impact of falling objects. Sharp objects or obstructions and elevated walking and working surfaces common to many laboratories may be hazardous to personnel.

Corridors and exit paths that are narrow or obstructed may provide inadequate means of egress in case of fire or other emergency.

Electrical Hazards

Electrically operated laboratory equipment may provide a source of ignition, due to heated surfaces (such as hot plates or heating mantles),

FIGURE 41.1. *Biological hazards present special fire protection problems.*

overheated equipment (such as an overloaded motor turning an agitator shaft), or faulty wiring on a piece of equipment. Even though fire protection personnel may not be able to assess compliance with NFPA 70, the *National Electrical Code*, they should look for actual or potential damage to wiring, power cords, electric motors, etc. They should inquire about experience with equipment overheating and power failure that could cause fire. They should also assess the likelihood of flammable gases and vapors being ignited by heated equipment, either in the course of routine laboratory operations or due to a spill or leakage.

According to NFPA 45, *Standard on Fire Protection for Laboratories Using Chemicals*, laboratories are not classified as hazardous locations in accordance with Article 500 of the *National Electrical Code*. However, extraordinary conditions may require use of what is commonly termed "explosion-proof" electrical equipment. Examples include the interiors of walk-in laboratory hoods and pilot plant equipment where the volume of flammable liquids is exceptionally high.

Static electricity generated by pouring liquids, especially nonpolar solvents, should be dissipated by bonding metal dispensing and receiving containers together or to a suitable ground. Information can be found in NFPA 77, *Recommended Practice on Static Electricity*. Special measures should be taken to control static electricity if flammable concentrations of gases or vapors may be present and can be ignited.

Biological Hazards

Biological hazards may exist where animals are used in research, such as in areas of nutrition or toxicology. Biological hazards may exist on a more subtle basis where microorganisms are present in specimens tested in research or industrial processes. Special fire protection and detection systems are needed to protect animal colonies or biological cultures.

Radiation Hazards

Sources of ionizing radiation include radioisotopes and any equipment emitting ionizing radiation. Nonionizing radiation includes lasers, high intensity light, infrared, ultraviolet, and microwave radiation. Ionizing radiation is not likely to cause a fire, but radioisotopes can be released as a result of a fire, thus posing a danger to occupants of the laboratory, fire fighting personnel, and the general public. The fire hazard due to nonionizing sources is primarily electrical in nature; however, some sources may be intense enough to ignite combustibles in the path of the radiation. Fire protection should be provided to protect sources of radiation from fire exposure that may damage equipment or release radiation.

Hypobaric and Hyperbaric Hazards

Hypobaric and hyperbaric hazards will occur only in high altitude or deep sea research or in chambers where atmospheric pressures are created for such research. The hazard involved in hypobaric facilities is the danger of anoxia or asphyxiation due to low partial pressure of oxygen. Hyperbaric facilities involve increased flammability of materials.

Thermal Hazards

Hazards due to high temperatures may be common in laboratories, but are usually so well controlled that they present little risk of causing a fire. Such sources of heat include ovens, furnaces, hot oil baths, molten salt baths, burners and torches, heating jackets, steam baths, and similar equipment. High temperatures can also be produced by overheated bearings, friction, arcing contacts, and exothermic chemical reactions. Precautions must be taken if combustible materials or flammable liquids or vapors can come in contact with heat sources whose temperatures are at or above ignition temperatures of the materials. Ignition can be prevented by ventilation, avoiding direct contact between combustible or flammable materials and heated equipment, or inerting of heated sources.

Cryogenic or supercooled liquids are not likely to present fire hazards unless the fluids are flammable gases. Cryogenic fluids that are at a lower temperature than the boiling point of oxygen can condense oxygen, thus leading to an oxygen-enriched atmosphere; cryogenic fluids that can condense oxygen include liquid nitrogen and liquid hydrogen. There may be an explosion hazard if cryogenic vessels rupture due to severe pressure buildup.

Chemical Hazards

Hazards involving chemicals are likely to be found in most laboratories. (Combustion hazards are discussed in the next section, and Chapter 31 discusses "Chemical Processes.") Chemical hazards are usually indicated on container labels or described in data sheets for specific chemicals. One excellent source of data on many common hazardous chemicals is NFPA 49, *Hazardous Chemicals Data*. NFPA 49 contains information on storage procedures, fire fighting tactics, and fire and health hazards.

Reactive chemicals: The category of reactive chemicals includes strong oxidizers, organic peroxides, and strong reducing agents. Some chemicals, such as perchlorates, benzoyl peroxide, or picric acid, are self-reactive when exposed to heat, friction, or shock. Strong oxidizers may not be self-reactive, but most will react vigorously or even explosively if mixed with organic materials. Some reactive chemicals may undergo spontaneous decomposition unless they are refrigerated or inhibited. Others may react violently with water or at elevated temperatures. Pyrophoric chemicals react spontaneously with ambient air.

It is not uncommon to store glass containers of incompatible materials in close proximity. A hazardous chemical reaction could result if such chemicals are intermixed due to spillage or breakage of glass containers. For information on hazardous combinations of chemicals, see NFPA 49 and NFPA 491M, *Manual of Hazardous Chemical Reactions*.

Fire protection personnel should carefully assess the hazards inherent in the storage and use of chemicals. Concentration and temperature may markedly affect the degree of hazard. For example, an explosion of a perchloric acid solution in a Los Angeles industrial plant many years ago generated extreme concern over the use of perchloric acid in laboratories. Perchloric acid that is dilute or used at ambient laboratory temperatures presents little hazard and does not become self-reactive unless made anhydrous. When heated at concentrations of 70 to 72 percent, however, perchloric acid boils off. The vapors can condense in laboratory hood exhaust ducts where they can then combine with organic material to form shock-sensitive organic perchlorates. Laboratories which routinely use perchloric acid for certain procedures will usually dedicate one or more laboratory hoods exclusively to such use. These hoods have special features detailed in NFPA 45.

Assessing the fire, explosion, and reactivity hazards of laboratory chemicals should be done by both fire protection personnel and laboratory personnel.

Corrosive, irritating, and toxic chemicals: Most chemicals possess some degree of toxicity, many are irritants, and some are corrosive or destructive to living tissue. Such chemicals pose hazards to laboratory and fire fighting personnel alike. It is prudent to estimate the potential hazards such chemicals present during emergency situations and cleanup operations. Consideration must also be given to the possibility that hose streams and automatic sprinkler discharge may spread spilled chemicals and cause pollution in the local environment.

Fire Hazards

Fire hazards are present in most laboratories, from the materials used, to combustible construction, furnishings, equipment, or books and papers. Flammable liquids are used in most laboratories for cleaning, preserving, dissolving or as a reactant. Many laboratories use flammable gas for burners or other heating equipment. Some laboratories use flammable solids for reactants or stains.

Flammable and combustible liquids are so commonly used that they warrant special emphasis. (See also Chapter 37, "Handling and Storage of Flammable and Combustible Liquids.") Although flammable liquids present a greater hazard because their flashpoints are below 100°F (37.7°C), combustible liquids can pose the same hazard if they are heated above their flashpoints or are atomized by spraying, dispersing, or otherwise increasing the surface-to-mass ratio. Spilling the liquids on fabric or clothing is an example of the latter.

Fire tests have shown that ignition of a 1-gal (3.78 L) spill of a flammable liquid can produce ceiling-level temperatures of 869°F (465°C) within 1 min. Ceiling-level temperatures can reach 1,400°F (760°C) within 2 min and can continue to climb if the fire is fed by other combustibles.[2] Heat release can reach 10,000 Btu per min per sq ft [113 626 (kJ/min)/m^2], and a 1-gal (3.78 L) spill can cover 20 sq ft (1.86 m^2) or more.[3] A 1-gal (3.78°C) spill in an unventilated area can create an explosive atmosphere in a volume as large as 2,000 cu ft (56.6 m^3). Overpressures from such an explosion can reach 100 psig (690 kPa).

THE SAFEGUARDS

Hazard control measures for laboratories must be selected wisely in order to be effective. It is very important to evaluate proposed control measures before they are required or implemented. The objective is to avoid creating other hazards, causing great interference with laboratory activities, or costing more than the possible benefits. For example, a poorly placed fire extinguisher can obstruct traffic and the use of space, and a fire door without a view panel can cause pedestrian traffic injuries and damage to equipment. Arbitrarily limiting the quantity of flammable or reactive chemicals used in a laboratory to a one- or two-day supply might be appropriate in a production area, but is likely to interfere with laboratory operations or create additional hazards due to transportation and handling. On the other hand, judiciously limiting quantities of such chemicals is now being recognized as a means of improving the utilization of expensive laboratory space and a step toward reducing waste disposal problems.

Consensus standards cannot always be tailored precisely to the needs of each individual laboratory. Standards are further limited by the compromises necessary to achieve a consensus, lack of adequate data upon which to base requirements,[4] and the speed with which technology moves relative to development of standards and regulations.

A basic problem in controlling laboratory hazards is the recognition and careful estimation of the hazards. Quite naturally, those who approach the problem tend to base decisions on their individual training and experience. Laboratory personnel often have had little training or experience with laboratory hazards and fires. Outside inspection personnel from insurance companies, fire departments, or regulatory agencies often have little understanding of laboratory procedures. Through a cooperative effort, hazards can be accurately estimated and reasonable and effective control measures can be developed.

Limiting Fuel

Limiting the fuel available in a laboratory is accomplished by excluding or protecting large glass containers of flammable chemicals, limiting total quantities of such chemicals, placing as much as possible in safety containers, and protecting as many as possible of the necessary flammable chemicals from early involvement in a fire by storage in appropriate flammable liquids storage cabinets or storage rooms.

General ventilation in laboratories and the small quantities of flammable liquids usually handled prevent flammable concentrations of vapors under most circumstances. If needed, special ventilation can be provided to keep vapor concentrations below the flammable range.

While it is desirable to limit the size of glass containers of flammable chemicals because of the serious hazards cited above, it is not always possible to exclude glass containers of 1-gal (3.78 L) size. NFPA 30 permits 1-gal (3.78 L) glass containers for combustible liquids and Class IC flammable liquids (flash point 73°F [22.8°C] or higher). NFPA 30 has an exception allowing 1-gal (3.78 L) glass containers for Class IA and IB liquids if glass is necessary for corrosion-resistance or liquid purity such as ACS analytical reagent grade or higher. NFPA 45 also permits 1-gal (3.78 L) glass containers for Class IA and IB liquids if the required purity would be adversely affected by storage in metal or if the liquid would cause excessive corrosion of the metal container. Most laboratory reagents that are Class IA and IB are commonly supplied in 1-gal (3.78 L) glass containers.

Safety cans: Where safety cans will not cause contamination, they should be used. The primary purpose of the safety can is to prevent breakage of the container, to limit spills from a dropped or knocked-over container, and to prevent rupture and sudden release of the contents due to a BLEVE (Boiling Liquid Expanding Vapor Explosion). Safety cans are available in terne plate (an alloy of lead and tin), stainless steel, and high-density polyethylene. They are also available with Teflon® linings. Safety cans usually have both UL (Underwriters Laboratories Inc.) and FM (Factory Mutual) listings. Flash arrester screens are required for FM approval.

The requirement that all flammable solvents be stored in safety cans is not one the author considers wholly practicable, not only because of the expense and hazard of transferring from glass containers to safety cans, but also because safety cans do not close tightly enough to retain extremely volatile solvents, such as ethyl ether. The major objection to the use of safety cans is that many solvents are contaminated by such storage and become unusable for certain analytical procedures. A possible alternative is to protect 1-gal (3.78 L) glass containers from impact by protective carrying containers or a protective resilient coating over the glass.

Quantity limits: The total amount of flammable liquids and gases in a laboratory should be limited to a quantity sufficient for convenient operation. NFPA 30 sets a limit of 25 gal (94.6 L) of Class IA liquids and 120 gal (454.2 L) of Classes IB through IIIA liquids in a single fire area, where use of these materials is incidental to the principal occupancy. NFPA 45 also establishes limits based on construction, fire area, separation from other fire areas and nonlaboratory spaces, presence of automatic sprinkler protection, and whether flammable and combustible liquids are stored in safety cans or storage cabinets. NFPA 45 also sets quantity limitations on compressed and liquefied flammable gases. Locating as many cylinders as possible outside the laboratory is preferred for firesafety as well as for improved utilization of laboratory space.

Storage cabinets: Storage cabinets are recognized as a means of providing limited insulation of flammable and combustible liquids and thereby limiting the fuel available in the first 10 min of a fire. The performance test set by NFPA 30 for an approved storage cabinet is that the internal temperature shall not exceed 325°F (162.7°C) when the cabinet has been exposed to standard fire test temperatures for 10 min (1,000°F [537.7°C] at 5 min and 1,300°F [704.4°C] at 10 min). More severe tests by the Los Angeles Fire Department in 1959 showed that 1-in. (25.4 mm) plywood cabinets provide greater insulation than conventional double-walled metal cabinets.[5] Both double-walled metal cabinets and 1-in. (25.4 mm) plywood cabinets meet the requirements of NFPA 30, although plywood construction provides greater fire protection at less cost. Providing cabinets convenient to the point of use will reduce the hazards of carrying bottles and transporting solvents.

Ventilation of storage cabinets is recommended only where required by local ordinance, or where there is some specific reason. Ventilation will require a steel duct and an adequate exhaust fan discharging to a safe location. The NFPA Technical Committee on General Storage of Flammable Liquids considers that connecting vents to storage cabinets reduces the limited fire protection provided by such cabinets because a single-walled duct will transmit heat faster than a double-walled cabinet.[6] Fires do not usually occur inside of cabinets because there is no source of ignition within the cabinet.

If greater protection than that provided by a storage cabinet is needed for working quantities of flammable liquids in the laboratory, a tight closet can be used if it is more fire resistant than the "ten minute" cabinet. For example, a closet with gypsum board or plastered walls and ceiling, concrete floor, and a 1¾-in. (44.4 mm) solid wood bonded core door would be more fire resistant than a metal or plywood cabinet. Such a closet would not have to meet all the requirements of a flammable liquids storage room, as defined in NFPA 30, if the contents were limited to the maximum storage capacity permitted for a storage cabinet.

Limiting Oxidizers

Atmospheric oxygen will always be present in the laboratory, but it can be excluded from laboratory equipment, if necessary, by inerting with a gas such as argon, nitrogen, or carbon dioxide in a concentration great enough to prevent combustion. Other oxidizers can be kept from combustible material by separate storage or by storage in an outer container.

Hazardous accumulations of peroxides in ethers and other peroxidizable compounds can be prevented by systematically dating containers when opened and disposing of any unused contents within recommended time limits, unless it is economical to test for and remove the peroxides.[7] Recommended storage limits vary from three to twelve months, depending on the compound.

Limiting Ignition Sources

Although laboratories usually have a variety of possible ignition sources such as motors, switches, ovens, hot plates, and gas burners, laboratories are not ordinarily considered hazardous locations requiring explosion proof electrical equipment and wiring installation. The unclassified status of laboratories is based on experience, general ventilation and the small quantities of flammables usually present. "No Smoking" regulations are frequently adopted in laboratories to control mobile sources of ignition, to prevent contamination of samples and equipment, and to limit hand-to-mouth ingestion of chemicals.

Refrigerated storage: The interiors of refrigerated compartments, however, are considered hazardous locations if used for storage of flammable liquids such as ethyl ether. Ethyl ether, pentane, and isopentane are examples of flammable liquids that cannot be cooled below their flash points in an ordinary refrigerator, freezer, coldroom or environmental chamber. Tests have shown that an open container of ethyl ether in the storage compartment of a refrigerator can generate a flammable concentration in as little as one hour.[8]

Basically, there are three types of laboratory refrigeration which can safely be used if needed for refrigerated storage of extremely flammable liquids such as ether and pentane: explosion-proof, flammable materials storage refrigerators, and modified domestic models. Explosion-proof refrigeration equipment is designed to protect against explosions from both internal and external sources of flammable vapors; it is designed for installation in a hazardous location. Such refrigeration equipment is typically found in pilot plants or laboratories where all electrical equipment is required to be explosion-proof.

The design concept of flammable materials refrigeration equipment is predicated on the typical laboratory environment. The intent is to eliminate ignition sources within the refrigerated compartment and to locate controls and compressors outside of the compartment. To reduce the potential for ignition of floor-level vapors, the controls and compressor may be located above the refrigerated compartment. Additional design features (self-closing door, compartment threshold) are intended to limit damage should an explosion occur.

It has been possible to modify a domestic refrigerator for use in a laboratory; however, it is not recommended and is not usually economical. Modifications include replacing any positive mechanical latch with a magnetic latch; removing all electrical controls and switches from the storage compartment and relocating them to the exterior of the equipment, preferably the top portion; removing all heaters and arcing devices; and sealing all openings in the storage compartment. These procedures can only be used on equipment with manual defrost. Self-defrosting equipment cannot be successfully modified.

Regardless of type used, each refrigerated storage compartment in a laboratory should be clearly labeled to show whether it can safely store flammable liquids.

Intrinsically safe and ventilated equipment: If laboratory operations generate sufficient concentrations of flammable vapors or gases to present serious ignition hazards, there are alternatives to explosion-proof electrical equipment. One approach is to use electrical equipment that is intrinsically safe, equipment that will not generate enough energy, in operation or failure, to ignite flammable gases or vapors. Another approach is to enclose the electrical components of the apparatus and ventilate or purge the enclosure with inert gas or fresh air at a pressure slightly above atmospheric in order to prevent flammable gases or vapors from reaching sources of ignition. Operation of the apparatus can be interlocked with the purge system so that the apparatus will not operate unless the system is operating. A variation of this second approach is to capture flammable gases or vapors at their source, thereby preventing a buildup of flammable concentrations.

In laboratories where quantities of flammable liquids and gases susceptible to ignition by static electricity are used, measures to control static electricity are needed. Such measures have been a routine requirement in hospital operating rooms where flammable anesthetics are used. Maintaining humidity at or above 50 percent and providing conductive shoes or grounding straps have also been standard practice in industrial process areas where static electricity is a likely source of ignition and fire. Bonding pouring operations and transfers from metal container to metal container should be standard practice to dissipate the static electricity generated by the contact and separation of liquids as well as particulate material. Although pouring small quantities from glass container to glass container usually doesn't generate enough static electricity to ignite vapors, particularly those of polar solvents, the generation of static electricity can be prevented by reducing the free-fall distance of the liquid, for example, pouring into a funnel that reaches the bottom of the receiving container. As an added precaution, such pouring should be done under controlled ventilation or in a hood to reduce the buildup of flammable vapors.

Permits for hot work: Finally, laboratories should establish a "hot work permit" system in order to exclude unexpected ignition sources from laboratory areas where flammable vapors may be present. Such a system should be used for controlling welding, cutting, and other maintenance or construction operations which may introduce ignition sources into areas where combustible or flammable materials may be present. (See Chapter 24 for a complete discussion of "Welding and Cutting.")

Fire and Smoke Detection

The early detection of fire and smoke will assist in the prompt evacuation of laboratory personnel and the summoning of the fire department or the plant emergency brigade. Automatic sprinkler system or some other automatic extinguishing system with an alarm provides good protection. If a system is installed for the detection of smoke, products of combustion, or infrared or ultraviolet radiation, the selection should be based upon complete information, i.e., the character of the fire hazards and the limitations and performance characteristics of the detection system. Selection of a system that is too sensitive or not sensitive enough could create adverse reactions to fire protection systems as well as cause unnecessary expense.

Limiting the Spread of Fire

The primary reasons for limiting firespread in laboratories are protection of personnel and property. The requirements for protecting personnel are set forth in NFPA 101, the *Life Safety Code*, which requires the enclosure and protection of vertical openings and exit paths and sets the number of routes of egress and maximum travel distances to stairs and other protected exit paths. The requirements for the protection of property, based on fire probabilities and maximum acceptable losses, should be set by management.

In some laboratories existing construction can be upgraded at minor expense to increase the separation effectiveness and thereby reduce the spread of fire and consequent damage. For example, laboratory doors of solid core wood with ordinary glass view panels will have greatly increased fire resistance if wired glass in steel frames is substituted for the ordinary glass which will quickly break under fire exposure.

Extinguishing Fires

Available records emphasize the effectiveness of automatic sprinkler systems in controlling laboratory fires.[9] Although sprinkler systems may not extinguish fires in most flammable liquids, they do provide cooling effects on uninvolved containers, and do reduce air temperatures so that trained personnel can approach and extinguish the fire.

Where many sprinklers may operate due to the rapid heat release of a large flammable liquid spill fire, consideration should be given to the installation of the "on-off" type of sprinkler that automatically shuts off when temperatures fall below the initial operating temperature.

The type of portable fire extinguisher recommended for a laboratory depends on the quantity of flammable liquids (Class B hazard) which may become involved in the fire, the value and sensitivity of electrical equipment (Class C hazard) which may be damaged by fire or extinguishing agents, and the presence of fire hazards which may require special extinguishers or extinguishing agents. Combustible metals (Class D hazard) require extinguishing agents developed specifically for combustible metals, such as the proprietary agents G-1 and Met-L-X powders. The preferred extinguishing agent for electrical equipment is generally carbon dioxide, considering cost, safety of use, and cleanup.

Dry chemical extinguishers, vaporizing gas extinguishers, or Aqueous Film Forming Foam (AFFF) fire extinguishers are generally required where flammable liquids are used, stored, or likely to be spilled. If there is the usual likelihood of fires involving both Class A (ordinary combustible solids) and Class B (flammable liquids) materials, the extinguishers chosen should have A and B ratings. (NFPA 10, *Portable Fire Extinguishers*, gives guidance on the selection and installation of portable fire extinguishers as well as an explanation of the classification and rating system used to describe the relative effectiveness of different types of portable extinguishers.)

In order to meet minimum NFPA requirements for fire extinguishers for flammable liquids in laboratories, it is necessary to specify extinguishers with at least a 20-B rating. Since the various B ratings are related to the area of burning liquid the extinguisher can put out, it is generally best to use the highest available B-rated extinguisher that can be handled easily. While it is possible to buy wheeled carbon dioxide fire extinguishers with B ratings, such extinguishers are cumbersome and have a limited range that renders them ineffective against flammable liquid spill fires.

For special hazards it is possible to provide packaged extinguishing systems using halogenated agents, carbon dioxide, or dry chemical. It is also possible to provide explosion suppression systems for the protection of personnel and equipment.

Evacuation of Occupants

The safe evacuation of occupants requires that a fire alarm system be heard clearly in all working and occupied areas and that clear and unobstructed means of egress exist. In many laboratories, emergency lighting will be required for occupants to reach and use the corridors and exits.

NFPA 45 specifies that means of egress shall comply with the exit requirements of NFPA 101, *Life Safety Code*, for general purpose industrial buildings. Educational buildings shall comply with the exit requirements for educational occupancies. Exit requirements for general purpose industrial buildings require a 44-in. (1.1 m) minimum width for corridors, exits, and exit discharge and a maximum occupant load of 100 persons per 22-in. (0.5 m) unit of exit width. Travel distance to the nearest of two exits shall not exceed 100 ft (30.4 m), or 150 ft (45.7 m) in a completely sprinklered building. (See NFPA 101, *Life Safety Code*, for additional details.)

NFPA 45 specifies that, if an alarm system is required to warn building occupants, the alarm shall also alert a plant fire brigade or public fire department. (See Chapter 3, "Plant Emergency Training and Organization.") This NFPA standard also calls for the development of procedures for laboratory emergencies: alarm actuation, evacuation, emergency shutdown procedures for equipment, and provision for fire fighting action that includes plans for fire control operations.

Summary

With national attention focusing again on the needs for research and development to achieve continuing technological and scientific innovation, firesafety in and fire protection for laboratories is becoming increasingly important. Conservation of resources will require close cooperation between laboratory management and fire protection and safety personnel in order to find economical and effective ways to achieve firesafety without significantly interfering with laboratory testing, analysis, study, development, or research.

BIBLIOGRAPHY

REFERENCES CITED

1. Steere, N. V., "Identifying Multiple Causes of Laboratory Accidents and Injuries," (Chapter 1), *Safety in the Chemical Laboratory*, vol. 3, Division of Chemical Education, Springfield, PA, 1974.
2. *Operation School Burning — Number 2*. National Fire Protection Association, Boston, MA, 1961.
3. Factory Mutual Engineering Corporation, *Handbook of Industrial Loss Prevention*, Second Edition, McGraw-Hill, New York, NY, 1967.
4. Steere, N. V., "Research Needed for Laboratory Safety Standards," (Chapter 19), *Safety in the Chemical Laboratory*, vol. 3, Division of Chemical Education, Springfield, PA, 1974.
5. Steere, N. V., "Fire Protected Storage for Records and Chemicals," (Chapter 14), *Safety in the Chemical Laboratory*, vol. 1, Division of Chemical Education, Springfield, PA, 1967.
6. Woodworth, M. E., personal communication to Veterans Administration, March 28, 1978.
7. Jackson, H. F., et al., "Control of Peroxidizable Compounds," (Chapter 30), *Safety in the Chemical Laboratory*, vol. 3, Division of Chemical Education, Springfield, PA, 1974.
8. Brewer, R. D., III, unpublished thesis, 1976, University of Illinois Medical Center, Chicago, IL.
9. Factory Mutual Engineering Corporation, "Laboratory Fires and Explosions," *Loss Prevention Data* 10-19, Factory Mutual System, Norwood, MA.

NFPA CODES, STANDARDS, AND RECOMMENDED PRACTICES

Reference to the following NFPA Codes, Standards, and Recommended Practices will provide further information on the safeguards for laboratories discussed in this chapter. (See the latest *NFPA Codes and Standards Catalog* for availability of current editions of the following documents.)

NFPA 10, *Portable Fire Extinguishers*.
NFPA 13, *Installation of Sprinkler Systems*.
NFPA 30, *Flammable and Combustible Liquids Code*.
NFPA 45, *Fire Protection for Laboratories Using Chemicals*.
NFPA 49, *Hazardous Chemicals Data*.

LABORATORIES

NFPA 70, *National Electrical Code.*
NFPA 77, *Recommended Practice on Static Electricity.*
NFPA 101, *Life Safety Code.*
NFPA 325M, *Fire Hazard Properties of Flammable Liquids, Gases, and Volatile Solids.*
NFPA 491M,*Manual of Hazardous Chemical Reactions.*

Additional Reading

Bartels, A. L., "Safeguards to Prevent Ignition by Hot Surfaces," *Electrical Review London*, vol. 203, no. 4 (1978), pp. 29-31.

Ellingson, A. C. and C. A. Trauth, Jr., *Approach to Incorporating Proven Quality Assurance, Reliability, and Human Factors Principles in Fire Protection Programs*, Sandia Labs, Albuquerque, NM, 1978. (available from NTIS).

Fire Protection Guide on Hazardous Materials, National Fire Protection Association, Quincy, MA, 1978.

Flash Point Index, 9th ed., National Fire Protection Association, Quincy, MA, 1978.

Fuscaldo, Anthony A., ed., *Laboratory Safety: Theory and Practice*, Academic Press, New York, NY, 1980.

Green, Michael F., and Amos Tark, *Laboratory Safety*, Macmillan, New York, NY, 1978.

Henry, Martin F., ed., *Flammable and Combustible Liquids Code Handbook*, National Fire Protection Association, Quincy, MA, 1978.

King, P. W. and J. Magid, *Industrial Hazard and Safety Handbook*, Butterworth Publishers, Inc., Woburn, MA, 1979.

Magison, E. C., *Electrical Instruments in Hazardous Locations*, 3rd ed., Instrument Society of America, Pittsburgh, PA, 1978.

McKinnon, G. P., ed., *Fire Protection Handbook*, Fifteenth Edition, National Fire Protection Association, Quincy, MA, 1981.

Pipitone, D., ed., *Safe Storage of Laboratory Chemicals*, John Wiley & Sons, NY, 1984.

"Procedures for Working with Substances that Pose Hazards Because of Flammability or Explosibility," in *Prudent Practices for Handling Hazardous Chemicals in Laboratories*, National Academy Press, Washington, DC, 1981.

Sittig, M., *Handbook of Toxic and Hazardous Chemicals*, Noyes Publications, Park Ridge, NJ, 1981.

Spindel, W., *Prudent Practices for Handling Hazardous Chemicals in Laboratories*, National Science Foundation, Washington, DC, 1981.

Steere, N., *Handbook of Laboratory Safety*, 2nd ed, CRC Press, Inc., West Palm Beach, FL, 1971.

Stevens, A., "Flammable Liquids — Why the Hazard?", *Journal of Chemical Education*, vol. 56, no. 3 (March 1979), pp. 119A-124A.

Talbot, G., "Static Electricity," *Fire*, 70 (871), 1978, pp. 397-398.

Welby, P. and K. R. Dickinson, "Monitoring Work Areas for Explosive and Toxic Hazards," *Chemical Engineering*, vol. 83, no. 22 (1976), pp. 139-145.

42

Boiler-Furnaces

O. W. Durrant

Boiler-furnaces use controlled combustion to generate steam to power machinery or provide heat required by industrial processes. Basically, there are two types of boilers — watertube and firetube. In watertube units, water passes through tubes surrounded by hot combustion gases and is converted to steam. In firetube units, however, the hot combustion gases pass through tubes that are immersed in circulating water, which is converted to steam. Though this chapter discusses watertube boilers, the safety principles outlined are also applicable to firetube boilers. Fuel burning systems and related control equipment for single- and multiple-burner industrial and public utility boiler-furnaces for fuels burned in suspension and on fluidized beds are discussed here. Solid fuel burning units, such as stokers, are not.

THE COMBUSTION PROCESS

Combustion may be defined as the rapid chemical combination of oxygen with the combustible elements of a fuel. There are three combustible

---

O. W. Durrant, retired from Babcock & Wilcox, Barberton, OH, is now a consultant.

elements of significance to this chapter — carbon, hydrogen, and sulfur. Sulfur is usually of minor importance as a source of heat, but can be of major significance in corrosion and pollution problems.

Carbon and hydrogen, when burned to completion with oxygen, unite according to the following:

$C + O_2 = CO_2 + 14{,}100$ Btu/lb (or 32 797 kJ/kg) of carbon
$2H_2 + O_2 = 2H_2O + 61{,}100$ Btu/lb (or 142 119 kJ/kg) of hydrogen

Air is the usual source of oxygen for boiler-furnaces. These combustion reactions are exothermic, and the heat released is about 14,100 Btu/lb (32 797 kJ/kg) of carbon burned and 61,100 Btu/lb (142 119 kJ/kg) of hydrogen burned. The objective of efficient combustion is the release of all available heat while minimizing losses from combustion imperfections and superfluous air.

The combination of combustible elements and compounds of a fuel with all the oxygen requires a *temperature* high enough to ignite the constituents, *turbulence* or mixing, and sufficient *time* for complete combustion. These factors are oftened referred to as the "three T's" of combustion.

The combustion process in a boiler-furnace results from a continual introduction of fuel and air in a flammable mixture. It is necessary to maintain control not only of the flow rates of both fuel and air, but also of the fuel-air ratio, at both the point of ignition and for the entire burner, as well as the source of ignition. If any one of the inputs is interrupted or becomes irregular, ignition can be lost and the controlled flammable mixture can become an uncontrolled explosive mixture.

Figure 42.1 shows schematically how fuel and air are introduced and burned in a furnace through a modern high capacity burner. The fuel and air are injected in separate impinging streams. Normally the total burner air is limited by some means at the point of introducing the fuel to provide a flammable mixture zone, often fuel rich, to ensure stable ignition, assuming there is sufficient ignition energy. The mixing and burning of the remaining fuel and air takes place in the furnace. With a turbulent, inherently stable burner, ignition always occurs at the burner and can be designed to be as stable with a fuel-rich ratio as with an air-rich ratio. With ignition established at the burner, combustion continues as the air and fuel continue to mix in the furnace and become flammable until all of the oxygen is consumed (fuel-rich ratio) or until all the fuel is consumed (oxygen-rich ratio). The combustion process, therefore, proceeds toward completion through a wide range of air-fuel ratios, with the final result dependent on the total ratio of air to fuel over the entire path of the flame.

FUELS

Natural gas, oil, and pulverized coal are the most commonly used fuels to fire boiler-furnaces. However, there are also alternative fuels, such as

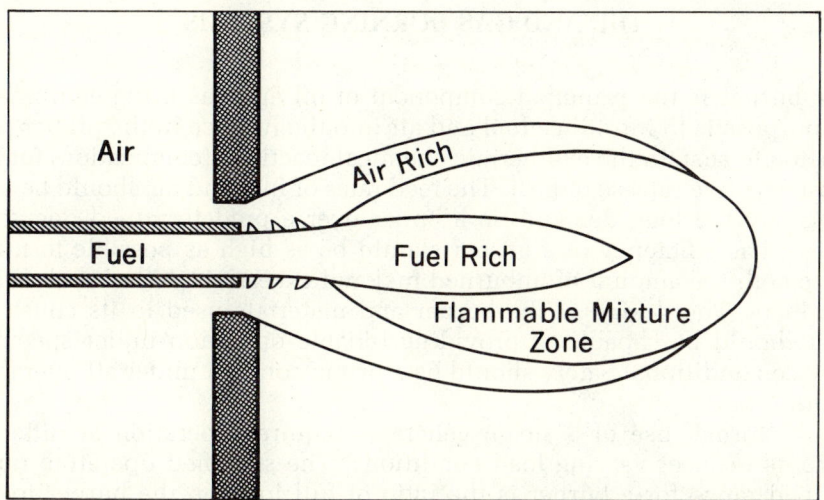

FIGURE 42.1. *Schematic of burner flame conditions. (Babcock & Wilcox)*

FIGURE 42.2. *Installation for combination oil and gas firing in a large electric utility boiler. (Babcock & Wilcox)*

crushed coal, methanol, or liquefied petroleum gas; and supplementary fuels, such as waste gases or flammable liquids in chemical plants and refineries, which may demand special precautions in addition to those required by the more common fuels. Sometimes more than one fuel is used; Figure 42.2 shows combination gas and oil burners on a larger public utility boiler-furnace.

Chapter 42

OIL AND GAS BURNING SYSTEMS

The burner is the principal component of oil and gas firing equipment. Its purpose is to introduce fuel and air into the furnace in the proper proportion to sustain the exothermic chemical reactions (combustion) for the most effective release of heat. The feed rates of fuel and air should be able to satisfy the load demand on a boiler over a predetermined operating range. The efficiency of a burner should be as high as possible to minimize both the amount of unburned fuel and excess air in the combustion products. The design of the burner and materials used in its construction should be capable of providing reliable operation under specified service conditions. Safety should be a prime concern under all operating conditions.

Normal use of a steam generator requires operation at different outputs to meet varying load conditions. The specified operating range or load range for a burner is the ratio of full load on the burner to the minimum load at which the burner must be capable of stable ignition and reliable operation. For example, with a boiler of 100,000 lb/hr (45 359 kg/hr) capacity (steam delivered), a load range of four-to-one on the burners means that the unit can produce stable ignition and complete combustion at any load from 100,000 lb/hr (45 359 kg/hr) down to 25,000 lb/hr (11 340 kg/hr) without necessitating a change in the number of burners in operation.

It is necessary to supply more than the theoretical air quantity to assure complete combustion of the fuel in the combustion chamber. The amount of excess air provided should be just enough to burn the fuel completely in order to minimize the heat loss in stack gases. The excess air normally required with oil and gas, expressed as a percent of theoretical air, is given in Table 42.1.

Continuity of service is enhanced by designing the furnace and arranging the burners to minimize slagging and fouling of heat-absorbing surfaces for the normal range of fuels burned.

The most frequently used burner for gas and oil is the circular burner shown in Figure 42.3. Normally, the capability of an individual circular burner is limited to 190-200 million Btu/hr (55 689 to 58 620 kW). The tangentially displaced doors built into the air register provide the turbulence necessary to mix the fuel and air and to produce stable flames. While the fuel is introduced to the burner in a fuel rich mixture in the center, the direction and velocity of the air and the fuel dispersion pattern thoroughly mix the fuel with the combustion air. The circular burner may be modified with an inner register to limit air in the primary ignition zone to limit formation of oxides of nitrogen (NO_x). The principle and performance are similar to those shown for the dual register coal burner (Figure 42.9).

BOILER FURNACES

TABLE 42.1. *Usual Amount Excess Air Supplied to Fuel-burning Equipment*

Fuel	Type of Furnace or Burners	Excess Air (% by weight)
Pulverized coal	Completely water-cooled furnace for slag-tap or dry-ash removal	15–20
	Partially water-cooled furnace for dry-ash removal	15–40
Crushed coal	Cyclone furnace, pressure or suction	10–15
Fuel oil	Oil burners, register-type	5–10
	Multifuel burners and flat-flame	10–20
Acid sludge	Cone- and flat-flame-type burners, steam atomized	10–15
Natural, coke-oven, and refinery gas	Register-type burners	5–10
	Multifuel burners	7–12
Blast-furnace gas	Intertube nozzle-type burners	15–18

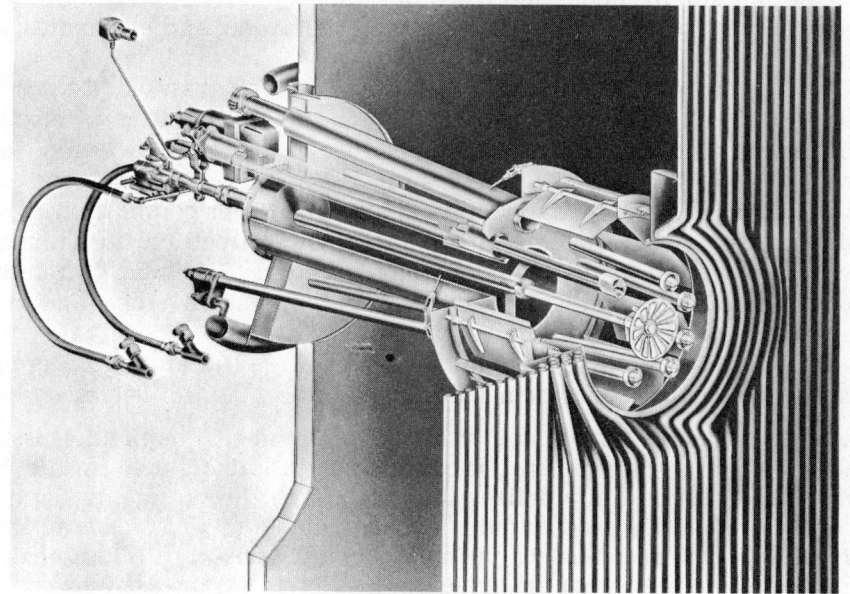

FIGURE 42.3. *Circular register for gas and oil firing. (Babcock & Wilcox)*

Oil Burners

To burn fuel oil at the high rates demanded of modern boiler units, it is necessary that the oil be atomized; that is, introduced into the furnace as

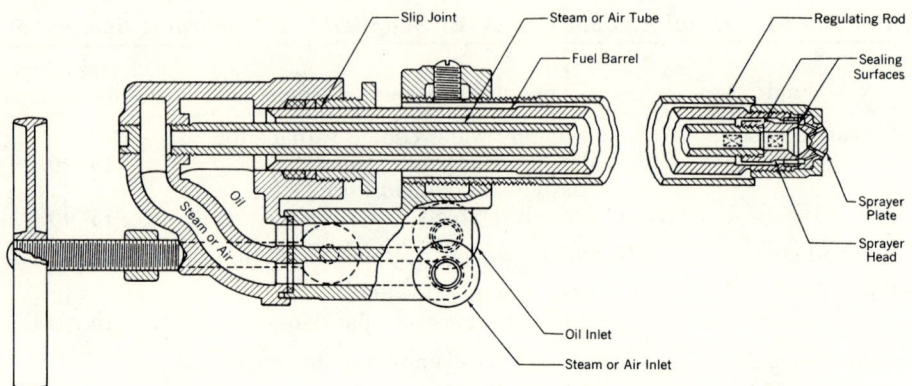

FIGURE 42.4. Steam (or air) oil atomizer assembly. (Babcock & Wilcox)

a fine mist. This exposes a large amount of fuel particle surface to air, which ensures prompt ignition and rapid combustion. There are several types of oil atomizers or vaporizers, but the discussion here is limited to the two most popular types — steam or air atomizers, and mechanical atomizers.

For proper atomization, oil of grades heavier than No. 2 must be heated to reduce their viscosities to 135 to 150 SUS.* Steam or electric heaters are used to raise the oil temperature to the required level — approximately 135°F (57.2°C) for No. 4 oil, 185°F (85°C) for No. 5, and 200 to 220°F (104.4°C) for No. 6. With certain oils, better combustion is obtained at somewhat higher temperatures than are required for atomization. However, oil temperature must not be raised to the point where vapor binding occurs in the pump that is supplying the oil, since this could cause flow interruptions followed by loss of ignition.

It is also important that oil be free of acid, grit, and other foreign matter that might clog or damage burners or their control valves.

Steam or Air Atomizers: The steam or air atomizer (Figure 42.4) is the most widely used. In general, it operates by producing a steam-fuel (or air-fuel) emulsion that atomizes the oil through the rapid expansion of the steam when released into the furnace. The atomizing steam must be dry because moisture causes pulsations, which can lead to loss of ignition. Where steam is not available, moisture-free compressed air can be substituted.

The steam atomizer performs more efficiently over a wider load range than other types. It normally atomizes the fuel properly down to 20 percent of rated capacity; in some instances, steam atomizers have been successfully operated at 5 percent of capacity. Frequently these ex-

---

*Abbreviation for Saybolt Universal Seconds. The efflux time in seconds (SUS) of 60 ml (approx 2 oz.) of sample flowing through a calibrated Universal orifice in a Saybolt viscometer under specified conditions.

BOILER FURNACES

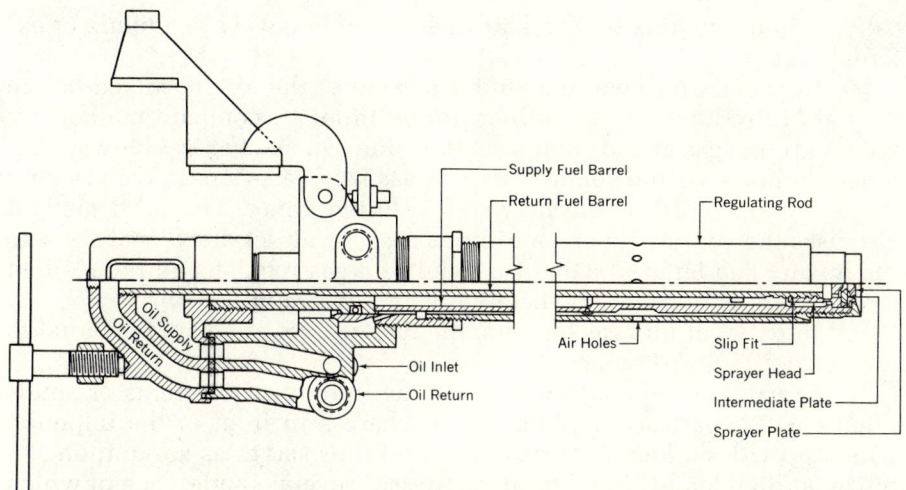

FIGURE 42.5. Mechanical return-flow oil atomizer assembly. (Babcock & Wilcox)

tremes in range cannot be fully utilized, because the temperature of the combustion space falls to a level insufficient to complete the combustion process adequately despite the excellent quality of atomization.

Mechanical Atomizers: In the mechanical atomizer, the pressure of the fuel itself is used as the means for atomization. Many forms have been developed. Those with rotating mechanical parts that are close to the furnace have lost favor, especially for multiburner boilers, because of the excessive maintenance required to keep them operational.

Return flow atomizers (Figure 42.5) are used for many marine installations and some stationary units where the use of steam is inappropriate, impractical or uneconomic. Oil pressure required at these atomizers for maximum capacity ranges from 600 to 1,000 psi (4137 to 6895 kPa), depending on capacity, load range, and fuel.

Mechanical atomizers are available in sizes up to 180 million Btu/hr (52 758 kW) input, about 10,000 lbs (4536 kg) of oil per hr. The acceptable operating range may be as much as ten to one or as little as three to one, depending on the maximum oil pressure used for the system, the furnace configuration, air temperature, and burner throat velocity. Return flow atomizers are ideally suited for standard grades of fuel oil where it is desired to meet load variations without changing sprayer plates or cutting burners in and out of service.

Natural Gas Burners

Natural gas is an ideal fuel for a burner, since it requires no preparation to be suitable for rapid and intimate mixing with combustion air. Basically, gas burners mix fuel and air in either of two ways — premix or external mix.

In premix burners, fuel and a portion of the air are mixed before they are introduced to the burner nozzle input. A common method involves mixing gas and air in the suction side of a mechanical blower. Another method uses the venturi effect. A gas jet creates a negative pressure at the air input orifice and draws air into the mixer. The third method also uses the venturi effect, but, in this case, an air jet draws fuel gas into the mixer. The latter system also requires a gas regulator to reduce fuel gas pressure to atmospheric before gas and air are mixed.

In external mix gas burners, the fuel and air are mixed external to the nozzle.

In external mix gas burners having individual elements or spuds (Figure 42.3), part or all of the gas discharges in front of the impeller, which provides a local fuel rich zone and thus serves as an ignition stabilizer at high loads. Each burner comprises several spuds, each of which is a gas pipe with multiple holes at the end to discharge gas for ignition at the burner throat. Each spud is fitted with a larger diameter flame holder to provide a local fuel rich zone to stabilize ignition at low inputs. Fuel ports are relatively large to minimize plugging in service. Gas spuds are located so that oil will not impinge on them when provision is made for firing multiple fuels. Because of its unusually good ignition stability, even under severe variations in airflow, the multiple spud-type burner is expected eventually to replace most natural gas element designs now used on circular-type burners.

With the proper selection of control equipment, a multifuel-fired furnace with a multiple spud-type burner is capable of changing from one fuel to another without a drop in load or boiler pressure. It is also capable of simultaneous firing of natural gas and oil in the same burner. The control system's primary function is to balance fuels, in the proper relationship with burner air flow to achieve complete and efficient combustion. This type of element is designed for use with natural gas or other gaseous fuels containing at least either 70 percent methane, 70 percent propane, or 25 percent hydrogen by volume. The element is also designed for a maximum input of 173 million Btu per hr (50 706 kW) per burner.

To provide safe operation, ignition of a gas burner should remain close to the burner impeller(s) throughout the full range of allowable gas pressures, not only with normal airflows, but also with abnormally high airflows. Ideally, it should be possible, at the minimum load, to pass full load airflow through the burner and as much as 25 percent in excess of theoretical air at full load without loss of ignition. With this range of airflow, it is not likely that ignition can be lost, even momentarily, during some upset in airflow due to improper operation or error.

PIPING AND CONTROL DEVICES

It is essential that boiler-furnace fuels be as free as possible from contamination and be under control to ensure combustion safety. The design and reliability of fuel-handling systems are, therefore, important factors in minimizing the risk of explosion and hostile fire.

Oil-Fired Systems

Fuel supply equipment should be designed and sized to ensure a continuous, steady flow of fuel that will meet all the operating requirements of the unit. This includes coordinating the burner header pressure regulator (Figure 42.6), burner shutoff valves, and associated piping volume to prevent fuel pressure transients that might exceed stable flame burner limits as a result of cutting burners in or out of service.

Fill and recirculation lines should be connected to storage tanks below the liquid level to avoid excessive evaporation and the generation of static electrical charges in free-falling fuel.

Fuel oils may contain abrasive, corrosive, or waxy contaminants, which may clog filters, produce wear, or otherwise damage oil-burning equipment. Strainers, fillers, traps, and sumps are some of the devices that may be used to remove harmful contaminants from oil burning systems.

It is good practice to route piping and locate valves to minimize their exposure to physical damage and temperature extremes that may alter fuel viscosity or pressure.

Burner shutoff valves should be located as close to the burner as possible to minimize the volume of oil that may remain in the burner line downstream of the valve or that may drain into the furnace following an emergency trip or burner shutdown. Positive means are provided to prevent oil from leaking into an idle burner.

Fuel oil must be delivered to the burners at a specific temperature and pressure to ensure proper atomization. Adequate recirculation provisions must be made for controlling the viscosity of the oil at the burners for initial light-off and subsequent operation. It is important that these systems prevent excessively hot oil from entering the pumps; otherwise, the pumps may become vapor-bound and interrupt the supply of fuel to the burners.

Positive means are needed to prevent fuel oil from entering the burner header system through the recirculating valve, particularly from the fuel supply system of another boiler. Check valves have proved unreliable for this function in heavy oil service.

Provisions should also be required for clearing or scavenging the passages of an atomizer into the furnace.

Chapter 42

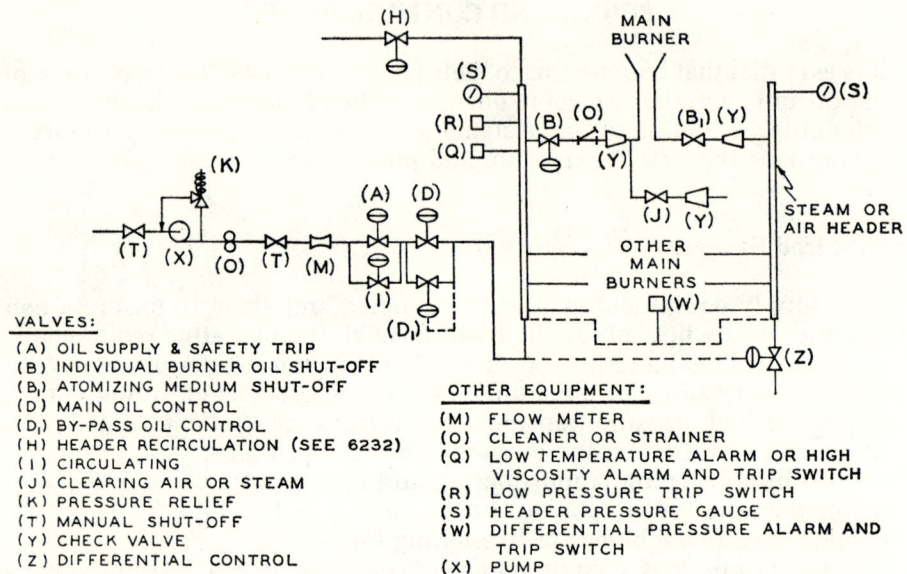

FIGURE 42.6. Schematic of typical main oil burner — steam or air atomizing.

Gas-Fired Systems

Natural gas fuel supply systems must be able to provide a continuous, steady flow of fuel that is adequate for all operating requirements of the unit and within specified fuel header pressure limits for the burners. The system (Figure 42.7) includes a burner header pressure regulating valve and a bypass or minimum pressure regulating valve. The bypass valve comes into use automatically for start-up and low flow operation to avoid operating with a header pressure below the minimum pressure prescribed for burner stability.

The portion of the fuel supply system shown in Figure 42.7 located outside the boiler room is arranged to prevent excessive fuel gas pressure in the burner supply system, even in the event of failure of the main supply constant gas pressure regulators. Usually this is accomplished by providing full relieving capacity vented to a safe location. Where full relieving capacity is not installed, a high supply gas pressure trip may be provided.

The system should have positive means to prevent gas from leaking into an idle furnace. Piping should be vented upstream of the last shutoff valve in any line to a burner or igniter. Provision should also be made in the gas piping to permit testing for leaks and for their subsequent repair. These provisions should include means for making tightness tests of the main gas supply safety shutoff valves and the main burner gas safety shutoff valves.

Vents should be located so that there is no possibility of vented

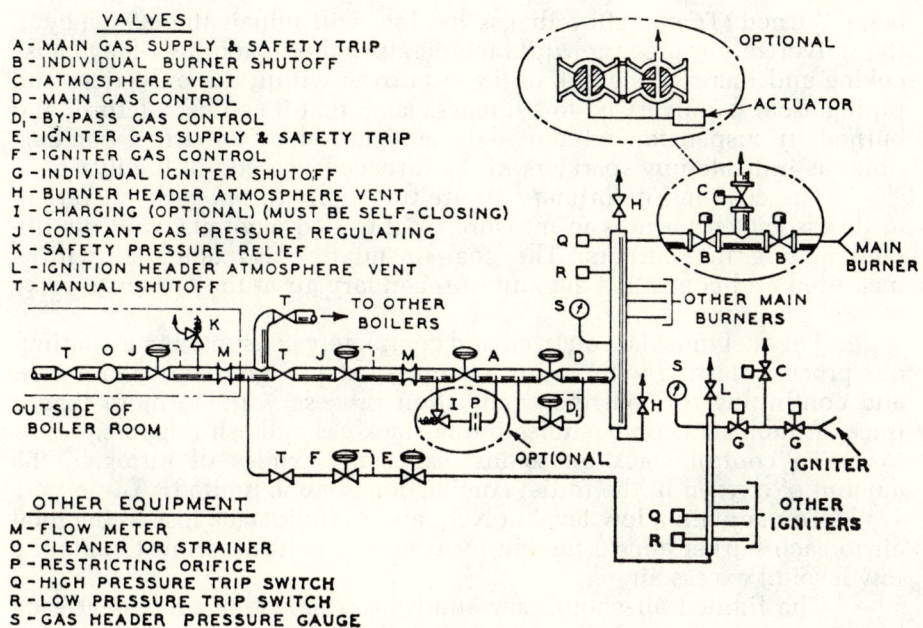

FIGURE 42.7. *Typical fuel supply system for natural gas-fired multiple burner boiler-furnace.*

gases entering the boiler room, adjacent buildings, or air intakes to ventilating systems. The vents should be placed high enough that escaping gas will not be a fire hazard. Header vent lines should be run independently, and the igniter vent subsystem should be run independent of the burner vent subsystem. There must be no cross-connection between venting systems of different boilers.

PULVERIZED COAL SYSTEMS

There are several possible arrangements for pulverized-coal-fired systems. One of the variables is the location of the primary air fan in relation to the air heater and pulverizer. The combination shown in Figure 42.8 is known as a "hot" system and includes individual primary air fans for each pulverizer located downstream of the air heaters. Often the blowers are located ahead of the pulverizer as shown in the figure. Sometimes, the air fans are exhausters located after the pulverizer.

Pulverized-coal-fired systems capable of burning a wide range of coal fuels have several necessary functions. Stored coal must be transferred to the pulverizer in measured and controlled quantities. Hot air mixed with tempering cooler air is blown into the pulverizer to dry the coal. The temperature of the air-coal stream at the outlet of the pulverizer is maintained between specific limits determined by the type of coal being burned. Temperature that is too low will impair the efficiency of the pulverization process, while temperature that is too high may cause coking and increase the risk of fire occurring within the pulverizer and piping. Coal is pulverized to a fineness such that it can be volatized and burned in suspension without delayed ignition or unburned combustibles as indicated by sparklers in the furnace or carbon in the flyash.

The coal and its primary air are transported from each pulverizer to its associated burners in measured, controlled amounts, well distributed among the burners. The coal-air mixture is combined with a measured and controlled amount of secondary air at the burner to serve several purposes:

The fuel must be volatized and completely consumed in a continuous process starting with an initial zone of stable ignition at the burner and continuing through the combustion process with no more than a trace of unburned combustibles in the stack gas and ash hoppers.

To control stack emissions, especially oxides of nitrogen, the amount of oxygen in the initial combustion zone is limited.

To maintain a low level of NO_x, and to limit stack losses, the total air to each burner and to the burner zone of the furnace must contain a low level of excess air.

The limited air should surround the combustion process at each burner to minimize reducing atmosphere in contact with tube surfaces.

Stable ignition is necessary at each burner to ensure continuous combustion and a means of flame detection at each burner.

Sufficient furnace water wall surface between burners is necessary both to minimize the generation of thermal NO_x by cooling the combustion zone, and to cool ash particles below the temperature that causes slag deposits on furnace or superheater surfaces.

Products of combustion are transported from the furnace through platen and pendant superheater and reheater surfaces, through the convection pass including the economizer surface, and through the air heater to accomplish the necessary heat transfer before being transported to the air pollution control system(s) by the induced draft(ID).

Achieving the environmental limitation on NO_x levels requires a low turbulence burner with the air and fuel well distributed to minimize the available oxygen in the initial ignition zone and excess air in the furnace. To achieve the conditions discussed in the preceding paragraphs, burner designs similar to that shown in Figure 42.9, and a compartmented burner system such as that shown in Figure 42.10 were developed.

Figure 42.9 shows the operating principles of a dual register pulverized coal burner to achieve low levels of oxides of nitrogen emissions. The burner, as its name implies, has two registers to proportion the air between the fuel rich, initial ignition zone and the secondary combustion air zone. A third means of adjustment is the position of the swirler vanes to control the turbulence and, thus, the stability of the initial ignition zone. The turbulence and quantity of secondary air to the initial ignition zone is limited to that required for initial ignition and for

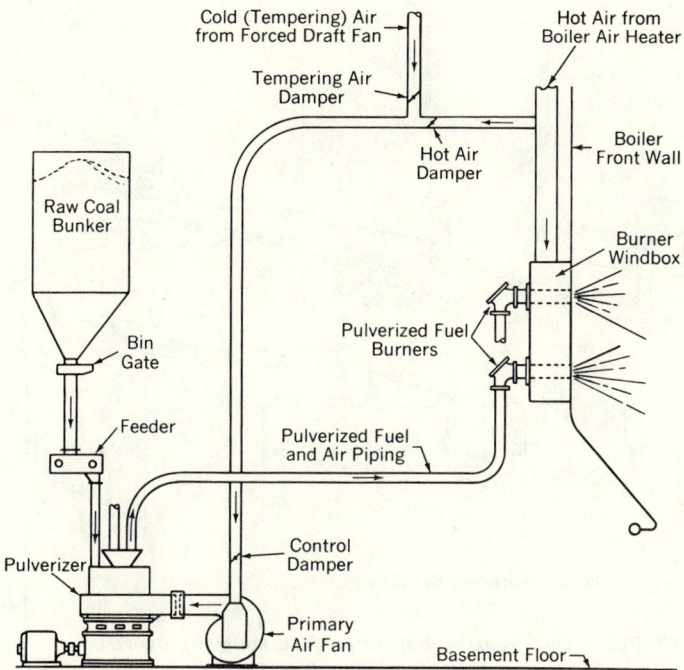

FIGURE 42.8. *Direct-firing system for pulverized coal. (Babcock & Wilcox)*

continuous combustion of the fuel. The remainder of the secondary combustion air is mixed with the fuel within the furnace to provide slow but efficient and continuous combustion until the fuel is completely consumed. A flow distribution device is located in the coal nozzle to ensure a uniform mix of the primary air and coal just prior to emission from the nozzle.

The compartmented burner system (Figure 42.10) provides each burner group served by its pulverizer with a separate windbox with secondary air measured and controlled at each end. Thus, combustion control has been modified from a total boiler basis to a per-pulverizer basis to achieve more precise and more flexible control of the combustion process.

Figure 42.10 also shows another fan-to-pulverizer combination of a pulverized coal system known as a "cold air" or "common primary" fan system. Here, the primary air fans blow (cold) air through (air) heaters to supply the common duct for hot primary air, and the common duct for cold primary air that bypasses the airheaters.

The large steam generating capacity of modern boilers coupled with poorer fuel quality have created the need for high capacity pulverizers capable of processing 50 tons (45.36 metric tons) or more of coal per hr. The pulverizer shown in Figure 42.11 is a roll-and-race-type that

Chapter 42

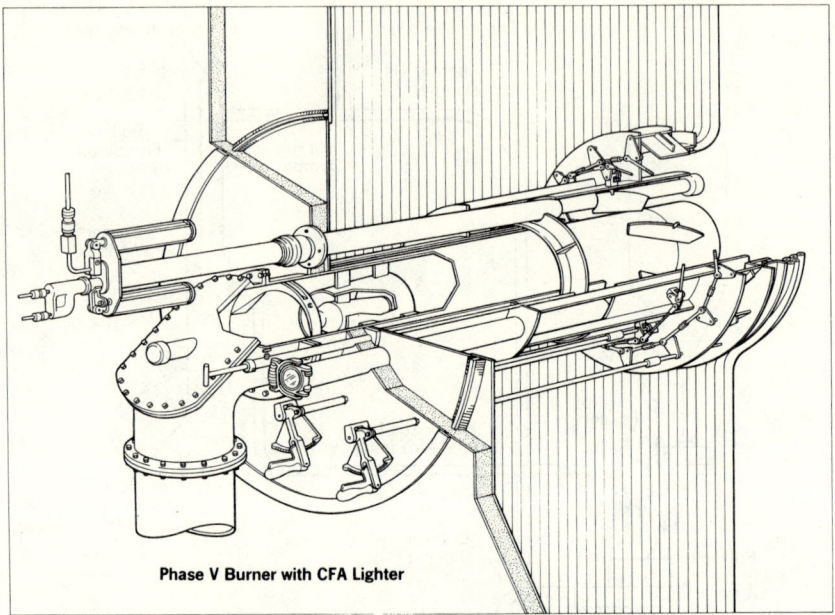

FIGURE 42.9. Dual register burner for NO_x control pulverized coal firing.

utilizes three large diameter rolls spaced equally around the mill. The rolls are mounted on axles; and, in turn, the roll assemblies are attached by a pivoted connection to a stationary overhead frame that keeps them in their roll path while permitting limited radial freedom of movement. Grinding pressure is supplied by springs that apply force to the axles of the rolls. The grinding ring rotates at low speed and is shaped to form a race in which the rolls run.

Circulation of coal is similar to that shown in Figure 42.11. Raw coal is fed into the mill either inside or outside of the grinding race. It immediately mixes with the partially ground coal that is circulated within the grinding zone by the airflow through the pulverizer. As the coal is reduced in size, the air carries it to the classifier. Fine coal, along with air, leaves the pulverizer through the outlet pipes. The oversize coal is returned to the grinding zone through a seal at the bottom of the classifier. Other types of grinders and mills are discussed in Chapter 34, "Grinding and Milling Operations."

FLUIDIZED BED COAL SYSTEMS

The rising cost of fuel, the requirements of air pollution laws to meet sulfur, NO_x, and particulate emission regulations have caused the utility and industrial power industry to develop more economical and efficient

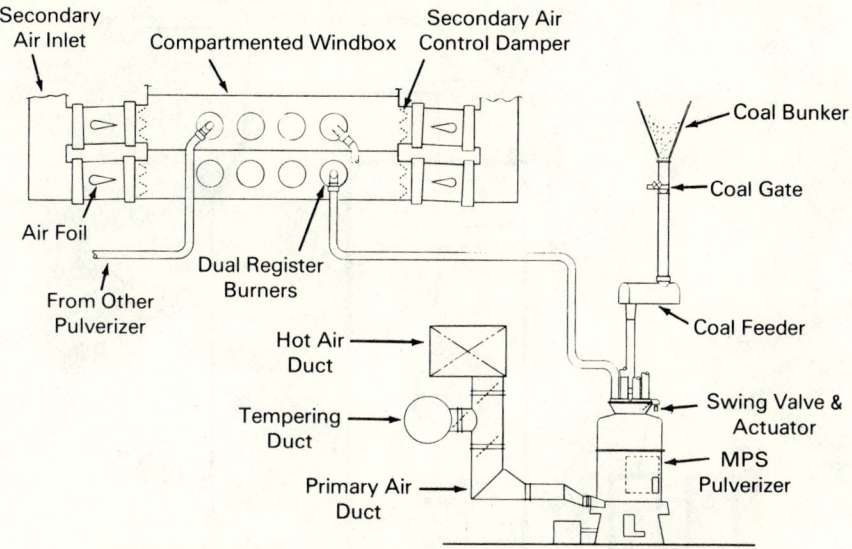

FIGURE 42.10. *Pulverized coal firing system. (Babcock & Wilcox)*

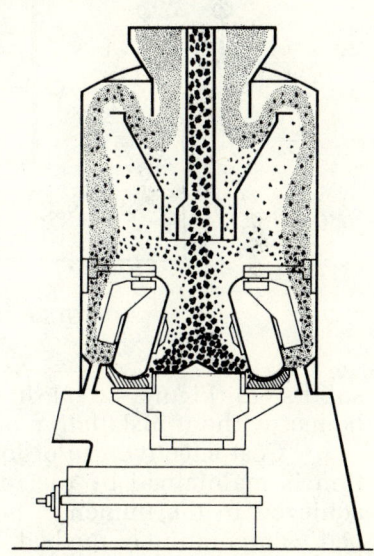

FIGURE 42.11. *MPS pulverizer coal recirculation. (Babcock & Wilcox)*

boiler systems. A recent technology developed to meet these challenges is the fluidized bed boiler. It is designed to meet federal sulfur and NO_x emissions regulations; it uses low-grade, lower-cost energy sources.

Fuel combustion is carried out in a fluidized bed in the presence of a bubbling bed of calcined limestone. Calcium oxide from the limestone reacts with sulfur dioxide produced in combination in a temperature range of 1500-1600°F (815 to 871°C) to form calcium sulfate in a dry

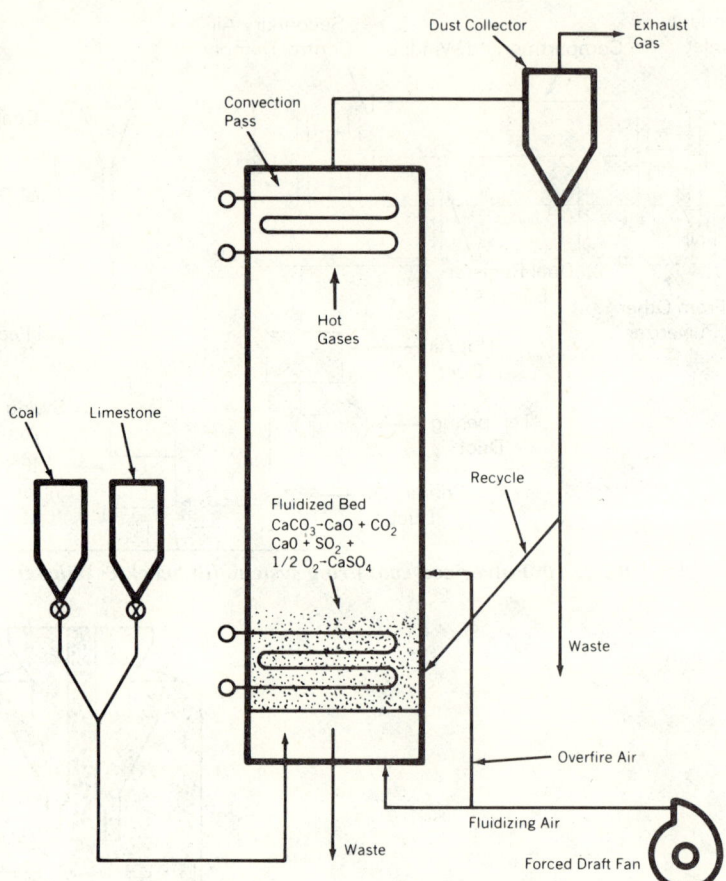

FIGURE 42.12. *Fluidized bed system.*

solid state (Figure 42.12). In this way, federal emission regulations can be met without installation of flue gas desulfurization equipment.

Cost-effective use of low grade fuels is possible because combustion is maintained by the residence time, turbulence and temperature achieved in the immense, hot mass of a bubbling, fluidized limestone bed. Temperature of the bed is below the softening temperature of ash so the ash remains in a solid condition throughout the boiler convection pass, superheater, reheater, and economizer. Problems of slagging and fouling are virtually eliminated from the bubbling, fluidized bed boiler.

The relatively low temperature of the combustion process in the range of 1500-1600°F (815 to 871°C) results in a much lowered production of NO_x.

The major difference between the bubbling, fluidized bed combustion of coal and pulverized coal combustion is the temperature at which combustion takes place. In Atmospheric Fluidized Bed Combustion

(AFBC), the temperature of the combustion environment is kept at 1550°F (843°C). This environment consists of a bed of solids (usually limestone) which are fluidized by air perculating upward from the bottom. Heat is constantly removed from the bed by a superheater surface within the bed, or boiler bank within the bed, to maintain this desired equilibrium temperature. The 1550°F (843°C) temperature, compared to 2500-3000°F (1371 to 1649°C) for pulverized coal boilers, is the source of three major advantages of AFBC over pulverized coal.

1. Sulfur dioxide (SO_2) absorption by a limestone bed is near optimum at this temperature, eliminating the need for a flue gas scrubber.
2. Nitrogen oxides (NO_x) emissions are kept well below federal emissions standards.
3. Slagging and fouling by the coal ash is prevented by being well below the fusion temperature of the ash.

Bubbles of flue gas are formed in the combustion process which rise through the bed causing highly turbulent mixing. Smaller limestone particles become entrained in the gas stream and are carried over through the convection pass. These smaller limestone particles are collected in multiclone dust collectors for recycling to the fluidized bed. Recyling of the unreacted limestone and unburned carbon for reinjection into the fluidized bed greatly increases efficiency of both sulfur capture and carbon burnup.

The integration of the fluidized bed combustion process with a boiler includes the boiler surface within the bed, superheater surface within the bed, waterwall enclosure to contain the bed, and a water-cooled air distributor plate to support the bed.

BOILER-FURNACE HAZARDS

Explosion and fire are the principal hazards of boiler-furnaces and their associated fuel supplies, pipes, ducts, and fans. Explosions are the result of the ignition of combustible mixtures of fuel and air that have accumulated in the confined spaces of the equipment. Generally, such accumulations are the result of a malfunction or operator error associated with an inadequate or improper purge or incorrect operation of the burner equipment.

A temporary loss of flame caused by an interruption in fuel or air delivery or in ignition energy may allow a combustible fuel-air mixture to accumulate in the furnace before ignition is re-established. Leaking fuel may collect in an idle furnace and ignite explosively when the burner is lighted. In multiburner units, the loss of flame at one or more burners may allow an explosive mixture to accumulate, only to be ignited by other burners, either while operating normally or while being lighted. A classic example of operator error is the failure to purge the furnace of combustible mixtures between repeatedly unsuccessful attempts to light off the burner.

Hazards of Oil Firing

Like other petroleum products, fuel oil is a complex blend of hydrocarbons having a wide variety of molecular weights, boiling, freezing, and flash points. When subjected to sufficiently high temperatures, fuel oils will partially decompose or volatize or both, thereby creating new and unpredictable liquid, gaseous, and solid fuels. While the refining process controls the properties of fuel oils within standard limits, the properties of crude oils can vary considerably. For example, crude oils contain volatile light ends, such as propane, butane, and pentane; fuel oils do not.

Historically, there were two grades of fuel oils — distillates and residuals. Residual fuel oils were the residues left after the distillation of crude oils to obtain distillate fuel oils. Modern refining practices involving the cracking process permit more and different types of finished products to be obtained from a single barrel of crude; thus, a wide range of fuel oils is now available. No. 1 and No. 2 fuel oil, and fuels commonly known as kerosenes, range oil, furnace oil, and diesel oil, may still be broadly classified as distillates, while No. 5 and No. 6 fuel oil may be referred to as residuals. Most power boiler fuel systems are designed for the heavier Grade 5 or 6 oil, although distillate fuels may be used for boilers designed with natural gas as the primary fuel. The use of crude oils, even those that have been given some treatment to remove the gassy light ends, is no longer a common practice. It is necessary to preheat the heavier oils (Grades 4, 5, and 6) to hold the viscosity of the oil flowing to the burners within acceptable limits that will ensure proper atomization.

If sludge is allowed to accumulate in storage tanks, it may be drawn into the burners and plug strainers or burner tips, causing a flameout. Water in the oil may also cause flameout.

If two shipments of fuel having widely different viscosities or specific gravities are stored in the same tank, a significant change in fuel input rate without a corresponding change in airflow may occur and impair the efficiency of combustion.

Combustion efficiency of mechanical atomizing burners may also be affected by a change in orifice size caused by wear. In burners operating with very little excess air, combustibles may collect in the furnace. Periodic flow testing and/or replacement of sprayer plates may be necessary. Unsafe operating conditions can also be created by the failure of installation or maintenance personnel to install a nozzle or sprayer plate in the burner assembly.

Very rapid transients in oil flow through operating burners can be caused by rapid operation of an oil supply valve, individual burner shutoff valves, or the regulating valve in the return oil line from the burner header. These uncontrolled changes in the fuel input to the furnace can introduce very hazardous conditions.

Oil flow to individual oil guns can be adversely affected by such conditions as variations in burner elevation, distance from the regulating valve, and pipe size, all of which can be hazardous on low pressure burners.

Hazards of Gas Firing

The primary hazards in gas-fired systems are gas leaks and the development of fuel-rich mixtures within the furnace or structural enclosures. Potentially hazardous accumulations are most likely to develop within buildings, particularly where gas piping is routed through confined areas that are not adequately ventilated.

Within the furnace it is possible for air-fuel ratios to be altered severely without producing any visible evidence at the burners, furnace, or stack, thus allowing the condition to progressively deteriorate. Combustion systems that respond to reduced boiler steam pressure or steam flow with an increase in their demand for fuel are potentially dangerous unless protected or interlocked to prevent the creation of a fuel-rich mixture and a loss of ignition.

Natural gas may be either wet or dry. A wet gas implies the presence of distillate, which, if carried into the burners, can result in momentary increases in fuel input and/or a flameout followed by possible reignition and explosion. For this reason, systems using wet gas require special attention.

Gases supplied from one or more sources can introduce unacceptable hazards if they have significant differences in volumetric heating value. With such variable supplies, it is necessary to provide instruments that are responsive to variations in heat value (e.g., specific gravity or heating value meters) and appropriate alarm and combustion compensation devices.

Discharges from relief valves or any other form of atmospheric vents can present a hazard unless special precautions are taken to prevent a source of ignition or reentry into the boiler room.

Burner pulsation is one of the most mystifying problems associated with gas firing and, to a lesser degree, with oil firing. When one or more burners on a large unit begin to pulsate, the action may become alarmingly violent, at times shaking the whole boiler. Adjusting only one burner may start or stop pulsation. At times, only minor burner adjustments eliminate pulsation. In other instances, it may be necessary to make physical alterations to the burners. Alterations may include modifying the gas ports, impinging gas streams on one another, or installing a device that changes the fuel-air mixing characteristics.

Chapter 42

Hazards of Coal Firing

Coal varies in size and in the amount of impurities it contains. Wide variations in the size of raw coal may cause erratic or uncontrolled feeding of the coal to the pulverizer. As delivered, coal may contain any number of foreign materials such as scrap iron, wood, rags, excelsior, or rock, which may interrupt coal feed, damage or jam equipment, or become a source of ignition within a pulverizer. Since as little as 0.05 oz (1.4 g) of pulverized coal per cu ft of air forms an explosive mixture and since a large boiler may burn 300 lbs (136 kg) or more of coal per second, an explosive mixture will develop quickly if a momentary flameout occurs.

A special hazard is the methane gas that is released from freshly crushed or pulverized coal and which may accumulate in enclosed spaces such as storage bins and within pulverizers and burner piping.

Fuel (finely divided coal) is conveyed through pipes from the pulverizer to the burners entrained in an air stream. To prevent the settling out of coal particles in the burner pipes, and potential preignition, it is necessary to maintain an air velocity that is sufficient to keep the pulverized coal in suspension.

Provisions should be made for cooling down and emptying the pulverizer and/or burner lines when the burners it supplies are shut down. This is to prevent spontaneous combustion and a possible explosion in the pulverizer or burner lines.

Pulverizer fires and explosions are serious hazards. A sudden and considerable increase in the temperature of the fuel-air mixture leaving the pulverizer or in the temperature of the outer casing are indications that a pulverizer fire has started. Such a fire may be caused by feeding burning fuel from the raw fuel bin into the pulverizer or by spontaneous combustion of fuel in the pulverizer or piping. If a fire occurs in any part of a pulverized fuel system, it should be looked upon as serious and should be dealt with promptly. The necessary steps for extinguishing the fire are the following:

1. Remove the pulverizer from service without emptying it to avoid creating an air-rich condition.
2. Isolate the pulverizer against air infiltration by closing all burner line valves and inlet air dampers, and sealing the air valve.
3. Admit steam or inert gas into the pulverizer through the connection provided for that purpose to smother the fire and to maintain an inert atmosphere until the coal has been removed and the housing has cooled.
4. Dump the coal stored in the pulverizer through the pyrite gates and into the sluice system.
5. Rotate the pulverizer without operating the primary air fan to complete the swirling and cleaning procedure.

A similar procedure should be followed for pulverizers that have been tripped full of subbituminous B, C, or lignite coal without immediate restart to avoid pulverizer explosions while out of service or when started up. Recommended emergency procedures should be reviewed with pulverizer manufacturers to include requirements peculiar to each design.

Meters that detect combustibles are not infallible when used with pulverized coal, as they measure only gaseous combustibles. Since the generation of combustibles requires both high temperature for volatilization and limited oxygen in the combustion zone, the lack of a meter indication of combustibles does not rule out the presence of unburned coal particles.

The limited operating experience of fluidized bed boilers has not yet identified any unique, hazardous condition. It is expected that the incidence of furnace explosions will be less than for pulverized coal primarily because of the larger size of crushed coal as compared to pulverized coal.

The furnace safety and burner interlock systems for pulverized coal are currently being applied to experimental and pilot plant installations of fluidized bed boilers. This practice will be continued with special emphasis on the requirements for continuous purge procedures and of interlocks for igniters and auxiliary burners. Compartment alarms and trips may be initiated by both high and low bed temperatures. For any installation the interlock requirements should be reviewed with the manufacturer and the consulting engineer for the project.

It is not the intent of this chapter to include startup, shutdown, or operating procedures or a detailed description of interlock and trip systems. This information is contained in the applicable NFPA standards listed at the end of this chapter.

Open Register Light-off or Continuous Purge Procedure

One important consideration is the use of the open register light-off or continuous purge procedure for all fuels. This procedure maintains airflows at or above the prescribed minimum of 25 percent of full load volumetric airflow during all operations of the boiler. Volumetric airflow is specified to recognize that weight flow rate must be increased at low air temperature to maintain air and gas velocities.

The open register purge rate or continuous purge procedure is based upon the concept that three basic operating conditions will significantly improve the margin of operating safety, particularly during startup. Those conditions are:

1. Minimum number of required equipment manipulations, thereby minimizing exposure to operating errors or equipment malfunction;
2. Means for establishing the desired fuel-rich condition at individual burners during light-off; and
3. An air-rich furnace atmosphere during light-off and warm-up by maintaining total furnace airflow at the same rate as that required for the furnace purge.

In its simplest form, the basic procedure that satisfies these three operating objectives is as follows:

- Place all or most of the burner air registers in a predetermined open position.
- Purge the furnace and boiler settings are with the burner air registers in that same predetermined position. The total airflow for purge must not be less than 25 percent of full load volumetric airflow.
- Light the first burner or group of burners without any change in airflow setting or in burner air register position.

Each boiler should be tested to determine whether any modifications are required to obtain satisfactory ignition or to satisfy other design limitations during light off and warm-up. For example, some boilers will be purged with the registers in the normal operating position. In this event, it may be necessary to momentarily close the registers of the burner being lighted to establish ignition. However, unnecessary modifications in basic procedure should be avoided, thereby satisfying to the greatest degree possible the three basic objectives set forth earlier, particularly that of keeping the number of equipment manipulations to a minimum.

FIRE AND EXPLOSION PROTECTION

The foundation of fire and explosion protection is prevention. Reliable equipment, good facility design, system monitoring and malfunction alarm instrumentation, operator training, and maintenance are key elements in preventing fires and explosions in boiler-furnace systems.

The entire system — boiler, furnace, fuel supply, air supply, vents, piping, and ducts — should be designed to specific parameters and for specific operating limits. At no time should these limits be exceeded.

Good facility design requires the boiler to be installed in a separate room or structure, preferably of noncombustible construction. Boiler-furnaces should be set on concrete floors or platforms that extend beyond the equipment for a distance of at least 4 ft (1.2 m) in each direction. If they must be set on combustible floors, there should be sufficient air circulation beneath the furnace to keep the temperature of the combustible floor below 160°F (71°C). Metal chimneys or smokestacks should not extend through combustible floors, ceilings, or walls. If such chimneys must be passed through combustible roofs, sufficient clearance and/or insulation should be provided to keep the temperature of the combustible materials below 160°F (71°C). Generally, minimum clearance is considered to be 18 in. (0.46 m).

A system of interlocks should be provided to prevent improper sequencing by operating personnel and to shut down operations if certain critical malfunctions occur. Audible and visual alarms serve to warn operators to take certain corrective steps; others may indicate what automatic functions have been performed to reduce a hazard. New boilers should not be fired until adequate safeguards have been installed and tested.

Safe operation cannot be ensured solely by equipment design and adherence to manufacturer's operating instructions. Knowledgeable and competent operators who understand the processes involved are also needed. Technical competence should be maintained by a continuing program of retraining so that increasing familiarity with the system does not tempt operating personnel to take "short cuts" in operating procedures or to bypass safety devices.

A program of preventive maintenance is needed to maintain the reliability of the equipment and its control devices. Poor preventive maintenance can lead to frequent corrective maintenance. Cleanliness and good housekeeping, especially where pulverized coal is the fuel, will also contribute to the prevention of fire and explosion.

Automatic sprinkler or water spray systems are practical forms of fire protection in combustible boiler rooms. In noncombustible boiler rooms where the potential for sustained fire may be slight, portable fire extinguishers or small hose lines may be adequate. Additional guidance on fire protection for fuel supply systems will be found in Chapter 37, "Handling and Storage of Flammable and Combustible Liquids," and in Chapter 38, "Handling and Storage of Industrial Gases."

BIBLIOGRAPHY

REFERENCES CITED

Factory Mutual Engineering Corporation, "Elements of Combustion, Controls, and Safeguards in Industrial Heating Equipment," *Loss Prevention Data* 6-0, September 1977, Factory Mutual System, Norwood, MA.

F. M. Eng. Corp., "Boiler-Furnaces Oil- or Gas-Fired Single Burner," *Loss Prevention Data* 6-4, July 1976.

F. M. Eng. Corp., Boiler-Furnaces Oil- or Gas-Fired Multiple Burner, *Loss Prev. Data* 6-5, May 1978.

McKinnon, G. P., ed., *Fire Protection Handbook*, Fifteenth Edition, National Fire Protection Association, Quincy, MA, 1981.

NFPA CODES, STANDARDS, AND RECOMMENDED PRACTICES

Reference to the following NFPA Codes, Standards, and Recommended Practices will provide further information on the safeguards for boiler-furnaces discussed in this chapter.

(See the latest *NFPA Codes and Standards Catalog* for availability of current editions of the following documents.)

NFPA 30, *Flammable and Combustible Liquids Code.*

NFPA 68, *Explosion Venting Guide.*

NFPA 85A, *Prevention of Furnace Explosions in Fuel Oil- and Natural Gas-Fired Single Burner Boiler-Furnaces.*

NFPA 85B, *Prevention of Furnace Explosions in Natural Gas-Fired Multiple Burner Boiler-Furnaces.*

NFPA 85D, *Prevention of Furnace Explosions in Fuel Oil-Fired Multiple Burner Boiler-Furnaces.*

NFPA 85E, *Prevention of Furnace Explosions in Pulverized Coal-Fired Multiple Burner Boiler-Furnaces.*

NFPA 85F, *Pulverized Fuel Systems, Installation and Operation.*

NFPA 85G, *Prevention of Furnace Implosions in Multiple Burner Boiler-Furnaces.*

ADDITIONAL READING

Durrant, O. W., "Design, Operation, Control and Modeling of Pulverized Coal Fired Boilers." Paper presented to Boiler-Turbine Modeling and Control Seminar, February 14-18, 1977, University of New South Wales, Sydney, Australia.

Durrant, O. W., and G. C. Krippene, "Combustion Principles and Processes for NO_x Control Natural Gas-Fired Utility Boilers." Presented to the American Gas Association, June 6, 1972, Atlanta, GA.

Durrant, O. W., and A. J. Zadiraka, "Control of Pulverized Coal-fired Utility Drum Boilers During Load Changes." Presented to American Power Conference, April 1981, Chicago, IL.

Graham and Trotman, Ltd., eds., *Boiler Operator's Handbook*, State Mutual Book and Periodical Service, Ltd., New York, NY, 1981.

Heil, T. C., and O. W. Durrant, "Designing Boilers for Western Coal." Presented to Joint Power Generation Conference, September, 1978, Dallas, TX.

Resource Systems International, *Boiler: Repair and Maintenance*, Reston Publishing Co., Reston, VA, 1982.

Resource Systems International, *Boiler Systems and Components I and II*, Reston Publishing Co., Reston, VA, 1982.

Shields, C., *Boilers*, McGraw-Hill, New York, NY, 1961.

Spring, Harry M., and Anthony L. Kohan, *Boiler Operator's Guide*, 2nd ed., McGraw-Hill, New York, NY, 1981.

Trinks, W., *Industrial Furnaces*, (vols. 1 and 2), John Wiley and Sons, New York, NY, 1967.

Walker, R. R., and A. J. Zadiraka, "Integrated Fuel and Air Control System for PC Firing." Presented to the American Power Conference, April 21-23, 1975, Chicago, IL.

Whitney, S. A., and J. W. Smith, "Industrial Fluidized Bed Design and Operation of the TVA Test Facility." Presented to 6th International Coal and Lignite Utilization Exhibition and Conference, Houston, TX, November 1983.

43

Fluid Power Systems

Henry Haggard

Fluid power systems are used for force multiplication (automobile lift), process actuation (injection molding), or control (aircraft flap and rudder control). A fluid is defined as anything that is capable of flowing; both gases and liquids are fluids. Fluid power may use air or an inert gas (pneumatic systems), a liquid (hydraulic systems), or a combination of A liquid and gas (hydropneumatic systems). For the purposes of this Handbook, the discussion will center on hydraulic systems.

FLUID POWER AND ITS APPLICATIONS

How Fluid Power Works

In its simplest form, a fluid power system consists of two cylinders of different sizes fitted with pistons and interconnected by pipes or hoses. When the system is filled with a hydraulic fluid, the pressure at a given

---

Henry Haggard, former Assistant Chief, Los Angeles City Fire Department, is now advisor, Fire Protection Administration, California State University, Los Angeles.

Chapter 43

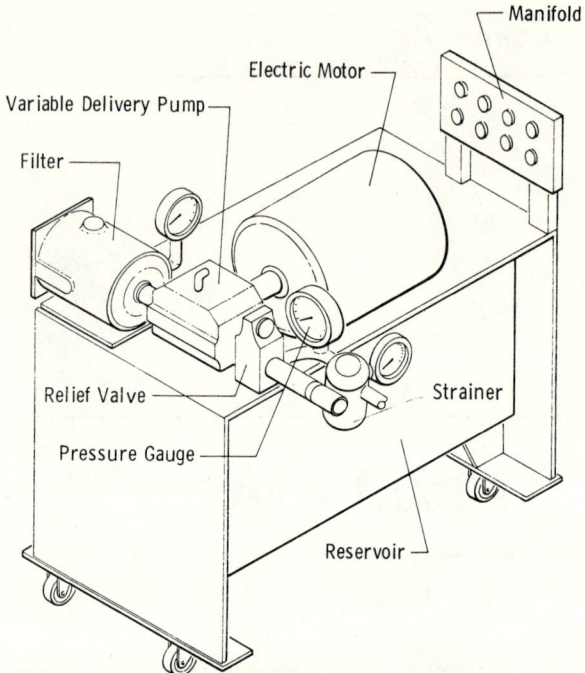

FIGURE 43.1. Mobile hydraulic test station.

elevation is the same at every point in the system. An increase in pressure at any point is transmitted throughout the system.

A force applied to the smaller of the two pistons increases the pressure in the system and produces a force at the larger piston that is directly proportional to the ratio of the surface areas of the two pistons. For example, if the area of the large piston is three times that of the small, the force developed at the large piston will be three times that applied to the small.

More complex fluid power systems may be equipped with pumps for supplying pressure and accumulators and reservoirs for storing hydraulic fluid. In addition, the system may also have fluid filters and a cooler. Fluid power systems may be controlled by manual, mechanical, pneumatic, hydraulic, or electrical means.

Uses of Fluid Power

Fluid power systems have been widely adopted for use in manufacturing and process machinery and for use in mobile equipment. Among the advantages of fluid power are efficient power transmission, substantial force multiplication, quick response, and flexibility in the arrangement of system components.

Some of the more common mobile applications of fluid power include actuating automotive power steering and power brakes, controlling the accessories of earth moving equipment, and multiplying force in materials handling equipment such as fork lift trucks.

The industrial uses of fluid power are many. It controls materials handling equipment, powers machine tools, and actuates automatic production line operations. It is used to bore, mill, shear, and grind. It powers a variety of hydraulic presses that form, straighten, bend, extrude, and stretch materials. It can control a delicate mechanical operation, or it can move a load of many tons.

CHARACTERISTICS OF HYDRAULIC FLUIDS

The wide range of fluid power system applications necessitates a variety of fluids to meet the requirements of any given system. Many different formulations of hydraulic fluids are available, and all of them represent some compromise from what might be considered an ideal fluid. Among the properties of importance in a hydraulic fluid are viscosity, viscosity index, pour point, compressibility, stability, lubricity, volatility, and compatibility.

There are four basic classes of hydraulic fluids — petroleum-base oils, water-glycol fluids, water-in-oil emulsions, and synthetic fluids. The latter include chlorinated hydrocarbons, phosphate esters, and blends of phosphate esters and mineral oil.

Viscosity

The most important property of a hydraulic fluid is viscosity or resistance to flow. Viscosity, unfortunately, is not a stable characteristic of any fluid; it varies with temperature fluctuations.

A fluid with a high viscosity usually causes a large pressure drop and results in low mechanical efficiency, sluggish operation, and high power consumption. Thus, high-pressure, high-precision systems suffer at low temperatures where viscosity is high. Under such conditions, the pumps that apply the necessary force to the liquid tend to cavitate or form vacuums in the fluid, causing irregular operation.

Low viscosity fluids present different problems. Although they do permit higher efficiency, they tend to increase wear and to lower volumetric efficiency. In addition, they tend to promote leakage, which can be hazardous. Lowering the viscosity increases the relative flow; thinner fluids flow more readily through small openings. Leakage may be more critically influenced by other factors such as operating pressure and temperature.

Viscosity Index

The readiness with which, or the extent to which, a fluid changes viscosity as temperature changes is called the viscosity index (VI). The ideal hydraulic fluid should have the same viscosity at all temperatures. Unfortunately, there is no ideal hydraulic fluid. Those fluids that come close to being the ideal are given a high viscosity index. A low viscosity index, on the other hand, indicates that a fluid's viscosity changes greatly with temperature variations. Petroleum oils, for example, have viscosity indexes ranging from 90 to 105, while viscosity indexes for glycols range from 160 to 200.

Pieces of mobile equipment that may be used outdoors in all types of weather require fluids with high viscosity indexes for their fluid power systems. The viscosity index can be raised by special additives, as is often the case with petroleum oils. Additives tend to be expensive and to lose their effectiveness under continued service unless carefully watched and replenished. They must be carefully selected for the particular system, as they may be incompatible with metals or seals.

Pour Point

Pour point, which appears to be a function of viscosity, also is affected by temperature. Pour point is relatively unimportant in a hydraulic fluid if the system operates in a warm environment. It does become significant in low ambient temperatures. The pour point of a fluid should be about 20°F (6.7°C) below the lowest temperature to be encountered.

Compressibility

Compressibility is the extent to which a fluid decreases in volume under pressure. Hydraulic fluids are relatively noncompressible but not completely so. As a general rule, compressibility is about 0.5 percent for each 1,000-psi (6895 kPa) increase in pressure up to 4,000 psi (27 580 kPa). Though normally of little significance, compressibility does affect the performance of fluid power systems in servo applications. It influences the system's gain or amplification and determines static rigidity. If the system is equipped with high-pressure positive displacement pumps, there is some loss in fluid volume and, thus, some power loss.

Stability

Ideally, hydraulic fluids should be stable and their characteristics should not change over long periods of use, but they do. There are several factors that cause fluids to change. Among them are mechanical stress, thermal degradation, oxidation, and hydrolysis. Mechanical stress from pump or valve action can cut polymer chemical chains and reduce the effectiveness of viscosity improvers. The resultant chemical changes may produce volatile compounds that dissipate the volume, corrosive materials that attack system components, or insoluble materials that cause excessive wearing of components.

Insoluble materials may be hard abrasive particles, sludge, or gum. Hard particles may accelerate wear in pumps, motors, and valves. Valve action will be slowed by the formation of sludges, gums, or varnish films. All of these insolubles can clog filters so that fluid flow is seriously reduced.

Thermal degradation, oxidation, and hydrolysis may contribute to the formation of corrosive agents, usually acidic. Corrosive agents widen the gaps between close-fitting parts and increase leakage. If they cause pitting, substantial localized loss of strength may result and the parts may fail under pressure.

Lubricity

Lubricity or lubricating ability is an essential characteristic of hydraulic fluids, because moving parts must be protected against wear. Lubricity is a natural quality of petroleum oils and is one of the reasons why such oils have long been used in fluid power systems. However, petroleum oils have some limitations as hydraulic fluids. They are flammable. Their long chain molecules tend to break down under high rates of shear; they are conducive to the formation of sludge and varnish; and they may show unwanted viscosity increases due to soluble oxidation products. Glycol-based fluids, on the other hand, do not have these limitations, and though their lubricating ability may not be as great as that of petroleum oils, it is sufficiently effective for many applications.

Other Qualities

Other qualities that govern the selection of a hydraulic fluid include volatility, foaming resistance, corrosion prevention, and compatibility. For example, natural rubber cannot be used for seals in systems using petroleum oils, and synthetic rubbers differ widely in their resistance to various fluids.

Zinc-coated surfaces may be attacked by rust preventive additives in nonpetroleum fluids, or they may react with oil to form metallic soaps. Hydraulic fluids always contain some air bubbles, which increase compressibility and result in erratic operation. The bubbles can also cause erosion of the pump. For these reasons, antifoaming agents may be added to the fluids.

Chapter 43

FIRE HAZARDS

While the operating temperatures of the hydraulic systems themselves are seldom high enough to present a fire hazard, the environment may provide ignition potentials for escaping hydraulic fluids. (See Chapter 37, "Handling and Storage of Flammable and Combustible Liquids.")

Flammability Factors

The elements that influence the flammability of hydraulic fluids are flash point, fire point, and autoignition temperature — in that order of importance. Flash point is the temperature at which a fluid will give off vapor that can be ignited by a flame or other ignition source. Fire point is the temperature at which flammable vapors are given off at a rate sufficient to support continuing combustion after the ignition source has been removed. Autoignition temperature is that temperature to which a fluid must be heated for it to ignite and burn without an external ignition source.

In actual practice, hydraulic fluids, including petroleum oils, have flash point temperatures more than twice the recommended system operating temperature, and autoignition temperatures are even higher. Thus, there is a comfortable margin of safety under normal operating conditions. However, in finely divided form such as mist or spray, liquids can be ignited at temperatures below their flash points. Thus, it is when the integrity of the system is impaired and leakage occurs near an external ignition source that the fire hazard becomes a reality.

Mobile Systems

Fluid power systems that are normally operated outdoors as part of mobile equipment often use petroleum-base hydraulic oils. Generally, the only potential ignition source is the exhaust system of the internal combustion engine. The hazard lies in leaking hoses or seals permitting pressurized oil to be sprayed on hot engine or exhaust surfaces and ignited.

Mobile materials handling equipment for industrial plants often moves from unheated storage sheds or yards to areas where process equipment is being operated at high temperatures. Under these conditions, the fire hazard becomes much greater because the ignition potential for leaking, sprayed, or spilled hydraulic fluid is much higher.

Fixed Systems

Some fixed systems may be less hazardous than any mobile system, but chances are that higher ambient temperatures and a greater number of potential ignition sources will be found within the confines of an industrial plant than will be found outdoors. Serious fires have been started when hydraulic fluid was sprayed on an ignition source located a considerable distance away.

The distance to which fluid may be sprayed and the extent of its dispersion depend in large measure on the size of the activating piston, its length of stroke, and the operating pressure. The size and shape of the leak will also affect distance and dispersion.

FIRE RESISTANCE OF HYDRAULIC FLUIDS

Since most fluid power systems are designed to operate in the 120 to 140°F (49 to 60°C) range, the inherent fire hazard may be considered small, even when petroleum-base oil is the hydraulic fluid. The major fire hazard is in the escape of the fluid from the system to an ignition source.

Of the four types of hydraulic fluids mentioned earlier, water-in-oil, water-glycol, and synthetic fluids can be classified as fire resistant. However, they differ widely in the degree of fire resistance and in other qualities. The synthetics, such as phosphate esters and chlorinated hydrocarbons, provide a high degree of fire resistance, but they are less efficient than other types at low temperatures. On the other hand, they are more stable than water-in-oil or water-glycol fluids.

Water-in-Oil Fluids

These are emulsions in which finely divided particles of water are dispersed throughout a continuous outer phase of oil. These emulsions can be formulated with a wide range of viscosities and are suitable for many applications. The viscosity of water-in-oil fluids increases with an increase in water content. Therefore, if the viscosity is too high for a particular application, it cannot be reduced by adding water to the emulsion. Normally, such fluids have a water content of about 40 percent. It is the water content that provides the fire resistance; therefore, it is important that the water content be maintained. When such fluids are sprayed on a hot surface, the conversion of the water to steam lowers the temperature of the surface and denies oxygen to the combustion process. Water-in-oil fluids are not recommended for use at operating temperatures in excess of 150°F (65.5°C) because water loss due to evaporation becomes significant, resulting in reduced viscosity and increased flammability. Some authorities contend that water-in-oil fluids should not be used in applications where magnesium is involved because of the metal's reactivity with water.

Water-in-oil emulsions are not chemically stable and tend to separate if allowed to stand for long periods or if subjected to freezing temperatures. Contamination by other hydraulic fluids, even other formulations of water-in-oil emulsions, may also result in the breakdown of the fluid, causing it to lose those qualities that make it effective.

Water-Glycol Fluids

Water-glycol fluids are mixtures of water, polyglycols, ethylene glycol, and additives to provide corrosion protection. Unlike water-in-oil emulsions, water-glycols are true chemical solutions and are quite stable. However, water may evaporate from the solution, especially at operating temperatures above 150°F (65.5°C). Decreasing water content produces increasing viscosity. Water-glycol fluids usually have a water content of 35 to 50 percent. These fluids also derive their fire resistance from the water they contain and are not recommended for use in systems involving magnesium.

Synthetic Fluids

Phosphate ester fluids: These are generally of the triaryl type and may be used alone or blended with mineral oil. They are characterized by low viscosity indexes and must have VI improvers added if they are to be effective at low temperatures. The additives tend to lower the shear resistance of the fluids so that, after continued use, the original viscosity is reduced. Phosphate ester fluids will burn at elevated temperatures, but they will not propagate flame and are self-extinguishing when the ignition source is removed. When sprayed on hot or molten metal, these fluids will ignite at temperatures between 1,400 and 1,500°F (760 and 815°C). Autoignition temperatures normally are in the 1,100 to 1,300°F (593 to 704°C) range.

Under most conditions, phosphate esters are excellent lubricants with lubricity equal to, or better than, the best petroleum hydraulic oils.

Chlorinated hydrocarbon fluids: Chlorinated hydrocarbon fluids have qualities and characteristics much the same as those of phosphate ester fluids. Both types will attack some of the materials used for seals, gaskets, and hoses. They will also damage certain paints. Substitution of fire resistant fluids in petroleum oil systems will result in impaired mechanical efficiency and extensive physical damage to system components.

FIRE PROTECTION

Since the primary hazard in fluid power systems is the accidental release, often under pressure, of the hydraulic fluid, fire protection begins with good maintenance and operating practices that are designed to prevent an incident from occurring. In the event that a fire does occur, certain fire suppression equipment should be available to minimize the loss.

FLUID POWER SYSTEMS

Maintenance and Operations

High-pressure leaks can be caused by the failure of hoses, gaskets, seals, valves, couplings, and unions. System components should be checked at least daily or, in the case of multiple shift operations, at least once each shift. Particular attention should be paid to moving parts and vibrations that could loosen couplings or abrade hoses.

Stores of hydraulic fluids in cans and drums should be separated from the rest of the plant by open space or by fire-resistive construction. Bulk storage of large quantities of fluid may need to be protected by automatic fire suppression equipment. Additional guidance on the Handling and Storage of Flammable and Combustible Liquids will be found in Chapter 37.

Any spillage, of course, should immediately be removed. If the fluid is petroleum oil, it may be pumped, skimmed, or absorbed, depending on the quantity involved. Spills of water-glycol fluids can be cleaned by flushing or mopping with water. Synthetic fire resistant phosphate ester and chlorinated hydrocarbon fluids can be cleaned up with the same methods and solvents used for petroleum oils. The same method can also be used for water-in-oil emulsions. No effort should be made to salvage or reclaim spilled hydraulic fluids, as they will be highly contaminated and harmful to the fluid power system.

In handling hydraulic fluids in the course of normal operations, care should always be taken to prevent the fluids from becoming contaminated.

Waste hydraulic fluids should not be mixed with the normal plant effluent, for each type requires a different treatment. The presence of large quantities of petroleum oil may present a fire hazard in a disposal system. Waste petroleum oil fluids should be collected in drums and disposed of in a suitable manner. Water-in-oil emulsions should be separated and the oil disposed of in the same manner as other waste oils. Synthetic fluids also can be treated as oily wastes. Water-glycols are readily handled by conventional sewage treatment processes, or they can be degraded by aeration and oxidation.

If any maintenance, repairs, or modifications to the system require cutting and welding operations, oil deposits should be cleaned up and the system depressurized before work is begun. The safety precautions discussed in Chapter 24 also should be taken under these conditions.

Suppression Equipment

Plants using petroleum oil as a hydraulic fluid should be equipped with automatic sprinklers. Hydraulic fluids escaping under pressure will become mixed with air and sprayed considerable distances. If the spray contacts an ignition source, the resulting flame will resemble a torch. Though sprinklers may be of little use against such a flame, they will afford a measure of protection to the structure and other combustibles housed in it. The most effective way to extinguish the oil flame is to cut off the fuel supply; an automatic switch is desirable for this purpose.

If the fluid power system is large or sprinkler discharge may be deflected away from some areas containing combustibles, a small hose with a spray nozzle is an important adjunct to the sprinkler system.

Portable fire extinguishers of types suitable for use against flammable liquid fires will provide effective protection for mobile systems and for fixed systems using small amounts of fluid, usually less than 100 gal (378 L).

Operators should be well trained in the operation of fluid power systems. They also should be schooled in the location, function, and operation of emergency shutoff devices, fire extinguishers, and fire hoses. See Chapter 3, "Plant Emergency Training and Organization," for a complete discussion.

BIBLIOGRAPHY

REFERENCES CITED

Factory Mutual Engineering Corporation, "Hydraulic Fluids," *Loss Prevention Data 7-98*, August 1975, Factory Mutual System, Norwood, MA.

McKinnon, G. P., ed., *Fire Protection Handbook*, Fifteenth Edition, National Fire Protection Association, Quincy, MA, 1981.

NFPA CODES, STANDARDS, AND RECOMMENDED PRACTICES

Reference to the following NFPA Codes, Standards, and Recommended Practices will provide further information on the safeguards for the fluid power systems and hydraulic fluids discussed in this chapter. (See the latest *NFPA Codes and Standards Catalog* for availability of current editions of the following documents.)

NFPA 10, *Portable Fire Extinguishers.*

NFPA 13, *Installation of Sprinkler Systems.*

NFPA 14, *Standpipe and Hose Systems.*

NFPA 24, *Installation of Private Fire Service Mains and their Appurtenances.*

ADDITIONAL READING

American Institute of Physics, *Hydraulic Devices*, McGraw-Hill, New York, NY, 1975.

Anders, James E., *Industrial Hydraulics Troubleshooting*, McGraw-Hill, New York, NY, 1983.

"Fluid Systems 1978: Looking Ahead Ten Years," *Machine Design*, vol. 50, no. 22 (Sept. 28, 1978), pp. 2-5.

"Fluids," *Machine Design*, vol. 50, no. 22 (Sept. 28, 1978), pp. 112, 115-116.

Giles, Ronald V., *Fluid Mechanics and Hydraulics*, McGraw-Hill, New York, NY, 1962.

Henrikson, K. H., "Fire Resistant Fluids in Mobile Equipment," 650671, Sept. 1965, Society of Automotive Engineers, New York, NY.

Hydraulic Handbook, 8th ed., Gulf Publishing Co., Houston, TX, 1983.

Hydraulic Handbook, 7th ed., State Mutual and Periodical Service, Ltd., New York, NY, 1981.

Johnson, Olaf A., *Fluid Power for Industrial Use: Hydraulics*, 2 vols., Krieger Publishing Co., Melbourne, FL, 1981.

Merritt, L. C., *Hydraulic Control Systems*, Wiley-Interscience, New York, NY, 1967.

Millett, W. H., "Fire Resistant Hydraulic Fluids," Sept. 1973, E. F. Houghton & Co., Philadelphia, PA. (Paper presented at the Twenty-ninth National Conference on Fluid Power.)

Pingree, Daniel, "Looking at Fire Hazards: Hydraulic Fluids," *Fire Journal*, vol. 59, no. 6 (Nov. 1965), p. 23.

Pippenger, John, and Tyler Hicks, *Industrial Hydraulics*, 3rd ed., McGraw-Hill, New York, NY, 1979.

Protheroe, A. R., "Fire Resistent Hydraulic Fluids: New Capabilities Expand Applications," *Hydraulics and Pneumatics*, vol. 31, no. 5 (1978), pp. 73-75.

Roberts, A. F., and F. R. Brookes, "Hydraulic Fluids: An Approach to High Pressure Spray Flammability Testing Based on Measurement of Heat Output," *Fire and Materials*, vol. 5, no. 3 (1981), pp. 87-92.

Stewart, Harry L., *Hydraulic and Pneumatic Power for Production*, 4th ed., Industrial Press Inc., New York, NY, 1977.

Standard Practice for the Use of Fire Resistant Fluids for Fluid Power Systems, USA Standard B93.5-1966, National Fluid Power Association, Thiensville, WI.

Sabersky, Rolf H., et. al, *Fluid Flow: First Course in Fluid Mechanics*, 2nd ed., Macmillan, New York, NY, 1971.

Snyder, C. E., A. A. Krawetz, T. Tovrog, "Determination of the Flammability Characteristics of Aerospace Hydraulic Fluids," *Lubrication Engineering*, vol. 37, no. 12 (Dec. 1981), pp. 704-714.

"Synthetic Fire-Resistent Hydraulic Fluids for General Industrial Use," *Hydraulic, Pneumatic and Mechanical Power*, April 1979, pp. 145-148.

Tuve, Richard L., *Principles of Fire Protection Chemistry*, National Fire Protection Association, Boston, MA, 1976.

Warren, S. M., and J. R. Kilner, *Fireproof Brake Hydraulic System*, Aeronautical Laboratories, Wright-Patterson Air Force Base, OH, 1981.

44

Refrigeration Systems

Emil E. Roy

Industrial refrigeration systems perform four major functions. They provide: low-temperature storage for food and other products that deteriorate at normal ambient temperatures; cooled air for the comfort of building occupants; low-temperature air or gas-air mixtures for critical processes, such as the grinding of spices, rubber, and resins and the fast freezing of foods; and they convert cryogenic gases to liquids for more efficient storage and transport.

HOW THE SYSTEM WORKS

Basically, all refrigeration systems take heat from one space or medium and transfer it to another. To accomplish the transfer, a material is needed that will readily absorb, transport, and release heat under controlled conditions. Such materials, or refrigerants, perform in both the liquid and gaseous states. When the liquid changes to a gas, it absorbs heat from its surroundings. Conversely, when the gas condenses back to

---

Emil E. Roy is Vice President, Loss Prevention Services, for Frank B. Hall & Company.

a liquid, it releases heat to its surroundings. The most effective refrigerant is a material that can vaporize and condense at the desired temperatures and pressures.

The refrigeration process is continuous. At the condenser, where it arrives in a high-pressure gaseous state, the refrigerant is condensed back to the liquid state, releasing heat in the process. It is then stored in a receiver or accumulator until it is needed. When temperature controls call for refrigeration, the liquid passes through an expansion valve into the evaporator where it vaporizes and absorbs heat. It is then carried back to the condenser.

Refrigeration systems require two pressure levels — low pressure in the evaporator and high pressure in the condenser. Pressure differentials are obtained either by mechanical compression or by heat absorption.

Figure 44.1 illustrates the relationship of a mechanical compressor to the evaporator and condenser coils. Mechanical compressors can be of the reciprocating type with horizontal, vertical, or "V" design (see Figure 44.2) or of the centrifugal type (see Figure 44.3). Regardless of the type, compressors should be made of materials that are compatible with the refrigerant used and equipped with the proper safeguards against overpressure and leakage. Typical safeguards include relief valves, purge valves, liquid separators, and oil separators.

Heat absorption system: The compression stage of a heat absorption system occupies the same relative position in the refrigeration system as does the mechanical compressor. In heat absorption systems, heat supplied by steam, hot water, or a gas flame raises the pressure of the refrigerant so that it will flow through the expansion valve to the evaporator. The refrigerant, lithium bromide for air conditioning applications or ammonia for low-temperature process applications, is vaporized and goes into solution with an absorbent, usually water. After it has passed through a heat exchanger, the solution enters a generator in which heat drives the refrigerant from the absorbent. The refrigerant is then purified and returned to the condenser for reuse (see Figure 44.4).

Compression systems: Most of the commercial and industrial refrigeration systems in use today are of the mechanical compressor type. This type permits a wider selection of refrigerants to be used than does the absorbent type.

In addition to being classified according to the type of compression stage employed, refrigeration systems are further differentiated as direct or indirect. In the direct type, the evaporator is in direct contact with the product or space to be cooled, or it is in air-circulating passages connected to the area or product. A household refrigerator is an example of a direct system.

An indirect system uses an intermediate heat transfer fluid, which is first cooled by the refrigerant and then circulated to the material or space. The intermediate fluid is called "brine" because early refrigerating systems used a mixture of sodium or calcium salt and water. Today,

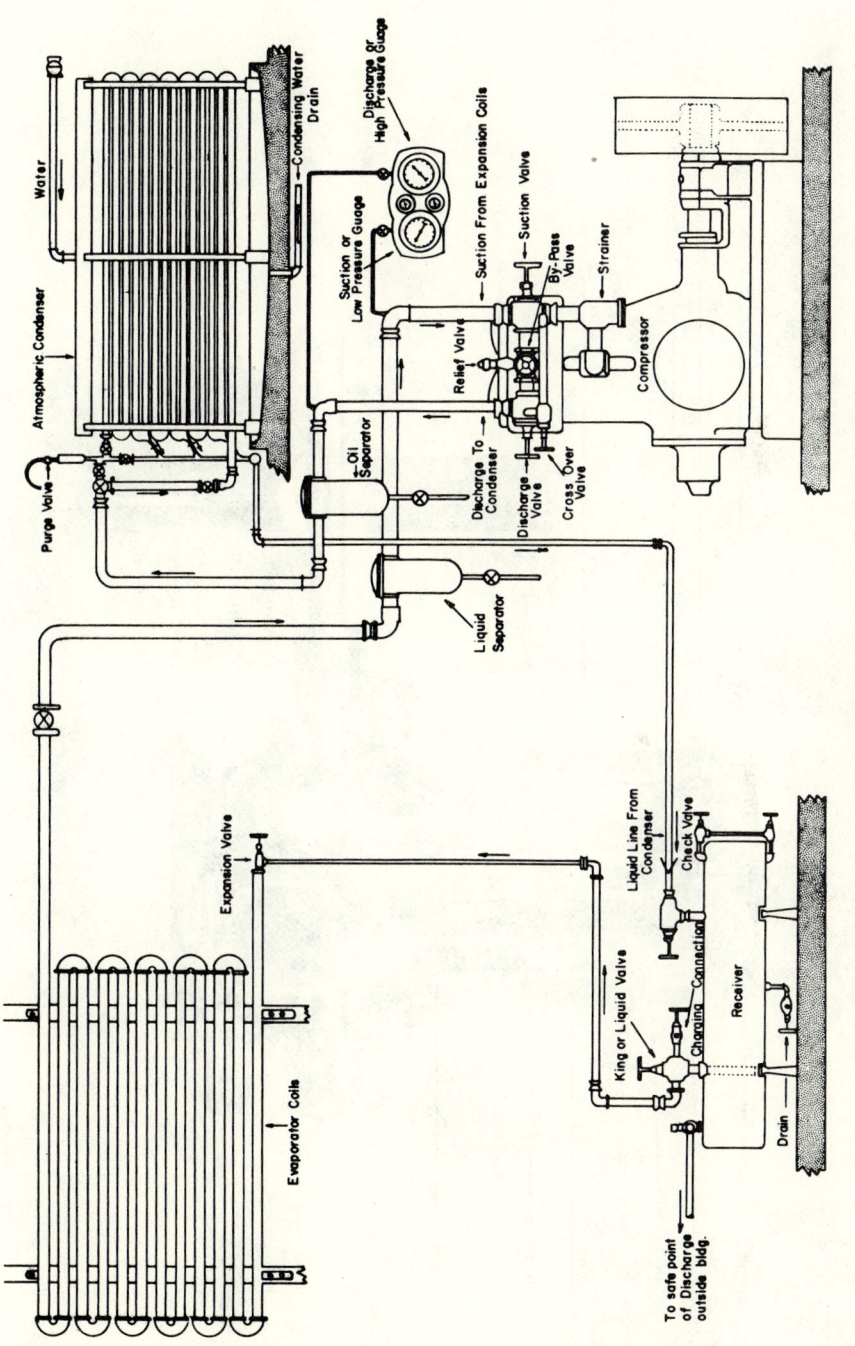

FIGURE 44.1. Typical ammonia refrigeration plant. (Factory Mutual System)

Chapter 44

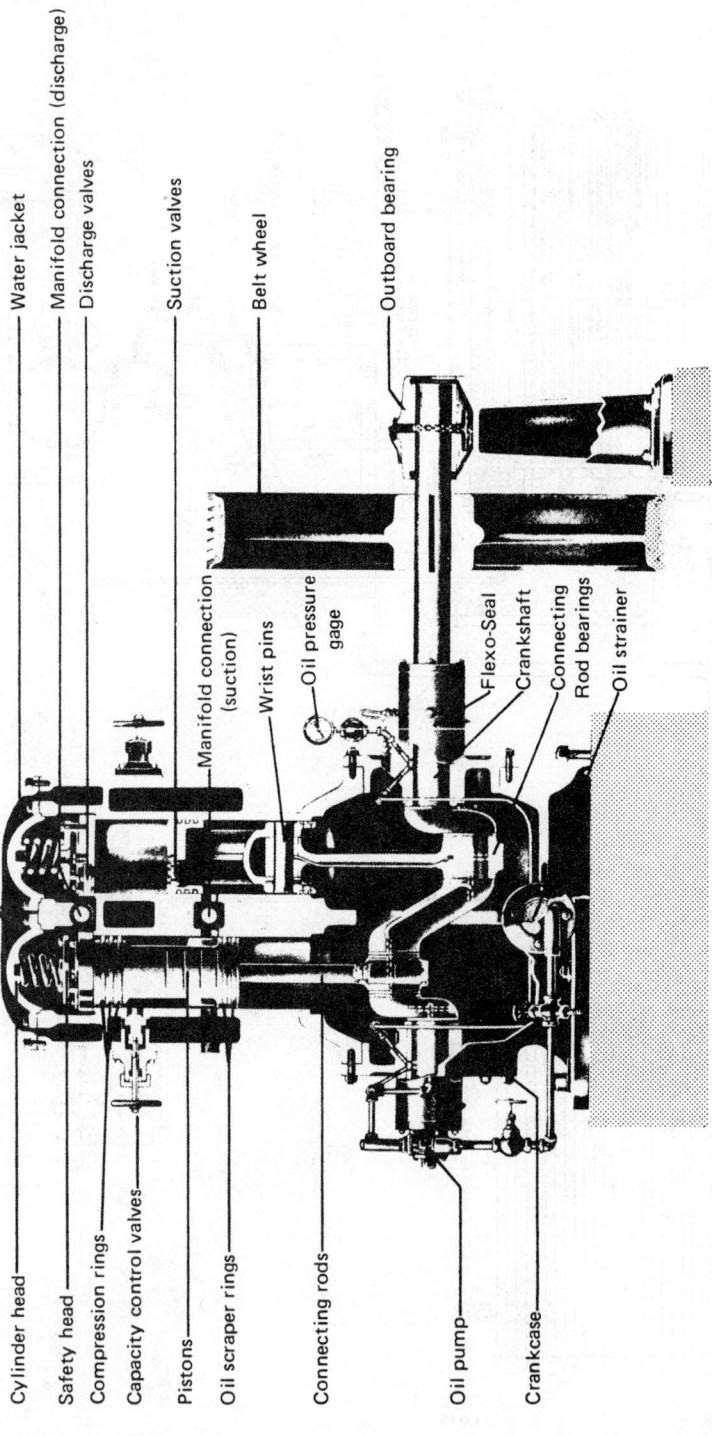

FIGURE 44.2. *Reciprocating compressor, vertical single acting. (Factory Mutual System)*

REFRIGERATION SYSTEMS

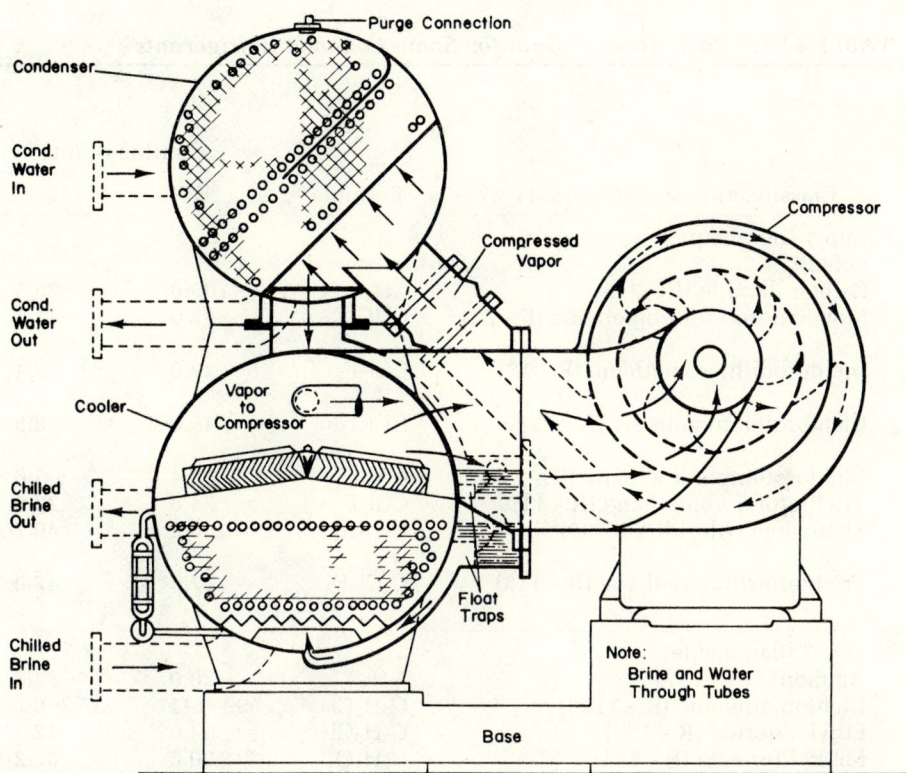

FIGURE 44.3. Cross-section of centrifugal refrigeration system showing refrigeration cycle. (Factory Mutual System)

organic chemical solutions, such as ethylene glycol, as well as brine, are employed in indirect systems.

CHARACTERISTICS OF REFRIGERANT GASES

One does not usually associate low temperatures with fire hazards, but refrigeration systems are of concern for two reasons. Some of the products used as refrigerants are flammable, and some are explosive in critical combination with air. Others are toxic and, if they escape during a fire, can cause injury or death or interfere with fire fighting operations.

Table 44.1 shows refrigerant gases divided into three classes according to their flammability. Those in Group 1 are either nonflammable or very weakly so. They also have rather low levels of toxicity and thus present very little hazard. However, all gases except oxygen present a hazard to life if they displace breathing air.

TABLE 44.1. *Basic Hazard Data for Some Common Refrigerants*

Classification of Refrigerants	Formula	Boiling point °F	Boiling point °C
Group 1 (nonflammable, except as noted):			
Carbon dioxide (R - 744)	CO_2	−109.0	−78.3
Monochlorodifluoromethane (R - 22)	$CHClF_2$	−42.0	−41.1
Dichlorodifluoromethane (R - 12)	CCl_2F_2	−22.0	−29.4
Dichlorofluoromethane (R - 21)	$CHCl_2F$	48.0	8.8
Dichlorotetrafluoroethane (R - 114)	$C_2Cl_2F_4$	38.4	3.6
Trichlorofluoromethane (R - 11)	CCl_3F	74.8	23.8
Methylene chloride (R - 30)	CH_2Cl_2	105.2	40.7
Trichlorotrifluoroethane (R - 113)	$C_2Cl_3F_3$	117.6	47.6
Group 2 (flammable):			
Ammonia (R - 717)	NH_3	−28.0	−33.3
Dichloroethylene (R - 1130)	$C_2H_2Cl_2$	99–141	37.2–60.6
Ethyl chloride (R - 160)	C_2H_3Cl	54.0	12.2
Methyl formate (R - 611)	$C_2H_4O_2$	90.0	32.2
Group 3 (highly flammable):			
Butane (R - 600)	C_4H_{10}	31.0	−0.6
Ethane (R - 170)	C_2H_6	−128.0	−88.9
Propane (R - 240)	C_3H_8	−44.0	−42.2
Ethylene (R - 1150)	C_2H_4	−154.8	−106.0

The refrigerants in the second group are both flammable and toxic. If they escape from the refrigeration system, they can reach critical concentrations in the surrounding air and produce combustion explosions, endangering lives and hindering fire fighting. They have relatively narrow ranges of flammability, and their upper limits may be reached before combustion is initiated. While Group 2 refrigerants in tightly closed refrigerated areas have been involved in combustion explosions, their toxicity is the greater problem. Ammonia systems require careful attention to the compatibility of the materials used. Moist ammonia reacts vigorously with copper and zinc and many brass and bronze alloys, rapidly eroding them.

Though the refrigerants in Group 3 do not present the same toxicity hazards as those in Group 2, they present much greater fire hazards. They generally have lower flammability limits and a wider range. Butane, ethane, and propane are all components of liquefied natural gas

TABLE 44.1. (continued)

Calculated density gas (air = 1)	Auto-ignition Temperature		Flammable Limits, Percent by Volume in Air		Toxicity
	°F	°C	Lower	Upper	
1.52		632			Slightly toxic
2.98	1,170		Very weakly flammable		Slightly toxic
4.17					Nontoxic for ordinary exposure
3.55	1,026	552	Very weakly flammable		Slightly toxic
5.89					Nontoxic
4.79					Slightly toxic
2.93	1,139	615	Very weakly flammable		Slightly toxic
6.46	1,256	680	Very weakly flammable		Slightly toxic
0.59	1,204	651	16.0	25.0	Toxic
3.35	856	458	9.7	12.8	Moderately toxic
2.22	966	519	3.8	15.4	Moderately toxic
2.07	869	465	5.0	23.0	Toxic
2.01	900–1,000	482–538	1.55	8.6	Slightly toxic
1.04	959	515	3.0	12.5	Slightly toxic
1.52	920–1,120	493–604	2.2	9.6	Slightly toxic
0.98	914	490	2.7	36.0	Slightly toxic

(LNG) and liquefied petroleum gas (LPG), and they present relatively the same fire and explosion hazards as do LNG and LPG. Ethylene, which is also a component of LPG, has a wide flammability range and a high burning velocity.

HAZARDS OF REFRIGERATION

Since most of the common refrigerants are at least slightly toxic, a leak in the system will be a hazard to health and life safety. The danger to people and products is much greater in a direct system than in an indirect system. However, the degree of hazard is dependent upon how completely the direct portion of the system is isolated from the space or product being cooled.

Leaks can occur for many different reasons. Pipe joints may

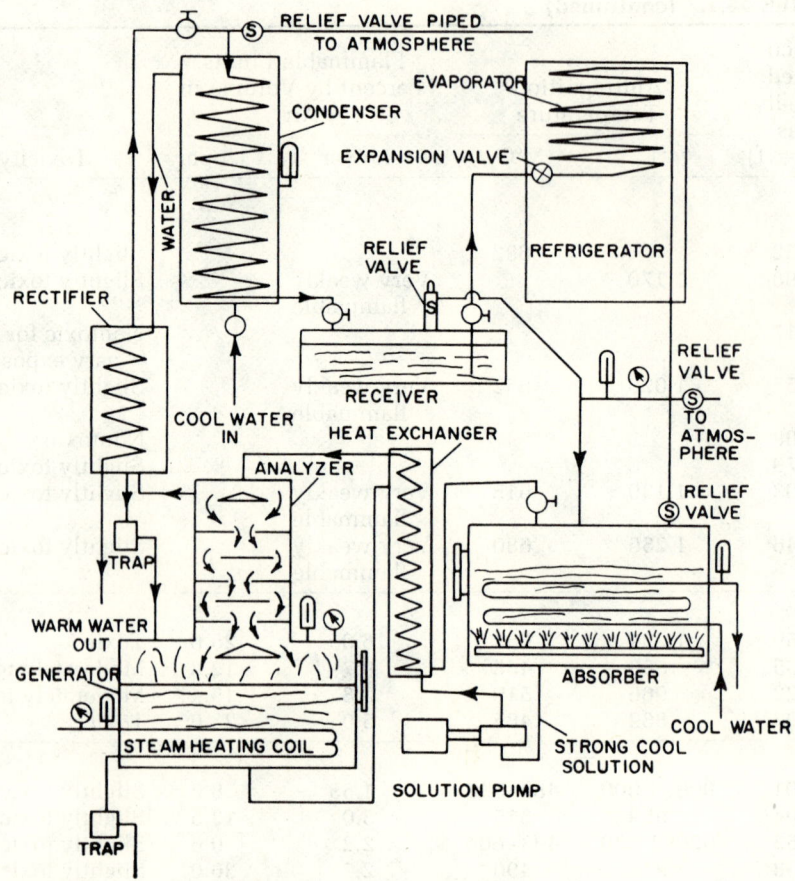

FIGURE 44.4. Heat absorption type refrigeration system. (National Safety Council)

weaken and fail because of excessive vibration. Dissimilar metals in piping and valves may set up an electrolytic action that destroys the integrity of the junction. Incompatibility between the metal and refrigerant may cause corrosion and failure. Mechanical impurities in the refrigerant can become lodged in valves, causing malfunctions. In compression systems, the introduction of a noncompressible liquid into the suction side of the compressor can cause the compressor to crack or burst. Faulty valves within the system may permit recompression of the gas to pressures high enough to cause failure.

Refrigerants are normally gases at atmospheric temperatures and pressures. Therefore, they mix readily with air. If they are flammable, there is the danger that they will reach flammable concentrations, ready to be ignited by some source of heat energy (see Figure 44.5).

If the surrounding volume of air is large enough and other condi-

FIGURE 44.5. *Result of an anhydrous ammonia explosion in Houston, Texas on December 11, 1983. (Houston Fire Department)*

tions are favorable, the refrigerant gas may be diluted sufficiently to minimize or eliminate the hazard. However, diffusion, in some cases, may not take place rapidly enough to avoid the hazard. Experience has shown that structures need not be completely filled with a flammable gas-air mixture for a combustion explosion to occur. Most such explosions occur with less than 25 percent of the enclosure occupied by the flammable mixture.

EMERGENCY CONTROLS AND PROCEDURES

Refrigerant gases present two basic hazards — toxicity and flammability — which can create nonfire and fire emergencies, respectively.

Control of the nonfire emergency is accomplished by directing, diluting, and dispersing the gas to prevent it from coming in contact with people. Simultaneously, steps should be taken to stop the flow of gas. Air, water, and steam are practical media for accomplishing dilution and dispersion. Water in the form of a spray from hoses or fixed nozzles is most commonly used for this purpose.

Control of fire emergencies generally consists of reducing the heat of the fire with water and, if possible, shutting off the gas supply. Water should be applied as a spray from hoses or fixed nozzles. Carbon dioxide, dry chemical, and halogenated extinguishing agents are useful in most instances. Ammonia is easily controlled with water, but suitable breathing apparatus and protective clothing should be worn.

With the exception of carbon dioxide, the refrigerants in Group 1 (see Table 44.1) are halogenated hydrocarbons, commonly known as halons. Related halons, most of which are bromine compounds, are used as fire extinguishing agents. Halon refrigerants are usually referred to by numbers (R-11, etc.) as is done in Table 44.1. They have varying characteristics and, therefore, have varying applications.

R-11 and R-4 are low-pressure refrigerants. This means that the pressure in the evaporator is below atmospheric pressure; thus, the pressure in the condenser is relatively low. For cooling water for air conditioning and industrial processing, R-11 is commonly used. For cooling brines to temperatures between 20°F and −20°F (−6 and −28°C), R-114 is usually employed. R-12 is used for very low temperatures down to −120°F (−84°C). Though classed as nonflammable, halon refrigerants can present some hazards. R-12, for example, can react vigorously with aluminum at elevated temperatures of about 1,200°F (−684°C).

These halogenated refrigerants are commonly used in centrifugal refrigeration systems. In such systems, the brine is cooled by passing it through tubes immersed in the refrigerant. The vaporized refrigerant is forced by the compressor into the condenser, where it is liquefied. At this point, an economizer partially cools the compressed refrigerant by partial evaporation, and the vapor goes to the second stage of the compressor. A purge recovery unit takes air or other noncondensable gas out of the refrigerant and discharges it to the atmosphere.

Refrigeration equipment should be installed with careful regard to the many variables that affect life safety and fire hazards. The occupancy should determine whether a direct or indirect system should be installed. Limitations should be imposed upon the quantity of refrigerant available, based upon the concentration that would be attained if the entire quantity were discharged into the area. The quantity, of course, should be determined by the toxicity and flammability of the refrigerant. Materials used in the system should be compatible with the type of refrigerant used, and the system should include the proper safety devices and relief valves. Electrical equipment and wiring should be properly grounded and shielded and all other possible ignition sources eliminated.

BIBLIOGRAPHY

REFERENCES

Factory Mutual Engineering Corporation, "Refrigeration," *Loss Prevention Data 7-13*, October 1975 (with revisions), Factory Mutual System, Norwood, MA.

McKinnon, G. P., ed., *Fire Protection Handbook*, Fifteenth Edition, National Fire Protection Association, Quincy, MA, 1981.

NFPA Codes, Standards, and Recommended Practices

Reference to the following NFPA Codes, Standards, and Recommended Practices will provide further information on the safeguards for refrigeration equipment discussed in this chapter. (See the latest *NFPA Codes and Standards Catalog* for availability of current editions of the following documents.)

NFPA 10, *Portable Fire Extinguishers.*
NFPA 12, *Carbon Dioxide Extinguishing Systems.*
NFPA 12A, *Halon 1301 Fire Extinguishing Systems.*
NFPA 12B, *Halon 1211 Fire Extinguishing Systems.*
NFPA 17, *Dry Chemical Extinguishing Systems.*
NFPA 30, *Flammable and Combustible Liquids Code.*
NFPA 325M, *Fire Hazard Properties of Flammable Liquids, Gases, Volatile Solids.*

Additional Reading

Factory Mutual Engineering Corp., "Refrigerated Warehouses," *Loss Prevention Data*, 8-29, May 1976.
"Fire and Explosion Hazards of Compressors," *Fire Prevention*, (122), 1977, pp. 19-21.
Henry, Martin F., *Flammable and Combustible Liquids Code Handbook*, National Fire Protection Association, Quincy, MA, 1978.
Loader, K., "Fire Protection of Cold Storage Warehouses," *Fire Prevention*, Feb. 1981, pp. 11-15.
McRae, H., "Ammonia Explosion Destroys Ice Cream Plant," *Fire Command*, April 1984, pp. 36-37.
Meacock, M. H., *Refrigeration Processes: A Practical Handbook on the Physical Properties of Refrigerants and their Applications*, Pergamon Press Inc., Elmsford, NY, 1974.
Miller, Rex, *Refrigeration and Air Conditioning Technology*, Bennett Publishing Co., Peoria, IL, 1982.
Number Designation of Refrigerants, ASHRAE 34, 1978, American Society of Heating, Refrigeration and Air Conditioning Engineers, Atlanta, GA.
Safety Code for Mechanical Refrigeration, ASHRAE 15, 1978, American Society of Heating, Refrigeration and Air Conditioning Engineers, Atlanta, GA.
"Sprinkler Protection of Small Cold Stores," *Fire Prevention*, vol. 143, no. 17 (August 1981).
Weiby, P. and K. R. Dickinson, "Monitoring Work Areas for Explosive and Toxic Hazards," *Chemical Engineering*, vol. 83, no. 22 (1976), pp. 139-145.

45

Air Moving Equipment

W. W. Schliestett

The term "air moving equipment," as used in this chapter, includes all blower and exhaust systems for removal and collection of dusts, vapors, and other waste materials. Also included are pneumatic systems for conveying process materials in finely divided form, dust collectors, and electrostatic precipitators.

The safe removal of dangerous materials such as dusts, flammable vapors, and corrosive fumes from industrial processes is vitally important to the health and safety of employees. Also, the protection of the environment from industrial wastes is of increasing public concern. Mechanized blower and exhaust systems (air moving equipment) are indispensable for these purposes.

Air moving systems generally consist of hoods or enclosures connected to duct work and fans which either exhaust directly to outdoors or to cyclone separators, dust collectors, or, in the case of pneumatic conveyors, transfer materials from one industrial process to another.

Air moving equipment, despite its firesafety advantages, can present serious fire and explosion hazards due to the potential of igniting

---

W. W. Schliestett is the President of Air Masters, Inc., Charlotte, NC.

Chapter 45

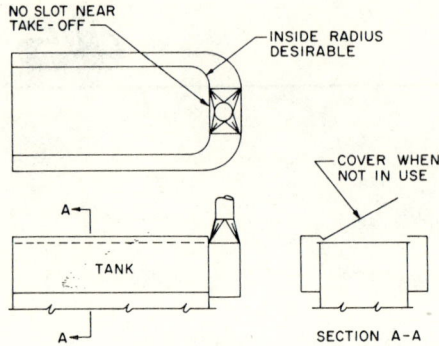

FIGURE 45.1. *Slot exhaust such as used for solvent degreasing tanks. (American Conference of Governmental Industrial Hygienists)*

flammable materials by sources such as sparks from fans or foreign materials in stock, hot bearings, electrical sparks, and spontaneous ignition.

There is also the possibility of fire spreading through ductwork from one area or building to another. Also, violent explosions in flammable vapors or dusts in exhaust systems or collectors have occasionally caused heavy loss of life as well as the destruction of property.

System Design

Fundamentally, air moving equipment is designed to carry away waste material given off from a manufacturing process or operation. This requires an air velocity ("capture velocity") at the place of generation of the material to be exhausted sufficient to overcome the other forces acting on the material. These forces involve the size and shapes of particles, their specific gravity, the gravitational force and inertia given the particles as a result of the process or operation, and outside effects such as air currents. Usual "capture velocities" are in the range of 50 to 2,000 fpm (15 to 610 m/min).

Ideally, equipment producing dusts or vapors should be completely enclosed so that the full benefit of the exhaust system can be utilized. Generally this is not possible and the design of a suitable hood, canopy, or other suction inlet is necessary. On a dip tank, for example, where vapors are to be exhausted, lateral slots around the periphery of the tank or a side hood are used where an overhead hood would interfere with the work (see Figures 45.1 and 45.2). If the process involves vapors that are lighter than air, or if it is a "hot" process and the advantage of a thermal up-draft can be used, a canopy or hood is generally very effective (see Figure 45.3). Grinding operations that can generate fines and dusts need hood arrangements, as shown in Figure 45.4, to capture the fine particles which may be combustible.

Complete removal of the material to be exhausted is the objective, eliminating pockets where vapors may accumulate and preventing deposits of condensed material in and around equipment. Dissimilar ma-

AIR MOVING EQUIPMENT

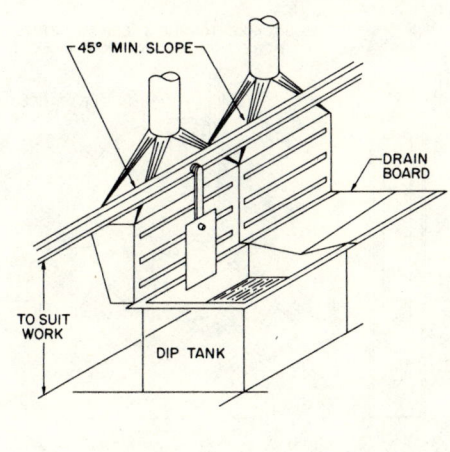

FIGURE 45.2. Side hood such as used for dip tanks. For best results the drainboard should be enclosed as a drying tunnel. (American Conference of Governmental Industrial Hygienists)

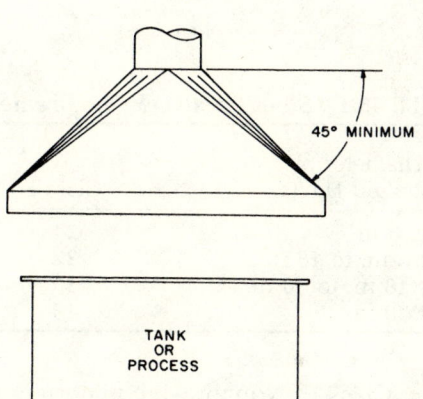

FIGURE 45.3. Canopy hood. (American Conference of Governmental Industrial Hygienists)

terials should not be handled through a single exhaust system where the intermingling or contact of one type of material with another would create a fire or explosion hazard in the duct system.

Ducts for air moving equipment are preferably made of sheet steel or another noncombustible material and should be of sufficient strength to remain rigid and withstand abrasion from the materials being carried. Special materials are required when corrosive materials are handled. Joints are made tight to avoid leakage, and laps are made with the inner lap in the direction of air flow. Fittings are designed to minimize friction losses and to avoid air pockets. Numerous hand holes or access panels are provided in order to facilitate inspection and cleaning.

Materials handled in air moving systems are divided into three classes by NFPA 91, *Blower Exhaust Systems for Dust, Stock, Vapor Removal*. They are:

Class I: Nonabrasive applications such as paint spray, woodworking, pharmaceutical and food products, discharge ducts from dust collectors.

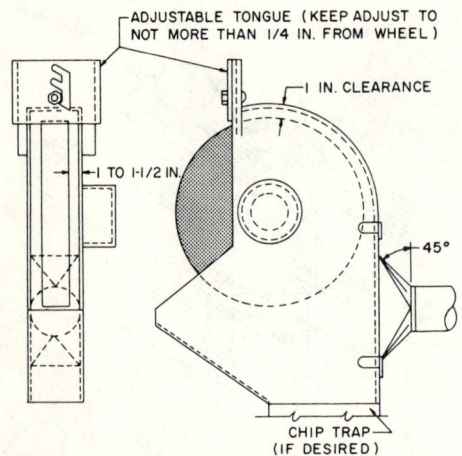

FIGURE 45.4. Grinder wheel hood. (1 in. = 25.4 mm) (American Conference of Governmental Industrial Hygienists)

TABLE 45.1 *Steel Ductwork Requirements*

Diameter of Straight Ducts	U.S. Standard Gage for Steel Duct		
	Class I	Class II	Class III
Up to 8 in.	24	22	20
Over 8 in. to 18 in.	22	20	18
Over 18 in. to 30 in.	20	18	16
Over 30 in.	18	16	14

Class II: Nonabrasive materials in high concentrations (such as for low pressure pneumatic conveying); moderately abrasive materials; and highly abrasive materials in light concentrations. Typical examples are conveying of chemicals and wood dusts; exhaust of foundry shakeouts and sand-handling systems, grain dusts; coal crushing, screening, and grinding; buffing and polishing.

Class III: All highly abrasive materials in moderate to heavy concentrations and moderately abrasive materials in heavy concentrations such as low pressure conveying of tobacco; exhaust systems from sand and grit blasting, abrasive cleaning operations, rock and ore screening, crushing dryers and kilns; fly ash from boiler stacks.

The class of material moved in a system governs the thickness of metal used in the construction of ducts of various sizes (see Table 45.1). For Class II and Class III materials, sheet metal elbows, wyes, and bends are made from metal at least two gages heavier than that used for straight ductwork of the same diameter. *Exception:* for No. 14 gage (4.6 mm) and heavier, all ductwork is of the same gage.

Ducts should run directly from the equipment or process served to the outside. Running ducts through firewalls, piercing floors, or extending them through attics and other concealed spaces is not good practice.

Clearance of ducts from combustible construction or combustible contents depends largely on the material passing through the duct. The diameter of the duct and the fire protection or insulation provided for combustible construction are also governing factors. Details on duct clearances are given in NFPA 91. If temperatures of exhaust gases run higher than 900°F (482°C), the duct should be lined with refractory material.

Design and Selection of Fans

The common types of fans used in blower and exhaust installations are centrifugal and axial flow. There are variations in design in each of these two general classifications (see Figure 45.5). A straight blade centrifugal fan is suitable for exhausting large particles of materials because it will not clog and the blades can withstand considerable abrasion. Propeller fans can move large volumes of air but only against small resistance. Such a fan is not suitable for long ducts where the friction loss may be high.

Consideration should be given to the following general fire protection features of fans: (1) noncombustible construction, (2) remote control to shut down the fan in case of fire, (3) accessibility for maintenance, and (4) structural quality to assure minimum wear and prevent distortion and misalignment. When exhausting flammable solids or vapors, there is a possibility of ignition from a spark when the fan hits the housing due to misalignment. This possibility is minimized by using a fan and fan housing of nonferrous materials.

Fire Extinguishing Systems

The number of fires that originate in and spread through ducts justifies the installation of automatic extinguishing systems in ducts and associated equipment where combustible deposits or accumulations occur, or where the duct itself is combustible. Automatic sprinklers generally furnish reliable and economical protection for air moving equipment, but carbon dioxide, dry chemical, and other types of extinguishing systems also have applications.

Manual Extinguishing Equipment

Portable fire extinguishers of appropriate type or small hose with spray nozzles are helpful where fires may occur in air moving equipment. This is true even where fixed pipe extinguishing systems have been provided.

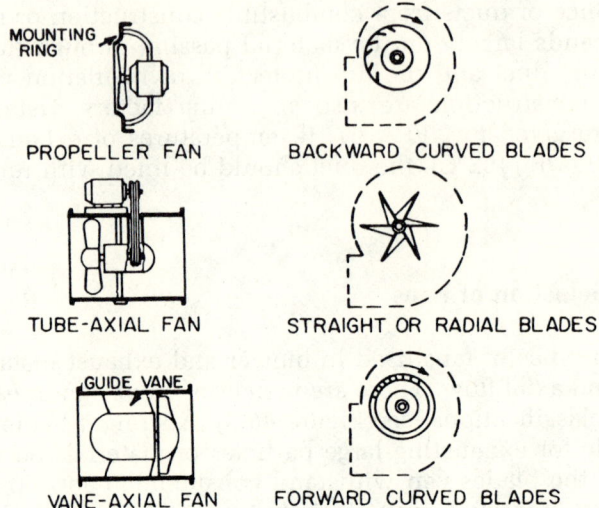

FIGURE 45.5. *Types of fans suitable for use in air moving systems. (American Conference of Governmental Industrial Hygienists)*

Explosion Venting

Explosion venting provides protection that prevents or at least minimizes damage to duct systems and dust collectors for materials that may form an explosive mixture with air. Vents in duct systems should lead to outdoors by the most direct practical route. See Figure 45.6, a suggested relief vent at a bend in an indoor duct.

Explosion Prevention

A continuous flow of inert gas may be used to protect enclosed grinding and dust removal equipment for materials such as pyrophoric metals and certain chemicals where ignition sources are difficult to control.

Explosion suppression systems are helpful for some situations where repeated explosions have occurred or are likely to occur. Details of suppression systems are given in NFPA 69, *Explosion Prevention Systems*.

SPECIAL CONSIDERATIONS FOR SPECIFIC USES

Flammable Vapors

Adequate ventilation is an essential safeguard against flammable liquid fires and explosions. With many flammable liquid processes, ventilation is also necessary to safeguard the health of employees. Usually the amount of ventilation required for safety to health is more than enough for fire and explosion safety. (See Chapter 37 for a complete discussion of Handling and Storage of Flammable and Combustible Liquids, and Chapter 25 for Spray Finishing and Powder Coating.)

It is important that process generated vapors be taken directly from the process to the outdoors. Processes generating flammable vapors are best located in cutoff rooms, along an outside wall to facilitate vapor removal and explosion venting. Hazardous flammable liquid processes should not be located in an upper story, above processes or storage which could be damaged by water used in extinguishing a fire.

Using a single system to exhaust flammable vapors as well as to exhaust particles from spark-producing processes is obviously dangerous. Similarly, drawing different vapors, not individually flammable but hazardous as a mixture, into a single exhaust system is extremely poor practice. For example, if perchloric acid vapors were drawn into a system that was exhausting organic materials, fire or explosion could result because perchloric acid acts as an oxidizing agent at elevated temperatures.

When flammable vapors cannot readily be picked up at a specific source, general ventilation, through a system of suction ducts with inlets to the room, may be employed. It is important that the ventilation system provide sufficient air movement to maintain the vapor concentration well below the lower explosive limit. A general rule is to provide 10,000 cu ft (283.2 m^3) of air (referred to 70°F or 21.1°C) for each gallon of solvent evaporated.

Where heavier-than-air vapors or mixtures are handled, exhaust openings are best located near the floor level. Conversely, for vapors or mixtures lighter than air, openings near the top of the room or enclosure will be more effective.

Provision of automatic fire protection is particularly important where flammable residues can accumulate, for example, in paint spray booths and exhaust ducts, and the exhaust stacks of ovens evaporating flammable solvents. Where automatic sprinklers are installed, sprinklers in horizontal ducts should be spaced no more than 12 ft (3.66 m) apart, with one sprinkler at the inlet side of the exhaust fan. One head should also be installed at the top of each vertical section of duct or stack. For exhaust stacks located outside of buildings, a sprinkler should be provided for each 30 ft (9.1 m) of elevation.

Automatic control is preferred for sprinklers in ducts and stacks. Where sprinklers are subject to freezing, they can be installed on a nonfreeze system or connected to an existing dry-pipe system if one is available. An alternate method is to provide open heads controlled by an automatic water control valve. Sprinklers or nozzles may be covered with thin paper bags to protect them from residues. The bags should be cleaned or replaced as necessary to ensure proper operation of the fusible elements.

Periodic cleaning of booths, fans, and ductwork is extremely important where buildup of flammable deposits can occur.

Chapter 45

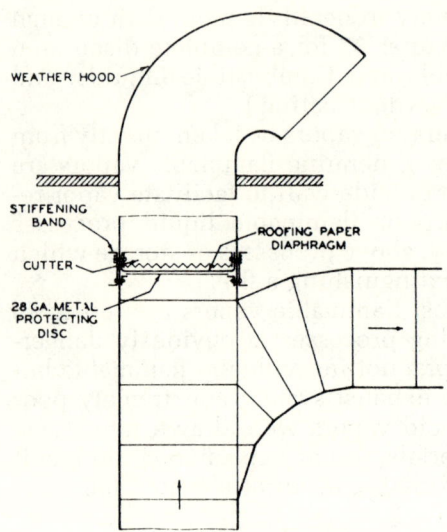

FIGURE 45.6. Suggested form of explosion relief vent suitable for use at duct turns. The thin metal protecting disc shown beneath the rupture diaphragm serves to prevent abrasion of the diaphragm, and upon operation of the device will blow free.

Corrosive Vapors

Frequently, corrosive vapors must be exhausted from a process. The degree of expected corrosion is the governing factor in the construction of the ducts. In some cases a heavier gage metal is sufficient. Other situations may be handled by a protective lining or stainless steel.

Plastic duct systems (Figure 45.7) are used only if the vapors are nonflammable and the plastic has a flame spread rating of 25 or less and a smoke developed rating of 50 or less. Good practice calls for automatic fire protection at the hood, canopy, or intake to plastic duct systems. Large and important plastic duct systems may also need protection within the ducts themselves.

Pneumatic Conveying and Collecting Systems

Pneumatic conveying systems are widely used for transporting dusts and other materials in finely divided form. Materials handled in this manner include grain and flour, textile fibers, wood chips and refuse, metal powders, chemicals, and many others. Most of these materials are combustible and easily ignited. Pneumatic systems for handling them present fire and explosion hazards which require special consideration. (See Chapters 6, 12, 17, and 31 for specific discussions of materials handed by these systems.)

Prevention of ignition sources: Sparks from fans are leading causes of fires and explosions. Systems conveying combustible dusts and fibers should be arranged so that the fan is on the clean air side of the collec-

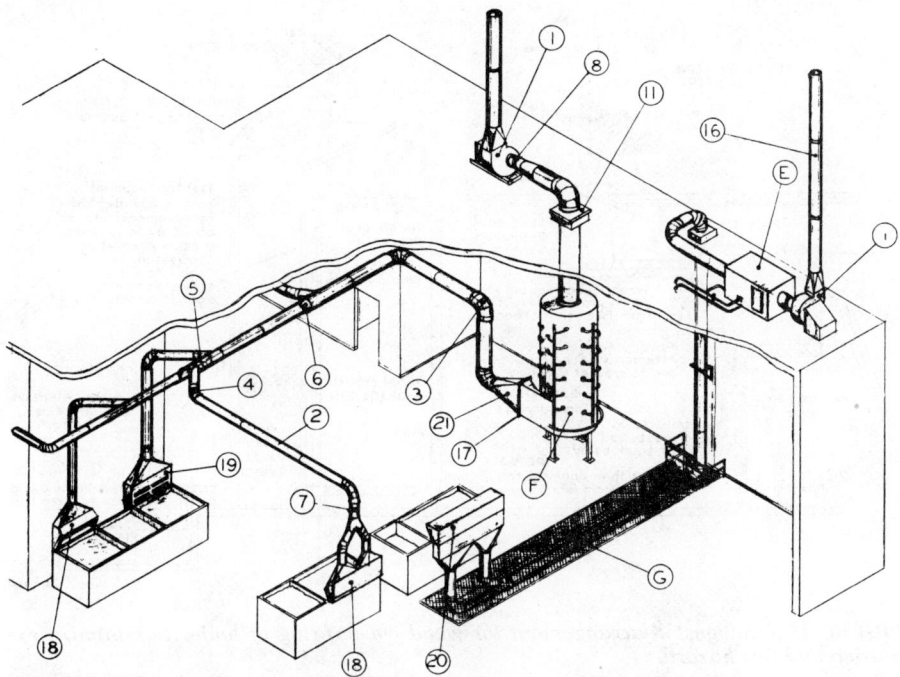

FIGURE 45.7. Exhaust system for one-story building occupied by various types of fume hoods with vertical-type fume scrubber and service trench.

tor. If this is not possible, nonsparking materials can be used for the fan and its housing. Fan bearings and motors are located outside of ducts unless they are designed for use in dusty atmospheres.

Magnetic separators, permanent magnet or electromagnet, are used where there is a possibility of ferrous materials entering the system.

Electrical ignition sources can be minimized by using electrical equipment suitable for Class II, Division 1 or 2 occupancies as outlined in Article 500 of NFPA 70, *National Electrical Code*, or by locating switches, motors, and other spark-producing equipment outside the combustible dust area.

The possibility of static sparks can be reduced by electrically bonding and grounding all parts of the conveying system.

Cutting and welding should not be permitted on ductwork or dust collectors unless collector screens are removed and equipment is thoroughly cleaned. (See Chapter 24 for a comprehensive discussion of welding and cutting.)

Separating and collecting equipment: Systems for collecting combustible dusts should be constructed of steel and located outside the building. Cyclones, cloth screen, and bag-type collectors are commonly used.

Chapter 45

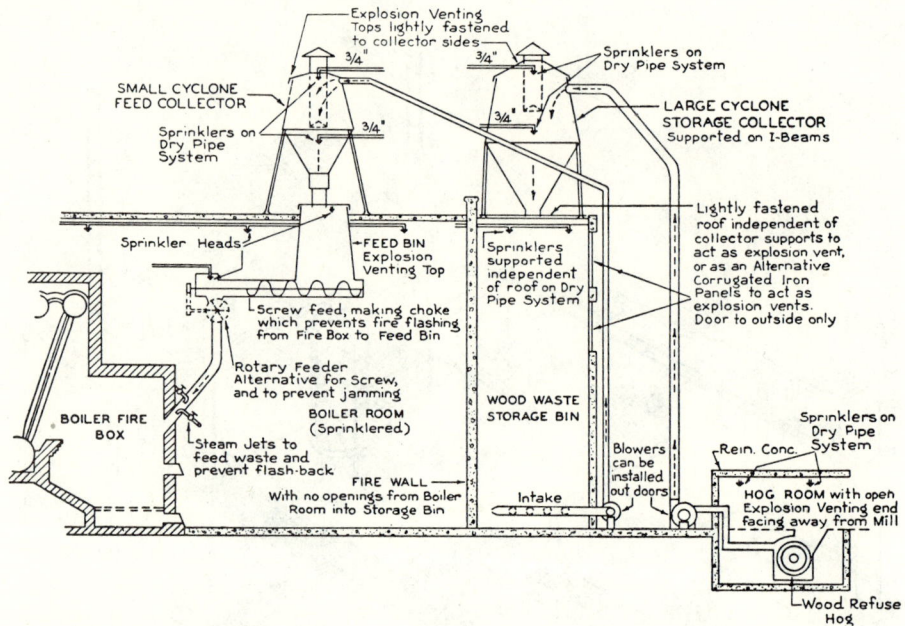

FIGURE 45.8. Suggested arrangement for wood waste firing of boiler to minimize explosion and fire hazard.

The material collected generally drops directly into hoppers, bins, or silos. Wood waste should not be delivered directly from cyclones or separators into the fireboxes of boilers or incinerators. A suggested arrangement for wood waste firing is shown in Figure 45.8.

The provision of explosion venting and the installation of automatic fire extinguishing systems are serious considerations with regard to separating and collecting equipment for combustible dusts. Typical examples of some collecting equipment and how sprinklers can be used in some of them are shown in Figures 45.9, 45.10, and 45.11.

Electrostatic Precipitators for Pollution Control

Electrostatic precipitators are extensively used in pollution control to remove fly ash from flue gases of pulverized coal fired boilers, impurities in blast furnace gas, dust impurities in the manufacture of phosphorus, and asphalt particles from asphalt impregnating processes.

The precipitators impart an electrical charge to particles in a gas or air stream that causes them to adhere to collector plates of an opposite charge. The particles are then removed by shaking or scraping.

Serious fires and explosions have occurred in electrostatic precipitators, usually due to arcing between collector plates igniting combustible residue accumulations. In precipitators using oil seals, leaking oil

AIR MOVING EQUIPMENT

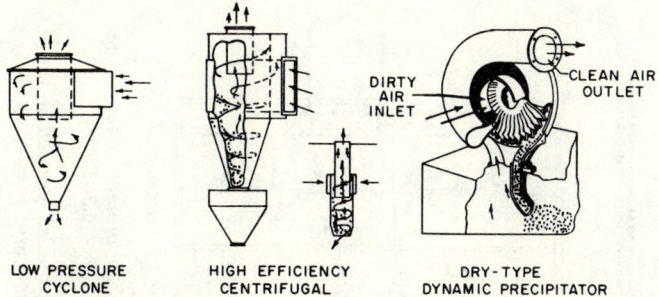

FIGURE 45.9. Examples of typical types of dust collecting equipment. (American Conference of Governmental Industrial Hygienists)

can be a factor. Malfunction of process burners can also result in explosive concentrations of gases which can be ignited by arcing in the precipitator.

Prevention of precipitator fires and explosions is largely a matter of good maintenance, particularly in the prevention of excessive buildup of particles on the collector plates.

Ventilation of Kitchen Cooking Equipment

Exhaust systems for restaurant cooking equipment, as found in industrial plant cafeterias, present a troublesome problem due to the condensation of grease in the interior of the ducts. These accumulations may be ignited by sparks from the cooking range itself, or more often by a small fire in overheated cooking oil or fat on the range surface. Details of a typical kitchen range exhaust system are shown in Figure 45.12.

The following is a summary of the factors to be considered in designing a good duct system for commercial-type cooking equipment.

1. Design the system to minimize grease accumulations, with a minimum air flow of 1,500 fpm (457.2 m/min) through any duct.
2. Arrange ducts with ample clearance from combustibles.
3. Use ducts not lighter than No. 16 Manufacturer's Standard Gage steel or No. 18 Manufacturer's Standard Gage stainless steel. Make seams and joints liquid-tight with a continuous external weld.
4. Have no connections with any other ventilating or exhaust system.
5. Lead ducts directly outside the building.
6. Provide openings for inspection and cleaning.
7. Provide grease filters or other grease removal devices.

Automatic fire extinguishing systems furnish desirable protection for kitchen exhaust hoods and ducts. Detailed recommendations for safe-

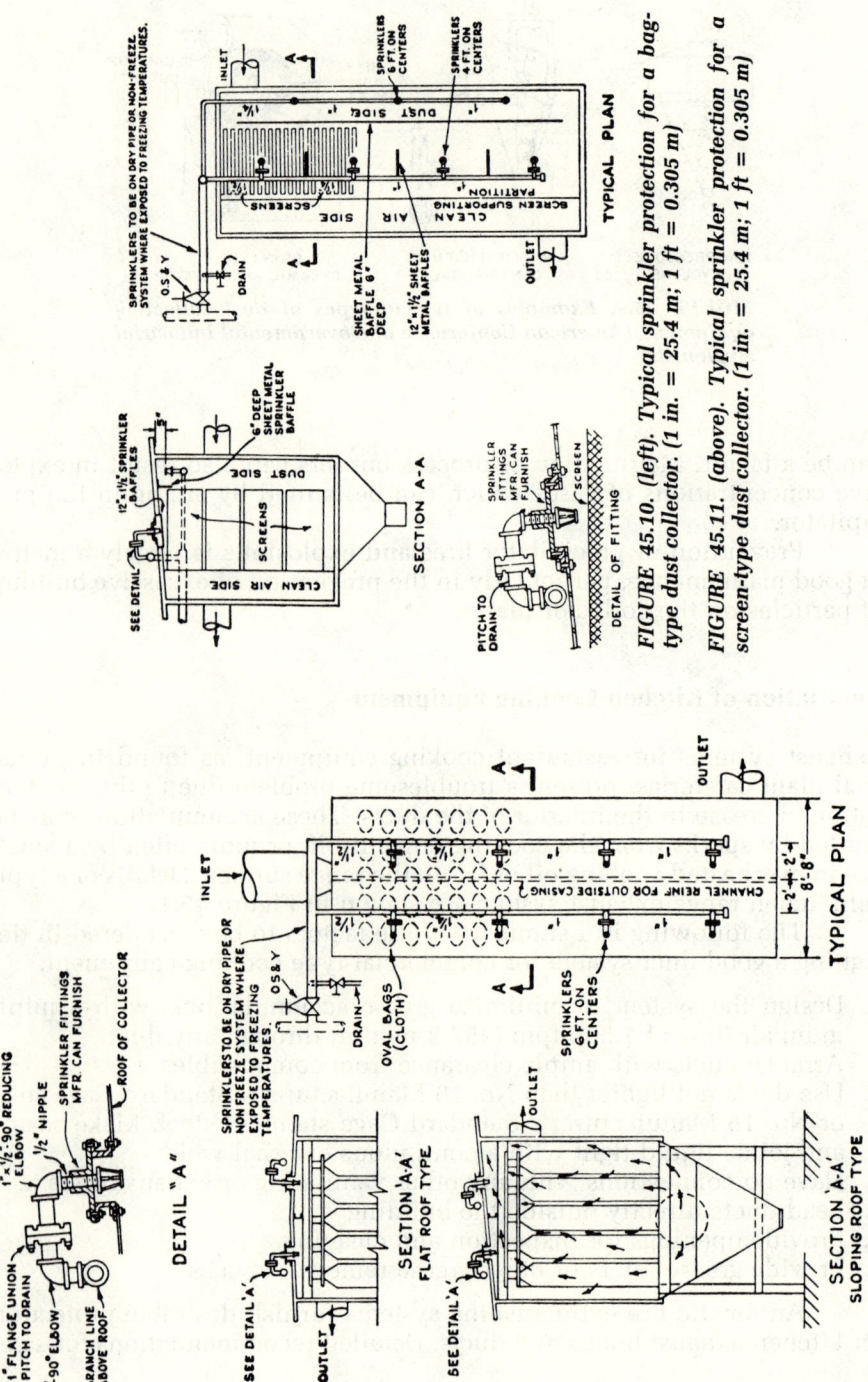

FIGURE 45.10. (left). Typical sprinkler protection for a bag-type dust collector. (1 in. = 25.4 m; 1 ft = 0.305 m)

FIGURE 45.11. (above). Typical sprinkler protection for a screen-type dust collector. (1 in. = 25.4 m; 1 ft = 0.305 m)

AIR MOVING EQUIPMENT

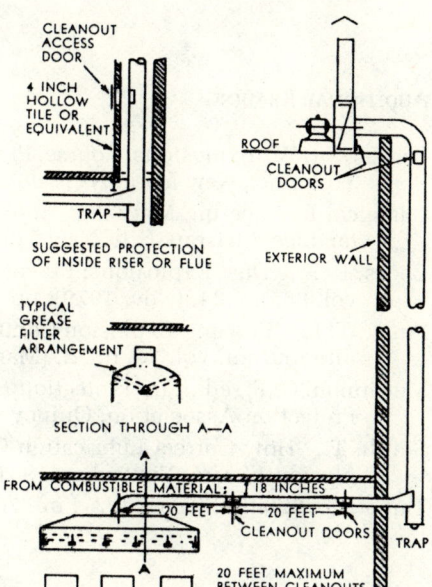

FIGURE 45.12. *A typical kitchen range exhaust system arrangement, showing vertical riser outside the building. When it is necessary to locate the riser inside the building, it should be enclosed in a masonry shaft. (1 in. = 25.4 mm)*

guarding commercial cooking equipment are contained in NFPA 96, *Vapor Removal from Commercial Cooking Equipment*.

BIBLIOGRAPHY

NFPA CODES, STANDARDS, AND RECOMMENDED PRACTICES

Reference to the following NFPA Codes, Standards, and Recommended Practices will provide further information on the safeguards for air moving equipment discussed in this chapter.(See the latest *NFPA Codes and Standards Catalog* for availability of current editions of the following documents.)

NFPA 13, *Installation of Sprinkler Systems*.
NFPA 68, *Explosion Venting Guide*.
NFPA 69, *Explosion Prevention Systems*.
NFPA 70, *National Electrical Code*.
NFPA 77, *Recommended Practice on Static Electricity*.
NFPA 91, *Blower and Exhaust Systems for Dust, Stock and Vapor Removal or Conveying*.
NFPA 96, *Vapor Removal of Smoke and Grease Laden Vapors from Commercial Cooking Equipment*.
NFPA 650, *Pneumatic Conveying Systems for Handling Combustible Materials*.

Additional Reading

Bartknecht, W., *Explosions: Course, Prevention, Protection*, Springer-Verlag, New York, Inc., New York, NY, 1981.

Chemical Engineering Magazine, *Safe and Efficient Plant Operation and Maintenance*, McGraw-Hill, New York, NY, 1980.

Cocks, R. E., "Dust Explosions: Prevention and Control," *Chemical Engineering*, vol. 86, no. 24, (Nov. 1979), pp. 94-101.

Frank, T. E., "Fire and Explosion Control in Bag Filter Dust Collection Systems," *Fire Journal*, vol. 75, no. 2, (March 1981), pp. 75-80, 94.

McKinnon, G. P., ed., *Fire Protection Handbook*, Fifteenth Edition, National Fire Protection Association, Quincy, MA, 1981.

Smith, T., "How Correct Lubrication Can Minimize Risk of Air Compressor Explosions," *Fire*, November, 1981, p. 341.

Talbot, G., "Static Electricity," *Fire*, 70(871), 1978, pp. 397-398.

46

Materials Handling Systems

John K. Bouchard

Almost everything which is manufactured has to be moved from place to place within the plant as the material progresses from raw stock to finished product to storage warehouse. Efficient handling of materials is therefore essential, and highly specialized equipment has been developed for the purpose.

Cranes, industrial trucks, and mechanical stock conveyors of various types are among the most widely used materials handling equipment. They can introduce serious fire and explosion hazards unless care is given to their selection, installation, protection, operation, and maintenance. This chapter describes the basic safeguards for industrial trucks and mechanical conveyors, including protection of conveyor openings. Safeguarding cranes against damage by fire and windstorms is also discussed. Pneumatic conveyors are discussed in Chapter 45, "Air Moving Equipment."

John K. Bouchard is Assistant Division Director, General Engineering, on the staff of the NFPA.

Chapter 46

INDUSTRIAL TRUCKS

Industrial trucks are available in many special designs to suit the type of load to be handled. Lift trucks of the fork or squeeze clamp types are the most common. They may be propelled by electric motors, or by gasoline, diesel, or LP-Gas engines. Use of industrial trucks introduces the dangers of fire, explosion, water damage, and mechanical damage. They should be selected and operated in accordance with the hazards of the locations where they are to be operated. NFPA 505, *Powered Industrial Trucks*, lists thirteen different designations for industrial trucks and tractors. In summary they are:

Electric-Powered Trucks

Four types of electric-powered trucks are available. Each type is specially designed for use in locations with hazards ranging from ordinary to severe.

Type E: These trucks have the minimum necessary safeguards and are for use in ordinary hazard areas.

Type ES: These trucks have additional safeguards to prevent emission of sparks from the electrical system and to limit surface temperatures. They are recommended for areas where easily ignitible fibers are stored or handled (except in process of manufacturing).

Type EE: These trucks have the electric motor and all other electrical equipment completely enclosed and are recommended for use in hazardous locations other than those that require Type EX.

Type EX: These trucks are of explosion-proof-type (Class 1, Group D) or dust-tight-type (Class II, Group D) construction and are recommended for areas where there are likely to be explosive mixtures of flammable vapors or combustible dusts during normal operations.

Gasoline-Powered Trucks

Two types of gasoline-powered trucks are available.

Type G: These trucks have the minimum necessary safeguards and are recommended for use in areas of light fire hazard.

Type GS: These trucks have additional safeguards in the electrical, fuel, and exhaust systems. They are recommended for occupancies where there are readily ignited combustible materials.

Diesel-Powered Trucks

Three types of diesel-powered trucks are available.

Type D: This type is considered comparable in hazard to the Type G gasoline-powered truck.

Type DS: This type is considered comparable in hazard to the Type GS gasoline-powered truck.

Type DY: These trucks are equipped with additional safeguards which make them less hazardous than Type DS gasoline-powered trucks. Surface and exhaust gas temperatures are limited to a maximum of 325°F (162.8°C), there is no electrical system, and other safeguards are provided to minimize the fire hazard normally associated with internal combustion engines.

LP-Gas-Powered Trucks

LP-Gas-powered trucks present the same general hazards as gasoline-powered trucks plus the additional hazards associated with a combustible gas stored at high pressure.

Two types of LP-Gas-powered trucks are available: *Type LP* and *Type LPS*. They are considered comparable in fire hazard to Types G and GS gasoline-powered trucks, respectively.

Dual-Fuel Trucks

Two types of dual-fuel trucks are available. They are equipped to be operated on either gasoline or LP-Gas, and they are designated either *Type G/LP*, units comparable in fire hazard to Types G and LP; or *Type GS/LPS*, units that require the same safeguards against the hazards of exhaust, fuel, and electrical systems as do Types GS and LPS.

Operation and Maintenance

An axiom for the safe use of industrial trucks, particularly in hazardous areas, is using only trucks approved for that type of service. A system of markings has been developed to identify the different types of approved trucks and the areas where they may be used. Figure 46.1 illustrates the uniform system of markings recommended by the NFPA. The markings have distinctive shapes and lettering. The appropriate marking is affixed to both sides of the truck so that its type is easily identified.

As shown in Figure 46.2, markers of corresponding shape are posted at the entrances to hazardous areas. Thus, when a truck approaches a hazardous area, the operator or a nearby supervisor can make

Chapter 46

FIGURE 46.1. Markers used to identify the various types of approved industrial trucks. The markers for LPS, GS, DS, ES, and GS/LPS are 4 in. (102 mm) square. The width of others is 5 in. (127 mm).

FIGURE 46.2. Building markers for posting at entrances to hazardous areas. The width of each is 11 in. (280 mm). Note that the shape of each marker corresponds to the shape of the truck markers in Figure 46.1.

a quick check to ascertain whether the truck should be driven into the area by comparing markings.

Table 46.1 summarizes the various hazardous locations that may be encountered (as defined by Article 500 of NFPA 70, *National Electrical Code*), gives some representative occupancies where the various locations will be found, and identifies the types of trucks permitted in each location.

Use of trucks should be restricted to trained, competent personnel. Careless operation can cause severe damage in many forms, including fire, explosion, sprinkler pipe breakage, and building damage.

The possibility of breakage to sprinkler piping can be reduced by maintaining a clearance of at least 3 ft (0.91 m) below sprinklers and by providing substantial guards at exposed risers and connections near passageways. A good source for safe operating practices is the *American National Standard Safety Code for Powered Industrial Trucks*, ANSI B56.1-1969.

The greatest potential fire source for gasoline-, diesel-, and LP-Gas-powered trucks are fuel leaks or spills which are ignited by the heat of the engine, the ignition system, or other electrical equipment. This danger is reduced somewhat for diesel trucks because of the higher flash point of diesel fuel. However, it increases with LP-Gas trucks which have the additional hazard of combustible gas under high pressure. Leaking LP-Gas, which is heavier than air, tends to gravitate toward low spots or pits.

Electrically-powered trucks have experienced comparatively few fires. Nevertheless, electrical short circuits, hot resistors, and exploding batteries have been known to be ignition sources.

Recharging and Refueling

Refueling and battery recharging operations should be performed in specified, well-ventilated areas (outdoors where practicable) away from manufacturing and storage areas.

Trucks using gasoline or diesel fuel should be refueled only from approved dispensing pumps in safe locations away from ignition sources. Many fires have occurred from spills during emergency refueling in manufacturing or storage areas. See Chapter 39, "Liquefied Petroleum Gases at Industrial Plants," for information on the hazards and safeguards associated with the fueling of LP-Gas trucks.

Maintenance

It is important to establish a system of regularly scheduled preventive maintenance. This system should include periodic checks of fuel lines, carburetor, fuel tanks, lubrication, electrical equipment, and safety devices. A special location away from manufacturing and storage areas should be provided for servicing and repairing trucks.

CONVEYING SYSTEMS

Belt Conveyors

Belt conveyors provide an efficient means of transporting bulk and packaged materials. While belt conveyors have become indispensable to the operation of modern industry, they are vulnerable to fire. Belt conveyors present two principal hazards — that of the material being carried, and that of the belt itself. As is the case with conveyors of any type, there is the danger of transmitting a fire from one building or area to another. Every year, a number of serious conveyor fires cause large losses to prop-

TABLE 46.1. Recommended Types of Trucks for Various Occupancies (Factory Mutual)

Location	Typical Occupancies	Types of Trucks[b,c] Approved and Listed
Indoor or outdoor locations containing materials of ordinary fire hazard	Grocery warehouse Cloth storage Paper manufacturing and working Textile processes except opening, blending, bale storage, and other Class III locations Bakery Leather tanning Foundries and forge shops Sheet-metal working Machine-tool occupancies	Electrical—Type E Gasoline—Type G Diesel—Type D LP-Gas—Type LP Dual-fuel—Type G/LP
Class I, Division 1,[a] Locations in which explosive concentrations of flammable gases or vapors may exist under normal operating conditions or where accidental release of hazardous concentrations of such materials may occur simultaneously with failure of electrical equipment	Few areas in this division in which trucks would be used	Electrical—Type EX[e] Gasoline, diesel, LP-Gas, and dual-fuel—not recommended for this service
Class I, Division 2,[a] Locations in which flammable liquids or gases are handled in closed systems or containers from which they can escape only by accident or locations in which hazardous concentrations are normally prevented by positive mechanical ventilation	Paint mixing, spraying, or dipping Storage of flammable gases in cylinders Storage of flammable liquids in drums or cans Solvent recovery Chemical processes using flammable liquids Paper and cloth coating using flammable solvents in closed equipment Rubber-cement mixing	Electrical—Type EE Diesel—Type DY Gasoline, diesel Types D & DS, LP-Gas, and dual-fuel—not recommended for this service

MATERIALS HANDLING SYSTEMS

Class/Division	Examples	Type of Truck
Class II, Division 1,[a] Locations in which explosive mixtures of combustible dusts may be present in the air under normal operating conditions, or where mechanical failure of equipment might cause such mixtures to be produced simultaneously with arcing or sparking of electrical equipment or in which electrically conductive dusts may be present	Grain processing Starch processing Starch molding (candy plants) Wood-flour processing	Electrical—Type EX preferred Type EE[d] Diesel—Type DY Gasoline, diesel Types D & DS, LP-Gas, and dual-fuel—not recommended for this service
Class II, Division 2,[a] Locations in which explosive mixtures of combustible dusts are not normally present or likely to be thrown into suspension through the normal operation of equipment but where deposits of such dust may interfere with the dissipation of heat from electrical equipment or where such deposits may be ignited by arc or sparks from electrical equipment	Storage and handling of grain, starch, or wood flour in bags or other closed containers Grinding of plastic molding compounds in tight systems Feed mills with tightly enclosed equipment	Electrical—Type EE preferred, Type ES Gasoline—Type GS[d] Diesel—Type DS[d] LP-Gas—Type LPS[d] Dual-Fuel—Type GS/LPS[d]
Class III, Division 1,[a] Locations in which easily ignitable fibers or materials producing combustible flyings are handled, manufactured, or used	Opening, blending, or carding of cotton or cotton mixtures Cotton gins Sawing, shaping, or sanding areas in wood-working plants Preliminary processes in cordage plants	Electrical—Type EE preferred, Type ES[d] Diesel—Type DY Gasoline, diesel Types D & DS, LP-Gas, dual-fuel not recommended for this service
Class III, Division 2,[a] Locations in which easily ignitable fibers are stored or handled (except in process of manufacture)	Storage of textile and cordage fibers Storage of excelsior, kapok, or Spanish moss	Electrical—Type ES Gasoline—Type GS[d] Diesel—Type DS[d] LP-Gas—Type LPS[d] Dual-fuel—Type GS/LPS[d]

[a] Hazardous location as classified in Art. 500 National Electrical Code and Sec. 32 Canadian Electrical Code. See Data Sheet 5-1.
[b] Type G (gasoline), Type D (diesel), Type LP (LP-Gas), and Type G/LP (gasoline and LP-Gas) trucks are considered to have comparable fire hazard.
[c] Types GS (gasoline), Type DS (diesel), Type LPS (LP-Gas), and Type GS/LPS (gasoline and LP-Gas) trucks are considered to have comparable fire hazard.
[d] Acceptable if kept clean and well maintained.
[e] Electrical truck Type EX is presently listed by Underwriters' Laboratories only.

erty and serious interruptions to production. The following examples are typical.

At a Canadian mine, a rubber belt carrying iron ore was ignited by a spark from a welding operation. About 500 ft (152 m) of belting, conveyor motors, and the mostly noncombustible enclosure was damaged with loss of $260,000.

At an Ohio manufacturing plant, spontaneous ignition was the probable cause of a coal fire on a conveyor belt in a 700-ft (213 m) long inclined metal enclosure. The belt was entirely destroyed and the enclosure seriously damaged. The loss was $100,000.

At a California grain elevator, smoldering fire in grain dust set fire to a belt conveyor used to transfer grain from silos to ships. About 600 ft (183 m) of belting, conveyor machinery, and the unsprinklered enclosure was damaged. The loss was $60,000.

Conveyor belts are generally constructed with built-up plies of rubber or synthetic plastic materials laminated to a woven fabric base, which may be made of cotton, rayon or nylon. Some are made of fire retardant materials which make them difficult to ignite, but tests conducted by the Factory Mutual Research Corporation with conveyor belting that met the fire retardancy specifications of the U.S. Bureau of Mines showed that, once ignited, the belt burned completely and released heat rapidly.

The material conveyed by the belt frequently adds to the fire hazard. Combustible bulk materials include wood chips, wood waste, coal, sulfur, and grain. Materials in cardboard cartons or other combustible packaging moving on conveyors can contribute to fires, even when the material within the packaging is noncombustible.

Fire Causes in Conveyors

Friction between the conveyor belt and a roller or other object is another common cause of fire. Conveyors should be inspected regularly to detect belt slippage and/or defective rollers. All moving parts should be lubricated according to a definite schedule.

Careless cutting or welding operations on the conveyor or its housing is another leading fire cause. Hot globules of molten metal could ignite combustibles on the belt, accumulations of oil soaked debris around it, or the belt itself. If cutting or welding must be done, suitable precautions should always be taken. These include removing accumulations of combustibles, covering the belt with a flame resistant tarpaulin, and having a fire watcher stand by with an extinguisher or small hose.

Discharging excessively hot materials from kilns, ovens, or furnaces onto the belt is a fire hazard at some plants. Heat sensing and alarm devices help to prevent such fires. Electrical short circuits, spontaneous ignition, smoking, and incendiarism are other causes of conveyor fires.

Fire Protection

Loss experience shows that automatic sprinkler or water spray protection is generally needed for important belt conveyors, particularly if they are enclosed. Controls should be provided to shut down the conveyor when sprinklers operate.

    Hose stream protection to reach all portions of the conveyor is also needed. This protection can be provided by 1½-in. (38.1 mm) hose connections to sprinkler lines, or from hydrants.

Miscellaneous Mechanical Conveyors

Chain conveyors equipped with hooks and roller conveyors are very commonly used in assembly lines for various products.

    Screw conveyors, pan, and bucket conveyors have specialized uses for handling hot or molten materials. Belt conveyors are not generally suitable for handling materials over 150°F (65.5°C).

Bucket elevators: These elevators are used in nearly all bulk processing plants to convey loads vertically, and they are susceptible to the same fire hazards as other mechanical conveyors. The same precautions are taken for temperature, dust control, protection of openings, and elimination of friction and overheating. Elevators are best enclosed in substantial dust-tight casings, preferably of noncombustible construction, extending, without reduction in size, through the roof and fitted with light, weatherproof covers designed to lift readily and relieve explosion pressure within.

    Preferably, bucket elevators should be installed close to an outside wall of the building. Short, direct vents through the wall should be provided at 20-ft (6.1 m) intervals on tall elevator legs. Dust-tight doors should be provided for access to head and boot pulleys. There should be ample clearance around elevator boots for cleaning and oiling. Elevators should be safeguarded against overheating or choking by automatic releases actuated by overloading or reduced operating speed.

Protection of Conveyor Openings

It is undesirable from a fire hazard standpoint to run a conveyor through a fire wall, or, for that matter, through any wall or floor, because the opening provides a means of transmitting fire from one building or area to another. Alternative arrangements can sometimes be made to prevent such openings, particularly in major fire walls. In one-story buildings, for example, it is sometimes possible to run the conveyor through the roof, over the wall, and down on the other side in an inverted V-housing arranged to vent fire to the atmosphere (Figure 46.3).

Fire doors: When conveyors must pass through fire walls, the openings

Chapter 46

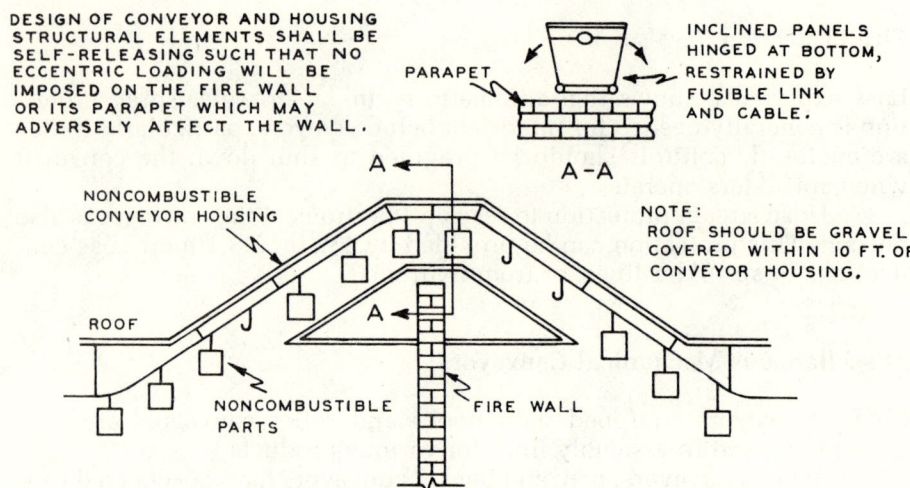

FIGURE 46.3. A conveyor carried over a fire wall.

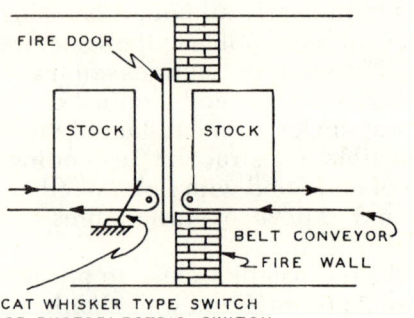

FIGURE 46.4. Protection of openings where a belt conveyor can be interrupted. an electromagnetic fire door release interlocks with switch which stops the conveyor with properr spacing of stock to prevent obstruction to door closer.

should be protected by an arrangement of fire doors. Careful design of controls is needed to stop the conveyor at a position where stock will not block the doors from closing.

Many ingenious methods have been designed for protecting openings for conveyors of different types.

Figures 46.4 through 46.6 illustrate various conveyor designs and/or programming devices which will minimize or eliminate the possibility of fire door obstruction by the conveyor or conveyed stock. The illustrations show only the basic concepts; proper performance depends on conservative design, good workmanship, inspection, and maintenance. Guidelines to observe are:

1. Select a design that is as simple and direct-acting as possible. Emphasis should be on "fail-safe" operation.
2. Program the sequence of operating steps and interlocks so that obstructions (conveyor, conveyorized material, etc.) to the fire door clo-

MATERIALS HANDLING SYSTEMS

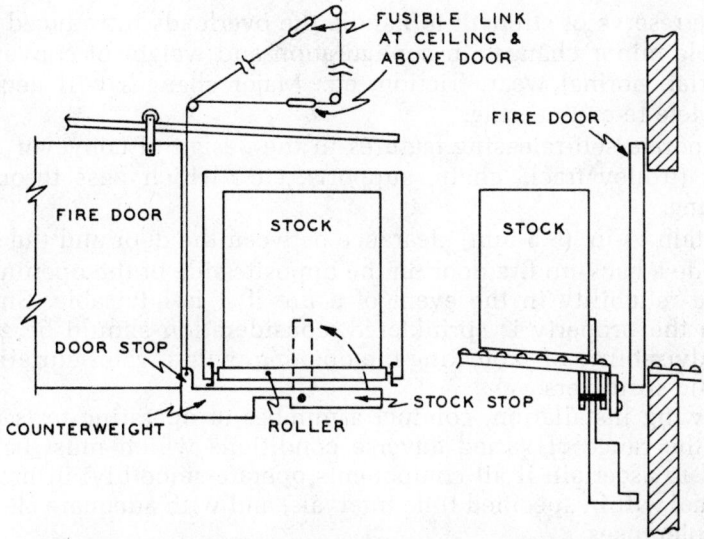

FIGURE 46.5. *A method of stopping stock on a gravity roller conveyor. The release of the counterweight rotates the stock stop to the vertical position and releases the door when the stock interval is correct. The short section of the hinged conveyor is also released to allow for the passage of the door.*

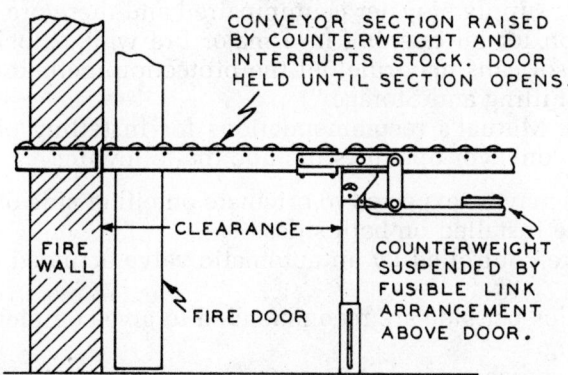

FIGURE 46.6. *A counterweight hinged section of a roller conveyor.*

sure are positively and permanently (until manually reset) removed from the door's path before it is released to close.
3. Design structural and mechanical components, linkages, clearances, etc., in a conservative manner. Counterweights, springs, and other operating forces (uninterruptible by initial fire stages) must have an

943

ample reserve of strength to handle the overloads introduced by reasonable minor changes in configuration and weight of conveyorized material, normal wear, friction, etc. Major changes will necessitate complete re-engineering.
4. Incorporate self-releasing features in the design of conveyor components (trolley track, chain, supports, etc.) which pass through the opening.
5. Maintain ⅜-in. (9.5 mm) clearance between the door and the sill.
6. Provide a back-up fire door on the opposite side of the opening to increase reliability in the event of a fire if it is advisable. Similarly, when the property is sprinklered, consideration should be given to the advisability of protecting the opening with a water curtain of automatic sprinklers.
7. Following installation, conduct a number of operating tests that reflect the range of varied adverse conditions which must be anticipated to ascertain if all components operate smoothly, in proper sequence, within specified time intervals, and with adequate clearances and tolerances.
8. Close all fire doors during inoperative periods. Routine closure stimulates emergency operation and provides a regular inspection of the fire door.

Water spray protection: Water spray protection was developed by Factory Mutual as a means of protecting conveyor openings in fire partitions or floors when fire doors are impractical. It is a substitute for fire doors only when the supply of water is unimpaired and therefore is not acceptable protection for an opening in a major fire wall. "Rocketing" items such as arerosol cans may rule out the protection method. (See Chapter 35, "Aerosol Filling and Storage.")

Factory Mutual's recommendations for installing of water spray protection at conveyor openings include the following:

1. Where fire may be expected to originate on either side of the opening, nozzles are installed on both sides.
2. Nozzles are controlled by an automatic valve actuated by a heat detector.
3. Four nozzles per side are recommended to give complete coverage of the opening.
4. Water discharge rates between 2 and 4 gpm/sq ft [81.5 and 163 (L/min)/m^2] or more, depending upon the height of the opening and unfavorable draft effects, are considered desirable.
5. Nozzles are located at an angle not more than 30° between the center line of nozzle discharge and a line perpendicular to the plane of the opening.
6. All communicating openings to the fire area are protected in a standard manner to prevent the nozzle counterdraft from forcing air from the fire area into other areas.

Conveyor openings through floors may also be protected by this

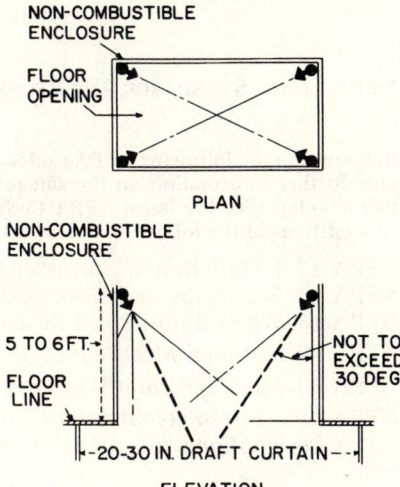

FIGURE 46.7. *Spray nozzle protection for floor openings. (1 ft equals 0.305 m; 1 in. equals 25.4 mm.)*

method, provided an enclosure is constructed around the conveyor from the floor to or slightly beyond the spray nozzles, and draft curtains extend 20 to 30 in. (0.51 to 0.76 m) below and around the floor opening (see Figure 44.7).

The effectiveness of this protection system is, of course, dependent upon rapid detection and appropriate interlocks between the detection system and the machinery.

CRANES

Cranes are essential to the operation of facilities where heavy materials must be lifted or moved about. Cranes which move along rails include overhead traveling, gantry, tower, and bridge cranes. There are also mobile cranes mounted on vehicles.

Overhead traveling cranes are used either indoors or outdoors. Gantry, tower and bridge cranes are mainly for outdoor use. Outdoor cranes are susceptible to damage by high winds unless a positive means of anchorage, in addition to the usual operating or service brakes, is provided. Large cranes which move along rails are generally provided with automatic or manual rail clamps. Other means of anchorage are crane traps, wedges, and cables.

Preferably, crane operators' cabs should be of noncombustible construction. Consideration should be given to egress arrangements for the operator's safety in a fire emergency. The cabs should be kept free of oily waste, rubbish and other combustibles. All electrical equipment should be in conformance with NFPA 70, *National Electrical Code*. It should be securely mounted and kept in good repair.

BIBLIOGRAPHY

NFPA Codes, Standards, and Recommended Practices

Reference to the following NFPA Codes, Standards, and Recommended Practices will provide further information on the safeguards for materials handling systems discussed in this chapter. (See the latest *NFPA Codes and Standards Catalog* for availability of current editions of the following documents.)

NFPA 13, *Installation of Sprinkler Systems.*
NFPA 14, *Standpipe and Hose Systems.*
NFPA 15, *Water Spray Fixed Systems.*
NFPA 70, *National Electrical Code.*
NFPA 80, *Fire Doors and Windows.*
NFPA 505, *Firesafety Standard for Powered Industrial Trucks Including Type Designations and Areas of Use.*

Additional Reading

Block, Richard A., *Crane Operation and Preventive Maintenance*, Marine Education Textbooks, Houma, LA, 1983.

Crane Safety and Operation, Signs of the Times Publishing Co., Cincinnati, OH, 1971.

Factory Mutual Engineering Corporation, "Fire Protection for Belt Conveyors," *Loss Prevention Data* 7-11, August 1972, Factory Mutual System, Norwood, MA.

F. M. Eng. Corp., "Industrial Trucks," *Loss Prevention Data* 7-39, April 1975.

F. M. Eng. Corp., "Protection of Openings," *Loss Prevention Data* 1-23, August 1976.

Kuchta, J. M., M. J. Sapko, and F. J. Perzak, "Improved Fire Resistance Test Method for Belt Materials," *Fire Technology*, vol. 17, no. 2 (May 1981), pp. 120-130.

McKinnon, G. P., ed., *Fire Protection Handbook*, Fifteenth Edition, National Fire Protection Association, Quincy, MA, 1981.

"Moving Fire: Fire Hazards of Belt Conveyors," *Record*, vol. 54, no. 6 (1977), pp. 18-21.

Patterson, C. B., "Powered Industrial Trucks: Appraising Their In-Plant Fire Safety," *Fire Journal*, vol. 66, no. 5 (Sept. 1972), pp. 103-104.

47

Electrical Installations in Industrial Locations

B. W. Whittington

Industrial electrical systems containing distribution and utilization segments are not normally considered to be high fire risk areas. However, in any location where electrical energy is distributed and used, components and connections that are not properly selected, located, and maintained can become potential fire propagating points. Thus, this chapter discusses the electrical equipment and equipment arrangements in industrial locations that have a potential for fire incidence. It also includes recommendations for minimizing this potential and discusses how proper maintenance and attention to good housekeeping can be significant deterrents to fire incidence.

POTENTIAL FOR FIRE INCIDENCE

Field observations at industrial locations and articles from trade magazines indicate that failures of electrical connections are the causes of

---

B. W. Whittington, P.E., a recently retired Electrical Engineering Consultant from Union Carbide Corporation, now has his own consulting firm in Charleston, West Virginia. He is a Fellow of the Institute of Electrical Engineers (IEEE) and a Past President of the Industry Applications Society of IEEE.

many equipment burnouts and fires. Many of these equipment failures result from improper terminations, poor workmanship, different characteristics of dissimilar metals, improper binding screws or splicing devices, environmental and atmospheric conditions present at the facility, and airborne contaminants from other industrial manufacturing establishments in the vicinity. Therefore, opportunities for trouble in electrical installations at industrial locations can be said to be many, since any moderate-sized electrical system will contain literally thousands of splices, terminals, and connection points. Generally, these consist of copper or aluminum and, in most instances, some of each.

Copper is usually considered to be more "forgiving" than aluminum in that minor procedural mistakes in making connections and terminations are less likely to result in failure. Aluminum, which is becoming more widely used because of attractive economics, requires more care in the preparation of terminations and splices. However, new product and material designs that provide for increased levels of safety of aluminum wire terminations have recently been developed by the electrical industry. Some of the fundamental precautions to be followed when using aluminum conductors are covered later in this chapter in the section titled "Failures and Faults from Mechanical Connectors."

Units having concentrated heat-emitting areas include: (1) switching devices, such as fused switches and circuit breakers; (2) distribution equipment in the form of transformers, panelboards, switchboards, and motor controls; and (3) utilization devices, typified by lighting, resistance and induction heaters, and motors. Historically, the design quantities of heat generated from these devices has been based on an economic balance involving the cost of materials and electrical energy, the heat-resistant qualities of the required insulating materials, and marketplace pricing. In many instances — particularly during the past several decades of relatively cheap electrical energy and intense competitive-pricing pressures — much equipment design has evolved into a concept of decreased material content, decreased space for cooling, and increased operating temperatures. Higher operating temperatures, when combined with high ambient temperatures and the added heat from poorly made terminations and splices, have resulted in increased fire potential problems in all types of electrical systems.

Selection of Equipment

The potential for trouble represented by the aforementioned factors must be recognized when selecting equipment and placing it in locations where ambient temperature and loading will cause the total temperature rise of the equipment components not to be excessive. The negative effects of such a situation can be cumulative. Very high ambient temperatures combined with temperatures emanating from a maximum loading condition and a poorly made electrical connection are ingredients that can lead to electrical failure -electrical failure with its associated potential for destructive area fires.

Government and marketplace pressures, in their concern for energy conservation, are causing a slow evolution in equipment and systems designs toward lower operating temperatures and decreased potential as fire sources. This evolution is accelerated or slowed dependent upon the cyclic effect of recessionary periods, spot energy shortages, or costs. During this evolution, designers and specifiers should be extremely selective when selecting equipment. Apply the equipment with consideration to its operating temperature and potential fire source.

Additionally, designer and specifier attention should be directed toward specifying distribution equipment where insulating materials on cabling and equipment components does not readily support combustion. Some materials burn for an extended time (seconds) in the presence of an electrical arc of low magnitude, and will continue to burn after the arc has been removed. Tests have shown that for some materials, 5 amp at 1,300 V is sufficient for ignition in about 5 sec. Where protective devices, fuses, and circuit breakers rapidly remove the faulted circuit from the supply, combustion temperatures are not generally attained.

Where interruption is delayed because of design considerations or failure of the protective device, combustion is likely to occur in some insulation materials. This is particularly true in large industrial systems where many steps of ground-fault relaying are required to bring about coordination and proper isolating of a segment of ground fault. Several seconds may elapse before the protection device clears the area faulted, thus permitting the potential for insulation combustion.

Minimizing fire potential through equipment selection: Minimizing the fire potential requires careful consideration by the designer or specifier in the selection of the distribution equipment and its component insulation materials. National testing laboratories are generally the best and most accurate source for checking for this type of information.

Certain components of electrical distribution systems inherently operate at higher temperatures. Some of these components are:

1. Service entrance equipment, which uses current-limiting fuses as the fault-interrupting device.
2. Dry-type transformers used for reducing distribution voltages for lighting or small motors.
3. Enclosed switchboards, panelboards, and motor control centers.
4. Utilization devices in the form of resistance and induction heaters, electric motors, and lighting components.

Special care should be exercised in making secure, minimal-resistance connections to these components, locating them in lower ambient temperatures whenever possible, and never allowing combustible materials to be stored against, or immediately adjacent to, these units.

MAIN SUBSTATION EQUIPMENT

In large industrial locations where load requirements may dictate service entrance equipment at transmission or subtransmission voltages, the service conductors may be uninsulated aluminum or copper bus conductors suspended from, or placed on, porcelain or epoxy supports. In such instances, amperage is not generally the criteria used for selecting conductors. Usually, the determining criteria are the supplying system's maximum fault current capability and the resultant conductor and support system required to withstand the substantial mechanical forces generated by downstream faults.

Main Service Entrance Devices

The main service entrance device is likely to be a free-standing oil, air, or gas-insulated circuit breaker where the dominant design consideration is based on interrupting needs. As in all electrical connections, care should be taken to properly terminate and splice the connections in order to avoid high resistance and, subsequently, overheated junction points. While the open-air, high-voltage environment does not have the combustibles usually found in indoor locations with associated fire potential on connection failures, unnecessary power outages can be avoided.

Fire Hazards in Main Outdoor Substations

Fire hazards in a main outdoor substation derive mainly from the character of the insulating liquids in the transformers and circuit breakers. Improper application of the main incoming line circuit breakers — such as the use of the units on systems with short-circuit fault duties exceeding equipment ratings — could result in equipment tank failure if the unit is required to interrupt a full system fault. If the insulating liquid is flammable, it may be ignited by the failure, resulting in equipment destruction and possibly the destruction of adjacent units or buildings. Statistically, the likelihood of such an event is remote and damage is normally confined to the equipment specifically involved; however, circumstances may warrant the consideration of special fire protection measures. Such protection may be achieved by: (1) spacing (physical isolation), (2) diking or curbing measures sufficient to contain the quantity of possible flaming fluid from the faulted unit, or (3) the use of water spray equipment directed upon the circuit breaker or transformer units. Fire-resistant insulating fluids in the form of silicone oil or specially formulated synthesized hydrocarbon oils are available for nonarc extinguishing units such as transformers. Any of these alternatives — spacing, diking or curbing, water spray protection, or fire-resistant fluids — may provide the degree of protection required or desired. In any case, proper housekeeping is mandatory in order to keep combustibles away from these units.

SERVICE ENTRANCE EQUIPMENT

In smaller industrial installations, service entrance conductors are generally enclosed in conduit, routed either overhead or underground, and terminated in customer-owned equipment of circuit breaker or fused-switch design. Careful attention is required concerning the proper selection of the conductor size and the temperature rating of the equipment's insulation system for the load ampere requirement. NFPA 70, the *National Electrical Code*, specifies minimum requirements for the conductor size and its insulation material temperature rating based on the maximum amperage to be carried and the conductor grouping. Care should be exercised to allow for larger conductor size where ambient temperatures may be greater than those used in the *National Electrical Code* sizing calculations.

Achieving full conductor ampacity: To achieve full ampacity of the selected conductors, avoid routing the supply conduit near heat sources (such as steam mains) that will substantially raise the ambient temperature of the conduit and its conductors; such routing will result in a decrease in the rated ampacity of the supply circuit. It is extremely important, from both fire hazard and system failure viewpoints, to properly clean and firmly secure the conductor-terminal interface of the line and load terminals of the main service entrance equipment. Percentage loading of the conductor ampacity is generally higher at this point than in other segments of the electrical system, and special care is warranted.

Cleaning of conductors and terminals: Terminals and conductors should be thoroughly cleaned to remove any high-resistance oxides or other semi-insulating foreign material from the interface surfaces of the terminals and lugs. Full-circumference compression lugs are preferable for connecting the wire to the lug as this method minimizes the possibility for high resistance, and subsequently overheated, connections. Additional precaution should be taken in purchasing equipment containing adequate training space for permitting allowable radius bending of the terminating conductors, thereby giving full and nonstressed entrance into the terminals or lugs. Numerous electrical failures with their associated area fire potential are directly traced to overly sharp radius bends, terminations made without full conductor penetration, substantial mechanical stress because conductors are forced into lugs at an angle, and conductors pressed against sharp projections or points because of lack of adequate training space.

The contamination of insulators and barriers by waste products produced by nearby manufacturing processes causes tracking, arcing, and flashover which can result in the ignition of combustibles.

Equipment ratings: Most electrical equipment is rated for a given temperature rise at rated load over 40°C (104°F) base. The locating of service entrance equipment and conductors in areas of high ambient temperatures (such as furnace rooms, boiler rooms, or areas with poor ventilation) should be avoided. Derating of the equipment and conductors will be necessary for ambient temperatures exceeding 40°C (104°F), and connections and operating components that may already be operating at above-normal temperatures will be elevated closer to failure points. Better utilization of equipment ratings, a decreased number of electrical failures, and a decreased fire incidence number can be achieved by placing the equipment in well-ventilated rooms at normal ambient temperatures. This is of major importance in the designing of an electrical facility, and lack of attention to these requirements has been a major contributing factor to electrical failures and associated fires. The importance of adequate ventilation and not storing foreign materials, especially combustibles, in electrical equipment rooms cannot be overemphasized.

Equipment and components with short-circuit current capabilities: One of the more important criteria in the avoidance of electrical failures and associated fires is the selection of equipment and connecting components having short-circuit current capabilities compatible with the maximum fault current which can be furnished by the supplying utility. This short-circuit current capability consists of the ability of the interrupting devices (circuit breakers or fuses) to safely and effectively interrupt the flow of fault current and the ability of all equipment conductors and support systems on the supply side of the fault to mechanically and thermally withstand the forces exerted by that current from its initiation until its interruption.

THE DISTRIBUTION SYSTEM

While industrial fire hazards are the main emphases of this chapter, it is not possible to disassociate two other basic concerns, or ingredients, in the design of an industrial electrical system. These two concerns are *reliability* and *safety*.

Reliability and Safety Factors

Continuity of production in an industrial facility is dependent upon the reliability of the facility's electrical system and the components of that system. The inherent safety of the system design and its components is imperative, and this criteria can generally best be obtained by following available codes and standards.

All of the factors that contribute to increased reliability and safety must be carefully examined to achieve an acceptable industrial electrical

FIGURE 47.1. Damage caused by improper and inadequate installation of molded-case circuit breaker. (Union Carbide Corp.)

system design. Some plants can tolerate interruptions, while others require the highest degree of service continuity. Many other factors, in addition to reliability and safety, must be considered. These factors include simplicity of operation, maintenance, flexibility, and first cost. Although it is not within the scope of this chapter to comment on all of these factors, it is important to emphasize those most directly bearing on fire hazards with their associated impact on reliability and safety. The distribution system designer or specifier must select the voltage, system configuration, method of grounding, fault protective devices, and equipment sizes and types for effectively, adequately, and safely serving the electrical loads. All this must be done within an acceptable first cost.

System Grounding Configurations

One of the more important facets in the selection of a reliable and safe electrical system design is the choice of the system's grounding configuration. The system may be either solidly grounded wye, high-resistance wye, low-resistance wye, delta corner-grounded, or delta ungrounded.

The service continuity required and the degree of operational control available generally dictate the choice of the grounding arrangement. For many years a substantial number of industrial plant distribution systems have been operated delta ungrounded at one or more voltage levels. This was done to prevent an immediate outage on the initiation of a ground fault, the intent having been to achieve an additional degree of service continuity. This method of grounding is still found in many older industrial plants. An adequate detection system with an audible alarm must be employed to alert plant operating personnel to the initial ground condition. Such a detection system is required by both the National Electrical Code and U.S. Department of Labor regulations on employee safety. Immediate detection and isolating procedures are required to remove the grounded equipment prior to the occurrence of a ground fault on another phase. Personnel safety and equipment protection may be maintained at very acceptable levels when these procedures are quickly and properly accomplished.

High-resistance grounded-wye configurations: A grounding method rapidly gaining more acceptance in industrial plant electrical distribution systems requiring the highest degree of service continuity is the high-resistance grounded-wye configuration. This grounding method, like the ungrounded system, does not cause a service interruption on the initiation of a ground fault on one phase. Further, because of being intentionally grounded through a high resistance, it limits the amount of ground-fault current flow to just slightly above the capacitive-charging current of the affected system exposure. This magnitude of current (generally about 5 amp for the average system) greatly reduces the amount of heat generated in the fault path, which may reduce fire potential in the area of the ground fault. This design concept has been successfully used in low-voltage configurations and in some medium-voltage levels. It can and does provide increased service continuity, and minimizes arc damage to conductors and equipment. It should only be utilized where adequate monitoring and alarming equipment is used with associated quick response by plant maintenance personnel in detecting and removing the faulted area.

The high-resistance grounded system configuration and method of responding to the initial ground fault is an obvious advantage to continuous process facilities such as chemical processes, steel production, automotive assembly lines, etc. Affected equipment may come to an orderly shutdown which, in turn, minimizes production losses, causes minimal arc damage to equipment, affords the lowest possible degree of flash hazard to personnel, and lessens the fire hazard potential. It should be thoroughly emphasized that use of proper ground-detection devices and quick removal of phase-to-phase faulted areas are essential. Since neither the detection devices nor the required maintenance services are generally found in commercial facilities, high-resistance grounding is not recommended for such establishments. The use of more conventional, solidly grounded wye systems, with the removal of the ground fault upon initiation, is recommended.

Protection devices on grounded systems: While protection devices will be covered in more detail later in this chapter, their application on solidly grounded wye systems merits discussion here. In recent years, particularly on high-fault capacity, low-voltage 208Y/120 volt systems, there have been numerous reported cases of arcing fault burnouts in which severe damage to, or complete destruction of, electrical equipment has been caused by the energy released in the arc. Typically, the arcing fault becomes established between a phase and ground, or between phases and ground. The fault arc releases enormous amounts of energy based on amperes squared multiplied by time (I^2t), with heat so intense that it vaporizes copper or aluminum conductors and destroys the surrounding steel enclosures. Any combustibles stored in the immediate vicinity of the equipment would also be ignited.

It is characteristic of these arcing conditions that phase overcurrent protective devices do not operate quickly enough to remove the fault from the source since they may be of lower magnitude than the protective device setting, or too much time may lapse before clearing occurs. The designer or specifier must recognize that prevention of arcing fault burnouts, with the present state of protective system design, must rely on fast and sensitive detection of the arcing fault current accompanied by the removal of the faulted circuit within 10 to 20 cycles. This can be accomplished, due to recent refinements in ground fault relaying, by the additional use of a protective assembly utilizing a special design current transformer and a low burden solid-state relay. The sensing transformer encloses all phase conductors and responds only to ground fault current flow allowing it to quickly sense and cause removal of arcing grounds.

Choice of Configuration Type

Selection of either the high-resistance wye or the solidly grounded wye configuration should be carefully made, since no system has a completely reliable and universally applicable means of protection against low-level phase-to-phase or phase-to-ground arcing faults. The recommendation made earlier in this chapter concerning choosing insulation systems that do not readily support combustion becomes especially important under these conditions.

Additional precautions should be taken in the establishment of a low-impedance ground-fault return path. Arcing may occur in conduit or cable tray systems serving as ground fault return paths where the design has not effectively achieved low-impedance paths and low-resistance connections.

FIGURE 47.2. *Equipment Damage Resulting from Non-removal of Arcing Ground (Union Carbide Corporation)*

Classification of Distribution Systems

Electrical distribution systems in textile, chemical, petroleum, mining, grain handling, and other similar industrial locations require selective analysis as to the presence of potentially explosive gases, dusts, or fibers. Hazardous (classified) areas are defined in detail in Article 500 of NFPA 70, the *National Electrical Code*, and thorough knowledge of this material is required of the electrical designer working in any of the previously mentioned industrial locations.

Basically, there are three Classes, two Divisions, and various subgroups to be considered in the classification analysis. Each of the three Classes has two categories: Division 1 or Division 2. The three classes are:

1. Class I locations are defined as those in which flammable gases or vapors may be present in quantities sufficient to produce explosive or ignitible mixtures.
2. Class II areas cover locations in which concentrations of combustible dusts may be present.
3. Class III areas are concerned with the presence of an ignitible mixture of fibers.

Three essential simultaneous conditions create a hazardous (classified) area: (1) vapors, dusts, or fibers present in sufficient quantity to be explosive, (2) its proper mixture or suspension in air, and (3) a source of ignition.

Division 1 category is defined as a condition where preceding condition (1) — vapors, dusts, or fibers present in sufficient quantities to be explosive — may exist continuously, frequently, or on a periodic schedule along with the likelihood of condition (2) — its proper mixture or suspension in air — occurring. All that is needed for a hazardous situation is the occurrence of condition (3) — a source of ignition — and the mathematical probability of such an event is unacceptable for personnel and equipment protection. Division 2 category states that the combination of conditions (1) and (2) is likely to occur only during an abnormal event of the operating facility or a malfunction of equipment with a corresponding mathematical probability of event occurrence. It is judged acceptable with certain types of equipment. The different groups of vapors, dusts, or fibers are arranged as to temperature-ignition points and mechanical forces generated upon ignition.

"Blanket" and "selective" classification of areas: During the post World War II years when manufacturing facilities were rapidly expanded and new compounds and materials were being developed and processed, conservative approaches were used in the electrical classification of industrial areas. Classifications denoted as "blanket" classifications were used because knowledge of materials and compounds along with time-proven operating and event occurrence experience were not available or, at best, uncertain. Some installations — because of the conservative approach — were unnecessarily expensive; however, several decades of operating experience and favorable-event occurrence are now placing emphasis on liberalized uses of some electrical equipment, and the use of "selective" rather than "blanket" classification. Blanket classification, as the term implies, classifies a large given area around a known ignitible-mixture emitting point. Selective classification first takes into consideration the pressure of the contained mixture, its weight with respect to air, and then selects a more tightly defined area to be classified.

Blanket classification of areas is generally associated with the American Petroleum Institute's recommended practice for electrical installations, RP-500, A, B, and C, while selective classification may be typified by NFPA 497, *Classification of Class I Hazardous Locations for Electrical Installations in Chemical Plants*. In larger industrial facilities where electrical classification teams are a permanent part of the plant administration, selective classification under NFPA 497 results in a less expensive, first-cost electrical system with associated decreased maintenance and replacement costs — all attained with totally acceptable event statistics. As an example of equipment use changes, modifications have been made in the *National Electrical Code* during the last decade allowing the use of nonarmored PVC (polyvinylchloride) jacketed cables in cable trays in Class I, Division 2 areas — a practice that has been safely used and experience-proven by the chemical industry for approximately thirty years. This and other equipment use changes are cumulative and time proven. Together with the use of selective classification, they allow the electrical designer more flexibility in the classification of areas and use of equipment in the areas classified. Totally acceptable personnel safety and equipment protection from fire hazards are achieved at optimized costs.

SUPPORT AND CONNECTIONS OF SYSTEM CONDUCTORS

Electrical distribution system conductors may be installed in many different ways in an industrial plant. Each support system has characteristics that make it more suitable for certain conditions than for others, and the electrical designer has the responsibility for selecting the one that gives the optimum required reliability and safety for a specific process or operation within acceptable economic restraints.

Conductor Housing

The conductors may be housed in rigid metallic-conduit systems located either overhead or underground (those located underground are generally in a pit and encased in concrete duct systems), direct buried, placed in cable trays, or arranged in messenger-suspended aerial cables or open-wire conductors on insulators in above-surface configurations. Rigid metallic-conduit systems achieve the highest degree of mechanical protection available in aboveground conduit systems, but have the disadvantage of being the most expensive to install. Alternates of thinner wall, intermediate-grade steel, aluminum, plastic, and fiber materials are currently being used to decrease the first cost. While they do offer some degree of flexibility in replacement of existing circuits, the event of external fire or a conductor fault may make it impossible to remove damaged conductors, thereby causing maximum cost and production loss in replacing both conduit and conductor. Also, during fires, toxic or corrosive fumes may be transmitted into an unaffected area, thereby causing unnecessary equipment damage. A surface fire under an overhead rigid-metallic conduit bank is a specific example. The use of this support method has its best application in industrial electrical systems of smaller size where maximum mechanical protection for the conductors is required, and where a minimal number of conductors in a common routing occurs.

Messenger-Suspended Aerial Cables and Cable Trays

The use of messenger-suspended aerial cables and cable trays is increasingly used in industrial wiring systems. Most aerial cable systems are replacing open-conductor wiring configurations, while cable tray support systems are being used as alternates for above- and below-ground rigid conduit systems. Aerial cables provide greater safety and reliability. They also require less surface space than open wiring, while the cable tray system offers low first cost, system flexibility in serving changing loads, quick accessibility for repair or addition of cables, and space saving where a large number of circuits using a common routing are involved. Generally, increased conductor ampacities are attainable with aerial and cable tray installations.

Direct Burial of System Conductors

Where the cost of providing overhead supports proves to be overly expensive, and where electrical loads are known precisely and are not subject to change, direct burial of the system conductors should be considered. This method offers maximum protection against mechanical damage from surface activities and protection against most surface fires. Cables used must be suitable for the purpose (resistance to moisture, crushing, and rodents), and protection must be provided against excavation activities.

Cable Tray System Fire Protection

Since the cable tray system is increasingly used, additional comments in its fire protection are warranted. Many of the cable tray systems, when passing through areas requiring mechanical protection from falling objects or possible flaming materials such as welding or process emissions, are provided with ventilated (or nonventilated) metallic covers. Barrier strips of suitable metal may also provide electromagnetic barriers where communication circuits are located adjacent to power conductors. National standards have been developed for constructing seals or fire stops when passing through walls or partitions, and fire protection is sometimes employed with the use of below-tray fire resistant barriers together with water spray systems. These fire protection systems have the purpose of protecting the integrity of the tray-housed circuit until the surface fire can be controlled, or of allowing time to effect orderly shutdown of the production units affected by the fire location.

Additional work has been done, and is continuing, on national testing laboratory listings of cable insulation systems that are nonflame propagating. As noted at the beginning of this chapter, cables selected for use in cable tray or other support systems should have insulation materials that are certified as nonflame propagating.

Cable Terminations (Splices)

While the selection of conductor support systems by the electrical designer is influenced by the impact of external fires, the point in the in-

FIGURE 47.3. Fire seals used to maintain the integrity of fire-rated walls, floors, partitions, or ceilings. (O.A./Gedney Co.)

FIGURE 47.4. A cable tray installation. (Husky Trough and Ladder)

dustrial electrical system known to be responsible for a substantial number of electrical fires is the cable termination, or splice. Thus, the subject of copper and aluminum conductor interfaces noted at the beginning of this chapter is continued here in further detail. There are two basic types of connectors used for making terminations of conductors: (1) mechanical, and (2) compression.

Mechanical-type connectors: Mechanical-type connectors obtain the pressure to attach the connector to the electric conductor from an integral screw, cone, or other mechanical part. It thus applies force and distributes it through the use of bolts or screws with sections suitable for the load current and mechanical forces encountered.

Compression connectors: Compression connectors are those in which the pressure to attach the connector to the electric conductor comes from an external device that changes the size and shape of the connector and conductor. Basically, it is a tube that has an inside diameter slightly larger than the outside diameter of the conductor, and a wall thickness suitable for load current and mechanical stresses. A connection is made by compressing the tube and conductor into an integral shape by means of a specially designed die or tool. Small units may be applied with a hand tool, while larger ones require a hydraulic compression unit.

Many industrial facilities are experiencing, particularly with the use of aluminum conductors, an increasing number of failures and faults from the mechanical connectors with the associated potential for equipment fires. These experience factors are causing an increased requirement and specification for the use of the compression connector. This is especially true when terminating and connecting aluminum conductors.

Considerations for use of aluminum conductors: The differences between copper and aluminum should be considered when specifying and using connectors for aluminum conductors. Aluminum develops a normal oxide coating which has a relatively high electrical resistance, and it also has a higher coefficient of expansion than copper. Under certain conditions corrosion is possible because aluminum is anodic to other commonly used metals, including copper, under the presence of moisture. Aluminum is subject to additional "creep," which is defined as continued deformation of the material under stress, and the use of an inadequate connector could cause relaxation of contact pressure with a resulting deterioration and failure of the connector. It is important to recognize the existence of the oxide film and remove it by abrading materials and, subsequently, apply a connector sealing compound designed to prevent the reoccurrence of the aluminum oxide.

National testing laboratory listings are available for aluminum connectors in both mechanical and compression types. The most satisfactory ones are specifically designed to prevent troubles from creep, the presence of oxide film, and the differences between coefficients of expansion between aluminum and other metals.

The connection of an aluminum connector to an aluminum or copper pad also requires the interface to be clean if plated. If not plated, the surface should be cleaned by wire brushing and then coated with a joint compound. In heavy duty applications where cold flow or creep may occur, Belleville washers are recommended and the aluminum should always be placed above the copper. Installers and maintenance personnel must be instructed and trained in the proper method of utilizing aluminum or copper and aluminum conductors and devices.

It is not possible in this one chapter to cover all the details of the problems that need to be considered concerning connection points. It is, however, important for the electrical designer and specifier to be fully aware of the details associated with potential problems, since these details are cumulative and have substantial negative impact on electrical system reliability and associated fire potential hazards.

UTILIZATION UNITS AND THEIR PROTECTIVE DEVICES

Most of the electrical energy used in industrial facilities is consumed or utilized in motors, lighting, and process-heating devices. There are many peripheral devices such as capacitors, reactors, ballasts, etc., that are — intentionally or unintentionally — energy-consuming devices, and there are some facilities that use substantial quantities of electrical energy in arc furnaces or similar devices. However, the three basic categories noted — motors, lighting, and process-heating devices (particularly electric motors) — constitute the majority of industrial electrical-system utilization-device fire source potential; these three basic categories are specifically covered in the following paragraphs.

Motors

Motors are the cause of many fires at industrial locations. The ignition of the motor insulation or of nearby combustible materials may be caused by sparks or arc from motor winding short circuits or grounds, or from improperly operating brushes. Bearings may overheat because of improper lubrication or motor overloading, and sometimes excessive bearing wear allows the rotor to rub on the stator. Use of individual drives on machines of several different types sometimes makes it necessary to install motors in locations and under conditions which are injurious to motor insulation. Dust that can conduct electricity may be deposited on the insulation, or deposits of textile fibers, etc., can prevent the normal dissipation of heat. Motors should be cleaned and lubricated regularly. All motor installations should comply with the requirements of the *National Electrical Code*, which includes special guidelines for motors in hazardous locations.

Motors, as with most electrical equipment, should be installed in areas having minimum dust, dirt, and other foreign materials and should be placed in ambient temperatures and ventilated in accordance with their design requirements. They should be chosen and placed to minimize their use in areas classified as "Hazardous" in accordance with Article 500 of the *National Electrical Code*. It is important, both from the standpoint of reliability and fire potential, to select the proper motor enclosure for the environment. Open type or drip-proof enclosures should only be used in areas free of dirt and contamination and where wind-driven rain or excessive moisture is not present. Many industrial facilities, because of the need to decrease the number of motor failures, are standardizing on TEFC (totally enclosed, fan-cooled) severe-duty motors — a motor enclosure which effectively prevents outside contaminated air from flowing across the motor windings and interior.

Design features for selection of motors: In medium-voltage motors, open types, TEFC (totally enclosed, fan-cooled), WPI (weather-protected one), and WPII (weather-protected two), enclosures are available and selection should be made on the basis of the reliability required and the environmental conditions. Two additional criteria in electric motor selection are: (1) insulation-class temperature rise, and (2) motor efficiency. On a national basis, several factors are causing pressures on motor manufacturers to design and build motors of higher efficiency and, in most cases, decreased operating temperatures. Many large industrial users have written specifications requiring lower temperature rise, increased efficiency and power factor, and longer bearing life. All these elements work cumulatively to help lower operating temperatures, increase operating reliability, and provide a corresponding decrease in motor failures and associated fire incidence.

Each motor served must also have a source of supply with its associated controller and protective device(s). These controllers, in either low- or medium-voltage class, may take the form of magnetic contactors, circuit breakers, or combinations thereof. They may employ either fuses or overload relays working through circuit breaker or contactor mechanisms to achieve circuit interruption upon overload or fault conditions. In order to properly protect the motor against those overload conditions and to minimize electric arc damage during faults, the selection of the fuses or the settings of the overload relays should be strictly within the limits set by the *National Electrical Code*. One of the more abused criteria in motor overload protection selection or replacement of existing units is oversizing. It follows that oversizing is also responsible for many unnecessary motor failures, repair, and fire statistics. As noted in the first part of this chapter, it is of utmost importance to locate motor controllers using thermally sensitive overload devices in areas of normal to low ambient temperatures.

Lighting Fixtures and Devices

While lighting fixtures and devices are not statistically known to be significant as a fire source in industrial locations, care should be exercised in the selection of fixtures for hazardous (classified) locations. The selection of fixtures for explosion-proof requirements should be guided by certification by a testing laboratory which leaves the designer or specifier small error potential. When selecting nonexplosionproof fixtures for Class I, Division 2 areas, the rule of the *National Electrical Code* that restricts the operating temperature of the lamp to 80 percent of the ignition temperature of the flammable vapor present should always be followed. The fixture should then be properly maintained to assure its original integrity.

Protective devices: The protective device is the most important part of the industrial plant electrical distribution system when overloads, sustained overvoltages, and electrical faults occur. The complexity of the electrical distribution systems vary greatly from very small industrial plants to very large ones. For example, small industrial plants may have a simple system with low-voltage protection only, while a large industrial plant may incorporate an intricate network of high-voltage transmission, medium- and low-voltage distribution substations, uninterruptible power sources, and in-plant generation.

The basic requirements for increased electrical system reliability have been previously discussed in this chapter, and the prompt and positive action of the protective device occupies a vital role in achieving those requirements. These protective devices consist primarily of two basic types: (1) the relay-actuated circuit breaker, and (2) the fuse. They provide the intelligence and initiate the action which enables circuit-switching equipment to respond to abnormal or dangerous system conditions.

The relays used previously were primarily of the electromagnetic induction type, with a recent substantial increase in the use of solid-state units. In smaller and low-voltage equipment, overload relays with bimetallic elements are used for thermal-overload protection.

The most commonly used fuses are a variation of low and high voltage and current-limiting and noncurrent-limiting types. A fuse may be defined as "an overcurrent protective device with a circuit-opening fusible part that is heated and severed by the passage of overcurrent through it."

Most industrial facilities employ a combination of relays and fuses, using relays and circuit breakers where a closer degree of protection is required and fuses where a lesser degree of protection is acceptable but where lower first cost is the criteria. The basic protection afforded is for overload, short-circuit, and ground fault, with many other sophisticated relays available for differential, reverse current, under or overfrequency, and undervoltage protection. Each, when sensing high overloads or fault currents, has the purpose of removing only the circuit faulted, accomplishing this quickly enough to allow continued operation of nonaffected circuits and minimal mechanical stress on equipment carrying the fault current from its source to the fault location. In a complex system, coordination of the applied relays and/or fuses is a design requirement which is necessary in order for the interrupting equipment to operate selectively and remove only the faulted portion of the system. Fast, selective operation of the relay or fuse in removing an overload or fault is a positive step in reducing industrial plant fire incidence statistics.

Process Heating Devices

The generation of heat with electricity and electric heating systems for ovens and furnaces are often the cause of fires at industrial facilities. Most electrical heating appliances for heating small areas and for other small heating jobs involve resistor heating elements which are usually one or more metal-alloy wires, nonmetallic carbon rods, or printed circuits. Resistor heating is used in radiators, unit heaters, convectors, central hot water systems, central warm air heating systems, and paneltype radiant heat installations for walls, floors, and ceilings.

Industrial furnaces generally employ transformers which may be either the dry type, fire-resistive-fluid-insulated, or oil-insulated type. The *National Electrical Code* requires oil-insulated transformers of a total rating exceeding 75 kVA to be located in a fire-resistive vault. Oil-filled circuit breakers that control arc furnaces are subjected to unusually severe duty and, unless frequently inspected and properly maintained, may fail with disastrous results. Circuit breakers on circuits operated at more than 600 V and which are used to control oil-filled transformers are located outside the transformer vaults. Vents on high-voltage circuit breakers should be piped outdoors. Electric arc furnace circuits are also subject to high surge voltages, due to the nature of the operation, which can cause failure of the arc circuit breakers used. In these cases, shunt capacitors are installed in the circuit to prevent these high-voltage surges.

Inductive and dielectric heat-generating equipment employing high-frequency alternating currents is in use in many heat treating processes. In order to eliminate both the personal and fire hazards of such equipment, construction and installation should comply with the special requirements of the *National Electrical Code*. Other hazards of electric furnaces are similar to those of furnaces employing other means of heating.

Types of electric heating systems for ovens and furnaces: Electric heating systems for ovens and furnaces are of five types: (1) resistance, (2) infrared, (3) induction, (4) arc, and (5) dielectric.

1. *Resistance heat* is produced by the flow of current through a resistive conductor. Resistance heaters may be of *open*-type with bare heating conductors or *insulated sheath*-type with heater conductors covered by a protecting sheath which may be filled with electrical insulating material.
2. *Infrared heat* is transmitted as electromagnetic waves from incandescent lamps with filaments that operate at temperatures lower than the filament temperature of ordinary incandescent lamps so that most of the radiation occurs in the infrared part of the spectrum. These waves pass through air and transparent substances but not opaque objects and therefore release their heat energy to these objects.
3. *Induction heat* is developed by currents induced in the charge. Induc-

FIGURE 47.5. A liquid-immersed transformer filled with a high fire point nonpropagating liquid with a fire point in excess of 300°C (572°F). (Square D Co.)

tion heaters have an electric coil surrounding the oven space, and heating is by electric currents induced in the work, being processed.
4. *Arc heat* is developed by the passage of an electric current between either a pair of electrodes or between electrodes and the work causing an arc which releases energy in the form of heat.
5. *Dielectric heat* is developed in dielectric materials when exposed to an alternating electric field. The frequencies used are generally higher (in the order of three megacycles or more) than those in induction heating. This type of heater is useful for heating materials which are commonly thought of as being nonconductive.

Electric systems can be arranged so that the processing operation does not require an oven enclosure. The use of "ovenless" or unenclosed heating systems can employ lamps, resistance-type electric elements, or infrared heaters in the vaporization of flammable, toxic, or corrosive liquids and their thermal decomposition products. Enclosures around "ovenless" systems are advisable to prevent flammable, toxic, or corrosive vapors from escaping into the general area, and to help provide better ventilation and safeguards for personnel. However, heating systems having an energy input of under 100 kw (kVa) may be excluded if adequate area ventilation is provided (Article 500 of the *National Elec-*

trical Code gives guidance for electrical installations in hazardous locations).

All parts of heaters operating at elevated temperatures within an oven or furnace and all other energized parts are protected to prevent contact by persons, as well as to prevent accidental contact with materials being processed, and contact with drippage from the materials.

ELECTRICAL MAINTENANCE AND HOUSEKEEPING

An effective electrical system equipment preventive maintenance program can be defined as a program that enhances safety and also reduces equipment failure to a minimum consistent with good economic judgment. Industrial fire hazard statistics have the potential of being significantly reduced by electrical designer and specifier selection and arrangement of proper equipment, accurate area classification, and a well-designed protective system. The potential will only be realized if plant management establishes an effective, on-going program of preventive maintenance, a reasonable approach to keeping the environment clean of combustibles, and measures for retaining the integrity of fire preventive devices.

Reliability maintenance (RM) or electrical preventive maintenance (EPM) programs have the purpose of reducing hazards to life and property that result from failure or malfunctions of industrial electrical systems and equipment. A well-planned and executed EPM program has additional substantial benefits in the reduction of costly and unplanned industrial electrical system shutdowns and the prevention of substantial losses in production. Often, the latter point is the one that persuades management to establish an EPM program. Statistics are available indicating that, in a two-year monitored period (1967-68), one-half of the losses associated with electrical equipment failures might have been prevented by an effective EPM program.

The purchase of electrical system equipment generally includes, as part of the purchase, a recommendation from the manufacturer on proper maintenance required to achieve maximum integrity and life service. These recommendations should always be followed when establishing an EPM program. As noted in NFPA 70B, *Electrical Equipment Maintenance*, there are four basic steps in the planning and development of such a program. They are:

1. Compile a listing of all plant equipment and systems.
2. Determine which of these are most important, critical, and costly in terms of repair and production loss.
3. Develop a system for documenting and maintaining exactly what needs to be done.
4. Train internal personnel for the required work, or arrange for an external contract with a company that has the necessary expertise.

The success of such a program is dependent upon the continuing support of management and the technical competence and administrative skill of that management. Properly executed, it will result in electrical distribution system operation with minimal fire hazards, equipment failures, and production losses. Additionally, and most important, it will, with supplemental training, afford maximum safety to all electrical system operating personnel.

BIBLIOGRAPHY

REFERENCES CITED

Factory Mutual Engineering Corporation, *Loss Prevention Data*, 1972 ed., Factory Mutual System, Norwood, MA, ch. 5-1, pp. 1-7.

Fire Protection Manual for Hydrocarbon Processing Plants, 1964 ed., Gulf Publishing Company, Houston, TX, pp. 175-177.

McClung, L. B., and Whittington, B. W., "Ground Fault Tests on High-Resistance Grounded 13.8 kV Electrical Distribution System of Modern Large Chemical Plant-11," *IEEE Transactions On Industry Applications*, vol. 1A-10, no. 5, September/October 1974, pp. 601-617.

Recommended Practice for Classification of Areas for Electrical Installations in Petroleum Refineries, RP-500A, American Petroleum Institute, Washington, DC.

Recommended Practice for Classification of Areas for Electrical Installations at Drilling Rigs and Production Facilities on Land and on Marine Fixed and Mobile Platforms, RP-500B, American Petroleum Institute, Washington, DC.

Recommended Practice for Classification of Areas for Electrical Installations at Petroleum and Gas Pipeline Transportation Facilities, RP-500C, American Petroleum Institute, Washington, DC.

Recommended Practice for Electric Power Distribution for Industrial Plants, 1976 ed., Institute of Electrical and Electronic Engineers, New York, NY, pp. 29-38, 73-124, 273-309, 319-350.

Recommended Practice for Grounding of Industrial and Commercial Power Systems, 1972 ed., Institute of Electrical and Electronic Engineers, New York, NY, pp. 13-24.

NFPA CODES, STANDARDS, AND RECOMMENDED PRACTICES

Reference to the following NFPA Codes, Standards, and Recommended Practices will provide further information on the safeguards for electrical installations discussed in this chapter. (See the latest *NFPA Codes and Standards Catalog* for availability of current editions of the following documents.)

NFPA 49, *Hazardous Chemicals Data.*

NFPA 61B, *Prevention of Fires and Explosions in Grain Elevators and Facilities Handling Bulk Raw Agricultural Commodities.*
NFPA 70, *National Electrical Code.*
NFPA 70B, *Electrical Equipment Maintenance.*
NFPA 70E, *Electrical Safety Requirements for Employee Workplaces.*
NFPA 86A, *Ovens and Furnaces: Design, Location, and Equipment.*
NFPA 86B, *Industrial Furnaces: Design, Location, and Equipment.*
NFPA 86C, *Industrial Furnaces Using Special Processing Atmospheres.*
NFPA 496, *Purged and Pressurized Enclosures for Electrical Equipment.*
NFPA 497, *Classification of Class I Hazardous Locations for Electrical Installations in Chemical Plants.*
NFPA 497M, *Classification of Gases, Vapors and Dusts for Electrical Equipment in Hazardous (Classified) Locations.*

Additional Reading

Bartknecht, W., *Explosions: Course, Prevention, Protection*, Springer-Verlag New York, Inc., New York, NY, 1981.

Electrical Instruments in Hazardous Locations, National Fire Protection Association, Quincy, MA, 1978.

Electrical Safety in Hazardous Environments, IEE Conference Publication Ser. No. 218, Peregrinus, England, 1982.

Factory Mutual Engineering Corporation, *Handbook of Industrial Loss Prevention*, 2nd ed., "Process Furnaces," and "Industrial Ovens and Dryers," McGraw-Hill, New York, NY, 1967, pp. 37-1 to 37-15 and 40-1 to 40-32.

Fire and Electricity, National Fire Protection Association, Quincy, MA, 1982.

Magison, E. C., *Electrical Instruments in Hazardous Locations*, 3rd ed., Instrument Society of America, Pittsburgh, PA, 1978.

McKinnon, G. P., ed., *Fire Protection Handbook*, Fifteenth Edition, National Fire Protection Association, Quincy, MA, 1981.

Nailen, R. L., "Hazardous Area Electrical Equipment," *Fire Engineering*, vol. 131, no. 6 (1978), pp. 32,34,36.

Osborn, Richard W., ed., *Tapping in to the NEC*, National Fire Protection Association, Quincy, MA, 1982.

48

Industrial Storage Practices

Martin M. Brown

An industrial plant official describing his fire hazards concluded by saying ". . . and the big north building is just used for storage." Stored goods, while admittedly passive starters of fires, can react in a monumental way. One of the largest industrial fires in history was at an automotive parts warehouse in West Germany, October 1977. It caused over $100 million in damage despite automatic sprinkler protection. A major factor was the temporary storage of canned oil in cartons 20 ft (6.1 m) high in the aisles. In 1983 the K Mart Corporation's 1.2 million sq ft (11 500 m^2) Distribution Center in Falls Township, Pennsylvania was destroyed when a fire involving palletized storage of petroleum-liquid-based aerosol containers overwhelmed the sprinkler systems (see Figure 48.1). The loss estimate was in excess of $100 million.

Martin M. Brown, P.E., a Fire Protection Consultant based in White Plains, New York, is a member of the NFPA General Storage Committee and a past president of the Society of Fire Protection Engineers.

Chapter 48

FIGURE 48.1. *An extremely fast developing fire spread through aerosol storage in the K Mart Corporation's Distribution Center in Falls Township, PA. Rocketing aerosol cans spread the fire throughout the warehouse, overtaxing the fire protection systems and resulting in early roof collapse and broken sprinkler piping (Bucks County Courier Times).*

The Problems with Goods in Storage

The potential for damage is higher in warehouses than in other areas of an industrial plant. The damage potential can be attributed to several general conditions:

High accumulation of value in a single fire area: Unless the plant has extremely expensive process equipment, as in utility power generating or continuous chemical process plants, the tendency is for the warehouse portion to have the greatest accumulation of valuable contents. It will often be the key to the maximum probable loss. Even with successful fire control, water and smoke damage can be excessive when vulnerable commodities are present, such as in the food industry.

Combustibility of packaging: Many stored items, such as machine or motor parts, are only moderately combustible or noncombustible; however, they are often cushioned in foamed plastic within their containers, for safe handling and shipping. Other more traditional cushioning materials include shredded paper, excelsior, and straw. Much of the volume of a storage area is made up of such fast-burning cushioning, although it is not readily visible.

High piling: Storing goods to great heights is the norm, with heights of 25 ft (7.6 m) common. Mechanical stacking equipment and narrow aisles, and occasional rack storage of 50 to 100 ft (15.2 to 30.4 m) in height, are highly conducive to fast vertical fire spread. Automatic sprinkler systems that could well cope with most storage occupancies through the years did their job well because water could reach the fire. They now have to be modified to meet the challenge of high stacking.

Horizontal air spaces: Unsprinklered areas that foster fast fire spread exist within high-piled palletized and rack storage out of reach of traditional fire extinguishing methods.

Fire protection criteria for a plant, including water supply and distribution, are often dictated by the needs of the portion of the plant used for storage. Fortunately, fires are less frequent in warehouses than in manufacturing areas, but that is offset by the high loss potential per fire.

Location of Storage

Most industrial plants have buildings for storing raw materials, finished products, or goods in an intermediate stage of production. Some materials, covered or resistant to the weather, are kept outdoors. Good protection is needed in areas designated for warehousing and for areas of incidental storage which are characteristic of large operations. Off-premise warehouses may be owned, or rented with little plant management control over their operation, maintenance, or protection. They can have high value, be critical to the plant operation, or both. When such plant storage is in the same area with storage owned by others, the overall hazard is governed by the storage with the greatest hazard.

Scope of This Chapter

This chapter deals with commodities in general, and classifications which have undergone fire protection studies that are not covered elsewhere in this book. It does not cover flammable liquids (Chapter 37), industrial gases (Chapter 38), dusty commodities (Chapter 6), oxidizing materials, or certain materials for which NFPA standards for safe practices in storage and handling have been developed, e.g., pyroxylin plastic, magnesium, titanium, ammonium nitrate, or explosive materials (see Bibliography).

FIRE CAUSES IN STORAGE OCCUPANCIES

Ignition sources that can be occasionally tolerated elsewhere should be eliminated in a storage area because of the high potential for destruction. Warehouse fires often start when no one is immediately present. The main causes or contributing factors are:

Bad housekeeping: This factor is an almost essential partner to ignition in storage fires. The packing or unpacking of goods usually requires that a quantity of loose combustible substances such as foamed polystyrene beads, cocoons of foamed plastic, shredded paper, excelsior, and straw, be present. Combustible bits such as baled fibers, loosely paper-wrapped furniture, or rolls of carpeting sometimes break off from some stored goods. Labels or tags may become disattached, and damaged containers may spill out their contents. Ways to safely handle and remove packing materials and loose materials that accumulate at floor level fall within the area of good housekeeping practices (see Chapter 50).

Industrial trucks (forklifts): These vehicles, powered by either LP-Gas (usually propane), gasoline, diesel oil, or electric batteries, are used to move stock both horizontally and vertically in warehouses. All are capable of starting fires if not properly maintained. In addition, gasoline spills during refueling, usually while engines and exhaust pipes are still hot, have caused fires. The importance of fueling these trucks outside the warehouse is magnified by the economic impracticality of installing sprinkler protection that will handle a liquid spill fire in typical rack storage. More information on safe practices in the care and use of industrial trucks will be found in Chapter 37, "Handling and Storage of Industrial Gases," Chapter 46, "Materials Handling Systems," and in Reference 1.

Welding and cutting: Repair or replacement of steel racks or overhead structures will sometimes necessitate "hot work." A multimillion dollar fire in rolled paper was started by hot sparks falling from overhead craneway repair work. A formal written permit system is used in many industrial plants to control this hazard. The use of mechanical fastenings and mechanical saws or cutting wheels is sometimes recommended in place of welding and cutting. Fixed rack structures are usually designed to facilitate removal or repair of damaged sections without resorting to flame cutting or welding in the storage area. (See Chapter 24 for a complete discussion of "Welding and Cutting.")

Smoking: Smoking should be controlled. Managers sometimes accept the argument that a curtailment of freedom to smoke will cause workers to sneak "smokes" in an unsafe, combustible space behind stock, or that outside truckers cannot be stopped from smoking when they enter the shipping area. The control of smoking requires a sincere management effort to enforce observance of permissible and prohibited smoking areas.

Arson: Storage areas are often chosen as the starting place of incendiary

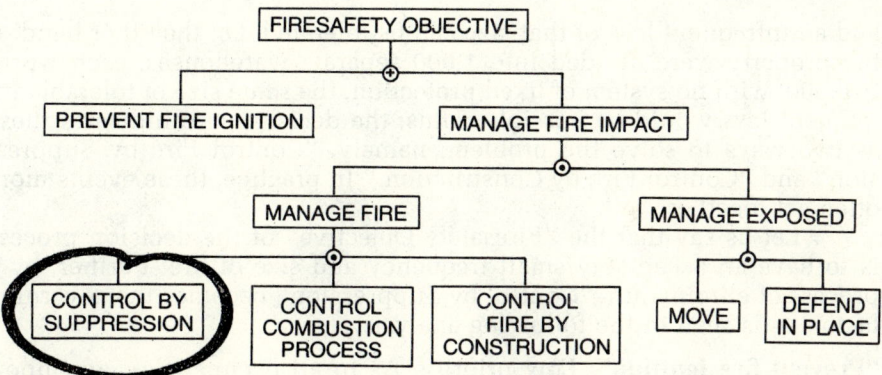

FIGURE 48.2. A decision tree showing "Control by Suppression" as the governing factor (see text for analysis).

fires, because the perpetrator usually doesn't need to bring his own "plant" (easily ignitible fuel). Warehouses have fewer employees than manufacturing areas, and those workers may not question a visitor, who may be an unknown important official from the front office. The arsonist can then find a concealed place for the crime. With arson an ever increasing threat, basic security measures are needed in all high value areas. A good security program will include controlled access of visitors, adequate fences, locked gates, good exterior lighting, patrols or visits during idle hours, locked warehouse doors, well-maintained windows, screening of new employees, and visual identification such as badges.

FIRE PROTECTION SYSTEMS

To see the factors involved in achieving the firesafety objective for a warehouse, a "decision tree" may be a useful analytical tool. (Figure 48.2 is an abbreviated display of a decision tree.) The decision tree offers a method of graphically showing the various events that can influence the achievement of an objective. Each event can succeed or fail, and there is, for any situation, a probability of success for each event. If we had numerical values for all the probabilities, they could be inserted in the tree and the overall probability of reaching the objective calculated. Even without the numbers, we can see which factors are economically or physically feasible for each case and how they interact. For example, "Prevent Fire Ignition" and "Manage Fire Impact" (Figure 48.2) involve an either-or situation. If fire prevention could be a 100 percent probability, there would be no need for fire management.

Let us consider an interesting but wholly improbable situation: A warehouse with good fire prevention is valued at $10,000,000. If it is adequately protected by automatic sprinklers, a $10,000 loss is possible, and an infrequent loss of that size can be tolerated. On the other hand, if the property were divided into 1,000 separate warehouses, each worth $10,000, with no system of fixed protection, the same size of tolerable infrequent loss would be possible. Thus, the decision tree indicates these as two ways to solve the problem; namely, "Control Fire by Suppression" and "Control Fire by Construction." In practice, these events more often act together.

Let us say that the "Firesafety Objective" of the decision process is to have an acceptably small frequency and size of fire. Further, by a process of elimination, "Control by Suppression" becomes the governing factor, as is seen in the following analysis:

"Prevent fire ignition": Low priority. As in all occupancies, an important factor, but in warehouses the frequency of fires is already low, and more success in fire prevention is secondary to controlling those fires which inevitably do occur. Prevention plays a relatively minor role compared to "Manage Fire Impact" except, perhaps, in a small isolated warehouse where the latter is not economically feasible.

"Manage exposed": Low priority. Moving, or defending in place, the contents of warehouses is usually difficult because of the very nature of the contents. At best it serves merely to lessen, to a degree, the impact of a large fire. People inside do need full consideration, but they are only slightly endangered, and few in number. Fire fighters can be severely exposed, but not if "Manage Fire," another main section of the decision tree, is successful.

"Control combustion process": Low priority. Arrangements of combustibles to diminish their fire severity are sometimes practical, but would most likely inhibit the operations and the economical use of the space.

"Control fire by construction": Low practicality. Compartmentation is effective in reducing the extent of an uncontrolled fire and was at one time a key element in warehouse design. It now plays a secondary role, generally to separate storage from other occupancies. A French report on four large warehouse fires commented,". . . these fires . . . give once more evidence of the need for the compartmentation of large open plan areas even if this can sometimes hamper storage handling - procedures." In one of the four fires, the building was equipped with automatic sprinklers, but while all of them operated, there was still extensive damage, apparently because the system was not adequate for the hazard. Good walls can themselves stop the fire, or at least enable fire fighters to make a stand, but the relentless trend has been to erect ever larger continuous areas for effective use of materials handling equipment. In reality, the forces against compartmentation are overwhelming, except when a huge value is subject to a single fire. Then, an underwriting or risk management concept may dictate subdivision by one or more structurally independent fire walls.

For reasons given in the preceding analysis, the greatest burden of success in achieving the firesafety objective in warehouses lies with

"Suppress Fire." Fire in storage can have a devastating effect, so we focus on ways to control and extinguish it. The other factors in the decision tree also play their parts, but fire suppression is the core of the problem. Therefore, much of this chapter is devoted to fire suppression.

FIRE CONTROL IN WAREHOUSES

Successful fire suppression encompasses detection, control, and extinguishment. In today's state of the art, automatic sprinklers are heavily relied upon for control; thus, it would seem the problem narrows down to using a good sprinkler system to do the job. Failures of supposedly good systems have been known to occur, with fire sweeping through an entire fire area, opening all sprinklers. There are, of course, reasons for these failures, which emphasize the need for careful design and the maintenance of conditions preconceived in the design. Some examples of sprinklered warehouse fire disasters reinforce those points:

1962. Roll paper warehouse, Florida, 60 ft (18.3 m) high with 30 ft (9.14 m) high unbanded paper on end. Total loss, $3,000,000.

Oct. 1972. Nylon yarn warehouse, Luxembourg. About 9200 m² (99,000 sq ft) in area, 10 m (33 ft) high, the building contained yarn in boxes and on aluminum reels, all polypropylene shrink-wrapped, partly on wood pallets, partly in racks. Sprinklers were at the roof and at half-height in the rack storage. Sprinklers operated, but were quickly overpowered, apparently because of insufficient in-rack sprinkler coverage.

May 12, 1977: Supermarket warehouse, Paris, France. The building was about 17 000 m² (183,000 sq ft) in area, one story high, and contained multiple racks for pallets with good aisles. Sprinklers were under the roof and in a criss-cross pattern at three levels inside the racks. In addition to foodstuffs, storage included wines and liqueurs, oils, cosmetics, aerosols, and some rubber tires. Palletized storage was up to 9 m (30 ft) high, and the goods were covered with plastic sheets. Fire spread rapidly, helped by a pile of empty cartons, and smoke was abundant. Burning flammable liquids flowed along the floor, and numerous explosions were heard, probably from aerosol cans. Sprinklers kept most of the steel structure of the roof and racks from collapsing, but could not stop the fire. Apparently the system was not suitable to cope with this occupancy.

Oct. 20, 1977: Automotive parts depot, Merkenich (Cologne), West Germany (see Figure 48.3). A variety of combustible and noncombustible articles in wire mesh baskets were piled just under 20 ft (6 m) high, either one on top of the other or in double row racks. Water supplies could deliver 4,000 gpm (15.1 m³/min) at 82 psi (565 kP). Sprinkler density at the origin of the fire was 0.46 gpm/sq ft [18 (L/min)/m²] over a 5,000-sq ft (465 m²) area. The sprinkler protection, meeting "highly protected risk" standards of an American insurer, was in service, yet fire destroyed

FIGURE 48.3. Sprinklered automotive parts warehouse in West Germany that suffered a $100 million loss. It was one of the largest industrial fires in history. (Copyright by Verufsfeurwehr der Stadt Köln)

800,000 sq ft (74 322 m^2) of the property. While several unfortunate things happened, the most adverse was the temporary storage of motor oil in cans in cardboard cartons in the 8- to 9-ft (2.4 to 2.7 m) wide aisles, stored 20 ft (6.1 m) high, and totaling 151 988 liters (40,151 gals). The cans had spouts sealed with an aluminum foil strip stuck on by adhesive, making the oil available to feed the fire when heated. The decision to permit this temporary storage nullified the capability of the sprinkler protection, which depended upon open aisles.

In the United States a number of fires in storage of foamed polyurethane and polystyrene have overpowered automatic sprinklers, either because the material was temporarily piled higher than planned, or because the system design did not fully compensate for the severity of the hazard. Recent fire testing has developed improved standards for adequate sprinkler protection.

Fire Control — A Definition for Warehouses

"Control" is the critical part of the fire suppression process, and can be defined as that which: 1) keeps a fire from spreading beyond an acceptable maximum, 2) keeps the building structure from burning or collapsing, and 3) maintains both conditions for a long enough time to extinguish the fire successfully or permit manual extinguishment.

Terms in this definition are further explored:

"Acceptable maximum." For most types of storage this term is defined as an acceptable maximum area. It is usually related to the size of area in which the design of the sprinkler system anticipates sprinklers will operate in a fire (the "demand area"). That area usually extends beyond the fire itself, and its limits are flexible, being inversely related to the density of the discharge. If there are no sprinklers, an acceptable maximum is the volume of fire that can be controlled by other means and tolerated as a loss.

"Keep the building structure from burning or collapsing." Fire tests have shown that the degree to which certain types of structures can resist the effects of fire can be anticipated as a function of the maximum ceiling temperature generated in the fire.[2]

"Time to extinguish the fire." Enough water or other extinguishing agent must be available for the time needed to put out the fire. Good operating conditions of the extinguishing equipment involved (sprinkler systems, hand hose, etc.) and prompt and efficient manual fire fighting must be present. It also involves such items as the size of the fire service water storage tank, adequacy of hydrants, watchman service, central station supervisory service, quality of fire brigade, and public fire department response. For example, two hours of water supply is a typical minimum needed to keep fire in check by sprinklers while other fire fighting and suppression activities proceed.

Fire Control System

Buildings of combustible construction need automatic fire control, and invariably that means automatic sprinklers. Valuable commodities, even if entirely noncombustible, should not be endangered by a fast-spreading destructive roof fire resulting from some temporary unexpected fire condition in the building.

In buildings that are not combustible, an automatic means of control is needed unless the total value is small and expendable, or the contents are essentially noncombustible or subdivided into fire-resistive compartments or both. Major dependence is placed on automatic sprinkler systems, with the occasional use of automatic flooding high expansion foam systems for special applications, such as storage of roll paper or rubber tires. Foam is generally a complement to the sprinklers.

The general rules of NFPA 13, *Sprinkler Systems*, apply to the installation of systems in warehouses where storage piles are low, but as most warehouses have storage higher than 12 ft (3.6 m) on pallets and racks, some supplemental requirements apply. Information on these requirements are found in the NFPA storage standards, i.e., NFPA 231, *Indoor General Storage*; NFPA 231C, *Rack Storage of Materials*; NFPA 231D, *Storage of Rubber Tires*; NFPA 231E, *Recommended Practice for the Storage of Baled Cotton*; and NFPA 231F, *Storage of Roll Paper*. Two conditions are carefully analyzed in determining what constitutes appropriate sprinkler protection for high storage areas. They are the *commodities themselves* and the *storage arrangements* for them. Both are essential components of the analysis leading to good sprinkler protection for the storage arrangement in question.

CLASSIFICATION OF COMMODITIES

Significant fire hazard properties of a commodity include its heat of combustion, rate of heat release, and rate of flame spread. However, commodities are often complex items whose fuel content, arrangement, shape, and form affect their performance in a fire. A packaged commodity must be considered as a whole, since that is the way it burns. For ease in applying protection requirements, stored commodities are divided into general hazard classifications based upon the behavior of typical items in each classification. The classifications are:

Class I: These commodities are essentially noncombustible products on combustible pallets, in ordinary corrugated cartons with or without single thickness dividers, or in ordinary paper wrappings with or without pallets. Examples of Class I products include metal parts, empty cans, stoves, washers, dryers, metal cabinets, glass products and noncombustible food stuffs.

Class II: These commodities are Class I products in slatted wooden crates, solid wooden boxes, multiple thickness paperboard cartons, or equivalent combustible packaging material with or without pallets. Examples of Class II products include light bulbs and Class I products in small cartons or small packages placed in ordinary paperboard cartons.

Class III: These commodities are wood, paper, natural fiber cloth, or comparatively slow burning plastics (Group C) or products thereof with or without pallets. These products may contain a limited amount of Group A or B plastics. Metal bicycles with plastic handles, pedals, seats, and tires are an example of a commodity with a limited amount of plastics. Examples of Class III products include books, plastic-coated food containers, shoes, luggage, furniture, and combustible foods.

Class IV: These commodities are Class I, II, or III products containing an appreciable amount of Group A plastics in ordinary corrugated cartons and Class I, II, or III products in corrugated cartons with Group A plastic packing, with or without pallets. An example of plastic packing material is a metal typewriter in a foamed plastic cocoon in an ordinary carton. Examples of Class IV products include small appliances, typewriters, and cameras with plastic parts; plastic backed tapes and nonviscose synthetic fabrics or clothing; telephones; vinyl floor tiles; wood or metal frame upholstered furniture or mattresses with plastic covering and/or padding; plastic padded metal bumpers and dashboards; insulated conductor and power cable on wood or metal reels or in cartons; inert solids in plastic containers; and building construction insulating panels of polyurethane sandwiched between nonplastic material.

Plastic Commodities

Unmodified plastics are placed in one of three groups (A, B, or C) according to their rate of burning, with those having the highest rate falling in the "A" group (the "bad actors" such as expanded polyurethane and polystyrene fall in this group). Criteria for rating plastics are based upon miscellaneous large- and small-scale tests and recognize that additives to unmodified plastics can change their burning characteristics.

Essentially, plastics that fall in Group B and "free-flowing" Group A plastics (those in the form of powder, pellets, flakes, or random small objects that can fall out of their containers, fill flue spaces, and create a smothering effect on the fire) are the same as Class IV storage. Group C plastics are essentially the same as Class III storage. Group A plastics are further subclassified according to whether the material is nonexpanded or expanded (foamed, cellular) and also whether it is cartoned or exposed (exposed plastics can become involved in a fire more quickly than those encased in cartons).

The plastic groupings are helpful in identifying the degree of hazard contributed by plastics to the different storage classifications. For example, a limited amount of Group A or B plastics, such as represented by plastic parts on metal bicycles, are tolerable in Class III storage and do not appreciably influence the degree of fire protection required. Judgment is needed in determining whether an amount of plastic is a "limited amount" suitable for inclusion in a Class III commodity. Similar judgments have to be made for the other classes of commodities and for the influence the different plastic groupings would have on protection requirements.

The Classification of Plastics in NFPA 231 follows:

Group A

    ABS (Acrylonitrile-Butadiene-Styrene Copolymer)
    Acrylic (Polymethyl Methacrylate)
    Acetal (Polyformaldehyde)
    Butyl Rubber
    EPDM (Ethylene-Propylene Rubber)
    FRP (Fiberglass Reinforced Polyester)
    Natural Rubber (if expanded)
    Nitrile Rubber (Acrylonitrile-Butadiene Rubber)
    PET (Thermoplastic Polyester)
    Polybutadiene
    Polycarbonate
    Polyester Elastomer
    Polyethylene
    Polypropylene
    Polystyrene
    Polyurethane
    PVC (Polyvinyl Chloride - highly plasticized, e.g., coated fabric, unsupported film)

SAN (Styrene Acrylonitrile)
SBR (Styrene-Butadiene Rubber)

Group B

Cellulosics (Cellulose Acetate, Cellulose Acetate Butyrate, Ethyl Cellulose)
Chloroprene Rubber
Fluoroplastics (ECTFE -Ethylene-Chlorotrifluoroethylene Copolymer; ETFEEthylene-Tetrafluoroethylene Copolymer; FEP - Fluorinated Ethylene-Propylene Copolymer)
Natural Rubber (not expanded)
Nylon (Nylon 6, Nylon 6/6)
Silicone Rubber

Group C

Fluoroplastics (PCTFE - Polychlorotrifluoroethylene; PTFE Polytetrafluoroethylene)
Melamine (Melamine Formaldehyde)
Phenolic
PVC (Polyvinyl Chloride - rigid or lightly plasticized, e.g., pipe, pipe fittings)
PVDC (Polyvinylidene Chloride)
PVF (Polyvinyl Fluoride)
PVDF (Polyvinylidene Fluoride)
Urea (Urea Formaldehyde)

British Standards — Commodity Classes

British Fire Offices' Committee (F.O.C.) sprinkler system rules[4] also incorporate four general classes of goods which, when stored in heights exceeding 4 m (13 ft), are classified by lists of examples which include the following:

1. Category I: Carpets, clothing, groceries.
2. Category II: Baled waste paper, plastics (nonfoamed).
3. Category III: Foamed plastics, rubber goods, idle wood pallets.
4. Category IV: Scrap foamed plastics.

No useful correlation between NFPA storage standards and British standards is feasible. NFPA standards are somewhat more stringent, and more broadly based on full-scale tests. The F.O.C. rules are mentioned here because they are used in a number of countries as an alternate to NFPA standards, and in a few countries as the prevailing standard supported by the local insurance industry.

Determining the Classification

The classification (Class I, II, III, IV or Plastics A, B or C) for a commodity is based upon the most severe hazard in the storage area. Miscellaneous types of storage, such as in hardware or automotive parts warehouses, have commodities in several classes but generally belong in Class IV because there is an appreciable amount of Class IV storage. It is frequently better to assign a next higher class than is present if there is any probability of the higher class being introduced later. The difference in protection cost at the outset could be small compared to the cost of upgrading the protective system at a later date.

Higher hazard commodities can sometimes be segregated to avoid influence on the overall hazard class. Thus, exposed foamed plastics or labeled flammable liquid drums can be stored in special rooms having strong internal protection and good fire separation from the main area. Minor quantities in small containers, such as cans of cigarette lighter fluid in a grocery warehouse or small propane cylinders on do-it-yourself torches in a hardware warehouse, tend, however, to intrude unnoticed by warehouse personnel. This condition probably occurs because they are awkward to identify and segregate during normal operations. It is usual practice to accept nominal quantities of such items in small containers, without classifying the entire storage as flammable liquids or flammable gases (which are generally considered to be more severe than the classes above and beyond the scope of this chapter). Vigilance is in order to see that quantities of such items continue to be limited.

Flammable aerosols, that is, flammable liquids in small pressurized spray cans, have recently demonstrated a shocking ability to rocket about while flaming, spreading fire in scattered locations, thereby overtaxing sprinkler systems. The Factory Mutual System has conducted full scale fire tests to observe the effects of flammable aerosols and has issued in-house standards for their protection including physical cut-offs, low piling and copious sprinklers within storage racks. Since aerosols find their way into many warehouses, their presence requires special study.

STORAGE ARRANGEMENT

From a fire suppression viewpoint, storage is divided into four main categories: bulk storage, solid piling, palletized storage, and rack storage. Bins and narrow shelves are also of importance, but they are usually employed in stock rooms holding moderate quantities of items for direct use. The differences between the four categories that affect the fire behavior and difficulty of fire control are basically in the nature of horizontal and vertical air spaces or "flues" that the storage configurations create.

Bulk Storage

Bulk storage consists of piles of unpackaged materials, in a loose, free-flowing condition, such as powder, granules, pellets or flakes. The materials will be found in silos, bins, tanks, or in large piles on the floors of storage buildings (see Chapter 6).

With conveyor equipment, such as belt conveyors, air fluidizing through ducts, and bucket conveyors ("legs"), the material is agitated during certain stages of its movement. If prone to developing airborne dust, a combustible material presents an explosion hazard, notably in grain storage facilities. Conveyor belts are generally combustible, and, together with the commodity, can burn in inaccessible places such as at elevations high above the floor or in tunnels. Automatic sprinklers are frequently needed in the housings around conveying equipment.

Fires in large piles tend to burrow and require prolonged soaking to be reached. Spontaneous ignition fires may start in the interior of piles, and may be difficult to locate if internal heating has not been continually monitored by immersed heat sensors. The question of whether to install automatic sprinklers over large piles in noncombustible buildings is a difficult one,[5] but good hose stream access around the perimeter is always important.

Solid Piling

Solid piling consists of cartons, boxes, bales, bags, etc., in direct contact with each other to the full dimensions of each pile. Air spaces, or flues, exist only where there is imperfect contact, or where a pile is close to but not touching another pile. Stacking is by hand or by lift trucks using side clamps or the usual prongs when bales or other packages can be pried up from below without serious damage in the absence of pallets (see Figure 43.4).

Compared with palletized and rack storage, solid piling gives fire the least opportunity for fire development, and gives water application the greatest chance to be effective. Still, where outer surfaces possess rapid flame spread properties, high piling, starting at about 15 ft (4.5 m), can present a severe hazard. Movement of stock during a fire to get at the burning interior is more difficult.

Solid piling of general classes of goods up to 15 ft (4.5 m) high requires sprinkler protection complying with the NFPA sprinkler standards. However, for plastics over 5 ft (1.5 m) high, and other piling between 15 and 30 ft (4.5 and 9 m) high, NFPA 231, *Indoor General Storage*, gives comprehensive protection requirements except that for plastics in Group A (except free-flowing) the standard only applies up to 25 ft (7.6 m) height.

FIGURE 48.4. An example of solid piling without pallets. Lack of air spaces such as found in palletized or rack storage gives the least opportunity for fire spread; however, it appears this warehouse is not sprinklered. (Clark Industrial Truck Div.)

Palletized Storage

Palletized storage consists of unit loads mounted on pallets. The load is usually an approximate cube, about 3 to 4 ft (0.9 to 1.2 m) in height, made up of a single package or block of packages presenting a top surface that can sustain the weight of additional pallet loads on top of it without crushing the packages or forming too unstable a pile. Each pallet is about 4 in. (102 mm) high, usually of wood, but some are made of metal, plastic, expanded plastic, or cardboard (see Figure 48.5). It accepts the prongs of a forklift truck, and can also be an integral part of a sizable shipping container. The height of palletized storage is limited by the stackability of the packaged goods, usually 30 ft (9.1 m) at most.

Horizontal air spaces are formed by the pallets themselves. These spaces significantly assist the spread of fire within the pile, because they are out of reach of water from sprinklers. The airspaces are usually continuous in one direction for the entire width of a pile. An effort was made about 1967 to promote fire-stopped wood pallets that accept lift truck prongs from one side only. This method could have relieved the palletized storage protection problem.

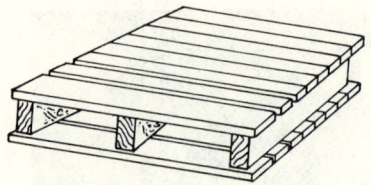

FIGURE 48.5. A conventional wood pallet.

TABLE 48.1. Protection for Indoor Storage of Idle Pallets of Wood or Nonexpanded Polyethylene

Height of Pallet Storage		Sprinkler Density Requirements		Area of Sprinkler Demand			
				sq ft		m²	
ft	m	gpm/sq ft	L·min/m²	286°F	165°F	141°C	74°C
up to 6	up to 1.8	0.20	8.1	2000	3000	186	279
6 to 8	1.8 to 2.4	0.30	12.2	2500	4000	232	372
8 to 12	2.4 to 3.6	0.60	24.4	3500	6000	325	556
12 to 20	3.6 to 6.1	0.60	24.4	4500	—	418	—

When vertical flues are very narrow — less than 6 in. (152 mm) — air movement and flaming is restricted. Fire tests of plastics storage using wider flues — 12 in. (0.3 m) — showed enhanced air movement and flaming through a pile. These particular dimensions may not be as pertinent for storage other than plastics.

Sprinkler system requirements for palletized storage up to 12 ft (3.65 m) high are contained in NFPA 13, *Sprinkler Systems*. However, for plastics over 5 ft (1.5 m) high, and other goods within heights above 12 ft (3.65 m) and up to 30 ft (9.1 m), NFPA 231, *Indoor General Storage*, gives requirements, except for plastics in Group A.

Idle Pallets

Piles of wood or plastic pallets introduce a severe fire condition. After pallets are used for a short time they dry out and their edges can become frayed and splintered. In this condition, they can ignite easily from a small ignition source. Even with sprinklers operating, the undersides of the pallets provide a dry area on which a fire can grow and expand. The process of fire jumping to other pallets in the pile continues until the fire bursts through the top of the stack. Once this occurs, there is a strong updraft of flame, and very little water can reach the base of the fire. A 6-ft (1.8 m) storage height is considered the maximum in which such a fire can be kept within manageable limits by ordinary hazard sprinkler protection. For higher piles of pallets, higher sprinkler system discharge densities are needed (see Table 48.1).

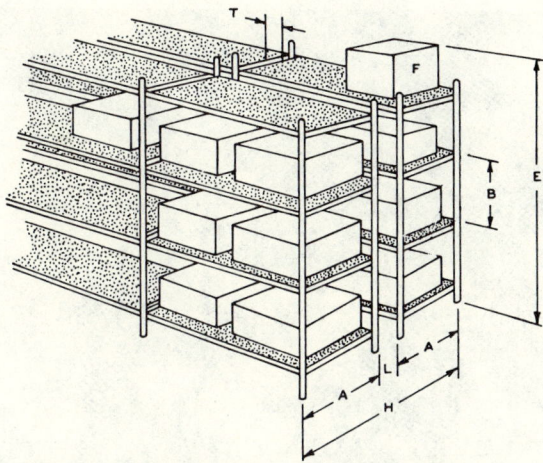

FIGURE 48.6. *Double row rack with solid shelves. Legend: A — shelf depth, B — shelf height, E — storage height, F — commodity, H — rack depth, T — transverse flue space, L — longitudinal flue space.*

Rack Storage

A storage rack consists of a structural framework into which are placed unit loads, generally on pallets (see Figures 48.6 and 48.7). The height of storage racks is, potentially, limited only by the vertical reach of the materials handling machinery which, like the racks themselves, can be designed for great heights. While most racks are roughly 25 ft (7.6 m) high, an appreciable number are higher, and in fully automated warehouses racks are sometimes up to 100 ft (30 m) and have even exceeded that height. Some of these systems adapt their steel racks as a framework for supporting the exterior walls and roof of the building (see Figure 48.8). As in any important building, this structural system should be adequate enough to withstand snow loads and wind pressure.

In addition to open single and double row racks (see Figure 48.8), there are numerous variations (see Figures 48.9 and 48.10). Portable racks approximate pallets except that the upper pallet does not rest on the storage below, but leaves an air space between (see Figure 48.11).

Automated materials handling machinery for high racks can operate in narrow aisles, such as 4-ft (1.2 m) widths across which fire can readily jump (see Figure 48.12). In racks, horizontal air spaces about 1 ft (0.3 m) in depth are under each tier of supports which provide clearance for handling. Since racks are seldom completely filled, the random vacancies also contribute to the air spaces. Vertical flues exist at upright rack members and between unit loads. They are a mixed blessing, allowing upward fire spread, but making possible some sprinkler water penetration into the usually open-style racks.

In high automated rack storage, even if the contents damage is not

Chapter 48

FIGURE 48.7. Typical double row rack storage under 25 ft (7.6 m) in height with commodities on wood pallets. (UNARCO Materials Storage)

severe, the fact that the materials handling mechanisms must be accurately aligned in order to function makes them vulnerable to the effects of rack distortion by heat, possibly leading to the impairment of the entire warehouse operation.

Fortunately, racks are a permanent type of installation which can support sprinkler piping. Otherwise, the combination of horizontal air spaces sheltered from overhead sprinkler water discharge, great height, and narrow aisles would pose an impossible problem for a conventional sprinkler system installed at the ceiling level.

With the acceleration of rack storage in the 1960s, there was a dangerous shortage of fire data upon which to base standards. In 1967, in a unique act of research cooperation, certain large insurance organizations, rack manufacturers, the fire protection equipment industry, and industrial users joined together to organize the Rack Storage Fire Protection Committee, which developed and financed a program of full-scale fire tests. Data from the next few years of tests were turned over to an NFPA committee and a standard (NFPA 231C, *Rack Storage of Materials*) was developed for storage racks up to 25 ft (7.6 m) high, based upon automatic sprinkler protection. (For rack storage up to 12 (3.7 m) ft high, provisions of the NFPA general sprinkler standard still apply.) Further test results enabled amendments to NFPA 231C to provide for any storage height above 25 ft (7.6 m), making liberal use of in-rack automatic sprinklers.

Meanwhile, fire research in England has led to the development of an interesting method: a system of small vertical deluge sprinkler system zones that are intended to quickly put water on a low level fire *and* the

FIGURE 48.8. *High steel storage rack as structural support for roofs and walls of a storage facility. (Clark Handling Systems Div.)*

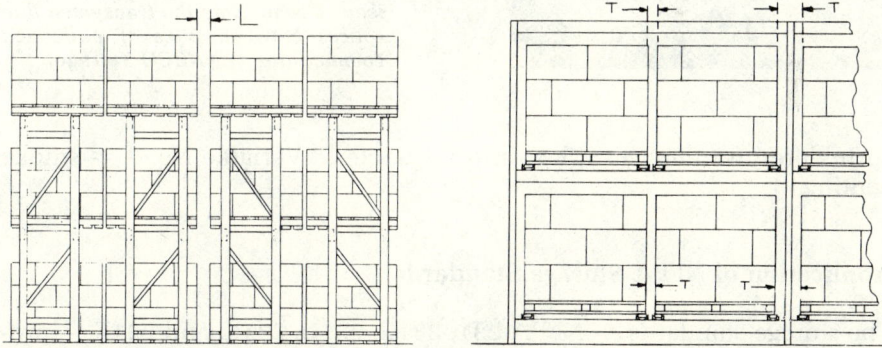

FIGURE 48.9. *Typical multiple row racks showing locations of longitudinal (L), and transverse (T) flue spaces. On the left is the end view; on the right is the aisle view.*

goods above it, preventing vertical fire spread.[6] This has apparent advantages, and has been considered in the United States.

In Germany, in addition to sprinklers at all levels, a system of early warning zoned fire detectors has been promoted to pneumatically operate smoke and heat vents in the roof and alert fire fighters, for added protection. Not all fire protection engineers agree that this is desirable in

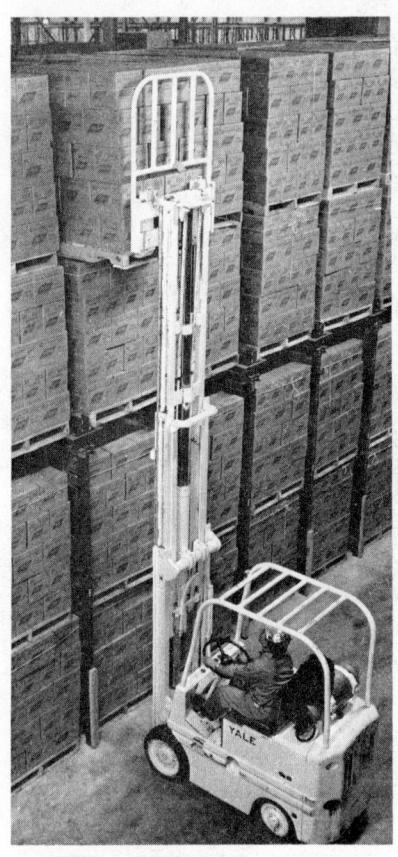

FIGURE 48.10. Multiple rack palletized storage. Note the transverse flue spaces between piles of palletized commodities. (UNARCO Storage)

sprinklered warehouses. (See, in this chapter, "Variable No. 9 — smoke venting.")

Application of NFPA Storage Standards

The storage standards — NFPA 231, 231C, 231D, 231E, and 231F — provide some flexibility in planning protection for storage occupancies instead of merely stipulating minimum requirements for protection; the user is offered alternatives. For example, there are curves representing a range of sprinkler discharge densities and the areas they would cover for a variety of storage conditions. Any point chosen from a curve gives the density-area criteria for the particular conditions represented by the curve (higher densities of discharge over smaller areas can result in less damage and are preferred). Often the planner is not bound by one curve, but can alter conditions so that a more favorable curve can be used. For example, consider a case (see Figure 48.13) where Class IV commodities

INDUSTRIAL STORAGE PRACTICES

 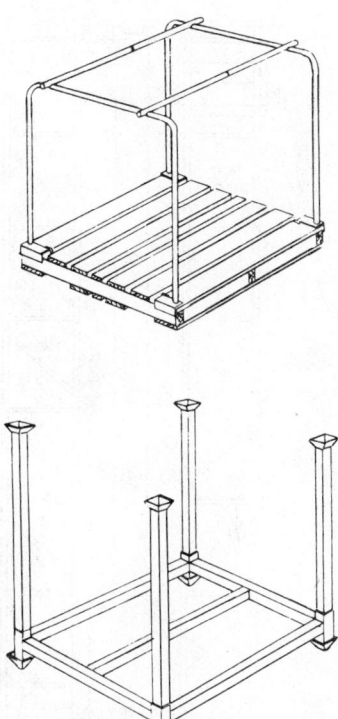

FIGURE 48.11. Typical portable racks. Note (at right) the air space between the top of the piled commodities and the rack above. (Equipment Co. of America)

are stored 20 ft (6.1 m) high in double row racks without sprinklers located in the racks. (See also "Water Supply" in this chapter.) There are 8-ft (2.4 m) aisles and 286 °F (141°C) ceiling sprinklers. A point on Curve E may be selected, such as 0.415 gpm/sq ft (16.9 [L/min]/m^2) of ceiling sprinkler discharge density over 4,000 sq ft (372 m^2) of sprinkler operating area. But, by installing one level of in-rack sprinklers at about the mid-height of the racks, Curve A may be substituted, requiring only 0.27 gpm/sq ft (13.2 [L/min]/m^2) over 4,000 sq ft (372 m^2). This effects a reduction in the water supply needed of about 640 gpm (2423 L/min) for ceiling sprinklers, while the added in-rack sprinkler demand would be only about 190 gpm (719 L/min). The cost of the in-rack sprinkler installation should be balanced against lower water supply cost *and* the reduced fire damage expected.

Other choices are available, including different in-rack sprinkler arrangements for rack storage over 25 ft high (7.6 m).

Chapter 48

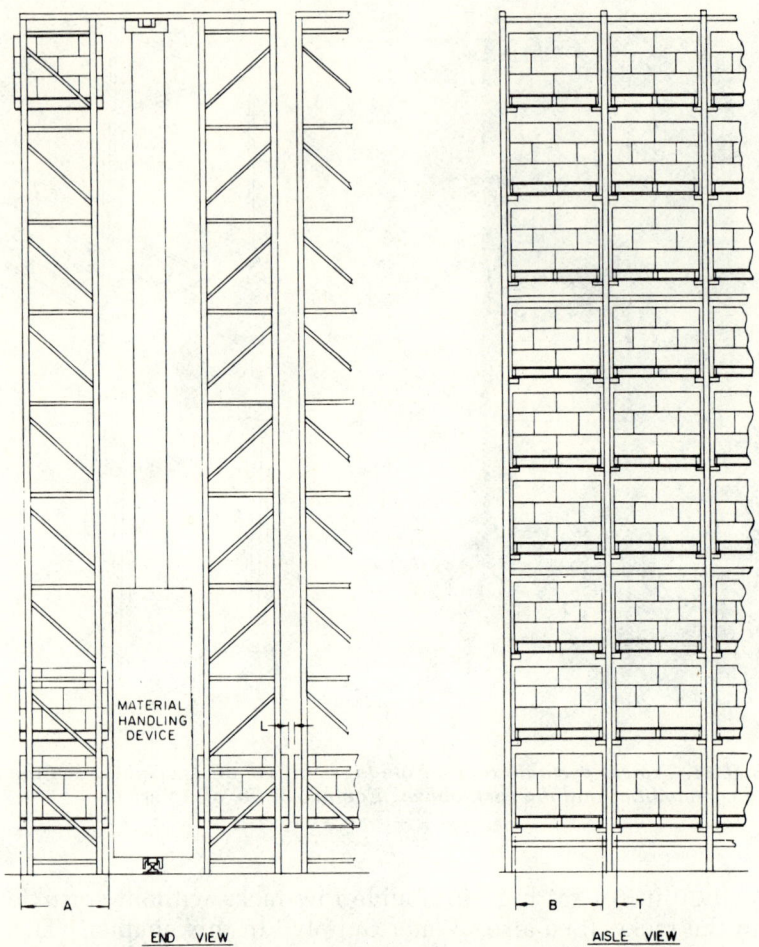

FIGURE 48.12. *A typical automatic storage material-type rack. The material handling device must operate in a narrow space. Legends: A — load depth, B — load width, L — longitudinal space, T — transverse flue space.*

The Influence of Variables

The NFPA storage standards are excellent guides, but they do not cover all situations. It is sometimes necessary to improvise by extrapolation or interpolation. While intended mostly for planning new installations, the recommendations and requirements in the standards assist in upgrading substandard installations and in judging the deficiency of protection so that the cost of an improvement (and options if any) can be weighed against the seriousness of the deficiency.

Before improvisation upon the basic protection requirements can be attempted, the variables that can be considered must be understood.

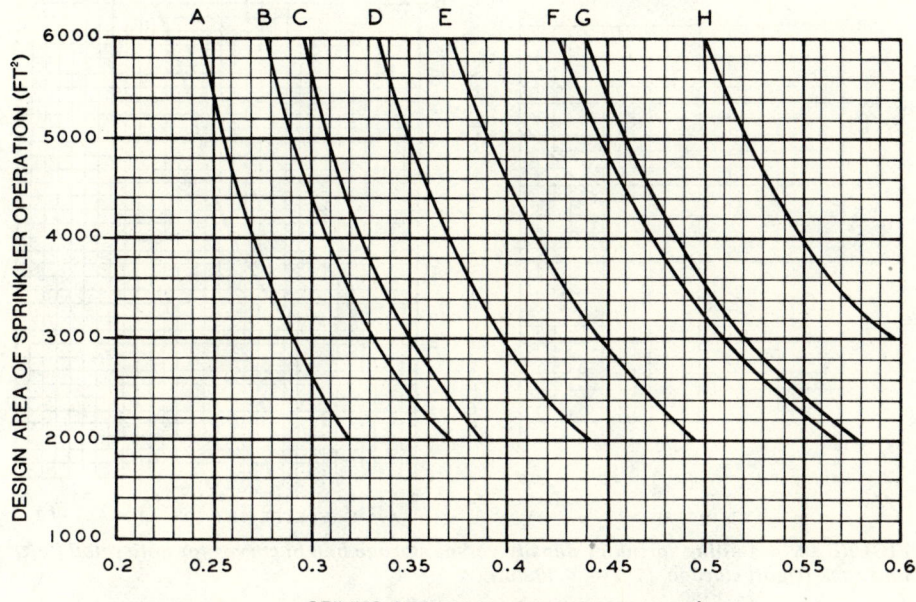

FIGURE 48.13. Typical sprinkler system design curves for rack storage of materials. These curves apply to systems protecting 20-ft (6.1 m) high double row racks holding Class IV nonencapsulated commodities on conventional pallets. Other curves govern other storage conditions and configurations as covered in the NFPA storage standards. [1 ft = 0.305 m; 1 sq ft = 0.0929 m^2; 1 gpm/ft^2 = 40.746 [L/min]/m^2; 5/9 °F−32 = °C]

Curve Legend
A—8 ft aisles with 286°F ceiling sprinklers and 165°F in-rack sprinklers
B—8 ft aisles with 165°F ceiling sprinklers and 165°F in-rack sprinklers
C—4 ft aisles with 286°F ceiling sprinklers and 165°F in-rack sprinklers
D—4 ft aisles with 165°F ceiling sprinklers and 165°F in-rack sprinklers

Curve Legend
E—8 ft aisles with 286°F ceiling sprinklers
F—8 ft aisles with 165°F ceiling sprinklers
G—4 ft aisles with 286°F ceiling sprinklers
H—4 ft aisles with 165°F ceiling sprinklers

Armed with an appreciation of these variables, the evaluation of existing protection and improvisation in original planning or upgrading of protection can be properly employed. The author considers eleven variables and each is discussed in the following paragraphs:

Variable No. 1 — storage height: Other than the fire properties of the commodities themselves, probably no other condition has a more profound influence on the progress of fire in a storage occupancy and difficulty of fire control than storage height. In rack storage tests, fire intensity was seen to vary with the square of the storage height. As an example (without in-rack sprinklers), a 10 percent increase in rack storage

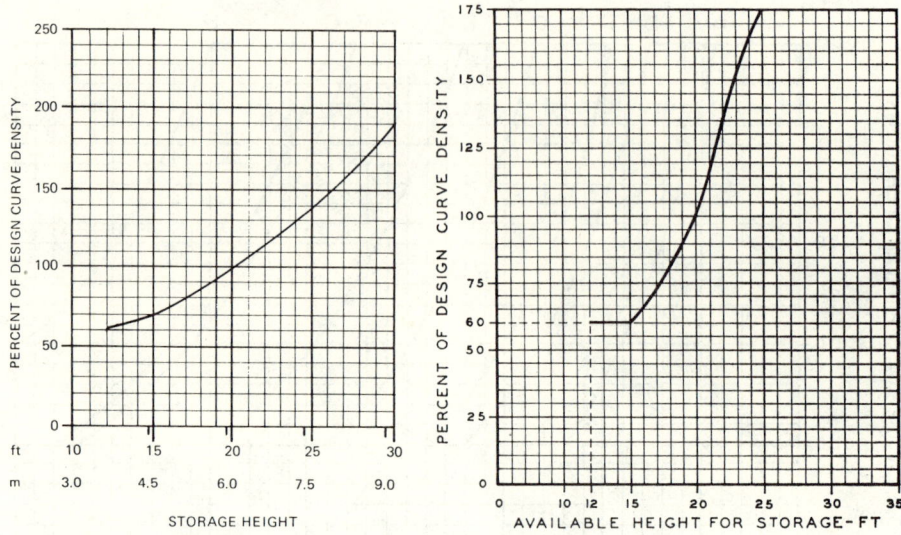

FIGURE 48.14. *Ceiling sprinkler density versus storage height curves for both piled (left) and rack (right) storage. (1 ft = 0.305 m)*

height from 20 to 22 ft (6.1 to 6.7 m) requires a 35 percent increase in sprinkler discharge density (see Figure 48.14).

Although increasing height causes greater increase in fire intensity, all heights up to 15 ft (4.6 m) in solid piles or 12 ft (3.65 m) in palletized and rack storage are treated the same, i.e., palletized storage 5 ft (1.5 m) high requires the same level of sprinkler discharge as that required for storage 12 ft (3.65 m) high except for plastics. This example illustrates the principle that a safer condition must sometimes be grouped with the more hazardous conditions to allow for the effects of any future changes that may influence the degree of hazard.

Because of perennial uncertainties in the future use of buildings, it is safe practice to use the maximum height available for storage for protection calculations rather than the estimated maximum storage height. This would be to about 18 in. (0.46 m) below the sprinkler deflectors. Of course, with permanent racks, the building height can be disregarded. (Exception: see Variable No. 5.) For racks over 25 ft (7.6 m) in height, as height increases there is no change in the ceiling sprinkler requirements, because fire control for all but the highest storage is essentially by in-rack sprinklers. In such high rack storage, ceiling sprinklers primarily protect the ceiling structure and only the top portion of the storage.

Variable No. 2 — aisle width: Wide aisles help to get water on the fire, retard transfer of fire from one pile to another, and permit access for fire fighting and salvage. For rack storage up to 25 ft (7.6 m) high, the width of aisles is a factor in determining sprinkler density for aisle widths of 4 and 8 ft (1.2 and 2.4 m); for aisle widths between 4 and 8 ft (1.2 and 2.4

m) a direct linear interpolation between curves can be made (see Figure 48.13).

An old, but at times useful, rule for determining the aisle width needed to restrict fire jumping across aisles called for width to be at least one-half the pile height. Palletized storage piles should be not more than 50 ft (15.2 m) wide between major aisles, so that small hose streams can penetrate 25 ft (7.6 m) to the center of a pile. Although the solid piled and palletized storage standard does not give "credit" for wide or frequent aisles, the presence of aisles is assumed, and the more and wider the better. Fortunately, aisles are usually necessary for efficient operation in conventional warehouses, with 8-ft (2.4 m) width the normal minimum for maneuvering a lift truck. Where there are frequent aisles, approximately every 10 ft (3.04 m), it is generally agreed that a somewhat lower ceiling sprinkler density is needed over solid or palletized storage than indicated in NFPA 231.

Double row rack storage needs aisles 4 ft (1.2 m) or more in width to achieve the protection intended by NFPA 231C, *Rack Storage of Materials*. Should the aisles be, for instance, 3½ ft (1.07 m) or narrower, the storage would have to be classified as "multiple row racks," requiring a higher ceiling sprinkler density. Storage in the aisles in the huge automotive parts warehouse converted a double row rack configuration to multiple row racks, thereby making the sprinkler design inadequate.

Variable No. 3 — in-rack sprinklers: General. Within racks, there are sizable unsprinklered areas that defy the ability of ceiling sprinklers. In-rack sprinklers were conceived to provide a degree of wetting for those concealed areas, thus interrupting the otherwise free spread of fire horizontally and vertically. The higher the rack, the more essential the in-rack sprinklers become. In-rack sprinklers are of ordinary temperature −130 to 170°F (54.4 to 76.7°C) — rating, have ½-in. (12.7 mm) orifices, and are either pendant or upright, with water shields to avoid cooling by water spray from any in-rack sprinklers that may be above. Since the standard sprinkler was originally designed for ceiling use, covering a given area with a fairly uniform spray, and not for the limited spray coverage in the flue spaces of a rack, it is likely that special designs for this purpose will evolve.

For rack storage containing solid or slatted shelves of any height, sprinklers are needed beneath each shelf unless transverse flue spaces at least 6 in. (152 mm) wide are maintained. (An exception is single row racks up to 25 ft in height.) Otherwise, the rules for open racks apply (see Figure 48.6).

In-rack sprinklers are not needed if an automatic high expansion foam system is used in conjunction with ceiling sprinklers.

Storage up to 25 ft (7.6 m). For rack storage arrangements without solid shelves up to 20 ft (6.1 m) high, in-rack sprinklers are optional for most conditions, but their presence appreciably reduces the ceiling sprinkler density requirement. If in-rack sprinklers are installed at one level only, they are located at the tier level nearest to one-half to two-thirds of the storage height and are arranged longitudinally in the center of the rack. They become practical when storage height is at least 15 ft (4.6 m).

For the higher range of storage heights under 25 ft (7.6 m), one level of in-rack sprinklers is usually mandatory in double row racks, and in multiple row racks except that two levels are needed when Class IV storage is over 20 ft (6.1 m) high.

Sometimes racks are introduced into an existing building having an ordinary hazard ceiling sprinkler system. Installing one or more levels of in-rack sprinklers may relieve the necessity of upgrading the ceiling sprinkler system.

Storage over 25 ft (7.6 m). Ceiling sprinkler discharge can effectively wet exposed storage at the aisles only for the upper 15 ft (4.6 m) or so of the rack; thus, for rack storage over 25 ft (7.6 m) high, the lower faces of the rack are protected by "face" sprinklers that are located in the racks, located at transverse flues and positioned within 18 in. (0.46 m) of the aisle. They protect against vertical spread of fire over the face of the storage and assist in controlling fire in the racks. Various arrangements of face sprinklers and centrally located in-rack sprinklers have been devised (see Figure 48.15 for examples of arrangements for Class I, II, and III commodities), including the use of horizontal barriers. The arrangements prescribed can be repeated indefinitely as pile height increases.

Horizontal barriers are solid barriers, such as sheet metal or wallboard, covering the entire length and width of the rack, including all flue spaces. Their effect, when equipped with in-rack sprinkler coverage below, is to reduce the number of centrally located in-rack sprinklers needed. The barriers have the same restrictive effect on vertical fire spread as a full complement of face and central in-rack sprinklers at the same level. Figure 48.14 shows how the presence of barriers can reduce the number of sprinklers.

Variable No. 4 — temperature rating of ceiling sprinklers: Generally, in storage occupancies of low height, ordinary temperature — 135 to 170°F (57.2 to 76.7°C) — sprinklers are used, subject to higher temperature ratings for high ambient temperature conditions.* Some engineers recommend intermediate — 175 to 225°F (79.4 to 107°C) — or high temperature — 250 to 300°F (121 to 149°C) — sprinklers to reduce the number of sprinklers operating unnecessarily beyond the actual fire when the commodity may be expected to produce a rapid rate of heat release, which is often the case. For high storages a distinctly lower ceiling sprinkler system discharge density and design area of sprinkler operation are allowed by the standards when the temperature ratings recommended for a particular storage and protection configuration are used.

In palletized or solid piled storage, or rack storage up to 25 ft

---

*Automatic sprinklers have six different temperature ratings: ordinary, 135 to 170°F (57.2 to 76.7°C); intermediate, 175 to 225°F (79.4 to 107°C); high, 250 to 300°F (121 to 149°C); extra high, 325 to 375°F (163 to 191°C); very extra high, 400 to 475°F (204 to 246°C); and ultra high, 500 to 575°F (260 to 302°C).

INDUSTRIAL STORAGE PRACTICES

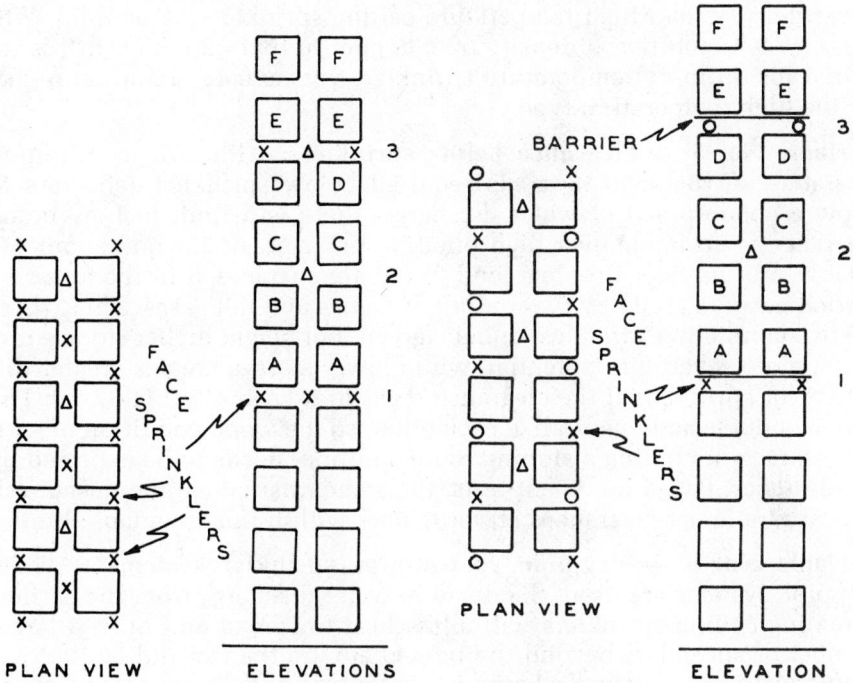

NOTES:
1. Sprinklers labeled 1 required when loads labeled A or B represent top of storage.
2. Sprinklers labeled 1 and 2 required when loads labeled C or D represent top of storage.
3. Sprinklers labeled 1 and 3 required when loads labeled E or F represent top of storage.
4. For storage higher than represented by loads labeled F, the cycle defined by notes 2 and 3 is repeated, with stagger as indicated.

FIGURE 48.15. *In-rack sprinkler arrangement for Class I, II, or III commodities with height of storage over 25 ft (7.6 m). Note how the presence of horizontal barriers within the racks (right) reduces the number of in-rack sprinklers required (left). The symbols O, △ or X indicate sprinklers installed in a vertical or horizontal alternating or "staggered" arrangement on the sprinkler piping.*

(7.6 m), high temperature sprinklers can yield a 40 percent reduction of sprinkler operating area, as compared with ordinary or intermediate temperature sprinklers. Somewhat surprisingly, the situation is reversed for rack storage over 25 ft (7.6 m). In a series of fire tests, in-rack sprinklers operated prior to ceiling sprinklers, indicating that in the over-25-ft (7.6 m) racks, in-rack sprinklers convert what would normally be a rapidly developing fire, from the standpoint of ceiling sprinklers, to a slower developing fire with a lesser degree of heat release. This result differed from tests involving 20-ft-(6.1 m) high storage; in those tests ceiling sprinklers operated before the in-rack sprinklers. Therefore, in over-25-ft (7.6 m) rack storage, ordinary temperature ceiling sprinklers require a distinctly lower density than high temperature ceiling sprinklers. [Example: With Class IV commodities, a density of 0.35 gpm/sq ft [14.2(L/min)/m^2] is required for ordinary temperature sprinklers versus 0.45 [18.3(L/min)/m^2] for the high temperature type.]

Variable No. 5 — clearance below sprinklers: Although a minimum clearance of 18 in. (0.46 m) is required below sprinkler deflectors to allow a good spread of water discharge, there is a limit to how much clearance is desirable over high piled or rack storage. Beyond about 4½ ft (1.37 m), the finer droplets tend to remain suspended in the air or be carried away laterally by the updraft of flames and hot gases. Thus, there are two conflicting variables; under a given roof height higher storage creates a more severe fire potential, while lower storage creates greater difficulty for sprinklers if the clearance starts to exceed 4½ ft (1.37 m). The two may not exactly neutralize each other, so the worst condition of variable storage, including a sloping roof condition, needs to be explored by the designer. Based on recent tests, the standards reduce the design criteria where lesser clearances are provided, within the acceptable limits.

Variable No. 6 — dry-pipe vs wet-pipe sprinkler systems*: Where dry-pipe systems are used, the delay in water discharge from the earliest operating ceiling sprinklers will allow heat to spread and open a larger number of sprinklers beyond the immediate fire than would be the case with a wet-pipe system. To compensate for this, the design area of sprinkler operation is usually increased by 30 percent for dry-pipe systems. The density should be selected so that the design area, after the 30 percent increase, does not exceed the upper area limit given in the design curve, such as the curves in Figure 48.12.

Variable No. 7 — pile stability: Instability of piles is often undesirable, fostering collapse into the aisles, providing a bridge for fire to cross over, impeding fire fighting operations, and possibly endangering building walls. However, tests show that, in some cases, sprinklers are more effective after some collapse, spillage of contents, or leaning of stacks across flue spaces has taken place. The favorable effect of a collapse in these situations is evidently in the choking off of the upward flow of fire in flue spaces.

Pile stability, under fire conditions, is difficult to judge, but some guidelines are available. Compartmented cartons containing stiff cardboard dividers were found to be stable under test fire conditions, while noncompartmented cartons tended to be unstable in the same tests. Storage on pallets within height limits suitable for the commodity to withstand compressive forces, and items held in place by materials which do not deform readily under fire conditions are both examples of

---

*In a wet-pipe sprinkler system the piping to which the automatic sprinklers are attached contains water under pressure at all times. In a dry-pipe system the piping contains air or nitrogen under pressure; when a sprinkler operates the pressure is reduced, a "dry-pipe valve" is opened, and water enters the piping to flow out of any opened sprinklers.

INDUSTRIAL STORAGE PRACTICES

stable storage. If there are leaning stacks, crushed bottom cartons, or reliance on combustible bands for stability, pile instability under fire conditions can be predicted.

The sometimes beneficial effects of pile collapse should not, generally, lead one to purposely plan for it. Racks, in particular, should have adequate cross-bracing, anchorage at floors, and tight connections, as rack collapses tend to be chain reactions.

Variable No. 8 — encapsulation: Encapsulation is a method of packaging consisting of a plastic sheet, such as polyethylene, completely enclosing the sides and top of a pallet load containing a commodity package or packages. This covering apparently deflects the water from fire that has penetrated a pallet load from below. This is likely to happen until the fire gains sufficient intensity to overcome the surface cooling of the sprinkler water. For this reason, in-rack sprinklers are quite necessary. Tests show that a higher ceiling sprinkler density is needed for rack storage at all heights, even with in-rack sprinklers, where there is encapsulation. The adverse effect of encapsulation has not yet been appreciably noted in palletized storage pile tests. Standards require no difference in the ceiling sprinkler density, with or without encapsulation, in non-rack storage. To convert an encapsulated pallet load to a nonencapsulated load, one need only remove the top covering portion of plastic.

Variable No. 9 — smoke venting: Smoke removal is important to manual fire fighting and overhaul (mopping up). Ideally, ventilation operations in a well-protected warehouse should be deferred until automatic sprinkler operation has reduced all temperatures in the building to ambient, which should occur within 30 min of ignition. However, this often is difficult advice for fire fighters to follow. Temperatures observed at doorways or atop insulated roofs will not portray all conditions in the building. Also, the fire officer in command will probably not know whether the sprinkler system is entirely adequate for the type of occupancy in the building at the time of the fire. To merely pump into the sprinkler system and stand by, waiting for the building to cool down, represents a gamble. Hence, fire fighters will most likely make holes in the roof, and enter the building to put small or large hose streams on the fire.

Automatic roof vents: There is a running controversy on the subject of automatic roof vents, which, when actuated by temperature, will open, allowing smoke and hot gases to escape near the seat of the fire. Tests show they are quite effective in unsprinklered buildings. On the other hand, the operation of automatic sprinklers tends to interrupt the operation of the vents, and to spoil the stack effect that causes the hot smoke to escape. In fact, the initial effect of sprinkler operation is often to beat down smoke, increasing the obscuration of vision within the building.

The NFPA storage standards do not call for smoke removal facilities because the fire tests upon which they are based did not employ smoke venting. However, the standards do make the comment that venting through eaveline windows, doors, monitors, or gravity or mechanical exhaust systems is essential to smoke removal after control of the fire is achieved. Some warehouses are of extremely large area, which makes dependence on eaveline windows or doors difficult. Automatic roof smoke vents usually have a manual release, operable from above the roof and/or at floor level, that can be used during a fire.

Proponents and opponents of automatic roof vents in sprinklered warehouses disagree as to their advantages, but several points of general concurrence can be made:

1. If there is excellent and frequent prefire planning, a fire fighting organization can be sufficiently fortified with knowledge of the occupancy and sprinkler system capability to develop with confidence a strategy of waiting for sprinklers to cool things down before ventilating and attacking the fire. This requires the full cooperation and assistance of warehouse management.
2. Roof vents equipped for convenient manual operation will be of assistance to fire fighters when they decide to ventilate.
3. If roof vents are equipped with mechanical ventilation instead of depending on the stack effect, the smoke cooled by sprinkler water can be removed sufficiently to make the building more tenable for fire fighters.
4. Roof vents should be considered when designing a building. In an existing warehouse they involve cutting openings in the roof deck and covering, reinforcement, and flashing to preserve weather resistance. Then, they tend to be economically unfeasible.
5. Smoke or no smoke, the sprinkler system must be kept in operation during manual fire fighting and overhaul.

Variable No. 10 — high expansion foam: While automatic high expansion foam extinguishing systems can be an independent means of fire suppression, there is a reluctance to use them as the sole means of automatic fire control. They are expensive, complicated (relative to sprinkler systems), do not protect the roof structure until foam reaches that level, involve the entire contents of a protected area regardless of how small the fire may be, and present a problem of foam residue removal after discharge. However, they have found acceptance as a partner to automatic sprinklers for certain high challenge storage occupancies, such as rubber tires, roll paper, and exposed plastics, where excessive foam contact is not a serious problem.

A high expansion foam system uses a series of foam makers at ceiling level. When its fire detection system actuates it, the foam system causes, in each foam-maker, a spray of water mixed with special foam concentrate to strike a screening while a fan blows through it. A large volume of bubbles of uniform size is produced, which cascades down into the warehouse area, gradually filling it up. The foam has an expansion ratio of up to 1,000 to 1, flashes to steam on contact with burning material, and engulfs other material, keeping it from burning. Building doors automatically close when the system trips. NFPA 11A, *Medium and High*

Expansion Foam Systems, contains the requirements for installation of these systems.

When high expansion foam systems are used in combination with ceiling sprinkler systems, the sprinkler discharge density may be reduced to one-half the density otherwise specified, but not less than 0.15 gpm/sq ft (6.1 [L/min]/m^2) for Class I through IV solid piled or palletized commodities, including idle pallets and plastics; or not less than 0.24 gpm/sq ft (9.8 [L/min]/m^2) for rubber tire storage. In rack storage, in-rack sprinklers are not required when high expansion foam is installed, and ceiling density can be reduced to 0.2 gpm/sq ft (8.1 [L/min]/m^2) for Class I, II, and III commodities and to 0.25 gpm/sq ft (10.2 [L/min]/m^2) for Class IV.

Variable No. 11 — sprinkler orifice size: Any required sprinkler discharge density can be had with any sized automatic sprinkler that is normally used — ½-in. (12. 7 mm) or larger orifice — by coordinating the sprinkler spacing and pressure available at the sprinkler. In the piping system design, however, when the higher values of density are needed, a larger than ½-in. (12.7 mm) orifice can be beneficial. Full-scale fire tests using densities of 0.40 gpm/sq ft (16.3 [L/min]/m^2) and higher give better results with large orifice — $^{17}/_{32}$ in. (13.5 mm) — sprinklers and 70- to 100-sq ft (6.5 to 9.3 m^2) spacing, than when using ½-in. (12.7 mm) orifice sprinklers at 50-sq ft (4.6 m^2) spacing. Substituting the $^{17}/_{32}$-in. (13.5 mm) for a ½-in. (12.7 mm) sprinkler results in a delivery of 40 percent more water under the same pressure, with less of the weakly penetrating fine spray.

The NFPA standard for indoor general storage does not specify orifice size, and the NFPA rack storage and rubber tire standards are silent as to ceiling sprinkler orifice size; therefore, the choice is up to the user. Designers usually take advantage of the economy and improved performance of the $^{17}/_{32}$-in. (13.5 mm) orifice when density is to be about 0.40 gpm/sq ft (16.36 [L/min]/m^2) or higher. The extra large orifice sprinkler developed by Factory Mutual Research and now available commercially has a 0.64-in. (16.25 mm) orifice. It generally provides even better protection for fast-burning, high heat release storage fires, by delivering a greater amount of water as large drops which better penetrate a strong updraft of fire. It will probably be covered in standards as a special option, for certain high challenge storage fires only.

Requirements have been cited from various NFPA storage standards merely to illustrate some of the statements included under "Variables" and to show how they enter into warehouse fire suppression system design. However, the standards themselves contain the comprehensive requirements, graphs, tables, and explanatory indices that should be pursued in detail as applicable.

FIRE PROTECTION FOR WAREHOUSES

While automatic sprinklers are considered the main line of fire defense in storage occupancies, they are by no means the only method of protection. Hose systems, portable extinguishers, and manual fire fighting operations also have a role in protecting storage occupancies. Each is discussed below.

Automatic Sprinkler Systems

For warehouse ceiling sprinkler systems, the most important design consideration is, of course, a required "Sprinkler Discharge Density" at a corresponding "Design Area of Sprinkler Operation" (see Figure 48.13). The widespread use of hydraulically calculated sprinkler systems has fostered loop and grid systems in which practically every length of pipe is used to almost its ultimate practical carrying capacity in meeting design requirements. In the older "tree" layout of piping employing a pipe schedule or hydraulic design, the portion nearest the riser has appreciably more carrying capacity than it requires in order to satisfy the minimum need at the hydraulically remote portions of the system. Economy in pipe sizing and labor costs is achieved by using a modern hydraulically calculated grid-type design layout, with considerably smaller size pipes than those permitted by older pipe schedules.

Unfortunately, such calculated systems have little room for error in estimating storage conditions, or to compensate for future changes, which are endemic to many warehouses. Should a public water supply that is the sole supply to a property deteriorate, the system won't deliver the promised density/area. Should the occupancy change (higher storage, racks replacing palletized piles, a more hazardous commodity, etc.), the sprinkler system becomes substandard and cannot be economically upgraded.

The "tree" system can be reinforced easily by running an auxiliary feed main to the far end of the system, looping crossmains, or by similar "tricks." To upgrade a grid-type hydraulically designed system sometimes requires superimposing a second complete piping layout under the first. For this reason, the practice of meeting the design parameter by choice of a tree system or by purposely overdesigning a grid system is a far-sighted option.

Water Supply

The fire service water supply at a warehouse must satisfy design requirements for ceiling sprinklers, in-rack sprinklers (if any), hydrants, and standpipe and hose systems for indoor use. An adequate water supply will have the capacity to supply these protective system components and still have enough pressure remaining in the mains to guarantee sufficient operating pressure at the sprinklers.

Multiplying the needed ceiling sprinkler density by the design area of sprinkler operation produces a flow requirement in gallons per minute to which about 10 percent should be added to compensate for unavoidable inequalities in the sprinkler system. To the ceiling sprinkler demand is added the in-rack sprinkler demand, if any, based upon operation of a number of the most hydraulically remote in-rack sprinklers. That number depends upon the number of sprinkler levels and the class of commodity. For one level only, in racks with Class I, II, or III commodities, six sprinklers are assumed. For more than one level of Class IV commodities the number is fourteen. In-rack sprinklers in storage up to 25 ft (7.6 m) need at least 15 psi (103 kPa) and have ½-in. (12.7 mm) orifices, so each will discharge a minimum of 22 gpm (83.3 L/min). Thus, 22 gpm (83.3 L/min) is the multiplier for the number of in-rack sprinklers assumed to operate. For rack storage over 25 ft (7.6 m), 30 gpm (113.5 L/min) per operating in-rack sprinkler is the multiplier.

After adding ceiling and in-rack sprinkler water demands, another 500 gpm (1893 L/min) is added for large and small hose streams without altering the residual pressure requirement, the assumption being that they will be used concurrently with sprinkler system operation.

The total water supply estimate can be expressed as a flow rate at a corresponding minimum residual pressure. That pressure is affected by the sprinkler system piping design, but in general at least 30 to 50 psi (207 to 345 kPa) pressure is needed at the ceiling level distribution point or top of riser. When greater pressure is available, more economic system design is possible, except that for standard orifice sprinklers, it is not desirable to have more than 60 psi (414 kPa) at the remote end sprinkler, as it is believed to cause excessive ineffective ultra-fine spray in the system discharge.

The water supply has to be available for a sufficient time, usually two hours, for full fire control by sprinklers, manual fire fighting that may be necessary, and "mop-up" operations. The duration estimate presumes prompt automatic application of an adequate water supply, prompt alarm, good hydrant accessibility, and reasonable response by a fire fighting organization of acceptable ability. The total time required affects the size of fire service water tank that may be required, or figures in the evaluation of the water storage capability in a small town public water supply distribution system.

The reliability requirement for a water supply at a high value warehouse, for example, can lead to a need for a multiple supply; i.e., a good fire pump-suction tank private supply may back up an adequate public supply, or there may be two similar private supplies. Such redundancy is based on good judgments; obviously it is not prudent for a $50,000,000 warehouse to depend entirely on a connection to a single city water main, no matter how powerful that supply is, because of the possibility of its being out of service for awhile.

The layout of yard mains, tanks, pump houses, power supply to pumps, hydrants, and valves warrants qualified fire protection engineering assistance in the planning stage. In correcting an existing water supply deficiency, alternate methods can be considered, and the variables discussed earlier in this chapter offer such alternatives. Adding a booster fire pump may be an economical alternative when many sprinkler systems on the property need, and will benefit from, the higher pressure. On the other hand, if just one sprinkler system out of many is deficient, it may be more economical to achieve adequacy by upgrading the piping layout in that system only, or by replacing ceiling sprinklers there with ones of a different temperature rating, or by installing in-rack sprinklers in that area, or by lowering pile heights in that area (not a good long-range solution), or a combination of methods.

Large and Small Hose

Warehouses of moderate area close to city hydrants present no special problem. But when the small dimension of a warehouse exceeds about 200 ft (61 m), hose lays from public hydrants to the far side of the building can be a serious problem. Private hydrants then become important around the perimeter with access doors to the warehouse near them. Private hydrant spacing of 250 ft (76 m) is common for such situations. Guidance on number and location of hydrants will be found in NFPA 24, *Installation of Private Fire Service Mains and Their Appurtenances.* Indoor small hose is recommended in high piled or rack storage buildings for use by occupants on incipient fires. Their relatively good reach and long lasting supply make them more effective than portable fire extinguishers on high storage fires, but extinguishers are also needed as an alternate. In many countries, axial feed reels of plastic hose are used instead of the accordian-fold, linen hose or woven jacket, lined hose common in the United States. They provide simple, readily available tools for incipient fires.

Portable Fire Extinguishers

For selecting types and sizes of fire extinguishers and locating them properly, guidance is provided by NFPA 10, *Portable Fire Extinguishers.* But one caution is in order: There is a tendency to provide dry chemical (ABC type) fire extinguishers everywhere because they are all-purpose units. Such extinguishers have a limited range, or reach, and time of discharge. For example, the 20-lb ABC unit (about as heavy as can be well handled by the average person) has a maximum effective horizontal range of about 20 ft (6 m) and discharges within 25 seconds. It is rated 3-20A compared with 2-A for a 2½ gal (9.5 L) water-type unit, but the latter lasts for a minute with an effective range of up to 40 ft (12 m), providing an advantage in most high warehousing situations, assuming Class BC units (for flammable liquid and electrical fires) are also provided. Laboratory tests upon which extinguisher ratings are based do not necessarily consider high piled storage fires.

Fire Fighting

Even with adequate fire control by sprinklers or other mechanical means, final manual extinguishment of storage fires is not always a simple procedure. Sprinkler discharge tends to decrease visibility by the presence of the downpour and the smoke driven down from the ceiling (see Variable No. 9 — smoke venting — discussed previously). In the past, some ill-advised fire officers upon arrival ordered sprinklers shut off to assist in getting at the seat of the fire; however, *the necessity of working with sprinklers operating, is now, fortunately, the recognized standard procedure.*

Other manual fire extinguishing problems involve getting at residual fire in the upper reaches of high racks, gaining access through aisles blocked by collapsed storage, and removing burning items from the building, such as fiber bales or rubber tires that cannot be fully extinguished while in the original, still smoldering pile. Any planned system of fire protection should simplify the manual fire fighting tasks. Automatic alarms to the fire department from operating sprinkler systems, as well as adequate hydrants, emergency access around the clock, audible/visual indication of which sprinkler system is operating, and clear identification of sprinkler valves, hydrants, interior small hose stations, and siamese pumper connections, all contribute to successful extinguishment. Prefire inspections and planning should be encouraged.

Plant Emergency Organization

A warehouse, like any other industrial property, needs some organization for emergencies. Even if it is merely the procedure in a small building for notifying the public fire department and using a fire extinguisher, who is to do it and how it is to be done must be preplanned (see Chapter 3). Larger properties frequently warrant inplant assignments for communication, using extinguishers and small hose, checking the open condition of valves, operating the fire pumps, advising responding fire fighters, salvage, and more, all duties of a private fire brigade. (NFPA 27, *Private Fire Brigades*, gives guidance on the organization, duties, and training of brigades.)

Chapter 48

SPECIAL OCCUPANCIES

Rubber Tires

Fires in rubber tires, while infrequent, are very hot, smoky, and difficult to control, and, finally, to extinguish. Many full-scale fire tests have shown the need for high sprinkler densities to tame the fire and protect the building structure.

Tires are piled solid, in compact portable racks referred to as pallets, and in racks. Since a tire is not packaged, it contributes its own circular air space to the storage, and these form considerable horizontal or vertical flues, depending on whether tires are "on side" or "on tread." The interiors of tire carcasses can flame vigorously, mostly out of reach of sprinkler water. Final extinguishment involves the laborious application of water to individual tires, usually by removal from the building.

NFPA 231D, *Storage of Rubber Tires*, calls for strong sprinkler densities varying with the method of storage and height. As an example, 20-ft (6.1 m) high palletized storage or fixed rack storage with regular pallets requires a 0.6 gpm/sq ft (24.4 [L/min]/m^2) sprinkler density over 5,000 sq ft (464 m^2) if ordinary temperature sprinklers are used, or 3,000 sq ft (279 m^2) with high temperature sprinklers. If a high expansion foam system is installed, the density can be reduced to 0.3 gpm/sq ft (12.2 [L/min]/m^2). Or, with one line of in-rack sprinklers in rack storage, 0.4 gpm/sq ft (16.3 [L/min]/m^2) would suffice.

In combination with sprinklers, high expansion foam is very effective on rubber tires, with minimal water damage or contamination. To obtain full penetration to the inner reaches of the tires, it is advisable to allow an additional hour of foam soaking time after sprinklers are shut off, maintaining the foam level.

Chapter 9, "Rubber Products," contains additional information on problems associated with rubber tire storage.

Roll Paper

Roll paper is stored on side or, more commonly, on end, and one might guess that the latter arrangement would permit easier paths for sprinkler water to penetrate the vertical air spaces and slow the fire. Experience shows otherwise.

One interesting fire in 1962 involved unbanded rolls of Kraft paper piled on end up to 30 ft (9.1 m) high. Spaces between stacks were about 12 in. (0.3 m), which would be desirably wide flues in palletized storage. However, in this configuration it led to rapid fire growth, as it allowed the paper to peel freely from the rolls and continuously feed fast-burning fuel to the fire (started by hot metal falling from a cutting torch). The wet-pipe sprinkler system, fed by what was then considered a good water supply, failed to control the fire, destroying the warehouse in a $3,000,000 loss. In 1966 another fire in similar storage up to 32 ft (9.75

m) high with up to 15 in. (0.38 m space) between stacks opened 387 sprinklers in the dry-pipe system, but, supported by a strong 6,000-gpm (22.7 m³/min) total water supply, sprinklers did help control the fire.

In storage on end, peeling or delamination during a fire is a major problem which can be somewhat abated by metal banding, steel baling wire applied tightly by hand, fire-retardant-treated tight paper wrappers covering ends as well as sides, or close spacing between columns of rolls. Effective close spacing would occur where stacks or columns are less than 4 in. (101 mm) apart.

NFPA 231F, *Standard for the Storage of Roll Paper*, prescribes sprinkler density-area requirements according to the arrangement of rolls, whether banded or unbanded, height of piles, and weight of paper, i.e., heavy weight, medium weight or light weight. Light weight paper and tissue are not covered. The standard allows reduced density-area parameters when automatic high expansion foam systems are used (see Variable No. 10 — high expansion foam — earlier in this chapter).

Chapter 16, "Printing and Publishing," and Chapter 14, "Pulp and Paper Mills," contains further information on hazards associated with rolled paper storage.

Carpet Storage

Rugs or carpeting in rolls are generally stored in racks that accommodate 12- to 15-ft (3.6 to 4.6 m) long rolls; therefore, racks are at least that deep, and sometimes placed back-to-back forming a total width of 24 to 30 ft (7.3 to 9.1 m) between aisles. Of course, small carpet pieces and rolls may be in cartons and stored in regular solid or palletized piles or conventional racks.

Long carpet rolls would deflect if not supported along their length; hence, solid or slatted shelves are used in the wide racks. An exception would be for carpets individually encased in strong paperboard tubes that have enough bending strength to be supported in open-style racks without shelving. Note that in-rack sprinklers are needed under each solid or slatted shelf in racks unless there are transverse flue spaces at least 6 in. wide (Variable No. 3 — in-rack sprinklers). With carpet racks having as many as 10 tiers, the amount of sprinklers and piping for in-rack sprinklers can be prohibitive. Therefore, protection focuses on providing a minimum 2-in. (51 mm) wide transverse flue at vertical supports, which are normally 10 to 12 ft (3 to 3.7 m) apart, and providing a level of in-rack sprinklers, including face sprinklers for the wider racks, located about every third tier for storage over 12 ft (3.65 m) high. Tiers vary in height from 24 to 40 in. (0.61 to 1 m), so the sprinkler installation is a tight squeeze (see Figure 48.16). Some engineers recommend vertical transverse barriers at every third set of vertical supports, in preference to relying upon 2-inch (51 mm) wide transverse flues.

With back-to-back racks, a 12-in. (0.3 m) space should be provided between vertical members, creating a longitudinal flue. If rolls tend

FIGURE 48.16. Racks for the storage of carpeting on slatted shelves. The closeness of the tiers to each other plus the depth of the tiers complicate problems of providing in-rack sprinkler protection. (Clark Industrial Truck Div.)

to project into this space, toe plates or other means can be fastened to the racks to stop the rolls physically at the rear face. Unless there is a large quantity of carpet with attached expanded (foam) rubber backing, fire does not tend to spread rapidly in normal carpet rolls along the length of the rack (see Figure 48.17). Nevertheless, since the racks often extend for several hundred feet, a valuable restriction to fire spread is to provide clear spaces 8 ft (2.4 m) wide about 100 ft (30 m) apart along the length of the rack. This serves as an assist to in-rack sprinklers, or as a fire stop where no in-rack sprinklers are used — storage under 12 ft (3.65 m) high.

Carpet padding: Padding or backing is generally stored in carpet warehouses. It is usually made of expanded plastic, expanded rubber, or jute. These materials are much faster burning than the carpeting, per se, and should be segregated for special arrangement and protection. Adequate ceiling sprinkler density in the warehouse as a whole will probably not be adequate for the padding unless the storage is kept low, such as less than 5 ft (1.5 m) for expanded rubber or plastic.

FIGURE 48.17. Test fire in $^{80}/_{20}$ polyester/nylon tufted carpet with foam rubber backing. Fire spread very rapidly. With woolen carpets there was no fire spread. (Fire Research Station — Crown Copyright Reserved)

Records Protection

Data recorded and stored on paper, cards, plastic film, etc., can be essential to production, sales, and service. There have been destructive records storage fires, even in fire-resistive buildings. While not industrial in nature, the fire at the Military Personnel Records Center near St. Louis, Mo., in 1973 was interesting in that fire severity caused serious structural damage to the reinforced concrete building. It also provided an example of how vital records that have been wet can be successfully salvaged, using a vacuum heat recycling process.[9] Filing and records rooms in buildings often have a high fire loading* because of the trend toward "open files." Steel file cabinets have been credited with limiting the fire load, represented by paper files, to 10 percent of that for exposed paper in open files.

For purposes of evaluating protection, records are classified in standards (NFPA 75 and 232) as follows: vital records, important records, and useful records. Where warranted, Class 1 records are kept in

---

*The expected maximum density of combustible material in a given fire area, usually stated in psf of ordinary combustibles.

fire-resistive safes. Large fire-resistive vaults, as found in banks and some industrial locations, are used for vital or important combustible records; however, there is a tendency to give them too much credit as safe havens against common perils, including fire and water damage, even though unsprinklered. This is particularly true if the vault is poorly located, such as in a basement area where the vault could be exposed to a long term "cooking effect" in a major fire from burning debris filling the basement around the vault.

Baled Fibers

The main hazard of baled cotton and other fibers of vegetable origin comes from the surfaces of the bales that have a multitude of exposed minute fibers. Fire flashes quickly over these vertical surfaces, as well as over any loose particles on the floor or lint on overhead piping and structure. Fire on one side of an automatic-closing fire door can progress through the open doorway on loose floor scraps before the door can automatically close. Housekeeping is therefore especially important.

Extra high piling creates intolerable fire control conditions, and insurance companies were at one time able to urge that cotton bales be stored in special warehouses having low head room in each story.

Fire tends to penetrate between and into bales, requiring removal of burning bales from the building for extinguishment. Considerable smoke is emitted and complicates fire fighting. Certain fibers, such as jute, are likely to swell when wet, so they should be piled with regard to stability in a fire, and with clearance from walls.

NFPA 231E, Recommended Practice for the Storage of Baled Cotton, limits heights of baled storage to 15 ft (4.6 m) and pile sizes to 700 bales, with a 12-ft (3.65 m) main aisle and 4-ft (1.2 m) cross aisles, in maximum fire divisions of 10,000 basles.[10] It calls for large operating areas for sprinklers.

For wool, which is slower burning than most fibers, the operating area can be less.

Ordinary dry chemical (BC type) fire extinguishers are very effective in knocking down surface fires on bales; however, they should be backed up by water spray from small hose or garden hose or water-type extinguishers equipped with spray nozzle attachments, to extinguish smaller fires that may have penetrated into the bales preferably using "wet water" (a chemical agent additive to increase water's penetrating and spreading ability).

Chapter 17, "Textile Manufacturing," has additional information on storage and handling baled fibers, particularly cotton.

Baled Wastes

Paper: Like baled fibers, baled waste paper is stored in solid piles into which fire tends to burrow. When paper is finely shredded, fire can flash over the surface of bales, much as happens with baled fiber. Waste paper becomes mushy and difficult to handle when wet; in fact, the integrity of the bales slowly disappears as the hose stream application progresses, making removal of burning bales difficult. Smoke can also be a troublesome problem. Baled paper is sometimes stored very high because of its low unit value, but beyond 30 ft (9 m) in height there is no fire protection standard that applies. The Factory Mutual standard calls for 0.25 gpm/sq ft (10.2 [L/min]/m^2) sprinkler density over 3,000 sq ft (279 m^2) for 20-ft (6.1 m) high storage using wet-pipe systems.[11] With dry-pipe systems, distinctly greater areas are specified.

Baled cloth scraps and rags: These rank between baled fibers and baled waste paper as to fire spread and control, but much depends on the type of fabric and whether it is clean or soiled.

Pesticides

Being poisonous to insects and small animals, pesticides are harmful, even fatal to humans when ingested or breathed as products of combustion. They are stored as solutions in flammable or combustible liquids, as powders or granules in combustible packages, or compressed gases for fumigation. Special attention must be given to the personnel hazard in addition to the commodity's combustibility.

Pesticide fires can endanger persons fighting the fire or standing nearby, and pollute water or soil in the area by poisonous run-off from fire fighting. It has been recommended that in isolated locations fire fighters allow a storage shed to burn itself out without any attempt to apply fire extinguishing water, and for all persons to stand off on the windward side. This, however, presumes no interexposure with high value contents, no compressed fumigant gases that could explode under fire exposure, no fertilizers that are explosive, and no inhabited buildings nearby.

Automatic sprinkler protection is advisable, but with provision for safe accumulation and disposal of water run-off. Fire fighting strategy should include plans to use protective clothing and respiratory equipment, to use spray nozzles rather than straight streams to reduce water run-off, to use specialized medical back-up services, and to avoid container breakup and pesticide dispersal.

NFPA 43D, *Storage of Pesticides in Portable Containers*, gives general requirements for both inside and outside storage of pesticides.

Chapter 48

SPECIAL STORAGE FACILITIES

Piers and Wharves

Warehouses often form the superstructure of piers and wharves. They may be referred to as "sheds," which, on piers, can project great distances over the water. They have the same general characteristics as other warehouses, but with more problems. Among them are: reduced accessibility for land-based fire department vehicles; the frequently combustible nature of the substructure, which can be exposed to floating burning debris or flammable liquids; the danger of water supply mains freezing; and the hazard of ships colliding with the pier. On the favorable side, commodities tend to be low-piled because of the temporary transfer nature of the storage (ship to shore or vice versa).

At the Luckenbach Pier in Brooklyn, N.Y. on December 3, 1956, fire and an explosion in stored Cordeau detonant fuse for explosives killed ten people and caused $7.6 million in damage. It illustrated the need for better limitations on hazardous materials, in this case contiguous with ordinary combustibles.

NFPA 87, *Piers and Wharves*, calls for a water supply for both hose streams and sprinkler systems to be available for at least 4 hrs — about double the usual duration. To protect the underside of a combustible substructure, sprinklers should be either standard pendent sprinklers in the upright position, or old-style sprinklers in the upright position, the purpose being to wet the undersurface.

It is desirable to limit the area between transverse fire walls in a pier making these walls continuous with substructure fire walls which in turn should extend to low water. They are spaced not more than 450 ft (137 m) apart.

Storage Garages

Motor vehicles, including automobiles and trucks, in "dead storage" are housed in storage garages or outdoors. Because of their generally fast-burning upholstery and gasoline or diesel oil content, fires in individual vehicles can be severe, but the overall building fire loading is moderate, evidently because the considerable amount of metal in the vehicles absorbs heat, and the average weight of combustibles per square foot is not high.

Storage garages and enclosed parking structures (not open air parking structures) need automatic sprinkler protection if located in basements, in buildings over 50 ft (15.2 m) in height having combustible roof/floor assemblies, or inside or below a building used for another purpose (typically high rise apartments, hotels, or offices). In the latter case, an alternative to sprinklers is a supervised ionization-type smoke detection system together with a mechanical ventilating system capable of exhausting smoke. Dispensing of flammable liquid fuels from underground tanks is sometimes a feature inside such garages, but the pumps should be above grade and within 20 ft (6.1 m) of an outside door on a floor sloped to pitch toward the door, and cut off from the remaining part of the garage by 2-hour partitions, a requirement of NFPA 88A, *Parking Structures.*

Refrigerated Storage

Coolers and freezer storage buildings are primarily for foodstuffs, although certain antibiotics, pharmaceuticals, and unstable chemicals are kept under refrigeration. To date, fire loss experience in refrigerated warehouses appears to have been very good during normal occupancy. It is during construction or temporary shutdown for repair that some devastating fires have occurred, usually from cutting and welding (see Chapter 24, "Welding and Cutting"). The housekeeping is generally excellent, and smoking mostly nonexistent because smoke odor clings to frozen food packages. Nevertheless, the potential for large loss is present in the containers, pallets, dunnage, waxy paper containers, and electrical equipment.

Insulation at walls and ceilings can spread fire quickly. Present practice involves prefabricated sandwich-panels with insulation cores, lining sheets of expanded polystyrene, or foamed-in-place expanded polyurethane. Older warehouses, and some recent ones, use cork or other cellulose-type insulation, or noncombustible insulation such as fibrous glass or expanded glass block. When an expanded plastic lining is used, it should be covered with an approved thermal barrier, which can be portland cement plaster on lath, gypsum plaster wallboard, fire retardant plywood, or an approved fire-resistive inorganic spray material; a few of these are now listed for the purpose by Underwriters Laboratories and the Factory Mutual System. When a nonplastic combustible lining is used, it too should be covered by a thermal barrier if the occupancy does not require sprinklers.

The same sprinkler system design used for a nonrefrigerated building is appropriate for this occupancy. However, some feel that the commodity hazard classification can be lowered one step in freezer areas because of ice in the packages. Dry-pipe systems have been used successfully for freezer buildings, but to combat the tendency for an ice plug to form inside the feed pipe just beyond the point where it penetrates the boundary wall of the freezer area, dry air is used in the piping. This requires not only taking the air from inside the freezer area, but putting it through a high efficiency dryer. Even so, priming water in the relatively warm dry-pipe valve tends to continually feed water vapor into the air so that, after a period of years, it will accumulate as ice at the point that the feed pipe enters the freezer area. The NFPA sprinkler standard calls for a removable feature in the piping at that point, and periodic internal examinations.

To keep water from entering sprinkler piping inside the freezer area should the dry-pipe valve trip prematurely, a combined deluge/dry-pipe valve system can be used. In this arrangement, water pressure under the dry-pipe valve is maintained through a -in. (3.2 mm) orifice in a bypass, while the main water supply to the dry-pipe valve is held back by a deluge valve that is electrically tripped by a separate fire detection system in the freezer.

Isolated Storage Buildings

Some high value warehouses are needed for only a few years in construction projects in remote areas. Such a warehouse can at times contain much electrical and electronic equipment, wire, cables, fixtures, pipe fittings, etc. Also there are experimental or mining sites that may have a large warehouse. These remote areas are often without public water mains or fire departments, making adequate private protection more costly. Factors have to be weighed, such as the criticality of the building relative to the project mission and the time for replacement of vital contents. Standards and building codes do not rigidly apply. A suitable decision could be to separate unprotected warehousing into small units, or to provide fully adequate private protection, or to effect a compromise such as protective systems with a limited water supply. Along with any compromised physical protection should be increased vigilance regarding watchman service, housekeeping, maintenance, and smoking control.

Underground Storage

Large underground warehouses can be a boon to energy conservation, as rock caverns have a nearly constant temperature. In the United States, some caverns have been used for compressed gas storage and some old mines for records storage. There also have been feasibility studies for permanent storage of high-level radioactive wastes from nuclear power plants.
    In general, the problems of underground structures include: exits, venting of smoke and heat, and access for fire fighting. Automatic sprinkler protection is vital to protect combustible storage and associated personnel, and can be installed under constructed floors and ceilings.

Outdoor Storage

Certain combustible commodities are typically stored outdoors, including most classes of lumber, coal and sulfur in bulk piles, roll paper, forest products, wood chips (in Michigan in 1975, a fire in a huge wood chip pile burned for several weeks), logs, vehicles, and various commodities covered with protective sheets of weather-resistant membranes.

INDUSTRIAL STORAGE PRACTICES

The disadvantages of outdoor storage include the absence of automatic sprinkler protection and the effect of wind on the spread of fire. Even without wind, storage containing large quantities of wood or paper tends to produce flying brands by convection, hastening fire spread. In the arrangement of the storage, important emphasis should be given to spacing between piles, access for fire fighting and hydrants, and not the least, spacing from important buildings or structures to minimize the interexposure.

Commodities subject to spontaneous heating, such as bituminous coal, require special handling, including compacting or internal temperature monitoring or both. A fire lasted about one year in an incident in a 180-ft by 100-ft by 40-ft (55 by 30 by 12 m) maximum height coal pile although fought intermittently with a large hose stream.

For fire fighting, major reliance is placed on available hydrants. Monitor nozzles of much greater fire extinguishing power than hand-held hose streams can be beneficial, as for pulpwood piles and foamed plastic storage.

Outdoor storage does have some protection advantages, though generally they are outweighed by disadvantages. Fire fighters can approach a hot, smoky fire safely from the windward direction, and may have greater freedom to apply hose streams to the tops of the storage.

Early fire detection is largely dependent upon watchman tours. Ultraviolet radiation sensors, although sometimes falsely alarmed by a small match flame, have been used successfully in certain applications. A number of NFPA standards are available covering specific types of outdoor storage.

Air-Supported Structures

For warehousing in temporary or remote locations, such as at construction sites, or as a low-cost adjunct to more permanent conventional types of buildings, air-supported structures of considerable size are sometimes used, based on the convenience and economics of the risk involved (see Figures 48.18 and 48.19). These structures consist of a plastic-coated fabric envelope resembling a balloon that is kept in a rigid condition by low positive air pressure within. When used for warehousing such as palletized or rack storage, they may be equipped with loading dock arrangements, such as air-locks with electric roll-up doors. Structures under 150 ft (46 m) in width or diameter may comply with NFPA 102, which deals with wind resistance, strength, load distribution, and pressurization of air-supported structures. There is a trend toward using a cable harness net system to encapsulate the envelope in all directions, increasing stability during wind load.

These structures do not support overhead piping or wiring, and so are not able to satisfy the usual need for automatic fire control systems. Therefore, they are not suitable for long-term combustible storage of high value, or storage of high hazard. When there are compelling reasons to

FIGURE 48.18. A typical air-supported structure that is suitable for warehousing purposes. (Air-Tech Industries, Inc.)

FIGURE 48.19. Interior of air-supported warehouse — 100 ft by 300 ft (30 by 91 m) — showing rack storage. (Air-Tech Industries, Inc.)

use them, it is preferable to have a number of separate smaller units rather than one large structure.

Construction

Any important building (warehouse or otherwise) merits durable construction that also resists wind and snow loads. Many types meeting these criteria are combustible or vulnerable to high temperatures, but the automatic sprinklers virtually always needed to protect the storage contents compensate for these hazards.

Indeed, problems arise mainly when "there is nothing to burn" in a building. Then, one must consider the type of covering on a steel deck roof, plastic panels in walls and insulation in roofs or walls. Certain expanded plastic insulated core panel walls should be used only with sprinklers.[12] But even with automatic sprinklers, the use of expanded plastics in close contact with the top of a metal deck roof is inadvisable because of its low melting point. (Exception: fire retarded polyurethanes which do not melt but char and vaporize in place.)

Buildings used for both manufacturing and warehousing should have a good barrier wall (preferably a true fire wall) between these components because the greater activity (hazard) in the former can expose the latter's high values to the damaging effects of a fire. Incidental adjoining areas such as boiler, machinery, or service rooms should be cut off by fire partitions.

Steel columns that are within storage racks over 15 ft high (4.6 m) in which there are no in-rack enclosed sprinklers need to be fireproofed, protected by one or two sidewall sprinklers, or need to be under high density ceiling sprinkler discharge in accordance with the rack storage standard.

BIBLIOGRAPHY

REFERENCES CITED

1. Patterson, C. B., "Powered Industrial Trucks: Appraising Their In-Plant Fire Safety," *Fire Journal*, vol. 66, no. 5 (Sept. 1972), pp. 103-104.
2. Thompson, N. J., Chapter 4, "Travel of Combustion Products," *Fire Behavior and Sprinklers*, National Fire Protection Association, Boston, MA, 1964, pp. 49-56.
3. Factory Mutual Engineering Corporation, "Solid, Palletized, and Rack Storage of Plastics," *Loss Prevention Data* 8-9, Factory Mutual System, Norwood, MA.
4. *Rules of the Fire Offices' Committee for Automatic Sprinkler Installations* London, England, 29th Edition, 1973.
5. F. M. Eng. Corp., "Bulk Storage of Raw Sugar," *Loss Prevention Data* 8-26, March 1967.
6. Nash, B., "A Non-Electric Resetting Zoned Sprinkler System for High-Racked Storages," *Fire Prevention Science and Technology*, no. 18 (December 1977), Fire Protection Association, London, England, pp. 14-20.
7. F. M. Eng. Corp., "Flow and Pressure Requirements for Pipe Schedule System Sprinkler Demand," *Loss Prevention Data* 2-77.
8. F. M. Eng. Corp., "Roll Paper Storage," *Loss Prevention Data* 8-21.
9. "Federal Fire Council Recommended Practice No. 2, Salvaging and Restoring Records Damaged by Fire and Water," 1963, Federal Fire Council, Washington, DC.

Chapter 48

10. F. M. Eng. Corp., "Baled Fiber Storage," *Loss Prevention Data* 8-7.
11. F. M. Eng. Corp., "Storage of Baled Waste Paper," *Loss Prevention Data* 8-22.
12. F. M. Eng. Corp., "Walls, Insulated Core," *Approval Guide*.

NFPA CODES, STANDARDS, AND RECOMMENDED PRACTICES

Reference to the following NFPA Codes, Standards, and Recommended Practices will provide further information on the safeguards for industrial storage practices discussed in this chapter. (See the latest *NFPA Codes and Standards Catalog* for availability of current editions of the following documents.)

NFPA 10, *Portable Fire Extinguishers*.
NFPA 11A, *High Expansion Foam Systems*.
NFPA 13, *Installation of Sprinkler Systems*.
NFPA 24, *Installation of Private Fire Service Mains and Their Appurtenances*.
NFPA 27, *Private Fire Brigades*.
NFPA 30, *Flammable and Combustible Liquids Code*.
NFPA 40E, *Storage of Pyroxylin Plastic*.
NFPA 43A, *Storage of Liquid and Solid Oxidizing Materials*.
NFPA 43C, *Storage of Gaseous Oxidizing Materials*.
NFPA 43D, *Storage of Pesticides in Portable Containers*.
NFPA 44A, *Manufacture, Storage and Transportation of Fireworks*.
NFPA 46, *Recommended Safe Practice for Storage of Forest Products*.
NFPA 48, *Magnesium Storage, Handling and Processing*.
NFPA 51B, *Fire Prevention in Use of Cutting and Welding Processes*.
NFPA 58, *Storage and Handling of Liquefied Petroleum Gases*.
NFPA 59, *Storage and Handling of Liquefied Petroleum Gases at Utility Gas Plants*.
NFPA 59A, *Liquefied Natural Gas, Production, Storage, and Handling*.
NFPA 75, *Electronic Computer/Data Processing Equipment*.
NFPA 82, *Incinerators, Waste and Linen Handling Systems and Equipment*.
NFPA 87, *Piers and Wharves*.
NFPA 88A, *Parking Structures*.
NFPA 102, *Tents, Grandstands and Air-Supported Structures Used for Places of Assembly*.
NFPA 204M, *Smoke and Heat Venting*.
NFPA 231, *Indoor General Storage*.
NFPA 231C, *Rack Storage of Materials*.
NFPA 231D, *Storage of Rubber Tires*.
NFPA 231E, *Recommended Practice for the Storage of Baled Cotton*.
NFPA 231F, *Storage of Roll Paper*.
NFPA 232, *Protection of Records*.
NFPA 232AM, *Archives and Record Centers*.
NFPA 481, *Titanium Handling and Storage*.

NFPA 490, *Storage of Ammonium Nitrate.*

NFPA 495, *Manufacture, Transportation, Storage and Use of Explosive Materials.*

Additional Reading

"Concern Over Warehouse Fires Leads to Tests on Aerosols," *Fire*, January 1981, pp. 410-412.

Deacon, F. C., "Designing Fire Protection to Limit Monetary Loss," *SFPE Technology Report No. 80-2*, Society of Fire Protection Engineers, Boston, MA, 1980.

"Designing to Limit Loss in High-Rise Rack Warehouses," *Kemper Group Report*, vol. 10, no. 3 (Sept. 1981), pp. 2-9.

Goring, G., "Sprinkler Protection of Storage Risks," *Fire Protection*, vol. 8, no. 2 (June 1981), pp. 20-25.

Harrington, J. L. and R. B. Hopkinson, *Rack Storage Protection*, Worcester Polytechnical Institute, Worcester, MA, 1977.

Herzog, G. R., "Management of Flammable Liquid Storage Tank Fires, *International Fire Chief*, vol. 48, no. 2 (Feb. 1981), pp. 26-27.

McKinnon, G. P., ed., *Fire Protection Handbook*, Fifteenth Edition, National Fire Protection Association, Quincy, MA, 1981.

Pignato, J. A., J. J. Hottinger, T. W. Berger, "One Way to Protect Loading Racks," *Fire Journal*, vol. 47, no. 2 (March 1980), pp. 56-59.

Young, R. A. and P. Nash, "The Fire Protection of Modern High Bay Storages," *Fire Prevention Science and Technology*, no. 18 (December 1977), pp. 4-13.

49

Industrial Waste Control

Lawrence G. Doucet

The term "industrial waste" generally describes wastes emanating from manufacturing facilities, processing plants, factories and the like. However, it essentially has no meaning without further definition, because it encompasses an almost infinite variety of materials.

It has been estimated that more than 350 million tons of industrial waste are generated in the United States every year. A large portion of these wastes are considered hazardous. A recent survey by the U.S. Environmental Protection Agency (EPA) determined that in 1981 about 150 million tons of hazardous industrial waste were generated, and that the largest 150 generators alone produced about 35 billion gallons of hazardous waste. These quantities are projected to increase steadily at an annual rate of 5 to 10 percent. The typical, average waste generation rates of a few major industries are given in Table 49.1. The approximate distribution of hazardous wastes generated by some major industries are shown in Figure 49.1.

Disposal of such tremendous waste quantities is a matter of great public concern in light of recent revelations of improper or indiscrimi-

---

Lawrence G. Doucet is a principal of Doucet & Mainka, P.C., Consulting Engineers, Peekskill, NY.

TABLE 49.1. *Industrial Solid-Waste Production Rates*[1]

SIC* Code	Industry	Waste Production Rate (tons/employee/year)
201	Meat processing	6.2
2033	Cannery	55.6
2037	Frozen foods	18.3
Other 203	Preserved foods	12.9
Other 20	Food processing	5.8
22	Textile-mill products	0.26
23	Apparel	0.31
2421	Sawmills and planning mills	162.0
Other 24	Wood products	10.3
25	Furniture	0.52
26	Paper and allied products	2.00
27	Printing and publishing	0.49
281	Basic chemicals	10.00
Other 28	Chemical and allied products	0.63
29	Petroleum	14.8
30	Rubber and plastic	2.6
31	Leather	0.17
32	Stone, clay	2.4
33	Primary metals	24
34	Fabricated metals	1.7
35	Nonelectrical machinery	2.6
36	Electrical machinery	1.7
37	Transportation equipment	1.3
38	Professional and scientific instruments	0.12
39	Miscellaneous manufacturing	0.14

*Standard Industries Classification

nate dumping practices at numerous locations. It has been estimated that there are more than 30,000 hazardous waste dumpsites throughout the United States, of which more than 2,000 present significant health and environmental problems. The EPA has documented hundreds of cases whereby indiscriminate or improper hazardous waste disposal has severely damaged the environment and endangered life.

To protect against further hazardous waste mismanagement, the Congress enacted the Resource Conservation and Recovery Act of 1976 (RCRA). Under this Act, the EPA has instituted a complex set of regulations for controlling hazardous wastes from generation through final disposal. This is known as "cradle-to-grave" control. To deal with dumpsite cleanup costs and liabilities not covered by RCRA, the EPA has enacted legislation known as the "Superfund" bill. The Toxic Substances Control Act (TSCA) of 1976 was established to regulate chemicals not controlled under other regulations. This Act, for example, specifies disposal requirements for the highly toxic polychlorinated biphenyl compounds (PCBs).

For most industries, particularly those generating hazardous

INDUSTRIAL WASTE CONTROL

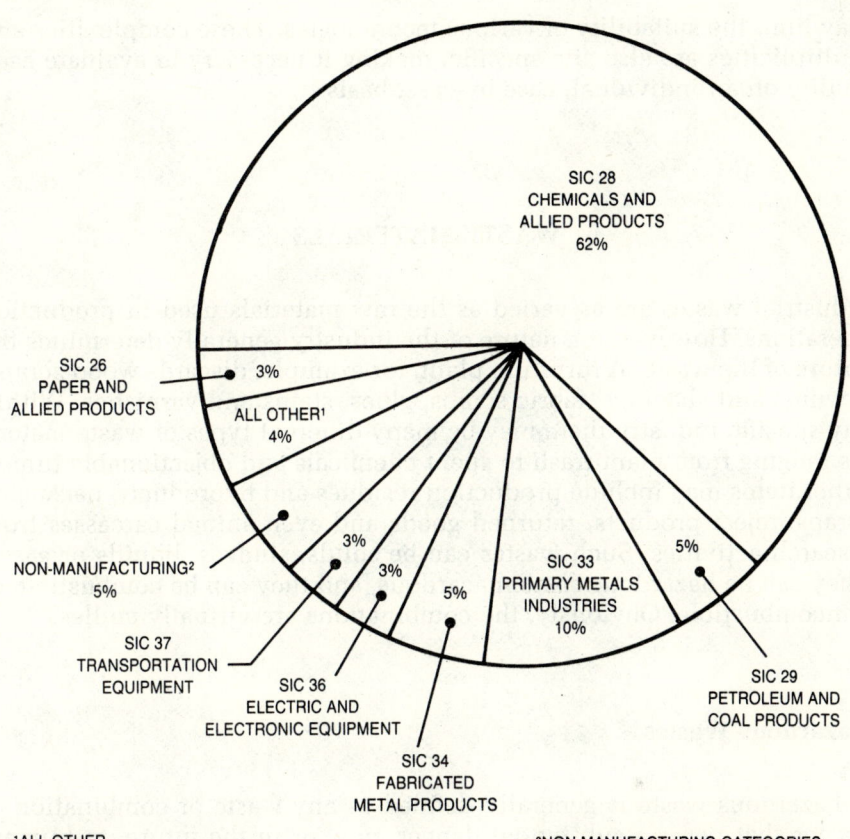

FIGURE 49.1. *Pie chart showing the distribution of hazardous waste generation by Standard Industrial Classification (SIC) code in 1980.*[2]

wastes, the evaluation and selection of waste management and disposal systems is difficult and complex. First of all, there are numerous options and alternate technologies available, each with differing benefits, risk factors and costs. Secondly, the forms and properties of waste materials vary widely, and greatly impact the feasibilities and cost-effectiveness of the alternate technologies. Thirdly, regulatory requirements and restrictions may limit the suitability of various technologies. These complexities and multiplicities are also site specific, making it necessary to evaluate each facility on an individual, case-by-case, basis.

WASTE MATERIALS

Industrial wastes are as varied as the raw materials used in production operations. However, the nature of the industry generally determines the nature of the waste. A furniture plant, for example, discards wood scraps, sawdust and shavings, fabric scraps, glues, stains and varnishes. Within any specific industry there may be many different types of waste materials ranging from plant trash to spent chemicals and objectionable fumes. Other items may include production residues and byproducts, packaging scraps, reject products, returned goods and even animal carcasses from research activities. Such wastes can be solids, sludges, liquids or gases. They can be hazardous or nonhazardous, and they can be combustible or noncombustible. Obviously, the combinations are virtually endless.

Hazardous Wastes

A hazardous waste is generally defined as any waste or combination of wastes that poses a substantial danger, now or in the future, to human, plant or animal life, and which, therefore, cannot be handled or disposed of without special precautions. General categories of hazardous waste are toxic chemical, flammable, corrosive, reactive, explosive, radioactive, and biological. These can contaminate surface and ground waters as well as the ambient atmosphere. They can poison, burn, maim, blind, and kill people and other living organisms. Some are nondegradable and persist in nature indefinitely. Some may accumulate in living things, and some may work their way into the food chain. Also, some may catch fire or explode at normal temperatures and pressures when exposed to air or water, or by being jarred or dropped.

Currently, the EPA, under RCRA, defines and regulates wastes that are hazardous because of either ignitibility, corrosivity, reactivity, or toxicity. The regulations not only identify specific hazardous criteria, but also list specific materials, constituents, and sources generating wastes with these hazardous characteristics. Radioactive waste materials are regulated by the Nuclear Regulatory Commission (NRC). Biological, or infectious, waste materials are usually regulated by state and local environmental agencies and health departments.

Hazardous Characteristics

Ignitibility: Liquid wastes are considered ignitibly, or flammably, hazardous if their flash point is less than 140°F (60°C). These mainly consist of contaminated organic solvents, but they may include oils, plasticizers, complex organic sludges, and off-specification chemicals.

Nonliquid wastes are considered ignitibly hazardous if they are capable, under standard temperatures and pressures, of causing fires through friction, absorption of moisture, or through spontaneous chemical changes. These include pyrophoric materials such as phosphorous and aluminum alkyls, which ignite when exposed to air.

Organic vapors and fumes may be flammably hazardous if released in combustible concentrations. These may travel considerable distances, reach an ignition source and ignite. NFPA 86A, *Ovens and Furnaces*, identifies specific dilution requirements for the safe venting of such gases.

Corrosivity: Corrosive wastes can severely damage skin and other living tissues, as well as construction materials such as metals. Wastes are defined as corrosively hazardous if their pH values are 2 or less, 12 or more, or if they corrode steel at a rate greater than one-quarter inch per year. These wastes generally comprise strong acids and alkalis.

Reactivity: Hazardous wastes characterized as reactive are normally unstable and readily undergo violent changes without detonation. Some reactive wastes are capable of detonation or explosive decomposition at standard temperatures and pressures. Also, some reactive wastes react violently with water, and some form potentially explosive mixtures or toxic vapors if mixed with water. Reactive materials, such as sodium, potassium and aluminum alkyls, react violently with water and burn fiercely. Strong oxidizers in contact with organic materials may cause rapid combustion or explosion.

Toxicity: Toxic wastes poison, or produce injury, upon contact with or through accumulation in or on the body of a living organism. This may occur by inhalation or through contaminated food or water. It is both a short-term and long-term problem. Most toxic wastes contain either metals, such as arsenic, barium, cadmium or copper, or synthetic organics, such as various pesticides. RCRA specifies extraction procedure (EP) testing to determine whether such materials are present in significant concentrations to be a toxic hazard.

Explosives: These mainly consist of obsolete ordnance and manufacturing wastes from commercial explosives industries. They may also consist of wastes from the manufacturing of various propellants and pyrotechnics.

Radioactivity: Most radioactive wastes consist of conventional, nonradioactive materials contaminated with radionuclides in concentrations ranging from parts per billion to as much as 50 percent. Biological hazards from radioactive materials are due to the effects of penetrating and ionizing radiation rather than chemical toxicity. The long-term hazard associated with each waste is not necessarily proportional to the nominal level of radioactivity, but rather to the specific toxicity and decay rate of each radionuclide. Radioactive industrial wastes are primarily generated from biomedical research activities. These generally comprise items such as animal carcasses and liquid scintillation counting vials (LSC) with tracer level, or low-level, concentrations of radioactivity.

Biohazard: Wastes characterized as biohazards contain pathogens capable of producing infection. A pathogen is any disease-producing microorganism or material, and includes bacteria, fungi, viruses, viroids, rickettsiae and protozoa. Industrial wastes which are biohazards are almost exclusively generated from biomedical research activities.

Waste Characterization

Waste stream characterization provides the starting point for designing waste handling and disposal systems. This involves identification of average waste composition and properties, deviations from these averages, and the forecasting of likely changes.

A characterization program usually begins with a detailed inventory and classification of all waste stream sources, components and constituents. These are then organized into various groupings, or categories, to facilitate further evaluations. Typical groupings include form, such as solids, sludges and liquids, combustible vs. noncombustible, and hazardous vs. nonhazardous. The waste classification system shown in Table 49.2. has proven particularly useful to the author.

The next step in a characterization program usually involves determination of the physical and chemical properties of the waste materials, or groups of waste materials. Physical properties of concern typically include physical state, size, shape, weight, volume, temperature, viscosity and pressure. Chemical properties typically include elemental constituents, toxicity, corrosivity, explosivity, flash point, heat of combustion and products of combustion. Table 49.3. is a comprehensive checklist of parameters which may be significant in a waste characterization program.

Depending on the specific waste, sufficient characterization data may be readily available in published literature or engineering handbooks. When such data is not available or inadequate, laboratory testing and analytical work may be required. Waste testing methodologies are typically in accordance with American Society for Testing and Materials (ASTM) procedures. However, hazardous waste sampling and analysis procedures usually comply with "Test Methods for Evaluating Solid Waste — Physical/Chemical Methods," SW-846, as issued by the EPA.

WASTE MANAGEMENT SYSTEMS

Waste management generally comprises the collection, internal transport, interim storage and final disposal of waste materials. A great variety of alternate processes and equipment are available for providing these functions as part of a total waste management system. Depending upon the specific types, forms, quantities and hazardous characteristics of the waste materials, management systems can range from a few simple processes to a complex combination of many processes.

Waste handling generally refers to those functions associated with the movement of wastes after creation, excluding storage, processing, treatment, and final disposal. It includes collection of wastes at points of generation, transport of the materials, and unloading from the transport system. These systems can range in sophistication from fully manual to fully automatic operation.

Waste handling systems and equipment include conveyors, chutes, carts and transport vehicles, elevators and lifts, as well as pumps and piping for liquids and ducts and blowers for gases.

Waste storage includes the interim containment of accumulated materials prior to subsequent handling, processing, treatment or disposal. Storage, depending on the waste form, can be in loose, compacted, or other processed form; in bulk or individual containers such as drums or cartons.

Waste processing includes those functions which prepare or alter the waste by changing its shape, size, uniformity, or consistency. Such processes are usually provided to facilitate other handling, storage, treatment, or disposal processes. Waste processing systems and equipment include compactors, shredders, crushers, pulpers, pulverizers, baggers and encapsulators, extruders, and dewatering devices.

Systems for storing and handling liquid chemical wastes often require special designs and safety provisions. Problems of concern typically include viscosity, freeze protection, abrasion, corrosion, slagging, and adverse chemical reactions, such as spontaneous ignition, rapid heat release, foaming, precipitation, solidification, and vaporization of low boiling point compounds. Safety provisions include spill and runoff containment, static electricity prevention, flame arrestors, and systems for handling the vapors or fumes vented from the storage tanks. In addition, special precautions are usually provided to assure that incompatible wastes are segregated to prevent undesirable reactions.

For safely handling fumes or vapors from industrial ovens and furnaces, dilution air is required such that the combustible concentrations are below 50 percent, and usually less than 25 percent, of lower explosive limits (LEL).

Waste handling systems also comprise equipment and devices for loading or charging wastes into treatment systems and processes. For example, mechanical loading devices may be part of a system for charging solid wastes into an incinerator. Such devices not only provide for a regulated loading rate, but they also provide a continual seal between the

TABLE 49.2. Waste Classification System[3]

Class	Remarks	Example
I. Solid wastes		
A. Putrescibles		
Household garbage		
Vegetable and fruit processing wastes		
Animal manure		
Dead animals		
Meat, poultry and seafood processing wastes		
Others, not elsewhere classified[1]		
B. Bulky combustibles	Bulky is defined to mean a material of a size to present problems by jamming in a compaction truck hopper, an incinerator feed chute, or other such problems in handling or disposal. Its dimensions may not be defined, except by the size and nature of handling or disposal equipment.	
Wood		Timbers, pallets, cross-ties
Paper and paper products		Large cardboard packing boxes, box-car linings
Cloth		Filter cloths, mattress
Plastics		Styrofoam logs, polystyrene sheets, garden hose
Rubber		Belting, tires
Leather		Conveyor belting
Yard and street wastes		Tree limbs
C. Bulky noncombustibles		
Metal		Drums, bedsprings
Mineral		Carboys, bathroom fixtures
D. Small combustibles	Small is defined to mean a piece of waste material of a small enough size to give no danger of jamming equipment, or otherwise causing problems because of its size. It refers to the size of a piece of the material, not the size of the delivered load.	
Wood		
Paper and paper products		
Cloth		Gloves
Leather		Shoes
Plastics		Milk cartons
Rubber		Galoshes, butyl rubber crumb

INDUSTRIAL WASTE CONTROL

E.	Yard and street wastes	Street sweeping, leaves
	Small noncombustibles	Cans
	Metal	Bottles
	Mineral	Furnace ashes
	Ashes	Solvent drums
F.	Non-empty cans, bottles and drums	Bottles of wastes from laboratories
		This functional class will require precise definition of the contents by one of the other classes. It must be considered to establish the quantity of material delivered in this manner instead of in bulk form.
G.	Gas cylinders	Oxygen, acetylene
H.	Powders and dusts	Pesticides, grain dusts, chemical powders, coal dusts
	Organic	
	Metallic inorganic	
	Non-metallic inorganic	
	Explosive	
I.	Pathological wastes	
	Cloth, paper and plastic	
	Animal and human wastes	
	Instruments and utensils	
J.	Sludges	Materials which are solid in appearance, but are wet, either from water or liquid organics. The solids portion is classified.
	Chlorinated	
	Brominated	
	Fluorinated	
	Acid	
	Alkaline	
	Water-reactive (unhydrolyzed)	
	Air-reactive	May represent a material highly reactive with the moisture in air, or with the oxygen in the air.
	Putrescible	May represent the wet form of any of the materials listed under the Functional Class of the same name, but also includes such materials as wastewater treatment plant sludges.
	Miscellaneous organic	Refers to metals in the uncombined form. Inorganic compounds.
	Metallic inorganic	Particles of metal in oil.
	Non-metallic inorganic	Filter cakes, $CaCO_3$ precipitate
K.	Demolition and construction	This functional class will be subdivided into the Analytical classes shown for bulky and small, combustibles and non-combustibles. It is included as a separate class to quantify materi-

TABLE 49.2. *(Continued)*

Class	Remarks	Example
	als from this source, and to recognize that a delivery may contain a wide mixture of large and small combustibles and non-combustibles.	
L. Abandoned vehicles		
M. Radiological wastes	This functional class will require further definition by one of the other classes. It must be considered to define this type of contamination to refuse, and to consider the special measures involved.	
II. Liquid wastes		
A. Wastewaters	Waste liquids composed almost entirely of water, but containing contaminants in low enough concentration (usually much less than 1.0 percent) to be handled through a sewer system to a waste-water treatment plant. This definition is proposed for exclusion purposes, since these wastes are not considered normally as refuse.	
B. Contaminated waters	Waters containing contaminants in a concentration too high, or of a nature, that handling through a wastewater treatment plant is too practical.	
Chlorinated		
Brominated		
Fluoridated		
Acid		
Alkaline		
Putrescibles		Blood, grease
Insoluble oils		
Soluble oils		
Toxic organics		
Toxic inorganics		
Soluble metals		
Others, (not elsewhere classified)†		
C. Liquid organics	Liquid at all ambient temperatures.	Many solvents
Chlorinated		
Brominated		
Fluoridated		
Sulfurated		

INDUSTRIAL WASTE CONTROL

Acid
Alkaline
Water reactive (unhydrolyzed)
Shock reactive
Toxic and hazardous
Soluble metals
Others, (not elsewhere classified)† .... Pesticides

D. Tars — Stiff materials which are semi-solid at low ambient temperatures.

Chlorinated .... Chlorine-substituted hydrocarbons
Brominated .... Bromine-substituted hydrocarbons
Fluorinated .... Fluorine-substituted hydrocarbons
Sulfurated .... Sulfur-substituted hydrocarbons
Acid .... Low pH, corrosive solvents
Alkaline
Water-reactive .... Unhydrolyzed materials which react violently
Chemically reactive
Self-reactive (monomers)
Toxic and hazardous
Soluble metals .... Sodium, calcium
Others (not elsewhere classified)†

E. Slurries — Liquid materials which contain solids, but which readily flow or pump. The liquid and solid materials both are classified.

Organic in water .... Lime slurry
Inorganic in water
Organic in a liquid organic .... Metallic sodium in oil
Inorganic in a liquid organic

III. Gaseous wastes — This classification is restricted to those gaseous materials which have been or might become the responsibility of a disposal group or department to treat, burn, or otherwise alter before discharge to the atmosphere.

A. Odorous .... Mercaptans, H$_2$S
B. Combustible particulate
   Solids
   Mists
C. Organic vapors .... Volatile solvents
D. Acid gases .... SO$_2$, HCl

†Other (not elsewhere classified)—Represents the "cleanest" of material and is not intended as a catchall grouping.

1031

TABLE 49.3. *General, Physical, and Chemical Parameters of Possible Significance in the Characterization of Solid Wastes.*[4]

General parameters

Compositional weight fractions
  A. Domestic, commercial and
      institutional
    Paper (broken into subcategories)
    Food waste
    Textiles
    Glass and other ceramics
    Plastics
    Rubber
    Leather
    Metals
    Wood (limbs, sawdust)
    Bricks, stones, dirt, ashes
  B. Other municipal
    Dead animals
    Street sweepings
    Catch-basin cleanings
  C. Agricultural
    Field
    Processing
    Animal raising
  D. Industrial
  E. Mining/metallurgical
  F. Special
    Radioactive
    Munitions, etc.
    Pathogenic
    Moisture
Process weight fractions
  Combustible
  Compostable
  Processable by landfill
  Salvageable
  Having intrinsic value

Physical parameters

Total wastes
  Size
  Shape
  Volume
  Weight
  Density
  Density stratification
  Surface area
  Compaction
  Compactability
  Temperature
  Color
  Odor
  Age
  Radioactivity
  Physical state
    Total solids
    Liquid
    Gas
Solid wastes
  Soluble (%)
  Suspendable (%)
  Combustible (%)
  Volatile (%)
  Ash (%)
    Soluble (%)
    Suspendable (%)
  Hardness

Particle characteristics
  Size distribution
    "shape"
    "surface"
    "porosity"
    "sorption"
    "density"
    "aggregation"
Liquid wastes
  Turbidity
  Color
  Taste
  Odor
  Temperature
  Viscosity data
    Specific gravity
    Stratification
  Total solids (%)
    Soluble (%)
    Suspended (%)
    Settlable (%)
  Dissolved oxygen
  Vapor pressure
  Effect of shear rate
  Effect of temperature
  Gel formation
Gaseous wastes
  Temperature
  Pressure

Volume
Density
Particulate (%)
Liquid (%)

Chemical parameters

General
- pH
- Alkalinity
- Hardness (CaCO$_3$)
- MBAS (methylene-blue active substances)
- BOD (biochemical oxygen demand)
- COD (chemical oxygen demand)
- Rate of availability of nitrogen
- Rate of availability of phosphorus
- Crude fiber
- Organic (%)
- Combustion parameters
  - Heat content
  - Oxygen requirement
  - Flame temperature
  - Combustion products (including ash)
  - Flash point
  - Ash-fusion characterization
  - Pyrolysis characterization
- Toxicity
- Corrosivity
- Explosivity
- Other safety factors
- Biological stability
- Attractiveness to vermin

Inorganic and elemental
- Moisture content
- Carbon
- Hydrogen
- (P$_2$O$_5$ and phosphate)
- Sulfur content
- Alkali metals
- Alkaline-earth metals
- Precious metals

Heavy metals
  especially Mercury
  Lead
  Cadmium
  Copper
  Nickel
Toxic materials
  Chromium
  especially Arsenic
  Selenium
  Beryllium
  Asbestos
Eutrophic materials
  Nitrogen
  Potassium
  Phosphorus

Organic
- Soluble (%)
- Protein nitrogen
- Phosphorus
- Lipids
- Starches
- Sugars
- Hemicelluloses
- Lignins
- Phenols
- Benzene oil
- ASB (alkyl benzene sulfonate)
- CCE (carbon chloroform extract)
- PCB (Polychlorinated biphenyls)
- PNH (Polynuclear hydrocarbons)
- Vitamins (e.g., B-12)
- Insecticides (e.g., Heptochlor, DDT, Dieldrin, etc.)

furnace and ambient surroundings during loading to prevent flame and smoke exfiltrations. Table 49.4. gives the applicability of various waste management and disposal methods for different waste classes.

TREATMENT AND DISPOSAL SYSTEMS

Because of the high costs and liabilities typically associated with waste treatment and disposal, initial efforts are usually directed to reducing waste quantities. Techniques include reuse and recycling of materials which would otherwise be wastes, changes in manufacturing and production operations, and substitutions of raw materials to others generating less waste. Also, substitutions and reductions in the usage of hazardous chemicals may result in reduced hazardous waste disposal requirements.

A recent study for the EPA reported 43 potentially feasible waste treatment processes. These processes include physical treatment, chemical treatment, biological treatment, and thermal treatment. They provide such functions as volume reduction, separations of waste stream components, destruction and detoxification of hazardous constituents. Often several treatment processes are linked in series to provide a required degree of treatment. Residues from these processes usually require disposal, such as land burial or deep well injection.

Selection of treatment processes depend on the type, form and quantities of waste, required performance to satisfy environmental requirements, and overall system economics. Table 49.5. summarizes the currently available hazardous waste treatment and disposal processes discussed below:

Physical Treatment

Physical treatment processes generally provide for separation of waste stream components or phases. These are particularly useful for concentrating specific hazardous constituents within dilute waste mixtures, thus reducing the overall quantities requiring disposal as a hazardous waste. These also provide for the recovery of specific materials in resource recovery operations.

The following physical treatment processes are potentially feasible for wastes:

Adsorption
Centrifugation
Dialysis
Electrodialysis
Electrolysis
Electrophoresis

INDUSTRIAL WASTE CONTROL

TABLE 49.4. Waste Management and Disposal Methods for Different Waste Classes¹

Waste Classification Utilization Chart	Potential Hazards				Management Methods															
	Toxicity	Explosiveness/ Flammability	Pathogenicity	Radioactivity	Std. Storage	Spec. Storage	Std. Coll. Vehicle	Spec. Coll. Vehicle	Std. Pipeline	Spec. Pipeline	Compaction	Grinding	Pulping (w/H$_2$O)	Composting	Incineration	Incineration (Liquid)	Incineration (Spec.)	Chemical Treat. or Alteration	Sanitary Landfill	Ocean Disposal‡
Solid wastes																				
Putrescible			X			X	X	X			X	X	X	X	X		X		X	
Bulky Combustible		X				X		X				X			X		X		X	
Bulky Non-Combustible					X		X				X	X							X	
Small Combustible		X			X	X	X	X			X	X	X		X		X		X	
Small Non-Combustible					X	X	X	X			X	X							X	
Non-Empty Cans, Bottles, Drums*	X				X	X		X									X		X	
Gas Cylinders		X				X		X									X	X	X	
Powders and Dusts			X	X		X		X									X	X	X	
Pathological Wastes			X			X		X						X	X		X	X	X	X
Sludges	X					X		X				X			X		X		X	X
Demolition and Construction	X					X		X			X	X							X	
Abandoned Vehicle								X											X	
Radiological Waste				X		X		X									X	X	X	
Liquid wastes§																				
Waste Waters	X			X					X	X						X		X		X
Contaminated Waters†	X		X			X		X	X	X						X		X		X
Liquid Organics†	X	X				X		X	X	X						X	X	X		
Tars	X					X		X	X	X						X	X	X		
Slurries†						X		X	X	X										
Gaseous wastes§																				
Odorous	X					X		X									X	X		
Combustible Particulate	X	X				X		X									X	X		
Organic Vapors	X	X				X		X									X	X		
Acid Gases	X	X						X									X	X		

*For contents treatment see proper classification; this class for container only.
†Pipeline selection dependent upon specific contaminant.
‡Ocean disposal as locally permitted or approved—not recommended procedure.
§Final disposal of these wastes requires an initial transformation to the solid class (except as noted).

TABLE 49.5. Hazardous Waste Treatment and Disposal Processes[5]

Process	Functions Performed[†]	Types of Waste[‡]	Forms of Waste[§]	Resource Recovery Capability
Physical treatment:				
Carbon sorption	VR, Se	1, 3, 4, 5	L, G	Yes
Dialysis	VR, Se	1, 2, 3, 4	L	Yes
Electrodialysis	VR, Se	1, 2, 3, 4, 6	L	Yes
Evaporation	VR, Se	1, 2, 5	L	Yes
Filtration	VR, Se	1, 2, 3, 4, 5	L, G	Yes
Flocculation/setting	VR, Se	1, 2, 3, 4, 5	L	Yes
Reverse osmosis	VR, Se	1, 2, 4, 6	L	Yes
Ammonia stripping	VR, Se	1, 2, 3, 4	L	Yes
Chemical treatment:				
Calcination	VR	1, 2, 5	L	
Ion exchange	VR, Se, De	1, 2, 3, 4, 5	L	Yes
Neutralization	De	1, 2, 3, 4	L	Yes
Oxidation	De	1, 2, 3, 4	L	
Precipitation	VR, Se	1, 2, 3, 4, 5	L	Yes
Reduction	De	1, 2	L	
Thermal treatment:				
Pyrolysis	VR, De	3, 4, 6	S, L, G	Yes
Incineration	De, Di	3, 5, 6, 7, 8	S, L, G	Yes
Biological treatment:				
Activated sludges	De	3	L	No
Aerated lagoons	De	3	L	No
Waste stabilization ponds	De	3	L	No
Trickling filters	De	3	L	No
Disposal/storage:				
Deep-well injection	Di	1, 2, 3, 4, 6, 7	L	No
Detonation	Di	6, 8	S, L, G	No
Engineered storage	St	1, 2, 3, 4, 5, 6, 7, 8	S, L, G	No
Land burial	Di	1, 2, 3, 4, 5, 6, 7, 8	S, L	No
Ocean dumping	Di	1, 2, 3, 4, 7, 8	S, L, G	No

[†]Functions: VR, volume reduction; Se, separation; De, detoxification; Di, disposal; and St, storage.

[‡]Waste types: 1, inorganic chemical without heavy metals; 2, inorganic chemical with heavy metals; 3, organic chemical without heavy metals; 4, organic chemical with heavy metals; 5, radiological; 6, biological; 7, flammable; and 8, explosive.

[§]Waste forms: S, solid; L, liquid; and G, gas.

Filtration
Flocculation
Flotation
Freeze Crystallation
Freeze Drying
Freezing
High Gradient Magnetic Separation
Reserve Osmosis
Stripping
Ultrafiltration
Zone Refining

Chemical Treatment

Chemical treatment processes are particularly useful for the detoxification of hazardous wastes. They also provide for the separation of specific waste stream components. However, they are generally limited to liquid forms of waste materials.
   The following chemical treatment processes are potentially feasible for wastes:

Chemical Oxidation
Chemical Reduction
Hydrolysis
Liquid-Liquid Solvent Extraction
Neutralization
Ozonation
Photolysis

Bilogical Treatment

Biological treatment processes utilize microorganisms for the decomposition of organic compounds in the waste. These are applicable to aqueous waste streams with solid or solvent organics. Composting is a biological treatment process used for solid wastes.
   The following biological treatment processes are potentially feasible for wastes:

Activated Sludge
Aerated Lagoons
Anaerobic Digestion
Composting
Enzyme Treatment
Trickling Filters
Stabilization Ponds

Thermal Treatment

Thermal treatment processes include incineration and pyrolysis. Incineration is high temperature oxidation, or combustion, while pyrolysis is thermal decomposition without the addition of oxygen. Thermal treatment processes are suitable for all forms of waste materials.

Thermal treatment processes such as incineration provide the greatest weight and volume reduction of all treatment processes. In addition, they can destroy or detoxify hazardous organics, sterilize infectious materials, and provide for waste heat recovery. Some industries utilize thermal treatment processes in conjunction with production operations for the conversion, reclamation, and reuse or recovery of residuesand byproducts. Examples include the use of distillation for spent solvent reclamation and hydrochloric acid recovery from the incineration of chlorinated hydrocarbons in conjunction with a quench/neutralization system.

Incineration

Incineration is an engineered process. Variables most affecting system design, construction and operation include waste combustibility and heating values, operating temperatures, furnace retention times for the waste and products of combustion, turbulence within the combustion zone, and emission requirements. There are many basic types of industrial incinerators. Each is generally suitable for specific types, forms, and quantities of waste; and each has limitations, advantages, and disadvantages.

The standard, basic types of industrial incinerators include:

Open Burning: Basically involves detonation of explosive wastes in a remote, open areas on a flat, gravel base.

Open-Pit or Air Curtain Destructor: Single chamber, above or below ground, with an open top for burning explosive or reactive materials, such as nitrocellulose. A high velocity air stream across the open top provides a "curtain" of air to promote combustion turbulence within the pit.

Multiple Chamber Incinerator: Consists of a primary and one or more secondary chambers, and operates with high excess combustion air — typically 200 to 300 percent. Wastes are loaded into the primary chamber for combustion, and most combustibles entrained in the flue gases are burned in the secondary chambers. Usually limited to solid wastes, but hearth designs can accommodate sludges. Liquid wastes can be burned in suspension.

Controlled Air Incinerator: A two-stage combustion process: wastes in the primary chamber, or first-stage, are burned under starved air, or oxygen deficient conditions; most smoke and volatiles from the first-stage are combusted in the secondary chamber under excess air conditions — typically 100 to 200 percent. Usually limited to solid wastes and limited quantities of sludges. Liquid wastes can be burned in suspension.

Rotary Kiln Incinerator: Wastes loaded into slowly rotating, horizontally inclined, cylindrical chamber. Rotation provides tumbling and maximum turbulence for optimum combustion. Most combustibles entrained in the flue gases are burned in secondary chambers. Highly versatile system suitable for most organic wastes, including solids and sludges. Liquid wastes can be burned in suspension or loaded in drums or containers.

Multiple Hearth Incinerator: Wastes move slowly downward through vertically stacked hearths within vertically oriented, cylidrical chamber. Rotating rabble arms with plow blades move wastes across and downward through each hearth level. Usually limited to organic sludges, but can accommodate granulated solid wastes. Liquids and gases can be injected between various hearths.

Fluidized Bed Incinerator: Wastes injected into hot agitated bed of granular particles which are suspended within cylindrical chamber by a high pressure blower. Rapid combustion and heat transfer between wastes and bed. Usually limited to organic sludges, but can accommodate granulated solid wastes. Liquids and gases can be injected into the bed.

Liquid Injection Incinerators: This refers generically to those systems designed to burn liquid wastes in suspension within a combustion chamber. Depending on heating value, wastes can be injected through a nozzle or combusted through a burner system.

Gaseous Waste Incinerators: This refers to those systems designed for the combustion of fumes, vapors, odors and the like. Three basic types include:

1. Flares: open burning of combustible gases which are either near or above their lower explosive limits (LEL) of concentration or above their upper explosive limits (UEL) of concentration.
2. Thermal Burners: combustion of dilute, or low combustible, gases in the presence of a burner flame within a chamber.
3. Catalytic Burners: combustion of dilute gases which are preheated, exposed to a catalyst material and oxidized at relatively lower temperatures. Catalyst, subject to damage or suppression from gas stream contaminants.

Other types of industrial incinerators which are either highly specialized, very unique, or still under development include:

Molten-Salt Incineration
Wet Air Oxidation or Zimmerman Process
Watergrate Furnaces
Infrared or Radiant Heat Incineration
Plasma Destruction

High-Temperature Fluid-Wall Reactor (HTFW) or Thagard Process
Supercritical Fluid Technology or MODAR Process
Critical Fluid System (CFS)

Pyrolysis

As indicated, pyrolysis is a thermal treatment process similar to incineration except that decomposition occurs in the absence of oxygen. Usually heat is added externally, and temperatures are maintained much lower than typical incineration temperatures.

Pyrolysis basically transforms organic materials into solids, liquids and gaseous organics of much simpler structure. These can be subsequently burned for heat recovery, while inorganic materials, such as potentially slagging salts, remain with the solid ash. Rotary kilns, multiple hearth incinerators, and special rotary hearth and batch designs can all be operated under a pyrolytic mode.

Other Technologies

Other thermal treatment technologies and variations to conventional incineration used for waste combustion include:

Firing in conventional boilers
Cement kilns
Other industrial furnaces or ovens
At-sea incineration
Mobile incineration

Steam sterilization, or autoclaving, is a thermal treatment process for biologically hazardous or infectious wastes. Autoclave equipment is designed to kill pathogens in waste by exposing them to relatively high temperatures — between 240 and 280°F (116 to 138°C) — for periods ranging from several minutes to several hours. Retention times are dependent on steam temperatures and pressures, types and forms of the waste, quantities of waste, and specific type and design of the autoclave system.

Table 49.6. summarizes the major incineration technologies, and Table 49.7 shows the applicability of various incineration technologies to different waste types.

ULTIMATE DISPOSAL

Disposal is basically the final treatment or disposition of waste materials or residues from other treatment processes. For example, even with in-

TABLE 49.6. Summary of Organic Waste Incinerators[6]

Type	Process Principle	Application	Combustion Temp.	Residence Time
Rotary kiln	Slowly rotating cylinder mounted at slight incline to horizontal. Tumbling action improves efficiency of combustion.	Most organic wastes; well suited for solids and sludges; liquids and gases.	810–1650°C (1500–3000°F)	Several seconds to several hours
Multiple hearth	Solid feed slowly moves through vertically stacked hearths; gases and liquids fed through side ports and nozzles.	Most organic wastes, largely in sewage sludge; well suited for solids and sludges; also handles liquids and gases.	750–980°C (1400–1800°F)	Up to several hours
Liquid injection	Vertical or horizontal vessels; wastes atomized through nozzles to increase rate of vaporization	Limited to pumpable liquids and slurries (750 SSU or less for proper atomization).	650–1650°C (1200–3000°F)	0.1 to 1 sec
Fluidized bed	Wastes are injected into a hot agitated bed of inert granular paticles; heat is transferred between the bed material and the waste during combustion.	Most organic wastes; ideal for liquids, also handles solids and gases.	750–870°C (1400–1600°F)	Seconds for gases and liquids; longer for solids
Pyrolysis	Thermal decomposition in the absence of oxygen; transforms organic materials into solids, liquids and gaseous organic materials of simpler structure.	Primary as fuel; useful with solids and sludges; potential for resource recovery of breakdown products.	480–810°C (900–1500°F)	Normally 12–15 min

TABLE 49.7. Matrix for Matching Waste Type With Incineration Processes[7]

Waste Type	Rotary Kiln*	Multiple Hearth*	Fluidized Bed*	Liquid Incinerator	Catalytic Combustor	Multiple-Chamber Incinerator	Wet-Air Oxidation	Molten-Salt Incinerator
Solids								
Granular homogeneous	X	X	X					
Irregular bulky (Pallets, etc.)	X					X		
Low melting point (Tars, etc.)	X		X	If material can be melted and pumped				
Organic compounds with fusible ash constituents	X	X						X
Gases								
Organic vapor laden				X	X			
Liquids								
High organic strength aqueous wastes, often toxic	If equipped with auxiliary liquid injection nozzles			X			X	
Organic liquids	If equipped with auxiliary liquid injection nozzles		X	X				X
Solids/liquids								
Waste contains halogenated aromatic compounds (2,200°F minimum)	X		X	If liquid				X
Aqueous organic sludges	Provided waste does not become sticky upon drying	X					X	

*Suitable for pyrolysis operation

cineration, as much as 20 percent of the total mass remains as ash and/or sludges from the air pollution control system which needs disposal. Alternate disposal processes include land burial, deep well injection and ocean dumping.

Land Burial

Two types of land burial include sanitary landfills and secure, or industrial landfills. These are engineered systems as opposed to indiscriminate dumping. Sanitary landfills are used for the disposal of nonhazardous solid wastes. Wastes are covered daily with earth to minimize health vector problems, blowing of debris and open burning. However, unless specially designed, sanitary landfills have the potential for surface and groundwater pollution from leaching, as well as air pollution from gas venting.

Secure landfills for hazardous wastes are regulated under RCRA. These require special liner materials to prevent leaching, special caps to prevent surface exposures, venting systems for handling gases and monitoring and test wells for verifying the integrity of the liner structure.

Deep Well Injection

Deep well injection involves the injection of liquid wastes into an underground reservoir which has been identified as geologically secure. Since a prime concern is to protect all usable underground water, the reservoir must be located below potable water aquifers and isolated by thick, nearly impermeable strata, such as shale, limestone or dolomite. Average well depths are about 5,000 ft, with some as deep as 10,000 ft. Most are located in the southwest where the geology favors this disposal technology.

Ocean Dumping

Sewage sludges, industrial wastes and explosives have been dumped at sea for years in designated and approved locations. The U.S. Army Corps of Engineers and the U. S. Coast Guard are basically responsible for the sea transportation and disposal areas for industrial wastes. Of course, the primary concern with this type of disposal is the threat to marine life and coastal areas from improperly sealed containers of hazardous wastes.

Table 49.8. compares some of the alternate hazardous waste treatment and disposal technologies.

TABLE 49.8. Comparison of Some Hazard Reduction Technologies[a]

	Disposal		Treatment		
	Landfills and Impoundments	Injection Wells	Incineration and Other Thermal Destruction	Emerging High-Temperature Decomposition[b]	Chemical Stabilization
Effectiveness: How well it contains or destroys hazardous characteristics	Low for volatiles, questionable for liquids; based on lab and field tests	High, based on theory, but limited field data available	High, based on field tests, except little data on specific constituents	Very high, commercial-scale tests	High for many metals, based on lab tests
Reliability issues:	Siting, construction, and operation. Uncertainties: long-term integrity of cells and cover, liner life less than life of toxic waste	Site history and geology; well depth, construction and operation	Monitoring uncertainties with respect to high degree of DRE; surrogate measures, PICs, incinerability[d]	Limited experience Mobile units; onsite treatment avoids hauling risks Operational simplicity	Some inorganics still soluble Uncertain leachate test, surrogate for weathering
Environmental media most affected	Surface and ground water	Surface and ground water	Air	Air	Ground water
Least compatible wastes[c]	Liner reactive; highly toxic, persistent, and bioaccumulative	Reactive; corrosive; highly toxic, mobile, and persistent	Highly toxic and refractory organics, high heavy metals concentration	Some inorganics	Organics
Costs: Low, Mod, High	L–M	L	M–H (Coincin. = L)	M–H	M
Resource recovery: potential	None	None	Energy and some acids	Energy and some metals	Possible building material

[a] Wastes listed do not necessarily denote common usage.
[b] Molten salt, high-temperature fluid wall, and plasma arc treatments.
[c] Wastes for which this method may be less effective for reducing exposure, relative to other technologies.
[d] DRE = destruction and removal efficiency. PIC = product of incomplete combustion.

INDUSTRIAL WASTE CONTROL

CODES, REGULATIONS AND STANDARDS

The proper and safe management and disposal of industrial wastes require in-depth knowledge of the material properties, alternate treatment technologies and applicable regulations. There are also various codes and standards dealing with waste handling and disposal. The more significant of these are listed below.

Federal Codes and Regulations Applicable to Waste Management and Disposal

Resource Conservation and Recovery Act of 1976 (RCRA), Subtitle C, Hazardous Waste Regulations, 10 CFR 40.
Nuclear Regulatory Commission (NRC), "Standards for Protection Against Radiation," 10 CFR 20.
Toxic Substances Control Act of 1976 (TSCA)
Clean Air Act of 1963 (CAA), including:
- National Ambient Air Quality Standards (NAAQS)
- National Emission Standards for Hazardous Air Pollution (NESHAP)
- Prevention of Significant of Deterioration (PSD)
- Nonattainment Regulations (NA)

Federal Pesticide Control Act of 1972; also known as Federal Insecticide, Fungicide and Rodenticide Act (FIFRA)
Oil Pollution Act of 1961
Federal Water Pollution Control Act of 1948 (FWPCA)
Occupational Safety and Health Act of 1970 (OSHA), including National Institute of Occupational Safety and health (NIOSH)
Hazardous Substances and Hazardous Waste Response, Liability and Compensation Act; also known as "Superfund"

General Standards Dealing With Waste Materials and Their Management

NFPA 43, *Liquid and Solid Oxidizing Materials*
NFPA 48, *Magnesium Storage and Processing*
NFPA 49, *Hazardous Chemicals Data*
NFPA 321, *Basic Classification of Flammable and Combustible Liquids*
NFPA 325M, *Fire Hazard Properties of Flammable Liquids, Gases and Volatile Solids*
NFPA 481, *Titanium Production, Handling and Storage*
NFPA 482M, *Zirconium Production, Processing and Handling*
NFPA 491M, *Hazardous Chemical Reactions*
EPA Industrial Environmental Research Laboratory, "Test

Methods for Evaluating Solid Waste- Physical/Chemical Methods," SW-846, Revision B, July, 1981

EPA Industrial Environmental Research Laboratory, "Sampling and Analysis Methods for Hazardous Waste Incineration," First Edition, Contract 68-02-3111, February, 1982.

U.S. Department of Health and Human Services- Public Health Service: Center for Disease Control and National Institutes of Health, "Biosafety in Microbiological and Biomedical Laboratories," Draft, March, 1983.

General Standards Dealing With Waste Treatment and Disposal Systems

NFPA 30, *Flammable and Combustible Liquids Code*
NFPA 86A, *Ovens and Furnaces: Design, Location and Equipment*
NFPA 82, *Incinerators, Waste and Linen Handling Systems and Equipment*
NFPA 801, *Facilities Handling Radioactive Materials*
EPA Office of Solid Waste, "Draft Manual for Infectious Wate Management," SW-957, September, 1982
Incinerator Institute of America, "Incinerator Standards," 1968

General Standards Dealing with Incineration Burners and Equipment

NFPA 31, *Installation of Oil Burning Equipment*
NFPA 54, *National Fuel Gas Code*
UL 296, *Oil Burners*
UL 372, *Primary Safety Controls for Gas and Oil Fired Appliances*
UL 795, *Gas Burners*

The above codes and standards are not all inclusive. Other agencies, such as Industrial Risk Insurers (IRI), American National Standards Institute (ANSI) and American Society for Testing and Materials (ASTM), may have standards applicable to specific waste management and disposal operations. In addition, state and local regulations and requirements may be more stringent than federal regulations or standards.

BIBLIOGRAPHY

REFERENCES CITED

1. Wilson, D. G., *Handbook of Solid Waste Management*, Van Nostrand Reinhold Co., NY, 1977.
2. *Hazardous Waste Generation and Commercial Hazardous Waste Management*, U.S. EPA, SW-894, December 1980, Page III-3.
3. State-Wide Comprehensive Solid-Waste Management Study, New York State Department of Health, Albany, NY, 1970.
4. Ulmer, N. S., *Physical and Chemical Parameters and Methods for Solid-Waste Classification*, Open File Progress Report RS-03-68-17, U.S, EPA, 1970.
5. *Report to Congress - Disposal of Hazardous Wastes*, U.S. EPA, SW-115, 1974.
6. Shen, T. T., Chen, M. and Lauber, J., "Incineration of Toxic Chemical Wastes," *Pollution Engineering*, pp. 45-50, October 1978.
7. Hitchcock, D., "Solid Waste Disposal: Incineration," *Chemical Engineering*, May 21, 1979, pp. 185-194.
8. *Technologies and Management Strategies for Hazardous Waste Control*, Congressional Office of Technology Assessment, OTA-M-197, March 1983.

NFPA CODES, STANDARDS, AND RECOMMENDED PRACTICES

Reference to the following NFPA Codes, Standards, and Recommended Practices will provide further information on the control of hazardous wastes discussed in this chapter. (See the latest *NFPA Codes and Standards Catalog* for availability of current editions of the following documents.)

NFPA 30, *Flammable and Combustible Liquids Code.*
NFPA 31, *Installation of Oil Burning Equipment.*
NFPA 43, *Oxidizing Materials, Storage of Liquid and Solid.*
NFPA 48, *Magnesium Storage, Handling and Processing.*
NFPA 49, *Hazardous Chemicals Data.*
NFPA 54, *National Fuel Gas Code.*
NFPA 82, *Incinerators, Waste and Linen Handling Systems and Equipment.*
NFPA 86A, *Ovens and Furnaces: Design, Location and Equipment.*
NFPA 321, *Basic Classification of Flammable and Combustible Liquids.*
NFPA 325M, *Fire Hazard Properties of Flammable Liquids, Gases, and Volatile Solids.*
NFPA 481, *Titanium Production, Processing, Handling and Storage.*
NFPA 482, *Zirconium Production, Processing, Handling and Storage.*
NFPA 491M, *Manual of Hazardous Chemical Reactions.*
NFPA 801, *Facilities Handling Radioactive Materials.*

Additional Reading

Brown, D. A. G., "Chemical Companies Know Their Risks," *Fire Engineers Journal*, 41(121), March 1981, pp. 30-32.

Chemical Engineering, *Industrial Waste Water and Solid Waste Engineering*, McGraw-Hill, 1980.

Controlling Hazardous Wastes - Research Summary, U.S EPA, May, 1980.

Conway, Richard A. and Richard D. Ross, *Handbook of Industrial Waste Disposal*, Van Nostrand Reinhold, New York, NY, 1980.

Disposing of Small Batches of Hazardous Waste, U.S. EPA, SW-562, 1976.

Doucet, L. G., *Waste Incineration - Selection, Costing and Implementation*, Vanderbilt University CEE, 1980.

Edwards, B. H., and Coghlan-Jordan, K., *Emerging Technologies for the Control of Hazardous Waste*, Noyes Data Corp., NJ, 1983.

Engineering Handbook for Hazardous Waste Incineration, U.S. EPA, SW-889, September, 1981.

"Firms Avidly Seeking New Hazardous Waste Treatment Routes," *Chemical Engineering*, News Feature, September 6, 1982, pp. 53-57.

Hackman III, E., *Toxic Organic Chemical Destruction and Waste Treatment*, Noyes Data Corp., NJ, 1978.

Handbook of Key Federal Regulations and Criteria for Multimedia Environmental Control, U.S. EPA, August, 1979.

Hazardous Wastes, U.S. EPA, SW-138, 1975.

Hazardous Waste Management Guide, J. J. Keller and Associates, Inc., Neenah, WI, 1983.

Incineration in Hazardous Waste Management, U.S. EPA, SW-141, 1975.

Kiang, Y. H., and Metry, A. A., *Hazardous Waste Processing Technology*, Ann Arbor Science, MI, 1982.

Paulson, E. G., "How to Get Rid of Toxic Organics," *Chemical Engineering*, Deskbook Issue, October 17, 1977, pp. 21-26.

Pavoni, J. L., Heer Jr., J. E., and Hagerty, D. J., *Handbook of Solid Waste Disposal*, Van Nostrand Reinhold Co., New York, NY, 1975.

Prudent Practices for the Disposal of Chemicals from Laboratories, Committee on Hazardous Substances of National Research Council, National Academic Press, Washington, DC, 1983.

Sax, I, *Dangerous Properties of Industrial Materials*, Van Nostrand Reinhold Co., New York, NY, 1975.

Vance, Mary, *Industrial Waste Disposal: A Bibliography*, Vance Bibliographies, Monticello, IL, 1982.

"Second Chance for Pyrolysis," *Chemical Engineering*, News Feature, December 13, 1982, pp. 41-43.

Sittig, M., *Incineration of Industrial Hazardous Waste and Sludges*, Noyes Publishing Co., Park Ridge, NJ, 1981.

———., *Landfill Disposal of Hazardous Waste and Sludges*, Noyes Publishing Co., Park Ridge, NJ, 1979.

Worthy, W., "Hazardous Treatment Technology Grows," *C&EN*, March, 1982, pp. 10-16.

50

Industrial Housekeeping Practices

Kathleen M. Robinson

Good housekeeping is the care and maintenance of property and the provision of equipment and service. It is basic to firesafety and should be a major concern in every type of industrial occupancy, from the simplest workshop to the most complex industrial facility.

A good housekeeping program concerns itself with the less complex aspects of operational tidiness and order, waste control, and the regulation of such personal practices as smoking which, without reasonable controls, could lead to hazardous conditions.

Poor housekeeping contributes to loss potential by increasing fire and explosion hazards in several ways:

1. It provides more places for a fire to start.
2. It creates a greater continuity of combustibles that makes it easier for fire to spread.
3. It provides a greater combustible loading for the initial fire to feed upon.
4. It creates the potential for flash fires or dust explosions when layers of lint or dust are allowed to accumulate.

Kathleen M. Robinson is Editor of *The Sentinel*, published by Industrial Risk Insurers, Hartford, CT.

5. It increases the potential for spontaneous ignition.

In addition to the increased hazard, poor housekeeping can have a negative effect on production. Quality proves hard to maintain when the workspace is crowded and messy. Efficiency suffers because people normally tend to work faster and more accurately if their surroundings are clean. Thus, good housekeeping will not only prevent fires but can improve production and employee morale as well.

Good Housekeeping Theory

The degree of effort and attention needed for proper housekeeping is influenced, of course, by the type of buildings and the overall size of the facility involved. But most significant are the specific occupancies of the facility. Some processes produce more waste, leakage, and vapors than others, thus contributing to the extent of housekeeping problems. In addition, the acceptable level of cleanliness varies from occupancy to occupancy. What is satisfactory in a foundry would probably not be tolerable in an office building. And the cleanliness of the average office would hardly be satisfactory for an electronic "clean room."

Proper housekeeping does not just happen. It requires the leadership and wholehearted support of management and the cooperation of all employees. Management cannot merely decree that good housekeeping is a desired goal. It must place the responsibility and authority for achieving that goal with a committee (or individual). The committee, in turn, must inspect, evaluate, and, with management, finally define the desired cleanliness levels for the various sections of the facility. It must also establish the frequency of periodic inspections and devise a report form and distribution list. At the same time, the committee must identify the responsibilities of the individual workers towards both the general area and their immediate workspaces.

In small industrial, commercial, or office occupancies, the services of professional cleaning and maintenance firms may be contracted for, thus placing the immediate responsbility for good housekeeping with the managers of the hired individuals or crews. Nevertheless, plant management is still ultimately responsible for auditing the work being performed and making sure that the contractor complies with the plant's established goals.

Where proper housekeeping does not exist, it is usually because inadequate attention is paid to, or inadequate action taken in, one or more of the following areas.

Communication. Management must obviously publicize its commitment to good housekeeping and the delegation of its authority to the committee or individuals to whom overall responsibility has been assigned. This publicity must be reinforced periodically and recognition given whenever notable improvement or outstanding performance takes place.

Besides the scheduled inspection reports, the housekeeping committee must be able to meet with management periodically to review performance, to revise goals, and to offer and substantiate recommendations for major expenditures. The committee should also encourage feedback from employees in the form of suggestions and constructive criticism. Simpler or more efficient housekeeping methods are often more readily identified by the worker who uses them than by management, and action on the proffered suggestions emphasize management's sincere desire to achieve the publicized goals.

Equipment. Housekeeping efforts should not founder from lack of necessary tools or equipment. This includes the tools normally used by maintenance personnel, such as brooms, dust pans, mops, and vacuum cleaners, as well as other items, such as those that encourage the proper disposal of trash. The simple step of putting a sufficient number of easily accessible wastebaskets or trash receptacles at points of need can cut down on the amount of waste deposited on the floor or in the product.

In some production and handling areas, dust, lint, and other waste may be produced constantly. In these cases, vacuum pick-up stations tied into an exhaust and collection system may be needed at specific points of waste generation. For area cleaning, powered floor sweepers or rail-mounted traveling cleaners may be warranted. Plant sections in which large quantities of scrap or discarded packing materials continuously accumulate may need not only large trash containers, but also motorized equipment that can be emptied or replaced frequently.

Layout and Storage. Overcrowding is a major impediment to proper housekeeping. Blocked or restricted aisles limit access and, in so doing, hamper efficient cleaning and trash pick-up. Lack of sufficient workspace and storage capacity leads to inefficient operations, to an inability to create order, and finally to worker frustration. The creative use of racks, shelving, and bins is often a rewarding answer.

With its negative influence on good housekeeping, disorganized and haphazard storage is usually a detriment to effective fire protection, as well. Fire extinguishers, small hose stations, and extinguishing system control valves can become blocked and inaccessible while other fire equipment, such as fire doors, may be made inoperable.

Environment. Equally as important as the natural environment is the artificial environment created in the workplace. The control of process fumes, vapors, dusts, and flyings required today for the well-being of the worker has also had a beneficial impact on overall cleanliness in many work areas.

Adequate lighting, now recognized as a prerequisite for quality and high productivity, also helps improve housekeeping. So do the light-colored walls and floors commonly used to make lighting more effective and surroundings more pleasant. They make spills, leaks, and waste accumulations highly visible, and visible waste is more likely to be cleaned up than waste hidden by poor lighting and dark surroundings.

Manpower. Too many housekeeping programs fail because they depend solely on manpower to achieve their goals. They ignore other factors, such as automated equipment and increased employee awareness, that could help keep the facility clean.

Nevertheless, manpower is an important part of any housekeeping program and it cannot be overlooked. People will always be needed to make sure that things are being done as they should be, and management must recognize that fact if established goals are to be met.

BUILDING CARE AND MAINTENANCE

The three basic requirements for good housekeeping are proper layout and equipment, correct materials handling and storage, and cleanliness and order. Any facility that implements these basics has laid the foundation for good housekeeping. Using them, the facility can develop special housekeeping practices to deal with its own specific problems.

The care and maintenance of buildings requires special housekeeping practices. These are particularly noteworthy because they either introduce fire hazards into or reduce the fire danger to buildings.

Floors

The general care, treatment, cleaning, and refinishing of floors may present a fire hazard if flammable solvents or finishes are used or if combustible residues are produced in quantity. Many fires have resulted from the use of gasoline to clean floors, for example. In general, cleaning or finishing compounds containing solvents with flash points below room temperature are too dangerous for ordinary use, except in very small quantities. The magnitude of the hazard depends on the conditions of use and the precautions taken. Many cleaning compounds presenting little or no hazard are listed by fire testing laboratories.

Sweeping compounds: Compounds used for sweeping floors generally consist of sawdust or some other combustible material treated with oil. Such compounds are hazardous, the degree of danger depending upon the characteristics of the oil. The use of sawdust or similar materials to absorb oil spillage increases the fire hazard unnecessarily since noncombustible oil-absorptive materials are available for this purpose.

Floor oils: Compounds containing oils and low-flash-point solvents are a hazard, particularly when freshly applied. In addition, component oils may be subject to spontaneous heating. To reduce the fire hazard, suitable attention must be given to the safe storage of oily mops, sponges, and wiping rags in metal or other noncombustible containers. Any combustible oil used to excess increases the combustibility of the floor. Oil-soaked floors, the product of years of use, also show increased combustibility.

Floor waxes: Low-flash-point solvents are hazardous, especially when used with electric polishers. In such instances, ignition might result from friction and sparking. Water emulsion waxes are preferable.

Furniture polishes: Furniture polishes containing oils subject to spontaneous heating become hazardous when rags that are saturated with these polishes are not disposed of properly. Such oil-soaked rags should be placed in metal or other noncombustible containers.

Flammable cleaning solvents: Flammable cleaning solvents need not be used since a number of nonhazardous cleaning agents are available. These relatively safe materials are stable and have high flash points and low toxicity. There are several commercial stable solvents available which have flash points ranging from 140 to 190°F (60 to 88°C) and have a comparatively low degree of toxicity. Safe materials are available for most of the preceding purposes.

Dust and Lint

A necessary procedure in many industrial occupancies is the removal of combustible dust and lint accumulations from walls, ceilings, and exposed structural members. Unless this procedure is performed safely, as by vacuum cleaners or air moving (blower and exhaust) systems, this procedure may present a fire or explosion hazard. In some cases, vacuum cleaning equipment must be equipped with dust-ignition-proof motors to assure safe operation in dust-laden atmospheres.

Care should be taken not to dislodge into the atmosphere any appreciable quantities of combustible dust or lint which might ignite or form an explosive mixture with air. A lot of work can be eliminated by applying suction at locations where dust may escape from processing machinery and conveying the aspirated dust to safely located collectors. Blowing down dust with compressed air may create dangerous dust clouds, and such cleaning should be done only when other methods cannot be used and after all possible sources of ignition have been eliminated. In most localities, it is possible to obtain the services of reliable professional industrial cleaning specialists to remove dust accumulations safely.

Exhaust Ducts and Related Equipment

The exhaust ducts from the hoods over cooking ranges, such as those found in plant cafeterias, present troublesome problems because grease condenses inside the ducts and on exhaust equipment. Grease accumulations may be ignited by sparks from the range or, more often, by small fires in overheated cooking oil or fat. Without these grease accumulations in the hood and duct, stove top fires can often be extinguished or allowed to burn out without causing appreciable damage. Fires occur frequently in frying because cooking oils and fats are heated to their flash points and may reach their self-ignition temperatures when accidentally overheated or spilled on the hot stove top.

Grease removal devices: All exhaust systems for kitchen cooking equipment must be equipped with a grease removal device. These include such items as grease extractors, grease filters, or special fans designed to remove grease vapors effectively and provide a fire barrier. Grease filters, including frames, and other grease removal devices should be made of noncombustible materials.

Ducts: There is no practical method for preventing all kitchen duct fires, but the danger can be minimized through a combination of precautions as outlined in NFPA 96, *Vapor Removal from Cooking Equipment*. It is good practice to clean hoods, grease removal devices, fans, ducts, and associated equipment frequently. The exhaust system should be inspected daily or weekly, depending on its use, to determine if grease or other residues are accumulating in it.

Clean ducts are essential to firesafety, but they often remain dirty because cleaning them is a difficult and unpleasant job. One source of help is a commercial firm that specializes in this sort of work. In any case, never try burning the grease out; it is a dangerous practice, even though duct systems installed according to NFPA standards are designed to withstand burnout.

In cleaning the exhaust system, avoid using flammable solvents or other flammable cleaning aids. Do not start the cleaning process until all electrical switches, detection devices, and extinguishing system supply cylinders have been turned off or locked in a "shut" position. This will prevent both the exhaust fan and the fire extinguishing system (if the exhaust duct is equipped with one) from actuating accidentally. Once the cleaning process is completed, the switches and other controls should be returned to normal operating position.

Satisfactory cleaning results have been obtained with a powder compound consisting of one part calcium hydroxide and two parts calcium carbonate. This compound saponifies the grease or oily sludge (converts it to soap), thus making it easier to remove and clean. The process requires proper ventilation. Another cleaning method is to loosen the grease with steam and then scrape the residue out of the duct. This has proven to be quite effective.

Spraying duct interiors with hydrated lime after cleaning is a fire prevention method used commercially. This procedure tends to saponify the grease and may facilitate subsequent cleaning, but it does not provide permanent fire retardency.

INDUSTRIAL OCCUPANCY AND PROCESS HOUSEKEEPING

Housekeeping programs developed for industrial occupancies and processes must give special consideration to disposal of rubbish, control of smoking habits, housekeeping hazards, and lockers and cupboards, where and as applicable.

Disposal of Rubbish

The proper handling and disposal of rubbish is an integral part of the housekeeping process, and its success depends primarily upon having and observing a satisfactory routine. The proper and regular disposal of combustible waste products is of the utmost importance.

In both industrial and commercial properties, the removal of combustible waste products at the end of each workday or at the end of each work shift is a common practice. In some properties, more frequent waste disposal is necessary. In others, the collection, storage, and disposal routine vary with the nature of the property use. In all cases, however, an adequate program for dealing with this problem is a firesafety essential. Keeping a place tidy also depends on providing enough wastebaskets, bins, cans, and other proper containers so that building users will find tidiness convenient. See Chapter 49, "Industrial Waste Control," for further information on good practices to follow in disposing of industrial wastes.

Receptacles: Noncombustible containers should be used for the disposal of waste and rubbish. This is true even of such small receptacles as ashtrays and wastebaskets and applies, of course, to the larger units found in industrial properties. Industrial waste barrels should be made of metal and equipped with a fitted cover. Care should be taken to avoid mixing waste materials where such mixing introduces hazards of its own.

Plastic wastebaskets of varying sizes are readily available and are popular because they are quiet, attractive, and scratch- and dent-resistant. Not all plastic baskets have the same burning characteristics, however. Some melt and burn readily, adding fuel to the fire and creating a comparatively serious fire exposure problem by collapsing and spilling their burning contents. This is also true of many plastic liners used for wastebaskets and receptacles. Other baskets may contribute relatively little fuel to the fire while maintaining their shape fairly well. If a plastic basket is to be used, the buyer should keep in mind that some are superior to others. Several manufacturers have been concerned enough to make marked improvements in the fire behavior of their products, and

Chapter 50

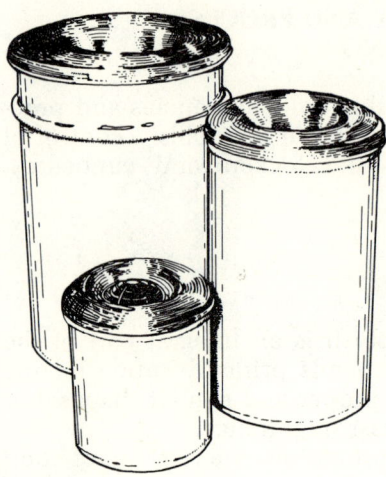

FIGURE 50.1. *Waste containers designed to snuff out accidental fires in their contents and to limit external surface temperatures to no more than 175°F (80°C) above room temperature. (Justrite Mfg. Co.)*

the prospective buyer should look for information concerning such superiority in the manufacturer's literature.

Segregation of waste: It is not good housekeeping practice to dump all manner of dry waste down refuse chutes or to place it in a common bin or storage receptacle. This cannot be done safely. For example, combustible metal dusts and metal powders dumped into chutes may explode. Mercury batteries and pressurized containers, such as the ubiquitous aerosol can, may also explode when incinerated or mixed with rubbish which is subsequently burned. Precautions should be taken to keep combustible items separate from each other and from noncombustible items.

Control of smoking habits: Smoking may be a difficult problem to handle, particularly because of the personal factors involved. It is a habit that is hard to break, even though it may put the smoker in direct conflict with the firesafety regulations and production standards in effect in various areas. When these areas contain flammable liquids or dusty and linty atmospheres, self-preservation alone makes smoking control relatively easy. However, control of smoking is also required in less obvious places, such as shipping and receiving areas, with their large quantities of loose packing materials, and storage areas, which may have high-piled concentrations of combustible materials. In these areas, carefully planned smoking regulations are necessary.

Smoking regulations should be specific as to location and, preferably, time. Areas in which smoking is permissible, as well as those in which it is limited or prohibited entirely, must be clearly marked by appropriate signs that leave no question as to what is allowed where.

In addition to sensible regulations, smoking control also requires adequate receptacles for spent smoking materials.Properly designed ashtrays are essential to safe smoking. They should be made of noncombustible materials, with grooves or snuffers that hold cigarettes securely.

FIGURE 50.2. Examples of signs permitting or forbidding smoking in designated areas. (Factory Mutual System)

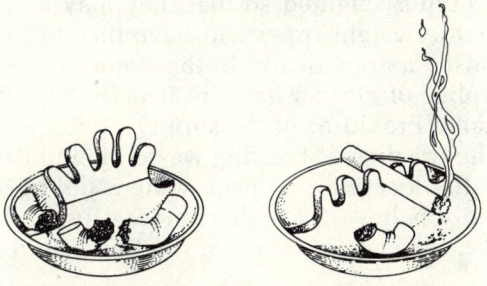

FIGURE 50.3. Two types of well-designed ash trays.

Their sides should be steep enough to force smokers to place cigarettes entirely within the ashtray. In industrial buildings, large containers of sand are often used to conveniently and safely extinguish and dispose of spent smoking materials.

Improperly designed ashtrays may constitute a hazard, particularly if they allow a lit cigarette or cigar to fall or roll away. A lighted butt may too easily come in contact with combustible materials and start a fire under certain circumstances.

The contents of ashtrays must be disposed of carefully because a live butt may well be mixed in with apparently innocuous ashes. If lighted smoking materials were to be dumped into an ordinary wastebasket, it could set paper or some other piece of combustible rubbish on fire. To prevent this from happening, reserve special covered metal containers for discarded smoking materials only.

Chapter 50

Industrial Housekeeping Hazards

Some industrial occupancies have special housekeeping problems inherent to the nature of their operations. For these particular problems, specific planning and arrangements are necessary.

Clean waste and rags: Clean cotton waste or wiping rags are generally considered to be mildly hazardous, chiefly because they are readily flammable when not baled and there is always the likelihood that dirty waste may become mixed with them. The presence of dirty waste or small amounts of certain oils may lead to spontaneous heating. Reclaimed waste is considered somewhat more hazardous than new waste. It is common practice to handle clean waste in the same manner as dirty waste, although the fire hazard is relatively small.

Large supplies of clean waste are best kept in bins made entirely of metal or of wood lined with metal and provided with covers that are normally kept closed. Several bins may be provided where the supplies are large or where different kinds of waste are kept. The covers on such bins should be counterweighted so that they may be readily raised and lowered. The counterweight ropes can have fusible links to ensure that the covers are closed automatically in the event of fire.

Local supplies of clean waste are usually kept in small, properly marked waste cans. Providing local supply points for clean waste can help eliminate the practice of keeping waste in clothes lockers, drawers, benches, and similar locations. If clean waste is put in such places, workers may mistakenly believe that other, more oily waste is also allowed there when, in fact, the combination of the two may result in fire.

Coatings and lubricants: Paints, grease, and similar combustibles are widely used at industrial occupancies, and a good housekeeping program will make sure that their combustible residues are collected and disposed of safely. Nonsparking tools are recommended for cleaning spray booths and associated exhaust fan blades and ducts to avoid possible ignition of combustible residues. The discharge of vapors from spray booths should be so arranged that the vapors are conducted directly to the outside and the residues accumulate safely.

Particular care must be taken to keep sprinklers free from deposits. A thin coating of grease placed on sprinklers and cleaned frequently is one satisfactory method. Another is to enclose each sprinkler in a light paper or plastic bag that is changed daily. See NFPA 13A, *Care and Maintenance of Sprinkler Systems*, for more definitive information on the maintenance of sprinklers and sprinkler systems.

Drip pans: Drip pans are essential at many locations, notably under motors, machines using cutting oils, and bearings. They are also used with borings and turnings that may contain oil. Drip pans should be made of noncombustible material and contain an oil-absorbing compound. At many industrial occupancies, commercial oil-absorbing compounds con-

INDUSTRIAL HOUSEKEEPING PRACTICES

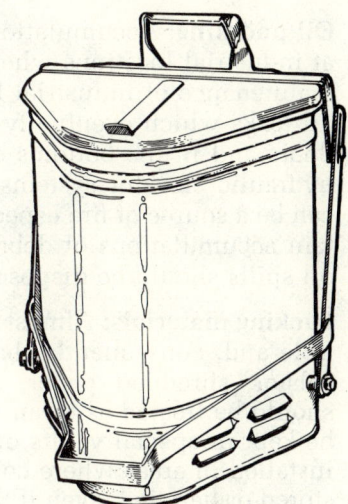

FIGURE 50.4. *A portable waste can which is equipped with a self-closing cover. This type of can can be used to store oily waste materials, particularly if they are subject to spontaneous heating. (The Protectoseal Co.)*

sisting largely of diatomaceous earth are used instead of sawdust or sand. The regular removal of oil-soaked material is recommended.

Flammable liquid spills: Flammable liquid spills may be anticipated wherever such products are handled or used. At industrial occupancies, some means of coping with these spills must be kept on hand. These include a supply of suitable absorptive material and special tools to help limit the spill. Workers should understand and promptly take the steps needed to cut off sources of ignition, ventilate the area, and safely dissipate any flammable vapors.

Flammable liquids waste disposal: The disposal of combustible liquid waste often presents a troublesome problem. Waste liquids, such as automobile crankcase fluids, must never be drained into sewers but placed in metal drums until they can be disposed of safely. In some cities, there are firms that make a specialty of collecting waste petroleum products and keeping them for further use, such as coating driveways and race tracks. Waste products must not be burned in oil burners unless the burner and accessories have been designed or properly adapted to handle such liquids. Many fire departments like to receive waste oils for use in training fire fighters to handle combustible liquid fires.

Oily waste: Oily wiping rags, sawdust, lint, clothing, and other items are highly dangerous, particularly if they contain oils subject to spontaneous heating. To dispose of all such materials in ordinary quantities, a standard waste can that has been tested and approved or listed by recognized testing laboratories is best. For large amounts, heavy metal barrels with covers are ideal. Good practice calls for cans containing oily waste to be emptied daily and for wiping rags to be kept in covered metal containers until they can be laundered.

Oil puddling: Accumulations of oil can present a housekeeping problem at industrial locations where a considerable amount of oil is used. Poor maintenance of industrial hydraulic elevator installations may result in oil leaks which eventually form puddles on the elevator machine room floors and in the bottoms of hoistway pits. Although most oils used in hydraulic elevator systems have high flash points, any combustible oil can be a source of fire especially when it is found in puddles which contain accumulations of debris. Puddled oil and materials used to absorb oil spills should be disposed of in metal barrels.

Packing materials: Almost all packing materials used today are combustible and, consequently, hazardous. Plastic pellets and rigid forms, excelsior, shredded paper, sawdust, burlap, and other such materials should be treated as clean waste. However, large quantities may have to be kept in special vaults or storerooms. Automatic sprinklers should be installed in areas where considerable quantities of packing materials are stored or handled, even if the balance of the building is not so protected.

Used or waste packing materials and the crating materials from receiving and shipping rooms must be removed and disposed of as promptly as possible in order to minimize the danger of fire. Ideally, the packing and unpacking processes should be conducted in an orderly manner so that excessive quantities of packing materials do not become strewn about the premises.

Lockers and Cupboards

Many industrial facilities provide their employees with lockers in which to put their personal belongings. These lockers may present a fire hazard if they are untidy or are used as general storerooms for such waste material as oily rags and cloths or paint-smeared clothing. These items may ignite spontaneously or they may be accidentally ignited by matches and imperfectly extinguished pipes and cigars that employees inadvertently leave in their lockers.

Wooden lockers can ignite and spread fire. Metal ones are preferred and they should be inspected regularly. Metal lockers may contain a fire if they are of solid construction, fronts and bottoms, partitions and backs. Backs and partitions made of expanded metal or wire screen may allow fire to spread unchecked and should not be used.

Lockers that are arranged in two tiers, one upon the other, are generally unsatisfactory. They do not hold clothes without mussing them and it often becomes the habit of those using such lockers to keep their clothes outside the locker or to throw them haphazardly into the locker, thus increasing the danger of spontaneous heating when such clothing is spotted with oil or paint.

Some lockers are provided with mechanical exhaust ventilation. Where this is true, NFPA 91, *Blower and Exhaust Systems Standard*, should be followed to avoid spreading a fire that originates in a locker.

Industrial plants that furnish and wash their employees' protective clothing may use a system of wire baskets, one per employee, suspended from the ceiling by a small chain running over a pulley instead of lockers. This method has proved successful in maintaining cleanliness and thereby reducing the fire hazard.

Where automatic sprinklers are installed, lockers must have expanded metal or screen tops to enable water from the sprinklers to reach the contents. Paper can be pasted on the tops to keep dust out. Sloping tops are also advisable to help prevent both fires and the accidents. Material cannot be placed on top of a locker designed this way.

Wooden supply cupboards constitute a fire hazard in places like machine and paint shops where woodwork becomes oil- or paint-soaked and where clothes or oily waste may be left in them. Wooden cupboards should be inspected regularly to make certain that they are always clean. The ideal cupboards for tools and similar items are made entirely of steel.

OUTDOOR HOUSEKEEPING PRACTICES

Good housekeeping practices are as essential out-of-doors at industrial occupancies as they are indoors. Failure to comply with good housekeeping practices out-of-doors may threaten the security of exposed structures and goods stored outside. The accumulation of rubbish and waste and the growth of tall grass and weeds adjacent to buildings or stored goods are probably the most common hazards. A regular program for policing the grounds is essential.

Grass and Weed Control

Tall grass, dry weeds, and bushes around buildings, along highways, on railroad properties, and along the streets of large industrial complexes present a definite fire hazard. To reduce this hazard, those responsible for maintaining these properties have always tried to control or destroy such vegetation.

One way to get rid of unwanted vegetation is to apply a chemical solution which poisons the weeds. Among the chemicals used are chlorate compounds, particularly sodium chlorate. Unfortunately, those who use chlorate compounds do not always realize that they are oxidizing agents. When these compounds come in contact with combustible materials, they virtually prime those materials for a fire or explosion. During hot periods in the summer, large numbers of fires have resulted from the use of sodium chlorate solutions on dry grass and weeds. Fires have also been reported in buildings and other structures in which such solutions have been spilled. Chlorate compounds spilled on clothing may present a danger to personnel, too.

There are weed killers that are not toxic and do not pose a fire hazard. Calcium chloride and agricultural borax, applied dry or in solution, are effective nonhazardous weed killers, as are various proprietary solutions. A number of commercial chemical weed killers, such as ammonium sulfamate, have little or no fire hazard and only a slight toxic hazard. Sodium arsenite and other compounds containing arsenic are efficient herbicides, but they are poisonous and not generally recommended.

The amounts of various chemicals needed to effectively kill weeds and the duration of their effect vary depending on the weed-killing agent used, the character of the vegetation, the climate, and the soil. Manufacturers' directions indicate the proper amounts to be used under various conditions.

Another way to remove vegetation is to burn it. This may be done only where environmental regulations permit outdoor burning and must be carefully controlled. Adequate fire extinguishing equipment must be readily available at all times.

Using this method, grass and weeds are usually cut down, collected in piles, and ignited. When the grass is too damp to propagate fire easily, flame-throwing torches may be used on the piles. However, these torches introduce a hazard if not carefully operated.

In fact, all burning introduces a hazard. Grass fires frequently spread out of control and ignite nearby buildings. To avoid this hazard, controlled burning should only be done at certain times of the year and then under the direct supervision of the fire department.

The fire department issues fire permits to help them control burning. These permits provide the department with an opportunity to educate the public in safe burning. They also help the fire department limit burning to nonhazardous periods of the year and better manage burning done during the hazardous periods.

At many industrial occupancies, pest control is an important housekeeping function. Information on effective pest control operations and fumigation is contained in NFPA 57, *Standard on Fumigation*.

Outdoor Storage

Goods stored outdoors should be properly separated from buildings of combustible construction and from other combustible storage which might constitute an exposure hazard. These separations should be maintained by the housekeeping staff, who must see to it that they are never blocked, even temporarily, by such things as contractors' shacks, discarded crates, pallets, or other combustibles. Obstructed aisles could hamper fire fighting operations if the need for them ever arises. Passageways between storage piles should also be unobstructed and clear of combustibles.

Proper housekeeping also requires that smoking in outdoor storage areas be controlled. Suitable signs should be posted and large noncombustible receptacles should be provided for the disposal of smoking materials before entering a "no smoking" area.

Outdoor Rubbish Disposal

Combustible waste materials stored outdoors to await subsequent disposal as rubbish should be placed not less than 20 ft (6 m), and preferably 50 ft (15 m), from buildings and at least 50 ft (15 m) from public highways and sources of ignition, such as incinerators. It should also be enclosed with a secure noncombustible fence of adequate height.

The most satisfactory solution to the rubbish disposal problem is regular public collection. Burning rubbish is generally unsafe and is not permitted in built-up urban areas.

If rubbish must be burned outdoors, it should be done in the early morning or at night because the night moisture reduces the chance that sparks will ignite combustibles in the surrounding area. This is the reasoning behind certain fire department and forestry service regulations that limit outdoor burning to certain days or times of day. Of course, there are some times when outdoor burning is strictly prohibited. Most parts of the United States and Canada experience days when things are so dry that any burning is dangerous.

Even when rubbish is not burned outdoors, it can present a fire hazard. All rubbish dumps, even landfills, are susceptible to fire. And sparks and flying brands from dump fires can carry the fire long distances. This is also true of sparks and brands produced by bonfires and incinerators which lack adequate spark arresters.

BIBLIOGRAPHY

NFPA CODES, STANDARDS, AND RECOMMENDED PRACTICES

Reference to the following NFPA Codes, Standards, and Recommended Practices will provide further information on good industrial housekeeping practices discussed in this chapter. (See the latest *NFPA Codes and Standards Catalog* for availability of current editions of the following documents.)

NFPA 13A, *Care and Maintenance of Sprinkler Systems.*
NFPA 91, *Blower and Exhaust Systems for Dust, Stock, Vapor Removal.*
NFPA 96, *Vapor Removal from Commercial Cooking Equipment.*

Chapter 50

ADDITIONAL READING

"How to Apply Good Housekeeping," *The Handbook of Property Conservation*, ch. 23, Factory Mutual System, Norwood, MA, 1973, pp. 189-193.

McKinnon, G. P., ed., *Fire Protection Handbook*, Fifteenth Edition, National Fire Protection Association, Quincy, MA, 1981.

Morrow, L. C., ed., "Sanitation and Housekeeping," *Maintenance Engineering Handbook*, 2nd ed., McGraw-Hill, New York, NY, 1957, pp. 14-1 to 14-71.

Index

Index

A

Absorption, in chemical processes, 662
Acetylene
  as industrial gas, 798
  as welding fuel, 502
  working pressure, 816
Additives, for paints and coatings, 146
Adiabatic heat, 802
Adsorption, in chemical processes, 662
Aerosols
  classification, 732
  containers for, 729
  filling plants, 734
  fire behavior, 733
  propellants, 729
  safeguards, 159, 734
  storage, 737
  warehousing, 739
Air-handling systems
  for nuclear power plants, 77
Air moving equipment
  corrosive vapors, 926
  ducts, 921
  electrostatic precipitators, 928
  explosion prevention, 924
  fans, 923
  fire extinguishing systems, 923
  flammable vapors, 924
  pneumatic conveyors, 926
  purposes, 919
  system design, 920
Air-supported structures, 1015
Alarm systems, 18
Alpha radiation, 745
Aluminum
  explosion hazards, 237
  finishing, 237
  fire control, 243
  fire hazards, 237
  life safety, 246
  machinability, 640
  raw materials, 234
  refining, 235
  safeguards, 242
Ammonia, as industrial gas, 799
Animal oils
  process hazards, 125
  safeguards, 141
  storage, 126
Atomizers, for spray finishing, 521

B

Ball clay, 377
Bauxite, refining, 235
Beta radiation, 745
Bindery, 340
Bleaching, in pulp and paper making, 291, 296
BLEVE,
  conditions for, 771
  in laboratories, 862
Blowers, see *Air moving equipment*
Boiler furnaces
  coal-burning, 884
  combustion process, 871
  continuous purge, 891
  controls, 879
  explosion and fire protecition, 892
  fire tube, 871
  fuels, 872
  gas-burning, 874
  hazards, 44, 887
  light-off, 891
  natural gas burners, 877
  oil-fired, 51, 874
  piping for, 879
  pulverized coal, 881
  water tube, 871
Brazing, 504
Browns Ferry Nuclear Power Plant, 70
Building design fundamentals, 14
Bulkheads, shipyard,
  fire hazards, 456
Butane
  as industrial gas, 799
  in liquefied petroleum gas, 828

C

Carbide, as fuel in nuclear power plants, 74
Carbon dating, 748
Carbon dioxide, as industrial gas, 804
Card machines, textiles, 357
Carpet, storage of, 1007
Carpeting, manufacture of, 365
Chemical processes
  crushing and grinding, 658
  drying, 663
  fire protection, 666
  fluid flow, 655
  hazards, 664
  heat transfer, 656
  mixing, 657
  raw material handling, 655
  reactors, 652
  separation, 658
Chemicals, for pulp and paper, hazards, 291
Chemical salts,
  melting points, 617
China clay, 377

Index

Chlorine, as industrial gas, 801
Clay products,
  drying, 380
  fire hazards, 387
  forming, 378
  glazing, 385
  kiln firing, 381
  production, 377
  raw materials, 376
  safeguards, 390
  types, 375
Clinch River Breeder Reactor Plant, 74
Coal
  fire hazards, 43
  fuel for electric generating plants, 40
  safeguards, 48
  storage, 41
Coal-burning systems
  for boiler furnaces, 881
  fluidized bed, 884
Coal dust, in mine explosions, 212
Coal mining
  strip, 195
  underground, 202
Coaters, for dipping and coating
  curtain, 556
  flow, 554
  roll, 556
Coating and dipping
  applications, 553
  design considerations, 566
  equipment, 554
  fire protection, 567
  hazard reduction, 558
  ignition sources, 563
  process hazards, 556
  thermal spraying, 509
Coatings
  architectural, 143
  fluid, 521
  manufacture of, 143
    fire control, 162
    fire hazards, 155
    production process, 155
    product storage, 154
    raw materials, 145
  product, 144
  special purpose, 144
Combing, of textiles, 359
Combustible liquids, *see also Flammable liquids*
  classification, 768
  defined, 767
  hazards, 769
Combustion, spontaneous,
  of grains, 102

Commodities
  British classes, 982
  classification, 980
  plastics, 981
Compression heating, 802
Computer centers
  fire detection, 849
  fire hazards, 847
  fire suppression, 851
  housekeeping, 852
  importance, 843
  production process, 846
  protection, 848
  raw materials, 845
  safeguards, 847
Computers, hazards of, 1, 847
Construction, firesafety during, 5
Containers
  cryogenic liquids, 823
  liquefied gas, 823
Control room, electrical,
  protection of, 64
Conveyors
  belt,
    fire causes, 940
    fire hazards, 937
    fire protection, 941
  grain, 103
  mechanical, 941
  pneumatic, 926
Cooling towers
  hazards, 44
  for nuclear power plants, 77
  protection for, 61
Copper, machinability of, 640
Cotton
  baled, 350
  storage, 1010
Cranes, materials handling, 945
Curtain coaters, 556
Cutting, *see also Welding*
  arc, 501
  electrical, 496
  machine tool, 629
  oxy-fuel gas, 505
  personnel protection, 516
  safeguards, 509
  thermal, 495
Cutting fluid, 637

D

Data processing, *see Computer centers*
Dehydrators, agricultural, *see Dryers, agricultural*
Diapers, disposable
  fire hazards, 329

production, 329
  raw materials, 328
  safeguards, 329
Dipping, see Coating
Dip tanks, for dipping and coating, 554
Drawing, of textiles, 360
Drills, exit, 18
Dryers, agricultural
  construction, 703
  extinguishing equipment, 705
  heating, 700
  installation, 703
  types, 700
Ducts, air moving, 922
Dust explosions
  conditions for, 719
  elements of, 109
  inerting, 724
  prevention, 723
  suppression, 725
  venting, 724
Dusts
  collecting equipment, 927
  combustible, 472
  control of, in grain handling 107
  explosion characteristics, 710
  explosibility index, 708
  as fire hazard in plastics, 403

E

Edible oil, processing, 127
Egress, means of,
  design requirements, 16
Electrical systems, industrial
  cable tray protection, 959
  conductors
    aluminum, 961
    support for, 958
  connections, 958
  distribution system
    classification, 956
    grounding, 953
    reliability, 952
    selection, 955
  equipment selection, 948
  fire potential, 947
  maintenance, 967
  protective devices, 962
  service entrance equipment, 951
  substations, 950
Electric generating plants
  boilers, 53
  design, 40, 45
  extinguishing systems, 56
  fire protection, 48
  fuels, 40

hazards, 40
nuclear, 69
pollution control, 43
safeguards, 45
turbine generators, 54
ventilation, 53
water supply, 47
Electricity, production, 42
Electricity, static,
  in plastics production, 408
Electrocoating, in motor vehicle assembly, 422
Elevators, grain
  bucket, 104
  fire control, 120
  liners for, 105
Emergency action plan
  chart, 30
  criteria, 21
  need for, 8
  OSHA requirements, 24
  procedures, 24
Ethylene
  as industrial gas, 800
  in plastics products, 399
Evacuation drills, 18
Evacuation plans, 25
Exhaust systems, see Air moving equipment
Exit drills, 18
Exit facilities, design of, 16
Explosion
  possibility of, 2
  prevention, in air moving equipment, 924
  severity, of materials, 708
  venting, for air moving equipment, 924
Explosions
  chemical, 664
  combustion, 771
  detonation, 771
  dust, 109
  in flammable liquids, 774, 777
  grain,
    in handling, 102
    ignition sources, 112
    location, 113
    safeguards, 115
  mine,
    ignition sources, 213
    underground, 211
  molten metal, 239
Explosive limits, of grain dusts, 110
Explosive materials, classification, 709
Extinguishers, use of, 26
  training for, 27

Index

Extinguishing systems,
  for air moving equipment, 923
  carbon dioxide, 57
  dry chemical, 58
  for flammable liquids, 789
  foam, 57
  halon, 28
  for turbine generators, 56
Extractors, for solvent extraction, 677

F

Fans, air moving, 923
Fat splitting, hazards of, 138
Fiberboard
  conveying systems, 263
  fire control systems, 265
  production, 258
  raw materials, 258
  safeguards, 264
Fire
  evaluating possibilities of, 2
  type and severity, 13
Fire brigade
  incipient fires, 22
  in motor vehicle assembly, 432
  organization of, 8
  OSHA statement, 31
  prefire plan for, 35
  requirements, 29
  selection, 31
  shipyards, 461
  structural, 22
  training, 33
Fire clay, 377
Fire control program, 2
Fire detection
  equipment, 1
  in laboratories, 865
Fire emergency, options for, 22
Fire prevention
  equipment, 1
  inspection for, 6
  program objectives, 2
Fire prevention manager, 2
Fire protection equipment,
  influence of, 14
Fire risk management, 1
  considerations for, 4
Firesafety
  during construction, 5
  planning for, 12
Fire severity, equated to fire loading, 13
Fission, 748
Flammable liquids, *see also* Combustible liquids
  classification, 768
  combustion explosions, 771
  defined, 767
  detonation explosions, 771
  dispensing, 780
  fire in, 770
  fire prevention, 784
  hazards, 769
  loss control, 784
  in motor vehicle assembly, 417
  storage, 772, 777
  transfer, 780
Flammable off-gases, 83
Flammable solvents, in plastics production, 405
Flexography, 340
Flint, in clay products, 377
Flow coaters, 554
Fluid power, *see also* Hydraulic fluids
  applications, 895
  defined, 895
  fire hazards, 900
  fire protection, 902
  hydraulic fluids, 897
  maintenance, 903
  uses, 896
Fluorine, as industrial gas, 802
Flux, in clay products, 377
Food processing
  building construction, 473
  combustible dusts, 472
  fire hazards, 470
  flammable liquids, 471
  gases, 471
  loss potential, 465
  processes, 467
  raw materials, 466
  safeguards, 473
  storage, 472
Fourdrinier (paper machine), 297
Fuel oil
  filters, 52
  fire hazards, 43
  heaters, 51
  pumps, 52
  safeguards, 51
  storage, 41
Furnaces, *see also* Ovens
  special atmosphere, 585
Furniture manufacture
  fire control, 285
  fire hazards, 281
  fire losses, 275
  flammable solvents, 279
  production process, 277
  raw materials, 276
  safeguards, 285
Fusion, 748

G

Gamma radiation, 745
Garages, as storage facilities, 1012
Gas-burning systems, 874
Gases
　flammable, 798
　flammable, off-gases, 83
　inert, 804
　oxidizing, 801
　refrigerant, 911
Gases, industrial
　container handling, 813
　defined, 795
　equipment for, 809
　filling densities, 815
　fire prevention, 820
　leak inspection, 818
　liquefied, 796, 798
　operating practices, 813
　properties, 796
　shipping, 806
　specific gravity, 815
　storage, 808
Generators,
　diesel, 81
　electric,
　　hazards of, 1
　hydrogen cooled, 84
　turbine, 44
Grain
　conveyors, 103
　drying, 106
　dust control, 107
　explosion hazard, 109
　fire hazard, 108
　production, 102
　spontaneous combustion, 102
　storage, 103
Grain elevators, see Elevators, grain
Grain explosions
　location, 113
　safeguards, 115
Graphite, as moderator in nuclear generator plants, 74
Gravure printing, 338
Grinding, see Milling
Grog, in clay products, 377
Guard service, 5

H

Halon systems, 28
Handicapped persons, arrangements for, 13
Hardboard
　conveying systems, 263
　fire control, 265
　production, 258
　raw materials, 258
　safeguards, 264
Heaters, fuel oil, 51
Heating, oxy-fuel gas, 505
Heat processing
　equipment, 571
　fire and explosion problems, 572
　operation, 582
Helium, for heat transfer in nuclear power plants, 74
Hexane, as extraction solvent, 675
Hose system, small, 22
Housekeeping, industrial, see Industrial housekeeping
Hydraulic fluids, see also Fluid power,
　classes of, 897
　compressibility, 898
　fire hazards, 900
　fire protection, 902
　fire resistance, 901
　synthetic, 902
　viscosity, 897
　water-glycol, 902
　water-in-oil, 901
Hydraulic systems, in plastics production, 408
Hydrocarbons, as aerosol propellants, 730
Hydrogen
　as generator coolant, 44, 84
　as industrial gas, 800
　manufacture of, 131
　in nuclear power plants, 77, 84
　storage, 59
Hydrogenation, of oils, 129

I

Ignition sensitivity, of materials, 708
Ignition sources, in laboratories, 863
Industrial gas equipment, 809
Industrial gases, see Gases, industrial
Industrial housekeeping
　building maintenance, 1052
　equipment, 1051
　grass and weed control, 1061
　hazards, 1058
　loss potential, 1049
　manpower, 1051
　outdoor storage, 1062
　rubbish disposal, 1055
　theory, 1050
Industrial occupancies,
　life safety in, 11, 17
Inspection, fire prevention, 6
　periodic, 8

Index

Insulation, electrical cable,
  in nuclear power plants, 82
Isobutane, as aerosol propellant, 730

K

Kaolin, 377
Kaolinite, 376
Kilns
  for clay products, 381
  lumber, see Lumber kilns
Kitchen equipment, ventilation for, 929
Knitting, textiles, 365
Kraft paper process, 293

L

Laboratories, industrial
  evacuation, 867
  fire control, 866
  hazards, 856
  intrinsically safe equipment, 864
  safeguards, 860
  smoke detection, 865
  storage cabinets, 862
Lecithin, 674
Letterpress, 334
Life, potential for loss of, 11
Life safety
  in facility planning, 4
  in industrial occupancies, 11
  in motor vehicle assembly, 433
  in nuclear power plants, 98
  occupancy classification, 15
  in paper processing, 310
  preplan, 12
  psychological factors, 13
Life Safety Code, NFPA 101, 15
Liquefied gas, defined, 796
Liquefied petroleum gas
  approved equipment, 833
  delivery, 839
  fire protection, 836
  industrial uses, 800, 827
  as life truck fuel, 837
  properties, 829
  standards, 830
  storage, 830
  valves for, 834
Lithography, 337
Loss,
  maximum foreseeable (MFL), 2
  maximum possible (MPL), 2
  maximum probable (MPL), 2
LPG, see Liquefied petroleum gas
LP-Gas, see Liquefied petroleum gas

Lumber
  drying, 255
  finishing, 256
  storage, 257
Lumber kilns
  construction, 697
  fire hazards, 699
  heat sources, 698
  safeguards, 699
  types, 694

M

Machine tools, see also Machining,
  uses, 627
Machining
  cutting fluids, 637
  cutting tool, 629
  electrical discharge, 632
  electrochemical, 633
  fire hazards, 644
  fire protection, 646
  metal powders, explosibility, 635
  milling machine, 632
  power requirements, 644
  processes, 628
  raw materials, 628
  safeguards, 645
  tool actuators, hydraulic, 638
Magnesium, machinability, 640
Manager, fire prevention and control, 2
MAPP, see Methylacetylene-propadiene,
  stabilized
Marine railway, 445
Materials handling
  belt conveyors, 937
  cranes, 945
  industrial trucks, 934
  mechanical conveyors, 941
  in motor vehicle assembly, 419
  systems, 933
Metal mining, underground, 204
Metal powders,
  explosibility of, 635
  ignition, 635
Metals,
  machinability, 639
  ferrous, 640
Methane, in mine explosions, 212
Methylacetylene-propadiene, stabilized as
  industrial gas, 801
Milk cartons
  fire control, 328
  fire hazards, 326
  production process, 326
  raw materials, 325
  safeguards, 328

Millers, grain, 101
Milling
  equipment, 712
  explosion hazards, 708
  hazards, 717
  materials, 707
Milling machine, 632
Mills, types of, 712
Minerals, U.S. production, 194
Mining
  explosion safeguards, 220
  fire detection, 222
  firesafety, 219
  fire suppression, 223
  methods, 194
  miner safety, 227
  surface, 194
  underground, 201
    hazards, 211
    nonmetal, 204
Monomer, 397
Motor vehicle assembly
  body construction, 421
  hazards, 426
  housekeeping, 430
  life safety, 433
  materials handling, 419
  production process, 420
  raw materials, 416
  safeguards, 436
  subassemblies, 418
MPS, see Methylacetylene-propadiene, stabilized

N

Natural gas, types, 827
Natural gas burners
  controls for, 880
  for boiler furnaces, 877
n-Butane, as aerosol propellant, 731
NFPA Life Safety Code, 15
Nickel, machinability, 642
Nitrocellulose, for coatings, 146
Nitrogen, as industrial gas, 805
Nuclear fuel, storage of, 71
Nuclear power plants
  air handling systems, 77
  combustibles in, 72
  combustible waste, 82
  construction considerations, 85
  coolant pump motors, 81
  cooling towers, 77
  diesel-driven generators, 81
  fire brigade, 97
  fire control system, 88
  fire hazards, 78

  fuel, 71
  hazardous materials, 80
  hydrogen in, 77, 84
  life safety provisions, 98
  reactors, 72
  safeguards, 84
  support systems, 75
  as users of radioactive materials, 744

O

Occupancies, industrial
  examples of, 16
  life safety in, 17
Offset printing, 337
Oil, animal
  process hazards, 125
  storage, 126
Oil burners
  for boiler furnaces, 875
  controls for, 879
  piping for, 879
Oil burning systems, 874
Oil, edible, processing, 127
Oil, fuel
  for electric generating plants, 41
  storage, 41
Oil quenching
  central oil system, 605
  fire protection, 607
  loss potential, 597
  material transfer, 601
  quench oils, 598
  quench tanks, 599
  safeguards, 606
  temperature control, 603
Oils, deodorization, 132
  hazards, 133
Oils, hydrogenation, 129
Oils, for paints and coatings, 147
Oilseeds
  production, 102
  solvent extraction, 673
Oil, vegetable, storage of, 126
Operations, continuity of, 4
OSHA Subpart L, 21
Ovens
  Class A, defined, 573
    operation, 582
  Class B, defined, 573
    operation, 584
  Class C, defined, 573
  Class D, defined, 574
    operation, 587
  construction, 576
  fire prevention, 591
  fuel hazards, 581

Index

handling systems, 574
heating systems, 571
location, 576
maintenance, 593
operator training, 593
safety controls, 590
special atmosphere, 585
Oxidizers, in laboratories, 863
Oxygen,
 as industrial gas, 802
 liquid, 803

P

Paints, *see also* Coatings
 fire control system, 162
 fire hazards, 155
 production process, 151
 product storage, 154
 raw materials, 145
 safeguards, 157
Panic, danger of, 13
Paper processing
 bleaching, 296
 chemicals, 291
 finishing, 299
 fire brigade, 306
 fire control, 305
 fire hazards, 300
 fuels, 292
 kraft process, 293
 paper machine, 297
 processes, 292
 raw materials, 290
 roll storage, 299, 1006
 safeguards, 305
Paper products
 milk carton, 325
 shipping containers, 316
 uses of, 289
Paper, roll, storage, 1006
Particleboard
 conveying systems, 263
 fire control, 265
 production, 258
 raw materials, 258
 safeguards, 264
Pesticides, storage, 1011
Picker room, textiles, 356
Piers, shipyard,
 fire hazard, 456
 and wharves,
  as storage facilities, 1012
Pigments, for paints and coatings, 146
Plastics
 examples, 395
 hazards, 396

in motor vehicle assembly, 417
storage, 981
terminology, 396
thermoplastics, 394
thermoset, 394
Plastics products
 dust, as fire hazard, 403
 ethylene in, 399
 fire control, 410
 fire hazards, 403
 flammable solvents, 405
 housekeeping, 409
 production process, 398
 raw materials, 398
 safeguards, 410
 storage, 409
Plenums, charcoal filter,
 in nuclear power plants, 82
Plutonium, as nuclear fuel, 71
Plywood
 fire hazards, 270
 production process, 266
 raw materials, 266
 safeguards, 271
Pollution control
 in electric generating plants, 43
Powder coating,
 electrostatic, 523, 541
 enclosures for, 539
 fire prevention, 547
 fire protection, 549
 fluidized bed, 541
 hazards, 547
 powder recovery, 539
 spray process, 536
Precipitators, electrostatic,
 for air pollution control, 928
Prefire plan, for fire brigade, 35
Printing
 bindery, 340
 fire hazards, 342
 flexography, 340
 gravure, 338
 letterpress, 334
 lithography, 337
 production process, 333
 raw materials, 322
 safeguards, 345
 screen, 339
 storage, 341
 thermography, 339
Production machines,
 hazards, 1
Propane
 as aerosol propellant, 730
 in liquefied petroleum gas, 828

Propellants, hydrocarbon,
  fire hazards, 731
Property, protection of, 4
Publishing, see Printing
Pulping chemicals, 291
Pumps
  fuel oil, 52
  turbine driven, 81

Q

Quenching oils, see Oil quenching

R

Rack storage, 987
Radiation, protection from, 753
Radioactive materials,
  characteristics, 745
  containment, 758
  contamination control, 757
  criticality, 759
  fire protection, 759
  hazards, 749
  labeling, 750
  life safety, 762
  in nuclear power plants, 744
  production, 749
  protection from, 753
  radiation machines, 750
  shielding, 755
  transportation, 750
  uses, 744
Radioactive waste, 69
Radioisotopes, 750
Reactor
  boiling water, 73
  high temperature gas (HTGR), 72
  liquified metal fast breeder, 74
  pressurized water, 72
Records protection, 1009
Refractories, 376
Refrigerated storage, 1013
Refrigeration systems
  compression, 908
  emergency control, 915
  gases for, 911
  hazards, 913
  heat absorption, 908
  operation, 907
Resins, for paints and coatings, 145
Roll coaters, 556
Roving, textiles, 360
Rubber products
  fire brigade, 191

fire control, 186
fire hazards, 179
process, 171
raw materials, 170
safeguards, 185
sprinkler systems, 189

S

Salt baths, molten
  applications, 611
  defined, 609
  hazards, 616
  quenching, 614
  safeguards, 619
  types, 610
Salts, chemical,
  melting points, 617
Sawmills,
  fire hazards, 253
  lumber sorters, 254
  raw materials, 251
  safeguards, 254
  saw building, 252
Screen printing, 339
Scrubbers, overspray, 535
Semiconductor manufacturing
  fire hazards, 486
  loss potential, 481
  loss prevention, 492
  production process, 482
  safeguards, 487
  special hazards, 489
Shale, in clay products, 377
Shielding, radioactive, 755
Shipbuilding, importance, 441
Shipping containers,
  fire control, 324
  production, 316
  production hazards, 320
  safeguards, 323
Shipyards
  construction basin, 445
  fire brigades, 461
  fire hazards, 443, 452
  fire loss, 443
  fire protection, 442
  marine railway, 445
  properties of, 444
  raw materials, 446
  safeguards, 460
  vessel construction, 449
  vessel repair, 451
  watch service, 462
  ways, 444, 452
Smoke detection, in laboratories, 865

Index

Soap making
  hazards, 138
  process, 136
Sodium,
  for heat transfer in nuclear power plants, 74
Soldering, hard, 504
Solvent extraction
  fire protection, 688
  hazards, 682, 685
  process, 675
  raw materials, 674
  solvents, 675
Solvents, flammable,
  in plastics production, 405
  for paints and coatings, 146
Spinning, textiles, 360
Spontaneous combustion
  of grains, 102
  in mines, 216
Sprinklers, automatic, life safety record, 15
Sprinkler systems
  for aerosol storage, 738
  for coal fires, 50
  in electric generating plants, 56
  for flammable liquids, 788
  in food processing, 474
  installation, 21
  in paper processing, 306
  in-rack, 995
  for radioactive materials, 760
  for storage occupancies, 1002
Spray booths, 529
Spray coating, see Spray finishing
Spray finishing
  continuous coater, 530
  decorating machine, 530
  enclosures for, 528
  fire protection, 546
  fluid coatings, 521
  fluid supply, 524
  hazards, 543
  heaters, 525
  open floor, 533
  overspray collectors, 533
  powder coatings, 523
  spray guns, 526
Spray guns
  dry powder, 537
  fluid, 526
Spray rooms, 532
Stairways, enclosed, 17
Standpipe system, Class II, 24
Storage
  hazards, 1
  problems, 972
Storage occupancies
  air-supported structures, 1015
  baled fibers, 1010
  baled waste, 1011
  bulk storage, 984
  carpet, 1007
  commodities classification, 980
  emergency organization, 1005
  fire causes, 974
  fire control, 977
  fire protection, 975, 1002
  garages, 1012
  hose systems, 1004
  isolated, 1014
  NFPA Standards, 990
  outdoor, 1014
  palletized, 985
  pesticides, 1011
  piers and wharves, 1012
  plastics, 981
  rack storage, 987
  records, 1009
  refrigerated, 1013
  roll paper, 1006
  rubber tires, 1006
  solid piling, 984
  sprinkler density, 993
  underground, 1014
Surfacing, metal, 505

T

Tanks, for flammable liquids, 772
Textile manufacture
  carpeting, 365
  cleaning, 365
  finishing, 365
  knitting, 365
  safeguards, 369
  spinning, 360
  weave room fires, 364
Textile mills
  electrical installations, 368
  fires in, 350
Textiles
  blending, 361
  card room, 357
  combing, 359
  cotton stock, 350
  drawing, 360
  opener rooms, 352
  picker room, 356
  production process, 351
  raw materials, 350
  roving, 360
  weaving, 363
Thermal spraying, 509

Thermography, 339
Tires, rubber
  making, 176
  storage, 184, 1006
Titanium, machinability of, 642
Transformers,
  oil-insulated, 44
  protection of, 62
Trucks, industrial
  identification for, 936
  for materials handling, 934
  operation and maintenance, 935
  types, 934, 938
Turbine generators, protection of, 54

U

Uranium
  machinability, 643
  as nuclear fuel, 71
  production, 744

V

Vegetable oil
  process hazards, 125
  processing safeguards, 141
  storage, 126
Ventilation
  for flammable liquids, 787
  for kitchen cooking equipment, 929
  for radioactive materials, 761
Vessel construction, 449
Vessel repair, 451
  fire hazards, 459
Viscosity
  of hydraulic fluids, 898
  index, 898

W

Warehousing, see also Storage occupancies
  fire control, 977
  fire protection, 1002
Waste, combustible
  in nuclear power plants, 82
Waste, hazardous
  characteristics, 1025
  disposal processes, 1036
  generation of, 1023
  hazard reduction, 1044
Waste, industrial
  classification, 1028
  codes, regulations and standards, 1045
  disposal systems, 1034
  hazard reduction, 1044
  incinerators, 1041
  management systems, 1027
  solid waste, 1022
  treatment
    biological, 1037
    chemical, 1037
    physical, 1034
    thermal, 1038
  types, 1042
Waste, radioactive, 69
Waste recovery
  in textile plants, 366
Wastes, storage of, 1011
Watch service,
  in shipyards, 462
Water supply
  for electric generating plants, 47
  for fire protection, 8
  for nuclear power plants, 88
Ways, shipyard, 444, 452
Welding
  braze, 504
  electrical, 496
  electroslag, 500
  flash, 500
  oxy-acetylene, 502
  oxy-fuel gas, 502
    equipment, 506
  personnel protection, 516
  precautions, 511
    special, 515
  resistance, 499
  safeguards, 509
  types, 495
Welding, arc,
  flux cored, 497
  gas metal, 497
  gas tungsten, 497
  plasma, 498
  shielded metal, 497
  submerged, 499
Whiteware, 375
Wood, as paper ingredient, 290
Wood processing
  fire control, 307
  life safety, 310
Wood products manufacture, 249

X

X-rays, 746

Z

Zinc, machinability, 642
Zirconium, machinability, 643

Don't "OK" another fire inspection report... without consulting this classic manual!

National Fire Protection Association
Batterymarch Park, Quincy, MA 02269
Telephone orders: (617) 770-3500

The NFPA Inspection Manual has been trusted by the firesafety profession for over 30 years. Now fully updated in a new 5th edition, this practical manual shows you how to identify and correct fire hazards. Features new, in-depth treatment of the most dangerous fire hazards, including gas, combustible dusts, chemicals, and explosives.

387 pp., pocket-sized for easy use in the field.

ORDER YOUR COPY TODAY!

Please send me _____ **NFPA Inspection Manuals** (SPP-11C) at $21.00 each (**NFPA members: $18.90**)
TOTAL: _____
☐ Check enclosed
☐ Please bill me, plus shipping and handling
NFPA member I.D. #

Name _____
Address _____
City _____ State _____ Zip _____
Charge to my   ☐ VISA   ☐ MASTERCARD
Card #

Signature _____ Exp. date _____

Looking for the answers to today's fire protection problems?

National Fire Protection Association
Batterymarch Park, Quincy, MA 02269
Telephone orders: (617) 770-3500

Stay on top of developments in fire protection techniques, systems and equipment all from one comprehensive source. **The fifteenth edition of the Fire Protection Handbook** features current information on every aspect of firesafety... facts you'll use again and again.

Approximately 1,400 pages, 20 sections on specific subjects and occupancies, 165 fact-packed chapters, and nearly 1,000 illustrations, photographs and diagrams!

QTY _____ **Fire Protection Handbook** (F16-FPH1581)
$60.00 each/**NFPA Members $54.00**
TOTAL $ _____
☐ Check enclosed
☐ Please bill me, plus shipping and handling
NFPA member I.D. #

Signature _____ Exp. date _____

Name _____
Address _____
City _____ State _____ Zip _____
Charge to my   ☐ VISA   ☐ MASTERCARD
Card #

NO POSTAGE NECESSARY IF MAILED IN THE UNITED STATES

BUSINESS REPLY MAIL
FIRST CLASS MAIL   PERMIT NO. 5347   QUINCY, MA

POSTAGE WILL BE PAID BY ADDRESSEE

NATIONAL FIRE PROTECTION ASSOCIATION
BATTERYMARCH PARK
QUINCY, MASSACHUSETTS 02269

NO POSTAGE NECESSARY IF MAILED IN THE UNITED STATES

BUSINESS REPLY MAIL
FIRST CLASS MAIL   PERMIT NO. 5347   QUINCY, MA

POSTAGE WILL BE PAID BY ADDRESSEE

NATIONAL FIRE PROTECTION ASSOCIATION
BATTERYMARCH PARK
QUINCY, MASSACHUSETTS 02269